T0210818

Preface

We are pleased to present the proceedings of the 25th International Conference on Medical Image Computing and Computer-Assisted Intervention (MICCAI) which – after two difficult years of virtual conferences – was held in a hybrid fashion at the Resort World Convention Centre in Singapore, September 18–22, 2022. The conference also featured 36 workshops, 11 tutorials, and 38 challenges held on September 18 and September 22. The conference was also co-located with the 2nd Conference on Clinical Translation on Medical Image Computing and Computer-Assisted Intervention (CLINICCAI) on September 20.

MICCAI 2022 had an approximately 14% increase in submissions and accepted papers compared with MICCAI 2021. These papers, which comprise eight volumes of Lecture Notes in Computer Science (LNCS) proceedings, were selected after a thorough double-blind peer-review process. Following the example set by the previous program chairs of past MICCAI conferences, we employed Microsoft's Conference Managing Toolkit (CMT) for paper submissions and double-blind peer-reviews, and the Toronto Paper Matching System (TPMS) to assist with automatic paper assignment to area chairs and reviewers.

From 2811 original intentions to submit, 1865 full submissions were received and 1831 submissions reviewed. Of these, 67% were considered as pure Medical Image Computing (MIC), 7% as pure Computer-Assisted Interventions (CAI), and 26% as both MIC and CAI. The MICCAI 2022 Program Committee (PC) comprised 107 area chairs, with 52 from the Americas, 33 from Europe, and 22 from the Asia-Pacific or Middle East regions. We maintained gender balance with 37% women scientists on the PC.

Each area chair was assigned 16–18 manuscripts, for each of which they were asked to suggest up to 15 suggested potential reviewers. Subsequently, over 1320 invited reviewers were asked to bid for the papers for which they had been suggested. Final reviewer allocations via CMT took account of PC suggestions, reviewer bidding, and TPMS scores, finally allocating 4–6 papers per reviewer. Based on the double-blinded reviews, area chairs' recommendations, and program chairs' global adjustments, 249 papers (14%) were provisionally accepted, 901 papers (49%) were provisionally rejected, and 675 papers (37%) proceeded into the rebuttal stage.

During the rebuttal phase, two additional area chairs were assigned to each rebuttal paper using CMT and TPMS scores. After the authors' rebuttals were submitted, all reviewers of the rebuttal papers were invited to assess the rebuttal, participate in a double-blinded discussion with fellow reviewers and area chairs, and finalize their rating (with the opportunity to revise their rating as appropriate). The three area chairs then independently provided their recommendations to accept or reject the paper, considering the manuscript, the reviews, and the rebuttal. The final decision of acceptance was based on majority voting of the area chair recommendations. The program chairs reviewed all decisions and provided their inputs in extreme cases where a large divergence existed between the area chairs and reviewers in their recommendations. This process resulted in the acceptance of a total of 574 papers, reaching an overall acceptance rate of 31% for MICCAI 2022.

In our additional effort to ensure review quality, two Reviewer Tutorials and two Area Chair Orientations were held in early March, virtually in different time zones, to introduce the reviewers and area chairs to the MICCAI 2022 review process and the best practice for high-quality reviews. Two additional Area Chair meetings were held virtually in July to inform the area chairs of the outcome of the review process and to collect feedback for future conferences.

For the MICCAI 2022 proceedings, 574 accepted papers were organized in eight volumes as follows:

- Part I, LNCS Volume 13431: Brain Development and Atlases, DWI and Tractography, Functional Brain Networks, Neuroimaging, Heart and Lung Imaging, and Dermatology
- Part II, LNCS Volume 13432: Computational (Integrative) Pathology, Computational Anatomy and Physiology, Ophthalmology, and Fetal Imaging
- Part III, LNCS Volume 13433: Breast Imaging, Colonoscopy, and Computer Aided Diagnosis
- Part IV, LNCS Volume 13434: Microscopic Image Analysis, Positron Emission Tomography, Ultrasound Imaging, Video Data Analysis, and Image Segmentation I
- Part V, LNCS Volume 13435: Image Segmentation II and Integration of Imaging with Non-imaging Biomarkers
- Part VI, LNCS Volume 13436: Image Registration and Image Reconstruction
- Part VII, LNCS Volume 13437: Image-Guided Interventions and Surgery, Outcome and Disease Prediction, Surgical Data Science, Surgical Planning and Simulation, and Machine Learning – Domain Adaptation and Generalization
- Part VIII, LNCS Volume 13438: Machine Learning – Weakly-supervised Learning, Machine Learning – Model Interpretation, Machine Learning – Uncertainty, and Machine Learning Theory and Methodologies

We would like to thank everyone who contributed to the success of MICCAI 2022 and the quality of its proceedings. These include the MICCAI Society for support and feedback, and our sponsors for their financial support and presence onsite. We especially express our gratitude to the MICCAI Submission System Manager Kitty Wong for her thorough support throughout the paper submission, review, program planning, and proceeding preparation process – the Program Committee simply would not have be able to function without her. We are also grateful for the dedication and support of all of the organizers of the workshops, tutorials, and challenges, Jianming Liang, Wufeng Xue, Jun Cheng, Qian Tao, Xi Chen, Islem Rekik, Sophia Bano, Andrea Lara, Yunliang Cai, Pingkun Yan, Pallavi Tiwari, Ingerid Reinertsen, Gongning Luo, without whom the exciting peripheral events would have not been feasible. Behind the scenes, the MICCAI secretariat personnel, Janette Wallace and Johanne Langford, kept a close eye on logistics and budgets, while Mehmet Eldegez and his team from Dekon Congress & Tourism, MICCAI 2022's Professional Conference Organization, managed the website and local organization. We are especially grateful to all members of the Program Committee for their diligent work in the reviewer assignments and final paper selection, as well as the reviewers for their support during the entire process. Finally, and most importantly, we thank all authors, co-authors, students/postdocs, and supervisors, for submitting and presenting their high-quality work which made MICCAI 2022 a successful event.

We look forward to seeing you in Vancouver, Canada at MICCAI 2023!

September 2022

Linwei Wang

Qi Dou

P. Thomas Fletcher

Stefanie Speidel

Shuo Li

Organization

General Chair

Shuo Li — Case Western Reserve University, USA

Program Committee Chairs

Linwei Wang	Rochester Institute of Technology, USA
Qi Dou	The Chinese University of Hong Kong, China
P. Thomas Fletcher	University of Virginia, USA
Stefanie Speidel	National Center for Tumor Diseases Dresden, Germany

Workshop Team

Wufeng Xue	Shenzhen University, China
Jun Cheng	Agency for Science, Technology and Research, Singapore
Qian Tao	Delft University of Technology, the Netherlands
Xi Chen	Stern School of Business, NYU, USA

Challenges Team

Pingkun Yan	Rensselaer Polytechnic Institute, USA
Pallavi Tiwari	Case Western Reserve University, USA
Ingerid Reinertsen	SINTEF Digital and NTNU, Trondheim, Norway
Gongning Luo	Harbin Institute of Technology, China

Tutorial Team

Islem Rekik	Istanbul Technical University, Turkey
Sophia Bano	University College London, UK
Andrea Lara	Universidad Industrial de Santander, Colombia
Yunliang Cai	Humana, USA

Clinical Day Chairs

Jason Chan The Chinese University of Hong Kong, China
Heike I. Grabsch University of Leeds, UK and Maastricht
 University, the Netherlands
Nicolas Padoy University of Strasbourg & Institute of
 Image-Guided Surgery, IHU Strasbourg,
 France

Young Investigators and Early Career Development Program Chairs

Marius Linguraru Children's National Institute, USA
Antonio Porras University of Colorado Anschutz Medical
 Campus, USA
Nicole Rieke NVIDIA, Deutschland
Daniel Racoceanu Sorbonne University, France

Social Media Chairs

Chenchu Xu Anhui University, China
Dong Zhang University of British Columbia, Canada

Student Board Liaison

Camila Bustillo Technische Universität Darmstadt, Germany
Vanessa Gonzalez Duque Ecole centrale de Nantes, France

Submission Platform Manager

Kitty Wong The MICCAI Society, Canada

Virtual Platform Manager

John Baxter INSERM, Université de Rennes 1, France

Program Committee

Ehsan Adeli Stanford University, USA
Pablo Arbelaez Universidad de los Andes, Colombia
John Ashburner University College London, UK
Ulas Bagci Northwestern University, USA
Sophia Bano University College London, UK
Adrien Bartoli Université Clermont Auvergne, France
Kayhan Batmanghelich University of Pittsburgh, USA

Hrvoje Bogunovic	Medical University of Vienna, Austria
Ester Bonmati	University College London, UK
Esther Bron	Erasmus MC, the Netherlands
Gustavo Carneiro	University of Adelaide, Australia
Hao Chen	Hong Kong University of Science and Technology, China
Jun Cheng	Agency for Science, Technology and Research, Singapore
Li Cheng	University of Alberta, Canada
Adrian Dalca	Massachusetts Institute of Technology, USA
Jose Dolz	ETS Montreal, Canada
Shireen Elhabian	University of Utah, USA
Sandy Engelhardt	University Hospital Heidelberg, Germany
Ruogu Fang	University of Florida, USA
Aasa Feragen	Technical University of Denmark, Denmark
Moti Freiman	Technion - Israel Institute of Technology, Israel
Huazhu Fu	Agency for Science, Technology and Research, Singapore
Mingchen Gao	University at Buffalo, SUNY, USA
Zhifan Gao	Sun Yat-sen University, China
Stamatia Giannarou	Imperial College London, UK
Alberto Gomez	King's College London, UK
Ilker Hacihaliloglu	University of British Columbia, Canada
Adam Harrison	PAII Inc., USA
Mattias Heinrich	University of Lübeck, Germany
Yipeng Hu	University College London, UK
Junzhou Huang	University of Texas at Arlington, USA
Sharon Xiaolei Huang	Pennsylvania State University, USA
Yuankai Huo	Vanderbilt University, USA
Jayender Jagadeesan	Brigham and Women's Hospital, USA
Won-Ki Jeong	Korea University, Korea
Xi Jiang	University of Electronic Science and Technology of China, China
Anand Joshi	University of Southern California, USA
Shantanu Joshi	University of California, Los Angeles, USA
Bernhard Kainz	Imperial College London, UK
Marta Kersten-Oertel	Concordia University, Canada
Fahmi Khalifa	Mansoura University, Egypt
Seong Tae Kim	Kyung Hee University, Korea
Minjeong Kim	University of North Carolina at Greensboro, USA
Baiying Lei	Shenzhen University, China
Gang Li	University of North Carolina at Chapel Hill, USA

Fuyong Xing	University of Colorado Denver, USA
Ziyue Xu	NVIDIA, USA
Yanwu Xu	Baidu Inc., China
Pingkun Yan	Rensselaer Polytechnic Institute, USA
Guang Yang	Imperial College London, UK
Jianhua Yao	Tencent, China
Zhaozheng Yin	Stony Brook University, USA
Lequan Yu	University of Hong Kong, China
Yixuan Yuan	City University of Hong Kong, China
Ling Zhang	Alibaba Group, USA
Miaomiao Zhang	University of Virginia, USA
Ya Zhang	Shanghai Jiao Tong University, China
Rongchang Zhao	Central South University, China
Yitian Zhao	Chinese Academy of Sciences, China
Yefeng Zheng	Tencent Jarvis Lab, China
Guoyan Zheng	Shanghai Jiao Tong University, China
Luping Zhou	University of Sydney, Australia
Yuyin Zhou	Stanford University, USA
Dajiang Zhu	University of Texas at Arlington, USA
Lilla Zöllei	Massachusetts General Hospital, USA
Maria A. Zuluaga	EURECOM, France

Reviewers

Alireza Akhondi-asl
Fernando Arambula
Nicolas Boutry
Qilei Chen
Zhihao Chen
Javid Dadashkarimi
Marleen De Bruijne
Mohammad Eslami
Sayan Ghosal
Estibaliz Gómez-de-Mariscal
Charles Hatt
Yongxiang Huang
Samra Irshad
Anithapriya Krishnan
Rodney LaLonde
Jie Liu
Jinyang Liu
Qing Lyu
Hassan Mohy-ud-Din

Manas Nag
Tianye Niu
Seokhwan Oh
Theodoros Pissas
Harish RaviPrakash
Maria Sainz de Cea
Hai Su
Wenjun Tan
Fatmatulzehra Uslu
Fons van der Sommen
Gijs van Tulder
Dong Wei
Pengcheng Xi
Chen Yang
Kun Yuan
Hang Zhang
Wei Zhang
Yuyao Zhang
Tengda Zhao

Yingying Zhu
Yuemin Zhu
Alaa Eldin Abdelaal
Amir Abdi
Mazdak Abulnaga
Burak Acar
Iman Aganj
Priya Aggarwal
Ola Ahmad
Seyed-Ahmad Ahmadi
Euijoon Ahn
Faranak Akbarifar
Cem Akbaş
Saad Ullah Akram
Tajwar Aleef
Daniel Alexander
Hazrat Ali
Sharib Ali
Max Allan
Pablo Alvarez
Vincent Andrearczyk
Elsa Angelini
Sameer Antani
Michela Antonelli
Ignacio Arganda-Carreras
Mohammad Ali Armin
Josep Arnal
Md Ashikuzzaman
Mehdi Astaraki
Marc Aubreville
Chloé Audigier
Angelica Aviles-Rivero
Ruqayya Awan
Suyash Awate
Qinle Ba
Morteza Babaie
Meritxell Bach Cuadra
Hyeon-Min Bae
Junjie Bai
Wenjia Bai
Ujjwal Baid
Pradeep Bajracharya
Yaël Balbastre
Abhirup Banerjee
Sreya Banerjee

Shunxing Bao
Adrian Barbu
Sumana Basu
Deepti Bathula
Christian Baumgartner
John Baxter
Sharareh Bayat
Bahareh Behboodi
Hamid Behnam
Sutanu Bera
Christos Bergeles
Jose Bernal
Gabriel Bernardino
Alaa Bessadok
Riddhish Bhalodia
Indrani Bhattacharya
Chitresh Bhushan
Lei Bi
Qi Bi
Gui-Bin Bian
Alexander Bigalke
Ricardo Bigolin Lanfredi
Benjamin Billot
Ryoma Bise
Sangeeta Biswas
Stefano B. Blumberg
Sebastian Bodenstedt
Bhushan Borotikar
Ilaria Boscolo Galazzo
Behzad Bozorgtabar
Nadia Brancati
Katharina Breininger
Rupert Brooks
Tom Brosch
Mikael Brudfors
Qirong Bu
Ninon Burgos
Nikolay Burlutskiy
Michał Byra
Ryan Cabeen
Mariano Cabezas
Hongmin Cai
Jinzheng Cai
Weidong Cai
Sema Candemir

Qing Cao
Weiguo Cao
Yankun Cao
Aaron Carass
Ruben Cardenes
M. Jorge Cardoso
Owen Carmichael
Alessandro Casella
Matthieu Chabanas
Ahmad Chaddad
Jayasree Chakraborty
Sylvie Chambon
Yi Hao Chan
Ming-Ching Chang
Peng Chang
Violeta Chang
Sudhanya Chatterjee
Christos Chatzichristos
Antong Chen
Chao Chen
Chen Chen
Cheng Chen
Dongdong Chen
Fang Chen
Geng Chen
Hanbo Chen
Jianan Chen
Jianxu Chen
Jie Chen
Junxiang Chen
Junying Chen
Junyu Chen
Lei Chen
Li Chen
Liangjun Chen
Liyun Chen
Min Chen
Pingjun Chen
Qiang Chen
Runnan Chen
Shuai Chen
Xi Chen
Xiaoran Chen
Xin Chen
Xinjian Chen

Xuejin Chen
Yuanyuan Chen
Zhaolin Chen
Zhen Chen
Zhineng Chen
Zhixiang Chen
Erkang Cheng
Jianhong Cheng
Jun Cheng
Philip Chikontwe
Min-Kook Choi
Gary Christensen
Argyrios Christodoulidis
Stergios Christodoulidis
Albert Chung
Özgün Çiçek
Matthew Clarkson
Dana Cobzas
Jaume Coll-Font
Toby Collins
Olivier Commowick
Runmin Cong
Yulai Cong
Pierre-Henri Conze
Timothy Cootes
Teresa Correia
Pierrick Coupé
Hadrien Courtecuisse
Jeffrey Craley
Alessandro Crimi
Can Cui
Hejie Cui
Hui Cui
Zhiming Cui
Kathleen Curran
Claire Cury
Tobias Czempiel
Vedrana Dahl
Tareen Dawood
Laura Daza
Charles Delahunt
Herve Delingette
Ugur Demir
Liang-Jian Deng
Ruining Deng

Yang Deng
Cem Deniz
Felix Denzinger
Adrien Depeursinge
Hrishikesh Deshpande
Christian Desrosiers
Neel Dey
Anuja Dharmaratne
Li Ding
Xinghao Ding
Zhipeng Ding
Ines Domingues
Juan Pedro Dominguez-Morales
Mengjin Dong
Nanqing Dong
Sven Dorkenwald
Haoran Dou
Simon Drouin
Karen Drukker
Niharika D'Souza
Guodong Du
Lei Du
Dingna Duan
Hongyi Duanmu
Nicolas Duchateau
James Duncan
Nicha Dvornek
Dmitry V. Dylov
Oleh Dzyubachyk
Jan Egger
Alma Eguizabal
Gudmundur Einarsson
Ahmet Ekin
Ahmed Elazab
Ahmed Elnakib
Amr Elsawy
Mohamed Elsharkawy
Ertunc Erdil
Marius Erdt
Floris Ernst
Boris Escalante-Ramírez
Hooman Esfandiari
Nazila Esmaeili
Marco Esposito
Théo Estienne

Christian Ewert
Deng-Ping Fan
Xin Fan
Yonghui Fan
Yubo Fan
Chaowei Fang
Huihui Fang
Xi Fang
Yingying Fang
Zhenghan Fang
Mohsen Farzi
Hamid Fehri
Lina Felsner
Jianjiang Feng
Jun Feng
Ruibin Feng
Yuan Feng
Zishun Feng
Aaron Fenster
Henrique Fernandes
Ricardo Ferrari
Lukas Fischer
Antonio Foncubierta-Rodríguez
Nils Daniel Forkert
Wolfgang Freysinger
Bianca Freytag
Xueyang Fu
Yunguan Fu
Gareth Funka-Lea
Pedro Furtado
Ryo Furukawa
Laurent Gajny
Francesca Galassi
Adrian Galdran
Jiangzhang Gan
Yu Gan
Melanie Ganz
Dongxu Gao
Linlin Gao
Riqiang Gao
Siyuan Gao
Yunhe Gao
Zeyu Gao
Gautam Gare
Bao Ge

Rongjun Ge
Sairam Geethanath
Shiv Gehlot
Yasmeen George
Nils Gessert
Olivier Gevaert
Ramtin Gharleghi
Sandesh Ghimire
Andrea Giovannini
Gabriel Girard
Rémi Giraud
Ben Glocker
Ehsan Golkar
Arnold Gomez
Ricardo Gonzales
Camila Gonzalez
Cristina González
German Gonzalez
Sharath Gopal
Karthik Gopinath
Pietro Gori
Michael Götz
Shuiping Gou
Maged Goubran
Sobhan Goudarzi
Alejandro Granados
Mara Graziani
Yun Gu
Zaiwang Gu
Hao Guan
Dazhou Guo
Hengtao Guo
Jixiang Guo
Jun Guo
Pengfei Guo
Xiaoqing Guo
Yi Guo
Yuyu Guo
Vikash Gupta
Prashnna Gyawali
Stathis Hadjidemetriou
Fatemeh Haghighi
Justin Haldar
Mohammad Hamghalam
Kamal Hammouda

Bing Han
Liang Han
Seungjae Han
Xiaoguang Han
Zhongyi Han
Jonny Hancox
Lasse Hansen
Huaying Hao
Jinkui Hao
Xiaoke Hao
Mohammad Minhazul Haq
Nandinee Haq
Rabia Haq
Michael Hardisty
Nobuhiko Hata
Ali Hatamizadeh
Andreas Hauptmann
Huiguang He
Nanjun He
Shenghua He
Yuting He
Tobias Heimann
Stefan Heldmann
Sobhan Hemati
Alessa Hering
Monica Hernandez
Estefania Hernandez-Martin
Carlos Hernandez-Matas
Javier Herrera-Vega
Kilian Hett
David Ho
Yi Hong
Yoonmi Hong
Mohammad Reza Hosseinzadeh Taher
Benjamin Hou
Wentai Hou
William Hsu
Dan Hu
Rongyao Hu
Xiaoling Hu
Xintao Hu
Yan Hu
Ling Huang
Sharon Xiaolei Huang
Xiaoyang Huang

Yangsibo Huang
Yi-Jie Huang
Yijin Huang
Yixing Huang
Yue Huang
Zhi Huang
Ziyi Huang
Arnaud Huaulmé
Jiayu Huo
Raabid Hussain
Sarfaraz Hussein
Khoi Huynh
Seong Jae Hwang
Ilknur Icke
Kay Igwe
Abdullah Al Zubaer Imran
Ismail Irmakci
Benjamin Irving
Mohammad Shafkat Islam
Koichi Ito
Hayato Itoh
Yuji Iwahori
Mohammad Jafari
Andras Jakab
Amir Jamaludin
Mirek Janatka
Vincent Jaouen
Uditha Jarayathne
Ronnachai Jaroensri
Golara Javadi
Rohit Jena
Rachid Jennane
Todd Jensen
Debesh Jha
Ge-Peng Ji
Yuanfeng Ji
Zhanghexuan Ji
Haozhe Jia
Meirui Jiang
Tingting Jiang
Xiajun Jiang
Xiang Jiang
Zekun Jiang
Jianbo Jiao
Jieqing Jiao

Zhicheng Jiao
Chen Jin
Dakai Jin
Qiangguo Jin
Taisong Jin
Yueming Jin
Baoyu Jing
Bin Jing
Yaqub Jonmohamadi
Lie Ju
Yohan Jun
Alain Jungo
Manjunath K N
Abdolrahim Kadkhodamohammadi
Ali Kafaei Zad Tehrani
Dagmar Kainmueller
Siva Teja Kakileti
John Kalafut
Konstantinos Kamnitsas
Michael C. Kampffmeyer
Qingbo Kang
Neerav Karani
Turkay Kart
Satyananda Kashyap
Alexander Katzmann
Anees Kazi
Hengjin Ke
Hamza Kebiri
Erwan Kerrien
Hoel Kervadec
Farzad Khalvati
Bishesh Khanal
Pulkit Khandelwal
Maksim Kholiavchenko
Ron Kikinis
Daeseung Kim
Jae-Hun Kim
Jaeil Kim
Jinman Kim
Won Hwa Kim
Andrew King
Atilla Kiraly
Yoshiro Kitamura
Stefan Klein
Tobias Klinder

Lisa Koch
Satoshi Kondo
Bin Kong
Fanwei Kong
Ender Konukoglu
Aishik Konwer
Bongjin Koo
Ivica Kopriva
Kivanc Kose
Anna Kreshuk
Frithjof Kruggel
Thomas Kuestner
David Kügler
Hugo Kuijf
Arjan Kuijper
Kuldeep Kumar
Manuela Kunz
Holger Kunze
Tahsin Kurc
Anvar Kurmukov
Yoshihiro Kuroda
Jin Tae Kwak
Francesco La Rosa
Aymen Laadhari
Dmitrii Lachinov
Alain Lalande
Bennett Landman
Axel Largent
Carole Lartizien
Max-Heinrich Laves
Ho Hin Lee
Hyekyoung Lee
Jong Taek Lee
Jong-Hwan Lee
Soochahn Lee
Wen Hui Lei
Yiming Lei
Rogers Jeffrey Leo John
Juan Leon
Bo Li
Bowen Li
Chen Li
Hongming Li
Hongwei Li
Jian Li

Jianning Li
Jiayun Li
Jieyu Li
Junhua Li
Kang Li
Lei Li
Mengzhang Li
Qing Li
Quanzheng Li
Shaohua Li
Shulong Li
Weijian Li
Weikai Li
Wenyuan Li
Xiang Li
Xingyu Li
Xiu Li
Yang Li
Yuexiang Li
Yunxiang Li
Zeju Li
Zhang Li
Zhiyuan Li
Zhjin Li
Zi Li
Chunfeng Lian
Sheng Lian
Libin Liang
Peixian Liang
Yuan Liang
Haofu Liao
Hongen Liao
Ruizhi Liao
Wei Liao
Xiangyun Liao
Gilbert Lim
Hongxiang Lin
Jianyu Lin
Li Lin
Tiancheng Lin
Yiqun Lin
Zudi Lin
Claudia Lindner
Bin Liu
Bo Liu

Chuanbin Liu
Daochang Liu
Dong Liu
Dongnan Liu
Fenglin Liu
Han Liu
Hao Liu
Haozhe Liu
Hong Liu
Huafeng Liu
Huiye Liu
Jianfei Liu
Jiang Liu
Jingya Liu
Kefei Liu
Lihao Liu
Mengting Liu
Peirong Liu
Peng Liu
Qin Liu
Qun Liu
Shenghua Liu
Shuangjun Liu
Sidong Liu
Tianrui Liu
Xiao Liu
Xingtong Liu
Xinwen Liu
Xinyang Liu
Xinyu Liu
Yan Liu
Yanbei Liu
Yi Liu
Yikang Liu
Yong Liu
Yue Liu
Yuhang Liu
Zewen Liu
Zhe Liu
Andrea Loddo
Nicolas Loménie
Yonghao Long
Zhongjie Long
Daniel Lopes
Bin Lou

Nicolas Loy Rodas
Charles Lu
Huanxiang Lu
Xing Lu
Yao Lu
Yuhang Lu
Gongning Luo
Jie Luo
Jiebo Luo
Luyang Luo
Ma Luo
Xiangde Luo
Cuong Ly
Ilwoo Lyu
Yanjun Lyu
Yuanyuan Lyu
Sharath M S
Chunwei Ma
Hehuan Ma
Junbo Ma
Wenao Ma
Yuhui Ma
Anderson Maciel
S. Sara Mahdavi
Mohammed Mahmoud
Andreas Maier
Michail Mamalakis
Ilja Manakov
Brett Marinelli
Yassine Marrakchi
Fabio Martinez
Martin Maška
Tejas Sudharshan Mathai
Dimitrios Mavroeidis
Pau Medrano-Gracia
Raghav Mehta
Felix Meissen
Qingjie Meng
Yanda Meng
Martin Menten
Alexandre Merasli
Stijn Michielse
Leo Milecki
Fausto Milletari
Zhe Min

Tadashi Miyamoto
Sara Moccia
Omid Mohareri
Tony C. W. Mok
Rodrigo Moreno
Kensaku Mori
Lia Morra
Aliasghar Mortazi
Hamed Mozaffari
Pritam Mukherjee
Anirban Mukhopadhyay
Henning Müller
Balamurali Murugesan
Tinashe Mutsvangwa
Andriy Myronenko
Saad Nadeem
Ahmed Naglah
Usman Naseem
Vishwesh Nath
Rodrigo Nava
Nassir Navab
Peter Neher
Amin Nejatbakhsh
Dominik Neumann
Duy Nguyen Ho Minh
Dong Ni
Haomiao Ni
Hannes Nickisch
Jingxin Nie
Aditya Nigam
Lipeng Ning
Xia Ning
Sijie Niu
Jack Noble
Jorge Novo
Chinedu Nwoye
Mohammad Obeid
Masahiro Oda
Steffen Oeltze-Jafra
Ayşe Oktay
Hugo Oliveira
Sara Oliveira
Arnau Oliver
Emanuele Olivetti
Jimena Olveres

Doruk Oner
John Onofrey
Felipe Orihuela-Espina
Marcos Ortega
Yoshito Otake
Sebastian Otálora
Cheng Ouyang
Jiahong Ouyang
Xi Ouyang
Utku Ozbulak
Michal Ozery-Flato
Danielle Pace
José Blas Pagador Carrasco
Daniel Pak
Jin Pan
Siyuan Pan
Yongsheng Pan
Pankaj Pandey
Prashant Pandey
Egor Panfilov
Joao Papa
Bartlomiej Papiez
Nripesh Parajuli
Hyunjin Park
Sanghyun Park
Akash Parvatikar
Magdalini Paschali
Diego Patiño Cortés
Mayank Patwari
Angshuman Paul
Yuchen Pei
Yuru Pei
Chengtao Peng
Jialin Peng
Wei Peng
Yifan Peng
Matteo Pennisi
Antonio Pepe
Oscar Perdomo
Sérgio Pereira
Jose-Antonio Pérez-Carrasco
Fernando Pérez-García
Jorge Perez-Gonzalez
Matthias Perkonigg
Mehran Pesteie

Jorg Peters
Terry Peters
Eike Petersen
Jens Petersen
Micha Pfeiffer
Dzung Pham
Hieu Pham
Ashish Phophalia
Tomasz Pieciak
Antonio Pinheiro
Kilian Pohl
Sebastian Pölsterl
Iulia A. Popescu
Alison Pouch
Prateek Prasanna
Raphael Prevost
Juan Prieto
Federica Proietto Salanitri
Sergi Pujades
Kumaradevan Punithakumar
Haikun Qi
Huan Qi
Buyue Qian
Yan Qiang
Yuchuan Qiao
Zhi Qiao
Fangbo Qin
Wenjian Qin
Yanguo Qin
Yulei Qin
Hui Qu
Kha Gia Quach
Tran Minh Quan
Sandro Queirós
Prashanth R.
Mehdi Rahim
Jagath Rajapakse
Kashif Rajpoot
Dhanesh Ramachandram
Xuming Ran
Hatem Rashwan
Daniele Ravì
Keerthi Sravan Ravi
Surreerat Reaungamornrat
Samuel Remedios

Yudan Ren
Mauricio Reyes
Constantino Reyes-Aldasoro
Hadrien Reynaud
David Richmond
Anne-Marie Rickmann
Laurent Risser
Leticia Rittner
Dominik Rivoir
Emma Robinson
Jessica Rodgers
Rafael Rodrigues
Robert Rohling
Lukasz Roszkowiak
Holger Roth
Karsten Roth
José Rouco
Daniel Rueckert
Danny Ruijters
Mirabela Rusu
Ario Sadafi
Shaheer Ullah Saeed
Monjoy Saha
Pranjal Sahu
Olivier Salvado
Ricardo Sanchez-Matilla
Robin Sandkuehler
Gianmarco Santini
Anil Kumar Sao
Duygu Sarikaya
Olivier Saut
Fabio Scarpa
Nico Scherf
Markus Schirmer
Alexander Schlaefer
Jerome Schmid
Julia Schnabel
Andreas Schuh
Christina Schwarz-Gsaxner
Martin Schweiger
Michaël Sdika
Suman Sedai
Matthias Seibold
Raghavendra Selvan
Sourya Sengupta

Carmen Serrano
Ahmed Shaffie
Keyur Shah
Rutwik Shah
Ahmed Shahin
Mohammad Abuzar Shaikh
S. Shailja
Shayan Shams
Hongming Shan
Xinxin Shan
Mostafa Sharifzadeh
Anuja Sharma
Harshita Sharma
Gregory Sharp
Li Shen
Liyue Shen
Mali Shen
Mingren Shen
Yiqing Shen
Ziyi Shen
Luyao Shi
Xiaoshuang Shi
Yiyu Shi
Hoo-Chang Shin
Boris Shirokikh
Suprosanna Shit
Suzanne Shontz
Yucheng Shu
Alberto Signoroni
Carlos Silva
Wilson Silva
Margarida Silveira
Vivek Singh
Sumedha Singla
Ayushi Sinha
Elena Sizikova
Rajath Soans
Hessam Sokooti
Hong Song
Weinan Song
Youyi Song
Aristeidis Sotiras
Bella Specktor
William Speier
Ziga Spiclin

Jon Sporring
Anuroop Sriram
Vinkle Srivastav
Lawrence Staib
Johannes Stegmaier
Joshua Stough
Danail Stoyanov
Justin Strait
Iain Styles
Ruisheng Su
Vaishnavi Subramanian
Gérard Subsol
Yao Sui
Heung-Il Suk
Shipra Suman
Jian Sun
Li Sun
Liyan Sun
Wenqing Sun
Yue Sun
Vaanathi Sundaresan
Kyung Sung
Yannick Suter
Raphael Sznitman
Eleonora Tagliabue
Roger Tam
Chaowei Tan
Hao Tang
Sheng Tang
Thomas Tang
Youbao Tang
Yucheng Tang
Zihao Tang
Rong Tao
Elias Tappeiner
Mickael Tardy
Giacomo Tarroni
Paul Thienphrapa
Stephen Thompson
Yu Tian
Aleksei Tiulpin
Tal Tlusty
Maryam Toloubidokhti
Jocelyne Troccaz
Roger Trullo

Chialing Tsai
Sudhakar Tummala
Régis Vaillant
Jeya Maria Jose Valanarasu
Juan Miguel Valverde
Thomas Varsavsky
Francisco Vasconcelos
Serge Vasylechko
S. Swaroop Vedula
Roberto Vega
Gonzalo Vegas Sanchez-Ferrero
Gopalkrishna Veni
Archana Venkataraman
Athanasios Vlontzos
Ingmar Voigt
Eugene Vorontsov
Xiaohua Wan
Bo Wang
Changmiao Wang
Chunliang Wang
Clinton Wang
Dadong Wang
Fan Wang
Guotai Wang
Haifeng Wang
Hong Wang
Hongkai Wang
Hongyu Wang
Hu Wang
Juan Wang
Junyan Wang
Ke Wang
Li Wang
Liansheng Wang
Manning Wang
Nizhuan Wang
Qiuli Wang
Renzhen Wang
Rongguang Wang
Ruixuan Wang
Runze Wang
Shujun Wang
Shuo Wang
Shuqiang Wang
Tianchen Wang

Tongxin Wang
Wenzhe Wang
Xi Wang
Xiangdong Wang
Xiaosong Wang
Yalin Wang
Yan Wang
Yi Wang
Yixin Wang
Zeyi Wang
Zuhui Wang
Jonathan Weber
Donglai Wei
Dongming Wei
Lifang Wei
Wolfgang Wein
Michael Wels
Cédric Wemmert
Matthias Wilms
Adam Wittek
Marek Wodzinski
Julia Wolleb
Jonghye Woo
Chongruo Wu
Chunpeng Wu
Ji Wu
Jianfeng Wu
Jie Ying Wu
Jiong Wu
Junde Wu
Pengxiang Wu
Xia Wu
Xiyin Wu
Yawen Wu
Ye Wu
Yicheng Wu
Zhengwang Wu
Tobias Wuerfl
James Xia
Siyu Xia
Yingda Xia
Lei Xiang
Tiange Xiang
Deqiang Xiao
Yiming Xiao

Hongtao Xie
Jianyang Xie
Lingxi Xie
Long Xie
Weidi Xie
Yiting Xie
Yutong Xie
Fangxu Xing
Jiarui Xing
Xiaohan Xing
Chenchu Xu
Hai Xu
Hongming Xu
Jiaqi Xu
Junshen Xu
Kele Xu
Min Xu
Minfeng Xu
Moucheng Xu
Qinwei Xu
Rui Xu
Xiaowei Xu
Xinxing Xu
Xuanang Xu
Yanwu Xu
Yanyu Xu
Yongchao Xu
Zhe Xu
Zhenghua Xu
Zhoubing Xu
Kai Xuan
Cheng Xue
Jie Xue
Wufeng Xue
Yuan Xue
Faridah Yahya
Chaochao Yan
Jiangpeng Yan
Ke Yan
Ming Yan
Qingsen Yan
Yuguang Yan
Zengqiang Yan
Baoyao Yang
Changchun Yang

Chao-Han Huck Yang
Dong Yang
Fan Yang
Feng Yang
Fengting Yang
Ge Yang
Guanyu Yang
Hao-Hsiang Yang
Heran Yang
Hongxu Yang
Huijuan Yang
Jiawei Yang
Jinyu Yang
Lin Yang
Peng Yang
Pengshuai Yang
Xiaohui Yang
Xin Yang
Yan Yang
Yifan Yang
Yujiu Yang
Zhicheng Yang
Jiangchao Yao
Jiawen Yao
Li Yao
Linlin Yao
Qingsong Yao
Chuyang Ye
Dong Hye Ye
Huihui Ye
Menglong Ye
Youngjin Yoo
Chenyu You
Haichao Yu
Hanchao Yu
Jinhua Yu
Ke Yu
Qi Yu
Renping Yu
Thomas Yu
Xiaowei Yu
Zhen Yu
Pengyu Yuan
Paul Yushkevich
Ghada Zamzmi

Ramy Zeineldin

Dong Zeng

Rui Zeng

Zhiwei Zhai

Kun Zhan

Bokai Zhang

Chaoyi Zhang

Daoqiang Zhang

Fa Zhang

Fan Zhang

Hao Zhang

Jianpeng Zhang

Jiawei Zhang

Jingqing Zhang

Jingyang Zhang

Jiong Zhang

Jun Zhang

Ke Zhang

Lefei Zhang

Lei Zhang

Lichi Zhang

Lu Zhang

Ning Zhang

Pengfei Zhang

Qiang Zhang

Rongzhao Zhang

Ruipeng Zhang

Ruisi Zhang

Shengping Zhang

Shihao Zhang

Tianyang Zhang

Tong Zhang

Tuo Zhang

Wen Zhang

Xiaoran Zhang

Xin Zhang

Yanfu Zhang

Yao Zhang

Yi Zhang

Yongqin Zhang

You Zhang

Youshan Zhang

Yu Zhang

Yubo Zhang

Yue Zhang

Yulun Zhang

Yundong Zhang

Yunyan Zhang

Yuxin Zhang

Zheng Zhang

Zhicheng Zhang

Can Zhao

Changchen Zhao

Fenqiang Zhao

He Zhao

Jianfeng Zhao

Jun Zhao

Li Zhao

Liang Zhao

Lin Zhao

Qingyu Zhao

Shen Zhao

Shijie Zhao

Tianyi Zhao

Wei Zhao

Xiaole Zhao

Xuandong Zhao

Yang Zhao

Yue Zhao

Zixu Zhao

Ziyuan Zhao

Xingjian Zhen

Haiyong Zheng

Hao Zheng

Kang Zheng

Qinghe Zheng

Shenhai Zheng

Yalin Zheng

Yinqiang Zheng

Yushan Zheng

Tao Zhong

Zichun Zhong

Bo Zhou

Haoyin Zhou

Hong-Yu Zhou

Huiyu Zhou

Kang Zhou

Qin Zhou

S. Kevin Zhou

Sihang Zhou

Tao Zhou
Tianfei Zhou
Wei Zhou
Xiao-Hu Zhou
Xiao-Yun Zhou
Yanning Zhou
Yaxuan Zhou
Youjia Zhou
Yukun Zhou
Zhiguo Zhou
Zongwei Zhou
Dongxiao Zhu
Haidong Zhu
Hancan Zhu

Lei Zhu
Qikui Zhu
Xiaofeng Zhu
Xinliang Zhu
Zhonghang Zhu
Zhuotun Zhu
Veronika Zimmer
David Zimmerer
Weiwei Zong
Yukai Zou
Lianrui Zuo
Gerald Zwettler
Reyer Zwiggelaar

Outstanding Area Chairs

Ester Bonmati   University College London, UK
Tolga Tasdizen   University of Utah, USA
Yanwu Xu   Baidu Inc., China

Outstanding Reviewers

Seyed-Ahmad Ahmadi   NVIDIA, Germany
Katharina Breininger   Friedrich-Alexander-Universität
                       Erlangen-Nürnberg, Germany
Mariano Cabezas   University of Sydney, Australia
Nicha Dvornek   Yale University, USA
Adrian Galdran   Universitat Pompeu Fabra, Spain
Alexander Katzmann   Siemens Healthineers, Germany
Tony C. W. Mok   Hong Kong University of Science and
                       Technology, China
Sérgio Pereira   Lunit Inc., Korea
David Richmond   Genentech, USA
Dominik Rivoir   National Center for Tumor Diseases (NCT)
                       Dresden, Germany
Fons van der Sommen   Eindhoven University of Technology,
                       the Netherlands
Yushan Zheng   Beihang University, China

Honorable Mentions (Reviewers)

Chloé Audigier   Siemens Healthineers, Switzerland
Qinle Ba   Roche, USA

Meritxell Bach Cuadra	University of Lausanne, Switzerland
Gabriel Bernardino	CREATIS, Université Lyon 1, France
Benjamin Billot	University College London, UK
Tom Brosch	Philips Research Hamburg, Germany
Ruben Cardenes	Ultivue, Germany
Owen Carmichael	Pennington Biomedical Research Center, USA
Li Chen	University of Washington, USA
Xinjian Chen	Soochow University, Taiwan
Philip Chikontwe	Daegu Gyeongbuk Institute of Science and Technology, Korea
Argyrios Christodoulidis	Centre for Research and Technology Hellas/Information Technologies Institute, Greece
Albert Chung	Hong Kong University of Science and Technology, China
Pierre-Henri Conze	IMT Atlantique, France
Jeffrey Craley	Johns Hopkins University, USA
Felix Denzinger	Friedrich-Alexander University Erlangen-Nürnberg, Germany
Adrien Depeursinge	HES-SO Valais-Wallis, Switzerland
Neel Dey	New York University, USA
Guodong Du	Xiamen University, China
Nicolas Duchateau	CREATIS, Université Lyon 1, France
Dmitry V. Dylov	Skolkovo Institute of Science and Technology, Russia
Hooman Esfandiari	University of Zurich, Switzerland
Deng-Ping Fan	ETH Zurich, Switzerland
Chaowei Fang	Xidian University, China
Nils Daniel Forkert	Department of Radiology & Hotchkiss Brain Institute, University of Calgary, Canada
Nils Gessert	Hamburg University of Technology, Germany
Karthik Gopinath	ETS Montreal, Canada
Mara Graziani	IBM Research, Switzerland
Liang Han	Stony Brook University, USA
Nandinee Haq	Hitachi, Canada
Ali Hatamizadeh	NVIDIA Corporation, USA
Samra Irshad	Swinburne University of Technology, Australia
Hayato Itoh	Nagoya University, Japan
Meirui Jiang	The Chinese University of Hong Kong, China
Baoyu Jing	University of Illinois at Urbana-Champaign, USA
Manjunath K N	Manipal Institute of Technology, India
Ali Kafaei Zad Tehrani	Concordia University, Canada
Konstantinos Kamnitsas	Imperial College London, UK

Pulkit Khandelwal	University of Pennsylvania, USA
Andrew King	King's College London, UK
Stefan Klein	Erasmus MC, the Netherlands
Ender Konukoglu	ETH Zurich, Switzerland
Ivica Kopriva	Rudjer Boskovich Institute, Croatia
David Kügler	German Center for Neurodegenerative Diseases, Germany
Manuela Kunz	National Research Council Canada, Canada
Gilbert Lim	National University of Singapore, Singapore
Tiancheng Lin	Shanghai Jiao Tong University, China
Bin Lou	Siemens Healthineers, USA
Hehuan Ma	University of Texas at Arlington, USA
Ilja Manakov	ImFusion, Germany
Felix Meissen	Technische Universität München, Germany
Martin Menten	Imperial College London, UK
Leo Milecki	CentraleSupelec, France
Lia Morra	Politecnico di Torino, Italy
Dominik Neumann	Siemens Healthineers, Germany
Chinedu Nwoye	University of Strasbourg, France
Masahiro Oda	Nagoya University, Japan
Sebastian Otálora	Bern University Hospital, Switzerland
Michal Ozery-Flato	IBM Research, Israel
Egor Panfilov	University of Oulu, Finland
Bartlomiej Papiez	University of Oxford, UK
Nripesh Parajuli	Caption Health, USA
Sanghyun Park	DGIST, Korea
Terry Peters	Robarts Research Institute, Canada
Theodoros Pissas	University College London, UK
Raphael Prevost	ImFusion, Germany
Yulei Qin	Tencent, China
Emma Robinson	King's College London, UK
Robert Rohling	University of British Columbia, Canada
José Rouco	University of A Coruña, Spain
Jerome Schmid	HES-SO University of Applied Sciences and Arts Western Switzerland, Switzerland
Christina Schwarz-Gsaxner	Graz University of Technology, Austria
Liyue Shen	Stanford University, USA
Luyao Shi	IBM Research, USA
Vivek Singh	Siemens Healthineers, USA
Weinan Song	UCLA, USA
Aristeidis Sotiras	Washington University in St. Louis, USA
Danail Stoyanov	University College London, UK

Ruisheng Su	Erasmus MC, the Netherlands
Liyan Sun	Xiamen University, China
Raphael Sznitman	University of Bern, Switzerland
Elias Tappeiner	UMIT - Private University for Health Sciences, Medical Informatics and Technology, Austria
Mickael Tardy	Hera-MI, France
Juan Miguel Valverde	University of Eastern Finland, Finland
Eugene Vorontsov	Polytechnique Montreal, Canada
Bo Wang	CtrsVision, USA
Tongxin Wang	Meta Platforms, Inc., USA
Yan Wang	Sichuan University, China
Yixin Wang	University of Chinese Academy of Sciences, China
Jie Ying Wu	Johns Hopkins University, USA
Lei Xiang	Subtle Medical Inc, USA
Jiaqi Xu	The Chinese University of Hong Kong, China
Zhoubing Xu	Siemens Healthineers, USA
Ke Yan	Alibaba DAMO Academy, China
Baoyao Yang	School of Computers, Guangdong University of Technology, China
Changchun Yang	Delft University of Technology, the Netherlands
Yujiu Yang	Tsinghua University, China
Youngjin Yoo	Siemens Healthineers, USA
Ning Zhang	Bloomberg, USA
Jianfeng Zhao	Western University, Canada
Tao Zhou	Nanjing University of Science and Technology, China
Veronika Zimmer	Technical University Munich, Germany

Mentorship Program (Mentors)

Ulas Bagci	Northwestern University, USA
Kayhan Batmanghelich	University of Pittsburgh, USA
Hrvoje Bogunovic	Medical University of Vienna, Austria
Ninon Burgos	CNRS - Paris Brain Institute, France
Hao Chen	Hong Kong University of Science and Technology, China
Jun Cheng	Institute for Infocomm Research, Singapore
Li Cheng	University of Alberta, Canada
Aasa Feragen	Technical University of Denmark, Denmark
Zhifan Gao	Sun Yat-sen University, China
Stamatia Giannarou	Imperial College London, UK
Sharon Huang	Pennsylvania State University, USA

Anand Joshi	University of Southern California, USA
Bernhard Kainz	Friedrich-Alexander-Universität Erlangen-Nürnberg, Germany and Imperial College London, UK
Baiying Lei	Shenzhen University, China
Karim Lekadir	Universitat de Barcelona, Spain
Xiaoxiao Li	University of British Columbia, Canada
Jianming Liang	Arizona State University, USA
Marius George Linguraru	Children's National Hospital, George Washington University, USA
Anne Martel	University of Toronto, Canada
Antonio Porras	University of Colorado Anschutz Medical Campus, USA
Chen Qin	University of Edinburgh, UK
Julia Schnabel	Helmholtz Munich, TU Munich, Germany and King's College London, UK
Yang Song	University of New South Wales, Australia
Tanveer Syeda-Mahmood	IBM Research - Almaden Labs, USA
Pallavi Tiwari	University of Wisconsin Madison, USA
Mathias Unberath	Johns Hopkins University, USA
Maria Vakalopoulou	CentraleSupelec, France
Harini Veeraraghavan	Memorial Sloan Kettering Cancer Center, USA
Satish Viswanath	Case Western Reserve University, USA
Guang Yang	Imperial College London, UK
Lequan Yu	University of Hong Kong, China
Miaomiao Zhang	University of Virginia, USA
Rongchang Zhao	Central South University, China
Luping Zhou	University of Sydney, Australia
Lilla Zollei	Massachusetts General Hospital, Harvard Medical School, USA
Maria A. Zuluaga	EURECOM, France

Contents – Part VII

Outcome and Disease Prediction

Surgical Data Science

Surgical Planning and Simulation

Machine Learning – Domain Adaptation and Generalization

Image-Guided Interventions
and Surgery

Real-Time 3D Reconstruction of Human Vocal Folds via High-Speed Laser-Endoscopy

Jann-Ole Henningson[1]($\boxtimes$)(iD), Marc Stamminger[1](iD), Michael Döllinger[2](iD), and Marion Semmler[2](iD)

[1] Friedrich-Alexander-University Erlangen-Nuremberg, Erlangen, Germany
jann-ole.henningson@fau.de
[2] Division of Phoniatrics and Pediatric Audiology at the Department of Otorhinolaryngology, Head and Neck Surgery, University Hospital Erlangen, Friedrich-Alexander-University Erlangen-Nuremberg, 91054 Erlangen, Germany

Abstract. Conventional video endoscopy and high-speed video endoscopy of the human larynx solely provides practitioners with information about the two-dimensional lateral and longitudinal deformation of vocal folds. However, experiments have shown that vibrating human vocal folds have a significant vertical component. Based upon an endoscopic laser projection unit (LPU) connected to a high-speed camera, we propose a fully-automatic and real-time capable approach for the robust 3D reconstruction of human vocal folds. We achieve this by estimating laser ray correspondences by taking epipolar constraints of the LPU into account. Unlike previous approaches only reconstructing the superior area of the vocal folds, our pipeline is based on a parametric reinterpretation of the M5 vocal fold model as a tensor product surface. Not only are we able to generate visually authentic deformations of a dense vibrating vocal fold model, but we are also able to easily generate metric measurements of points of interest on the reconstructed surfaces. Furthermore, we drastically lower the effort needed for visualizing and measuring the dynamics of the human laryngeal area during phonation. Additionally, we publish the first publicly available labeled in-vivo dataset of laser-based high-speed laryngoscopy videos. The source code and dataset are available at henningson.github.io/Vocal3D/.

Keywords: Human vocal folds · Endoscopy · Structured light

1 Introduction

Human interaction is fundamentally based on the ability to communicate with each other [2]. Over the last decades, communication-based professions have

---

Supplementary Information The online version contains supplementary material available at https://doi.org/10.1007/978-3-031-16449-1_1.

L. Wang (Eds.): MICCAI 2022, LNCS 13437, pp. 3–12, 2022.
https://doi.org/10.1007/978-3-031-16449-1_1

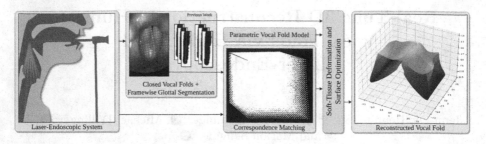

Fig. 1. Our pipeline for real-time reconstruction of vocal folds during phonation. First, we segment images into vocal fold and glottal area. Next, we estimate laser correspondences by taking the systems epipolar constraints into account. Lastly, we generate dense reconstructions by optimizing a parametric vocal fold model using a combined soft-tissue deformation and least squares surface fitting step.

increased drastically and up to 10% of the Western worlds workforce are now classified as heavy occupational voice users [22]. Hence, a lasting impairment of our oral expression is necessarily accompanied by severe social and economic limitations [5,14], increasing the significance of diagnosing laryngeal and voice-related disorders. Laryngeal disorders, impairing speech production, result from different causes ranging from functional abnormalities in the dynamic process as well as morphological alterations in the anatomical structures. Conventionally, human vocal folds have been examined using standard video endoscopy. However, when the dynamics of the vocal folds are of interest (i.e. the vocal folds during **phonation**), High-Speed Video Endoscopy (HSV) is generally used, as the frequency of the vibration necessitates a high temporal resolution of the recording camera [7,10,11,18]. Döllinger et al. [4] have shown that moving vocal folds have a significant vertical expansion in its motion, leading to the assumption that not only the lateral and longitudinal deformation of vocal folds during phonation is of interest, but also their vertical deformation. However, classical video-endoscopy and HSV-Imaging can not resolve the vertical deformation. In common literature, two systems are used for reconstructing the surface of the vocal folds during phonation. Either **a)** stereo-endoscopy systems [23,26], which suffer from a lack of feature points on the smooth tissue inside the larynx [25] or (as in our case) **b)** structured light supported endoscopy which projects a symmetric pattern onto the surface of the vocal folds [13,16,20,21], which can then be used for stereo triangulation. To reach clinical applicability, it is necessary that these systems work robustly without any human input. However, the methods proposed in [13,16,20,21] do require human input in form of a time-consuming manual labeling step. Thus, based upon a laser-based endoscopic high-speed video system [20], we propose the first fully-automatic real-time capable pipeline for reconstructing human vocal folds during phonation. An overview of our pipeline is shown in Fig. 1. Furthermore, we publish a dataset containing laser-endoscopy videos of 10 healthy subjects that can be used to drive further

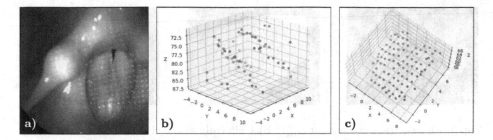

Fig. 2. a) Extracted local maxima lying on the surface of the vocal folds **b)** Reconstructed points by Mask Sweeping generated using Epipolar Constraints, **c)** Globally aligned and optimized triangulation using RHC. (Color figure online)

research in this area. We believe this work is taking a big step towards integrating 3D video endoscopy into the clinical routine.

2   Method

Our reconstruction pipeline can be properly divided into three significant parts, as shown in Fig. 1. It receives high-speed video images as input, that are taken with a calibrated laser-endoscopy system consisting of a high-speed video camera (4000 Hz, resolution 256 × 512) and an LPU projecting a symmetric grid of laser rays. A short video sequence (~100–200 ms) is recorded and passed to our pipeline. First, vocal folds and the glottal area are segmented and a region of interest is extracted. We use this segmentation to find a frame where the vocal folds are closed and quasi-planar. Secondly, we generate first laser ray correspondences in this frame by utilizing epipolar constraints given by the laser-endoscopy system. We globally align and refine these correspondences, using a novel RANSAC-based discrete hill climbing (RHC) approach. Third, by stereo triangulation, we generate per frame 3D points on the surface of the vocal folds. These are used to fit a novel 3D B-Spline model of the vocal folds, based on the M5 model [19] to generate realistically moving vocal folds. For step 1 we apply the technique of Koç et al. [12]. Steps 2 and 3 are presented in the following sections. Our pipeline is designed to be robust and real-time capable, such that immediate feedback is given to the user about the success of the recording.

2.1   Correspondence Matching via Epipolar Constraints

Based on a segmentation of the glottis and the vocal folds, we extract the laser points lying on the superior area of the vocal folds. To this end, we use dilatation filtering on the ROI of the vocal folds to find local maxima corresponding to the projected laser rays and weight the estimated local areas by their intensities for a sub-pixel accurate centroid calculation. Note that we only apply this part of the pipeline to the estimated frame in which the glottal area is minimal, as we

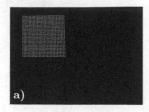 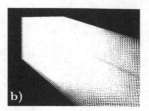 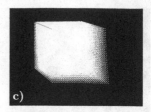

Fig. 3. a) Laser rays sampled at $u_{x,y}^{\infty,\infty}$ **b)** Uniformly sampled laser rays from $u_{x,y}^{0,\infty}$ **c)** Uniformly sampled laser rays from $u_{x,y}^{r,s}$ where $u_{0,0}^{r,s}$ is highlighted in green. By taking the epipolar constraints into account, the search space can be drastically reduced and laser correspondences can be found by sampling from $u_{x,y}^{r,s}$.

then use a temporal nearest neighbor search to label all of the remaining frames. The goal now is to find the proper correspondences between laser rays and their projected points in the image.

Epipolar Constraint-based Search Space Reduction. We assume that we have a calibration between high-speed camera and LPU and that their relative position is static. Thus, we can project each point on the laser ray $r_{x,y}(t) = o_l + t * d_{x,y}$ with distance t from the rays origin o_l to the point $u_{x,y}^t$ in image space, where $(x,y) \in [0 \dots M-1, 0 \dots N-1]$ are the indices of the ray's position in the laser grid. Thus, we can restrict the search space for the laser dots p_i to the epipolar line $u_{x,y}^{0,\infty} := u_{x,y}^0 \rightarrow u_{x,y}^\infty$. However, due to the high density of the lasergrid, we have to restrict the search space further, to be able to disambiguate all correspondences. In general, one can generate first depth estimations of fronto-parallel surfaces by measuring the laser points extent in image space. As a laser ray can be assumed to be collimated for close distances, the projected laser points circumference is inversely proportional to the surfaces depth. In our setup however, the image resolution does not allow for such a depth estimation. Instead, we use the observation of Semmler et al. [21], that states that an endoscopes working distance is in between 50 mm to 80 mm above the vocal folds, so we can confine the search space to this depth range. We refer to the projection of the reduced search space as $u_{x,y}^{r,s}$. A visual representation of this is given in Fig. 3. Note that, as the LPU projects a symmetric laser grid onto the vocal folds surface, neighboring search spaces overlap. Thus, a disambiguation is still necessary.

Estimating Laser Grid Correspondences. Let $P = \{p_i\}$ be the set of extracted local laser points. We then want to generate initial laser point - laser ray mappings, i.e. we need to know which p_i corresponds to which laser ray $r_{x,y}$. To this end, we rasterize the line $u_{x,y}^{r,s}$ for each ray $r_{x,y}$ into the image, whereas the mask of $u_{x,y}^{r,s}$ is directly dependent on the radius of the projected laser dots. Whenever a probable dot p_i is hit, we map it to $r_{x,y}$ and remove the local maximum from P. Note that in this stage, we will still have several wrong correspondences. After this initialization, we can use stereo triangulation to reconstruct the 3D world coordinates of the laser points. As can be seen in Fig. 2, the triangulated points

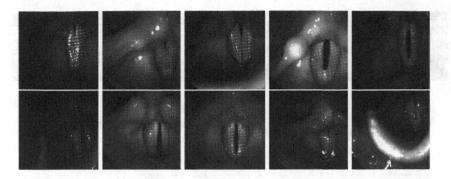

Fig. 4. The set of diverse videos contained in the labeled HLE dataset.

do not have any global alignment, due to overlapping search spaces. Thus, many correspondences are mislabeled by a small offset. To globally align the correspondences, we randomly select a single local maximum $p_i \in P$ and use a recursive grid-based search to label the remaining ones. Based on the selected starting point, all of the consecutively labeled local laser points may now be mislabeled by a discrete static offset. To find the correct labeling we propose RHC, a method specifically designed for labelling symmetric laser grids.

RANSAC-based Hill Climbing for discrete Correspondence Optimization. In this step of the pipeline, the found correspondences are globally aligned, but based on the chosen starting point for the grid-based search, the correspondences may only be close to their optimal solution. Thus, we search a static grid-based offset $a, b \in \mathbb{Z}$ such that the reprojection error $\mathcal{E}(x, \tilde{x})$ is minimized, i.e. $\min_{a,b} \sum_{p_i \in P} \mathcal{E}(p_i, r_{x+a, y+b})$. Where $\mathcal{E}(x, \tilde{x})$ is the Euclidean reprojection error of point x in image space and the reprojection of the intersection between the laser ray $\tilde{x}$ and the camera ray stemming from x. Instead of just naively brute-forcing a global optimum for a and b that might be inaccurate due to outliers, we propose a RANSAC-based [8] recursive hill climbing algorithm to find optimal parameters a and b. RHC optimizes the labels in such a manner that a local grid-based minimum is found. First, to be robust against outliers, e.g. local maxima stemming from specular reflections, we take a random subset $\tilde{P} \subseteq P$ of local maxima and their corresponding labels $\tilde{x}, \tilde{y} \in \tilde{X} \subseteq X$ and calculate their reprojection error $e = \mathcal{E}(\tilde{P}, r_{\tilde{x},\tilde{y}})$. We then calculate the reprojection error of the labels inside the 4-neighborhood of $r_{\tilde{x},\tilde{y}}$ and recurse in the direction of the smallest error $\hat{e} = \arg\min_{a,b} \mathcal{E}(\tilde{P}, r_{\tilde{x}+a, \tilde{y}+b})$. If $e < \tilde{e}$, we stop the recursion, otherwise, we repeat this process with $e = \mathcal{E}(\tilde{P}, r_{\tilde{x}+a, \tilde{y}+b})$ until a local minimum has been found. This algorithm is then repeated for different subsets $\tilde{P}$, until a convergence criterion or the maximum amount of iterations has been reached. Next, we update all of the labels based on the discrete labeling-offset a, b that produced the smallest reprojection error.

Finally, as the videos are of very short duration, we can assume the camera to be almost static. Thus, we can use a temporal nearest neighbor search on

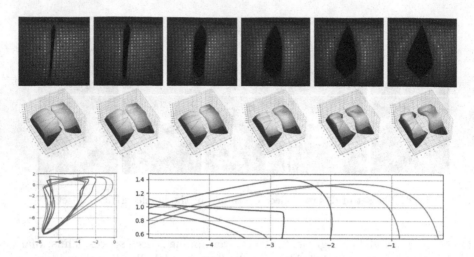

Fig. 5. First Row: Input to our pipeline. Second Row: Framewise reconstruction of a vocal fold model and visualization of the geodesic curvatures to measure. Third Row: Acquired cross sections and geodesic curvatures.

consecutive frames to label all of the remaining images and compute frame-wise point clouds of the superior vocal fold surface using stereo triangulation.

2.2 Surface Reconstruction

Goal of this step is to generate dense moving vocal folds to better guide practitioners in diagnosing laryngeal disorders, while simultaneously improving comprehensibility and plausibility of the data. To achieve real-time performance for reconstructing dense vocal fold models, we extend the M5 Model by Scherer et al. [19] to 3D using B-splines, such that we can reduce the amount of parameters to optimize. Formally, a B-spline surface is a piecewise polynomial function, where a surface point parameterized by (u, v) is defined by $\mathbf{S}(u, v) = \sum_{i=0}^{n} \sum_{j=0}^{m} N_i^p(u) N_j^q(v) \mathbf{P}_{ij}$. Here, $N_i^p(u)$ is the i-th polynomial basis function, where p is the degree in u direction and $\mathbf{P}_{ij}$ the set of control points building the surfaces convex hull. In case of NURBS and B-Splines, they are commonly computed using the Cox-De-Boor recursion formula [17]. Given the piecewise definition of the 2D-based M5 model in [19], we propose a B-Spline surface based M5 vocal fold model (BM5). To generate the BM5 we first subdivide each piece of the parametric function n times and generate points p_i lying on the M5's surface. We define the knots $u \in U$ to be uniformly sampled in the interval $[0, 1]$. Next, we extrude the parametric surface in z-direction. We define the knots of the knot vector $v_i \in V$ similar to U. Lastly, we define the control points of the BM5 to be exactly the points P lying on the extruded surface.

Surface Optimization from Sparse Samples. Let T be the set of triangulated points $t_i \in \mathbb{R}^3$. We then fit a plane to the triangulated points t_i and project the

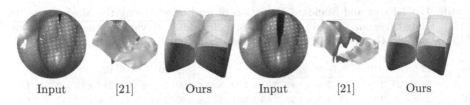

Input   [21]   Ours   Input   [21]   Ours

Fig. 6. Qualitative comparison of the approach by Semmler et al. [21] and our method using images of the HLE Dataset as input.

extreme points of the glottal midline as well as the glottal outline into the point cloud, generating $\hat{T}$. Next, we align T and $\hat{T}$ such that the glottal midline lies on the z-axis to generate a BM5 that lies directly below T. Next, we compute $\arg\min_k \|\hat{P}_{ij} - \hat{t}_k\|_2$, i.e. the nearest triangulated point $\hat{t}_k \in \hat{T}$ to every control point $\hat{P}_{ij}$ defining the superior surface of the vocal fold. We then use as-rigid-as-possible deformation [24] using the pre-computed mapping for each $\hat{P}_{ij}$ and $\hat{t}_k$ while restricting the deformation of the inferior control points of the BM5. Next, we optimize $\arg\min_P \mathcal{L}(\mathbf{S}(u, v), t_i)$, i.e. minimize the distance between specific points on the parameterized surface $\mathbf{S}(u, v)$ and their nearest neighbor t_i, where $\mathcal{L}(\mathbf{S}(u, v), t_i)$ is the metric to be minimized.

3   Results

We implemented our pipeline in Python, using the NumPy [9], PyTorch [15] and OpenCV [1] libraries. We achieve real-time performance (around 25 fps) on an Intel Core i7-6700K CPU and NVidia Quadro RTX 4000 GPU. For optimizing the BM5, we use the NURBS-Diff module proposed by Prasad et al. [3]. We use the approach by Koç et al. [12] for segmenting the vocal folds and glottal area. Lastly, for calibration we use the method proposed in [21], the systems error measures can be inferred from that work as well.

We propose the **human laser-endoscopic (HLE)** dataset, a labeled dataset consisting of 10 in-vivo monochromatic recordings using a laser-based endoscopic recording setup of 10 healthy subjects (Fig. 4). The videos were recorded using a 4000FPS camera with a spatial resolution of 256×512 pixels and labeled manually using [21], whereas a 18 by 18 symmetric laser pattern is projected into the laryngeal area [20]. Subjects were ordered to make a sustained /i/ vowel during recording. The dataset consists of high-quality to lower-quality recordings containing slight camera movement and under-exposed imagery. An excerpt of the HLE Dataset is given in Fig. 4.

RHC Evaluation. We evaluate the RHC algorithm on 21 videos of silicone M5 vocal folds under distinct viewing angles ($-15°$ to $+15°$) and distances (50 mm to 80 mm), in which a 31 by 31 laser grid is projected onto the vocal fold model [21]. We are interested in the general labelling error of our Mask Sweeping algorithm, the global alignment pass and the RHC algorithm. To show the validity of each

Table 1. L1 Error and Standard Deviation of grid offsets using the Mask Sweeping (MS), Global Alignment (GA) and RHC step for different viewing angles of silicone M5 vocal folds. Ground-truth data was generated manually using [21].

α	$-15°$	$-10°$	$-5°$	$0°$	$5°$	$10°$	$15°$	Average
MS	1.54 ± 1.25	3.16 ± 1.97	3.84 ± 1.81	3.31 ± 1.39	3.34 ± 1.39	3.21 ± 1.63	2.86 ± 1.78	3.26 ± 1.60
GA	2.29 ± 1.29	3.15 ± 1.65	2.60 ± 1.70	2.65 ± 1.41	2.31 ± 1.50	2.54 ± 1.44	3.30 ± 1.57	2.57 ± 1.51
RHC	1.49 ± 1.29	1.67 ± 1.27	1.39 ± 1.26	1.55 ± 0.54	1.26 ± 0.93	1.85 ± 1.33	1.36 ± 1.07	1.51 ± 1.10

step, we compute the averaged per label L1-Norm for 20 randomly generated regions of interest lying inside the laser pattern per video. As we use a grid-based labelling, the L1-Norm shows the general accuracy of our method. For example, in case of a 31 by 31 laser grid, we need to discern 961 different labels. If an algorithm now estimates (n, m) for every label, whereas $(n+1, m)$ would be correct, the averaged L1-Norm is one. The ground-truth labels were generated using the semi-automatic approach proposed in [21] for a direct quantitative comparison between the methods. The results can be seen in Table 1. It shows that the Mask Sweeping algorithm finds solutions close to a local optimum. However, as can be seen, globally aligning the points generally reduces the error, while RHC drastically minimizes the offset of the laser grid. We also tested RHC with an 8- and 12-neighborhood look-up, but couldn't observe any differences in labelling accuracy. In general, our approach works robustly in cases where the laser grid's edge is included in the ROI. However, in cases where the labelling is misaligned, RHC finds a mapping that lies on the epipolar lines $u_{x,y}^{r,s}$, thus still finding a solution capable for visualization purposes as the relative proportions of the triangulation stay intact.

Surface Reconstruction. As there does not exist any real-world ground-truth data for vocal folds during phonation, we're showing image based comparisons of our reconstructions using a silicone based M5 vocal fold model and the HLE dataset. We set the anchor weight of the ARAP algorithm to 10^5 and the iterations to 2. In the surface fitting step we use the Chamfer Distance [6] as a loss function, as we want to maximize the point cloud similarity between the points lying on the superior surface of the vocal folds and the triangulated points. We refine the control points for 5 iterations with a learning rate of 0.5. In Fig. 6 we show a qualitative comparison of the approach by Semmler et al. [21] and our method. It can be seen that in previous works, only the superior surface of the glottis could be reconstructed, while our method also takes the deformation of the inferior part of the vocal folds into account. In Fig. 5 a reconstructed glottal opening cycle is visualized depicting the temporal coherence of our method, as well as measurements taken of its geodesic curvature over time.

4   Conclusion

In this work, we proposed the first fully automatic pipeline for reconstructing dynamic vocal folds during phonation based on a laser-endoscopy system that records high-speed videos. We achieve this through highly specialized algorithms for correspondence estimation between symmetric laser grids and their projection in image space. These algorithms enable the triangulation of thousands of frames in a matter of seconds. Unlike other approaches, we do not only visualize and measure the upper surface of the vocal fold, but instead use a parametric reinterpretation of the M5 vocal fold model for a dense surface reconstruction, that also takes the inferior part of the vocal folds into account. Based on an ARAP-deformation and Least Squares Optimization, we can generate visually appealing reconstructions of vocal folds in real-time. Furthermore, we proposed a dataset that can be used to drive further research in this area.

Acknowledgement. We thank **Florian Güthlein** and **Bernhard Egger** for their valuable feedback. This work was supported by Deutsche Forschungsgemeinschaft (DFG) under grant STA662/6-1 (DFG project number: 448240908).

References

1. Bradski, G.: The OpenCV Library. Dr. Dobb's Journal of Software Tools (2000)
2. Cummins, F.: Voice, (inter-)subjectivity, and real time recurrent interaction. Front. Psychol. **5** (2014). https://doi.org/10.3389/fpsyg.2014.00760, https://www.frontiersin.org/article/10.3389/fpsyg.2014.00760
3. Deva Prasad, A., Balu, A., Shah, H., Sarkar, S., Hegde, C., Krishnamurthy, A.: Nurbs-diff: a differentiable programming module for NURBs. Comput.-Aided Des. **146**, 103199 (2022)
4. Döllinger, M., Berry, D.A., Berke, G.S.: Medial surface dynamics of an in vivo canine vocal fold during phonation. J, Acoust. Soc. Am. **117**(5), 3174–3183 (2005). https://doi.org/10.1121/1.1871772
5. Faap, R., Ruben, R.: Redefining the survival of the fittest: communication disorders in the 21st century. Laryngoscope **110**, 241–241 (2000). https://doi.org/10.1097/00005537-200002010-00010
6. Fan, H., Su, H., Guibas, L.: A point set generation network for 3d object reconstruction from a single image (2016)
7. Fehling, M.K., Grosch, F., Schuster, M.E., Schick, B., Lohscheller, J.: Fully automatic segmentation of glottis and vocal folds in endoscopic laryngeal high-speed videos using a deep convolutional LSTM network. PLoS ONE (2) (2020). https://nbn-resolving.de/urn/resolver.pl?urn=nbn:de:bvb:19-epub-87208-2
8. Fischler, M.A., Bolles, R.C.: Random sample consensus: a paradigm for model fitting with applications to image analysis and automated cartography. Commun. ACM **24**(6), 381–395 (1981). https://doi.org/10.1145/358669.358692
9. Harris, C.R., et al.: Array programming with NumPy. Nature **585**(7825), 357–362 (2020)
10. Kist, A., Dürr, S., Schützenberger, A., Döllinger, M.: Openhsv: an open platform for laryngeal high-speed videoendoscopy. Sci. Rep. **11** (2021). https://doi.org/10.1038/s41598-021-93149-0
11. Kist, A., et al.: A deep learning enhanced novel software tool for laryngeal dynamics analysis. J. Speech Language, Hearing Res. **64**, 1–15 (2021). https://doi.org/10.1044/2021_JSLHR-20-00498
12. Koc, T., Çiloglu, T.: Automatic segmentation of high speed video images of vocal folds. J. Appl. Math. **2014** (2014). https://doi.org/10.1155/2014/818415

13. Luegmair, G., Mehta, D., Kobler, J., Döllinger, M.: Three-dimensional optical reconstruction of vocal fold kinematics using high-speed videomicroscopy with a laser projection system. IEEE Trans. Med. Imaging **34** (2015). https://doi.org/10. 1109/TMI.2015.2445921

14. Merrill, R.M., Roy, N., Lowe, J.: Voice-related symptoms and their effects on quality of life. Ann. Otol. Rhinol. Laryngol. **122**(6), 404–411 (2013). https://doi.org/10. 1177/000348941312200610, https://doi.org/10.1177/000348941312200610, pMID: 23837394

15. Paszke, A., et al.: Pytorch: an imperative style, high-performance deep learning library. In: Wallach, H., Larochelle, H., Beygelzimer, A., d' Alché-Buc, F., Fox, E., Garnett, R. (eds.) Advances in Neural Information Processing Systems, vol. 32, pp. 8024–8035. Curran Associates, Inc. (2019). http://papers.neurips.cc/paper/9015-pytorch-an-imperative-style-high-performance-deep-learning-library.pdf

16. Patel, R., Donohue, K., Lau, D., Unnikrishnan, H.: In vivo measurement of pediatric vocal fold motion using structured light laser projection. J. Voice Off. J. Voice Found. **27**, 463–472 (2013). https://doi.org/10.1016/j.jvoice.2013.03.004

17. Piegl, L., Tiller, W.: The NURBS Book. Springer, Berlin (1995)

18. Schenk, F., Urschler, M., Aigner, C., Roesner, I., Aichinger, P., Bischof, H.: Automatic glottis segmentation from laryngeal high-speed videos using 3d active contours (2014)

19. Scherer, R.C., Shinwari, D., De Witt, K.J., Zhang, C., Kucinschi, B.R., Afjeh, A.A.: Intraglottal pressure profiles for a symmetric and oblique glottis with a divergence angle of 10 degrees. J. Acoust. Soc. Am. **109**(4), 1616–1630 (2001). https://doi. org/10.1121/1.1333420, https://asa.scitation.org/doi/abs/10.1121/1.1333420

20. Semmler, M., Kniesburges, S., Birk, V., Ziethe, A., Patel, R., Döllinger, M.: 3d reconstruction of human laryngeal dynamics based on endoscopic high-speed recordings. IEEE Trans. Med. Imag. **35**(7), 1615–1624 (2016). https://doi.org/10. 1109/TMI.2016.2521419

21. Semmler, M., et al.: Endoscopic laser-based 3d imaging for functional voice diagnostics. Appl. Sci. **7** (2017). https://doi.org/10.3390/app7060600

22. Snyder, T., Dillow, S.: Digest of education statistics, 2010. nces 2011–015. National Center for Education Statistics (2011)

23. Sommer, D.E., et al.: Estimation of inferior-superior vocal fold kinematics from high-speed stereo endoscopic data in vivo. J. Acoust. Soc. Am. **136**(6), 3290–3300 (2014). https://doi.org/10.1121/1.4900572, https://doi.org/10.1121/1.4900572

24. Sorkine, O., Alexa, M.: As-rigid-as-possible surface modeling, pp. 109–116 (2007). https://doi.org/10.1145/1281991.1282006

25. Stevens Boster, K., Shimamura, R., Imagawa, H., Sakakibara, K.I., Tokuda, I.: Validating stereo-endoscopy with a synthetic vocal fold model. Acta Acustica Unit. Acust. **102**, 745–751 (2016). https://doi.org/10.3813/AAA.918990

26. Tokuda, I., et al.: Reconstructing three-dimensional vocal fold movement via stereo matching. Acoust. Sci. Technol. **34**, 374–377 (2013). https://doi.org/10.1250/ast. 34.374

Self-supervised Depth Estimation in Laparoscopic Image Using 3D Geometric Consistency

Baoru Huang[1,2]([✉]), Jian-Qing Zheng[3,4], Anh Nguyen[5], Chi Xu[1,2],
Ioannis Gkouzionis[1,2], Kunal Vyas[6], David Tuch[6], Stamatia Giannarou[1,2],
and Daniel S. Elson[1,2]

[1] The Hamlyn Centre for Robotic Surgery, Imperial College London, London, UK
Baoru.Huang18@imperial.ac.uk
[2] Department of Surgery and Cancer, Imperial College London, London, UK
[3] The Kennedy Institute of Rheumatology, University of Oxford, Oxford, UK
[4] Big Data Institute, University of Oxford, Oxford, UK
[5] Department of Computer Science, University of Liverpool, Liverpool, UK
[6] Lightpoint Medical Ltd., Chesham, UK

Abstract. Depth estimation is a crucial step for image-guided intervention in robotic surgery and laparoscopic imaging system. Since per-pixel depth ground truth is difficult to acquire for laparoscopic image data, it is rarely possible to apply supervised depth estimation to surgical applications. As an alternative, self-supervised methods have been introduced to train depth estimators using only synchronized stereo image pairs. However, most recent work focused on the left-right consistency in 2D and ignored valuable inherent 3D information on the object in real world coordinates, meaning that the left-right 3D geometric structural consistency is not fully utilized. To overcome this limitation, we present M3Depth, a self-supervised depth estimator to leverage 3D geometric structural information hidden in stereo pairs while keeping monocular inference. The method also removes the influence of border regions unseen in at least one of the stereo images via masking, to enhance the correspondences between left and right images in overlapping areas. Extensive experiments show that our method outperforms previous self-supervised approaches on both a public dataset and a newly acquired dataset by a large margin, indicating a good generalization across different samples and laparoscopes.

Keywords: Self-supervised monocular depth estimation ·
Laparoscopic images · 3D geometric consistency

Supplementary Information The online version contains supplementary material available at https://doi.org/10.1007/978-3-031-16449-1_2.

L. Wang (Eds.): MICCAI 2022, LNCS 13437, pp. 13–22, 2022.
https://doi.org/10.1007/978-3-031-16449-1_2

1   Introduction

Perception of 3D surgical scenes is a fundamental problem in computer assisted surgery. Accurate perception, tissue tracking, 3D registration between intra- and pre-operative organ models, target localization and augmented reality [8,15] are predicated on having access to correct depth information. Range finding sensors such as multi-camera systems or LiDAR that are often employed in autonomous systems and robotics are not convenient for robot-assisted minimally invasive surgery because of the limited port size and requirement of sterilization. Furthermore, strong 'dappled' specular reflections as well as less textured tissues hinder the application of traditional methods [16]. This has led to the exploration of learning-based methods, among which fully convolutional neural networks (CNNs) are particularly successful [1,22].

Since it is challenging to obtain per-pixel ground truth depth for laparoscopic images, there are far fewer datasets in the surgical domain compared with mainstream computer vision applications [2,9]. It is also not a trivial task to transfer approaches that are based on supervised learning to laparoscopic applications due to the domain gap. To overcome these limitations, view-synthesis methods are proposed to provide self-supervised learning for depth estimation [10,15], with no supervision via per-pixel depth data. Strong depth prediction baselines have been established in [3,4,11]. However, all of these methodologies employed left-right consistency and smoothness constraints in 2D, *e.g.* [3,5,7], and ignored the important 3D geometric structural consistency from the stereo images.

Recently, a self-supervised semantically-guided depth estimation method was proposed to deal with moving objects [13], which made use of mutually beneficial cross-domain training of semantic segmentation. Jung *et al.* [12] extended this work by incorporating semantics-guided local geometry into intermediate depth representations for geometric representation enhancement. However, semantic labels are not common in laparoscopic applications except for surgical tool masks, impeding the extension of this work. Mahjourian *et al.* [17] presented an approach for unsupervised learning of depth by enforcing consistency of ego-motion across consecutive frames to infer 3D geometry of the whole scene. However, in laparoscopic applications, the interaction between the surgical tools and tissue creates a dynamic scene, leading to failure of local photometric and geometric consistency across consecutive frames in both 2D and 3D. Nevertheless, the 3D geometry inferred from left and right synchronized images can be assumed identical, allowing adoption of 3D- as well as 2D-loss.

In this paper, we propose a new framework for self-supervised laparoscopic image depth estimation called M3Depth, leveraging not only the left-right consistency in 2D but also the inherent geometric structural consistency of real-world objects in 3D (see Sect. 2.2 for the 3D geometric consistency loss), while enhancing the mutual information between stereo pairs. A U-Net architecture [20] was employed as the backbone and the network was fed with only left image as inputs but was trained with the punitive loss formed by stereo image pairs. To cope with the unseen areas at the image edges that were not visible in both cameras, blind masking was applied to suppress and eliminate outliers and give more emphasis to feature correspondences that lay on the shared vision field. Extensive experiments on both a public dataset and a new experimental dataset demonstrated the effectiveness of this approach and a detailed ablation study indicated the respective positive influence of each proposed novel module on the overall performance.

2   Methodology

2.1   Network Architecture

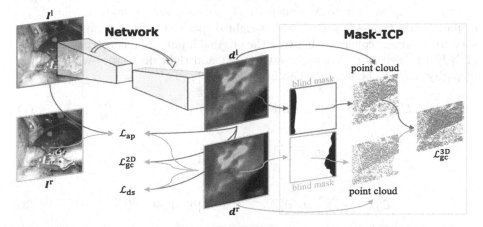

Fig. 1. Overview of the proposed self-supervised depth estimation network. ResNet18 was adopted as the backbone and received a left image from a stereo image pair as the input. Left and right disparity maps were produced simultaneously and formed 2D losses with the original stereo pair. 3D point clouds were generated by applying the intrinsic parameters of the camera and iterative closest point loss was calculated between them. Blind masks were applied to the 2D disparity maps to remove outliers from areas not visible in both cameras.

Network Architecture. The backbone of the M3Depth followed the general U-Net [20] architecture, *i.e.* an encoder-decoder network, in which an encoder was employed to extract image representations while a decoder with convolutional layer and upsampling manipulation was designed to recover disparity maps at the original scale. Skip connections were applied to obtain both deep abstract features and local information. To keep a lightweight network, a ResNet18 [6] was employed as the encoder with only 11 million parameters. To improve the regression ability of the network from intermediate high-dimensional features maps to disparity maps, one more ReLU [18] activation function and a convolutional layer with decreased last latent feature map dimension were added before the final sigmoid disparity prediction. Similar to Monodepth1 [3], in M3Depth, the left image $\boldsymbol{I}^{l}$ of a stereo image pair $(\boldsymbol{I}^{l}, \boldsymbol{I}^{r} \in \mathbb{R}_{+}^{h \times w \times 3})$ was always the input and the framework generated two distinct left and right disparity maps $\boldsymbol{d}^{l}$, $\boldsymbol{d}^{r} \in \mathbb{R}_{+}^{h \times w}$ simultaneously, $i.e.$ $\mathcal{Z} : \boldsymbol{I}^{l} \mapsto (\boldsymbol{d}^{l}, \boldsymbol{d}^{r})$. Given the camera focal length f and the baseline distance b between the cameras, left and right depth maps $\boldsymbol{D}^{l}, \boldsymbol{D}^{r} \in \mathbb{R}_{+}^{h \times w}$ could then be trivially recovered from the predicted disparity, $(\boldsymbol{D}^{l}, \boldsymbol{D}^{r}) = bf/(\boldsymbol{d}^{l}, \boldsymbol{d}^{r})$. h and w denote image height and width. Full details of the architecture are presented in the supplementary material.

Image Reconstruction Loss in 2D. With the predicted disparity maps and the original stereo image pair, left and right images could then be reconstructed by warping the counter-part RGB image with the disparity map mimicking optical flow [3,14]. Similar to Monodepth1 [3], an appearance matching loss $\mathcal{L}_{ap}$, disparity smoothness loss $\mathcal{L}_{ds}$ and left-right disparity consistency loss $\mathcal{L}_{lr}^{2D}$ were used to encourage coherence between the original input and reconstructed images $(\boldsymbol{I}^{l*}, \boldsymbol{I}^{r*})$ as well as consistency between left and right disparities while forcing disparity maps to be locally smooth.

$$\mathcal{L}_{ap}^{r} = \frac{1}{N} \sum_{i,j} \frac{\gamma}{2}(1 - \text{SSIM}(I_{ij}^{r}, I_{ij}^{r*})) + (1 - \gamma)\|I_{ij}^{r} - I_{ij}^{r*}\|_{1} \qquad (1)$$

$$\mathcal{L}_{lr}^{2D(r)} = \frac{1}{N} \sum_{ij} |\mathbf{d}_{ij}^{r} + \mathbf{d}_{ij+\mathbf{d}_{ij}^{r}}^{l}| \qquad (2)$$

$$\mathcal{L}_{ds}^{r} = \frac{1}{N} \sum_{ij} |\partial_{x}(\mathbf{d}_{ij}^{r})|e^{-|\partial_{x}I_{ij}^{r}|} + |\partial_{y}(\mathbf{d}_{ij}^{r})|e^{-|\partial_{y}I_{ij}^{r}|} \qquad (3)$$

where N is the number of pixels and γ was set to 0.85. Note that 2D losses were applied on both left and right images but only equations for the right image are presented here.

2.2   Learning 3D Geometric Consistency

Instead of using the inferred left and right disparities only to establish a mapping between stereo coordinates and generate reconstructed original RGB input images, a loss function was also constructed that registered and compared left and right point clouds directly to enforce the 3D geometric consistency of the whole scene. The disparity maps of the left and right images were first converted to depth maps and then backprojected to 3D coordinates to obtain left and right surgical scene point clouds $(\boldsymbol{P}^{l}, \boldsymbol{P}^{r} \in \mathbb{R}^{hw \times 3})$ by multiplying the depth maps with the intrinsic parameter matrix $(\boldsymbol{K})$. The 3D consistency loss employed Iterative Closest Point (ICP) [17,21], a classic rigid registration method that derives a transformation matrix between two point clouds by iteratively minimizing point-to-point distances between correspondences.

From an initial alignment, ICP alternately computed corresponding points between two input point clouds using a closest point heuristic and then recomputed a more accurate transformation based on the given correspondences. The final residual registration error after ICP minimization was output as one of the returned values. More specifically, to explicitly explore global 3D loss, the ICP loss at the original input image scale was calculated with only 1000 randomly selected points to reduce the computational workload.

2.3   Blind Masking

Some parts of the left scene were not visible in the right view and vice versa, leading to non-overlapping generated point clouds. These areas are mainly located at the left edge of the left image and right edge of the right image after rectification. Depth and image pixels in those area had no useful information for learning, either in 2D and 3D. Our experiments indicated that retaining the contribution to the loss functions for such pixels and voxels degraded the overall performance. Many previous approaches solved this problem by padding these areas with zeros [3] or values from the border [4], but this can lead to edge artifacts in depth images [17].

To tackle this problem, we present a blind masking module $\mathcal{M}^{l,r}$ that suppressed and eliminated these outliers and gave more emphasis to correspondences between the left and right views (Fig. 1). First, a meshgrid was built with the original left image pixel coordinates in both horizontal $\mathcal{X}_{\mathrm{grid}}$ and vertical $\mathcal{Y}_{\mathrm{grid}}$ directions. Then the $\mathcal{X}_{\mathrm{grid}}$ was shifted along the horizontal direction using the right disparity map $\boldsymbol{d}^r$ to a get a new grid $\mathcal{X}'_{\mathrm{grid}}$, which was then stacked with $\mathcal{Y}_{\mathrm{grid}}$ to form a new meshgrid. Finally, grid sampling was employed on the new meshgrid with the help of the original left image coordinates, from which the pixels that were not covered by the right view for the current synchronized image pair were obtained. By applying the blind masking on the depth maps for the stereo 3D point cloud generation, a 3D alignment loss was obtained as follows.

$$M_{ij}^{l,r} = \begin{cases} 1, \ (\mathbf{d}_{ij}^{l,r} + \mathcal{X}_{ij}) \in \{\mathcal{X}\} \\ 0, \ (\mathbf{d}_{ij}^{l,r} + \mathcal{X}_{ij}) \notin \{\mathcal{X}\} \end{cases} \tag{4}$$

$$\boldsymbol{P}^{l,r} = \mathrm{backproj}(\mathbf{d}^{l,r}, \boldsymbol{K}, \boldsymbol{M}^{l,r}) \tag{5}$$

$$\mathcal{L}_{\mathrm{gc}}^{3D} = \mathrm{ICP}(\boldsymbol{P}^l, \boldsymbol{P}^r) \tag{6}$$

2.4   Training Loss

Pixel-wise, gradient-based 2D losses and point cloud-based 3D losses were applied to force the reconstructed image to be identical to the original input while encouraging the left-right consistency in both 2D and 3D to derive accurate disparity maps for depth inference. Finally, an optimization loss used a combination of these, written as:

$$\begin{aligned} \mathcal{L}_{\mathrm{total}} &= (\mathcal{L}_{2D}^r + \mathcal{L}_{2D}^l) + \mathcal{L}_{lr}^{3D} \\ &= \alpha_{\mathrm{ap}}(\mathcal{L}_{\mathrm{ap}}^r + \mathcal{L}_{\mathrm{ap}}^l) + \alpha_{\mathrm{ds}}(\mathcal{L}_{\mathrm{ds}}^r + \mathcal{L}_{\mathrm{ds}}^l) + \alpha_{lr}^{2D}(\mathcal{L}_{lr}^{2D(r)} + \mathcal{L}_{lr}^{2D(l)}) + \beta\mathcal{L}_{\mathrm{gc}}^{3D} \end{aligned} \tag{7}$$

where α_* and β balanced the loss magnitude of the 2D and 3D parts to stabilize the training. More specifically, α_{ap}, α_{ds}, α_{lr}^{2D} and β were experimentally set to 1.0, 0.5, 1.0 and 0.001.

3 Experiments

3.1 Dataset

M3Depth was evaluated on two datasets. The first was the *SCARED* dataset [1] released at the MICCAI Endovis challenge 2019. As only the ground truth depth map of key frames in each dataset was provided (from structured light), the other depth maps were created by reprojection and interpolation of the key frame depth maps using the kinematic information from the *da Vinci* robot, causing a misalignment between the ground truth and the RGB data. Hence, only key-frame ground truth depth maps were used in the test dataset while the remainder of the RGB data formed the training set but with the similar adjacent frames removed.

To overcome the *SCARED* dataset misalignment and improve the validation, an additional laparoscopic image dataset (namely *LATTE*) was experimentally collected, including RGB laparoscopic images and corresponding ground truth depth maps calculated from a custom-built structured lighting pattern projection. More specifically, the gray-code detection and decoder algorithm [23] were used with both original and inverse patterns. To remove the uncertainty brought by occlusions and uneven illumination conditions, we used a more advanced 3-phase detection module, in which sine waves were shifted by $\pi/3$ and $2\pi/3$ and the modulation depth $\mathcal{T}$ was calculated for every pixel. Pixels with modulation depth under $\mathcal{T}$ were defined as uncertain pixels, and the equation for calculating the modulation is written as Eq. 8. This provided 739 extra image pairs for training and 100 pairs for validation and testing.

$$\mathcal{T} = \frac{2\sqrt{2}}{3} \times \sqrt{(\mathcal{I}_1 - \mathcal{I}_2)^2 + (\mathcal{I}_2 - \mathcal{I}_3)^2 + (\mathcal{I}_1 - \mathcal{I}_3)^2} \tag{8}$$

where $\mathcal{I}_1, \mathcal{I}_2, \mathcal{I}_3$ denotes original modulation and modulations after shifts.

Table 1. Quantitative results on the *SCARED* dataset. Metrics labeled with blue headings mean *lower is better* while those labeled with red mean *higher is better*.

Method	Abs Rel	Sq Rel	RMSE	RMSE log	$\delta < 1.25$	$\delta < 1.25^2$	$\delta < 1.25^3$
Mono2 [4]	1.100	74.408	56.548	0.717	0.102	0.284	0.476
PackNet3D [5]	0.733	37.690	32.579	0.649	0.288	0.538	0.722
Mono1 [3]	0.257	20.649	33.796	0.404	0.696	0.837	0.877
M3Depth	**0.116**	**1.822**	**9.274**	**0.139**	**0.865**	**0.983**	**0.997**

3.2   Evaluation Metrics, Baseline, and Implementation Details

Evaluation Metrics. To evaluate depth errors, seven criteria were adopted that are commonly used for monocular depth estimation tasks [3,4]: mean absolute error (Abs Rel), squared error (Sq Rel), root mean squared error (RMSE), root mean squared logarithmic error (RMSE log), and the ratio between ground truth and prediction values, for which the threshold was denoted as δ.

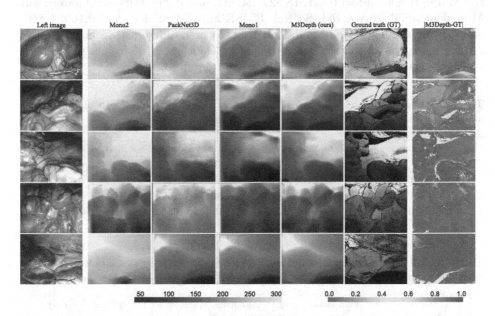

Fig. 2. Qualitative results on the *SCARED* dataset with error map of M3Depth. The depth predictions are all for the left input image. M3Depth generated depth maps with high contrast between the foreground and background and performed better at distinguishing different parts of the scene, reflecting the superior quantitative results in Table 1.

Baseline. The M3Depth model was compared with several recent deep learning methods including Monodepth [3], Monodepth2 [4], and PackNet [5], and both quantitative and qualitative results were generated and reported for comparison. To further study the importance of each M3Depth component, the various components of M3Depth were removed in turn.

Implementation Details. M3Depth was implemented in PyTorch [19], with an input/output resolution of 256×320 and a batch size of 18. The learning rate was initially set to 10^{-5} for the first 30 epochs and was then halved until the end. The model was trained for 50 epochs using the Adam optimizer which took about 65 h on two NVIDIA 2080 Ti GPUs.

4   Results and Discussion

The M3Depth and other state-of-the-art results on the *SCARED* and *LATTE*
dataset are shown in Table 1 and Table 2 using the seven criteria evaluation
metrics. M3Depth outperformed all other methods by a large margin on all seven
criteria, which shows that taking the 3D structure of the world into consideration
benefited the overall performance of the depth estimation task. In particular,
the M3Depth model had 0.141, 18.827, 24.522, and 0.265 units error lower than
Monodepth1 [3] in Abs Rel, Sq Rel, RMSE and RMSE log, and 0.169, 0.146,
and 0.12 units higher than Monodepth1 [3] in three different threshold criteria.
Furthermore, the average inference time of M3Depth was 105 frames per second,
satisfying the real-time depth map generation requirements.

Table 2. Quantitative results on *LATTE* dataset. Metrics labeled with blue headings
mean *lower is better* while labeled by red mean *higher is better*.

Method	Abs Rel	Sq Rel	RMSE	RMSE log	$\delta < 1.25$	$\delta < 1.25^2$	$\delta < 1.25^3$
Mono2 [4]	1.601	306.823	87.694	0.913	0.169	0.384	0.620
PackNet3D [5]	0.960	357.023	259.627	0.669	0.135	0.383	0.624
Mono1 [3]	0.389	57.513	99.020	0.424	0.268	0.709	0.934
M3Depth	**0.236**	**21.839**	**57.739**	**0.245**	**0.665**	**0.893**	**0.969**

Table 3. Ablation study results on the *SCARED* dataset. Metrics labeled with blue
headings mean *lower is better* while those labeled with red mean *higher is better*.

Method	3GC	Blind masking	Abs Rel	Sq Rel	RMSE	RMSE log	$\delta < 1.25$	$\delta < 1.25^2$	$\delta < 1.25^3$
Mono1 [3]	✗	✗	0.257	20.649	33.796	0.404	0.696	0.837	0.877
M3Depth w/ mask	✓	✗	0.150	3.069	13.671	0.249	0.754	0.910	0.956
M3Depth	✓	✓	**0.116**	**1.822**	**9.274**	**0.139**	**0.865**	**0.983**	**0.997**

Detailed ablation study results on the *SCARED* dataset are shown in Table 3
and the impact of the proposed modules, 3D geometric consistency (3GC) and
blind masking were evaluated. The evaluation measures steadily improved when
the various components were added. More specifically, the addition of the blind
masking further boosted the 3GC term, which shows the importance and necessity of removing invalid information from areas that are not visible to both cameras. We note that more quantitative results can be found in our supplementary
material.

Qualitative results comparing our depth estimation results against prior work
using the *SCARED* dataset are presented in Fig. 2. As the sample images shows,
the application of temporal consistency encouraged by the 3D geometric consistency loss can reduce the errors caused by subsurface features, and better
recover the real 3D surface shape of the tissue. Furthermore, depth outputs
from M3Depth show better results along the boundaries of objects, indicating
the effectiveness of the proposed 3GC and blind masking modules.

5   Conclusion

A novel framework for self-supervised monocular laparoscopic images depth estimation was presented. By combining the 2D image-based losses and 3D geometry-based losses from an inferred 3D point cloud of the whole scene, the global consistency and small local neighborhoods were both explicitly taken into consideration. Incorporation of blind masking avoided penalizing areas where no useful information exists. The modules proposed can easily be plugged into any similar depth estimation network, monocular and stereo, while the use of a lightweight ResNet18 backbone enabled real time depth map generation in laparoscopic applications. Extensive experiments on both public and newly acquired datasets demonstrated good generalization across different laparoscopes, illumination condition and samples, indicating the capability to large scale data acquisition where precise ground truth depth cannot be easily collected.

References

1. Allan, M., et al.: Stereo correspondence and reconstruction of endoscopic data challenge. arXiv:2101.01133 (2021)
2. Geiger, A., Lenz, P., Stiller, C., Urtasun, R.: Vision meets robotics: the kitti dataset. Int. J. Robot. Res. **32**(11), 1231–1237 (2013)
3. Godard, C., Mac Aodha, O., Brostow, G.J.: Unsupervised monocular depth estimation with left-right consistency. In: Proceedings of the IEEE Conference on Computer Vision and Pattern Recognition, pp. 270–279 (2017)
4. Godard, C., Mac Aodha, O., Firman, M., Brostow, G.J.: Digging into self-supervised monocular depth estimation. In: Proceedings of the IEEE/CVF International Conference on Computer Vision, pp. 3828–3838 (2019)
5. Guizilini, V., Ambrus, R., Pillai, S., Raventos, A., Gaidon, A.: 3d packing for self-supervised monocular depth estimation. In: Proceedings of the IEEE/CVF Conference on Computer Vision and Pattern Recognition, pp. 2485–2494 (2020)
6. He, K., Zhang, X., Ren, S., Sun, J.: Deep residual learning for image recognition. In: Proceedings of the IEEE Conference on Computer Vision and Pattern Recognition, pp. 770–778 (2016)
7. Huang, B., et al.: Simultaneous depth estimation and surgical tool segmentation in laparoscopic images. IEEE Trans. Med. Robot. Bion. **4**(2), 335–338 (2022)
8. Huang, B., et al.: Tracking and visualization of the sensing area for a tethered laparoscopic gamma probe. Int. J. Comput. Assist. Radiol. Surg. **15**(8), 1389–1397 (2020). https://doi.org/10.1007/s11548-020-02205-z
9. Huang, B., Zheng, J.Q., Giannarou, S., Elson, D.S.: H-net: unsupervised attention-based stereo depth estimation leveraging epipolar geometry. In: Proceedings of the IEEE/CVF Conference on Computer Vision and Pattern Recognition, pp. 4460–4467 (2022)
10. Huang, B., et al.: Self-supervised generative adversarial network for depth estimation in laparoscopic images. In: de Bruijne, M., Cattin, P.C., Cotin, S., Padoy, N., Speidel, S., Zheng, Y., Essert, C. (eds.) MICCAI 2021. LNCS, vol. 12904, pp. 227–237. Springer, Cham (2021). https://doi.org/10.1007/978-3-030-87202-1_22

11. Johnston, A., Carneiro, G.: Self-supervised monocular trained depth estimation using self-attention and discrete disparity volume. In: Proceedings of the IEEE/CVF Conference on Computer Vision and Pattern Recognition, pp. 4756–4765 (2020)
12. Jung, H., Park, E., Yoo, S.: Fine-grained semantics-aware representation enhancement for self-supervised monocular depth estimation. In: Proceedings of the IEEE/CVF International Conference on Computer Vision, pp. 12642–12652 (2021)
13. Klingner, M., Termöhlen, J.-A., Mikolajczyk, J., Fingscheidt, T.: Self-supervised monocular depth estimation: solving the dynamic object problem by semantic guidance. In: Vedaldi, A., Bischof, H., Brox, T., Frahm, J.-M. (eds.) ECCV 2020. LNCS, vol. 12365, pp. 582–600. Springer, Cham (2020). https://doi.org/10.1007/978-3-030-58565-5_35
14. Lipson, L., Teed, Z., Deng, J.: Raft-stereo: multilevel recurrent field transforms for stereo matching. In: 2021 International Conference on 3D Vision (3DV), pp. 218–227. IEEE (2021)
15. Liu, X., et al.: Dense depth estimation in monocular endoscopy with self-supervised learning methods. IEEE Trans. Med. Imaging 39(5), 1438–1447 (2019)
16. Luo, W., Schwing, A.G., Urtasun, R.: Efficient deep learning for stereo matching. In: Proceedings of the IEEE Conference on Computer Vision and Pattern Recognition, pp. 5695–5703 (2016)
17. Mahjourian, R., Wicke, M., Angelova, A.: Unsupervised learning of depth and egomotion from monocular video using 3D geometric constraints. In: Proceedings of the IEEE conference on Computer Vision and Pattern Recognition, pp. 5667–5675 (2018)
18. Nair, V., Hinton, G.E.: Rectified linear units improve restricted boltzmann machines. In: Icml (2010)
19. Paszke, A., et al.: Automatic differentiation in pytorch (2017)
20. Ronneberger, O., Fischer, P., Brox, T.: U-net: convolutional networks for biomedical image segmentation. In: Navab, N., Hornegger, J., Wells, W.M., Frangi, A.F. (eds.) MICCAI 2015. LNCS, vol. 9351, pp. 234–241. Springer, Cham (2015). https://doi.org/10.1007/978-3-319-24574-4_28
21. Rusinkiewicz, S., Levoy, M.: Efficient variants of the ICP algorithm. In: Proceedings Third International conference on 3-D Digital Imaging and Modeling, pp. 145–152. IEEE (2001)
22. Tran, M.Q., Do, T., Tran, H., Tjiputra, E., Tran, Q.D., Nguyen, A.: Light-weight deformable registration using adversarial learning with distilling knowledge. IEEE Trans. Med. Imaging 41, 1443–1453 (2022)
23. Xu, Y., Aliaga, D.G.: Robust pixel classification for 3d modeling with structured light. In: Proceedings of Graphics Interface 2007, pp. 233–240 (2007)

USG-Net: Deep Learning-based Ultrasound Scanning-Guide for an Orthopedic Sonographer

Kyungsu Lee[1], Jaeseung Yang[1], Moon Hwan Lee[1], Jin Ho Chang[1],
Jun-Young Kim[2], and Jae Youn Hwang[1(✉)]

[1] Daegu Gyeongbuk Institute of Science and Technology, Daegu, South Korea
{ks_lee,yjs6813,moon2019,jhchang,jyhwang}@dgist.ac.kr
[2] Daegu Catholic University Medical Center, Daegu, South Korea

Abstract. Ultrasound (US) imaging is widely used in the field of medicine. US images containing pathological information are essential for better diagnosis. However, it is challenging to obtain informative US images because of their anatomical complexity, which is significantly dependent on the expertise of the sonographer. Therefore, in this study, we propose a fully automatic scanning-guide algorithm that assists unskilled sonographers in acquiring informative US images by providing accurate directions of probe movement to search for target disease regions. The main contributions of this study are: (1) proposing a new scanning-guide task that searches for a rotator cuff tear (RCT) region using a deep learning-based algorithm, i.e., ultrasound scanning-guide network (USG-Net); (2) constructing a dataset to optimize the corresponding deep learning algorithm. Multidimensional US images collected from 80 patients with RCT were processed to optimize the scanning-guide algorithm which classified the existence of RCT. Furthermore, the algorithm provides accurate directions for the RCT, if it is not in the current frame. The experimental results demonstrate that the fully optimized scanning-guide algorithm offers accurate directions to localize a probe within target regions and helps to acquire informative US images.

Keywords: Ultrasound imaging · Deep learning · Probe navigation

1 Introduction

A rotator cuff tear (RCT) is a common injury that causes shoulder pain and leads to limited movement of the shoulder joint [20]. Its prognosis is aided by age, history of physical trauma, and the dominant arm. For efficient surgical

---

K. Lee and J. Yang—Contributed equally.

Supplementary Information The online version contains supplementary material available at https://doi.org/10.1007/978-3-031-16449-1_3.

L. Wang (Eds.): MICCAI 2022, LNCS 13437, pp. 23–32, 2022.
https://doi.org/10.1007/978-3-031-16449-1_3

planning, it is necessary to diagnose the precise size, shape, and location of tears in the supraspinatus tendon [6,23]. Hence, noninvasive imaging techniques such as ultrasound (US) and magnetic resonance imaging (MRI) have been widely utilized. However, owing to its limitations, such as the time-consuming scanning process and high medical cost, MRI is not easily accessible to patients. Contrarily, US imaging is widely utilized in clinics because of its diagnostic efficiency using real-time dynamic capture, low cost, and wide availability [8,9,19].

In ultrasonography using computer-aided diagnosis (CAD), deep learning algorithms have been widely employed because of their high diagnostic accuracy [5,11,13,15,16,18,22,24]. In CAD, a sonographer obtains US images by moving a probe and a deep learning network is used to detect diseases in these images. However, because the interpretation of US images is difficult, sonographers, even the experienced ones, have to constantly adjust and localize the probe at the target disease regions [3,7,17]. Therefore, despite the availability of a high-accuracy deep learning model, precise diagnosis is challenging because of the difficulty associated with the localization of the ultrasound probe. To address the aforementioned issues, recent studies have suggested deep learning techniques to guide the movements of probe for localization at the standard scan plane *Scanning-Guide* [7,14]. In particular, Droste *et al.* [7] proposed a deep learning algorithm to accurately predict the movements of the probe toward the standard scan plane by learning the recorded inertial measurement unit (IMU) signals of motion. However, the proposed algorithm exhibits limitations with regard to dataset construction, such as the considerable amount of time needed for generating the ground truth, noise in the IMU signals, and the need for IMU calibration in each trial. Additionally, Li *et al.* [14] proposed a scanning-guide algorithm that enables autonomous navigation toward the standard scan planes by optimizing image quality using deep reinforcement learning. Despite the accurate guidance toward the standard scan plane aided by the proposed deep learning models, unskilled sonographers face difficulties searching for target disease regions. Therefore, a more advanced scanning-guide algorithm that facilitates guidance toward the exact target regions is needed for better diagnosis.

In this study, we propose a deep learning-based scanning-guide algorithm to guide the probe toward the exact region of the target disease (RCT) without using external motion-tracking sensors (e.g., an IMU). In particular, we propose a novel scanning-guide task to search for target disease regions using a corresponding ultrasound scanning-guide network (USG-Net); this network will guide the probe toward the target region without the need for external motion sensors. For more effective guidance, USG-Net utilizes 2D and 3D US images that contain anatomical representations. Hence, we adopt a generative model to predict the 3D volume based on 2D US images. In addition, we develop an automatic dataset construction method and train the proposed network using the constructed dataset. Our data construction method involves the creation of randomly sliced 2D US images and the calculation of the corresponding ground truths. The experimental results illustrate the high accuracy of the USG-Net and demonstrate the feasibility of our proposed network, which has the potential for use in practical ultrasonic diagnosis.

2   Methods

In this section, we introduce the USG-Net and the dataset construction method we used in this study. Figure 2(b) shows an overview of the proposed system. The location of RCT was estimated using the captured 2D US image, and the direction of the next probe movement was provided to the sonographer.

2.1   Dataset Construction

Using a 3D US imaging system [12] developed for the acquisition of 3D shoulder images, skilled sonographers captured 3D US images of RCT from 80 patients. The data were collected in accordance with the Declaration of Helsinki, the protocol was approved by the Ethics Committee of the Institutional Review Board of Daegu Catholic University Medical Center (MDCR-20-007), and the clinical trials for this study were in accordance with the ethical standards. Two expert surgeons annotated the position of the RCT region using 3D US images. 2D planes were sliced from 3D volumetric US images with constraints because it is essential to exclude image planes that cannot be obtained using ultrasound imaging. As shown in Fig. 1, two parallel lines were randomly selected from the bottom and top faces. Along these lines, the 3D US image was sliced into a 2D US image, and two ground truths of RCT existence (G_p) and directions toward the RCT regions (G_d) were computed using the following procedure. The Euclidean distance between the 2D plane and points of RCT was calculated in a 3D vector space. Note that the distance is zero when the target is in the sliced 2D US image, i.e., $G_p = 1$. Otherwise, the distance is nonzero, i.e., $G_p = 0$. In addition, the direction from the 2D plane to the closest point was projected on the top face and categorized into nine directions, namely, LU, L, LD, D, RD, R, RU, U, and S, which correspond to the following probe movements: left-up,

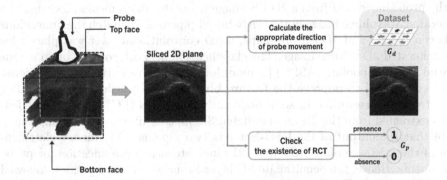

Fig. 1. Description of a proposed dataset construction method. The yellow line on the top face and the pink line on the bottom face are parallel to each other. The 2D plane is sliced using two lines. The dataset consists of the sliced 2D plane, the information about the existence of RCT, and the direction of the probe movement. (Color figure online)

left, left-down, down, right-down, right, right-up, up, and stop, respectively. The predicted direction of probe movement was determined using the obtained US image, which is susceptible to variation owing to the rotation of the probe in Euclidean coordinates.

2.2 Anatomical Representation Learning

During ultrasound imaging, probe movement is determined using anatomical knowledge. Because a 3D US image contains more information than a 2D US image, USG-Net aims to predict 3D anatomical structure information based on the latter, known as anatomical representation learning. To learn 3D anatomical representations, USG-Net adopts a generative model that predicts a 3D volume image from a 2D US image. Hence, USG-Net adopts the DEAR-3D network [21], which is a prediction model based on Wasserstein generative adversarial networks (GAN) [1]. Because it is difficult to generate a large 3D volume image using a single 2D slice, USG-Net generates a thin 3D volume image. In the plane where the 2D US image was sliced on the captured 3D image, voxels with a depth of D were generated in the direction perpendicular to the sliced 2D plane, and $(H + p, W + p)$ voxels were generated in the direction parallel to the sliced 2D plane. H, W, and p represent the height, width, and padding size, respectively. The generated volume forms a small 3D US image of size $(H + p, W + p, D)$. However, USG-Net aims to pre-train the generator in a GAN-based pipeline because the generator generates dummy 3D volume images in the early epochs, and the generative model has properties such as a vanishing gradient or easily occurring mode collapse.

2.3 USG-Net Architecture

As shown in Fig. 2(a), USG-Net classifies the existence of RCT and the direction of the probe movement from a 2D US image using the anatomical knowledge of a generated 3D volume image in the GAN-based pipeline. To match the dimensions between the 2D and 3D feature maps, a 3D convolutional layer was placed for the generated 3D volume image. The classification network consists of an atrous spatial pyramid pooling (ASPP) [4] module for a large receptive field and dense connectivity [10] to improve the feature blurring and vanishing gradient after repeated feed-forwarding. In addition, a fully connected (FC) layer receives features extracted from the 2D and predicted 3D images after concatenating them. Note that the output of the USG-Net has two pipelines: (1) binary classification of the existence of an RCT and (2) nine-categorical classification for probe movements. Hence, the penultimate FC layer branches as two pipelines, followed by softmax layers. Furthermore, after pre-training, USG-Net is trained without the discriminator because the GAN-based model consumes a significant amount of training time and requires a large memory allocation.

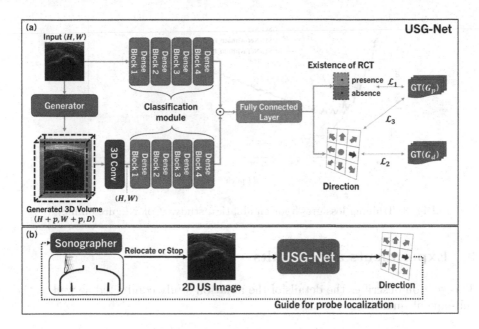

Fig. 2. (a) Overview of USG-Net architecture. (b) Overview of a proposed system. USG-Net guides the direction of probe movement until RCT is observed.

2.4 Loss Function

The loss functions for USG-Net were designed based on the cross-entropy in $\mathcal{L}_1$ and $\mathcal{L}_2$. Let P_p and G_p be the predicted existences of the RCT and the corresponding ground truth, respectively. Similarly, let P_d and G_d be the predicted directions of the probe movement and the corresponding ground truth, respectively. To train the USG-Net, cross-entropy losses are utilized in $\mathcal{L}_1$ using P_p and G_p, and $\mathcal{L}_2$ using P_d and G_d. Furthermore, the correlations ($\mathcal{L}_3$) between G_p and P_d are constrained. When G_p is 1 or 0, the probability of P_d should be more or less dominant in S. Therefore, the knowledge obtained from the classification of the existence of RCT is transferred into another pipeline, such that it is utilized to determine the direction of the probe movement. In summary, the loss function $\mathcal{L}_{total}$ for the network is as follows:

$$\mathcal{L}_{total} = -\left[(1 - G_p)\left\{ log(1 - P_p) + log(1 - P_d{}^s) \right\} + G_p \left(log P_p + log P_d{}^s \right) + \sum_t G_d{}^t log P_d{}^t \right] \quad (1)$$

where $G_p \in \{0, 1\}$ such that 1 indicates that RCT exists. $P_p \in \mathbb{R}$ is in the range of $[0, 1]$. In $G_d{}^t$ and $P_d{}^t$, t represents the direction of probe movement in the nine classes. For instance, $P_d{}^s$ indicates that the next probe movement is classified as S.

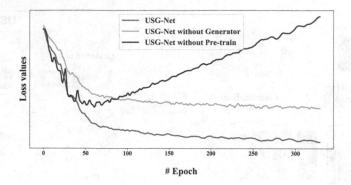

Fig. 3. Training loss graph for an ablation study. (Color figure online)

3   Experiments and Results

This section describes the details of the ablation study conducted as well as the subsequent quantitative analysis.

Ablation Study. To validate the training strategies associated with the generator, we performed an ablation study of the USG-Net for three cases: (1) our proposed network (USG-Net); (2) USG-Net excluding the generator (the trained predictor for 3D anatomical representations); (3) USG-Net using the generator without pretraining. As shown in Fig. 3, the USG-Net with the generator converged to a smaller loss value than without it. In contrast, when the generator was accompanied but not pre-trained, the loss diverged. Therefore, as a small loss convergence value was expected and training could be performed stably, we adopted the pretrained generator (blue line) as the training strategy for the USG-Net.

Experimental Results. To our knowledge, the current study is the first to propose such a method. As such, we did not have any deep-learning models of this type to compare to our proposed method. However, we performed a quantitative analysis of the nine direction classes using 5-fold cross-validation [2].

Table 1 shows the measured evaluation scores for various metrics. The classification performance of the S (stop) class tends to be higher than other direction classes. In addition, standard deviations are minimum between direction classes. The results demonstrate that the USG-Net can successfully predict the path and allow the sonographer to navigate an ultrasound probe toward the target region, thereby detecting RCT. For the USG-Net, high performance in predicting the S class implies that the anatomical representations of shoulder and the synthetic representations of RCT have been successfully learned during the optimization process using the cross-entropy loss between the existence of RCT and the direction of probe movement, as expressed using Eq. 1.

Table 1. Classification performance (Mean ± Standard deviation) of each direction class.

Direction	Sensitivity	Specificity	Precision	Recall	F1 Score	Cohen Kappa
LU	0.81 ± 0.06	0.98 ± 0.01	0.81 ± 0.06	0.81 ± 0.06	0.81 ± 0.06	0.79 ± 0.07
L	0.78 ± 0.10	0.98 ± 0.01	0.80 ± 0.06	0.78 ± 0.10	0.79 ± 0.08	0.76 ± 0.08
LD	0.79 ± 0.05	0.97 ± 0.01	0.78 ± 0.08	0.79 ± 0.05	0.78 ± 0.06	0.76 ± 0.07
D	0.76 ± 0.06	0.97 ± 0.01	0.73 ± 0.05	0.76 ± 0.06	0.75 ± 0.05	0.72 ± 0.06
RD	0.76 ± 0.10	0.97 ± 0.01	0.64 ± 0.08	0.76 ± 0.10	0.69 ± 0.09	0.67 ± 0.10
R	0.76 ± 0.05	0.97 ± 0.01	0.80 ± 0.05	0.76 ± 0.05	0.78 ± 0.05	0.75 ± 0.05
RU	0.75 ± 0.08	0.97 ± 0.01	0.79 ± 0.04	0.75 ± 0.08	0.77 ± 0.06	0.74 ± 0.06
U	0.78 ± 0.06	0.97 ± 0.01	0.76 ± 0.03	0.78 ± 0.06	0.77 ± 0.04	0.74 ± 0.05
S	0.81 ± 0.03	0.98 ± 0.01	0.84 ± 0.04	0.81 ± 0.03	0.83 ± 0.04	0.80 ± 0.04
Average	0.78 ± 0.05	0.97 ± 0.01	0.78 ± 0.05	0.78 ± 0.05	0.78 ± 0.05	0.75 ± 0.05

Furthermore, to analyze the direction predicted by our USG-Net, a confusion matrix for each direction class is shown in Fig. 4(a). Overall, our proposed network achieved an average classification accuracy of 77.86% for the nine classes. Specifically, the classes with large errors were around the true positives that partially included the error directions, indicating that the predicted directions were not completely incorrect (e.g., predicted: R, ground truth: RD). Despite the strict classification criteria, Cohen's kappa values were over 0.7, and the area under the curve (AUC) values were over 0.8 in all the classes, indicating the potential of our proposed network for use as an effective tool for practical diagnosis in clinics (Table 1, Fig. 4(b)). Moreover, statistical verification of the Delong's test reported that the p-value was less than 0.01.

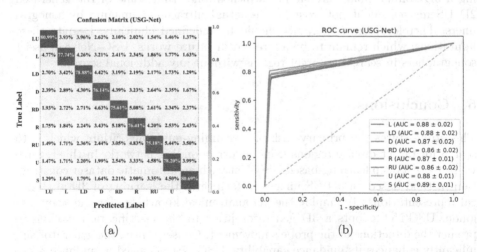

(a)   (b)

Fig. 4. (a) Confusion matrix between 9 direction classes. (b) ROC curve of each direction class.

4   Discussion and Future Work

Dataset Construction. We further propose a novel dataset construction method, involving the generation of 2D US images through random slicing of 3D US images and computation of the ground truth via vector calculation. We also propose a method to generate a dataset for training the scanning-guide algorithm. Notably, in addition to the custom-developed 3D US imaging system for data acquisition, other 3D US imaging systems can also be employed.

3D Anatomical Representation. Because the human body has three-dimensional structures, a 3D anatomical representation is more effective for scanning-guides. However, previous studies predicted the probe's direction using 2D images with motion sensor signals (e.g., IMU). By contrast, the proposed USG-Net utilizes a generative model to process 3D anatomical information without motion sensor signals. Hence, the generative model significantly enhances the classification and guiding performances.

Limitations. (1) In this study, we primarily focused on the development of new methodologies for US scanning-guides based on 3D US images. Although the experimental results were satisfactory for use in RCT diagnosis, validation by expert sonographers is required for the application of USG-Net in clinics. (2) USG-Net was applied to diagnose RCT. However, because the use of the proposed network is not specified in RCT diagnosis, it can be utilized for other diseases (e.g., calcific tendinitis and breast cancer). (3) The constraints for slicing a 3D volume image are simple, which implies that some of the generated 2D US images might not exist in the actual ultrasound imaging by sonographers. Therefore, the constraints should be redefined clinically. Despite these limitations, which remain to be addressed in future works, USG-Net can assist sonographers in searching target regions without any additional sensors.

5   Conclusions

This study states the primary task of a scanning-guide algorithm, which is to search for target disease regions using a corresponding network. In this regard, we propose a deep learning-based USG-Net with an automatic dataset construction method based on 3D US images to facilitates the learning of 3D anatomical representations. To employ the 3D anatomical knowledge for the scanning-guide, USG-Net adopts a 3D generator prior to the classification module to predict the directions of the probe's movement. Consequently, the generator significantly enhances its guidance capability. USG-Net achieved a guidance accuracy of 77.86%. The high evaluation performance demonstrates the effectiveness of the proposed network for this task, suggesting its potential as a novel tool for assisting sonographers in the diagnosis of various diseases using ultrasound imaging.

Acknowledgements. This work was supported by the Technology Innovation Program (No. 2001424) funded By the Ministry of Trade, Industry & Energy (MOTIE, Korea) and the Korea Medical Device Development Fund grant funded by the Korea government (the Ministry of Science and ICT, the Ministry of Trade, Industry and Energy, the Ministry of Health & Welfare, the Ministry of Food and Drug Safety) (Project Number: RS-2020-KD000125, 9991006798).

References

1. Arjovsky, M., Chintala, S., Bottou, L.: Wasserstein generative adversarial networks. In: International Conference on Machine Learning, pp. 214–223. PMLR (2017)
2. Arlot, S., Celisse, A.: A survey of cross-validation procedures for model selection. Stat. Surv. **4**, 40–79 (2010)
3. Baumgartner, C.F., et al.: Sononet: real-time detection and localisation of fetal standard scan planes in freehand ultrasound. IEEE Trans. Med. Imaging **36**(11), 2204–2215 (2017). https://doi.org/10.1109/TMI.2017.2712367
4. Chen, L.C., Papandreou, G., Kokkinos, I., Murphy, K., Yuille, A.L.: Semantic image segmentation with deep convolutional nets and fully connected CRFs. arXiv preprint arXiv:1412.7062 (2014)
5. Chiang, T.C., Huang, Y.S., Chen, R.T., Huang, C.S., Chang, R.F.: Tumor detection in automated breast ultrasound using 3-d cnn and prioritized candidate aggregation. IEEE Trans. Med. Imaging **38**(1), 240–249 (2018)
6. Dalton, S.: The conservative management of rotator cuff disorders (1994)
7. Droste, R., Drukker, L., Papageorghiou, A.T., Noble, J.A.: Automatic probe movement guidance for freehand obstetric ultrasound. In: Martel, A.L., et al. (eds.) MICCAI 2020. LNCS, vol. 12263, pp. 583–592. Springer, Cham (2020). https://doi.org/10.1007/978-3-030-59716-0_56
8. Fenster, A., Parraga, G., Bax, J.: Three-dimensional ultrasound scanning. Interface Focus **1**(4), 503–519 (2011)
9. Gee, A., Prager, R., Treece, G., Berman, L.: Engineering a freehand 3D ultrasound system. Pattern Recogn. Lett. **24**(4–5), 757–777 (2003)
10. Huang, G., Liu, Z., Van Der Maaten, L., Weinberger, K.Q.: Densely connected convolutional networks. In: Proceedings of the IEEE Conference on Computer Vision and Pattern Recognition, pp. 4700–4708 (2017)
11. Huang, Q., Huang, Y., Luo, Y., Yuan, F., Li, X.: Segmentation of breast ultrasound image with semantic classification of superpixels. Med. Image Anal. **61**, 101657 (2020)
12. Lee, M.H., Kim, J.Y., Lee, K., Choi, C.H., Hwang, J.Y.: Wide-field 3d ultrasound imaging platform with a semi-automatic 3d segmentation algorithm for quantitative analysis of rotator cuff tears. IEEE Access **8**, 65472–65487 (2020)
13. Lei, Y., et al.: Ultrasound prostate segmentation based on multidirectional deeply supervised v-net. Med. Phys. **46**(7), 3194–3206 (2019)
14. Li, K., et al.: Autonomous navigation of an ultrasound probe towards standard scan planes with deep reinforcement learning. In: 2021 IEEE International Conference on Robotics and Automation (ICRA), pp. 8302–8308. IEEE (2021)
15. Looney, P., et al.: Fully automated, real-time 3d ultrasound segmentation to estimate first trimester placental volume using deep learning. JCI Insight **3**(11), e120178 (2018)
16. Ouahabi, A., Taleb-Ahmed, A.: Deep learning for real-time semantic segmentation: application in ultrasound imaging. Pattern Recogn. Lett. **144**, 27–34 (2021)

17. Prevost, R., et al.: 3D freehand ultrasound without external tracking using deep learning. Med. Image Anal. **48**, 187–202 (2018)
18. Shin, S.Y., Lee, S., Yun, I.D., Kim, S.M., Lee, K.M.: Joint weakly and semi-supervised deep learning for localization and classification of masses in breast ultrasound images. IEEE Trans. Med. Imaging **38**(3), 762–774 (2018)
19. Song, X., et al.: Cross-modal attention for MRI and ultrasound volume registration. In: de Bruijne, M., et al. (eds.) MICCAI 2021. LNCS, vol. 12904, pp. 66–75. Springer, Cham (2021). https://doi.org/10.1007/978-3-030-87202-1_7
20. Tempelhof, S., Rupp, S., Seil, R.: Age-related prevalence of rotator cuff tears in asymptomatic shoulders. J. Shoulder Elbow Surg. **8**(4), 296–299 (1999)
21. Xie, H., Shan, H., Wang, G.: Deep encoder-decoder adversarial reconstruction (dear) network for 3d ct from few-view data. Bioengineering **6**(4), 111 (2019)
22. Xue, C., et al.: Global guidance network for breast lesion segmentation in ultrasound images. Med. Image Anal. **70**, 101989 (2021)
23. Yamamoto, A., et al.: Prevalence and risk factors of a rotator cuff tear in the general population. J. Shoulder Elbow Surg. **19**(1), 116–120 (2010)
24. Zhou, Y., et al.: Multi-task learning for segmentation and classification of tumors in 3D automated breast ultrasound images. Med. Image Anal. **70**, 101918 (2021)

Surgical-VQA: Visual Question Answering in Surgical Scenes Using Transformer

Lalithkumar Seenivasan[1] , Mobarakol Islam[2] , Adithya K Krishna[3] ,
and Hongliang Ren[1,4,5(✉)]

[1] Department of Biomedical Engineering, National University of Singapore,
Singapore, Singapore
lalithkumar_s@u.nus.edu, ren@nus.edu.sg
[2] Biomedical Image Analysis Group, Imperial College London, London, UK
m.islam20@imperial.ac.uk
[3] Department of ECE, National Institute of Technology, Tiruchirappalli, India
108118004@nitt.edu
[4] Department of Electronic Engineering, Chinese University of Hong Kong,
Hong Kong, Hong Kong
hlren@ee.cuhk.edu.hk
[5] Shun Hing Institute of Advanced Engineering, Chinese University of Hong Kong,
Hong Kong, Hong Kong

Abstract. Visual question answering (VQA) in surgery is largely unexplored. Expert surgeons are scarce and are often overloaded with clinical
and academic workloads. This overload often limits their time answering questionnaires from patients, medical students or junior residents
related to surgical procedures. At times, students and junior residents
also refrain from asking too many questions during classes to reduce dis-
ruption. While computer-aided simulators and recording of past surgical
procedures have been made available for them to observe and improve
their skills, they still hugely rely on medical experts to answer their
questions. Having a Surgical-VQA system as a reliable 'second opinion' could act as a backup and ease the load on the medical experts in
answering these questions. The lack of annotated medical data and the
presence of domain-specific terms has limited the exploration of VQA
for surgical procedures. In this work, we design a Surgical-VQA task
that answers questionnaires on surgical procedures based on the surgical
scene. Extending the MICCAI endoscopic vision challenge 2018 dataset
and workflow recognition dataset further, we introduce two Surgical-
VQA datasets with classification and sentence-based answers. To perform
Surgical-VQA, we employ vision-text transformers models. We further
introduce a residual MLP-based VisualBert encoder model that enforces
interaction between visual and text tokens, improving performance in
classification-based answering. Furthermore, we study the influence of
the number of input image patches and temporal visual features on the
model performance in both classification and sentence-based answering.

L. Seenivasan and M. Islam are co-first authors.

1   Introduction

Lack of medical domain-specific knowledge has left many patients, medical students and junior residents with questions lingering in their minds about medical diagnosis and surgical procedures. Many of these questions are left unanswered either because they assume these questions to be thoughtless, or students and junior residents refrain from raising too many questions to limit disruptions in lectures. The chances for them finding a medical expert to clarify their doubts are also slim due to the scarce number of medical experts who are often overloaded with clinical and academic works [6]. To assist students in sharpening their skills in surgical procedures, many computer-assisted techniques [2,17] and simulators [14,18] have been proposed. Although the systems assist in improving their skills and help reduce the workloads on academic professionals, the systems don't attempt to answer the student's doubts. While students have also been known to learn by watching recorded surgical procedures, the tasks of answering their questions still fall upon the medical experts. In such cases, a computer-assisted system that can process both the medical data and the questionnaires and provide a reliable answer would greatly benefit the students and reduce the medical expert's workload [20]. Surgical scenes are enriched with information that the system can exploit to answer questionnaires related to the defective tissue, surgical tool interaction and surgical procedures.

With the potential to extract diverse information from a single visual feature just by varying the question, the computer vision domain has seen a recent influx of vision and natural language processing models for visual question answering (VQA) tasks [15,23,27]. These models are either built based on the long short-term memory (LSTM) [5,21] or attention modules [20,22,29]. In comparison to the computer vision domain, which is often complemented with massive annotated datasets, the medical domain suffers from the lack of annotated data, limiting the exploration of medical VQA. The presence of domain-specific medical terms also limits the use of transfer learning techniques to adapt pre-trained computer-vision VQA models for medical applications. While limited works have been recently reported on medical-VQA [20] for medical diagnosis, VQA for surgical scenes remains largely unexplored.

In this work, **(i)** we design a Surgical-VQA task to generate answers for questions related to surgical tools, their interaction with tissue and surgical procedures (Fig. 1). **(ii)** We exploit the surgical scene segmentation dataset from

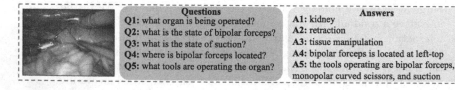

Fig. 1. Surgical-VQA: Given a surgical scene, the model predicts answer related to surgical tools, their interactions and surgical procedures based on the questionnaires.

the MICCAI endoscopic vision challenge 2018 (EndoVis-18) [3] and workflow recognition challenge dataset (Cholec80) [25], and extend it further to introduce two novel datasets for Surgical-VQA tasks. **(iii)** We employ two vision-text attention-based transformer models to perform classification-based and sentence-based answering for the Surgical-VQA. **(iv)** We also introduce a residual MLP (ResMLP) based VisualBERT ResMLP encoder model that outperforms VisualBERT [15] in classification-based VQA. Inspired by ResMLP [24], cross-token and cross-channel sub-modules are introduced into the VisualBERT ResMLP model to enforce interaction among input visual and text tokens. **(v)** Finally, the effects on the model's performance due to the varied number of input image patches and inclusion of temporal visual features are also studied.

2   Proposed Method

2.1   Preliminaries

VisualBERT [15]: A multi-layer transformer encoder model that integrates BERT [9] transformer model with object proposal models to perform vision-and-language tasks. BERT [9] model primarily processes an input sentence as a series of tokens (subwords) for natural language processing. By mapping to a set of embeddings (E), each word token is embedded $(e \in E)$ based on token embedding e_t, segment embedding e_s and position embedding e_p. Along with these input word tokens, VisualBERT [15] model processes visual inputs as unordered visual tokens that are generated using the visual features extracted from the object proposals. In addition to text embedding from BERT [9], it performs visual embedding (F), where, each visual token is embedded $(f \in F)$ based on visual features f_o, segment embedding f_s and position embedding f_p. Both text and visual embedding are then propagated through multiple encoder layers in the VisualBERT [15] to allow rich interactions between both the text and visual tokens and establish joint representation. Each encoder layer consists of an (i) self-attention module that establishes relations between tokens, (ii) intermediate module and (iii) output module consisting of hidden linear layers to reason across channels. Finally, the encoder layer is followed by a pooler module.

2.2   VisualBERT ResMLP

In our proposed VisualBERT ResMLP encoder model, we aim to further boost the interaction between the input tokens for vision-and-language tasks. The VisualBERT [15] model relies primarily on its self-attention module in the encoder layers to establish dependency relationships and allow interactions among tokens. Inspired by residual MLP (ResMLP) [24], the intermediate and output modules of the VisualBERT [15] model are replaced by cross-token and cross-channel modules to further enforce interaction among tokens. In the cross-token module, the inputs word and visual tokens are transposed and propagated forward, allowing information exchange between tokens. The resultant is then transposed back to allow per-token forward propagation in the cross-channel module. Both cross-token and cross-channel modules are followed by element-wise summation with a skip-connection (residual-connection), which are layer-normalized. The cross-token output (X_{CT}) and cross-channel output(X_{CC}) is theorised as:

$$X_{CT} = Norm(X_{SA} + (A((X_{SA})^T))^T) \tag{1}$$
$$X_{CC} = Norm(X_{CT} + C(GeLU(B(X_{CT})))) \tag{2}$$

where, X_{SA} is the self-attention module output, A, B and C are the learnable linear layers, $GeLU$ is the GeLU activation function [12] and NORM is the layer-normalization.

2.3 VisualBert ResMLP for Classification

Word and Visual Tokens: Each question is converted to a series of word tokens generated using BERT tokenizer [9]. Here, the BERT tokenizer is custom trained on the dataset to include surgical domain-specific words. The visual tokens are generated using the final convolution layer features of the feature extractor. A ResNet18 [11] model pre-trained on ImageNet [8] is used as the feature extractor. While the VisualBERT [15] model uses the visual features extracted from object proposals, we bypass the need for object proposal networks by extracting the features from the entire image. By employing adaptive average pooling to the final convolution layer, the output shape (s) is resize to s = [batch size x n x n x 256], thereby, restricting the number of visual tokens

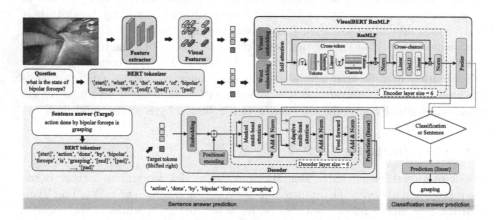

Fig. 2. Architecture: Given an input surgical scene and questions, its text and visual features are propagated through the vision-text encoder (VisualBERT ResMLP). **(i) classification-based answer:** The encoder output is propagated through a prediction layer for answer classification. **(ii) Sentence-based answer:** The encoder is combined with a transformer decoder to predict the answer sentence word-by-word (regressively).

(patches) to n^2. **VisualBERT ResMLP Module:** The word and visual tokens are propagated through text and visual embedding layers, respectively, in the VisualBert ResMLP model. The embedded tokens are then propagated through 6 layers of encoders (comprising self-attention and ResMLP modules) and finally through the pooling module. Embedding size = 300 and hidden layer feature size = 2048 are set as VisualBert ResMLP model parameters. **Classification-based answer:** The output of the pooling module is propagated through a prediction (linear) layer to predict the answer label.

2.4   VisualBert ResMLP for Sentence

Word and Visual Tokens: In addition to tokenizing the words in the questions and visual features in the image as stated in Sect. 2.3, the target sentence-based answer is also converted to a series of word tokens using the BERT tokenizer. **VisualBERT ResMLP encoder:** The VisualBERT ResMLP model follows the same parameters and configuration as stated in Sect. 2.3. **Sentence-based answer:** To generate the answer in a sentence structure, we propose to combine the vision-text encoder model (VisualBERT [15] or VisualBERT ResMLP) with a multi-head attention-based transformer decoder (TD) model. Considering the sentence-based answer generation task to be similar to an image-caption generation task [7], it is positioned as a next-word prediction task. Given the question features, visual features and the previous word (in answer sentence) token, the decoder model predicts the next word in the answer sentence. During the training stage, as shown in Fig. 2, the target word tokens are shifted rights and, together with vision-text encoder output, are propagated through the transformer decoder layers to predict the corresponding next word. During the evaluation stage, the beam search [28] technique is employed to predict the sentence-based answer without using target word tokens. Taking the '[start]' token as the first word, the visual-text encoder model and decoder is regressed on every predicted word to predict the subsequent word, until a '[end]' token is predicted.

3   Experiment

3.1   Dataset

Med-VQA: A public dataset from the ImageCLEF 2019 Med-VQA Challenge [1]. Three categories (C1: modality, C2: plane and C3: organ) of medical question-answer pairs from the dataset are used in this work. The C1, C2 and C3 pose a classification task (single-word answer) for 3825 images with a question. The C1, C2 and C3 consist of 45, 16 and 10 answer classes, respectively. The train and test set split follows the original implementation [1].

EndoVis-18-VQA: A novel dataset was generated from 14 video sequences of robotic nephrectomy procedures from the MICCAI Endoscopic Vision Challenge 2018 [3] dataset. Based on the tool, tissue, bounding box and interaction annotations used for tool-tissue interaction detection tasks [13], we generated two versions of question-answer pairs for each image frame. **(i) Classification (EndoVis-18-VQA (C)):** The answers are in single-word form. It consists of 26 distinct answer classes (one organ, 8 surgical tools, 13 tool interactions and 4 tool locations). **(ii) Sentence (EndoVis-18-VQA (S)):** The answers are in a sentence form. In both versions, 11 sequences with 1560 images and 9014 question-answer pairs are used as a training set and 3 sequences with 447 images and 2769 question-answers pairs are used as a test set. The train and test split follow the tool-tissue interaction detection task [19].

Cholec80-VQA: A novel dataset generated from 40 video sequences of the Cholec80 dataset [25]. We sampled the video sequences at 0.25 fps to generate the Cholec80-VQA dataset consisting of 21591 frames. Based on tool-operation and phase annotations provided in the Cholec80 dataset [25], two question-answer pairs are generated for each frame. **(i) Classification (Cholec80-VQA (C)):** 14 distinct single-word answers (8 on surgical phase, two on tool state and 4 on number of Tools). **(ii) Sentence (Cholec80-VQA (S)):** The answers are in a sentence form. In both versions, 34k question-answer pairs for 17k frames are used for the train set and $9K$ question-answer pairs for 4.5k frames are used for the test set. The train and test set split follows the Cholec80 dataset [25].

3.2   Implementation Details

Both our classification and the sentence-based answer generation models[1] are trained based on cross-entropy loss and optimized using the Adam optimizer. For Classification tasks, a batch size $= 64$ and epoch $= 80$ are used. A learning rate $= 5 \times 10^{-6}$, 1×10^{-5} and 5×10^{-6} are used for Med-VQA [1], EndoVis-18-VQA (C) and Cholec80-VQA (C) dataset, respectively. For the sentence-based answer generation tasks, a batch size $= 50$ is used. The models are trained for epoch $= 50$, 100 and 51, with a learning rate $= 1 \times 10^{-4}$, 5×10^{-5} and 1×10^{-6} on Med-VQA [1], EndoVis-18-VQA (S) and Cholec80-VQA (S) dataset, respectively.

4   Results

The performance of classification-based answering on the EndoVis-18-VQA (C), Cholec80-VQA (C) and Med-VQA (C1, C2 and C3) datasets are quantified based on the accuracy (Acc), recall and F-score in Table 1. It is observed that our proposed encoder (VisualBERT ResMLP) based model outperformed the current medical VQA state-of-the-art MedFuse [20] model and marginally outperformed the base encoder (VisualBERT [15]) based model in almost all datasets. While

---

[1] github.com/lalithjets/Surgical_VQA.git.

Table 1. Performance comparison of our VisualBERT ResMLP based model against MedFuse [20] and VisualBERT [15] based model for classification-based answering.

Dataset	MedFuse [20]			VisualBert [15]			VisualBert ResMLP		
	Acc	Recall	Fscore	Acc	Recall	Fscore	Acc	Recall	Fscore
Med-VQA (C1)	0.754	0.224	0.140	**0.828**	**0.617**	**0.582**	**0.828**	0.598	0.543
Med-VQA (C2)	0.730	0.305	0.303	**0.760**	0.363	0.367	0.758	**0.399**	**0.398**
Med-VQA (C3)	0.652	0.478	0.484	0.734	0.587	0.595	**0.736**	**0.609**	**0.607**
EndoVis-18-VQA (C)	0.609	0.261	0.222	0.619	**0.412**	0.334	**0.632**	0.396	**0.336**
Cholec80-VQA (C)	0.861	0.349	0.309	0.897	**0.629**	0.633	**0.898**	0.627	**0.634**

Table 2. k-fold performance comparison of our VisualBERT ResMLP based model against VisualBERT [15] based model on EndoVis-18-VQA (C) dataset.

Model	1st Fold		2nd Fold		3rd Fold	
	Acc	Fscore	Acc	Fscore	Acc	Fscore
VisualBert [15]	0.619	0.334	0.605	0.313	0.578	0.337
VisualBert ResMLP	**0.632**	**0.336**	**0.649**	**0.347**	**0.585**	**0.373**

the improvement in performance against the base model is marginal, a k-fold study (Table 2) on EndoVis-18-VQA (C) dataset proves that the improvement is consistent. Furthermore, our model (159.0M) requires 13.64% lesser parameters compared to the base model (184.2M).

For the sentence-based answering, BLEU score [16], CIDEr [26] and METEOR [4] are used for quantitative analysis. Compared to the LSTM-based MedFuse [20] model, the two variants of our proposed Transformer-based model

Table 3. Comparison of transformer-based models ((i) VisualBERT [15] + TD and (ii) VisualBERT ResMLP + TD) against MedFuse [20] for sentence-based answering.

Model	EndoVis-18-VQA (S)				Cholec80-VQA (S)			
	BLEU-3	BLEU-4	CIDEr	METEOR	BLEU-3	BLEU-4	CIDEr	METEOR
MedFuse [20]	0.212	0.165	0.752	0.148	0.378	0.333	1.250	0.222
VisualBert [15] + TD	**0.727**	**0.694**	5.153	**0.544**	**0.963**	**0.956**	**8.802**	**0.719**
VisualBert ResMLP + TD	0.722	0.691	**5.262**	0.543	0.960	0.952	8.759	0.711

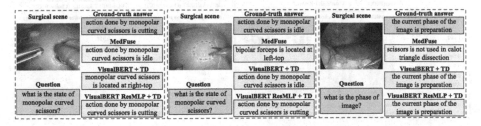

Fig. 3. Comparison of sentence-based answer generation by MedFuse [20], Visual-BERT [15] + TD and VisualBERT ResMLP + TD models.

(VisualBERT [15] + TD and VisualBERT ResMLP + TD) performed better on both the EndoVis-18-VQA (S) and Cholec80-VQA (S) dataset (Table 3). However, when compared within the two variants, the variant with the VisualBERT [15] as its vision-text encoder performed marginally better. While the cross-patch sub-module in VisualBERT ResMLP encoder improves performance in classification-based answering, the marginal low performance could be attributed to its influence on the self-attention sub-module. To predict a sentence-based answer, in addition to the encoder's overall output, the adaptive multi-head attention sub-module in the TD utilizes the encoder's self-attention sub-module outputs. By enforcing interaction between tokens, the cross-patch sub-module could have affected the optimal training of the self-attention sub-module, thereby affecting the sentence generation performance. It is worth noting that the VisualBERT ResMLP encoder + TD model (184.7M) requires 11.98% fewer parameters compared to the VisualBERT [15] + TD model (209.8M) while maintaining similar performances. Figure 3 shows the qualitative performance of sentence-based answering.

4.1 Ablation Study

Firstly, the performance of VisualBERT encoder and VisualBERT ResMLP encoder-based models for classification-based answering with a varying number of input image patches (patch size = 1, 4, 9, 16 and 25) is trained and studied. From Fig. 4 (a), it is observed that VisualBERT ResMLP encoder-based model generally performs better (in-line with observation in Table 1) than VisualBERT [15] encoder-based model, even with varied number of input patches. Secondly, from Fig. 4 (b), it is also observed that performances generally improved with an increase in the number of patches for classification-based

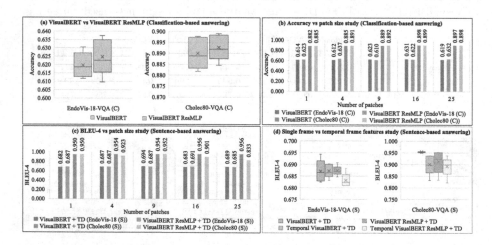

Fig. 4. Ablation study: For classification-based answering (a) Performance comparison of VisualBERT [15] vs VisualBERT ResMLP based model; (b) Accuracy vs patch size (number of patches) study. For sentence-based answering: (c) BLEU-4 vs patch size study; (d) Performance of single frame vs temporal visual features.

answering. However, the influence of the number of inputs patches on sentence-based answering remains inconclusive (Fig. 4 (c)). In most cases, the input of a single patch seems to offer the best or near best results.

Finally, the performance of models incorporated with temporal visual features is also studied. Here, a 3D ResNet50 pre-trained on the UCF101 dataset [10] is employed as the feature extractor. With a temporal size $= 3$, the current and the past 2 frames are used as input to the feature extractor. The extracted features are then used as visual tokens for sentence-based answering. Figure 4 (d) shows the model's performance on a varied number of input patches (patch size $= 1, 4, 9, 16$ and 25) from single frames vs. temporal frames on EndoVis-18-VQA (S) and Cholec80-VQA (S). It is observed that both transformer-based models' performance reduces when the temporal features are used.

5   Discussion and Conclusion

We design a Surgical-VQA algorithm to answer questionnaires on surgical tools, their interactions and surgical procedures based on our two novel Surgical-VQA datasets evolved from two public datasets. To perform classification and sentence-based answering, vision-text attention-based transformer models are employed. A VisualBERT ResMLP transformer encoder model with lesser model parameters is also introduced that marginally outperforms the base vision-text attention encoder model by incorporating a cross-token sub-module. The influence of the number of input image patches and the inclusion of temporal visual features on the model's performance is also reported. While our Surgical-VQA task answers to less-complex questions, from the application standpoint, it unfolds the possibility of incorporating open-ended questions where the model could be trained to answer surgery-specific complex questionnaires. From the model standpoint, future work could focus on introducing an asynchronous training regime to incorporate the benefits of the cross-patch sub-module without affecting the self-attention sub-module in sentence-based answer-generation tasks.

Acknowledgement. We thank Ms. Xu Mengya, Dr. Preethiya Seenivasan and Dr. Sampoornam Thamizharasan for their valuable inputs during project discussions. This work is supported by Shun Hing Institute of Advanced Engineering (SHIAE project BME-p1-21, 8115064) at the Chinese University of Hong Kong (CUHK), Hong Kong Research Grants Council (RGC) Collaborative Research Fund (CRF C4026-21GF and CRF C4063-18G) and (GRS)#3110167.

References

1. Abacha, A.B., Hasan, S.A., Datla, V.V., Liu, J., Demner-Fushman, D., Müller, H.: VQA-med: overview of the medical visual question answering task at imageclef 2019. clef2019 working notes. In: CEUR Workshop Proceedings, pp. 9–12. CEUR-WS.org <http://ceur-ws.org>. September

2. Adams, L., et al.: Computer-assisted surgery. IEEE Comput. Graph. Appl. **10**(3), 43–51 (1990)
3. Allan, M., et al.: 2018 robotic scene segmentation challenge. arXiv preprint arXiv:2001.11190 (2020)
4. Banerjee, S., Lavie, A.: Meteor: an automatic metric for MT evaluation with improved correlation with human judgments. In: Proceedings of the ACL Workshop on Intrinsic and Extrinsic Evaluation Measures for Machine Translation and/or Summarization, pp. 65–72 (2005)
5. Barra, S., Bisogni, C., De Marsico, M., Ricciardi, S.: Visual question answering: which investigated applications? Pattern Recogn. Lett. **151**, 325–331 (2021)
6. Bates, D.W., Gawande, A.A.: Error in medicine: what have we learned? (2000)
7. Cornia, M., Stefanini, M., Baraldi, L., Cucchiara, R.: Meshed-memory transformer for image captioning. In: Proceedings of the IEEE/CVF Conference on Computer Vision and Pattern Recognition, pp. 10578–10587 (2020)
8. Deng, J., Dong, W., Socher, R., Li, L.J., Li, K., Fei-Fei, L.: Imagenet: a large-scale hierarchical image database. In: 2009 IEEE Conference on Computer Vision and Pattern Recognition, pp. 248–255. IEEE (2009)
9. Devlin, J., Chang, M.W., Lee, K., Toutanova, K.: Bert: pre-training of deep bidirectional transformers for language understanding. arXiv preprint arXiv:1810.04805 (2018)
10. Hara, K., Kataoka, H., Satoh, Y.: Learning spatio-temporal features with 3D residual networks for action recognition. In: Proceedings of the IEEE International Conference on Computer Vision Workshops, pp. 3154–3160 (2017)
11. He, K., Zhang, X., Ren, S., Sun, J.: Deep residual learning for image recognition. In: Proceedings of the IEEE Conference on Computer Vision and Pattern Recognition, pp. 770–778 (2016)
12. Hendrycks, D., Gimpel, K.: Gaussian error linear units (gelus). arXiv preprint arXiv:1606.08415 (2016)
13. Islam, M., Seenivasan, L., Ming, L.C., Ren, H.: Learning and reasoning with the graph structure representation in robotic surgery. In: Martel, A.L., et al. (eds.) MICCAI 2020. LNCS, vol. 12263, pp. 627–636. Springer, Cham (2020). https://doi.org/10.1007/978-3-030-59716-0_60
14. Kneebone, R.: Simulation in surgical training: educational issues and practical implications. Med. Educ. **37**(3), 267–277 (2003)
15. Li, L.H., Yatskar, M., Yin, D., Hsieh, C.J., Chang, K.W.: Visualbert: a simple and performant baseline for vision and language. arXiv preprint arXiv:1908.03557 (2019)
16. Papineni, K., Roukos, S., Ward, T., Zhu, W.J.: Bleu: a method for automatic evaluation of machine translation. In: Proceedings of the 40th Annual Meeting of the Association for Computational Linguistics. pp. 311–318 (2002)
17. Rogers, D.A., Yeh, K.A., Howdieshell, T.R.: Computer-assisted learning versus a lecture and feedback seminar for teaching a basic surgical technical skill. Am. J. Surg. **175**(6), 508–510 (1998)
18. Sarker, S., Patel, B.: Simulation and surgical training. Int. J. Clin. Pract. **61**(12), 2120–2125 (2007)
19. Seenivasan, L., Mitheran, S., Islam, M., Ren, H.: Global-reasoned multi-task learning model for surgical scene understanding. IEEE Robot. Autom. Lett. **7**, 3858–3865 (2022)
20. Sharma, D., Purushotham, S., Reddy, C.K.: Medfusenet: an attention-based multimodal deep learning model for visual question answering in the medical domain. Sci. Rep. **11**(1), 1–18 (2021)

21. Sharma, H., Jalal, A.S.: Image captioning improved visual question answering. Multimedia Tools Appl. 1–22 (2021)
22. Sharma, H., Jalal, A.S.: Visual question answering model based on graph neural network and contextual attention. Image Vis. Comput. **110**, 104165 (2021)
23. Sheng, S., et al.: Human-adversarial visual question answering. In: Advances in Neural Information Processing Systems, vol. 34 (2021)
24. Touvron, H., et al.: Resmlp: feedforward networks for image classification with data-efficient training. arXiv preprint arXiv:2105.03404 (2021)
25. Twinanda, A.P., Shehata, S., Mutter, D., Marescaux, J., De Mathelin, M., Padoy, N.: Endonet: a deep architecture for recognition tasks on laparoscopic videos. IEEE Trans. Med. Imaging **36**(1), 86–97 (2016)
26. Vedantam, R., Lawrence Zitnick, C., Parikh, D.: Cider: consensus-based image description evaluation. In: Proceedings of the IEEE Conference on Computer Vision and Pattern Recognition, pp. 4566–4575 (2015)
27. Wang, Z., Yu, J., Yu, A.W., Dai, Z., Tsvetkov, Y., Cao, Y.: SIMVLM: simple visual language model pretraining with weak supervision. arXiv preprint arXiv:2108.10904 (2021)
28. Wiseman, S., Rush, A.M.: Sequence-to-sequence learning as beam-search optimization. arXiv preprint arXiv:1606.02960 (2016)
29. Zhang, S., Chen, M., Chen, J., Zou, F., Li, Y.F., Lu, P.: Multimodal feature-wise co-attention method for visual question answering. Inf. Fusion **73**, 1–10 (2021)

DSP-Net: Deeply-Supervised Pseudo-Siamese Network for Dynamic Angiographic Image Matching

Xi-Yao Ma[1,2], Shi-Qi Liu[1], Xiao-Liang Xie[1,2], Xiao-Hu Zhou[1],
Zeng-Guang Hou[1,2,3(✉)], Yan-Jie Zhou[1,2], Meng Song[1,2], Lin-Sen Zhang[1,4],
and Chao-Nan Wang[5]

[1] State Key Laboratory of Management and Control for Complex Systems,
Institute of Automation, Chinese Academy of Sciences, Beijing 100190, China
zengguang.hou@ia.ac.cn
[2] School of Artificial Intelligence, University of Chinese Academy of Sciences,
Beijing 100049, China
[3] CAS Center for Excellence in Brain Science and Intelligence Technology,
Beijing 100190, China
[4] University of Science and Technology Beijing, Beijing 100083, China
[5] Peking Union Medical College Hospital and Chinese Academy of Medical Science,
Beijing 100730, China

Abstract. During percutaneous coronary intervention (PCI), severe elastic deformation of coronary arteries caused by cardiac movement is a serious disturbance to physicians. It increases the difficulty of estimating the relative position between interventional instruments and vessels, leading to inaccurate operation and higher intraoperative mortality. Providing doctors with dynamic angiographic images can be helpful. However, it often faces the challenges of indistinguishable features between consecutive frames and multiple modalities caused by individual differences. In this paper a novel deeply-supervised pseudo-siamese network (DSP-Net) is developed to solve the problem. A pseudo siamese attention dense (PSAD) block is designed to extract salient features from X-ray images with noisy background, and the deep supervision architecture is integrated to accelerate convergence. Evaluations are conducted on the CVM X-ray Database built by us, which consists of 51 sequences, showing that the proposed network can not only achieve state-of-the-art matching performance of 3.48 Hausdorff distance and 84.09% guidewire recall rate, but also demonstrate the great generality to images with different heart structures or fluoroscopic angles. Exhaustive experiment results indicate that our DSP-Net has the potential to assist doctors to overcome the visual misjudgment caused by the elastic deformation of the arteries and achieve safer procedure.

Keywords: Dynamic angiography · Pseudo-siamese network · Deep supervision

L. Wang (Eds.): MICCAI 2022, LNCS 13437, pp. 44–53, 2022.
https://doi.org/10.1007/978-3-031-16449-1_5

1   Introduction

World Health Organization (WHO) reports that cardiovascular disease (CVD) is the world's leading cause of death [1]. Owing to its advantages including less trauma and better prognosis, percutaneous coronary intervention (PCI) is now the most common treatment for CVDs [12]. During PCI procedures, doctors manipulate interventional instruments such as guidewires to implement necessary treatments to the lesion under the guidance of angiographic images [8].

During PCI, interventional instruments are visible while vessels are invisible under X-ray. Doctors need to inject contrast medium to obtain the relative position of instruments to vessels. However, the angiographic process is ephemeral due to the strong blood pressure, so doctors can only get temporary information and have to perform "blindly" most of the time. Besides, it is infeasible to keep injecting contrast medium during the procedure, as the overdose will lead to acute renal failure to patients who already suffer from renal dysfunction [2,17]. Worse still, severe elastic deformation of the coronary arteries caused by cardiac movement can cause disturbance as shown in Fig. 1(a), which can easily lead doctors to make misjudgments and take inaccurate operations, increasing the risk of puncturing vascular wall with instruments and causing fatal hemorrhage.

One way to solve the problem is to provide X-ray fluoroscopic images with dynamic angiographic images in the same position of a cardiac contraction cycle. However, this task is extremely challenging for the following factors: (1) Vessel deformation between consecutive frames is nuanced, requiring strong discrimination ability of the model. (2) Different heart structures and fluoroscopic angles lead to poor robustness in model performance, as shown in Fig. 1(b). (3) It's difficult to extract representative features from X-ray images with low signal-noise-ratio (SNR). It is found a pivotal problem that restricts doctors from precise intervention procedure, while there are only a few studies investigating vessel matching problems before. Xu et al. [19] presented a system matching sclera vessels, which has higher SNR and fewer modalities, and [14,15] focus on the registration of the abdominal aorta that is hardly influenced by cardiac motion. None of the methods are suitable for this task.

To address the above-mentioned concerns, a deeply-supervised pseudo-siamese network is proposed. It is composed of two same yet independent con-

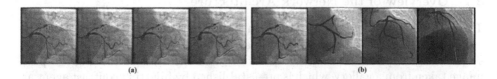

Fig. 1. Examples of angiographic images. (a) 4 consecutive frames of angiographic images. The guidewires in each image are marked in different colors, and they are all projected onto the fourth image. (b) Angiographic images with different heart structures and fluoroscopic angles. The main vessels are marked in red, and the guidewires are marked in cyan. (Color figure online)

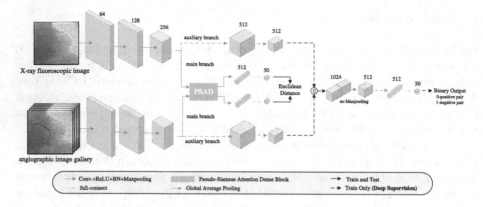

Fig. 2. The architecture of the proposed DSP-Net.

volutional neural networks (CNNs) to obtain information from two inputs separately. A pseudo siamese attention dense (PSAD) block is designed to extract features. Besides the main branch of the network, an auxiliary supervision branch is integrated to improve performance. Finally, a joint optimization strategy is adopted to calculate losses in the two branches respectively.

The main contributions can be summarized as follows:

- To the best of our knowledge, this is the first automatic approach exploring the task of dynamic angiographic image matching problem, showing great potential to achieve more efficient and safer procedure.
- The proposed DSP-Net processes X-ray fluoroscopic and angiographic images parallelly, achieving state-of-the-art performance on our medical image datasets, namely CVM X-ray Database.
- The designed PSAD block successfully distinguishes nuanced frames by learning the representative features and efficiently overcomes the noisy background, showing great generality to images from different sequences.

2 Method

2.1 Overview of the Network Architecture

The architecture of DSP-Net is shown in Fig. 2. It is composed of two identical sub-networks with two kinds of input images. One is the real-time X-ray fluoroscopic image captured during the operation, and the other is the angiographic image taken from a gallery, which is pre-established by injecting contrast agent at the beginning of the procedure. Note that parameters of those two sub-networks are not shareable, which show better performance when faced with inputs with great differences compared with siamese network.

The input pairs are first sent into several convolutional layers, where the first three blocks of VGG19 [16] are employed considering model size and computation

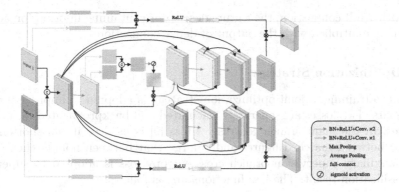

Fig. 3. The architecture of the PSAD block.

speed. For each sub-network, its feature map is sent into another two branches separately, namely the main branch and the auxiliary supervision branch. The designed PSAD block in the main branch extracts representative features via dense connection and fuses them. The attention structures added further refine those features. The auxiliary supervision branch consists of the last two convolutional blocks of VGG19, which can accelerate convergence and improve model performance. The main branch outputs the difference between two inputs as Eq. 1, while the auxiliary branch outputs a binary classification prediction.

$$D(X_1, X_2) = \|G(X_1) - G(X_2)\|_2 \qquad (1)$$

where X_1 and X_2 are the two inputs, G is the output of the main branch, and D calculates the Euclidean distance between the two feature vectors.

2.2 Pseudo Siamese Attention Dense Block

To highlight salient features, the designed PSAD block is utilized as is shown in Fig. 3. Concatenating two inputs to be a new one is shown effective by allowing transmissing information and compensating missing details for each other [3]. Therefore, the feature map pairs from last layers are first concatenated and then sent into convolutional layer to recover channel number. To highlight the informative regions, we apply average-pooling and max-pooling operations to the feature map along the channel axis. At last we adopt element-wise multiplication between the spatial map and the input map, and feed the result into a pseudo-siamese module with dense connection. The inputs are also added to the dense connection to complement information.

Besides finding the salient spatial regions, exploiting inter-channel relationship is also significant, so another channel attention module is applied. We first squeeze the spatial dimension of the input by simultaneously using average-pooling and max-pooling operations. The two feature vectors are delivered into a full-connection layer and then element-wisely added. With ReLU activation and another full-connection, the feature descriptor containing information across channels is multiplied with the output of dense connection.

2.3 Optimization Strategy

During the training, a joint optimization strategy is adopted to improve prediction results. Two losses are separately calculated and backpropagated in the two branches. In the main branch contrastive loss [9] is used as it can shorten the distance between matched figures and enlarge that between nonmatched ones. In the auxiliary supervision branch cross-entropy loss is adopted to supervise the classification result. The loss functions are as follows:

$$L_{main} = \sum_{i=1}^{P}(1 - Y^i)\frac{1}{2}(D(X_1^i, X_2^i))^2 + (Y^i)\frac{1}{2}\{\max(0, m - D(X_1^i, X_2^i))\}^2 \quad (2)$$

$$L_{aux} = -\sum_{i=1}^{P}Y^i log(\hat{Y}^i) + (1 - Y^i)log(1 - \hat{Y}^i) \quad (3)$$

where label Y is whether 0(match) or 1(nonmatch), P is the total number of the input data.

2.4 Evaluation Metrics

Hausdorff Distance (HD) is adopted to measure the maximal non-matched extent between guidewires and vessels, which is formulated as:

$$H(R_g^f, R_v^a) = \max\{h(R_g^f, R_v^a), h(R_v^a, R_g^f)\} \quad (4)$$

$$h(R_g^f, R_v^a) = \max_{a \in R_g^f} \min_{b \in R_v^a} \|a - b\| \quad (5)$$

$$h(R_v^a, R_g^f) = \max_{b \in R_v^a} \min_{a \in R_g^f} \|b - a\| \quad (6)$$

where R_g^f means the region of the guidewire of X-ray images, and R_v^a means the region of the vessel of angiographic images. The norm $\|.\|$ calculates Euclidean distance.

The ratio of the overlapped area between guidewires and vessels to the total area of the guidewires is considered as an evaluation metrics named guidewire recall rate (GRR). The range of GRR is 0 to 1, which is formulated as:

$$GRR = \frac{Area(R_g^f \cap R_v^a)}{Area(R_g^f)} \quad (Area(R_g^f) \leq Area(R_v^a)) \quad (7)$$

The larger GRR is, the better the result is. When GRR equals 1, it represents that the guidewires of X-ray images are all inside the vessels of angiographic images, implying the best matching performance.

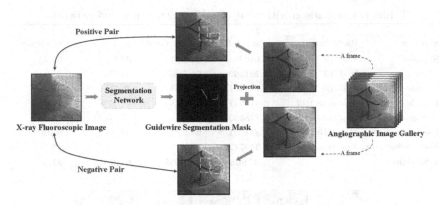

Fig. 4. The process of labeling. The guidewire tips of the angiographic images are marked in cyan, and the main vessels are in red. The segmentation result is in green. (Color figure online)

We sort the outputs of the main branch in the ascending order, and calculate the rank-1 and rank-3 HD and GRR to get comprehending evaluation results. Also, we count rank-n accuracy according to labels. Noted that since the images are all continuous frames derived from sequences, their morphological and positional differences are quite subtle, so we define that as long as there exists labeled matching images in the adjacent frames of the predicted one, the prediction is correct.

3   Experiments

3.1   Datasets and Setups

A new dataset is built by us named CVM X-ray Database. It consists of 51 sequences of coronary artery X-ray fluoroscopic images with 512×512 pixels. The images are labeled as pairs, which are either positive or negative. A positive pair implies that both angiographic images and X-ray images correspond to the same position during a cardiac contraction, i.e. they are matched, while a negative pair are nonmatched. The dataset includes 969 and 74 positive pairs in the training and testing set, which are independent from each other.

Detailed labeling process is shown in Fig. 4. First, a guidewire segmentation mask of the X-ray image is obtained by [20]. Then the masks are projected onto every angiographic image in the gallery. If the projected guidewire is inside the vessel, this pair will be labeled as positive, otherwise negative. As is highlighted in dotted yellow box, for positive pairs the two guidewire tips are almost coincided, while they are separated for negative pairs.

This model was implemented on the PyTorch library (version 1.7.1) with one NVIDIA TITAN Xp (12 GB). Adam optimizer was used with a learning rate of 0.0001. Moreover, the batch size is 10, and 300 epochs were used for training.

Table 1. Comparison with state-of-the-art structures and networks.

Structure	Backbone	Rank-1 HD↓	rank-1 GRR↑	rank-1 accu↑	FLOPs(M)
Siamese	VGG19	6.64035	62.03%	32.90%	62.55
TSNet [7]	VGG19	4.731560	64.71%	34.20%	361.11
Pseudo-Siamese	VGG19	5.713661	64.15%	36.96%	62.55
Pseudo-Siamese	Resnet18 [10]	6.479941	61.53%	54.17%	191.20
Pseudo-Siamese	RepVGG-A [6]	7.148590	69.69%	31.40%	15.72
Pseudo-Siamese	MobileNetV3 [11]	5.851081	60.93%	43.25%	**6.03**
Pseudo-Siamese	Xception [5]	7.343772	49.77%	34.00%	43.76
DSP-Net		**3.47908**	**84.09%**	**61.95%**	57.36

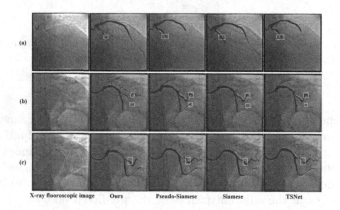

Fig. 5. Qualitative results of different approaches on CVM X-ray Database. The guidewires of X-ray images are marked in cyan and projected onto the predicted angiographic images. Main vessels are marked in red. Tips of guidewires are highlighted in yellow boxes. (Color figure online)

3.2 Comparison with State-of-the-Arts

To demonstrate the advantage of our DSP-Net, we compare it with two widely-used networks VGG19 [16] and Resnet18 [10], two lightweight networks Xception [5] and MobileNetV3 [11], and a recently proposed RepVGG-A [6]. Those backbones are combined with structures including siamese, pseudo-siamese and TSNet [7]. Table 1 clearly shows that DSP-Net outperforms other existing approaches in terms of rank-1 HD, rank-1 GRR and rank-1 accuracy, while possessing an acceptable amount of parameters.

The qualitative comparative results are shown in Fig. 5. Compared with other methods, DSP-Net can better capture the detailed information of the X-ray images and find the most accurate angiographic image where the guidewire of the X-ray image is located inside the vessels precisely. Take Fig. 5(c) as an example, if doctors are provided with angiographic images predicted by other methods, they may deliver the guidewire in the wrong direction as the yellow arrow shows, result in puncturing vascular wall. Also, as is shown, our model shows great generality and stability with different heart structures and fluoroscopic angles.

Table 2. Comparison with different modules added to pseudo-siamese dense block.

Description	rank-1 HD↓	rank-1 GRR↑	rank-1 accu↑	rank-3 accu↑
PSD	3.87286	82.44%	59.28%	91.00%
PSD+CBAM [18]	4.00592	84.03%	62.66%	86.72%
PSD+ASPP [4]	4.80341	78.22%	**66.93%**	91.00%
PSD+AA [13]	4.97830	79.36%	61.94%	90.07%
PSAD(Ours)	**3.47908**	**84.09%**	61.95%	**94.84%**

Table 3. Ablation study: experiment results on adding components to the baseline.

Deep supervision	PSAD	rank-1 HD↓	rank-1 GRR↑	rank-3 HD↓	rank-3 GRR↑	rank-1 accu↑	rank-3 accu↑
–	–	5.71366	64.15%	6.11309	64.62%	36.96%	78.06%
–	✓	5.43951	79.96%	5.40494	74.47%	40.00%	81.40%
✓	–	5.27489	58.07%	5.11292	56.86%	20.60%	44.80%
✓	✓	**3.47908**	**84.09%**	**4.54867**	**78.62%**	**61.95%**	**94.84%**

To verify the advantage of our proposed PSAD block, we compare it with three widely-used architectures, including CBAM [18], atrous spatial pyramid pooling (ASPP) [4], and Adaptive Attention (AA) [13]. As it demonstrates in Table 2, our PSAD achieves better performance in terms of rank-1 HD, rank-1 GRR and rank-3 accuracy. Particularly, compared with the original pseudo-siamese block with only dense connection, our approach improves rank-1 HD, rank-1 GRR and rank-1 accuracy by 0.39378, 1.65% and 2.67%, respectively.

3.3 Ablation Study

In order to comprehend how our proposed model works, we performed detailed ablation experiments to analyze the components by gradually adding them to the baseline, which is the pseudo-siamese network with VGG19 backbone. As shown in Table 3, PSAD block can bring improvements in every evaluation metrics over baseline, such as 0.27415 in rank-1 HD and 15.81% in rank-1 GRR. Noted that it enhances the rank-1 and rank-3 accuracy by 3.04% and 3.34% respectively. Adding deep supervision to the baseline also improved rank-1 and rank-3 HD. When both PSAD block and deep supervision are integrated, the model can observe an improvement of 24.99% and 16.78% over baseline in terms of rank-1 and rank-3 accuracy, and gain 2.23458 and 19.94% considering rank-1 HD and rank-1 GRR.

4 Conclusions

In this paper, we propose a novel framework DSP-Net to address the difficulties brought by cardiac movement during the procedure. By matching X-ray fluoro-scopic images with angiographic images, the model is able to provide doctors with dynamic reference images in PCI. Quantitative and qualitative evaluations have been conducted on our CVM X-ray Database to demonstrate that the approach achieves excellent performance. Extensive ablation studies prove the effectiveness of our proposed PSAD block and deep supervision architecture. By integrating these components, our DSP-Net achieves state-of-the-art performance. Specifically, the rank-1 HD and rank-1 GRR reach 3.47908 and 84.09% respectively, which is promising to assist doctors to overcome the visual misjudgment caused by the elastic deformation of the arteries and achieve safer procedure.

Acknowledgements. This work was supported in part by the National Natural Science Foundation of China under Grant 62073325, and Grant U1913210; in part by the National Key Research and Development Program of China under Grant 2019YFB1311700; in part by the Youth Innovation Promotion Association of CAS under Grant 2020140; in part by the National Natural Science Foundation of China under Grant 62003343; in part by the Beijing Natural Science Foundation under Grant M22008.

References

1. Cardiovascular diseases (cvds) (2021). https://www.who.int/news-room/fact-sheets/detail/cardiovascular-diseases-(cvds)
2. Briguori, C., Tavano, D., Colombo, A.: Contrast agent-associated nephrotoxicity. Prog. Cardiovasc. Dis. **45**(6), 493–503 (2003)
3. Chang, Y., Jung, C., Sun, J., Wang, F.: Siamese dense network for reflection removal with flash and no-flash image pairs. Int. J. Comput. Vision **128**(6), 1673–1698 (2020). https://doi.org/10.1007/s11263-019-01276-z
4. Chen, L.C., Papandreou, G., Kokkinos, I., Murphy, K., Yuille, A.L.: Deeplab: semantic image segmentation with deep convolutional nets, atrous convolution, and fully connected CRFs. IEEE Trans. Pattern Anal. Mach. Intell. **40**(4), 834–848 (2017)
5. Chollet, F.: Xception: deep learning with depthwise separable convolutions. In: Proceedings of the IEEE Conference on Computer Vision and Pattern Recognition, pp. 1251–1258 (2017)
6. Ding, X., Zhang, X., Ma, N., Han, J., Ding, G., Sun, J.: REPVGG: making VGG-style convnets great again. In: Proceedings of the IEEE Conference on Computer Vision and Pattern Recognition, pp. 13733–13742 (2021)
7. En, S., Lechervy, A., Jurie, F.: TS-NET: combining modality specific and common features for multimodal patch matching. In: Proceedings of the IEEE International Conference on Image Processing, pp. 3024–3028. IEEE (2018)
8. Grech, E.D.: Percutaneous coronary intervention. II: the procedure. BMJ **326**(7399), 1137–1140 (2003)
9. Hadsell, R., Chopra, S., LeCun, Y.: Dimensionality reduction by learning an invariant mapping. In: Proceedings of the IEEE Conference on Computer Vision and Pattern Recognition, vol. 2, pp. 1735–1742. IEEE (2006)
10. He, K., Zhang, X., Ren, S., Sun, J.: Deep residual learning for image recognition. In: Proceedings of the IEEE Conference on Computer Vision and Pattern Recognition, pp. 770–778 (2016)
11. Howard, A., et al.: Searching for mobilenetv3. In: Proceedings of the IEEE Conference on Computer Vision and Pattern Recognition, pp. 1314–1324 (2019)

12. Langer, N.B., Argenziano, M.: Minimally invasive cardiovascular surgery: incisions and approaches. Methodist Debakey Cardiovasc. J. **12**(1), 4 (2016)
13. Li, W., Liu, K., Zhang, L., Cheng, F.: Object detection based on an adaptive attention mechanism. Sci. Rep. **10**(1), 1–13 (2020)
14. Lv, J., Yang, M., Zhang, J., Wang, X.: Respiratory motion correction for free-breathing 3D abdominal MRI using CNN-based image registration: a feasibility study. Br. J. Radiol. **91**, 20170788 (2018)
15. Schulz, C.J., Böckler, D., Krisam, J., Geisbüsch, P.: Two-dimensional-three-dimensional registration for fusion imaging is noninferior to three-dimensional-three-dimensional registration in infrarenal endovascular aneurysm repair. J. Vasc. Surg. **70**(6), 2005–2013 (2019)
16. Simonyan, K., Zisserman, A.: Very deep convolutional networks for large-scale image recognition. arXiv preprint arXiv:1409.1556 (2014)
17. Tepel, M., Van Der Giet, M., Schwarzfeld, C., Laufer, U., Liermann, D., Zidek, W.: Prevention of radiographic-contrast-agent-induced reductions in renal function by acetylcysteine. N. Engl. J. Med. **343**(3), 180–184 (2000)
18. Woo, S., Park, J., Lee, J.Y., Kweon, I.S.: Cbam: convolutional block attention module. In: Proceedings of the European Conference on Computer Vision, pp. 3–19 (2018)
19. Xu, D., Dong, W., Zhou, H.: Sclera recognition based on efficient sclera segmentation and significant vessel matching. Comput. J. **65**, 371–381 (2020)
20. Zhou, Y.J., Xie, X.L., Zhou, X.H., Liu, S.Q., Bian, G.B., Hou, Z.G.: A real-time multi-functional framework for guidewire morphological and positional analysis in interventional x-ray fluoroscopy. IEEE Trans. Cogn. Dev. Syst. **13**, 657–667 (2020)

A Novel Fusion Network for Morphological Analysis of Common Iliac Artery

Meng Song[1], Shi-Qi Liu[1], Xiao-Liang Xie[1(✉)], Xiao-Hu Zhou[1],
Zeng-Guang Hou[1], Yan-Jie Zhou[1], and Xi-Yao Ma[1,2]

[1] Institute of Automation, Chinese Academy of Sciences, Beijing, China
xiaoliang.xie@ia.ac.cn
[2] University of Chinese Academy of Sciences, Beijing, China

Abstract. In endovascular interventional therapy, automatic common iliac artery morphological analysis can help physicians plan surgical procedures and assist in the selection of appropriate stents to improve surgical safety. However, different people have distinct blood vessel shapes, and many patients have severe malformations of iliac artery due to hemangiomas. Besides, the uneven distribution of contrast media makes it difficult to make an accurate morphological analysis of the common iliac artery. In this paper, a novel fusion network, combining CNN and Transformer is proposed to address the above issues. The proposed FTU-Net consists of a parallel encoder and a Cross-Fusion module to capture and fuse global context information and local representation. Besides, a hybrid decoder module is designed to better adapt the fused features. Extensive experiments have demonstrated that our proposed method significantly outperforms the best previously published results for this task and achieves the state-of-the-art results on the common iliac artery dataset built by us and two other public medical image datasets. To the best of our knowledge, this is the first approach capable of common iliac artery segmentation.

Keywords: Common iliac artery · Morphological analysis · Transformers · Convolution neural networks · Feature fusion

1 Introduction

Abdominal aortic aneurysm (AAA) has been the most common vascular disease of the abdominal aorta mainly manifested by local dilation and outward bulging under the action of various pathological factors (atherosclerosis, trauma, and infection), with a very high mortality rate of 85% to 90% [1]. Patients with AAAs are usually treated with endovascular aortic repair (EVAR), associated with superior perioperative outcomes, less surgical trauma, and shorter postoperative recovery compared with open repair [2]. In this procedure, a special bifurcated stent-graft with two iliac legs is placed over the abdominal aortic wall to isolate the high pressure arterial blood flow from the aneurysm wall and prevent rupture.

L. Wang (Eds.): MICCAI 2022, LNCS 13437, pp. 54–63, 2022.
https://doi.org/10.1007/978-3-031-16449-1_6

For the selection of stent, the shape and size of the iliac leg are important indicators as they determine whether the stent can be stably fixed in vessels with high pressure and high flow rate of blood while don't block the internal iliac artery and the external iliac artery. Also, the access diameter and shape of iliac artery determine the surgical approach [3,4]. Therefore, automatic morphological analysis of common iliac artery is vital for successful EVAR.

Nevertheless, the task is not straightforward for the following reasons: (1) Due to lesions in vessel, such as hemangioma, and specificity of different people, it is a challenge to discern the whole shape of vessels precisely; (2) The low contrast of vessels and background noise greatly interferes with the effectiveness of segmentation; (3) The edge pixels of vessels may be misclassified due to the uneven contrast agent, artifacts of the large intestine, and guidewire imaging.

There have been many studies on vessel segmentation [5,6]. It has been proved that CNN has the extraordinary ability of hierarchical learning [7]. However, due to the inherent bias, the lack of capturing global context information remains a problem. Recently, Transformer [8] has been applied to medical image segmentation. Chen et al. [12] proposed a U-shaped network where the Transformer acts as a feature extractor in the encoder. Zhang et al. [13] presented a fusion module to merge the features from CNN and transformer. However, scarce annotation of the medical dataset and large model remain challenges for the Transformer-based method.

To address the above-mentioned concerns, a novel Cross-fused architecture is proposed for common iliac artery segmentation, which consists of a parallel encoder, Cross-Fusion block, and a hybrid decoder. The CNN branch and transformer branch in the parallel encoder extract local details and global context information independently, and the Cross-Fusion block is proposed to merge local and global features from the same scale. Then, the hybrid decoder is designed to adapt the fused information and improve the stability of the model while ensuring performance. Moreover, skip-connection is applied between the Cross-Fusion block and decoder to integrate low-level semantic features. Additionally, deeply supervision method is adopted to speed up network convergence and solve the problem of scarce sufficient labeled data.

Our contributions can be summarized as follows: 1) to the best of our knowledge, this is the first automatic approach for morphological analysis of common iliac artery. 2) our proposed approach can address the difficulty of segmenting overlapping and severely deformed vessels to some extent, and the ends of the common iliac artery and the edge pixels can be more accurately segmented. 3) the proposed morphological analysis algorithm can obtain anatomical information of the common iliac artery: the minimum inner diameter, access diameter, and tortuosity.

2   Method

As shown in Fig. 1, the overall architecture of the proposed network consists of encoder, decoder, Cross-Fusion block, and skip connections. The original angiography images are sent into the parallel encoder, which consists of a CNN branch and a transformer branch, to extract local and global features respectively. Then, the features of the same scale extracted from both branches are fused by the Cross-Fusion block, where cross attention is applied to effectively fuse the information. Finally, the multi-scaled fused feature maps are transferred into the hybrid decoder by skip connections to recover the resolution from (H/32 × W/32) to (H × W). In order to avoid overfitting caused by the scarce annotation and speed up convergence, deeply supervision is applied in the network.

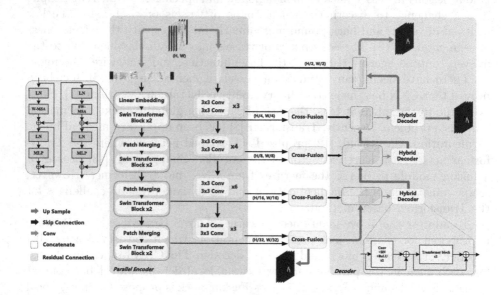

Fig. 1. Overview of the framework.

Parallel Encoder. The structure of the encoder consists of two independent branches in parallel. In the transformer branch, Swin Transformer block in [14] is applied as the feature extractor, which has lower computational complexity and better stability than other hierarchical transformers. In the CNN branch, ResNet34 is applied to extract local features, each of whose layers output feature maps of dimensions corresponding to the transformer branch.

Hybrid Decoder. To improve the stability and convergence speed of the model, a hybrid decoder block (refer to Fig. 1) that combines CNN and transformer is designed to insert into the traditional ASPP decoder structure. In this block, the feature maps first pass through an MLP block containing two consecutive 3×3 convolutions + BatchNorm + ReLU modules, then they require to get through a Transformer block. The residual connection is applied to obtain the output of a single decoder block, ensuring the stability of this structure in training, and preventing the network from overfitting.

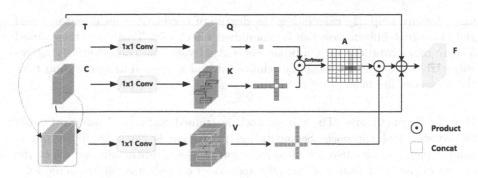

Fig. 2. Architecture of proposed cross-fusion module.

Cross-Fusion Block. Inspired by criss-cross attention in [15], a cross-fusion module is introduced to aggregate the extracted features from the CNN branch and transformer branch. As shown in Fig. 2, two kinds of feature maps **T** and **C** $\in \mathbb{R}^{C \times H \times W}$ are first sent two convolutional layers with 1×1 filters simultaneously to generate two feature maps **Q** and **K**, respectively, where $\{\mathbf{Q}, \mathbf{K}\} \in \mathbb{R}^{C' \times H \times W}$. C' is the number of channels, which is less than dimension C of the original feature map for dimension reduction. After obtaining **Q** and **K**, an attention map **A** is further generated by calculating the degree of correlation between **Q** and **K**. At each position **u** in the spatial dimension of **Q**, we can obtain a representation of all the dimensions of **Q** at that position **u**, expressed by vector $\mathbf{Q}_u \in \mathbb{R}^{C'}$. Meanwhile, the set $\mathbf{\Phi}_u \in \mathbb{R}^{(H+W-1) \times C'}$ expresses the horizontal and vertical features centered at the position in **K** corresponding to **u**. With $\mathbf{Q}_u$ and $\mathbf{\Phi}_u$, the correlation coefficient $\mathbf{\Psi}_{i,u}$ can be calculated using the following equation:

$$\Psi_{i,u} = Q_u \Phi_{i,u}^T, \tag{1}$$

where $\mathbf{\Phi}_{i,u} \in \mathbb{R}^{C'}$ represents the i-th element of $\mathbf{\Phi}_u$ and $\mathbf{\Psi}_{i,u}$ represents the correlation between the position vector **u** in **Q** and the element around the corresponding position in **K**, $i = [1, ..., H+W-1]$. The set of $\mathbf{\Psi}_{i,u}$ forms a feature map $\mathbf{\Psi} \in \mathbb{R}^{(H+W-1) \times (H \times W)}$, then a softmax operation is applied on $\mathbf{\Psi}$ over the channel dimension to calculate the attention map **A**. Then, we concatenate two original feature maps **T** and **C**, and apply another 1×1 convolutional layer on the connected feature map to generate **V**. For each position **u** in **V**, a collection of feature vectors that are in the same column and row at that position can be obtained, which is defined as $\mathbf{\Omega}_u$. And the same position of u in attention map **A**, a scalar value defined as $\mathbf{A}_u$, represents the attention score of original feature maps (**T** and **C**) at position **u**, and $\mathbf{A}_{i,u}$ represents the value of channel i. Therefore, with the feature map **V** and attention map **A**, the fused feature map can be calculated using the following formulation:

$$F_u = \Sigma_{i=0}^{H+W-1} A_{i,u} \Omega_{i,u} + T_u + C_u, \tag{2}$$

where F_u, T_u, C_u are vectors in **F**, **T**, **C** $\in \mathbb{R}^{C \times H \times W}$ respectively at position **u**. Then the original feature maps **T** and **C** are added to **F** to get the final fused feature map. By calculating the degree of correlation, local features and global context information can be combined effectively. Therefore, these fused feature representations have a broad contextual view and can selectively aggregate pixel-wise information. They achieve mutual gains and are more robust for the segmentation task.

Deeply Supervision. The full network is trained end-to-end using a Deeply Supervision [16] approach. Segmentation prediction is generated by a simple head, which contains two 3×3 convolution blocks. Additionally, we select the output of the last fusion module and the decoder output after all fusion modules to implement the deep supervision method. Dice loss is applied to supervise the above three outputs. The loss function can be shown as follows.

$$L_{totall} = \alpha L(R_1, G) + \beta L(R_2, G) + \gamma L(R_3, G), \tag{3}$$

where G is groundtruth, R_1, R_2, R_3 represent the three outputs, L represents Dice loss function and α, β, γ are adjustable hyperparameters.

Diagnosis Algorithm. To obtain the anatomical information of the CIA, we design a novel diagnosis algorithm. Firstly, we extract the boundary of the mask predicted by FTU-Net. Secondly, we obtain the centerline and skeleton of the vessels through the distance transform method. Then, each point on the centerline can make a tangent, and we take the line connecting the two points where this tangent intersects the boundary as the inner diameter of the blood vessel, to obtain the minimum inner diameter by comparison. Next, the bottom two pixels of the skeleton are chosen to calculate the access diameter. Finally, we perform polynomial fitting on the center line and calculate the curvature of this function. Because curves fitted by discrete points have the problem of curvature mutation, the curvature values that are equal or greater than 0.8 are taken out selectively, and the maximum value of the remains represents the largest tortuosity of the vessel.

3   Experiments and Results

Data Acquisition. CIA-DSA. A dataset is built which contains a total of 484 DSA images of the abdomen after contrast injection, collected from 100 patients at Peking Union Medical College Hospital. Each image has an average resolution of 1024×1024. Lableme is applied to annotate the common iliac artery, with a clinical expert to review and direct in the annotation process. All images and labels are resized into 512×512, and the dataset is randomly split into a training set and a testing set, containing 391 and 93 images respectively.

Kvasir. A public dataset for polyp segmentation [17] consists of 1000 images. We solely select 800 images for training and the remaining 200 images for testing, and the resolution of each image is resized into 512×512.

ISIC2017. The publicly available 2017 international Skin Imaging Collaboration [18] task 1 skin lesion segmentation dataset provides 2000 images for training, 150 images for validation, and 600 images for testing. Each image is resized into 512 × 512.

Table 1. Quantitative comparison results with state-of-the-art methods on CIA-DSA. The results of TransUNet and TransFuse are obtained by running code released by the authors.

Methods	Dice	mIoU	Recall	Precision
U-Net [19]	0.8999	0.8257	0.8944	0.9182
U-Net++ [20]	0.9024	0.8283	0.9041	0.9103
Attention U-Net [21]	0.8964	0.8185	0.9031	0.9003
TransUNet [12]	0.8989	0.8208	0.9019	0.9023
TransFuse [13]	0.9171	0.8519	0.9101	0.9298
FTU-Net	**0.9281**	**0.8680**	**0.9247**	**0.9344**

Table 2. Quantitative results on Kvasir and ISIC2017 datasets.

Methods	Kvasir			ISIC2017		
	Dice	mIoU	Precision	Dice	mIoU	Precision
U-Net++	0.8431	0.7574	0.8511	0.8030	0.7128	0.8866
Attention U-Net	0.8384	0.7550	0.8455	0.8016	0.7097	0.9011
TransUNet	0.8705	0.8007	0.8874	0.8498	0.7667	0.9027
TransFuse	0.8357	0.7835	0.8949	0.8425	0.7580	0.8977
FTU-Net	**0.9105**	**0.8548**	**0.9113**	**0.8620**	**0.7824**	**0.9288**

Implementation Details. Our proposed network was built in the PyTorch framework and trained using a single NVIDIA-A6000 GPU. The hyperparameters α, β and γ were set to 0.2, 0.3, 0.5 which can deliver maximum performance. Adam optimizer with an initial learning rate of 7e−5, weight decay of 0.5, and momentum of 0.999 was adopted for training. To get the optimal performance, the learning rate is multiplied by the factor of 0.9 every 30 epochs. Moreover, all models were trained within 200 epochs with a batch size of 4 on CIA-DSA and Kvasir datasets, and especially, the epoch number is set to 100 and batch size to 8 for the ISIC2017 dataset.

Comparison Experiments Results. To verify the effectiveness of our proposed approach in the common iliac artery segmentation task, main experiments are conducted on CIA-DSA by comparing FTU-Net with five previous SOTA methods. As shown in Table 1, it can be seen that our proposed approach achieves better performance than other existing networks on all indicators. It outperforms CNN-based methods by a large margin, and it observes an increase compared to the SOTA method TransFuse. It clearly demonstrates that our methods can better segment CIAs in DSA images.

Additionally, to demonstrate the generalization performance of our proposed architecture, we compared it with four widely used networks on Kvasir and ISIC2017 datasets. As shown in Table 2, our method can obviously improve the segmentation performance on both datasets. It achieves 91.05% in dice coefficient and 91.13% in precision on Kvasir, and it has an increase of 2.6% in precision compared to the best result with TransUNet on the ISIC2017 dataset. As can be seen in Fig. 3, our approach can accurately segment CIAs under various challenging scenarios, including severe deformation, locally obscured surface, and fuzzy boundary, whereas other methods usually lead to fracture and defect.

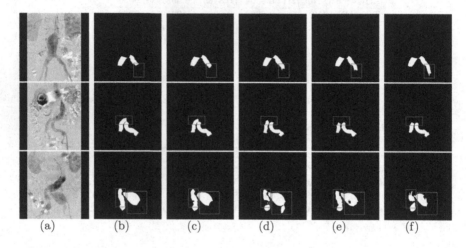

(a)   (b)   (c)   (d)   (e)   (f)

Fig. 3. Visualization results on CIA-DSA. (a) Original images. (b) Groundtruth. (c) Ours. (d) TransFuse. (e) TransUNet. (f) Attention U-Net. The red boxes represent regions that are prone to misclassification. (Color figure online)

Ablation Study. To evaluate the effectiveness of the components in our proposed network, an ablation study is conducted on the CIA-DSA dataset. We evaluate the function of each part by adding modules to the backbone. Comparing E.1 and E.2 with E.3, the combination of CNN and Transformer leads to better performance. Further, by comparing E.3 against E.4, we observe that the designed Cross-Fusion block can merge the local features and global context information well, which has an increase of 3.2%. Lastly, the hybrid decoder block can further improve network performance. It clearly demonstrates that with the backbone and composition setting, our approach can achieve better results (Table 3).

Table 3. Ablation study on Cross-Fusion module.

Index	Res34	Swin-S	Cross-Fusion	Hybrid Decoder	F1-Score	mIoU
E.1	√				0.8836	0.7952
E.2		√			0.8632	0.7645
E.3	√	√			0.8899	0.8036
E.4	√	√	√		0.9218	0.8637
E.5	√	√	√	√	**0.9281**	**0.8680**

Table 4. Qualitative validation of morphological algorithm.

Index	Mean Error	MSE
Min diameter	2.04	7.88
Access diameter	2.11	9.69
Curvature	0.2	-

Morphological Analysis. To evaluate the performance of the designed analysis algorithm, we add it after the inference of our proposed network. As shown in Fig. 4, it can be seen that our developed algorithm can obtain access diameter, the minimum inner diameter, and the maximum curvature of CIA with arbitrary shapes. These automatically calculated morphological information can help physicians select suitable stents and plan surgical procedures. As shown in the figure on the right, the large value 0.67 suggests that it is difficult for therapeutic surgical instruments to pass directly, thus the doctor may have to deliver the instrument from the right iliac artery and "climb the mountain" to the left side. To demonstrate the effectiveness of our proposed morphological algorithm, manually calibrated diameter and curvature by experts using Siemens commercial software are collected, and mean error and mean square error(MSE) between manual calibration and our results are calculated. As can be seen in Table 4, the results of the automatic algorithm can accurately help physicians to make surgical procedures and select medical instruments.

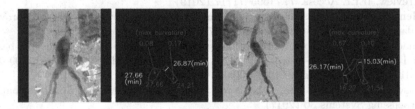

Fig. 4. Morphological analysis of CIAs. Green represents access diameter, white represents the minimum inner diameter, and blue represents the points with max curvature. (Color figure online)

4   Conclusion

In this paper, we introduce a novel fusion network combining CNN and Transformer and a novel diagnosis algorithm to address the challenging task of CIA morphological analysis. Quantitative and qualitative evaluations demonstrate that our proposed network achieves remarkable and stable improvements on three different types of datasets, achieving SOTA performance. Moreover, the diagnosis algorithm can obtain anatomical information automatically, which can help physicians for preoperative plan and stent-graft choice to reduce operation time and surgery cost. The experimental results indicate that our proposed approach has the potential to be applied in actual clinical scenarios.

Acknowledgements. This work was supported in part by the National Natural Science Foundation of China under Grant 62073325, and Grant U1913210; in part by the National Key Research and Development Program of China under Grant 2019YFB1311700; in part by the Youth Innovation Promotion Association of CAS under Grant 2020140; in part by the National Natural Science Foundation of China under Grant 62003343; in part by the Beijing Natural Science Foundation under Grant M22008.

References

1. Kent, K.C.: Abdominal aortic aneurysms. N. Engl. J. Med. **371**(22), 2101–2108 (2014)
2. Buck, D.B., et al.: Endovascular treatment of abdominal aortic aneurysms. Nat. Rev. Cardiol. **11**(2), 112 (2014)
3. Taudorf, M., et al.: Endograft limb occlusion in EVAR: iliac tortuosity quantified by three different indices on the basis of preoperative CTA. Eur. J. Vasc. Endovasc. Surg. **48**(5), 527–33 (2014)
4. Kim, H.O., et al.: Endovascular aneurysm repair for abdominal aortic aneurysm: a comprehensive review. Korean J. Radiol. **20**(8), 1247–1265 (2019)
5. Soomro, T.A., et al.: Deep learning models for retinal blood vessels segmentation: a review. IEEE Access. **7**, 71696–71717 (2019)
6. Meng, C., et al.: Multiscale dense convolutional neural network for DSA cerebrovascular segmentation. Neurocomputing **373**, 123–34 (2020)
7. Zhao, F., Chen, Y., Hou, Y., He, X.: Segmentation of blood vessels using rule-based and machine-learning-based methods: a review. Multimedia Syst. **25**(2), 109–118 (2017). https://doi.org/10.1007/s00530-017-0580-7
8. Vaswani, A., et al.: Attention is all you need. In: Advances in Neural Information Processing Systems 30 (2017)
9. Dosovitskiy, A., et al.: An image is worth 16x16 words: transformers for image recognition at scale. arXiv preprint arXiv:2010.11929 (2020)
10. Zheng, S., et al.: Rethinking semantic segmentation from a sequence-to-sequence perspective with transformers. In: 2021 IEEE/CVF Conference on Computer Vision and Pattern Recognition (CVPR), pp. 6881–6890. IEEE (2021)
11. Xie, E., et al.: SegFormer: simple and efficient design for semantic segmentation with transformers. In: Advances in Neural Information Processing Systems 34 (2021)

12. Chen, J., et al.: TransUNet: transformers make strong encoders for medical image segmentation. arXiv preprint arXiv:2102.04306 (2021)
13. Zhang, Y., Liu, H., Hu, Q.: TransFuse: fusing transformers and CNNs for medical image segmentation. In: de Bruijne, M., et al. (eds.) MICCAI 2021. LNCS, vol. 12901, pp. 14–24. Springer, Cham (2021). https://doi.org/10.1007/978-3-030-87193-2_2
14. Liu, Z., et al.: Swin transformer: Hierarchical vision transformer using shifted windows. In: 2021 IEEE/CVF International Conference on Computer Vision (ICCV), pp. 9992–10022 (2021)
15. Huang, Z., et al.: CCNet: criss-cross attention for semantic segmentation. In: 2019 IEEE/CVF International Conference on Computer Vision (ICCV), pp. 603–612. IEEE (2019)
16. Lee, C-Y., et al.: Deeply-supervised nets. In: Artificial Intelligence and Statistics, pp. 562–570 (2015)
17. Jha, D., et al.: Kvasir-SEG: a segmented polyp dataset. In: Ro, Y.M., et al. (eds.) MMM 2020. LNCS, vol. 11962, pp. 451–462. Springer, Cham (2020). https://doi.org/10.1007/978-3-030-37734-2_37
18. Codella, N.C., et al.: Skin lesion analysis toward melanoma detection: a challenge at the 2017 international symposium on biomedical imaging (ISBI), hosted by the international skin imaging collaboration (ISIC). In: 2018 IEEE 15th International Symposium on Biomedical Imaging (ISBI 2018), pp. 168–172. IEEE (2018)
19. Ronneberger, O., Fischer, P., Brox, T.: U-Net: convolutional networks for biomedical image segmentation. In: Navab, N., Hornegger, J., Wells, W.M., Frangi, A.F. (eds.) MICCAI 2015. LNCS, vol. 9351, pp. 234–241. Springer, Cham (2015). https://doi.org/10.1007/978-3-319-24574-4_28
20. Zhou, Z., Rahman Siddiquee, M.M., Tajbakhsh, N., Liang, J.: UNet++: a nested U-Net architecture for medical image segmentation. In: Stoyanov, D., et al. (eds.) DLMIA/ML-CDS -2018. LNCS, vol. 11045, pp. 3–11. Springer, Cham (2018). https://doi.org/10.1007/978-3-030-00889-5_1
21. Oktay, O., et al.: Attention U-Net: learning where to look for the pancreas. arXiv preprint arXiv:1804.03999 (2018)
22. Zhou, X.H., et al.: Learning skill characteristics from manipulations. IEEE Trans. Neural Netw. Learn. Syst. **PP**, 1–15 (2022)
23. Zhou, X.H., et al.: Surgical skill assessment based on dynamic warping manipulations. IEEE Trans. Med. Robot. Bionics **4**(1), 50–61 (2022)
24. Gui, M.J., et al.: Design and experiments of a novel Halbach-cylinder-based magnetic skin: a preliminary study. IEEE Trans. Instrum. Meas. **71**, 9502611 (2022)

Hand Hygiene Quality Assessment Using Image-to-Image Translation

Chaofan Wang[✉], Kangning Yang, Weiwei Jiang, Jing Wei,
Zhanna Sarsenbayeva, Jorge Goncalves, and Vassilis Kostakos

School of Computing and Information Systems, The University of Melbourne,
Melbourne, Australia
chaofanw@student.unimelb.edu.au

Abstract. Hand hygiene can reduce the transmission of pathogens and prevent healthcare-associated infections. Ultraviolet (UV) test is an effective tool for evaluating and visualizing hand hygiene quality during medical training. However, due to various hand shapes, sizes, and positions, systematic documentation of the UV test results to summarize frequently untreated areas and validate hand hygiene technique effectiveness is challenging. Previous studies often summarize errors within predefined hand regions, but this only provides low-resolution estimations of hand hygiene quality. Alternatively, previous studies manually translate errors to hand templates, but this lacks standardized observational practices. In this paper, we propose a novel automatic image-to-image translation framework to evaluate hand hygiene quality and document the results in a standardized manner. The framework consists of two models, including an Attention U-Net model to segment hands from the background and simultaneously classify skin surfaces covered with hand disinfectants, and a U-Net-based generator to translate the segmented hands to hand templates. Moreover, due to the lack of publicly available datasets, we conducted a lab study to collect 1218 valid UV test images containing different skin coverage with hand disinfectants. The proposed framework was then evaluated on the collected dataset through five-fold cross-validation. Experimental results show that the proposed framework can accurately assess hand hygiene quality and document UV test results in a standardized manner. The benefit of our work is that it enables systematic documentation of hand hygiene practices, which in turn enables clearer communication and comparisons.

Keywords: Hand hygiene · Handrub · Six-step hand hygiene technique · Healthcare-associated infections · Nosocomial infections

Supplementary Information The online version contains supplementary material available at https://doi.org/10.1007/978-3-031-16449-1_7.

1   Introduction

Healthcare-Associated Infections (HAIs) or nosocomial infections are a major patient-safety challenge in healthcare settings [22]. Appropriate hand hygiene is a simple and cost-efficient measure to avoid the transmission of pathogens and prevent HAIs [22]. However, research has found that hand hygiene quality in healthcare settings is generally unsatisfactory [17,18].

Typically, hand hygiene quality can be assessed by two methods: microbiological validation and Ultraviolet (UV) tests. Microbiological validation mainly uses samples from the fingertips (EN 1500) [14] or through the glove juice method (ASTM E-1174) before and after the World Health Organization (WHO) six-step hand hygiene technique. This approach evaluates hand hygiene quality in terms of bacteria count reduction [22]. Conversely, UV tests require subjects to use hand disinfectants mixed with fluorescent concentrates to perform the handrub technique, and then measure the skin coverage of the fluorescent hand disinfectants [4]. A strong correlation between the visual evaluation of UV tests and the degree of bacterial count reduction has been reported [6]. Compared to microbiological validation, UV tests can deliver an immediate and clearly visible result of skin coverage with hand disinfectants [19].

By assessing hand hygiene quality from UV tests, electronic hand hygiene monitoring systems could provide on-time intervention and periodic personalized hygiene education to Healthcare Workers (HCWs) to improve their hand hygiene practices [21]. The documented quality results can also be utilized to quantify hand hygiene technique effectiveness and provide corresponding improvement recommendations. However, subjects' hands come in diverse sizes and shapes, and their gestures and finger positions may differ across observations, resulting in difficulties in assessing hand hygiene quality and documenting its result through a standardized method that cannot be achieved by registration. Previous studies rely on manual annotation or traditional machine learning algorithms to analyze UV test results, which typically consider the presence, count, size, and/or location of the uncovered areas from the observations during the UV tests or the collected UV test images. Traditional machine learning algorithms can also summarize error distribution in terms of predefined hand regions. However, they lack the ability to further locate errors inside the hand regions or provide detailed morphology information [10,12,15]. Such information is crucial to enable consistent feedback and comparisons of hand hygiene quality. While manual annotation has been used to documented the size and location of uncovered areas on a normalized hand template, it is restricted to small sample sizes and lacks of standardized observational practices [4,5]. Thus, manual annotation can only be used for coarse-grained estimations of hand hygiene quality.

In this paper, we propose a novel deep learning-based framework to overcome these issues and evaluate hand hygiene quality on a large scale and in a standardized manner. Our contributions are twofold: 1. a method for segmenting hands from the background and classifying the hand areas covered with fluorescent hand disinfectants; 2. an approach for translating segmented hands into normalized hand templates to provide standardized high-resolution visualizations of hand hygiene quality.

2 Methods

Our framework includes two sub-models: 1. an Attention U-Net model localizes and segments hands from UV test images and identifies areas covered with fluorescent hand disinfectants (Fig. 1); and 2. a U-Net-based generator subsequently convert the segmented hands into normalized hand templates (Fig. 2).

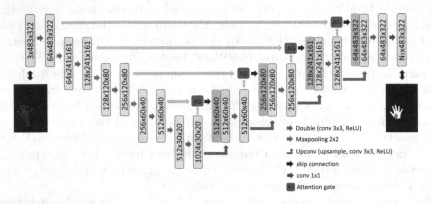

Fig. 1. Attention U-Net architecture. Input UV test images are progressively downsampled and upsampled by successive contracting path (left side) and expansive path (right side) to output images after hand segmentation and area classification. N_1 represents two classes, namely hand areas covered with fluorescent hand disinfectants (white) and uncovered hand areas (gray), and these two class images are then combined for better visualization. Attention gates highlight salient image regions and provide complimentary details to the upsampling network.

2.1 Hand Segmentation and Area Classification

Taking UV test images as input, we first apply the cascaded fully convolutional neural networks (i.e., U-Net [13]) with an attention gate (AG) mechanism [9] to capture spatial features and select informative feature responses. As shown in Fig. 1, this model contains three parts. Firstly, it has a contraction module consisting of alternating layers of convolution and pooling operators, which is used to capture local contextual information (like shapes or edges) progressively via multi-layers receptive fields and extract fine-grained feature maps. Secondly, it has a symmetrical expansion module where pooling operators are replaced by up-convolution operators, which is used to reconstruct and refine the corresponding hand segmentation and area classification images through successively propagating the learned correlations and dependencies to higher resolution layers. Moreover, it has multiple AGs connected to the correspondingly contracting and expansive paths to provide attention weights over the extracted different scales of feature maps in order to amplify feature responses in focus regions and simultaneously suppress feature responses in irrelevant background regions.

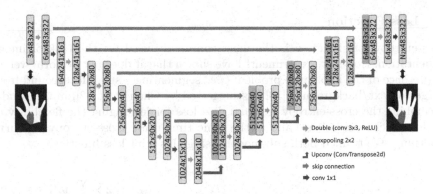

Fig. 2. U-Net-based generator architecture. Enlarged segmented hand images are translated into normalized hand templates via a fully convolutional network. N_2 represents two classes, namely hand areas covered with fluorescent hand disinfectants (white) and uncovered hand areas (gray), and these two class images are then combined for better visualization.

In this way, the finer details provided by each AG can supplement the corresponding upsampled coarse output through the skip layer fusion [7]. As defined by [9], for feature map x^l from a contracting convolutional layer l, we first computed the attention coefficients α^l by:

$$\alpha_i^l = \sigma_2(\psi^{\mathrm{T}}(\sigma_1(W_x^{\mathrm{T}} x_i^l + W_g^{\mathrm{T}} g_i + b_g)) + b_\psi) \qquad (1)$$

where x_i^l is the vector of each pixel i in x^l, g_i is a gating vector that is collected from the upsampled coarser scale and used for each pixel i to determine focus regions, σ_1 refers to the rectified linear unit (ReLU) activation function, σ_2 refers to the sigmoid activation function that is used to normalize the attention distribution, and W_x, W_g, ψ, b_g, and b_ψ are the trainable parameters.

2.2 Translation Between Segmented Hands and Hand Templates

To further convert the enlarged segmented hands into normalized hand templates that would be easy to compare, we built a U-Net-based generator. The main idea is to take advantage of the translation equivariance of convolution operation. Given a segmented hand image $X_s \in \mathbb{R}^s$, we seek to construct a mapping function $\phi : \mathbb{R}^s \to \mathbb{R}^t$ via a fully convolutional network to translate it into a normalized hand template. Unlike the Attention-based U-Net used for localization and segmentation, we do not introduce AGs into the generator architecture, since the self-attention gating is at the global scale, which lacks some of the inductive biases inherent to convolutional networks such as translation equivariance and locality [3].

2.3   Loss Function

Cross-entropy loss function is the most commonly used for the task of image segmentation. However, experiments have shown that it does not perform well in the presence of the imbalance problem (*i.e.*, segmenting a small foreground from a large context/background) [16]. As suggested by Chen *et al.* [2], in this study, we combine the cross-entropy loss and dice loss for leveraging the flexibility of dice loss of class imbalance and at the same time using cross-entropy for curve smoothing. We trained both sub-models using this joint loss function:

$$\mathcal{L} = -\frac{1}{N} \sum_{i=1}^{N} \beta(t_i \log p_i) + (1 - \beta)[(1 - t_i) \log(1 - p_i)] - \frac{2 \sum_{i=1}^{N} p_i t_i + \epsilon}{\sum_{i=1}^{N} (p_i + t_i) + \epsilon} \quad (2)$$

where p_i and t_i stand for pairs of corresponding pixel values of prediction and ground truth [8], N is *the number of samples* × *the number of classes*, ϵ is the added constant to avoid the undefined scenarios such as when the denominator is zero, β controls the penalization of FPs and FNs.

3   Experiments

3.1   Dataset Construction

To the best of our knowledge, there is no publicly available UV test-related dataset. Thus, we conducted a lab study with four tasks categories (*e.g.*, Shapes, Equally Split, Individual WHO Handrub Steps, Entire WHO Handrub Technique) to collect images with different skin coverage with fluorescent hand disinfectants (details in Appendix Fig. 2). The study protocol was reviewed and approved by the University of Melbourne's Human Ethics Advisory Group. From the lab study, we collected 609 valid UV test images for both hands and both sides (1218 images when separating left and right hands) from twenty-nine participants to evaluate the effectiveness of the proposed models.

For each of the 1218 images, we first labeled the ground truth for the hand segmentation and area classification image. Labeling the ground truth for hand segmentation consists of two steps: hand contour detection and wrist points recognition from the image taken under white light (Appendix Fig. 1b). Hand contour was recognized by the red-difference chroma component from the YUV system and Otsu's method for automatic image thresholding [11], while two authors manually marked wrist points. We acquired the ground truth for hand segmentation by cropping hand contour with wrist points. Then, we labeled hand areas covered with fluorescent hand disinfectants from the image taken under UV light (Appendix Fig. 1c). Since the fluorescent concentrate used in the experiment glows green under a UV lamp, we transferred these images to the Hue Saturation Value (HSV) color system and used the H channel to detect areas within the green color range with a threshold.

Regrading the ground truth for hand translation, manual translation between segmented hands and hand templates is impractical due to a lack of a standardized translation process. Instead, we decided to generate synthetic translation data by sampling triangles and trapezoids on the same relative positions in segmented hand and hand template pair to train the hand translation model, thereby obtaining the mapping information. We first split segmented hands and hand templates into 41 triangles based on the landmarks generated by MediaPipe (and manual labels for the standard hand templates) and finger-web points (convexity defects of hand contours) calculated by OpenCV [1,20,23]. Then for each of the 41 triangle pairs, we randomly sampled triangles or trapezoids within the triangle on the segmented hands and translated them into the corresponding positions within the triangle on the hand template through homography (shown in Fig. 4, and more examples can be found in our dataset repository) [1]. Furthermore, we resized the segmented hands to cover the image to remove irrelevant background and facilitate the training process.

3.2 Implementation Details

We implemented both the hand segmentation and area classification model and the hand translation model in Pytorch with a single Nvidia GeForce RTX 3090 (24GB RAM). The hand segmentation and area classification model was trained for 30 epochs, while 40 epochs were used for the translation model, and both models were trained with a batch size of 16. We resized the input images for both models to $3 \times 483 \times 322$ pixels (16% of the original image). We used RMSprop optimization with an initial learning rate of 10^{-5}, a weight decay of 10^{-8}, and a momentum of 0.9. On this basis, we applied the learning rate schedule: if the Dice score on the validation set not increased for 2 epochs, the learning rate would be decayed by a factor of 0.1. For data augmentation, we flipped images horizontally to increase the size of the dataset for the hand segmentation and area classification model, and we also performed rotation ($\pm 20°$) and resize ($\pm 5\%$) towards the dataset for hand translation model to increase its generalizability. Code and example dataset repository is available at https://github.com/chaofanqw/HandTranslation.

3.3 Evaluation Metrics

To evaluate the performance of both models, we conducted five-fold cross-validation. For each round, we retained six participants' data for testing, and the remaining participants' data were then shuffled and partitioned into the training and validation sets with a 90% and 10% breakdown respectively. The trained model with the highest Dice coefficient on the validation set was then evaluated on the test set, and the Dice coefficient and Intersection over Union (IOU) score across all five-folds were then averaged and reported.

Furthermore, to evaluate the performance for the hand segmentation and area classification model, we further compared its results with two other state-of-the-art segmentation models, namely U-Net and U-Net++, through the same five-fold cross-validation. Also, to visualize the real-life performance of the hand translation model, we employed the model to translate the segmented hands with different skin coverage (collected from the lab study) to hand templates.

4   Results

For the hand segmentation and area classification model, the Attention U-Net achieved the highest mean Dice coefficient (96.90%) and IOU score (94.02%). Meanwhile, U-Net++ achieved a comparable performance to Attention U-Net (Dice coefficient: 96.87%, IOU score: 93.95%), and both models outperformed U-Net (Dice coefficient: 96.64%, IOU score: 93.54%). Figure 3 provides the qualitative results of the hand segmentation and area classification over different hands, sides, and task categories.

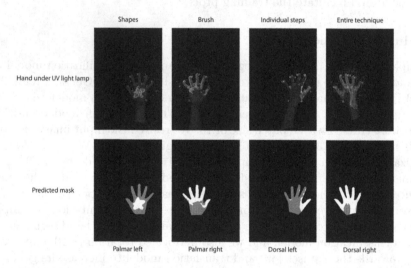

Fig. 3. Qualitative evaluation results over different hands, sides, and task categories. The upper row shows the original color images taken under UV light. The bottom row exhibits the predicted hand segmentation and area classification results, where black indicates background, gray indicates uncovered hand areas, and white indicates hand areas covered with hand disinfectants.

We further evaluated the Attention U-Net performance for each participant and task category. For the participant-wise model performance, the highest was seen by P30 (Dice coefficient: 97.96%, IOU score: 96.02%), while the lowest was seen by P2 (Dice coefficient: 95.68%, IOU score 91.75%). For the task-wise model performance, the highest was of the "Shapes" task (Dice coefficient: 97.07%, IOU score: 94.33%), while the lowest was of the "Individual WHO Handrub Steps" task (Dice coefficient: 96.89%, IOU score: 93.98%).

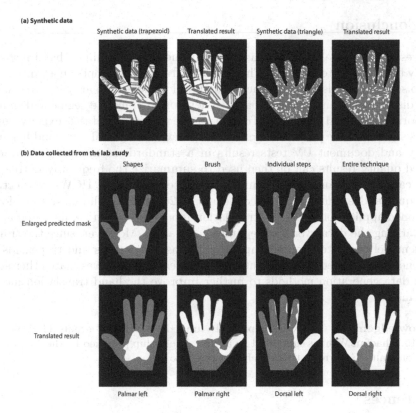

Fig. 4. Qualitative evaluation results over synthetic data and the lab study data. (a) The synthetic images with trapezoids or triangles within the segmented hands and the corresponding hand translation results. (b) The upper row shows the enlarged hand segmentation and area classification results from the lab study, and the bottom row exhibits the corresponding hand translation results.

For the hand translation model, the proposed system achieved the Dice coefficient of 93.01% and the IOU score of 87.34% on the synthetic dataset. Figure 4a provides qualitative results of the hand translation for the synthetic data. We then evaluated the trained model on the lab study data of segmented hands with different skin coverage. However, the model with the best performance on the synthetic dataset tends to overfit the shape of trapezoid and triangle and generates rough contours for the areas covered with hand disinfectants. Thus, to avoid overfitting, we chose to use the model trained with eight epochs for translating the lab study data to hand templates based on visual inspections (Fig. 4b).

5   Conclusion

We present an image-to-image translation framework to evaluate hand hygiene quality through UV test images. The proposed framework adopts an attention U-Net to segment hands from the background and classifies the areas covered with hand disinfectants and a U-Net-based generator to translate segmented hands into normalized hand templates. Trained on the presented dataset, experimental results show that the proposed framework can accurately evaluate hand hygiene quality and document UV tests results in a standardized manner. The documented quality results can be then used to summarize the frequently untreated areas caused by standardized hand hygiene techniques or HCWs' respective techniques and evaluate their effectiveness [4, 20]. Due to the nature of aforementioned application scenarios, translation model errors can be mitigated after summarizing the data over a large study population. Moreover, since the translation model tends to overfit sampling patterns of triangles and trapezoids of the generated synthetic dataset, future studies can aim to investigate other synthetic data generation methods to further improve the hand translation model performance.

Acknowledgement. This work is partially funded by NHMRC grants 1170937 and 2004316. Chaofan Wang is supported by a PhD scholarship provided by the Australian Commonwealth Government Research Training Program.

References

1. Bradski, G.: The OpenCV library. Dr Dobbs J. Softw. Tools (2000). https://doi.org/10.1111/0023-8333.50.s1.10
2. Chen, C., Dou, Q., Jin, Y., Chen, H., Qin, J., Heng, P.-A.: Robust multimodal brain tumor segmentation via feature disentanglement and gated fusion. In: Shen, D., et al. (eds.) MICCAI 2019. LNCS, vol. 11766, pp. 447–456. Springer, Cham (2019). https://doi.org/10.1007/978-3-030-32248-9_50
3. Dosovitskiy, A., et al.: An image is worth 16 × 16 words: transformers for image recognition at scale. arXiv preprint arXiv:2010.11929 (2020)
4. Kampf, G., Reichel, M., Feil, Y., Eggerstedt, S., Kaulfers, P.M.: Influence of rub-in technique on required application time and hand coverage in hygienic hand disinfection. BMC Infect. Dis. (2008). https://doi.org/10.1186/1471-2334-8-149
5. Kampf, G., Ruselack, S., Eggerstedt, S., Nowak, N., Bashir, M.: Less and less-influence of volume on hand coverage and bactericidal efficacy in hand disinfection. BMC Infect. Dis. (2013). https://doi.org/10.1186/1471-2334-13-472
6. Lehotsky, A., Szilagyi, L., Bansaghi, S., Szeremy, P., Weber, G., Haidegger, T.: Towards objective hand hygiene technique assessment: validation of the ultraviolet-dye-based hand-rubbing quality assessment procedure. J. Hospital Infection **97**(1), 26–29 (2017). https://doi.org/10.1016/j.jhin.2017.05.022, https://linkinghub.elsevier.com/retrieve/pii/S0195670117302943
7. Long, J., Shelhamer, E., Darrell, T.: Fully convolutional networks for semantic segmentation. In: Proceedings of the IEEE Conference on Computer Vision and Pattern Recognition, pp. 3431–3440 (2015)

8. Milletari, F., Navab, N., Ahmadi, S.A.: V-net: fully convolutional neural networks for volumetric medical image segmentation. In: 2016 Fourth International Conference on 3D Vision (3DV), pp. 565–571. IEEE (2016)

9. Oktay, O., et al.: Attention u-net: learning where to look for the pancreas. arXiv preprint arXiv:1804.03999 (2018)

10. Öncü, E., Vayısoğlu, S.K.: Duration or technique to improve the effectiveness of children' hand hygiene: a randomized controlled trial. Am. J. Infect. Control (2021). https://doi.org/10.1016/j.ajic.2021.03.012

11. Otsu, N.: Threshold Selection Method from Gray-Level Histograms. IEEE Trans. Syst. Man Cybern. (1979). https://doi.org/10.1109/tsmc.1979.4310076

12. Rittenschober-Böhm, J., et al.: The association between shift patterns and the quality of hand antisepsis in a neonatal intensive care unit: an observational study. Int. J. Nurs. Stud. (2020). https://doi.org/10.1016/j.ijnurstu.2020.103686

13. Ronneberger, O., Fischer, P., Brox, T.: U-Net: convolutional networks for biomedical image segmentation. In: Navab, N., Hornegger, J., Wells, W.M., Frangi, A.F. (eds.) MICCAI 2015. LNCS, vol. 9351, pp. 234–241. Springer, Cham (2015). https://doi.org/10.1007/978-3-319-24574-4_28

14. Standardization, E.C.: European Standard, EN1500:2013, CHEMICAL DISINFECTANTS AND ANTISEPTICS. HYGIENIC HANDRUB. TEST METHOD AND REQUIREMENTS (PHASE 2/STEP 2). European Committee for Standardization (2013)

15. Szilágyi, L., Lehotsky, A., Nagy, M., Haidegger, T., Benyó, B., Benyó, Z.: Steryhand: A new device to support hand disinfection. In: 2010 Annual International Conference of the IEEE Engineering in Medicine and Biology Society, EMBC 2010 (2010). https://doi.org/10.1109/IEMBS.2010.5626377

16. Taghanaki, S.A., et al.: Combo loss: handling input and output imbalance in multi-organ segmentation. Comput. Med. Imaging Graph. 75, 24–33 (2019)

17. Taylor, L.J.: An evaluation of handwashing techniques-1. Nursing times (1978)

18. Taylor, L.J.: An evaluation of handwashing techniques-2. Nursing times (1978)

19. Vanyolos, E., et al.: Usage of ultraviolet test method for monitoring the efficacy of surgical hand rub technique among medical students. J. Surg. Educ. (2015). https://doi.org/10.1016/j.jsurg.2014.12.002

20. Wang, C., et al.: A system for computational assessment of hand hygiene techniques. J. Med. Syst. 46(6), 36 (2022). https://doi.org/10.1007/s10916-022-01817-z

21. Wang, C., et al.: Electronic monitoring systems for hand hygiene: systematic review of technology (2021). https://doi.org/10.2196/27880

22. World Health Organization (WHO): WHO guidelines on hand hygiene in health care (2009)

23. Zhang, F., et al.: MediaPipe hands: on-device real-time hand tracking. arXiv preprint arXiv:2006.10214 (2020)

An Optimal Control Problem for Elastic Registration and Force Estimation in Augmented Surgery

Guillaume Mestdagh and Stéphane Cotin[(✉)]

Inria, Strasbourg, France
stephane.cotin@inria.fr

Abstract. The nonrigid alignment between a pre-operative biomechanical model and an intra-operative observation is a critical step to track the motion of a soft organ in augmented surgery. While many elastic registration procedures introduce artificial forces into the direct physical model to drive the registration, we propose in this paper a method to reconstruct the surface loading that actually generated the observed deformation. The registration problem is formulated as an optimal control problem where the unknown is the surface force distribution that applies on the organ and the resulting deformation is computed using an hyperelastic model. Advantages of this approach include a greater control over the set of admissible force distributions, in particular the opportunity to choose where forces should apply, thus promoting physically-consistent displacement fields. The optimization problem is solved using a standard adjoint method. We present registration results with experimental phantom data showing that our procedure is competitive in terms of accuracy. In an example of application, we estimate the forces applied by a surgery tool on the organ. Such an estimation is relevant in the context of robotic surgery systems, where robotic arms usually do not allow force measurements, and providing force feedback remains a challenge.

Keywords: Augmented surgery · Optimal control · Biomechanical simulation

1 Introduction

In the context of minimally-invasive surgery, abdominal organs are constantly subject to large deformations. These deformations, in combination with the limited visual feedback available to medical staff, make an intervention such as tumor ablation a complex task for the surgeon. Augmented reality systems have been designed to provide a three-dimensional view of an organ, which shows the current position of internal structures. This virtual view is superimposed onto intra-operative images displayed in the operating room.

Without loss of generality, we consider the case of liver laparoscopic surgery. In [8], the authors describe a full pipeline to produce augmented images. In

© The Author(s), under exclusive license to Springer Nature Switzerland AG 2022
L. Wang (Eds.): MICCAI 2022, LNCS 13437, pp. 74–83, 2022.
https://doi.org/10.1007/978-3-031-16449-1_8

their work, available data are intra-operative images provided by a laparoscopic stereo camera, and a biomechanical model of the organ and its internal structures, computed from pre-operative CT scans. During the procedure, a point cloud representing the current location of the liver surface is first extracted from laparoscopic images, and then the pre-operative model is aligned with the intra-operative point cloud in a non-rigid way.

The elastic registration procedure used to perform the alignment between pre- and intra-operative data is a widely studied subject. In particular, much attention has been given to choosing an accurate direct mechanical model for the liver and its surroundings. The liver parenchyma is usually described using a hyperelastic constitutive law (see [11] for an extensive review). Used hyperelastic models include Saint-Venant-Kirchhoff [6], neo-Hookean [12] or Ogden models [15]. Due to its reduced computational cost, the linear co-rotational model [14] is a popular solution when it comes to matching a real-time performance constraint [18,19,22]. Additional stiffness due to the presence of blood vessels across the parenchyma is sometimes also taken into account [7]. A less discussed aspect of the modeling is the interaction between the liver and its surroundings, which results in boundary conditions applied on the liver surface. Proposed approaches often involve Dirichlet boundary conditions where the main blood vessels enter the liver [18] or springs to represent ligaments holding the organ [15].

Another key ingredient to obtain accurate reconstructions is the choice of a registration procedure. In many physics-based registration methods in the literature, fictive forces or energies are added into the direct problem to drive the registration. Approaches based on the Iterative Closest Point algorithm introduce attractive forces between the liver and the observed point cloud and let the system evolve as a time-dependent process toward an equilibrium [8,19]. In [22], the authors model those attractive forces using an electrostatic potential. In [18], sliding constraints are used in the direct problem to enforce correspondence between the deformed liver model and the observed data. Such constraints are enforced by Lagrange multipliers that are also fictive forces. As the intra-operative observation is not a real protagonist of the physical model, those fictive forces do not reflect the true causes of displacements. This results in a poorly physically-consistent displacement field, regardless of the direct model. These methods cannot guarantee accurate registrations.

In this paper, we present a method to reconstruct a surface force distribution that explains the observed deformation, using the optimal control formalism. The problem formulation includes the choice of an elastic model (Sect. 2.1), a set of admissible forces and an objective function to minimize (Sect. 2.2). In particular, the set of admissible forces is specified by the direct model to ensure physically-consistent deformations. The adjoint method used to perform the optimization is described in Sect. 2.3.

Some existing works in the literature are concerned with reconstructing the physical causes of displacement. In [17], the effects of gravity and pneumoperitoneum pressure are taken into account to perform an initial intraoperative registration. In [20], the authors control the imposed displacement on parts of the liver boundary that are subject to contact forces, while a free boundary condition is applied onto other parts. A numerical method similar to our adjoint method is used in [10] to register a liver model onto a point cloud coupled with ultrasound data. While their method is very specific and tailored for linear elasticity, we present a more generic approach which is compatible with nonlinear models. Our method is relevant when a measurement of forces is needed, in particular when the surgeon interacts with the organ through a surgical robotic system without force sensor. We give an example of such a force estimation in Sect. 3.2.

2   Methods

In this section, we give details about the optimal control problem formulation. We first specify some notation around the direct model, then we introduce the optimization problem. In the last part, we describe the adjoint method, which is used to compute derivatives of the objective function with respect to the control.

2.1   Hyperelastic Model and Observed Data

The liver parenchyma in its pre-operative configuration is represented by a meshed domain Ω, filled with an elastic material. When a displacement field $\mathbf{u}$ is applied to Ω, the deformed mesh is denoted by $\Omega_{\mathbf{u}}$ and its boundary is denoted by $\partial\Omega_{\mathbf{u}}$. Note that the system state is fully known through the displacement of mesh nodes, stored in $\mathbf{u}$. The liver is embedded with a hyperelastic model. When a surface force distribution $\mathbf{b}$ is applied to the liver boundary, the resulting displacement $\mathbf{u_b}$ is the unique solution of the static equilibrium equation

$$\mathbf{F}(\mathbf{u_b}) = \mathbf{b}, \tag{1}$$

where $\mathbf{F}$ is the residual from the hyperelastic model. Note that $\mathbf{F}$ is very generic and may also account for blood vessels rigidity, gravity or other elements in a more sophisticated direct model.

The observed intra-operative surface is represented by a point cloud $\Gamma = \{y_1, \ldots, y_m\}$. We also define the orthogonal projection onto $\partial\Omega_{\mathbf{u}}$, also called the closest point operator,

$$p_{\mathbf{u}}(y) = \arg\min_{x \in \partial\Omega_{\mathbf{u}}} \|y - x\|.$$

Here, we consider that the orthogonal projection operator always returns a unique point, as points with multiple projections onto $\partial\Omega_{\mathbf{u}}$ represent a negligible subset of $\mathbb{R}^3$.

2.2   Optimization Problem

We perform the registration by computing a control $\mathbf{b}$ so that the resulting displacement $\mathbf{u_b}$ is in adequation with observed data. The optimization problem reads

$$\min_{\mathbf{b} \in B} \Phi(\mathbf{b}) \quad \text{where} \quad \Phi(\mathbf{b}) = J(\mathbf{u_b}) + R(\mathbf{b}), \tag{2}$$

where J measures the discrepancy between a prediction $\mathbf{u_b}$ and observed data, R is an optional regularization term, and B is the set of admissible controls.

The functional J enforces adequation with observed data and contains information such as corresponding landmarks between pre– and intra-operative surfaces. Many functionals of this kind exist in the literature, based on surface correspondence tools (see [21] and references therein). In this paper, we use a simple least-squares term involving the orthogonal projection onto $\partial\Omega_\mathbf{u}$, which reads

$$J(\mathbf{u}) = \tfrac{1}{2m} \sum_{j=1}^{m} \|p_\mathbf{u}(y_j) - y_j\|^2. \tag{3}$$

The functional $J(\mathbf{u})$ evaluates to zero whenever every point $y \in \Gamma$ is matched by the deformed surface $\partial\Omega_\mathbf{u}$. If $\mathbf{v}$ is a perturbation of the current displacement field $\mathbf{u}$, the gradient of (3) with respect to the displacement is defined by

$$\langle \nabla J(\mathbf{u}), \mathbf{v} \rangle = \tfrac{1}{m} \sum_{j=1}^{m} (r_j - y_j) \cdot \mathbf{v}(r_j), \tag{4}$$

where $r_j = p_\mathbf{u}(y_j)$ and $\mathbf{v}(r_j)$ is the displacement of the point of $\partial\Omega_\mathbf{u}$ currently at r_j under the perturbation $\mathbf{v}$.

The set of admissible controls B contains a priori information about the surface force distribution $\mathbf{b}$, including the parts of the liver boundary where $\mathbf{b}$ is nonzero and the maximal intensity it is allowed to take. The selection of zones where surface forces apply is critical to obtain physically plausible registrations. In comparison, a spring-based approach would result in forces concentrated in zones where the spring are fixed, which might disagree with the direct model.

2.3 Adjoint Method

We solve problem (2) using an adjoint method. In such a method, the only optimization variable is $\mathbf{b}$. Descent directions for the objective function Φ are computed in the space of controls, which requires to compute $\nabla\Phi(\mathbf{b})$. To differentiate $J(\mathbf{u_b})$ with respect to $\mathbf{b}$, we use an adjoint state $\mathbf{p_b}$ defined as the solution of the adjoint system

$$\nabla\mathbf{F}(\mathbf{u_b})^\mathrm{T} \mathbf{p_b} = \nabla J(\mathbf{u_b}). \tag{5}$$

A standard calculation (see for instance [1]) shows that

$$\mathbf{p_b} = \frac{\mathrm{d}}{\mathrm{d}\mathbf{b}} [J(\mathbf{u_b})] \qquad \text{and} \qquad \nabla\Phi(\mathbf{b}) = \mathbf{p_b} + \nabla R(\mathbf{b}).$$

The procedure to compute the objective gradient is summarized in Algorithm 1. For a given $\mathbf{b}$, evaluating $\Phi(\mathbf{b})$ and $\nabla\Phi(\mathbf{b})$ requires to solve the direct problem, and then to assemble and solve the adjoint problem, which is linear. In other words, the additional cost compared to a direct simulation is that of solving one linear system. The resulting gradient is then fed to a standard gradient-based optimization algorithm to solve (2) iteratively.

Algorithm 1: Computation of objective gradient using an adjoint method.

Data: Current iterate $\mathbf{b}$
Compute the displacement $\mathbf{u_b}$ by solving (1)
Evaluate $\nabla J(\mathbf{u_b})$ and $\nabla \mathbf{F}(\mathbf{u_b})$
Compute the adjoint state $\mathbf{p_b}$ by solving (5)
Result: $\nabla \Phi(\mathbf{b}) = \mathbf{p_b} + \nabla R(\mathbf{b})$

---

3 Results

Our method is implemented in Python (Numpy and Scipy), and we use a limited-memory BFGS algorithm [4] as an optimization procedure[1]. Our numerical tests run on an Intel Core i7-8700 CPU at 3.20 GHz with 16 GB RAM.

3.1 Sparse Data Challenge Dataset

To evaluate our approach in terms of displacement accuracy, we use the Sparse Data Challenge[2] dataset. It consists of one tetrahedral mesh representing a liver phantom in its initial configuration and 112 point clouds acquired from deformed configurations of the same phantom [3,5]. Once the registration is done, the final position of the mesh nodes for each point cloud is uploaded on the challenge website. Then the target registration error (TRE) is computed, using 159 targets whose positions are unknown to us.

Before we begin the elastic registration process, we use the standard Iterative Closest Point method [2] to perform a rigid alignment. Then we set a fixed boundary condition on a small zone of the posterior face to enforce the uniqueness of the solution to the direct elastic problem. As the liver mesh represents a phantom, no information about blood vessels is available and for this reason we just choose six adjacent triangles close to the center of the posterior face to apply the fixed displacement constraint. As specified by the challenge authors, the forces causing the deformation are contact forces applied onto the posterior face of the phantom, while the point cloud was acquired by observing the anterior surface. As a consequence, we label the anterior surface as the "matching surface", which is to be matched with the point cloud, while the posterior surface is labeled as "loaded surface", where the force distribution $\mathbf{b}$ is allowed to take nonzero values. Figure 1 shows the initial mesh with the matching and loaded surfaces in different colors, a point cloud and the liver mesh after rigid registration. In a clinical context, the surface in the field of view of the camera may be labeled as the matching surface while forces are allowed on the remaining hidden surface.

We solve problem (2) (with $R(\mathbf{b}) = 0$) using the adjoint method presented in Sect. 2.3. As only small deformations are involved in this dataset, we use a linear

---

[1] Code available at https://github.com/gmestdagh/adjoint-elastic-registration.
[2] See details and results at https://sparsedatachallenge.org.

Fig. 1. Left: Initial liver mesh with matching surface on top and loaded surface in the bottom. Right: Point cloud and liver mesh after rigid registration

elastic model for the liver phantom, with $E = 1$ and $\nu = 0.4$. The procedure is stopped after 200 iterations.

In Table 1, we reported the target registration error statistics for all datasets returned by the challenge website after submission of the results. Even for point clouds with a low surface coverage, the average target registration error stays below the 5 mm error that is usually required for clinical applications. In Fig. 2, our results are compared with other submissions to the challenge. We obtained the second best result displayed on the challenge website (the first one is that of the challenge organizers), which shows that our approach is competitive for registration applications. No information about methods used by other teams is displayed on the challenge website.

Table 1. Target registration error statistics (in millimeters) for all datasets, as returned by the website after submission.

Surface coverage	Average	Standard deviation	Median
20–28 %	3.54	1.11	3.47
28–36 %	3.27	0.85	3.19
36–44 %	3.13	0.82	3.13
All data sets	3.31	0.94	3.19

3.2 Force Estimation in Robotic Surgery

Estimating the force applied by a robotic arm onto an organ is necessary to avoid causing damage to living tissues. As certain standard surgical robots are not equipped with force sensors, many indirect methods based on image processing have been proposed [13]. Here we estimate a force in a context similar to [9], where the intra-operative point cloud is estimated from a laparoscopic camera.

We generate 5 synthetic test cases using a liver mesh of 3,046 vertices and a linear elastic model ($E = 20,000$ Pa, $\nu = 0.45$). Dirichlet conditions are applied at the hepatic vein entry and in the falciform region. A test case is a sequence of traction forces applied onto two adjacent triangles on the anterior surface of

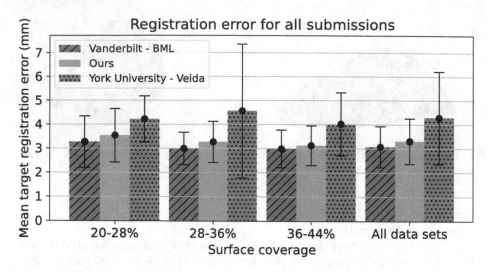

Fig. 2. Comparison of our TRE results with other participants to the challenge.

the liver to mimic the action of a robotic tool manipulating the liver. For each traction force, the resulting displacement is computed and a part of the deformed boundary is sampled to create a point cloud of 500 points. Each sequence consists of 50 forces, with a displacement of about 1 mm between two successive forces.

To avoid the inverse crime, we use a different mesh (3,829 vertices) for the registration, with the same linear elastic model as above. We allow the force distribution **b** to be nonzero only in a small zone (about 50 vertices) surrounding the triangles concerned by the traction force. In a clinical context, this approximate contact zone may be determined by segmenting the instrument tip on laparoscopic images. For a given traction force f_{true}, we solve the optimization problem with a relative tolerance of $5 \cdot 10^{-4}$ on the objective gradient norm and we compute the force estimation $f_{\text{est}} \in \mathbb{R}^3$ as the resultant of all the nodal forces of the reconstructed distribution **b**. Then we compute the relative error $\|f_{\text{est}} - f_{\text{true}}\| / \|f_{\text{true}}\|$. Figure 3 shows the original synthetic deformation and the reconstructed deformation and surface force distribution.

For each sequence, the observed point clouds are successively fed to the procedure to update the force estimation by solving the optimization problem. The optimization algorithm is initialized with the last reconstructed distribution, so that only a few iterations are required for the update. Table 2 shows the average error obtained for each sequence, together with the average execution time of the updates and the number of evaluations of the objective function.

The noise in the reconstructed distribution results in an intensity smaller than the reference, which is the main cause of an overall error of 10% in average. Note that the estimated force is proportional to the Young modulus of the model, which can be measured using elastography. According to [16], an error of 20% is typical for clinical elastographic measurements. In this context, the elastographic estimation is responsible for a larger part of the total error on the force estimation than our method.

Fig. 3. Synthetic deformation generated by a local force (left), reconstructed deformation using the point cloud (center) and zoom on the reconstructed force field (right). Nodal forces are summed to produce a resultant estimation.

Table 2. Average number of objective evaluations, execution time per update and relative error for each sequence.

Sequence	N. eval.	Update time	Relative error
Case 1	9.2	1.42 s	8.9%
Case 2	5.6	0.85 s	16.2%
Case 3	5.5	0.84 s	5.7%
Case 4	5.4	0.82 s	4.4%
Case 5	6.2	0.96 s	15.0%

4   Conclusion

We presented a formulation of the liver registration problem using the optimal control formalism, and used it with success in two different application cases. We showed that we can not only reconstruct an accurate displacement field, but also, in certain cases, give a meaning to the optimal force distribution returned by the procedure. By tuning the formulation parameters (namely the set of admissible controls), we easily added new hypotheses into the direct model without changing the code of our procedure. These numerical results highlight the relevance and the flexibility of an optimal control approach in augmented surgery. Due to its mathematical foundations, the optimal control formalism is probably an important step toward provable accuracy for registration methods. Current limitations in our work include the lack of results with hyperelastic models due to their high computational cost. In the next steps of our work, we intend to reduce computation times by initializing the optimization process with the output of a neural network.

References

1. Allaire, G.: Conception optimale de structures, Mathématiques & Applications (Berlin) [Mathematics & Applications], vol. 58. Springer-Verlag, Berlin (2007). https://doi.org/10.1007/978-3-540-36856-4
2. Besl, P.J., McKay, N.D.: Method for registration of 3-D shapes. In: Schenker, P.S. (ed.) Sensor Fusion IV: Control Paradigms and Data Structures, vol. 1611, pp. 586–606. International Society for Optics and Photonics, SPIE (1992)
3. Brewer, E.L., et al.: The image-to-physical liver registration sparse data challenge. In: Fei, B., Linte, C.A. (eds.) Medical Imaging 2019: Image-Guided Procedures, Robotic Interventions, and Modeling, vol. 10951, pp. 364–370. International Society for Optics and Photonics, SPIE (2019)
4. Byrd, R.H., Lu, P., Nocedal, J., Zhu, C.: A limited memory algorithm for bound constrained optimization. SIAM J. Sci. Comput. **16**(5), 1190–1208 (1995)
5. Collins, J.A., et al.: Improving registration robustness for image-guided liver surgery in a novel human-to-phantom data framework. IEEE Trans. Med. Imaging **36**(7), 1502–1510 (2017)
6. Delingette, H., Ayache, N.: Soft tissue modeling for surgery simulation. In: Computational Models for the Human Body, Handbook of Numerical Analysis, vol. 12, pp. 453–550. Elsevier (2004)
7. Haouchine, N., et al.: Impact of soft tissue heterogeneity on augmented reality for liver surgery. IEEE Trans. Vis. Comput. Graph. **21**(5), 584–597 (2015)
8. Haouchine, N., Dequidt, J., Peterlík, I., Kerrien, E., Berger, M., Cotin, S.: Image-guided simulation of heterogeneous tissue deformation for augmented reality during hepatic surgery. In: 2013 IEEE International Symposium on Mixed and Augmented Reality (ISMAR), pp. 199–208 (2013)
9. Haouchine, N., Kuang, W., Cotin, S., Yip, M.: Vision-based force feedback estimation for robot-assisted surgery using instrument-constrained biomechanical three-dimensional maps. IEEE Robot. Autom. Lett. **3**(3), 2160–2165 (2018)
10. Heiselman, J.S., Jarnagin, W.R., Miga, M.I.: Intraoperative correction of liver deformation using sparse surface and vascular features via linearized iterative boundary reconstruction. IEEE Trans. Med. Imaging **39**(6), 2223–2234 (2020)
11. Marchesseau, S., Chatelin, S., Delingette, H.: Nonlinear biomechanical model of the liver. In: Payan, Y., Ohayon, J. (eds.) Biomechanics of Living Organs, Translational Epigenetics, vol. 1, pp. 243–265. Academic Press, Oxford (2017)
12. Miller, K., Joldes, G., Lance, D., Wittek, A.: Total Lagrangian explicit dynamics finite element algorithm for computing soft tissue deformation. Commun. Numer. Methods Eng. **23**(2), 121–134 (2007)
13. Nazari, A.A., Janabi-Sharifi, F., Zareinia, K.: Image-based force estimation in medical applications: a review. IEEE Sens. J. **21**(7), 8805–8830 (2021)
14. Nesme, M., Payan, Y., Faure, F.: Efficient, physically plausible finite elements. In: Eurographics. Short papers, Dublin, Ireland, August 2005
15. Nikolaev, S., Cotin, S.: Estimation of boundary conditions for patient-specific liver simulation during augmented surgery. Int. J. Comput. Assist. Radiol. Surg. **15**(7), 1107–1115 (2020). https://doi.org/10.1007/s11548-020-02188-x
16. Oudry, J., Lynch, T., Vappou, J., Sandrin, L., Miette, V.: Comparison of four different techniques to evaluate the elastic properties of phantom in elastography: is there a gold standard? Phys. Med. Biol. **59**(19), 5775–5793 (sep 2014)
17. Özgür, E., Koo, B., Le Roy, B., Buc, E., Bartoli, A.: Preoperative liver registration for augmented monocular laparoscopy using backward-forward biomechanical simulation. Int. J. Comput. Assist. Radiol. Surg. **13**(10), 1629–1640 (2018)

18. Peterlík, I., et al.: Fast elastic registration of soft tissues under large deformations. Med. Image Anal. **45**, 24–40 (2018)
19. Plantefève, R., Peterlík, I., Haouchine, N., Cotin, S.: Patient-specific biomechanical modeling for guidance during minimally-invasive hepatic surgery. Ann. Biomed. Eng. **44**(1), 139–153 (2016)
20. Rucker, D.C., et al.: A mechanics-based nonrigid registration method for liver surgery using sparse intraoperative data. IEEE Trans. Med. Imaging **33**(1), 147–158 (2014)
21. Sahillioğlu, Y.: Recent advances in shape correspondence. Vis. Comput. **36**(8), 1705–1721 (2019). https://doi.org/10.1007/s00371-019-01760-0
22. Suwelack, S., et al.: Physics-based shape matching for intraoperative image guidance. Med. Phys. **41**(11), 111901 (2014)

PRO-TIP: Phantom for RObust Automatic Ultrasound Calibration by TIP Detection

Matteo Ronchetti[1(✉)], Julia Rackerseder[1], Maria Tirindelli[1,2],
Mehrdad Salehi[2], Nassir Navab[2], Wolfgang Wein[1], and Oliver Zettinig[1]

[1] ImFusion GmbH, München, Germany
`ronchetti@imfusion.com`
[2] Computer Aided Medical Procedures (CAMP), Technische Universität München,
Munich, Germany

Abstract. We propose a novel method to automatically calibrate
tracked ultrasound probes. To this end we design a custom phantom
consisting of nine cones with different heights. The tips are used as key
points to be matched between multiple sweeps. We extract them using a
convolutional neural network to segment the cones in every ultrasound
frame and then track them across the sweep. The calibration is robustly
estimated using RANSAC and later refined employing image based techniques. Our phantom can be 3D-printed and offers many advantages over
state-of-the-art methods. The phantom design and algorithm code are
freely available online. Since our phantom does not require a tracking target on itself, ease of use is improved over currently used techniques. The
fully automatic method generalizes to new probes and different vendors,
as shown in our experiments. Our approach produces results comparable
to calibrations obtained by a domain expert.

Keywords: Freehand ultrasound · Calibration · Phantom

1 Introduction

Ultrasound (US) imaging is a widely used medical imaging modality. Due to
its real-time, non-radiation-based imaging capabilities, it gains ever more pop-
ularity in the medical domain. By combining an ultrasound transducer with
a tracking system, 3D free-hand acquisitions become possible, further increasing the range of possible applications, for instance intra-operative navigation.
A tracking system consists of a tracking device, which is able to monitor the
position of a tracking target ("marker") in its own coordinate system [4]. The
marker is rigidly attached to a US probe, but regularly in an arbitrary fashion
relative to the image coordinate system. The geometrical relation mapping from
the image space into the marker's local coordinate system is commonly referred
to as (spatial) ultrasound calibration.

Accurate calibration often turns out to be a tedious task, it is therefore not
surprising that numerous approaches have been devised in the past to solve this

ⓒ The Author(s), under exclusive license to Springer Nature Switzerland AG 2022
L. Wang (Eds.): MICCAI 2022, LNCS 13437, pp. 84–93, 2022.
https://doi.org/10.1007/978-3-031-16449-1_9

(a)   (b)

Fig. 1. A photograph of the proposed 3D-printed phantom (a) and its corresponding labelmap (b), showing cones in blue and the base plate in red color. (Color figure online)

problem [20]. Many require additional tools or objects, which may also be tracked themselves. The simplest approach requires a tracked and pivot-calibrated stylus, which is held such that its tip is visible in the US image at multiple locations [21]. This technique has been improved by also digitizing points on the probe itself [25], or by enforcing stylus orientation through a phantom [27]. A majority of works utilize dedicated phantoms, i.e. objects with known geometric properties, for achieving the calibration [15]. Such phantoms include sphere-like objects [1], pairs of crossing wires [8], multiple point targets [19], planes [7], LEGO bricks [28], wires [2,3,5,23] and phantoms actively transmitting US echo back to the transducer [6]. Using crossing wires as a phantom, automatic robotic calibration using visual servoing has also been demonstrated [16]. Unfortunately, the described auxiliary control task is often not implementable on medical robotic systems as the required low-level joint control is encapsulated by the vendor. The aforementioned methods may be characterized by one or multiple disadvantages. First, manufacturing tolerances of phantoms and styluses (or the technique to calibrate them), as well as relative tracking between two tracking targets give rise to calibration errors. Phantoms may be tricky to assemble and geometrically characterize in code, for instances using the wires lead through the holes of 3D-printed objects. Finally, some approaches limit possible probe orientations, e.g. due to space between the wire phantom walls, which results in under-sampling of the six degrees of freedom, in turn leading to different in-plane and out-of-plane accuracies. In contrast to tool- or phantom-based calibration, image-based techniques work with arbitrary objects [26]. The idea resembles image registration as the similarity of US images acquired from different directions is iteratively maximized. This method has been shown to achieve high calibration accuracies for static objects but, just like registration problems, suffers from a limited capture range and hence needs to be initialized well.

   In this work, we propose a novel calibration technique that uses both feature- and image-based calibration on the same 3D-printed phantom. The phantom consists of nine cones and is compatible with a wide range of transducer designs and imaging depths. A machine learning model detects the cone tips in two sets of recorded US images. These are then matched and used to estimate a rigid calibration matrix, which is subsequently refined with an image-based method. The proposed method does not require any additional tracked tool or tracking on the phantom and no assembly other than 3D-printing, therefore it is faster and easier to use than existing approaches. In contrast to state of the art feature-based approaches, we use image-based refinement, which does not require tight manufacturing tolerances on the phantom. Furthermore, the usage of feature-based calibration as initialization for an image-based method removes the necessity of manual initialization, making the process fully automatic.

The CAD model for 3D-printing our phantom, the machine learning model for segmentation, as well as a reference implementation are freely available online[1].

2   Approach

The main idea behind the proposed method is to automatically detect multiple distinguishable keypoints in two sweeps, i.e. freehand US recordings slowly sweeping over the phantom, and match them to estimate the calibration matrix.

2.1   Phantom and Model Preparation

Phantom Design. We design a phantom composed of nine cones of different heights as depicted in Fig. 1a. The tips of the cones are used as keypoints, uniquely identifiable by their height from the base plane. To avoid confusion of cones of similar size, we make sure tips with comparable heights are not placed next to each other. Cone heights are spaced uniformly in the range [20 mm, 60 mm], a wider range would make cones more distinguishable but would also limit the range of usable imaging depths. We have found that using 9 cones is a good compromise between the number of keypoints and the number of mismatches due to similar cone heights. Our phantom design can be easily 3D-printed and produced in different scales depending on the image shape and depth.

Model Architecture and Training. A convolutional neural network [11,12, 17] is used to segment every frame of the sweeps, distinguishing background, cones and base plane. We follow the U-Net [22] architecture and make use of residual blocks [13], leaky ReLUs [18] and instance normalization [24]. The network does not make use of the temporal contiguity of frames but processes each frame independently. Augmentation on the training data is used to make the model as robust as possible against reflections and changes of probe. In particular, we re-scale every frame to $1mm$ spacing, crop a random 128×128 section, apply Cutout [9] with random intensity, add speckle noise at a resolution of 64×32 and Gaussian noise at full resolution.

---

[1] https://github.com/ImFusionGmbH/PRO-TIP-Automatic-Ultrasound-Calibration.

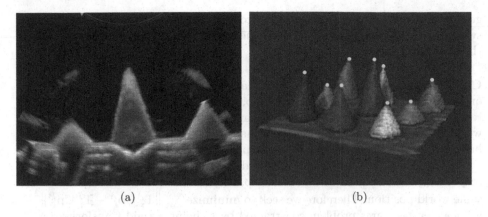

Fig. 2. Output of the segmentation model (a) and a compounded 3D labelmap with tip detections (b). Every *track* in (b) is visualized with a different color, distinguishing the different cones well. *See text for details.*

Data Labelling. We manually register tracked sweeps to the phantom's 3D label map (Fig. 1b) and use volume re-slicing to obtain 2D segmentation maps for every frame, Which allows us to quickly obtain a large dataset of labelled frames.

2.2   Calibration Method

The required input for the proposed method consists of two sets of freehand ultrasound frames, denoted A and B, from different principal directions (e.g. axial and sagittal), each covering ideally all phantom cones. The phantom must remain static throughout all recordings. We assume the image scale to be known and only consider a rigid calibration.

Tip Detection and Matching. After segmentation, every frame is processed to detect and track cones. First, a line is fitted to the base plane using RANSAC [10], if the number of inliers is too low the frame is skipped. Pixels located underneath the base and small connected components (< 25 mm^2) are discarded. The remaining connected components are considered cone detections and are tracked from frame to frame. We denote a list of detections corresponding to the same cone a *track*. Assignment of detections to tracks is done based on the area of the intersection of the bounding boxes. A new track is created every time a detection does not intersect with any existing track. Tracks are terminated if they do not receive new detections for more than 3 frames, as shown in Fig. 2b. After the last frame has been processed, tracks with less than 10 detections are discarded. For each detection in a track, we store the position of the highest point and its associated height. A mean smoothing filter is used to reduce the noise on height measurements. The point with maximum measured height is considered the tip of the cone track. Tips shorter than 1.5 cm or localized outside of the US geometry (when using convex or sector probes) are discarded. Tips from different sweeps are considered matches if their measured height difference is less than a threshold of 3 mm.

Calibration Estimation from Tip Matches. Given a point $\mathbf{p} = (x, y, 0, 1)^T$ on the i-th frame of an ultrasound sweep, its real world position is $\mathbf{q} = T_i C \mathbf{p} \in \mathbb{R}^4$, where T_i is the tracking matrix (mapping from the tracking target to the world origin) and C is the calibration matrix. Our goal is to estimate the calibration matrix C given n pairs of corresponding tip detections on frames.

Let $\mathbf{p}_i^A = (x_i^A, y_i^A, 0, 1)$ be the i-th tip detection on sweep A and p_i^B be the corresponding detection on the second sweep, these points should correspond to the same world position. Therefore we seek to minimize $\sum_i \left\| T_i^A C \mathbf{p}_i^A - T_i^B C \mathbf{p}_i^B \right\|^2$— a linear least-squares problem constrained by C being a rigid transformation. We manipulate the problem in order to bring it into the canonical least-squares formulation. The tracking and calibration matrices can be divided into blocks

$$
T = \left(\begin{array}{c|c} \Phi & \delta \\ \hline 0 & 1 \end{array} \right) \qquad C = \left(\begin{array}{c|c} R & t \\ \hline 0 & 1 \end{array} \right) \qquad R = \left(\begin{array}{ccc} | & | & | \\ \mathbf{u} & \mathbf{v} & \mathbf{w} \\ | & | & | \end{array} \right)
$$

where $\Phi, R \in \mathbb{R}^{3\times3}$ are rotation matrices, and $\mathbf{t}, \boldsymbol{\delta} \in \mathbb{R}^3$ are translation vectors. Note that, because $R^T R = I$ and $\det(R) = 1$, the column $\mathbf{w}$ must be $\mathbf{w} = \mathbf{u} \times \mathbf{v}$. Then:

$$
\sum_i \left\| T_i^A C \mathbf{p}_i^A - T_i^B C \mathbf{p}_i^B \right\|_2^2
$$

$$
= \sum_i \left\| (x_i^A \Phi_i^A - x_i^B \Phi_i^B) \mathbf{u} + (y_i^A \Phi_i^A - y_i^B \Phi_i^B) \mathbf{v} + (\Phi_i^A - \Phi_i^B) \mathbf{t} + \boldsymbol{\delta}_i^A - \boldsymbol{\delta}_i^B \right\|_2^2
$$

$$
= \left\| \underbrace{\left(\begin{array}{c|c|c} x_1^A \Phi_1^A - x_1^B \Phi_1^B & y_1^A \Phi_1^A - y_1^B \Phi_1^B & \Phi_1^A - \Phi_1^B \\ \vdots & \vdots & \vdots \\ x_n^A \Phi_n^A - x_n^B \Phi_n^B & y_n^A \Phi_n^A - y_n^B \Phi_n^B & \Phi_n^A - \Phi_n^B \end{array} \right)}_{=A} \left(\begin{array}{c} \mathbf{u} \\ \mathbf{v} \\ \mathbf{t} \end{array} \right) - \underbrace{\left(\begin{array}{c} \boldsymbol{\delta}_1^B - \boldsymbol{\delta}_1^A \\ \vdots \\ \boldsymbol{\delta}_n^B - \boldsymbol{\delta}_n^A \end{array} \right)}_{=d} \right\|_F^2
$$

We can then write the calibration problem as a constrained linear least squares problem with matrix of size $3n \times 9$:

$$
\underset{\mathbf{u},\mathbf{v},\mathbf{t}}{\arg\min} \left\| A \left(\begin{array}{c} \mathbf{u} \\ \mathbf{v} \\ \mathbf{t} \end{array} \right) - d \right\|_F^2 \qquad \text{subject to} \qquad \|\mathbf{u}\| = \|\mathbf{v}\| = 1 \text{ and } \langle \mathbf{u}, \mathbf{v} \rangle = 0. \quad (1)
$$

We solve the unconstrained problem, and project the solution $(\mathbf{u}', \mathbf{v}', \mathbf{t}')$ onto the constraints using SVD [14] obtaining $\mathbf{u}^*$ and $\mathbf{v}^*$. The translation $\mathbf{t}^*$ is then computed by solving the least squares problem while keeping $\mathbf{u} = \mathbf{u}^*$ and $\mathbf{v} = \mathbf{v}^*$ fixed. While it would be possible to iteratively improve the solution $(\mathbf{u}^*, \mathbf{v}^*, \mathbf{t}^*)$, for example using projected gradient, our approximation delivers results of sufficient quality for the image-based refinement to converge.

Table 1. Median fiducial pair distances of our method before and after image-based refinement, compared to the expert calibration.

	Initialization	Proposed		Expert
	Error	Error	Time	Error
Cephasonics 12 cm	0.88 mm	0.64 mm	17.3 s	**0.63 mm**
Cephasonics 15 cm	0.62 mm	**0.52 mm**	15.9 s	0.58 mm
Clarius linear	1.27 mm	**0.87 mm**	14.0 s	0.96 mm
GE 15 cm	1.28 mm	0.98 mm	24.4 s	**0.95 mm**

RANSAC Estimation. The tip matching procedure is subject to errors, therefore we use RANSAC [10] to robustly estimate the calibration matrix, disregarding the effect of erroneous matches. At every RANSAC iteration, we sample 4 pairs of tips and use them to produce a calibration hypothesis as described in the previous section. Although 3 pairs would be enough to solve Eq. 1, we observed that, because of noise and bad conditioning of the problem, it is beneficial to use 4 pairs. If the residual error of Eq. 1 is more than 100, the calibration hypothesis is immediately discarded. Otherwise the score of the hypothesis is computed by counting the numbers of inliers, i.e. pairs such that $\left\|T_i^A C \mathbf{p}_i^A - T_i^B C \mathbf{p}_i^B\right\|^2 \leq 1$ cm. The hypothesis with the highest score is refined by solving Eq. 1 again using all the inliers.

Image Based Refinement. The structure of our phantom produces ultrasound frames that exhibit many clear structures, therefore the calibration can be improved by using an image-based method. In particular, we use the method described in [26] on the same sweeps used to compute the calibration. This method refines the initial calibration to maximize the 2D normalized cross-correlation between frames of one sweep and in-plane reconstructions of all the slices of the other sweep.

Implementation Details. The algorithm is implemented as plugin to the ImFusion Suite 2.36.3 (ImFusion GmbH, Germany) using their SDK, including off-the-shelf implementations of the image-based calibration, volume reslicing, and PyTorch 1.8.0 integration for the machine learning model.

Limitations. Our algorithm needs the cone's tip and base to be visible in the same frame, in order to estimate its height. This limits the range of angles that the probe can cover and might limit the accuracy of the produced calibration. Our method is composed of multiple components which increase complexity of implementation. We mitigate this issue by releasing our source code. The proposed phantom design is not completely generic and would need to be adapted to specialized transducers for example end- or side-fire probes.

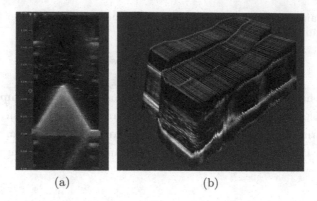

(a)   (b)

Fig. 3. A frame taken with the Clarius linear probe segmented by our model (a) and a 3D rendering of one of the sweeps used for the calibration of the Clarius probe (b). The image (a) exhibits different shape and content from frames taken with a convex probe (Fig. 2a). The sweeping motion (b) required to cover the whole phantom is unique to the linear probe and shows the generalization capabilities of our algorithm.

3   Experiments and Results

Dataset. We collect data using a variety of probes and tracking systems, to increase robustness of our method. Ultrasound frames are labelled as described in Sect. 2.1. Frames that do not show the base of the phantom or cones are discarded. We use the following hardware combinations: (1) Cicada research system (Cephasonics Inc., CA, USA) with convex probe tracked using a proprietary optical tracking system (Stryker Leibinger GmbH & Co KG, Germany). (2) LOGIQ E10S commercial system (GE Healthcare Inc., IL, USA) with a convex abdominal probe (GE C1-6) tracked using an Universal Robots, Denmark, UR5e robotic arm. (3) Clarius L7HD handheld portable linear probe (Clarius Mobile Health Inc., BC, Canada) and (4) Clarius C3 HD3 convex probe, both tracked using the same camera as in (1). We use sweeps from (1) and (2) as training data (2492 frames) except for two sweeps from each probe that are used for validation and testing (2095 frames). Sweeps from (3) and (4) are used exclusively for testing the entire calibration pipeline. By giving the same input sweeps as our method to a domain expert, we obtain a reference calibration. The expert manually initializes the calibration parameters, which consequently are optimized with the imaged-based calibration. These steps are potentially repeated until satisfactory overlap of sweep contents is achieved. We refer to the results as "Expert" calibration. For the evaluation, the cone tips are manually annotated in all sweeps by two users, and their positions after calibration in world coordinates are used as fiducials.

Results. We first evaluate whether the detected cone tips are correctly identified. The distance between cone tip annotations and detected tips are on average

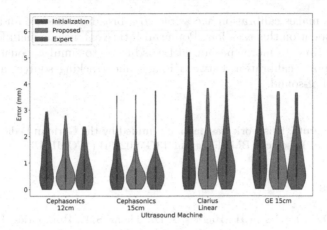

Fig. 4. Error distributions of fiducial pair distances of our method before (blue) and after image-based refinement (orange), and expert calibration with refinement (green) on data collected with different US systems. (Color figure online)

0.72 frames apart A few wrongly detected cones (later removed by RANSAC) cause the average distance to be 10.3 mm, but the median distance of 1.45 mm is well within the capture range of image-based refinement. The aim of our method is not to achieve better accuracy than current state-of-the-art methods, but rather match the results achieved by experienced users in a fraction of the time and without any human intervention. Therefore, we compare our results to the "Expert" calibration. Our calibration method is evaluated both with ("Proposed") and without the image-based refinement ("Initialization"). We compute the Euclidean distances between all $\binom{9}{2} = 36$ possible fiducial pairs and compare these to the ground truth distances of the phantom. Figure 4 depicts the distribution of errors obtained by different approaches on the datasets. It can be noticed that the results obtained by the proposed method achieve similar results to the expert calibration, both in average error and distribution shape. The comparably good results without image-based refinement show that our method can reliably produce an initialization within the narrow basin of attraction of the optimization method. Of particular interest are the results with the Clarius probe data, because frame geometry and sweep motion are different from any data used as training input (Fig. 3). Table 1 shows numerical results together with execution time of the Proposed method measured on a laptop running an Nvidia GTX1650. Runtime of our fully automatic method is in the order of seconds, while the manual approach takes several minutes.

4   Conclusion

In this paper, we proposed a fully automatic yet accurate approach to ultrasound calibration. We showed that our results are comparable to calibrations done by domain experts. Furthermore, the novel phantom design is easy to manufacture and use. This makes calibration accessible to a broader audience and keeps the overall time spent on this task low. Potential future extensions include phantom design adaptations to more specialized transducers, to simultaneously estimate also the temporal calibration between image and tracking source, and corrections of speed-of-sound.

Acknowledgment. This work was partially funded by the German Federal Ministry of Education and Research (BMBF), grant 13GW0293B ("FOMIPU").

References

1. Barratt, D.C., Davies, A.H., Hughes, A.D., Thom, S.A., Humphries, K.N.: Accuracy of an electromagnetic three-dimensional ultrasound system for carotid artery imaging. Ultrasound Med. Biol. **27**(10), 1421–1425 (2001)
2. Carbajal, G., Lasso, A., Gómez, Á., Fichtinger, G.: Improving n-wire phantom-based freehand ultrasound calibration. Int. J. Comput. Assist. Radiol. Surg. **8**(6), 1063–1072 (2013)
3. Chatrasingh, M., Suthakorn, J.: A novel design of N-fiducial phantom for automatic ultrasound calibration. J. Med. Phys. **44**(3), 191 (2019). http://www.jmp.org.in/text.asp?2019/44/3/191/266852
4. Chen, E.C., Lasso, A., Fichtinger, G.: External tracking devices and tracked tool calibration. In: Handbook of Medical Image Computing and Computer Assisted Intervention, pp. 777–794. Elsevier (2020)
5. Chen, T.K., Thurston, A.D., Ellis, R.E., Abolmaesumi, P.: A real-time freehand ultrasound calibration system with automatic accuracy feedback and control. Ultrasound Med. Biol. **35**(1), 79–93 (2009). https://doi.org/10.1016/j.ultrasmedbio.2008.07.004, https://linkinghub.elsevier.com/retrieve/pii/S0301562908003190
6. Cheng, A., Guo, X., Zhang, H.K., Kang, H.J., Etienne-Cummings, R., Boctor, E.M.: Active phantoms: a paradigm for ultrasound calibration using phantom feedback. J. Med. Imaging **4**(3), 035001 (2017)
7. Dandekar, S., Li, Y., Molloy, J., Hossack, J.: A phantom with reduced complexity for spatial 3-D ultrasound calibration. Ultrasound Med. Biol. **31**(8), 1083–1093 (2005)
8. Detmer, P.R., et al.: 3D ultrasonic image feature localization based on magnetic scanhead tracking: in vitro calibration and validation. Ultrasound Med. Biol. **20**(9), 923–936 (1994)
9. DeVries, T., Taylor, G.W.: Improved regularization of convolutional neural networks with cutout. arXiv preprint arXiv:1708.04552 (2017)
10. Fischler, M., Bolles, R.: Random sample consensus: a paradigm for model fitting with applications to image analysis and automated cartography. Commun. ACM **24**(6), 381–395 (1981)
11. Fukushima, K.: Neural network model for a mechanism of pattern recognition unaffected by shift in position-neocognitron. IEICE Tech. Rep. A **62**(10), 658–665 (1979)
12. Fukushima, K.: A self-organizing neural network model for a mechanism of pattern recognition unaffected by shift in position. Biol. Cybern. **36**, 193–202 (1980)

13. He, K., Zhang, X., Ren, S., Sun, J.: Deep residual learning for image recognition. In: Proceedings of the IEEE Conference on Computer Vision and Pattern Recognition, pp. 770–778 (2016)
14. Higham, N.J.: Matrix nearness problems and applications. In: Gover, M.J.C., Barnett, S. (eds.) Applications of Matrix Theory, pp. 1–27. Oxford University Press, Oxford (1989)
15. Hsu, P.W., Prager, R.W., Gee, A.H., Treece, G.M.: Freehand 3D ultrasound calibration: a review. Adv. Imaging Biol. Med. 47–84 (2009). https://doi.org/10.1007/978-3-540-68993-5_3
16. Krupa, A.: Automatic calibration of a robotized 3D ultrasound imaging system by visual servoing. In: Proceedings 2006 IEEE International Conference on Robotics and Automation, 2006. ICRA 2006, pp. 4136–4141. IEEE (2006)
17. Lecun, Y., Bottou, L., Bengio, Y., Haffner, P.: Gradient-based learning applied to document recognition. P. IEEE **86**(11), 2278–2324 (1998)
18. Maas, A.L., Hannun, A.Y., Ng, A.Y.: Rectifier nonlinearities improve neural network acoustic models. In: Proceedings ICML, vol. 30, p. 3. Citeseer (2013)
19. Meairs, S., Beyer, J., Hennerici, M.: Reconstruction and visualization of irregularly sampled three-and four-dimensional ultrasound data for cerebrovascular applications. Ultrasound Med. Biol. **26**(2), 263–272 (2000)
20. Mercier, L., Langø, T., Lindseth, F., Collins, D.L.: A review of calibration techniques for freehand 3-D ultrasound systems. Ultrasound Med. Biol. **31**(4), 449–471 (2005)
21. Muratore, D.M., Galloway, R.L., Jr.: Beam calibration without a phantom for creating a 3-D freehand ultrasound system. Ultrasound Med. Biol. **27**(11), 1557–1566 (2001)
22. Ronneberger, O., Fischer, P., Brox, T.: U-Net: convolutional networks for biomedical image segmentation. In: Navab, N., Hornegger, J., Wells, W.M., Frangi, A.F. (eds.) MICCAI 2015. LNCS, vol. 9351, pp. 234–241. Springer, Cham (2015). https://doi.org/10.1007/978-3-319-24574-4_28
23. Shen, C., Lyu, L., Wang, G., Wu, J.: A method for ultrasound probe calibration based on arbitrary wire phantom. Cogent Eng. **6**(1), 1592739 (2019)
24. Ulyanov, D., Vedaldi, A., Lempitsky, V.: Instance normalization: the missing ingredient for fast stylization. arXiv preprint arXiv:1607.08022 (2016)
25. Viswanathan, A., Boctor, E.M., Taylor, R.H., Hager, G., Fichtinger, G.: Immediate ultrasound calibration with three poses and minimal image processing. In: Barillot, C., Haynor, D.R., Hellier, P. (eds.) MICCAI 2004. LNCS, vol. 3217, pp. 446–454. Springer, Heidelberg (2004). https://doi.org/10.1007/978-3-540-30136-3_55
26. Wein, W., Khamene, A.: Image-based method for in-vivo freehand ultrasound calibration. In: Medical Imaging 2008: Ultrasonic Imaging and Signal Processing, vol. 6920, pp. 179–185. SPIE (2008)
27. Wen, T., Wang, C., Zhang, Y., Zhou, S.: A novel ultrasound probe spatial calibration method using a combined phantom and stylus. Ultrasound Med. Biol. **46**(8), 2079–2089 (2020)
28. Xiao, Y., Yan, C.X.B., Drouin, S., De Nigris, D., Kochanowska, A., Collins, D.L.: User-friendly freehand ultrasound calibration using Lego bricks and automatic registration. Int. J. Comput. Assist. Radiol. Surg. **11**(9), 1703–1711 (2016)

Multimodal-GuideNet: Gaze-Probe Bidirectional Guidance in Obstetric Ultrasound Scanning

Qianhui Men[1]([✉]), Clare Teng[1], Lior Drukker[2,3], Aris T. Papageorghiou[2], and J. Alison Noble[1]

[1] Institute of Biomedical Engineering, University of Oxford, Oxford, UK
qianhui.men@eng.ox.ac.uk
[2] Nuffield Department of Women's and Reproductive Health, University of Oxford, Oxford, UK
[3] Department of Obstetrics and Gynecology, Tel-Aviv University, Tel-Aviv, Israel

Abstract. Eye trackers can provide visual guidance to sonographers during ultrasound (US) scanning. Such guidance is potentially valuable for less experienced operators to improve their scanning skills on how to manipulate the probe to achieve the desired plane. In this paper, a multi-modal guidance approach (Multimodal-GuideNet) is proposed to capture the stepwise dependency between a real-world US video signal, synchronized gaze, and probe motion within a unified framework. To understand the causal relationship between gaze movement and probe motion, our model exploits multitask learning to jointly learn two related tasks: predicting gaze movements and probe signals that an experienced sonographer would perform in routine obstetric scanning. The two tasks are associated by a modality-aware spatial graph to detect the co-occurrence among the multi-modality inputs and share useful cross-modal information. Instead of a deterministic scanning path, Multimodal-GuideNet allows for scanning diversity by estimating the probability distribution of real scans. Experiments performed with three typical obstetric scanning examinations show that the new approach outperforms single-task learning for both probe motion guidance and gaze movement prediction. The prediction can also provide a visual guidance signal with an error rate of less than 10 pixels for a 224×288 US image.

Keywords: Probe guidance · Multimodal representation learning · Ultrasound navigation · Multitask learning

1 Introduction

Obstetric ultrasound (US) scanning is a highly-skilled medical examination that requires refined hand-eye coordination as the sonographer must look at a screen

Supplementary Information The online version contains supplementary material available at https://doi.org/10.1007/978-3-031-16449-1_10.

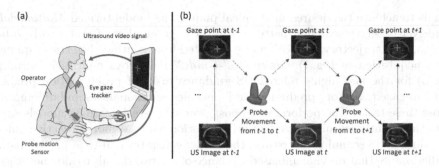

Fig. 1. Data acquisition and the correspondence between captured signals. (a) Overview of the multi-modality data acquisition in the clinical obstetric ultrasound scanning. (b) The unrolled US guiding process between the acquired US image, the probe motion signal, and the gaze signal.

and simultaneously manipulate a handheld probe. Computer-assisted scanning with probe motion guidance could improve the training process for non-specialists to develop their scanning skills, which has been increasingly investigated among researchers and clinicians [10,14,18]. Within the robotics field, work has focused on guiding operators to scan simple structures such as the liver [13], lumbar and vertebrae [11]. Such solutions are not feasible for obstetric scans because of the variety of fetal anatomy to be measured and the unpredictable fetal movement.

Previous studies in obstetric scanning guidance have proposed positioning the probe based on a behavioral cloning system [6] or landmark-based image retrieval [22]. In [6], different strategies are modeled for operators to either follow the next-action instruction or directly approach the anatomical Standard Plane (SP) [1]. Other work [18,19] deployed probe guidance signals to a robotic arm that is not practically applicable in a clinical environment. A common practice in these models is to treat probe guidance as an image-guided navigation problem. However, as multiple fetal anatomies can appear in a single US image, the gaze of the operator can provide instructive context about the likely next probe movement. Using gaze information to inform probe motion guidance has not been researched before now, and we explore this as the first aim of this work.

In addition to probe navigation, gaze information is also used as a guiding signal, usually in the form of gaze-point or saliency map (eye-tracking heat maps) prediction in US image or video. Cai et al. [2,3] leveraged visual saliency as auxiliary information to aid abdominal circumference plane (ACP) detection, and Droste et al. [5] extended it to various anatomical structures which is more applicable in real-time scanning guidance. Moreover, Teng et al. [17] characterized the visual scanning patterns from normalized time series scanpaths. Here, with the assumption that a sonographer will react to the next image inferred from their hand movement on probe, the second aim of this work is to explore whether probe motion is useful in guiding gaze.

In this work, we investigate how experienced sonographers coordinate their visual attention and hand movement during fetal SP acquisition. We propose the first model to provide useful guidance in both synchronized probe and gaze signals to achieve the desired anatomical plane. The model termed *Multimodal-GuideNet* observes scanning patterns from a large number of real-world probe motion, gaze trajectory, and US videos collected from routine obstetric scanning (data acquisition in Fig. 1). *Multimodal-GuideNet* employs multitask learning (MTL) for the two highly-related US guidance tasks of probe motion prediction and gaze trajectory prediction, and identifies commonalities and differences across these tasks. The performance boost over single-task learning models suggests that jointly learning gaze and probe motion leads to more objective guidance during US scanning. Moreover, the model generates real-time probabilistic predictions [8] that provide unbiased guidance of the two signals to aid operators.

2   Methods

Figure 1 outlines the principles of the approach. The probe orientation is recorded in 4D quaternions by an inertial measurement unit (IMU) motion sensor attached to the US probe, and the 2D gaze-point signal is captured by an eye-tracking sensor mounted on the bottom of the US screen. Given an US image starting at a random plane, its change in gaze between neighbour time steps, and its corresponding probe rotation, our multitask model *Multimodal-GuideNet* estimates the instructive next-step movements of both the gaze and probe for the SP acquisition. The two tasks: probe motion prediction and gaze shift prediction complement each other for more accurate US scanning guidance. The problem definition and network architecture are as follows.

2.1   Problem Formulation

Unlike previous US guidance models that only predict a fixed action, we regard the gaze and probe movements as random variables to account for inter- and intra-sonographer variation. For a more continuous prediction, the relative features are used from neighbour frames of these two modalities. Let $s_t = g_t - g_{t-1}$ be the shift of gaze point $g = (x, y)$ at time t and $r_t = q_{t-1}^* q_t$ be the rotation from the probe orientation $q = (q_w, q_x, q_y, q_z)$, where q^* is the conjugate. We make the assumption that the gaze shift s_t follows a bi-variate Gaussian distribution, i.e., $s_t \sim \mathcal{N}(\mu_t^s, \sigma_t^s, \rho_t^s)$, where μ_t^s and σ_t^s denote the mean and standard deviation respectively in 2D, and ρ_t^s is the correlation coefficient between x and y. Therefore, at every step, the model outputs a 5D vector for gaze estimation. Similarly, we achieve a 14D vector for probe rotation r_t which follows a multi-variate Gaussian distribution $r_t \sim \mathcal{N}(\mu_t^r, \sigma_t^r, \rho_t^r)$. The multitask objective for model training is to jointly minimize the negative log-likelihoods of the two learning tasks

$$\mathcal{L} = \sum_{t=t_0}^{T} \left(-\lambda_s \log(\mathbb{P}(s_t | \mu_t^s, \sigma_t^s, \rho_t^s)) - \lambda_r \log(\mathbb{P}(r_t | \mu_t^r, \sigma_t^r, \rho_t^r)) + \eta(1 - \|\mu_t^r\|^2)^2 \right),$$

$$(1)$$

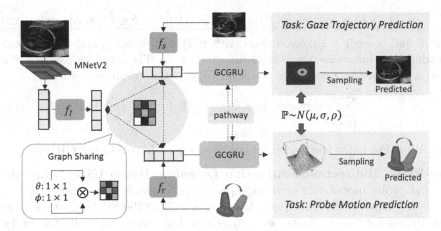

Fig. 2. Flowchart of *Multimodal-GuideNet* for a single time step. The two tasks share a modality-aware spatial graph from the three modalities.

where t_0 and T are the start and end indices for prediction. λ_s, and λ_r control the training ratio of the two tasks and are both set to 1. η is the weighting parameter for the quaternion prior to normalize $\boldsymbol{\mu}^r$ with $\eta = 50$.

2.2 Multimodal-GuideNet

To facilitate multitask learning, *Multimodal-GuideNet* constructs a lightweight graph shared among the three modalities. The network backbone is formed by graph convolutional Gated Recurrent Unit (GCGRU) [12] that automatically allocates useful dependencies within the graph at each time step. The designed lightweight spatial graph is also computationally efficient for online inference. Temporally, the gaze and probe dynamics complement each other through a bidirectional pathway. The entire multitask framework is presented in Fig. 2. To facilitate interactive learning within the graph structure, the input of each modality is embedded into an equal-sized 128-channel vector separately through a linear transformation block f_I, f_s, and f_r, each of which contains a fully-connected (FC) layer, a batch normalization (BN) layer, and a ReLU activation function. Before f_I, the grayscale US image is initially mapped to a flattened image representation $\boldsymbol{I}$ [6] with MobileNetV2 (MNetV2) [16].

Modality-Aware Graph Representation Sharing. To model spatial proximity at time t, we propose a common graph structure $G_t = (\mathcal{V}_t, \mathcal{E}_t)$ that is shared among the three modalities, where $\mathcal{V}_t = \{f_I(\boldsymbol{I}_t), f_s(\boldsymbol{s}_t), f_r(\boldsymbol{r}_t)\}$ is the vertex set with 3 nodes. $\mathcal{E}_t$ is the edge set specified by a 3×3 adaptive adjacency matrix $A_t + M_t$ with the first term indicating the spatial relationship within $\mathcal{V}_t$, and the second term a trainable adjacency mask [20] to increase graph generalization. Inspired by [21], the edge weight of any two nodes in A_t is formed by the affinity between the corresponding two modality features in the embedded space

$$A_t(j,k) = \text{softmax}(\theta(f_j(j_t))^T \phi(f_k(k_t))), \quad j,k \in \{I, s, r\} \tag{2}$$

where θ and ϕ are 1×1 convolutions with $\theta(x), \phi(x) \in \mathbb{R}^{256}$, and the *softmax* operation is to normalize the row summation of A_t. The message passed for s and r is therefore aggregated by one layer of a spatial graph convolution

$$\sum_{k \in \{I,s,r\}} \text{sigmoid}(A_t(j,k)f_k(k_t)W_j), \quad j \in \{s, r\} \tag{3}$$

where W_j is the input feature kernel specified for each gate in the GRU cell.

Gaze-Probe Bidirectional Adaptive Learning. During US scanning, the gaze and probe movements of sonographers are generally heterogeneous, i.e., they do not move at the same pace. Upon approaching the SP, the gaze is more prone to rapid eye movements between anatomical structures while the probe remains steady. We account for this effect by enclosing a bidirectional inverse adaptive *pathway* between hidden states of s and r in the time domain. Let h_t^s, h_t^r, $\tilde{h}_t^s$, $\tilde{h}_t^r$, and z_t^s, z_t^r refer to the hidden state, candidate activation, and update gate of s and r in GRU at time t respectively, we replace the original hidden state update $(1 - z_t) \odot h_{t-1} + z_t \odot \tilde{h}_t$ with:

$$h_t^s = \underbrace{\boldsymbol{\alpha}(1 - z_t^s) \odot h_{t-1}^s + \boldsymbol{\alpha} z_t^s \odot \tilde{h}_t^s}_{\text{update from gaze}} + \underbrace{(1 - \boldsymbol{\alpha})z_t^r \odot h_{t-1}^s + (1 - \boldsymbol{\alpha})(1 - z_t^r) \odot \tilde{h}_t^s}_{\text{inverse update from probe}}$$

$$h_t^r = \underbrace{\boldsymbol{\beta}(1 - z_t^r) \odot h_{t-1}^r + \boldsymbol{\beta} z_t^r \odot \tilde{h}_t^r}_{\text{update from probe}} + \underbrace{(1 - \boldsymbol{\beta})z_t^s \odot h_{t-1}^r + (1 - \boldsymbol{\beta})(1 - z_t^s) \odot \tilde{h}_t^r}_{\text{inverse update from gaze}}$$

$$\tag{4}$$

where $\boldsymbol{\alpha}, \boldsymbol{\beta}$ are the adaptive channel-wise weights for z_t^s and z_t^r, respectively, and $\odot$ is element-wise product. The number of hidden channels is set to 128 which is the same as $\boldsymbol{\alpha}$ and $\boldsymbol{\beta}$. With the proposed bidirectional pathway, the gaze and probe signals will adapt the domain-specific representation from each other to generate a more accurate scanning path. Other than the input operation for all gates (Eq. 3) and an adaptive hidden state (Eq. 4) for the output, we follow the operations in a standard GRU [4] to transfer temporal information.

3   Experiments

3.1   Data

The data used in this study were acquired from the PULSE (Perception Ultrasound by Learning Sonographic Experience) project [7]. The clinical fetal ultrasound scans were conducted on a GE Voluson E8 scanner (General Electric, USA) and the video signal was collected lossless at 30 Hz. The corresponding gaze tracking data was simultaneously recorded as (x, y) coordinates at 90 Hz with a Tobii Eye Tracker 4C (Tobii, Sweden). The probe motion was recorded with an IMU (x-io Technologies Ltd., UK) attached to the probe cable outlet as shown in Fig. 1(a). Approval from UK Research Ethics Committee was obtained for this study and written informed consent was also given by all participating sonographers and pregnant women. In total, there are 551 2nd and 3rd trimester scans carried out by 17 qualified sonographers. All three-modality data were downsampled to 6 Hz to reduce the time complexity.

3.2 Experimental Settings

The video frames were cropped to 224×288 and irrelevant graphical user interface information was discarded. To facilitate image representation learning, we pre-train MNetV2 with a large number of cropped US frames under the 14 SonoNet standard plane classifier [1] following the processing step of [6]. The clinical SP type is recognised automatically by Optical Character Recognition (OCR) with a total of 2121 eligible acquisitions labelled. For each acquisition, a multimodal data sample is selected 10s before the SP, which is the time for probe refinement. The raw gaze point is scaled to $(-0.5, 0.5)$ with the image center kept invariant, and the predicted μ_t^s is also normalized to the same range by *sigmoid* activation and a shift factor 0.5 before the minimization of multitask objective. The ratio of train:test is 4:1. In the training stage, we randomly select 32 continuous frames in each sample. The model is evaluated for three biometry SPs which are trans-ventricular plane (TVP), abdominal circumference plane (ACP), and femur standard plane (FSP) [15]. The AdamW optimizer is adopted with an initial learning rate of 1e−3 decayed by 1e−2 every 8 epochs. The whole network is first trained on all 14 classes of SPs for 20 epochs and separately fine-tuned for TVP, ACP, and FSP for 16 epochs.

3.3 Metrics and Baselines

We evaluate two probe scenarios: *Coarse Adjustment* where probe rotation angle to SP $\geq 10°$, and *Fine Adjustment* $\leq 10°$. The ratio of the two stages may vary from sample to sample and thus prediction performance is averaged among all frames in the same stage. For our method, we randomly sample 100 trajectories from the predicted distribution and average them as a final prediction $\hat{r}$ and $\hat{s}$. The two tasks are separately evaluated with different metrics: Probe movement is considered as correctly predicted if it is rotating towards the next target plane, i.e., $\angle(q_{t-1}\hat{r}_t, q_t) \leq \angle(q_{t-1}, q_t)$; The predicted gaze point $\hat{g}_t = g_{t-1} + \hat{s}_t$ is evaluated by pixel l_2 norm error. We compare our multitask model with two baselines and two single-task architectures: *Baseline (r)*, continuing the previous probe rotation at the current time step; *Baseline (g)*, using the previous gaze point at the current time step; *US-GuideNet* [6], single-task learning approach for probe guidance, where only probe motion is modeled and predicted from US video; *Gaze-GuideNet*, single-task learning approach for gaze prediction, where only gaze information is modeled and predicted from US video by discarding the probe stream from Multimodal-GuideNet.

4   Results and Discussion

4.1   Probe Motion Guidance

A detailed performance comparison for the probe guidance task is presented in Fig. 3. *Multimodal-GuideNet* achieves an overall consistent improvement over the single task-based *US-GuideNet* [6] for the two adjustment stages, which indicates that simultaneously learning the gaze patterns benefits the probe motion planning. The probe rotation for the femur (FSP) is difficult to predict when it gets close to the SP (at 0°). Different from a steady probe movement to achieve TVP and ACP, the probe manipulation close to FSP requires complicated twisting actions [15]. This also explains why incorporating gaze contributes more in the coarse adjustment (as at 30°) to locate the femur but not in the fine stage (as at 10°). Moreover, the flexible movements of fetal limbs increase the diversity in FSP detection, which explains why there is a slightly higher standard deviation observed for this plane. In the 5th subplot (w/ vs. w/o bi-path), the improvements indicate that the pathway between gaze and probe stabilizes the probe movement with a more accurate prediction, especially in fine adjustment.

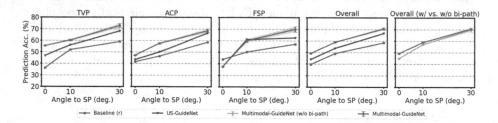

Fig. 3. Probe rotation accuracy (the higher the better) on the 3 evaluated standard planes and the overall prediction with ablations. The shaded area indicates the standard deviation of our model across all 100 samplings. *bi-path* signifies bidirectional pathway.

4.2   Gaze Trajectory Prediction

Figure 4 shows the prediction results for the gaze task. A common observation among all three planes is that for gaze prediction, the error of fine adjustment is generally larger than coarse adjustment. This is because in contrast to the fine-grained probe motion, eye gaze movement during that time is quite rapid, flitting between the observed anatomical structures. When comparing the three planes, the error ranges are the lowest for ACP and the highest for FSP especially for the fine adjustment. Since the key anatomical structures in ACP are relatively close to each other, the sonographer requires a smaller change in gaze. For FSP, sonographers switch focus between both femur ends which increases the uncertainty of the gaze state in the next time step. Comparing between methods, *Multimodal-GuideNet* reduces the error of *Gaze-GuideNet* for all cases, which demonstrates

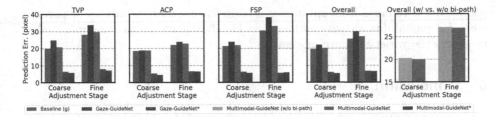

Fig. 4. Gaze prediction error (the lower the better) on the 3 evaluated standard planes and the overall prediction with ablations. The error of the best-generated gaze point that is closest to ground truth is reported in *Gaze-GuideNet** and *Multimodal-GuideNet**, respectively.

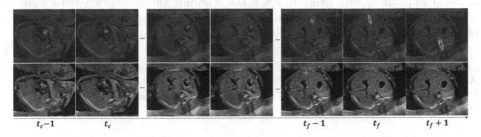

Fig. 5. Visualization of predicted saliency map (top row), gaze point (bottom row, red star), and corresponding ground truth gaze point (green star) for an ACP searching sequence. t_c and t_f are timestamps for coarse and fine adjustment, respectively. (Color figure online)

the effectiveness of multitask learning over single-task learning in gaze prediction. The bidirectional pathway also slightly improves the gaze prediction as compared in the 5th subplot. As a common evaluation in sampling-based generative models [9], we also report the performance of our best gaze point prediction among all samplings in *Gaze-GuideNet** and *Multimodal-GuideNet**. Their errors are within 10 pixels which shows the feasibility of the learned distribution in generating a plausible gaze trajectory. Practically, *Multimodal-GuideNet** could be useful when a precise gaze is needed such as when the sonographer focuses over a small range of underlying anatomical structure, and its improvement over *Gaze-GuideNet** indicates probe guidance could potentially help locate such a fixation point.

Figure 5 shows an example of predicted visual saliency and gaze point deduced from the generated gaze shift distribution. The predictions are highly accurate in all timestamps except for a significant gaze shift at frame t_f. However, the predicted saliency map at t_f correctly estimates the orientation of gaze shift. Saliency map-based numerical metrics are also evaluated in the supplementary material, where the multitask model generally outperforms the single-task one. In general, modeling the gaze information as a bi-variate distribution is technically advantageous over a saliency map-based predictor, as the problem complexity is reduced from optimizing a large feature map to only a few parameters for probability density estimation. The flexibility in gaze sampling also preserves the variety in gaze movements.

5 Conclusion

We have presented a novel multimodal framework for bidirectional guidance between probe motion and eye-tracking data in routine US scanning. We have explored multitask learning by jointly predicting the probe rotation and gaze trajectory from US video via a shared modality-aware graph structure. The performance gains over single-task predictions suggest that the two-modality signals complement each other to reach the scanning target, while ignoring any of them will lead to a biased guidance. The learned guidance signals with probability distributions also allow for diversity between individual scans in a practical environment.

Acknowledgements. We acknowledge the ERC (ERC-ADG-2015 694581, project PULSE), the EPSRC (EP/MO13774/1, EP/R013853/1), and the NIHR Oxford Biomedical Research Centre.

References

1. Baumgartner, C.F., et al.: SonoNet: real-time detection and localisation of fetal standard scan planes in freehand ultrasound. IEEE Trans. Med. Imaging **36**(11), 2204–2215 (2017)
2. Cai, Y., Sharma, H., Chatelain, P., Noble, J.A.: Multi-task SonoEyeNet: detection of fetal standardized planes assisted by generated sonographer attention maps. In: International Conference on Medical Image Computing and Computer-Assisted Intervention (MICCAI), pp. 871–879 (2018)
3. Cai, Y., Sharma, H., Chatelain, P., Noble, J.A.: SonoEyeNet: standardized fetal ultrasound plane detection informed by eye tracking. In: IEEE International Symposium on Biomedical Imaging (ISBI), pp. 1475–1478 (2018)
4. Cho, K., van Merrienboer, B., Gulcehre, C., Bougares, F., Schwenk, H., Bengio, Y.: Learning phrase representations using RNN encoder-decoder for statistical machine translation. In: Conference on Empirical Methods in Natural Language Processing (EMNLP) (2014)
5. Droste, R., et al.: Ultrasound image representation learning by modeling sonographer visual attention. In: International Conference on Information Processing in Medical Imaging, pp. 592–604 (2019)
6. Droste, R., Drukker, L., Papageorghiou, A.T., Noble, J.A.: Automatic probe movement guidance for freehand obstetric ultrasound. In: International Conference on Medical Image Computing and Computer-Assisted Intervention (MICCAI), pp. 583–592 (2020)
7. Drukker, L., et al.: Transforming obstetric ultrasound into data science using eye tracking, voice recording, transducer motion and ultrasound video. Sci. Rep. **11**(1), 1–12 (2021)

8. Graves, A.: Generating sequences with recurrent neural networks. arXiv:1308.0850 (2013)
9. Gupta, A., Johnson, J., Fei-Fei, L., Savarese, S., Alahi, A.: Social GAN: socially acceptable trajectories with generative adversarial networks. In: Proceedings of the IEEE Conference on Computer Vision and Pattern Recognition (CVPR), pp. 2255–2264 (2018)
10. Housden, R.J., Treece, G.M., Gee, A.H., Prager, R.W.: Calibration of an orientation sensor for freehand 3D ultrasound and its use in a hybrid acquisition system. Biomed. Eng. Online **7**(1), 1–13 (2008)
11. Li, K., et al.: Autonomous navigation of an ultrasound probe towards standard scan planes with deep reinforcement learning. In: IEEE International Conference on Robotics and Automation (ICRA), pp. 8302–8308 (2021)
12. Li, Y., Zemel, R., Brockschmidt, M., Tarlow, D.: Gated graph sequence neural networks. In: International Conference on Learning Representations (ICLR) (2016)
13. Mustafa, A.S.B., et al.: Development of robotic system for autonomous liver screening using ultrasound scanning device. In: IEEE International Conference on Robotics and Biomimetics (ROBIO), pp. 804–809 (2013)
14. Prevost, R., et al.: 3D freehand ultrasound without external tracking using deep learning. Med. Image Anal. **48**, 187–202 (2018)
15. Salomon, L.J., et al.: Practice guidelines for performance of the routine mid-trimester fetal ultrasound scan. Ultrasound Obstet. Gynecol. **37**(1), 116–126 (2011)
16. Sandler, M., Howard, A., Zhu, M., Zhmoginov, A., Chen, L.C.: MobileNetV2: inverted residuals and linear bottlenecks. In: Proceedings of the IEEE Conference on Computer Vision and Pattern Recognition (CVPR), pp. 4510–4520 (2018)
17. Teng, C., Sharma, H., Drukker, L., Papageorghiou, A.T., Noble, J.A.: Towards scale and position invariant task classification using normalised visual scanpaths in clinical fetal ultrasound. In: International Workshop on Advances in Simplifying Medical Ultrasound, pp. 129–138 (2021)
18. Toporek, G., Wang, H., Balicki, M., Xie, H.: Autonomous image-based ultrasound probe positioning via deep learning. In: Hamlyn Symposium on Medical Robotics (2018)
19. Wang, S., et al.: Robotic-assisted ultrasound for fetal imaging: evolution from single-arm to dual-arm system. In: Annual Conference Towards Autonomous Robotic Systems, pp. 27–38 (2019)
20. Yan, S., Xiong, Y., Lin, D.: Spatial temporal graph convolutional networks for skeleton-based action recognition. In: Thirty-Second AAAI Conference on Artificial Intelligence (2018)
21. Zhang, P., Lan, C., Zeng, W., Xing, J., Xue, J., Zheng, N.: Semantics-guided neural networks for efficient skeleton-based human action recognition. In: Proceedings of the IEEE Conference on Computer Vision and Pattern Recognition (CVPR), pp. 1112–1121 (2020)
22. Zhao, C., Droste, R., Drukker, L., Papageorghiou, A.T., Noble, J.A.: Visual-assisted probe movement guidance for obstetric ultrasound scanning using landmark retrieval. In: International Conference on Medical Image Computing and Computer-Assisted Intervention (MICCAI), pp. 670–679 (2021)

USPoint: Self-Supervised Interest Point Detection and Description for Ultrasound-Probe Motion Estimation During Fine-Adjustment Standard Fetal Plane Finding

Cheng Zhao[1]([✉]), Richard Droste[1], Lior Drukker[2], Aris T. Papageorghiou[2], and J. Alison Noble[1]

[1] Institute of Biomedical Engineering, University of Oxford, Oxford, UK
cheng.zhao@eng.ox.ac.uk
[2] Nuffield Department of Women's and Reproductive Health, University of Oxford, Oxford, UK

Abstract. Ultrasound (US)-probe motion estimation is a fundamental problem in automated standard plane locating during obstetric US diagnosis. Most recent existing recent works employ deep neural network (DNN) to regress the probe motion. However, these deep regression-based methods leverage the DNN to overfit on the specific training data, which is naturally lack of generalization ability for the clinical application. In this paper, we are back to generalized US feature learning rather than deep parameter regression. We propose a self-supervised learned local detector and descriptor, named USPoint, for US-probe motion estimation during the fine-adjustment phase of fetal plane acquisition. Specifically, a hybrid neural architecture is designed to simultaneously extract a local feature, and further estimate the probe motion. By embedding a differentiable USPoint-based motion estimation inside the proposed network architecture, the USPoint learns the keypoint detector, scores and descriptors from motion error alone, which doesn't require expensive human-annotation of local features. The two tasks, local feature learning and motion estimation, are jointly learned in a unified framework to enable collaborative learning with the aim of mutual benefit. To the best of our knowledge, it is the first learned local detector and descriptor tailored for the US image. Experimental evaluation on real clinical data demonstrates the resultant performance improvement on feature matching and motion estimation for potential clinical value. A video demo can be found online: https://youtu.be/JGzHuTQVlBs.

Keywords: Obstetric US · Probe motion · Local detector and descriptor

Supplementary Information The online version contains supplementary material available at https://doi.org/10.1007/978-3-031-16449-1_11.

1   Introduction

Motivation: Ultrasound (US) scan is an indispensable diagnostic tool in obstetrics care because of its safety, real-time nature and relatively low cost. However, obstetric US scanning is highly experienced-operator dependent. Automatic probe movement prediction for acquiring standard imaging planes might assist less-experienced users to perform scanning confidently. In the case of obstetric US scanning, it is typically not difficult for an operator to find a coarse approximate position for a standard imaging plane (searching); However, locating the probe position to accurately locate a diagnostically acceptable standard plane (fine-tuning) is not easy. Searching probe motion involves sharply translation and rotation, while fine-tuning probe motion is predominately rotation. Most of the time, probe movements in the fine-adjustment stage are rotation with very small translation. The proposed method focuses on the latter problem.

During the training of the trainee sonographer, the high-quality standard plane is stored in advance captured by expert sonographer. When close to the target plane of interest, it translates to US image-probe motion prediction between the current and target US image-pair. The predicted motion guides the trainee to capture the saved standard/target plane more efficiently to increase their practical scan ability. For the situation without pre-captured high-quality standard plane, the learned local feature can be inserted as a feature extraction frontend with a time-series model, e.g. LSTM and Transformer, for a series of motion predictions. The time-series predicted motions could guide the non-expert to choose the optimal action to approach the standard/target plane.

Related Work: Existing methods of US-probe motion estimation can be grouped into heuristic methods and data-driven deep regression methods. The former inferences the probe motion from elaborate hand-designed heuristics such as speckle decorrelation [2,7,14] extracted from the US image. The latter [6,8,12] takes advantage of the powerful non-linearity of DNN to regress the probe motion from the large-scale US-probe data. These learning-based regression methods leverage the DNN to overfit on the training data of some specific users. As mentioned in [17], the main limitation is naturally the lack of generalization ability in real clinical applications. So our proposed method is back to feature extraction and description. Due to the peculiar data structure, it is difficult to extract local features directly from the US image. Our target is to learn a general US local feature from the large-scale US-probe data recorded by experienced sonographers. The classic local features from computer vision such as SIFT [10], ORB [15] are not tailored for US images result in inferior performance of interest points detection and description on US image due to the domain variance between natural grey-style images and US images. Recently, the deep learned local feature [3,5,16] significantly improve the local feature extraction and matching on the natural images. However, these approaches are difficult to apply to US images due to the lack of geometry information (feature correspondence, known pose and depth) for the network training.

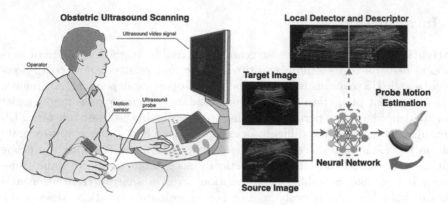

Fig. 1. The pipeline of the proposed approach.

Contribution: Instead of considering local feature extraction and probe motion estimation separately, we present a new way of jointly thinking about these two problems on US images. We present a self-supervised local feature, e.g. USPoint to predict keypoint locations, scores and descriptors for US-probe motion estimation during fine-adjustment standard plane finding. It is achieved by embedding a differentiable USPoint-based motion estimation inside the proposed network architecture, supervised only by automatically obtained ground truth (GT) probe motion information. The reason why we named it self-supervised learning is that similar to the stereo image or audio signals, the position information can be captured by the motion sensor naturally during clinical data collection. As Fig. 1 shows, US-probe motion estimation between the current and the target US images is achieved by learned local feature extraction and further deep motion estimation. Compared to conventional, hand-designed heuristics, our method learns priors over geometric transformations and regularities from data directly. Compared to conventional, deep parameter regression, our method is back to generalized US feature learning. In summary, our proposed network includes the following new features: I) a tailored US image-based local detector and descriptor are learned in a self-supervised manner without expensive human-annotation of local feature; II) the probe motion estimation is achieved by USPoint-based regression from the learned local features with attentional graph matches; III) these two tasks are jointly learned in a unified framework to enable collaborative learning for mutual benefit.

2   Methodology

Overview: As Fig. 2 shows, the proposed network architecture consists of four stages: feature encoder, local detector and descriptor, attentional graph matching and motion estimation. Stage one, feature encoder is composed of a series of convolution stacks, which transforms the US image to a dense high-dimensional feature representation. Stage two, local detector and descriptor, is a three-stream

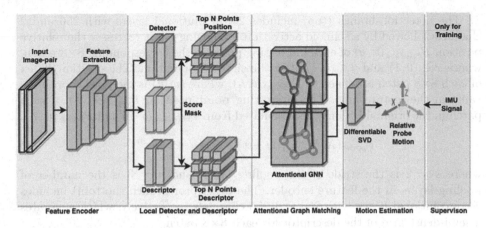

Fig. 2. The four-stage network architecture: feature encoder, local detector and descriptor, attentional graph matching and motion estimation. The IMU is only used to provide GT for training, while it is not required during inference.

architecture designed to jointly learn the pixel-wise position and feature vector of interesting points. It extracts the sparse feature representation of detected points from the dense feature representation of stage one. Stage three, attentional graph matching aggregates the feature representation of each interesting point from the other salient points, and further finds the correspondence according to the descriptors. Stage four, motion is estimated via the differentiable singular value decomposition (SVD) from the matched interesting point pairs. During training, the motion sensor (IMU) attached on the probe provides a motion signal as the supervision, while it is not required in the deployment. The complete network is trained in an end-to-end way without expensive human-annotations.

Feature Encoder: The feature encoder is a VGG-style [18] sub-network, which takes an US image as input and generates a dense, low spatial and high-dimensional feature representation. Each pixel in the feature map refers to an 8×8 pixel patch in the original US image. This dense feature representation is then fed to the local detector and descriptor to extract the sparse feature representation of each interesting point. The specific architecture and parameter are provided in the Supplementary Material.

Local Detector and Descriptor: As shown in Fig. 2, the local detector and descriptor jointly learn the position and feature vector of each interesting point, which is composed of three head-branches for score (middle), detection (top) and description (bottom). The score-branch (middle) includes 2 convolutional layers with 256 and 1 channels followed by a sigmoid activation which bounds score predictions S in the interval $(0, 1)$. It regresses the probability of interesting points in each 8×8 pixel patch, so that the top N interesting points can be selected from the original US image according to the scores.

The detection-branch (top) includes 2 convolutional layers with 256 and 2 channels followed by a sigmoid activation. Following [3], it regresses the relative position $P_{relative}(x, y)$ of each interesting point in the corresponding 8×8 patch, where $x \in (0, 1)$ and $y \in (0, 1)$ after the sigmoid operation. The position/index of each 8×8 patch is defined as $P_{patch}(w, h)$, where (w, h) is the pixel coordinate value in the feature map. The interesting point position $P_{point}(X, Y)$ of each patch in the original US image is calculated from $P_{relative}(x, y)$ and $P_{patch}(w, h)$,

$$P_{point}(X, Y) = ((w + x) * s^n, (h + y) * s^n), \tag{1}$$

where $s = 2$ is the stride of the pooling layer, and $n = 3$ is the number of pooling layers in the feature encoder. The descriptor-branch (bottom) includes 2 convolutional layers with 256 and 256 channels respectively, which generates a semi-dense grid of the descriptor for each 8×8 patch.

To achieve a real-time performance and reduce the computational memory, up-sampling back to the full resolution through deconvolution operation is abandoned. Instead, bi-cubic interpolation is used to interpolate descriptors according to the positions of interesting points. The top N interesting points are selected according to their detection scores S from the score branch. And their position (X, Y) and associated descriptor $D \in \mathbb{R}$ are further extracted by the detection and descriptor brunches respectively. This sub-network transforms dense feature representation to sparse feature representation for the next stage.

Attentional Graph Matching: Because the local feature description of each interesting point is extracted from a small patch in the original US image, the associated feature representation suffers from ambiguities due to lack of contextual cues. Inspired by [16], an Attentional Graph Neural Network (AGNN) [20] is employed to aggregate the contextual cues from the other interesting points to increase the distinctiveness. To integrate both position and appearance information of an interesting point $i, i \in N$ into the AGNN, the interesting point position information (X_i, Y_i, S_i) is embedded into a high-dimensional feature representation via a Multilayer Perception (MLP). This feature representation is further combined with the initial visual descriptor D_i to define a new feature vector F_i as,

$$F_i = D_i + MLP(X_i, Y_i, S_i). \tag{2}$$

The interesting points in both the source and the target images are considered as nodes of the complete graph. This graph includes two types of undirected edges, intra-image edges and extra-image edges. Intra-image edges connect the current interesting point i with all the other interesting points j in the same image, while extra-image edges connect the current interesting point i with all the other interesting points j in the other image. A message passing formulation [1] M is used for information propagation along both the intra-image and extra-image edges. The high-dimensional feature vector of each graph node is updated as,

$$F_i = F_i + MLP(F_i \oplus M_{i,j}^{intra}) + MLP(F_i \oplus M_{i,j}^{extra}), \tag{3}$$

where $\oplus$ refers to the concatenation. The intra-image message $M_{i,j}^{intra}$ and extra-image message $M_{i,j}^{extra}$ are calculated via the self-attention mechanism [19],

$$M_{i,j}^{intra} = \sum_{j \in (intra)} \text{softmax}(\frac{F_i \cdot F_j^T}{\sqrt{d}})F_j, \; M_{i,j}^{extra} = \sum_{j \in (extra)} \text{softmax}(\frac{F_i \cdot F_j^T}{\sqrt{d}})F_j. \tag{4}$$

Here d denotes the dimension of the feature vector. The AGNN can automatically build connections between a current interesting point and nearby or similar salient interesting points, because both position and appearance information are encoded into the feature representation of each graph node.

Further, the pairwise scores $\gamma_{i,j}$ are calculated via the similarity of matching the descriptor F_i^{source} and F_j^{target} of the interesting point i and j in the source and target images,

$$\gamma_{i,j} = \langle F_i^{source}, F_j^{target} \rangle, \forall (i,j) \in source \times target, \tag{5}$$

where $\langle \cdot, \cdot \rangle$ refers to inner product. The pair-matching is optimized by the Sinkhorn algorithm [4] to select the matched interesting point pairs.

Motion Estimation: Given the matched correspondence pairs $P_i^{source}, P_j^{target}$ with the pairwise score $\gamma_{i,j}$, the 3D relative motion between the source and the target US images is regressed by the differentiable SVD. Firstly, the matched pixel positions P are feed into a Transform Net [13] to obtain a matrix $T \in \mathbb{R}_{3\times3}$. Then the pixel coordinate values $P_i^{source}, P_j^{target}$ are transformed to three-dimensional values $p_i^{source}, p_j^{target}$ through $p = T \times P$. T is a dynamic and deterministic matrix, which is directly regressed by the Transform Net from data. Their centroid values $\bar{p}^{source}, \bar{p}^{target}$ are further calculated as,

$$\bar{p}^{source} = \frac{1}{N}\sum_{i=1}^{N} p_i^{source}, \;\; \bar{p}^{target} = \frac{1}{N}\sum_{j=1}^{N} p_j^{target}, \tag{6}$$

where N is the number of matched pairs. Then the cross-covariance matrix Σ is given as,

$$\Sigma = \sum_{i,j=1}^{N} (p_i^{source} - \bar{p}^{source})\gamma_{i,j}(p_j^{target} - \bar{p}^{target})^\top. \tag{7}$$

Σ is further decomposed by SVD, and then the relative motion between source and target US images is obtained as,

$$U, \Delta, V^\top = SVD(\Sigma), \; R = VU^\top, \; t = \bar{p}^{target} - R\bar{p}^{source}. \tag{8}$$

$U, V \in SO(3)$ and Δ is the diagonal but potentially signed. $R \in SO(3)$ is the rotation matrix, and is further transformed to a quaternion Q. $t \in \mathbb{R}^3$ is the translation vector.

Crucially, the motion estimation is differentiable, which backpropagates the motion error to the local detector, descriptor and matching sub-networks. This

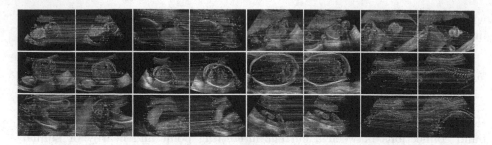

Fig. 3. Typical learned local detector, descriptor and matches for US image pairs. The color indicates the matching score (Jet colormap, blue = low, red = high). (Color figure online)

guarantees that the whole network is trainable in an end-to-end fashion. Most importantly, it builds a relationship between the local feature learning and the supervision signal from the IMU sensor, so that the local detector and descriptor are learned in a self-supervised way without expensive human-annotations. Moreover, the learned pairwise scores allow the matched interesting point pairs to contribute differently according to their confidences in the differentiable SVD. Thus, jointly learning the feature and motion enables collaborative learning between two sub-networks during backpropagation to minimise the loss function.

Loss Function: The US image pairs, i.e., source and target US images, are input to the network, and their relative 3D motion, i.e., translation t and quaternion Q, is predicted. The corresponding GT, i.e., translation t_{GT} and quaternion Q_{GT} measured by the IMU sensor is provided as the supervision for the network training. The loss function $\mathcal{L}$ is defined as,

$$\mathcal{L}(I^{source}, I^{target}) = \| t - t_{GT} \|_2 + \lambda(1 - |Q \cdot Q_{GT}|), \qquad (9)$$

where λ is the scale factor.

3   Experiments

Data Acquisition and Processing: The data used in this paper came from a large-scale multi-modal fetal US dataset *Anonymous* including synchronized US video and IMU signal. We consider the standard biometry planes of the 2nd and 3rd trimester, i.e. head circumference/trans-ventricular (HC/TV) plane, abdominal circumference (AC) plane and femur length (FL) plane. 47589 US image-pairs extracted from the video clips are then used for the proposed method training and testing, divided into five-fold cross validation. Please find more details on data processing and network training in the Supplementary Material.

Performance Evaluation: Typical learned local detector, descriptor and matches on the US image pairs are shown in Fig. 3. The colourized dots and lines refer to the detected interesting points and matched correspondences. The

Table 1. Performance comparison between the proposed method and baselines for local feature matching evaluation.

Method	MP (%)	MS (%)
SIFT [10] + FLANN [11]	45.06	6.58
SuperPoint [5] + SuperGlue [16]	66.57	33.16
Proposed Method	**74.72**	**53.98**

MP and MS refer to Matching Precision and Matching Score separately by percentage (%).

Table 2. Performance comparison between the proposed method and baseline for probe motion evaluation.

Method	50-frames		100-frames		150-frames		250-frames	
	PDP(%)	AO($°$)	PDP(%)	AO($°$)	PDP(%)	AO($°$)	PDP(%)	AO($°$)
Deep regression [12]	8.14	2.54	7.92	4.23	9.21	5.46	10.79	7.98
Proposed method	**4.76**	**1.99**	**4.83**	**2.91**	**5.46**	**3.23**	**6.27**	**5.45**

N-frames refers to the interval, i.e., number of frames between the US image pair. The average PDP and AO are evaluated by percentage (%) and degree ($°$).

Jet colourmap is used to represent their scores i.e. red refers to a high score while blue refers to a low score. We provide qualitative results on different anatomy planes such as the abdomen, heart, head, arm, foot, lip-nose and spine. Most of the matched correspondences maintain a good degree of global consistency with a few wrong lines crossing. The proposed method consistently predicts a high number of correct matches even within a lot of repeated texture. In order to demonstrate the robustness of learned local feature, the results for US image-pairs within different settings such as scale, crop and zoom settings are provided. We use four standard evaluation metrics: Matching Precision (MP), Matching Score (MS), Positional Drift Percentage (PDP) and Angular Offset (AO) used in the most popular benchmark datasets to analyse the experimental results. According to the metrics definitions, the MP, defined as (correct matches)/(predicted matches) is used for the learned descriptor evaluation, while the MS, defined as (correct matches)/(detected interesting points) is used for the learned detector evaluation. As GT for local feature evaluation, we manually annotated the correct matched correspondences on 300 US image pairs. It is very difficult to annotate the interesting points with matches directly in the US images. Our strategy is to manually select the corrected matches from the predicted matches by the proposed and baseline methods. In this case, manually annotating the interesting points in the US images is not required. For the probe translation evaluation, we use the positional drift percentage, defined as $\ell_2(\widehat{t} - t_{GT})/\ell_2(t_{GT})$, instead of the absolute positional error. Because the translation is usually very small during the US-probe fine-tuning, a small absolute positional error doesn't mean a good performance. For the probe rotation evaluation, we use the quaternion rather than Euler angle to represent 3D rotation because the Euler angle has a well-known Gimbal lock problem. Following PoseNet [9], the AO is defined as $2arccos(\widehat{Q} \cdot Q_{GT})$. The probe motion GT is directly obtained from the motion sensor (IMU) or US-Simulator. Note that the IMU and US-Simulator are only used to provide GT for network training and performance evaluation, while they are not required in practical inference.

As the baseline methods, we chose geometry-style SIFT [10] with FLANN [11], and learning-style SuperPoint [5] with SuperGlue [16] for the local feature matching comparison. We chose deep regression [12] method for the motion estimation comparison between two US images. The only difference is that we only use US image for the motion estimation for a fair comparison due to the proposed method only uses US image during inference. The original method [12] use both US image and IMU signal to predict the motion. From Table 1, we can see that the MP and MS of proposed method outperform both the geometry and learning based methods. The main reason is that our proposed local detector and descriptor are tailored for US image while the baselines are designed for natural images. When comparing with the MP and MS of the image-pairs from different anatomy standard planes, we did not find significant differences. From Table 2, it can be seen that the average PDP and AO of the proposed method outperform the baseline with different frame-intervals. For deep motion regression, elaborate matched local features provide more geoemtry cues than the original US images.

4   Conclusions

In this paper, we present a novel approach for jointly local feature extraction and motion estimation during the fine-adjustment stage of standard fetal plane finding in obstetric US scan. A novel local feature USPoint, including interest point detector, score and descriptor, tailored for US image is learned in a self-supervised manner. The probe motion is further estimated based on the learned local feature detection, description and matching. The collaborative learning between the local feature learning and motion estimation benefits both sub-networks. We will further verify the generalization ability of the USPoint applied in other US-based tasks in the next step. The future work will combine the US landmark retrieval [21] with this USPoint-based motion estimation to achieve a course-to-fine visual-assisted intervention [22] to guide US-probe movement.

Acknowledgments. We acknowledge the ERC (ERC-ADG-2015 694581, project PULSE), the EPSRC (EP/MO13774/1, EP/R013853/1), and the NIHR Biomedical Research Centre funding scheme.

References

1. Battaglia, P.W., et al.: Relational inductive biases, deep learning, and graph networks. arXiv preprint arXiv:1806.01261 (2018)
2. Chen, J.F., Fowlkes, J.B., Carson, P.L., Rubin, J.M.: Determination of scan-plane motion using speckle decorrelation: theoretical considerations and initial test. Int. J. Imaging Syst. Technol. **8**(1), 38–44 (1997)
3. Christiansen, P.H., Kragh, M.F., Brodskiy, Y., Karstoft, H.: Unsuperpoint: end-to-end unsupervised interest point detector and descriptor. arXiv preprint arXiv:1907.04011 (2019)
4. Cuturi, M.: Sinkhorn distances: lightspeed computation of optimal transport. In: Advances in Neural Information Processing Systems, pp. 2292–2300 (2013)
5. DeTone, D., Malisiewicz, T., Rabinovich, A.: Superpoint: self-supervised interest point detection and description. In: The IEEE Conference on Computer Vision and Pattern Recognition (CVPR) Workshops, June 2018
6. Droste, R., Drukker, L., Papageorghiou, A.T., Noble, J.A.: Automatic probe movement guidance for freehand obstetric ultrasound. In: Martel, A.L., et al. (eds.) MICCAI 2020. LNCS, vol. 12263, pp. 583–592. Springer, Cham (2020). https://doi.org/10.1007/978-3-030-59716-0_56
7. Gee, A.H., Housden, R.J., Hassenpflug, P., Treece, G.M., Prager, R.W.: Sensorless freehand 3D ultrasound in real tissue: speckle decorrelation without fully developed speckle. Med. Image Anal. **10**(2), 137–149 (2006)
8. Guo, H., Xu, S., Wood, B., Yan, P.: Sensorless freehand 3D ultrasound reconstruction via deep contextual learning. In: Martel, A.L., et al. (eds.) MICCAI 2020. LNCS, vol. 12263, pp. 463–472. Springer, Cham (2020). https://doi.org/10.1007/978-3-030-59716-0_44
9. Kendall, A., Grimes, M., Cipolla, R.: PoseNet: a convolutional network for real-time 6-DOF camera relocalization. In: Proceedings of the IEEE International Conference on Computer Vision, pp. 2938–2946 (2015)
10. Lowe, D.G.: Object recognition from local scale-invariant features. In: Proceedings of the Seventh IEEE International Conference on Computer Vision, vol. 2, pp. 1150–1157. IEEE (1999)
11. Muja, M., Lowe, D.: Flann-fast library for approximate nearest neighbors user manual. Computer Science Department, University of British Columbia, Vancouver, BC, Canada 5 (2009)
12. Prevost, R., et al.: 3D freehand ultrasound without external tracking using deep learning. Med. Image Anal. **48**, 187–202 (2018)
13. Qi, C.R., Su, H., Mo, K., Guibas, L.J.: PointNet: deep learning on point sets for 3D classification and segmentation. In: Proceedings of the IEEE Conference on Computer Vision and Pattern Recognition (CVPR), July 2017
14. Rivaz, H., Zellars, R., Hager, G., Fichtinger, G., Boctor, E.: 9C-1 beam steering approach for speckle characterization and out-of-plane motion estimation in real tissue. In: 2007 IEEE Ultrasonics Symposium Proceedings, pp. 781–784. IEEE (2007)
15. Rublee, E., Rabaud, V., Konolige, K., Bradski, G.: ORB: an efficient alternative to sift or surf. In: 2011 International Conference on Computer Vision, pp. 2564–2571. IEEE (2011)
16. Sarlin, P.E., DeTone, D., Malisiewicz, T., Rabinovich, A.: Superglue: learning feature matching with graph neural networks. In: The IEEE Conference on Computer Vision and Pattern Recognition (CVPR) (2020)

17. Sattler, T., Zhou, Q., Pollefeys, M., Leal-Taixe, L.: Understanding the limitations of CNN-based absolute camera pose regression. In: CVPR, pp. 3302–3312. IEEE (2019)
18. Simonyan, K., Zisserman, A.: Very deep convolutional networks for large-scale image recognition. arXiv preprint arXiv:1409.1556 (2014)
19. Vaswani, A., et al.: Attention is all you need. In: Advances in Neural Information Processing Systems, pp. 5998–6008 (2017)
20. Veličković, P., Cucurull, G., Casanova, A., Romero, A., Lio, P., Bengio, Y.: Graph attention networks. arXiv preprint arXiv:1710.10903 (2017)
21. Zhao, C., Droste, R., Drukker, L., Papageorghiou, A.T., Noble, J.A.: Visual-assisted probe movement guidance for obstetric ultrasound scanning using landmark retrieval. In: de Bruijne, M., et al. (eds.) MICCAI 2021. LNCS, vol. 12908, pp. 670–679. Springer, Cham (2021). https://doi.org/10.1007/978-3-030-87237-3_64
22. Zhao, C., Shen, M., Sun, L., Yang, G.Z.: Generative localization with uncertainty estimation through video-CT data for bronchoscopic biopsy. IEEE Robot. Autom. Lett. 5(1), 258–265 (2019)

Self-supervised 3D Patient Modeling with Multi-modal Attentive Fusion

Meng Zheng[1]([✉])(ID), Benjamin Planche[1](ID), Xuan Gong[2](ID), Fan Yang[1](ID), Terrence Chen[1], and Ziyan Wu[1](ID)

[1] United Imaging Intelligence, Cambridge, MA, USA
{meng.zheng,benjamin.planche,fan.yang,terrence.chen,ziyan.wu}@uii-ai.com
[2] University at Buffalo, Buffalo, NY, USA
xuangong@buffalo.edu

Abstract. 3D patient body modeling is critical to the success of automated patient positioning for smart medical scanning and operating rooms. Existing CNN-based end-to-end patient modeling solutions typically require a) customized network designs demanding large amount of relevant training data, covering extensive realistic clinical scenarios (*e.g.*, patient covered by sheets), which leads to suboptimal generalizability in practical deployment, b) expensive 3D human model annotations, *i.e.*, requiring huge amount of manual effort, resulting in systems that scale poorly. To address these issues, we propose a generic modularized 3D patient modeling method consists of (a) a multi-modal keypoint detection module with attentive fusion for 2D patient joint localization, to learn complementary cross-modality patient body information, leading to improved keypoint localization robustness and generalizability in a wide variety of imaging (*e.g.*, CT, MRI etc.) and clinical scenarios (*e.g.*, heavy occlusions); and (b) a self-supervised 3D mesh regression module which does not require expensive 3D mesh parameter annotations to train, bringing immediate cost benefits for clinical deployment. We demonstrate the efficacy of the proposed method by extensive patient positioning experiments on both public and clinical data. Our evaluation results achieve superior patient positioning performance across various imaging modalities in real clinical scenarios.

Keywords: 3D mesh · Patient positioning · Patient modeling

1 Introduction

The automatic patient positioning system and algorithm design for intelligent medical scanning/operating rooms has attracted increasing attention in recent years [14,15,17,32,34], with the goals of minimizing technician effort, providing superior performance in patient positioning accuracy and enabling contactless

---

Supplementary Information The online version contains supplementary material available at https://doi.org/10.1007/978-3-031-16449-1_12.

L. Wang (Eds.): MICCAI 2022, LNCS 13437, pp. 115–125, 2022.
https://doi.org/10.1007/978-3-031-16449-1_12

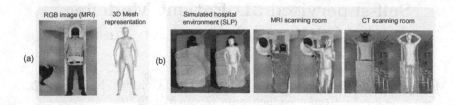

Fig. 1. (a) Mesh representation of a patient in MRI scanning room. (b) Failure cases of state-of-the-art mesh regressors (SPIN [19]) in challenging clinical scenarios, *e.g.*, simulated hospital environment [24], MRI and CT scanning rooms.

operation to reduce physical interactions and disease contagion between healthcare workers and patients. Critical to the design of such a patient positioning system, 3D patient body modeling in medical environments based on observations from one or a group of optical sensors (*e.g.*, RGB/depth/IR) is typically formulated as a 3D patient body modeling or pose estimation problem [2,16,18] in the computer vision field, defined as follows. Given an image captured from an optical sensor installed in the medical environment, we aim to automatically estimate the pose and shape—and generate a digital representation—of the patient of interest. Here we consider 3D mesh representations among several commonly used human representations (*e.g.*, skeleton, contour etc. [8]), which consist of a collection of vertices, edges and faces and contain rich pose and shape information of the real human body, as demonstrated in Fig. 1(a). The 3D mesh estimation of a patient can be found suitable for a wide variety of clinical applications. For instance, in CT scanning procedure, automated isocentering can be achieved by using the patient thickness computed from the estimated 3D mesh [17,21]. Consequently, there has been much recent work from both algorithm [6,9,18] as well as system perspectives [7,17].

State-of-the-art patient mesh estimation algorithms [38] typically rely on end-to-end customized deep networks, requiring extensive relevant training data for real clinical deployment. For example, training the RDF model proposed in [38] requires pairs of multi-modal sensor images and 3D mesh parameters (which are particularly expensive to create [25,26,28]). Moreover, conventional end-to-end 3D mesh estimation methods [16,19,38,39] assume a perfect person detection as preprocessing step for stable inference, *i.e.*, relying on an efficient person detection algorithm to crop a person rectangle covering the person's full body out of the original image. Hence, any error during this first person detection step propagates and further impacts the mesh estimation process itself (see Fig. 1(b)), and such detection errors are especially likely to occur when the target patient is *under-the-cover* (*i.e.*, occluded by sheets) or occluded by medical devices.

We thus propose a multi-modal data processing system that can (a) perform both person detection and mesh estimation, and (b) be trained over inexpensive data annotations. This system comprises several modules (*c.f.* Fig. 2). First, we train a multi-modal fused 2D keypoint predictor to learn complementary

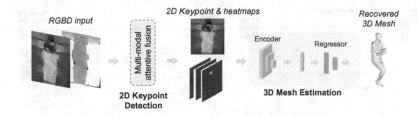

Fig. 2. Proposed framework to localize 2D keypoints and infer the 3D mesh.

patient information that may not be available from mono-modal sensors. We then process these 2D keypoints with a novel 3D mesh regressor designed to efficiently learn from inexpensively-produced synthetic data pairs in a self-supervised way. Besides technical contributions within each module, *e.g.*, cross-modal attention fusion for improved joint localization and self-supervised mesh regression (*c.f.* Sect. 2), we demonstrate the robustness and generalizability of our overall system over numerous imaging and clinical experiments (*c.f.* Sect. 3).

2   Methodology

(1) Multi-modal 2D Keypoint Detector with Attention Fusion. Most recent works in 2D pose estimation [4,10,35] are essentially single-source (*i.e.*, RGB only) architectures. Consequently, while they work reasonably in generic uncovered patient cases, they fail in more specific ones, *e.g.*, when the patient is covered by a cloth – a prevalent situation in numerous medical scanning procedures and interventions. As existing methods fail to ubiquitously work across imaging modalities and applications, we propose a multi-sensory data processing architecture that leverages information from multiple data sources to account for both generic as well as specialized scenarios (*e.g.*, cloth-covered patients). We first introduce how to individually train 2D keypoint detectors on single modalities (*e.g.*, RGB or depth), then how to learn complementary information from multiple modalities to improve detection performance and generalizability.

Given an RGB or depth person image, the 2D keypoint detection task aims to predict a set of N_J joint (usually predefined) locations of the person in the image space, which is typically achieved by learning a deep CNN network in most recent works. Here we adopt HRnet [35] as the backbone architecture which takes the RGB or depth image as input and outputs N_J 2D joint heatmaps, with the peak location of each heatmap $i = 1, ..., N_J$ indicating the corresponding joint's pixel coordinates, as illustrated in Fig. 3 (orange/blue block for RGB/depth).

While the training of RGB-based keypoint detector can leverage many publicly available datasets [12,22,29], the number of depth/RGBD image datasets curated for keypoint detection is much more limited, mainly due to lower sensor accessibility and image quality. Thus directly training depth-based keypoint detectors over such limited data can easily result in overfitting and poor detection performance during testing. To alleviate this, we propose to first utilize an

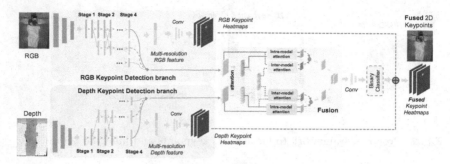

Fig. 3. Proposed RGBD keypoint detection framework with attention fusion. (Color figure online)

unsupervised pretraining technique [5] to learn a generalized representation from unlabeled depth images, which is proved to have better generalizability for downstream tasks like keypoint detection; and then to finetune with labeled training data for improved keypoint detection accuracy. This way, we can leverage a larger number of public depth or RGBD image datasets collected for other tasks for a more generic model learning. For further details w.r.t. the unsupervised pretraining for representation learning, we refer the readers to [5].

Color and depth images contain complementary information of the patient, as well as complementary benefits over different scenarios. *E.g.*, when the patient is covered by a surgical sheet (Fig. 1(b)), RGB features will be heavily affected due to the cover occlusion, whereas depth data still contain rich shape and contour information useful for patient body modeling. We seek to design an attentive multi-modal fusion network, to effectively aggregate complementary information across RGB and depth images by enforcing intra- and inter-modal attentive feature aggregation for improved keypoint detection performance. Specifically, we propose a two-branch score-based RGBD fusion network as shown in Fig. 3. In the proposed fusion network, we take the last stage features of the HRnet backbone from RGB and depth branches respectively, and forward them into a fusion module with intra-modal and inter-modal attentive feature aggregation, for a binary classification score prediction. This classifier aims to determine (*c.f.* output score) which modality (RGB or depth) results in the most reliable prediction, based on the prediction error from RGB and depth branches during training. For example, if the RGB prediction error is larger than depth prediction error, we set the classifier label to 0, and vice-versa (setting to 1 if RGB-based error is lower). After the binary classifier is learned, it will produce a probability score (range from 0 to 1) indicating the reliability of RGB and depth branch predictions, which is then utilized to weight keypoint heatmap predictions from each branch before their fusion. In this way, the proposed module is able to fuse complementary information from single modalities and learn enhanced feature representations for more accurate and robust keypoint detection.

(2) Self-supervised 3D Mesh Regressor. After producing the 2D keypoints, we aim to recover the 3D mesh representation of the patient for complete and dense patient modeling. Note that we use the Skinned Multi-Person Linear (SMPL) model [26], which is a statistical parametric mesh model of the human body, represented by pose $\theta \in \mathbb{R}^{72}$ and shape $\beta \in \mathbb{R}^{10}$ parameters. Unlike prior works [38] that require both images and the corresponding ground-truth 3D mesh parameters for training, our proposed method does not need such expensive annotations and relies only on synthetically generated pairs of 2D keypoint predictions and mesh parameters. Our method thus does not suffer from the biased distribution and limited scale of existing 3D datasets.

Specifically, to generate the synthetic training data, we sample SMPL pose parameters from training sets of public datasets (*i.e.*, AMASS [27], UP-3D [20] and 3DPW [37]), and shape parameters from a Gaussian distribution following [30]. We then render the 3D mesh given the sampled θ and β and project the 3D joints determined by the rendered mesh to 2D keypoint locations given randomly sampled camera translation parameters. The N_J 2D keypoint locations then can be formed into N_J heatmaps (as described in Sect. 2(1)) and passed to a CNN for θ and β regression. Here we use a Resnet-18 [11] as the baseline architecture for mesh regression. In our experiments, we extensively sampled the data points and generate 330k synthetic data pairs to train the mesh regressor. During testing, the 2D keypoint heatmaps inferred from the RGBD keypoint detection model (*c.f.* Sect. 2(1)) are directly utilized for 3D mesh estimation.

3   Experiments

Datasets, Implementation, and Evaluation. To demonstrate the efficacy of our proposed method, we evaluate on the public SLP dataset [24] (same train/test splits from the authors) and proprietary RGBD data collected (with approval from ethical review board) from various scenarios: computed tomography (CT), molecular imaging (MI), and magnetic resonance imaging (MRI).

To collect our proprietary MI dataset, 13 volunteers were asked to take different poses while being covered by a surgical sheet with varying covering areas (half body, 3/4 body, full body) and facial occlusion scenarios (with/without facial mask). We use 106 images from 3 subjects to construct the training set, and 960 images from the remaining 10 subjects as test set. For the dataset collected in MRI room, we captured 1,670 images with varying scanning protocols (*e.g.*, wrist, ankle, hip, *etc.*) and patient bed positions, with the volunteers being asked to show a variety of poses while being covered by a cloth with different level of occlusions (similar to MI dataset). This resulted in 1,410 training images and 260 testing ones. For our proprietary CT dataset, we asked 13 volunteers to lie on a CT scanner bed and exhibit various poses with and without cover. We collected 974 images in total. To test the generalizability of the proposed mesh estimator across imaging modalities, we use this dataset for testing only.

During training stage, we use all data from the SLP, MI and MRI training splits, along with public datasets COCO [22] and MPII [29] to learn our single

RGB keypoint detector, with the ground-truth keypoint annotations generated manually. For our depth detector, we pretrain its backbone over public RGBD datasets, *i.e.*, 7scene [31], PKU [23], CAD [36] and SUN-RGBD [33]. We then finetune the model over SLP, MI and MRI training data with keypoint supervision. We apply the commonly-used 2D mean per joint position error (MPJPE) and percentage of correct keypoints (PCK) [1] for quantifying the accuracy of 2D keypoints, and the 3D MPJPE, Procrustes analysis (PA) MPJPE [16] and scale-corrected per-vertex Euclidean error in a neutral pose (T-pose), *i.e.*, PVE-T-SC [30] (all in mm) for 3D mesh pose and shape evaluation.

3.1   2D Keypoint Prediction

(1) Comparison to State-of-the-art. In Table 1 (first row), we compare the 2D MPJPE of our keypoint prediction module with competing 2D detectors on the SLP dataset. Here, "Ours (RGB)" and "Ours (Depth)" refer to the proposed single-modality RGB and depth keypoint detectors, which achieve substantial performance improvement, including compared to the recent RDF algorithm of Yang *et al.* [38]. In Table 4, we compare the PCK@0.3 of proposed RGBD keypoint detector with off-the-shelf state-of-the-art 2D keypoint detector OpenPose [4] on SLP and MI datasets. We notice that our solution performs significantly better than OpenPose across different data domains, which demonstrates the superiority of the proposed method. We present more PCK@0.3 (torso) evaluations of the proposed multi-modal keypoint detector with attentive fusion in Table 5 on MRI and CT (cross-domain) dataset, proving the efficacy of the proposed multi-modal fusion strategy.

(2) Ablation Study on Multi-modal Fusion. Table 2 (A) contains results of an ablation study to evaluate the impact of utilizing single (RGB/depth) and multi-modal fused (RGBD) data for keypoint detection on SLP, MI, MRI and CT (cross-domain) data. We evaluated on different CNN backbones, *i.e.* HRNet [35] and ResNet-50 [11] (see supplementary material), and we observe consistent performance improvement across all datasets with multi-modal fusion, demonstrating the efficacy of our fusion architecture. See Fig. 4 for a qualitative illustration of this aspect.

(3) Ablation Study on Unsupervised Pretraining of Depth Keypoint Detector. To demonstrate the advantage of utilizing unsupervised pretraining strategy for generalized keypoint detection, another ablation study is performed w.r.t. our single depth-based detector on SLP and MI data, pretrained with varying number of unannotated data, then finetuned with a fix amount of labeled samples (SLP, MI and MRI data). We can see from Table 3 that the keypoint detection performance generally increases along with the quantity of pretraining data, proving the efficacy of the proposed unsupervised pretraining strategy.

3.2   3D Mesh Estimation

We next discuss the performance of our 3D mesh estimation module. To generate ground-truth 3D SMPL pose and shape annotations for all testing data, we

Table 1. Comparison on SLP [24] to existing methods, w.r.t. 2D keypoint detection and 3D mesh regression (modalities: "RGB" color, "T" thermal, "D" depth). Grey cells indicate numbers not reported in the references.

	Methods:	SPIN [19]	OP [4]	HMR [16]			RDF [38]			Ours		
				RGB	T	RGBT	RGB	T	RGBT	RGB	D	RGBD
2D	MPJPE (px)↓		293.8	37.2			36.6			17.1	14.2	13.2
2D	MPJPE (cm)↓		163.9	20.8			20.4			9.5	7.9	7.4
3D	MPJPE (mm)↓	236		155	149	143	144	138	137	123	118	115

Table 2. Ablation study and evaluation on different imaging modalities w.r.t. 2D keypoint detector (A) and 3D mesh regressor (B). († = cross-domain evaluation)

Data	(A) 2D detector ablation study			(B) 3D mesh regressor evaluation					
	2D MPJPE (px)↓			3D PA MPJPE (mm)↓			3D PVE-T-SC (mm)↓		
	RGB	D	RGBD	RGB	D	RGBD	RGB	D	RGBD
SLP	17.1	14.2	**13.2**	83.4	78.3	**77.3**	17.3	14.5	**13.3**
MI	13.0	13.6	**12.6**	97.0	101.5	**93.1**	22.9	26.6	**17.7**
MRI	7.7	15.6	**7.2**	103.1	99.3	**94.3**	19.8	17.8	**15.1**
CT†	23.3	22.5	**21.2**	110.9	107.5	**104.3**	17.3	20.2	**17.3**

Table 3. Impact of pretraining data (7scene [31], PKU [23], CAD [36], SUNRGBD [33]) on MPJPE accuracy (px) of proposed depth-based keypoint detector.

Pretrain datasets:	[31]	[31]+ [23]	[31]+ [23]+ [36]	[31]+ [23]+ [36]+ [33]
SLP	17.8	14.5	15.1	**13.5**
MI		14.5	13.6	13.8	**13.3**

Table 4. PCK@0.3 evaluation of our proposed 2D keypoint detector with competing methods on SLP (top) and MI (bottom).

Methods	R.Ak.	R.Kn.	R.H.	L.H.	L.Kn.	L.Ak.	R.Wr.	R.Eb.	R.Sh.	L.Sh.	L.Eb.	L.Wr.	Avg
OpenPose [4]	13.0	38.2	74.6	73.9	34.6	11.1	54.9	74.6	95.7	95.7	73.3	52.6	57.7
Proposed	98.4	98.4	100.0	100.0	99.6	98.2	92.5	97.2	99.9	99.3	96.1	94.7	97.9
Methods	R.Ak.	R.Kn.	R.H.	L.H.	L.Kn.	L.Ak.	R.Wr.	R.Eb.	R.Sh.	L.Sh.	L.Eb.	L.Wr.	Avg
OpenPose [4]	0.0	0.0	2.7	3.3	0.0	0.0	20.0	34.4	85.3	86.7	34.5	18.1	23.7
Proposed	97.6	99.3	99.9	99.7	97.2	95.4	91.6	97.8	100.0	99.8	98.7	92.5	97.5

apply an off-the-shelf 3D mesh predictor [13, 28] followed by manual refinement. Given this testing ground truth, we use the 3D MPJPE and PVE-T-SC metrics to quantify performance. Comparison to other 3D mesh estimation technique is shown in Table 1 (bottom row). Again, we observe substantial performance improvement in terms of 3D joint localization (*c.f.* 3D MPJPE) and per-vertex mesh accuracy (*c.f.* 3D PVE-T-SC) across a wide variety of cover conditions, despite purely relying on synthetic training data (whereas competitive methods require expensive 3D annotations). The proposed solution shines (on the SLP data) over the state-of-the-art, *e.g.*, recent method by Yang *et al.* [38] and one of

Table 5. PCK@0.3 (torso) evaluation of our proposed 2D keypoint detector on MRI and CT† (cross-validation: no CT training data used in model learning) testing data.

	R.Ak.	R.Kn.	R.H.	L.H.	L.Kn.	L.Ak.	R.Wr.	R.Eb.	R.Sh.	L.Sh.	L.Eb.	L.Wr.	Avg
MRI	97.0	98.7	99.4	99.4	98.7	98.5	96.8	98.1	99.4	99.4	98.7	96.8	98.4
CT†	91.3	93.2	93.9	94.0	92.5	91.3	84.9	88.6	93.6	93.9	88.9	86.2	91.0

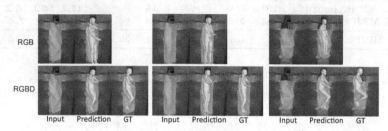

Fig. 4. Performance comparison between proposed RGB and RGBD model.

the most commonly used 3D mesh estimator SPIN [19]. Table 2 (B) further shows that these improvements are generally consistent across all imaging modalities. Qualitative mesh estimation results are shared in Fig. 5.

3.3   Automated Isocentering with Clinical CT Scans

To demonstrate the clinical value of the proposed method, we evaluate the isocentering accuracy in a clinical CT scanning setting. To do so, we mount an RGBD camera above the CT patient support and calibrate it spatially to the CT reference system. With the RGBD images captured by the camera, our proposed method can estimate the 3D patient mesh and compute the thickness of the target body part, which can then be used to adjust the height of the patient support so that the center of target body part aligns with the CT isocenter. We conducted this evaluation with 40 patients and 3 different protocols, and calculated the error based on the resulting CT scout scan as shown in Table 6. Compared to the currently deployed automated CT patient positioning system [3], our pipeline automatically aligns the center of target body part and scanner isocenter with mean errors of 5.3/7.5/8.1 mm for abdomen/thorax/head respectively vs. 13.2 mm median error of radiographers in [3], which clearly demonstrates the advantage of our proposed positioning workflow.

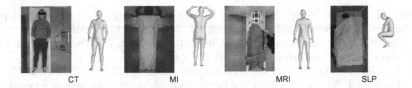

Fig. 5. Visualization of reconstructed mesh results on CT, MI, MRI and SLP.

Table 6. Evaluation on ISO-center estimation with clinical CT scans.

Protocol	Abdomen		Thorax		Head	
Error (mm)	Mean	STD	Mean	STD	Mean	STD
	5.3	2.1	7.5	2.9	8.1	2.2

4   Conclusion

In this work, we considered the problem of 3D patient body modeling and proposed a novel method, consisting of a multi-modal 2D keypoint detection module with attentive fusion and a self-supervised 3D mesh regression module, being applicable to a wide variety of imaging and clinical scenarios. We demonstrated these aspects with extensive experiments on proprietary data collected from multiple scanning modalities as well as public datasets, showing improved performance when compared to existing state-of-the-art algorithms as well as published clinical systems. Our results demonstrated the general-purpose nature of our proposed method, helping take a step towards algorithms that can lead to scalable automated patient modeling and positioning systems.

References

1. Andriluka, M., et al.: PoseTrack: a benchmark for human pose estimation and tracking. In: CVPR (2018)
2. Bogo, F., Kanazawa, A., Lassner, C., Gehler, P., Romero, J., Black, M.J.: Keep it SMPL: automatic estimation of 3D human pose and shape from a single image. In: ECCV (2016)
3. Booij, R., van Straten, M., Wimmer, A., Budde, R.P.: Automated patient positioning in CT using a 3D camera for body contour detection: accuracy in pediatric patients. Eur. Radiol. **31**, 131–138 (2021)
4. Cao, Z., Martinez, G.H., Simon, T., Wei, S., Sheikh, Y.A.: OpenPose: realtime multi-person 2D pose estimation using part affinity fields. IEEE Trans. Patt. Anal. Mach. Intell. (2019)
5. Chen, X., He, K.: Exploring simple Siamese representation learning. CVPR (2021)
6. Ching, W., Robinson, J., McEntee, M.: Patient-based radiographic exposure factor selection: a systematic review. J. Med. Radiat. Sci. **61**(3), 176–190 (2014)
7. Clever, H.M., Erickson, Z., Kapusta, A., Turk, G., Liu, K., Kemp, C.C.: Bodies at rest: 3D human pose and shape estimation from a pressure image using synthetic data. In: CVPR (2020)

8. Dang, Q., Yin, J., Wang, B., Zheng, W.: Deep learning based 2D human pose estimation: a survey. Tsinghua Sci. Technol. **24**(6), 663–676 (2019)
9. Georgakis, G., Li, R., Karanam, S., Chen, T., Košecká, J., Wu, Z.: Hierarchical kinematic human mesh recovery. In: ECCV (2020)
10. He, K., Gkioxari, G., Dollár, P., Girshick, R.: Mask R-CNN. In: ICCV (2017)
11. He, K., Zhang, X., Ren, S., Sun, J.: Deep residual learning for image recognition. In: CVPR (2016)
12. Johnson, S., Everingham, M.: Clustered pose and nonlinear appearance models for human pose estimation. In: BMVC (2010)
13. Joo, H., Neverova, N., Vedaldi, A.: Exemplar fine-tuning for 3D human pose fitting towards in-the-wild 3D human pose estimation. In: 3DV (2020)
14. Kadkhodamohammadi, A., Gangi, A., de Mathelin, M., Padoy, N.: Articulated clinician detection using 3D pictorial structures on RGB-D data. Med. Image Anal. **35**, 215–224 (2017)
15. Kadkhodamohammadi, A., Gangi, A., de Mathelin, M., Padoy, N.: A multi-view RGB-D approach for human pose estimation in operating rooms. In: WACV (2017)
16. Kanazawa, A., Black, M.J., Jacobs, D.W., Malik, J.: End-to-end recovery of human shape and pose. In: CVPR (2018)
17. Karanam, S., Li, R., Yang, F., Hu, W., Chen, T., Wu, Z.: Towards contactless patient positioning. IEEE Trans. Med. Imaging **39**(8), 2701–2710 (2020)
18. Kolotouros, N., Pavlakos, G., Black, M.J., Daniilidis, K.: Learning to reconstruct 3D human pose and shape via model-fitting in the loop. In: ICCV (2019)
19. Kolotouros, N., Pavlakos, G., Black, M.J., Daniilidis, K.: Learning to reconstruct 3D human pose and shape via model-fitting in the loop. In: ICCV (2019)
20. Lassner, C., Romero, J., Kiefel, M., Bogo, F., Black, M.J., Gehler, P.V.: Unite the people: closing the loop between 3D and 2D human representations. In: CVPR (2017)
21. Li, J., Udayasankar, U.K., Toth, T.L., Seamans, J., Small, W.C., Kalra, M.K.: Automatic patient centering for MDCT: effect on radiation dose. Am. J. Roentgenol. **188**(2), 547–552 (2007)
22. Lin, T., Maire, M., Belongie, S., et al.: Microsoft COCO: common objects in context. In: ECCV (2014)
23. Liu, C., Hu, Y., Li, Y., Song, S., Liu, J.: PKU-MMD: a large scale benchmark for continuous multi-modal human action understanding. arXiv:1703.07475 (2017)
24. Liu, S., Ostadabbas, S.: Seeing under the cover: a physics guided learning approach for in-bed pose estimation. In: MICCAI (2019)
25. Loper, M., Mahmood, N., Black, M.J.: MoSh: motion and shape capture from sparse markers. ACM Trans. Graph. **33**(6), 1–13 (2014)
26. Loper, M., Mahmood, N., Romero, J., Pons-Moll, G., Black, M.J.: SMPL: a skinned multi-person linear model. ACM Trans. Graph. **34**(6) (2015)
27. Mahmood, N., Ghorbani, N., Troje, N.F., Pons-Moll, G., Black, M.J.: AMASS: archive of motion capture as surface shapes. In: ICCV (2019)
28. Pavlakos, G., et al.: Expressive body capture: 3D hands, face, and body from a single image. In: CVPR (2019)
29. Pishchulin, L., et al.: DeepCut: joint subset partition and labeling for multi person pose estimation. In: CVPR, June 2016
30. Sengupta, A., Budvytis, I., Cipolla, R.: Synthetic training for accurate 3D human pose and shape estimation in the wild. In: BMVC (2020)
31. Shotton, J., Glocker, B., Zach, C., Izadi, S., Criminisi, A., Fitzgibbon, A.: Clustered pose and nonlinear appearance models for human pose estimation. In: CVPR (2013)

32. Singh, V., Ma, K., Tamersoy, B., et al.: DARWIN: deformable patient avatar representation with deep image network. In: MICCAI (2017)
33. Song, S., Lichtenberg, S.P., Xiao, J.: SUN RGB-D: A RGB-D scene understanding benchmark suite. In: CVPR (2015)
34. Srivastav, V., Issenhuth, T., Kadkhodamohammadi, A., de Mathelin, M., Gangi, A., Padoy, N.: MVOR: A multi-view RGB-D operating room dataset for 2D and 3D human pose estimation (2018)
35. Sun, K., Xiao, B., Liu, D., Wang, J.: Deep high-resolution representation learning for human pose estimation. In: CVPR (2019)
36. Sung, J., Ponce, C., Selman, B., Saxena, A.: Unstructured human activity detection from RGBD images. In: ICRA (2012)
37. Von Marcard, T., Henschel, R., Black, M.J., Rosenhahn, B., Pons-Moll, G.: Recovering accurate 3D human pose in the wild using IMUs and a moving camera. In: ECCV (2018)
38. Yang, F., et al.: Robust multi-modal 3D patient body modeling. In: MICCAI (2020)
39. Yin, Y., Robinson, J.P., Fu, Y.: Multimodal in-bed pose and shape estimation under the blankets. In: ArXiv:2012.06735 (2020)

SLAM-TKA: Real-time Intra-operative Measurement of Tibial Resection Plane in Conventional Total Knee Arthroplasty

Shuai Zhang[1,2], Liang Zhao[1(✉)], Shoudong Huang[1], Hua Wang[3(✉)], Qi Luo[3], and Qi Hao[2]

[1] Robotics Institute, University of Technology Sydney, Sydney, NSW, Australia
Liang.Zhao@uts.edu.au
[2] Department of Computer Science and Engineering, Southern University of Science and Technology, Shenzhen 518055, China
[3] Osteoarthropathy Surgery Department, Shenzhen People's Hospital, Shenzhen 518020, China
wanghuayisheng@sina.com

Abstract. Total knee arthroplasty (TKA) is a common orthopaedic surgery to replace a damaged knee joint with artificial implants. The inaccuracy of achieving the planned implant position can result in the risk of implant component aseptic loosening, wear out, and even a joint revision, and those failures most of the time occur on the tibial side in the conventional jig-based TKA (CON-TKA). This study aims to precisely evaluate the accuracy of the proximal tibial resection plane intra-operatively in real-time such that the evaluation processing changes very little on the CON-TKA operative procedure. Two X-ray radiographs captured during the proximal tibial resection phase together with a pre-operative patient-specific tibia 3D mesh model segmented from computed tomography (CT) scans and a trocar pin 3D mesh model are used in the proposed simultaneous localisation and mapping (SLAM) system to estimate the proximal tibial resection plane. Validations using both simulation and in-vivo datasets are performed to demonstrate the robustness and the potential clinical value of the proposed algorithm.

Keywords: TKA · Tibial resection · SLAM

This work was supported in part by the Australian Research Council Discovery Project (No. DP200100982), and in part by Shenzhen Fundamental Research Program (No: JCYJ20200109141622964).

Supplementary Information The online version contains supplementary material available at https://doi.org/10.1007/978-3-031-16449-1_13.

1   Introduction

Total knee arthroplasty (TKA) is considered to be the gold standard to relieve disability and to reduce pain for end-stage knee osteoarthritis. It is the surgery to replace a knee joint with an artificial joint. The number of TKA surgeries has increased at a dramatic rate throughout the world due to the growing prevalence of knee arthritis and the increased access to orthopaedic care. For example, the number of TKA procedures in USA is projected to grow to 1.26 million by 2030 [1]. Meanwhile, the frequency of TKA revisions has also increased and the most common reason is aseptic loosening caused by implant malalignment [2].

Regarding the optimal implant alignment, studies have confirmed that the distal femoral and proximal tibial should be resected perpendicular to their mechanical axes [3] (the standard accuracy requirement is less than 3° angle error). Otherwise, the inaccuracy of the distal femoral and proximal tibial resection will result in an increased risk of component aseptic loosening and early TKA failure, and those failures often occur on the tibial side [4]. However, as the most commonly used alignment guide device in the conventional jig-based TKA (CON-TKA), the extramedullary (EM) alignment guide for the proximal tibial resection shows a limited degree of accuracy [5]. Studies have shown that roughly 12.4% and 36% of patients who underwent CON-TKA procedures have the coronal and saggital tibial component angle errors more than 3°, respectively [6,7]. One main reason for causing the large errors is that no real-time feedback on the orientation of the proximal tibial resection is provided [5].

In this work, we propose a robust and real-time intra-operative tibial pre-resection plane measurement algorithm for CON-TKA which could provide real-time feedback for surgeons. The algorithm uses information from a pre-operative tibia CT scans, intra-operative 2D X-ray images, and a trocar pin 3D mesh model to estimate the poses of the pins and the X-ray frames. Based on these, the resection plane can be accurately calculated.

Our work has some relations to the research works on post-operative evaluation in orthopaedic surgeries [8–10], reconstruction of 3D bone surface model [11,12], and surgical navigation [13] using 2D-3D registration between pre-operative 3D CT scans and 2D X-ray images. Features (anatomical points, markers or edges), intensity or gradient information in 3D scans and X-ray images are used in the registration methods. However, these methods all have their own limitations. For example, optimising X-ray and implants separately which result in suboptimal results [8–10], requiring very accurate initialisation [11,12,14], relying on a custom-designed hybrid fiducial for C-arm pose estimation and tracking which results in large incision [13], and requiring the time-consuming generation of digitally reconstructed radiographs (DRRs) [14].

Some new techniques and devices have also been developed to improve the accuracy and reproducibility in maintaining proximal tibial resection, distal femoral resection, and implants alignment, including computer-assisted navigation system [15], robot-assisted surgery [16], patient-specific instrumentation [17], and accelerometer-based navigation devices [18]. However, they add many extra operative procedures and instruments, which increase the operation complexity, time, cost and even the risk of infections and blood clots. And no significant differences were found in mid-to-long term functional outcomes [19].

In summary, compared to existing methods, the key advantages of our proposed framework are (i) no external fiducials or markers are needed (ii) pin poses and X-ray poses are jointly optimised to obtain very accurate and robust intra-operative pre-estimation of the tibial resection plane, and (iii) it can be easily integrated clinically, without interrupting the workflow of CON-TKA.

2   Problem Statement

During a standard CON-TKA surgery, the tibial cutting block is first manually aligned using an EM tibial alignment guide and secured to the patient's tibia with a pair of trocar pins, then the proximal end of the tibia will be sawed off by an oscillating saw through the deepest portion of the trochlear groove on the tibial cutting block (refer to Fig. 1).

In the proposed framework, the aligned cutting block is removed before the execution of bone cutting and different views of X-ray images are captured for the proximal tibia with drilled pins on it using a C-arm X-ray device. The problem considered in this work is to use the pre-operative tibia mesh model, the pin mesh model, and the intra-operative X-ray observations to estimate the poses of the pins and X-ray frames w.r.t. the pre-operative tibia model intra-operatively.

Suppose N views of X-ray images are captured and $k \in \{1, ..., N\}$ is the index of the X-ray image, let $\mathcal{C}_k = \{\bar{\mathbf{p}}_{k,1}, ..., \bar{\mathbf{p}}_{k,\mathbb{N}_k}\}$ represent the edge observations of the tibia and fibula on the k-th X-ray image, and $\mathcal{C}_k^l = \{\bar{\mathbf{p}}_{k,1}^l, ..., \bar{\mathbf{p}}_{k,\mathbb{N}_k}^l\}$ represent the edge observations of the l-th pin on the k-th X-ray image, $\mathbb{N}_k$ is the number of tibia and fibula edge pixels extracted from the k-th X-ray image, $l \in \{1, 2\}$ denotes the index of pin drilled into the tibia (only two pins are used in the CON-TKA procedures) and $\mathbb{N}_{kl}$ is the number of pin edge pixels extracted from the l-th pin in the k-th X-ray image.

Suppose $\mathbf{p}_{k,i} = [u_{k,i}, v_{k,i}]^T$ is the ground truth coordinates of the i-th observed 2D edge point $\bar{\mathbf{p}}_{k,i}$, its corresponding 3D point from the pre-operative tibia model M_{tibia} is $\mathbf{P}_{k,i}^M \in \mathbb{R}^3$, then the observation model of the tibia edge point can be written as:

$$\bar{\mathbf{p}}_{k,i} = \mathbf{p}_{k,i} + w_{ki}, \ [\mathbf{p}_{k,i}^T, 1]^T \propto K(R_k^C \mathbf{P}_{k,i}^M + \mathbf{t}_k^C) \tag{1}$$

where w_{ki} is the zero-mean Gaussian noise with covariance matrix $\Sigma_{p_{ki}}$, K is the C-arm camera intrinsic matrix, $X_k^C = \{R_k^C, \mathbf{t}_k^C\}$ represents the rotation matrix and translation vector of the C-arm camera pose at which the k-th X-ray image is captured (in the frame of the pre-operative tibia model).

Similarly, suppose $\bar{\mathbf{p}}_{k,j}^l = \mathbf{p}_{k,j}^l + w_{kj}^l$ is the j-th observed 2D edge point, in which w_{kj}^l is the zero-mean Gaussian noise with covariance matrix $\Sigma_{p_{kj}^l}$, and $\bar{\mathbf{p}}_{k,j}^l$'s corresponding 3D point from the pin model M_{pin} is $\mathbf{P}_{k,j}^{M_l} \in \mathbb{R}^3$. Then, the observation model of the pin edge point is:

$$[(\mathbf{p}_{k,j}^l)^T, 1]^T \propto K(R_k^C(R_l^M \mathbf{P}_{k,j}^{M_l} + \mathbf{t}_l^M) + \mathbf{t}_k^C) \tag{2}$$

where $X_l^M = \{R_l^M, \mathbf{t}_l^M\}$ is the pose of the l-th pin relative to the pre-operative tibia model M_{tibia}. Thus (2) first transforms $\mathbf{P}_{k,j}^{M_l}$ from the pin model frame to the pre-operative tibia model frame by using the pin pose X_l^M, and then projects onto the k-th camera frame by using X_k^C.

Mathematically, the problem considered in this paper is, given the pre-operative 3D tibia model M_{tibia} and the pin model M_{pin}, how to accurately estimate the pin poses $\{X_l^M\}$, $l \in \{1,2\}$ from N views of 2D intra-operative X-ray observations, where the C-arm camera poses $\{X_k^C\}, k \in \{1, ..., N\}$ of the X-ray images are also not available.

3   Resection Plane Estimation

To obtain a more accurate and robust estimation, we formulate the problem as a SLAM problem where the pin poses and the C-arm camera poses are estimated simultaneously. Then, the proximal tibial pre-resection plane can be calculated using the estimated pin poses.

3.1   SLAM Formulation for Pin Poses Estimation

In the proposed SLAM formulation, the state to be estimated is defined as

$$\mathbf{X} \triangleq \left\{ \{X_k^C\}_{k=1}^N, \{X_l^M\}_{l=1}^2, \{\{\mathbf{P}_{k,i}^M\}_{i=1}^{N_k}\}_{k=1}^N, \{\{\{\mathbf{P}_{k,j}^{M_l}\}_{j=1}^{N_{kl}}\}_{l=1}^2\}_{k=1}^N \right\} \quad (3)$$

where the variables are defined in Sect. 2. The proposed SLAM problem can be mathematically formulated as a nonlinear optimisation problem minimising:

$$E = w_{rp}E_{rp} + w_{bp}E_{bp} + w_{mp}E_{mp} \quad (4)$$

which consists of three energy terms: Contour Re-projection Term E_{rp}, Contour Back-projection Term E_{bp} and Model Projection Term E_{mp}. w_{rp}, w_{bp} and w_{mp} are the weights of the three terms. Overall, the three energy terms are used together to ensure the accuracy and robustness of the estimation algorithm.

The Contour Re-projection Term penalises the misalignment between the contours of the outer edges of the tibia and the pins in the X-ray images and the contours re-projected by using the variables in the state in (3):

$$E_{rp} = \sum_{k=1}^N \left[\sum_{i=1}^{N_k} \|\bar{\mathbf{p}}_{k,i} - \mathbf{p}_{k,i}\|_{\Sigma_{p_{ki}}^{-1}}^2 + \sum_{l=1}^2 \sum_{j=1}^{N_{kl}} \|\bar{\mathbf{p}}_{k,j}^l - \mathbf{p}_{k,j}^l\|_{\Sigma_{p_{kj}^l}^{-1}}^2 \right] \quad (5)$$

where $\Sigma_{p_{ki}}$ and $\Sigma_{p_{kj}^l}$ are the covariance matrices of the 2D contour feature observations $\bar{\mathbf{p}}_{k,i}$ and $\bar{\mathbf{p}}_{k,j}^l$, respectively. $\mathbf{p}_{k,i}$ and $\mathbf{p}_{k,j}^l$ are given in (1) and (2).

The Contour Back-projection Term minimises the sum of squared coordinate distances (along the x, y and z axes, respectively) between the pre-operative tibia model M_{tibia} and the 3D back-projected contour points $\{\mathbf{P}_{k,i}^M\}$

in the state $\mathbf{X}$, and between the pin model M_{pin} and the 3D back-projected contour points $\{\mathbf{P}_{k,j}^{M_l}\}$ in the state $\mathbf{X}$:

$$E_{bp} = \sum_{k=1}^{N} \left[\sum_{i=1}^{\mathbb{N}_k} \left\| \mathbf{d}(\mathbf{P}_{k,i}^{M}, M_{tibia}) \right\|_{\Sigma_{P_{ki}^M}^{-1}}^2 + \sum_{l=1}^{2} \sum_{j=1}^{\mathbb{N}_{kl}} \left\| \mathbf{d}(\mathbf{P}_{k,j}^{M_l}, M_{pin}) \right\|_{\Sigma_{P_{kj}^{M_l}}^{-1}}^2 \right] \quad (6)$$

where $\mathbf{d}(\mathbf{P}_{k,i}^{M}, M_{tibia})$ represents the shortest coordinate distance from 3D point $\mathbf{P}_{k,i}^{M}$ to M_{tibia}, and $\mathbf{d}(\mathbf{P}_{k,j}^{M_l}, M_{pin})$ represents the shortest coordinate distance from $\mathbf{P}_{k,j}^{M_l}$ to M_{pin}, $\Sigma_{P_{ki}^M}$ and $\Sigma_{P_{kj}^{M_l}}$ are the corresponding covariance matrices.

The Model Projection Term penalises the misalignment between the observed contours in the intra-operative X-ray images and contours extracted from the projection of the pre-operative tibia model M_{tibia} and pin model M_{pin}:

$$E_{mp} = \sum_{k=1}^{N} \left[\sum_{m=1}^{\mathbb{N}_k^M} \left\| d(\tilde{\mathbf{p}}_{k,m}, \mathcal{C}_k) \right\|_{\Sigma_{P_{km}}^{-1}}^2 + \sum_{l=1}^{2} \sum_{n=1}^{\mathbb{N}_{kl}^M} \left\| d(\tilde{\mathbf{p}}_{k,n}^l, \mathcal{C}_k^l) \right\|_{\Sigma_{P_{kn}^l}^{-1}}^2 \right] \quad (7)$$

where $\{\tilde{\mathbf{p}}_{k,m}\}$, $m \in \{1, ..., \mathbb{N}_k^M\}$ are the 2D contour points representing the edge of the pre-operative tibia model projection, projected from M_{tibia} to the k-th X-ray image using X_k^C. $d(\tilde{\mathbf{p}}_{k,m}, \mathcal{C}_k)$ means the shortest Euclidean distance between $\tilde{\mathbf{p}}_{k,m}$ and the contour $\mathcal{C}_k$ of the k-th X-ray image. Similarly, $\{\tilde{\mathbf{p}}_{k,n}^l\}$, $n \in \{1, ..., \mathbb{N}_{kl}^M\}$ are the 2D contour points representing the edge of the l-th pin projection, projected from M_{pin} to the k-th X-ray image using X_k^C and X_l^M, $d(\tilde{\mathbf{p}}_{k,n}^l, \mathcal{C}_k^l)$ is the shortest Euclidean distance between $\tilde{\mathbf{p}}_{k,n}^l$ and the contour $\mathcal{C}_k^l$.

3.2 Resection Plane Estimation and the Approach Overview

The iterative Gauss-Newton (GN) algorithm is used to minimise the objective function (4). After the optimisation, the two pins are transformed into the frame

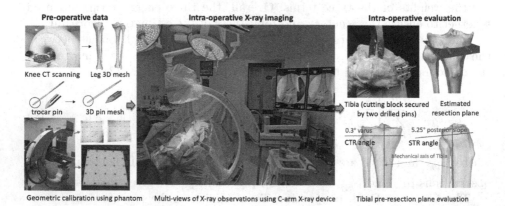

Fig. 1. The framework of proposed proximal tibial resection plane estimation.

of the pre-operative tibia model using the estimated poses, and the corresponding tibial resection plane (parallel to the plane fitted by the two paralleled trocar pins) is obtained by fitting all the vertices of the two transformed pin models. The coronal tibial resection (CTR) angle and the sagittal tibial resection (STR) angle of the estimated tibial resection plane are calculated w.r.t. the tibial mechanical axis and compared to the desired bone resection requirement before the surgeon cuts the proximal end of the tibia. Figure 1 shows the proposed framework which consists of pre-operative data preparation, intra-operative X-ray imaging, and intra-operative tibial resection plane estimation and evaluation.

4   Experiments

The proposed SLAM-TKA framework for real-time estimating the proximal tibial pre-resection plane is validated by simulations and in-vivo experiments. The pre-operative 3D tibia model is segmented from CT scans using the Mimics software, the 3D trocar pin mesh model is created using the Solidworks software. The minimum number of X-ray images are used both in simulations and in-vivo experiments, then $N = 2$. The C-arm X-ray device was calibrated using a calibration phantom, and the calibration dataset and code are made available in GitHub (https://github.com/zsustc/Calibration.git).

The proposed framework is compared to the feature-based 2D-3D registration "projection" strategy used in [8–10] and the "back-projection" strategy used in [11,12]. Four methods are compared, namely "projection (split)", "projection (joint) ", "back-projection (split)' and "back-projection (joint)". "split" means the poses of X-ray frames and pins are estimated separately, "joint" means all X-ray frames and pins are optimised together. We have implemented the compared methods ourselves since the source codes of the compared works are not publicly available. Given that there is practically no pixel-wise correspondences between the generated DRRs of the pin mesh model and X-ray images [14], currently the DRR-based 2D-3D registration method used in [13] is not compared.

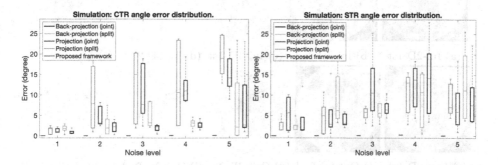

Fig. 2. Resection plane error for the simulation with five increasing noise levels.

4.1   Simulation and Robustness Assessment

To simulate the in-vivo scenario, zero-mean Gaussian noise with standard deviation (SD) of 2 pixels is added to the edge feature observations on the X-ray images. Five different levels of noise are added to the ground truth and used in the initialisation of state X, zero-mean Gaussian noise with SD of $\{0.1, 0.2, 0.3, 0.4, 0.5\}$ rad is added to the rotation angles, and noise with SD of $\{20, 40, 60, 80, 100\}$ mm is added to the translations and all the 3D edge points.

Ten independent runs are executed for each noise level to test the robustness and accuracy of the proposed algorithm and the compared methods. The absolute errors of CTR and STR angles (compared to the ground truth of the simulated resection plane) of the ten runs for the estimation of tibial resection plane are shown in Fig. 2. It shows that the proposed framework has the highest accuracy and robustness in terms of CTR and STR angles.

4.2   In-vivo Experiments

We use data from five CON-TKAs for evaluating the proposed algorithm. The doctor uses a stylus together with a touch screen to draw and extract the contour edges of the proximal tibia and pins from the intra-operative X-ray images, and only clear tibia and fibula contours are used as observations. The state variables $\{X_k^C\}$ are initialised by the C-arm rotation angles and translation measured using the device joint encoders, and $\{X_l^M\}$ are initialised by the pose of the ideal resection plane in the pre-operative tibia model according to the optimal tibial resection requirement. The state variables $\mathbf{P}_{k,i}^M$ are initialised by back-projecting their corresponding 2D contour observation points $\bar{\mathbf{p}}_{k,i}$ into the pre-operative tibia model frame using $\{X_k^C\}$ and the initial guesses for the point depths. And

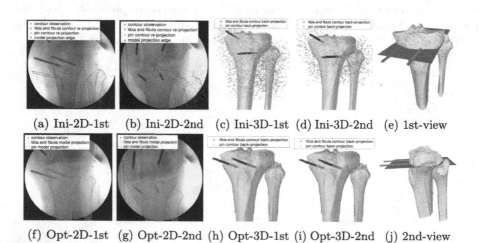

(a) Ini-2D-1st   (b) Ini-2D-2nd   (c) Ini-3D-1st   (d) Ini-3D-2nd   (e) 1st-view

(f) Opt-2D-1st   (g) Opt-2D-2nd   (h) Opt-3D-1st   (i) Opt-3D-2nd   (j) 2nd-view

Fig. 3. The initialisation and optimisation on the first in-vivo dataset. "Ini" and "Opt" are the abbreviations for initialisation and optimisation, respectively.

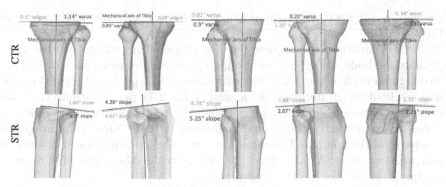

(a) 1st dataset (b) 2nd dataset (c) 3rd dataset (d) 4th dataset (e) 5th dataset

Fig. 4. Results from the proposed framework on the five in-vivo datasets.

the points depth can be initialised by using the distance between the patella of the patient and the X-ray tube from the C-arm X-ray device. Similar method is used to initialise the state variables $\mathbf{P}_{k,j}^{M_l}$.

Figure 3 shows the initialisation and optimisation results of the proposed algorithm on the first in-vivo dataset. Figure 3(a)–(d) show the tibia and pin contour re-projections onto the 2D X-ray images and contour back-projections into the pre-operative tibia model frame and the pin model frame using the initial values. It is obvious that the projected and back-projected contour features have large errors. After optimisation, referring to Fig. 3(f)–(i), the projection from the pre-operative tibia model and the pin model can fit well with the extracted observation, and the contour back-projection features are very close to the outer surface of the pre-operative tibia model and the pin model. Figure 3(e) and Fig. 3(j) show two different views of the estimated tibial resection plane.

Figure 4 shows the comparison between the estimated tibial resection plane angles (blue) using the proposed method and the ground truth (green) obtained from post-operative tibial CT scans for the five in-vivo datasets. The proposed method can achieve high accurate estimation of the tibial resection plane angles intra-operatively, and this can guarantee the accuracy of tibial resection by comparing the estimation to the clinical bone resection requirement (CTR and STR

Table 1. Resection plane estimation error in five in-vivo testings.

Invivo datasets	1		2		3		4		5	
Errors (°)	CTR	STR	CTR	STR	CTR	STR	CTR	STR	CTR	STR
Proposed framework	**1.64**	**1.15**	**1.44**	**−0.23**	**0.28**	**0.51**	**−1.13**	**0.79**	**0.76**	**0.94**
Projection (split)	3.72	6.15	3.43	−1.6	4.13	7.43	−4.21	−4.9	5.8	4.0
Projection (joint)	2.75	5.32	2.61	−1.78	2.24	8.46	−3.81	−7.11	6.23	3.51
Back-projection (split)	5.75	4.97	2.0	−2.89	5.58	8.04	−5.29	−5.45	6.75	4.67
Back-projection (joint)	3.51	5.73	3.16	−1.31	3.97	7.80	−4.84	−4.9	5.14	4.18

angles within $\pm 3°$ w.r.t. the mechanical axis) before the surgeon cut the proximal end of the tibia.

Table 1 summarises the accuracy of the proposed algorithm comparing to other four methods using the five groups of in-vivo datasets, and overall the proposed framework achieves the highest accuracy. The proposed algorithm converged in about 15–20 iterations with around 1.5–2.5 s per iteration. Thus, the computation time is short enough such that the tibial resection plane estimation can be completed without much influence on the CON-TKA procedure. In comparison, the compared methods use numerical optimisation tools such as simulated annealing (SA) algorithm [20] or the Nelder-Mead Downhill Simplex algorithm [21] and take hours to converge.

5   Conclusion

This paper presents SLAM-TKA, a real-time algorithm for reliably and intra-operatively estimating the tibial resection plane for CON-TKAs. It solves a SLAM problem using a patient-specific pre-operative tibia CT scans, a trocar pin mesh model and two intra-operative X-ray images. Simulation experiments demonstrate the robustness and accuracy of the proposed algorithm. In-vivo experiments demonstrate the feasibility and practicality of TKA-SLAM. In conclusion, the algorithm proposed in this paper can be deployed to improve tibial resection intra-operatively without the changing on the CON-TKA procedures. In the future, we aim to use a general tibia model together with different views of intra-operative X-ray images to recover the patient-specific tibia model, in order to replace the procedure of pre-operative tibia CT scanning and segmentation.

References

1. Gao, J., Xing, D., Dong, S., Lin, J.: The primary total knee arthroplasty: a global analysis. J. Orthop. Surg. Res. **15**, 1–12 (2020)
2. Pietrzak, J., Common, H., Migaud, H., Pasquier, G., Girard, J., Putman, S.: Have the frequency of and reasons for revision total knee arthroplasty changed since 2000? comparison of two cohorts from the same hospital: 255 cases (2013–2016) and 68 cases (1991–1998). Orthop. Traumatol. Surg. Res. **105**(4), 639–645 (2019)
3. Gromov, K., Korchi, M., Thomsen, M.G., Husted, H., Troelsen, A.: What is the optimal alignment of the tibial and femoral components in knee arthroplasty? an overview of the literature. Acta Orthop. **85**(5), 480–487 (2014)
4. Berend, M.E., et al.: The chetranjan ranawat award: tibial component failure mechanisms in total knee arthroplasty. Clin. Orthop. Relat. Res. (1976–2007) **428**, 26–34 (2004)
5. Iorio, R., et al.: Accuracy of manual instrumentation of tibial cutting guide in total knee arthroplasty. Knee Surg. Sports Traumatol. Arthroscopy **21**(10), 2296–2300 (2012). https://doi.org/10.1007/s00167-012-2005-7
6. Patil, S., D'Lima, D.D., Fait, J.M., Colwell, C.W., Jr.: Improving tibial component coronal alignment during total knee arthroplasty with use of a tibial planing device. JBJS **89**(2), 381–387 (2007)

7. Hetaimish, B.M., Khan, M.M., Simunovic, N., Al-Harbi, H.H., Bhandari, M., Zalzal, P.K.: Meta-analysis of navigation vs conventional total knee arthroplasty. J. Arthrop. **27**(6), 1177–1182 (2012)
8. Mahfouz, M.R., Hoff, W.A., Komistek, R.D., Dennis, D.A.: A robust method for registration of three-dimensional knee implant models to two-dimensional fluoroscopy images. IEEE Trans. Med. Imaging **22**(12), 1561–1574 (2003)
9. Kim, Y., Kim, K.-I., Choi, J.H., Lee, K.: Novel methods for 3D postoperative analysis of total knee arthroplasty using 2D–3D image registration. Clin. Biomech. **26**(4), 384–391 (2011)
10. Kobayashi, K., et al.: Automated image registration for assessing three-dimensional alignment of entire lower extremity and implant position using bi-plane radiography. J. Biomech. **42**(16), 2818–2822 (2009)
11. Baka, N., et al.: 2D–3D shape reconstruction of the distal femur from stereo x-ray imaging using statistical shape models. Med. Image Anal. **15**(6), 840–850 (2011)
12. Zheng, G., Gollmer, S., Schumann, S., Dong, X., Feilkas, T., González Ballester, M.A.: A 2D/3D correspondence building method for reconstruction of a patient-specific 3D bone surface model using point distribution models and calibrated X-ray images. Med. Image Anal. **13**(6), 883–899 (2009)
13. Otake, Y., et al.: Intraoperative image-based multiview 2D/3D registration for image-guided orthopaedic surgery: incorporation of fiducial-based c-arm tracking and GPU-acceleration. IEEE Trans. Med. Imaging **31**(4), 948–962 (2011)
14. Markelj, P., Tomaževič, D., Likar, B., Pernuš, F.: A review of 3D/2D registration methods for image-guided interventions. Med. Image Anal. **16**(3), 642–661 (2012)
15. Jones, C.W., Jerabek, S.A.: Current role of computer navigation in total knee arthroplasty. J. Arthrop. **33**(7), 1989–1993 (2018)
16. Hampp, E.L., et al.: Robotic-arm assisted total knee arthroplasty demonstrated greater accuracy and precision to plan compared with manual techniques. J. Knee Surg. **32**(03), 239–250 (2019)
17. Sassoon, A., Nam, D., Nunley, R., Barrack, R.: Systematic review of patient-specific instrumentation in total knee arthroplasty: new but not improved. Clin. Orthop. Relat. Res. **473**(1), 151–158 (2015)
18. Nam, D., Weeks, K.D., Reinhardt, K.R., Nawabi, D.H., Cross, M.B., Mayman, D.J.: Accelerometer-based, portable navigation vs imageless, large-console computer-assisted navigation in total knee arthroplasty: a comparison of radiographic results. J. Arthrop. **28**(2), 255–261 (2013)
19. Xiang Gao, Yu., Sun, Z.-H.C., Dou, T.-X., Liang, Q.-W., Li, X.: Comparison of the accelerometer-based navigation system with conventional instruments for total knee arthroplasty: a propensity score-matched analysis. J. Orthop. Surg. Res. **14**(1), 1–9 (2019)
20. Kirkpatrick, S., Gelatt, C.D., Jr., Vecchi, M.P.: Optimization by simulated annealing. Science **220**(4598), 671–680 (1983)
21. Nelder, J.D., Mead, R.: A simplex method for function minimization. Comput. J. **7**(4), 308–313 (1965)

Digestive Organ Recognition in Video Capsule Endoscopy Based on Temporal Segmentation Network

Yejee Shin[1], Taejoon Eo[1], Hyeongseop Rha[1], Dong Jun Oh[2], Geonhui Son[1], Jiwoong An[1], You Jin Kim[3], Dosik Hwang[1], and Yun Jeong Lim[2(✉)]

[1] School of Electrical and Electronic Engineering, Yonsei University, Seoul, Republic of Korea
[2] Department of Internal Medicine, Dongguk University Ilsan Hospital, Dongguk University College of Medicine, Goyang, Republic of Korea
drlimyj@gmail.com
[3] IntroMedic, Capsule Endoscopy Medical Device Manufacturer, Seoul, Republic of Korea

Abstract. The interpretation of video capsule endoscopy (VCE) usually takes more than an hour, which can be a tedious process for clinicians. To shorten the reading time of VCE, algorithms that automatically detect lesions in the small bowel are being actively developed, however, it is still necessary for clinicians to manually mark anatomic transition points in VCE. Therefore, anatomical temporal segmentation must first be performed automatically at the full-length VCE level for the fully automated reading. This study aims to develop an automated organ recognition method in VCE based on a temporal segmentation network. For temporal locating and classifying organs including the stomach, small bowel, and colon in long untrimmed videos, we use MS-TCN++ model containing temporal convolution layers. To improve temporal segmentation performance, a hybrid model of two state-of-the-art feature extraction models (i.e., TimeSformer and I3D) is used. Extensive experiments showed the effectiveness of the proposed method in capturing long-range dependencies and recognizing temporal segments of organs. For training and validation of the proposed model, the dataset of 200 patients (100 normal and 100 abnormal VCE) was used. For the test set of 40 patients (20 normal and 20 abnormal VCE), the proposed method showed accuracy of 96.15, F1-score@$\{50,75,90\}$ of $\{96.17, 93.61, 86.80\}$, and segmental edit distance of 95.83 in the three-class classification of organs including the stomach, small bowel, and colon in the full-length VCE.

Keywords: Video Capsule Endoscopy · Organ recognition · Temporal segmentation · Temporal convolutional networks

---

Y. Shin and T. Eo—These authors contributed equally to this work.

Supplementary Information The online version contains supplementary material available at https://doi.org/10.1007/978-3-031-16449-1_14.

1   Introduction

Video capsule endoscopy (VCE) is a noninvasive diagnostic modality that can provide direct mucosal visualization for whole gastrointestinal tracts [1]. In VCE, the patient gulps a capsule containing a camera-embedded device that passes through the gastrointestinal tracts to capture images that are transmitted to an external receiver. Although most VCE is used for the diagnosis of small bowel diseases in current clinical practice, esophageal, gastric, and colon capsule endoscopies have been developed to exploit the advantages of VCE. Because it is able to examine from mouth to anus using VCE, if automated reading for VCE is commercialized, VCE as a pan-endoscopy could be a breakthrough innovative screening diagnostic method of alimentary tract.

On the other hand, the biggest problem with the usage of VCE is the long reading time. Typical reading times vary between 30 and 120 min, which can be a tedious process for clinicians. The reading time is usually influenced by capsule's transit time to each organ and the experience of the reader [2–5]. To overcome this problem, complementary diagnostic tools based on artificial intelligence (AI) algorithms have been developed to shorten the interpretation time of VCE [6–10]. For example, if the capsule's transit time for each organ can be automatically measured, it can efficiently reduce the reading time [11]. In the previous work from [11], a 2D CNN-based algorithm was used to predict the duodenal transition time of the capsule, which demonstrates that the AI model can predict the transition time with the deviation time within 8 min. In this conventional method, however, only a transition point where the capsule enters the descending segment of the duodenum was detected, and the transition point between the small bowel and colon could not be detected. In addition, because only images captured from the stomach or small bowel were manually selected, this method cannot be directly applied to full-length VCE video.

The study aims to develop an automated organ recognition method for full-length VCE video based on a temporal segmentation network. For temporal locating and classifying organs in long untrimmed videos, MS-TCN++ [12] model consisting of temporal convolution layers [13, 14] is used. To improve the segmentation accuracy, input features for MS-TCN++ were extracted from the combination of two state-of-the-art networks for video feature extraction (i.e., TimeSformer [15] and I3D [16]). Consequently, the proposed algorithm temporally segments each full-length VCE video into three organs including the stomach, small bowel, and colon. To demonstrate the clinical efficacy of the proposed method, extensive experiments were performed using datasets of 200 patients. For the test set of 40 patients, accuracy, F1 score, and segmental edit distance were assessed.

2   Methods

2.1   Study Design (Fig. 1)

Preparation of Images for AI Training, Validation, and Test. The study protocol was approved by Dongguk University Ilsan Hospital Institutional Review Board DUIH 2022-01-032-001). We retrospectively included 260 patients who had undergone the CE examination on suspicion of suffering from small-bowel diseases at the Dongguk University Ilsan Hospital between 2002 and 2022. Data were obtained from a retrospective medical record review stored in Network Attached Storage (NAS) system. The examination was performed using an MiroCam device (MC1000W and MC1200, Intromedic Co. Ltd., Korea). The images were extracted in JPEG format using MiroView 4.0, with a matrix size of 320×320. Cases in which the capsule endoscope did not pass into the colon and cases in which organs could not be visually distinguished were excluded. For video labeling, two gastroenterologists specializing in VCE (Oh DJ and Lim YJ from Dongguk University Ilsan Hospital) independently performed image labeling for organ classification. Two gastroenterologists read the whole images from each WCE case manually and classified them as stomach, small bowel, and colon, and cross-checked with each other.

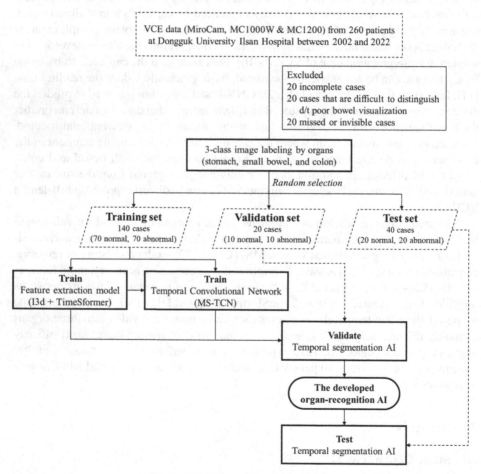

Fig. 1. Flowchart of the study design.

Distribution of Collected Images. The labeled datasets (n = 200) were then classified into training (n = 140, 70%), validation (n = 20, 10%), and test (n = 40, 20%) set. The training set contained 1,572,274, 8,721,126, and 7,809,614 frames for the stomach, small intestine, and colon, respectively. The validation set contained 136,266, 971,231, and 883,745 frames for stomach, small intestine, and colon, respectively. The test set contained 278,455, 2,046,571, and 1,541,448 frames for the stomach, small colon, and colon, respectively.

2.2 Proposed Organ Recognition Method

Our goal is to infer temporal segmentation for each video containing endoscopic frames $c_{1:T} = (c_1, \ldots, c_T)$ to localize where the capsule passed through digestive organs. We propose an organ recognition network for VCE. The proposed network consists of two main stages as shown in Fig. 2. Given the image frames $c_{1:T}$, we first obtain 2D feature maps that are extracted by a hybrid model. We introduce a new feature extractor architecture to outperform a single existed network. We finally use MS-TCN++ [12] to detect organs from multiple frames. More details are given below.

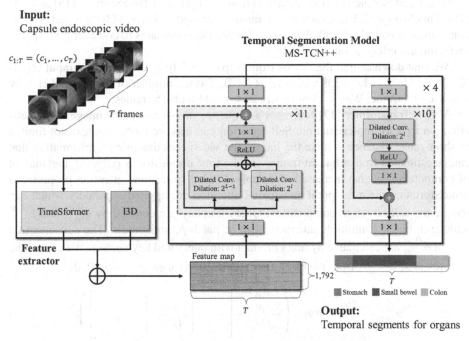

Fig. 2. The proposed organ recognition method for VCE

Temporal Segmentation Model. The model is based on sets of dilated temporal layers [13]. The temporal convolutional network (TCN) can expand the receptive field along the temporal direction. A dual-dilated layer uses two convolutional networks with two dilated factors to consider both small and large receptive fields at each layer. A multi-stage architecture stacked with TCN can gradually improve the prediction of the previous steps. We adopt the model with five stages. We utilize a cross-entropy loss for classification loss and mean square error for smoothing loss. The smoothing loss is used for reducing temporal segmentation error.

$$L_{cls} = \frac{1}{T}\sum_t^T - log(y_t) \tag{1}$$

$$L_{smoothing} = \frac{1}{T}\sum_t^T |log(y_t) - log(y_{t-1})|^2 \tag{2}$$

The weighting factors of 1 and 0.15 is used for classification loss and smoothing loss, respectively. We adopt Adam optimizer with a learning rate 0.00005.

Feature Extraction. It is very important to extract generalized features which can significantly affect the quality of the prediction results. Features for inputs of MS-TCN++ from each endoscopic video are extracted from the hybrid of TimeSformer [15] and I3D [16]. TimeSformer is based on a vision transformer model [17] and I3D is based on a 3D convolutional neural network that is suitable for video domains to consider both spatial and temporal information.

We first downsample the videos from 3 fps to 3/5 fps. A video as an input clip is $X \in \mathbb{R}^{3 \times T \times H \times W}$ where 3 is a channel of RGB, T is the number of frames, and H × W is a size of a frame. We use clips of size 224 × 224 with 15 frames.

Vision Transformer [17] leverages a sequence of patches in an image using self-attention to learn representation. Self-attention can capture global context not limited to short-range distance. Unlike the image, the video contains motion information that can be estimated from adjacent frames. As the time dimension is reflected, the amount of computation is also increased. We apply divided space-time attention proposed in TimeSformer, where temporal and spatial attention are applied separately which can reduce the self-attention computation. Query and key are denoted by $q_{(p,t)}^{(l,a)}$ and $k_{(p,t)}^{(l,a)}$ with each block l, multiple attention head a, patch p, and time t. The self-attention weight $\alpha_{(p,t)}^{(l,a)}$ are calculated by inner product with query and key vectors. We extract the features $W \in \mathbb{R}^{768 \times 1}$ from the block for classification token $(p, t) \rightarrow (0, 0)$.

$$\alpha_{(p,t)}^{(l,a)space} = softmax\left(\frac{q_{(p,t)}^{(l,a)T}}{\sqrt{D_h}} \cdot \left[k_{(0,0)}^{l,a} \left\{k_{(p',t)}^{(l,a)}\right\}_{p'=1,...,N}\right]\right) \tag{3}$$

$$\alpha_{(p,t)}^{(l,a)time} = softmax\left(\frac{q_{(p,t)}^{(l,a)^T}}{\sqrt{D_h}} \cdot \left[k_{(0,0)}^{l,a}\left\{k_{(p,t')}^{(l,a)}\right\}_{t'=1,...,T}\right]\right) \quad (4)$$

I3D is combined with a series of RGB images and optical flow charts corresponding to the video clip. RGB features are extracted from the existing image frame bundle and optical flow features are extracted from the gradient image frame bundle to which the Sobel filter is applied. I3D model is constructed with 3D convolution layers, max-pooling layers, Inception modules, and a 7×7 average pooling layer. Considering both temporal and spatial information, they are effective to be computed at the same speed. Therefore, the symmetric receptive field is applied to the last two max-pooling layers to control extracting speed of temporal and spatial information. The input clip X is passed through I3D to obtain output features $W \in \mathbb{R}^{1024 \times 1}$. Extracting optical flow feature $O \in \mathbb{R}^{2 \times T \times H \times W}$ where 2 is a channel of horizontal and vertical optical flow is also passed through I3D to obtain output feature $W \in \mathbb{R}^{1024 \times 1}$. The average value of RGB and optical flow is used. These processes have the effect of downsampling by 1/75. We adopt a target class of a median frame of the video clip.

To improve the performance, the I3D model pre-trained with ImageNet is finetuned by our VCE dataset, so that the model extracts a feature proper to the temporal segmentation task for organs. With an initial learning rate of 0.0001, we apply the Adam optimizer. The TimeSformer model pretrained with Kinetics-600 is also finetuned to extract better features that are more generalized to our tasks. We use SGD optimizer with an initial learning rate of 0.001 and a weight decay of 0.0001.

In terms of extracting features, the transformer model has the advantage of learning global context with weak inductive bias. Convolutional networks are effective in understanding relationships with adjacent pixels. We concatenate two features generated by TimeSformer and I3D model so that we take both advantages of the models and get more representative features for each video clip. We finally obtain features $W \in \mathbb{R}^{1792 \times 1}$.

Metrics. To evaluate how accurately organ is recognized, we calculate five metrics. We report the accuracy, precision, and recall for evaluating framewise performances. We use segmental edit distance (i.e., edit) and the segmental F1 score at intersection over union (IoU) ratio thresholds of 50%, 75%, and 90% (F1@{50, 75, 90}). While accuracy, precision, and recall are commonly used metrics for framewise classification, class imbalance and segmentation at wrong locations, not over segmentation, are not well reflected through those metrics. For these reasons, we adopt the segmental F1 score and edit distance as quantitative metrics for the temporal segmentation.

3 Results

We evaluate our proposed method for organ recognition through 20 normal and 20 abnormal cases. Train, valid, and test sets are selected randomly for each institution. Table 1 shows the results of the ablation study for different feature extraction methods. Since the number of classes is small and the overall area is well recognized, we set the F1 score at high overlapping thresholds of 50%, 75%, and 90%. Comparing before and after the finetuning of each feature extraction model, it can be seen that the latter one performs better. Compared with TimeSformer after finetuning, the accuracy of our method increased from 95.83 to 96.15, F1@{90} increased from 86.07 to 86.80, and the edit score is increased from 94.58 to 95.83. In the case of the I3D model, the overall scores are relatively lower than the others. Compared with the I3D model which is the combination of I3D and MS-TCN++ (baseline), scores are increased considerably, especially F1@{90} from 62.71 to 86.80. Consequently, the results demonstrate the effectiveness of our feature extraction method for temporal segmentation of organs.

Figure 3 shows the results of the above experiments in Table 1. The green, blue, and yellow areas depict the temporal ranges of stomach, small bowel, and colon, respectively. The red dotted line refers to the transition line where the organ changes. As shown in Fig. 3, the results of our proposed localization method are well-segmented along the temporal axis rather than the other models.

Table 1. Quantitative results for classifying three organs using different feature extraction models

Feature extraction		Fine-tuning	Accuracy	Precision	Recall	F1@{50,75,90}			Edit
I3D	TimeS-former								
✓ (baseline)			91.95	0.915	0.910	87.28	77.96	62.71	93.95
✓		✓	93.74	0.941	0.896	86.88	78.68	66.39	91.00
	✓		94.73	0.952	0.943	93.22	88.98	72.88	94.20
✓	✓		95.83	0.959	0.952	94.51	92.82	86.07	94.58
✓	✓	✓	**96.15**	**0.962**	**0.957**	**96.17**	**93.61**	**86.80**	**95.83**

*The highest metrics are bold-faced.

Although the proportion of the small bowel is low compared to the total video time, it shows that our method is well performed as in the case of Fig. 3(a, b). In the case of TimeSformer without I3D, most of the errors are found near organ transitions. In the case of I3D without TimeSformer, it is often observed that the temporal segments are not properly predicted as depicted in Fig. 3(a, b).

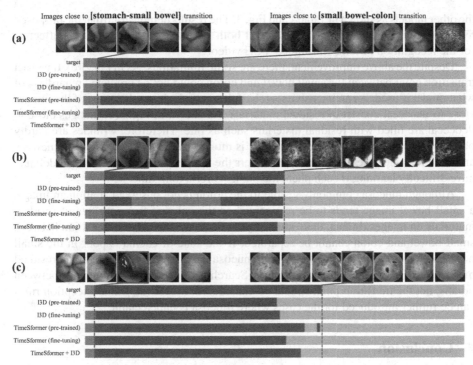

Fig. 3. Temporal segmentation results for organs in VCE. (a) a normal case, (b) an abnormal case (bleeding), and (c) an abnormal case (vascular). In (c), the proposed method failed to segment parts of ileum into the small bowel due to invisible ileocecal mucosa by bile and residual materials. (Color figure online)

The instability of the temporal segmentation of I3D may be caused by an inductive bias. In the case of the error at the front end of the small bowel, there are some common characteristics in the images. Observing the error at the end of the small bowel, vascular patterns and residual materials are often visible that are common features observed in the colon. Moreover, because the wrinkled parts are common features of the small bowel, if the wrinkled parts are invisible on the small bowel images, the model tended to misclassify the small bowel to the colon. The 3D convolutional neural network (3D CNN) has a relational inductive bias of locality and the properties above mentioned are important points in determining where to focus on [18, 19]. It demonstrates that extracted features are biased toward locally observed information such as residual materials and vascular patterns, therefore, it can cause misclassification of the I3D model. Unlike the 3D CNN, the transformer is based on MLP and patch-wise multiplication. It obtains a relatively weak inductive bias because it learns global large-range information. The temporal segmentation area was captured properly in the TimeSformer rather than the I3D model based on finetuning as shown in Fig. 3. It demonstrates that understanding the global context and long-range dependencies is effective in organ recognition. It can be interpreted that understanding both global and local contexts is important to generalize the temporal segmentation task for organs. As a result, the proposed method outperforms the other models as shown in Table 1 and Fig. 3. The results demonstrate that the proposed hybrid feature extraction which can extract both local and global contexts is effective for the recognition of long untrimmed VCE video.

Despite the above improvement, there are some limitations. As in Fig. 3(c), the exact transition point between the small bowel and colon could not be predicted regardless of what model is used. As an intensive search of this case, it shows that most of the images close to ileocecal consist of residual materials. As shown in Fig. 3(c), images close to ileocecal are filled with residual materials such as bile. Therefore, wrinkles are hardly observed in the small bowel. Such the case is more difficult to be classified framewise, and even if the deep-learning model considers the temporal information, the model may have difficulty learning if such phenomenon lasts for a long time.

Although mucosa commonly visible to the colon is not observed at all in an image, a small bowel image tends to be recognized as the colon if there are lots of residual materials on images. In fact, there are many clinical cases where the transition between small bowel and colon cannot be separated. If the state of enema is poor in the small bowel area close to ileocecal, the colonic mucosa is undetectable due to many residual materials such as bile, bleeding, and debris. Searching the valid transition range between organs, not the transition point should be considered. Moreover, frame exception rules can be further introduced for the cases where mucosa is undetectable.

4 Conclusion

To shorten the interpretation of VCE, automated organ recognition can play an important role in predicting the location of the wireless capsule endoscope and applications such as automated lesion detection [20–23]. Therefore, an automated organ recognition method for full-length VCE video should be developed, which is very helpful for reducing clinician's time costs by changing the manual marking process for organs into the automated manner. This study focuses on two main steps: (i) feature extraction of VCE and (ii) temporal segmentation. We propose a novel feature extractor consisting of TimsSformer and I3D that can exploit local and global contexts of long untrimmed VCE videos. Consequently, we show that the proposed method is highly effective for extracting features for organs and accurately segments the VCE video into organs.

Acknowledgements. This research was supported by a grant (grant number: HI19C0665) from the Korean Health Technology R & D project through the Korean Health Industry Development Institute (KHIDI) funded by the Ministry of Health & Welfare, Republic of Korea. And this research was supported by Basic Science Research Program through the National Research Foundation of Korea (NRF) funded by the Ministry of Science and ICT (2021R1C1C2008773), and Y-BASE R&E Institute a Brain Korea 21, Yonsei University. And this research was partially supported by the Yonsei Signature Research Cluster Program of 2022 (2022-22-0002), the KIST Institutional Program (Project No.2E31051-21-204), the Institute of Information and Communications Technology Planning and Evaluation (IITP) Grant funded by the Korean Government (MSIT) Artificial Intelligence Graduate School Program, Yonsei University (2020-0-01361).

References

1. Iddan, G., Meron, G., Glukhovsky, A., Swain, P.: Wireless capsule endoscopy. Nature **405**(6785), 417 (2000)
2. Zou, Y., Li, L., Wang, Y., Yu, J., Li, Y., Deng, W.: Classifying digestive organs in wireless capsule endoscopy images based on deep convolutional neural network. In: 2015 IEEE International Conference on Digital Signal Processing (DSP), pp. 1274–1278. IEEE (2015)
3. Flemming, J., Cameron, S.: Small bowel capsule endoscopy: indications, results, and clinical benefit in a university environment. Medicine **97**(14) (2018)
4. Magalhães-Costa, P., Bispo, M., Santos, S., Couto, G., Matos, L., Chagas, C.: Re-bleeding events in patients with obscure gastrointestinal bleeding after negative capsule endoscopy. World J. Gastrointest. Endosc. **7**(4), 403 (2015)
5. Nakamura, M., et al.: Validity of capsule endoscopy in monitoring therapeutic interventions in patients with Crohn's disease. J. Clin. Med. **7**(10), 311 (2018)
6. Leenhardt, R., et al.: A neural network algorithm for detection of gi angiectasia during small-bowel capsule endoscopy. Gastrointest. Endosc. **89**(1), 189–194 (2019)
7. Oki, T., et al.: Automatic detection of erosions and ulcerations in wireless capsule endoscopy images based on a deep convolutional neural network. Gastrointest. Endosc. **89**(2), 357–363 (2019)
8. Wu, X., Chen, H., Gan, T., Chen, J., Ngo, C.W., Peng, Q.: Automatic hookworm detection in wireless capsule endoscopy images. IEEE Trans. Med. Imaging **35**(7), 1741–1752 (2016)
9. Tsuboi, A., et al.: Artificial intelligence using a convolutional neural network for automatic detection of small-bowel angioectasia in capsule endoscopy images. Dig. Endosc. **32**(3), 382–390 (2020)
10. Hwang, Y., et al.: Improved classification and localization approach to small bowel capsule endoscopy using convolutional neural network. Dig. Endosc. **33**(4), 598–607 (2021)
11. Gan, T., Liu, S., Yang, J., Zeng, B., Yang, L.: A pilot trial of convolution neural network for automatic retention-monitoring of capsule endoscopes in the stomach and duodenal bulb. Sci. Rep. **10**(1), 1–10 (2020)
12. Li, S.J., AbuFarha, Y., Liu, Y., Cheng, M.M., Gall, J.: MS-TCN++: multi-stage temporal convolutional network for action segmentation. IEEE Trans. Pattern Anal. Mach. Intell. (2020)
13. Lea, C., Flynn, M.D., Vidal, R., Reiter, A., Hager, G.D.: Temporal convolutional networks for action segmentation and detection. In: proceedings of the IEEE Conference on Computer Vision and Pattern Recognition, pp. 156–165 (2017)
14. Farha, Y.A., Gall, J.: MS-TCN: multi-stage temporal convolutional network for action segmentation. In: Proceedings of the IEEE/CVF Conference on Computer Vision and Pattern Recognition, pp. 3575–3584 (2019)
15. Bertasius, G., Wang, H., Torresani, L.: Is space-time attention all you need for video understanding. arXiv preprint (3), 4 (2021). arXiv:2102.050952
16. Carreira, J., Zisserman, A.: Quo vadis, action recognition? A new model and the kinetics dataset. In: Proceedings of the IEEE Conference on Computer Vision and Pattern Recognition, pp. 6299–6308 (2017)
17. Dosovitskiy, A., et al.: An image is worth 16x16 words: transformers for image recognition at scale. arXiv preprint arXiv:2010.11929 (2020)
18. Greenspan, H., Van Ginneken, B., Summers, R.M.: Guest editorial deep learning in medical imaging: overview and future promise of an exciting new technique. IEEE Trans. Med. Imaging **35**(5), 1153–1159 (2016)
19. Jang, J., Hwang, D.: M3T: three-dimensional medical image classifier using multi-plane and multi-slice transformer. In: Proceedings of the IEEE Conference on Computer Vision and Pattern Recognition (2022)

20. Park, J., et al.: Recent development of computer vision technology to improve capsule endoscopy. Clin. Endosc. **52**(4), 328–333 (2019)
21. Park, J., et al.: Artificial intelligence that determines the clinical significance of capsule endoscopy images can increase the efficiency of reading. PLoS ONE **15**(10), e0241474 (2020)
22. Nam, J.H., et al.: Development of a deep learning-based software for calculating cleansing score in small bowel capsule endoscopy. Sci. Rep. **11**(1), 1–8 (2021)
23. Nam, S.J., et al.: 3D reconstruction of small bowel lesions using stereo camera-based capsule endoscopy. Sci. Rep. **10**(1), 1–8 (2020)

Mixed Reality and Deep Learning for External Ventricular Drainage Placement: A Fast and Automatic Workflow for Emergency Treatments

Maria Chiara Palumbo[1]([✉]), Simone Saitta[1], Marco Schiariti[2],
Maria Chiara Sbarra[1], Eleonora Turconi[1], Gabriella Raccuia[2], Junling Fu[1],
Villiam Dallolio[3], Paolo Ferroli[2], Emiliano Votta[1], Elena De Momi[1],
and Alberto Redaelli[1]

[1] Department of Electronics Information and Bioengineering,
Politecnico Di Milano, Milan, Italy
mariachiara.palumbo@polimi.it
[2] Department of Neurosurgery, Carlo Besta Neurological Institute
IRCCS Foundation, Milan, Italy
[3] Promev SRL, Lecco, Italy

Abstract. The treatment of hydrocephalus is based on anatomical landmarks to guide the insertion of an External Ventricular Drain (EVD). This procedure can benefit from the adoption of Mixed Reality (MR) technology. In this study, we assess the feasibility of a fully automatic MR and deep learning-based workflow to support emergency EVD placement, for which CT images are available and a fast and automatic workflow is needed. The proposed study provides a tool to automatically i) segment the skull, face skin, ventricles and Foramen of Monro from CT scans; ii) import the segmented model in the MR application; iii) register holograms on the patient's head via a marker-less approach. An ad-hoc evaluation approach including 3D-printed anatomical structures was developed to quantitatively assess the accuracy and usability of the registration workflow.

Keywords: Mixed reality · Deep learning · External ventricular drainage

1 Introduction

External Ventricular Drainage (EVD) represents the optimal surgical treatment for emergency cases of hydrocephalus [1]. This procedure consists in draining

M.C. Palumbo and S. Saitta—Equal first authorship.

Supplementary Information The online version contains supplementary material available at https://doi.org/10.1007/978-3-031-16449-1_15.

cerebrospinal fluid by inserting a catheter into the brain ventricles, positioning the tip on the Foramen of Monro (FoM). EVD placement is usually performed blindly, relying on paradigmatic anatomical landmarks localized freehand on the patient's head, thus carrying the risk of catheter malpositioning that can lead to hemorrhage, infections, brain tissue damage, multiple passes and reinsertion, entailing unnecessary brain trauma. Despite being a relatively simple procedure, a nearly 50% inaccuracy rate has been reported [1]. To address these issues, pre-operative images can provide helpful guidance during EVD placement and insertion [1,16], for instance, when combined with neuronavigation systems [10]. However, the high cost and significant encumbrance of these systems makes them hard to be adopted in the operating room. Mixed Reality (MR)-based solutions can represent a low-cost, easily portable guidance system for EVD placement. Previous studies have shown the feasibility of adopting MR based solutions for ventriculostomy procedures, providing visualization of internal brain structures [5], or to register the manually segmented holographic content on the patient's anatomy using marker based methods [3,9,13]. Nonetheless, these approaches do not represent adequate solutions for emergency cases where both automatic segmentation and quick registration need to coexist in a fast multistep routine. A first attempt to superimpose holograms on real world object using HoloLens2 Head Mounted Display (HMD) (Microsoft, Redmond, WA) and its depth camera via a marker-less approach, has been reported [7], resulting in low accuracy. The goal of this study was to assess the feasibility and accuracy of a MR and deep learning-based workflow to support EVD placement, in emergency procedures for which preoperative CT images are only available and a fast and automatic workflow is critical. The proposed study provides a tool to automatically i) segment the skull, skin, ventricles and FoM from computed tomography (CT) scans; ii) import the segmented model in the MR application; iii) register holograms on the patient's head via a marker-less approach.

2   Methods

The envisioned surgical workflow graphically depicted in Fig. 1 is divided into consecutive steps resulting in a final MR application specifically tailored for HoloLens2 (H2) device. The following sections provide a detailed description on the development of the automatic segmentation (Sect. 2.1), automatic registration (Sect. 2.2) and registration accuracy assessment (Sect. 2.3) methods.

2.1   Automatic Segmentation

3D CT scans of 200 subjects (including 41 cases of hydrocephalus) acquired between 2018 and 2022 were used as benchmark (pixel spacing ranging from 0.543×0.543 to $0.625 \times 0.625 \, mm^2$, slice thickness between 0.625 and 1.25 mm). The study was approved by the local ethical committee and informed consent was waived because of the retrospective nature of the study and the analysis of anonymized data.

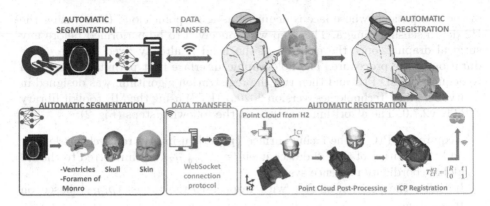

Fig. 1. Schematic representation of the developed workflow: the 3D CT images are automatically segmented to obtain the brain structures that will be sent to H2 device through an internet connection protocol. The user is then able to visualize the holographic models and set a path between the FoM and an entry point on the skin layer. The point cloud (PtC) of the skin surface of the patient is acquired using H2 depth camera and this data is used to estimate the transformation matrix $\mathbf{T}_{CT}^{H2}$.

Skin and Skull Segmentation was performed through i) automatic elimination of the mouth region in case of dental implant inducing artifacts; ii) thresholding (bone tissue: high-pass filter, threshold of 500 Hounsfield units (HU) [11]; skin: band-pass filter, lower and upper thresholds of -100 HU and 50 HU, respectively); iii) hole filling and smoothing.

Ventricle Segmentation was performed by a neural network (NN) based on a 3D UNET architecture, composed of an input layer of 16 filters, encoder and decoder branches of 5 resolution levels each, a bottleneck block and skip connections via concatenation [14]. The NN was trained and tested using all the CT volumes: the two lateral ventricles and the third ventricle were manually segmented by two experienced operators, guaranteeing that each acquisition was segmented by at least one user. The whole dataset was divided into training and test sets: 180 (90%) scans with their corresponding ground truth segmentations were randomly selected and used for NN training. A data augmentation routine was implemented to increase dataset diversity. Random rotations, mirroring and affine transformations were applied to the processed patches. A combination of Lovasz-Softmax loss and focal loss [12] was used to trained the NN to perform simultaneous segmentation of the lateral and third ventricles. The remaining 20 scans (10%), including 5 cases of hydrocephalus, were used for testing.

Detection of Foramen of Monro, which is the target for the tip of the catheter during EVD, was performed by a recursive thresholding approach [12].

2.2 Automatic Registration

The registration process was conceived to allow the end user to reliably superimpose the MR representation of the segmented skin model ($\mathcal{S}_{CT}$) on the head of the real patient, whose face is acquired as a 3D point cloud (PtC) using the H2 depth sensing camera. This step was conceived to be performed before any surgical draping, once the position of the head is already fixed for the entire duration of the procedure. The holographic interface that permits the surgeon to acquire the 3D PtC and then run the registration algorithm, was designed in Unity3D (Unity Technologies, version 2020.3.1LTS) using the MR toolkit library MRTK v2.7.0. The algorithm consisted of the following steps (Fig. 2):

- Acquire the PtC of the facing surface $\{p_{i,H2}\}$, using H2 research mode [15], and the position of the H2 device itself ($p_{camera,H2}$), both referred to the H2 global coordinate reference system.
- Initialize the position of $\mathcal{S}_{CT}$ with respect to the target $\{p_{i,H2}\}$ based on $p_{camera,H2}$.
- Apply an Hidden Point Removal algorithm [8] to $\mathcal{S}_{CT}$ to filter out points in the back of the head to avoid unnecessary computations.
- Extract $\{p_{i,H2}\}$ points belonging to the face through a DBSCAN density based clustering algorithm [6].
- Apply a Fast Global Registration algorithm [17] between the simplified $\mathcal{S}_{CT}$ and the cleaned $\{p_{i,H2}\}$, obtaining a first alignment, which is then refined via Local Refined Registration method based on a Point-to-Plane iterative closest point (ICP) algorithm [4] to obtain $\mathbf{T}_{CT}^{H2}$. All the described steps related to the PtC tweaking and registration were implemented using the open3D library [18].
- Send back $\mathbf{T}_{CT}^{H2}$ to H2 and apply it to the holographic model of $\mathcal{S}_{CT}$, thus allowing the surgeon to visualize the segmented model aligned with the patient face $\mathcal{S}_{H2} = \mathbf{T}_{CT}^{H2}\mathcal{S}_{CT}$.

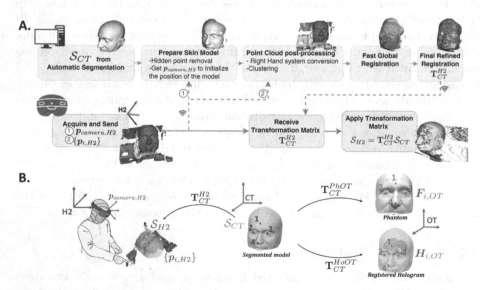

Fig. 2. A. Detailed representation of the registration workflow for the MR environment with B. The considered coordinate systems and transformation matrices.

2.3   Registration Accuracy Assessment

An experimental setup to evaluate the automatic registration accuracy was developed (Fig. 3A). The evaluation was divided into two different procedures to evaluate: 1. the accuracy of the hologram-to-phantom registration; 2. the targeting accuracy in a ventriculostomy-like procedure. For both tests, a 3D model of a patient with hydrocephalus was reconstructed from automated CT scan segmentation. The extracted skin model was modified to add four markers consisting in conical cavities on the skin surface at the left eye, right eye, nose and forehead, respectively. The modified model was 3D-printed and used to simulate a real patient's head. An optical tracker (OT) system (NDI Polaris Vicra) equipped with a trackable probe was used to acquire the position of the conical markers on the phantom, designed to accommodate the tip in a unique manner. The position of the markers was also reported in the holographic model where four red spheres of 1 mm of diameter were added to highlight each marker position.

To evaluate the accuracy of the hologram-to-phantom registration, an H2 expert user was asked to repeat the following procedure 10 times:

- Acquire the 4 markers position on the printed phantom in the OT space ($F_{i,OT}$, with $i = 1,..4$) (Fig. 3B).
- Wearing the H2 device execute the automatic registration and then, moving away the phantom, place the tip of the probe in correspondence of the 4 holographic marker to obtain their positions in the OT space ($H_{i,OT}$, with $i = 1,..4$). For this phase the probe is attached to a tripod which stabilizes the probe movement not limiting its orientation in the space (Fig. 3C).

The distance between the phantom and holographic model was assessed both by computing the Euclidean distance $e_{ij} = ||F_{ij,OT} - H_{ij,OT}||$ of the $i-th$ marker

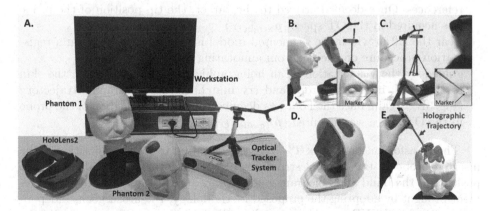

Fig. 3. A. Experimental setup showing the hardware and physical components needed to test the registration accuracy; user executing the first test is evaluating the position of the 4 marker on the printed phantom B., and on the registered holographic model C.; D. phantom used by neurosurgeons for the second test while E. trying to insert the OT probe to match the holographic trajectory.

and $j - th$ coordinate (x, y, z), but also providing an estimate of the error for each point of the skin surface and ventricle. To do this, a first transformation matrix $\mathbf{T}_{CT}^{PhOT}$ was computed to map $\mathcal{S}_{CT}$ with the 4 markers of the printed phantom $\mathbf{F}_{i,OT}$, and a second one $\mathbf{T}_{CT}^{HoOT}$ to map $\mathcal{S}_{CT}$ with the 4 markers of the registered hologram $\mathbf{H}_{i,OT}$ (Fig. 2B). A least-square solution based on the singular value decomposition algorithm [2] was applied to compute the transformation matrices. Finally, the point-to-point distance d_j with $j = 1, ..N$ vertices, was computed as:

$$d_j = ||\mathbf{T}_{CT}^{PhOT} \mathbf{m}_{j,CT} - \mathbf{T}_{CT}^{HoOT} \mathbf{m}_{j,CT}|| \tag{1}$$

with $\mathbf{m}_{j,CT}$ being the vertices of the segmented model of the skin and ventricles in the CT space. The root mean square error $(RMSE)$ was computed as:

$$RMSE = \sqrt{\frac{1}{N} \Sigma_{j=1}^{N} d_j^2} \tag{2}$$

To evaluate targeting accuracy, another phantom was 3D-printed from the same segmented model, provided with a large hole in proximity of a common entry point for a ventriculostomy procedure and with an internal box full of gelatin mimicking brain tissue (6% gelatin in water) (Fig. 3D). A membrane was placed on top of the hole entrance to reduce the bias experienced by the users when choosing the point of entrance. For this test, 7 experienced neurosurgeons were asked to:

- acquire the positions of the 4 conical markers on the phantom with the OT probe $\mathbf{F}_{i,OT}$.
- do a blind (i.e., without wearing the H2) ventriculostomy procedure using the OT probe as a catheter, trying to reach the FoM based on anatomical references. Once deemed arrived to the target, the tip position of the probe was acquired in the OT space $(\mathbf{p}_{blind,OT})$
- wear the H2 in which the segmented model is visualized to perform a registration procedure on the phantom annotating $\mathbf{T}_{CT}^{H2}$.
- guided by the visualization of an holographic path that goes from the skin to the FoM, insert the probe and try matching the holographic trajectory (Fig. 3E). Again, once the FoM was deemed reached, the position of the probe in the OT space was acquired $(\mathbf{p}_{Hguided,OT})$.

The transformation matrix $\mathbf{T}_{CT}^{PhOT}$ was computed to refer the position of the inserted probe to the internal structures of the brain, thus evaluating the tip position in the blind and hologram guided procedure. To quantify the improvement brought by adopting the proposed methodology, 7 neurosurgeons independently simulated EVD procedures on the gelatin phantom. Of note, none of the participant was familiar with the mixed reality device. The targeting accuracy improvement was computed as: $\Delta_\% = \left(\frac{|\mathbf{p}_{blind,OT} - \mathbf{p}_{Hguided,OT}|}{|\mathbf{p}_{blind,OT} - \mathbf{p}_{t,OT}|} \times 100 \right) \%$, where $\mathbf{p}_{t,OT}$ is the target position in OT space.

3 Results

3.1 Automatic Segmentation

Brain Ventricles Test Set Results. The trained NN was applied to run inference on the test set made of 20 CT scans, 5 of which presented hydrocephalus. Our automatic segmentation algorithm was able to generalize to new cases, extracting the 3D ventricle regions with accuracy (Fig. 4), requiring approximately 130 s per scan. For the lateral ventricles, mean Dice score and 95% Hausdorff distance were equal to of 0.88 and 14.4 mm, respectively. For the third ventricle, mean Dice score and 95% Hausdorff distance were equal to 0.62 and 19.4 mm. Considering the five test cases with hydrocephalus, a mean Dice score of 0.94 was obtained for the lateral ventricles and a mean value of 0.81 for the third ventricles, while 95% Hausdorff distances were equal to 18.3 mm and 7.48 mm, respectively.

For all test cases with hydrocephalus, the ventricle surface reconstruction accuracy was sufficient to automatically detect the FoM.

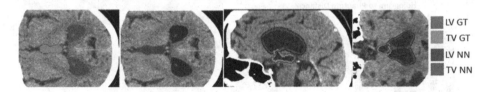

Fig. 4. Comparison of manual (ground truth, GT) segmentation in bright red (lateral ventricles, LV) and bright green (third ventricle, TV), vs. automatic segmentation obtained by the trained neural network (NN) in dark red (lateral ventricles, LV) and dark green (third ventricle, TV) on a test case. (Color figure online)

3.2 Registration Accuracy Assessment

Hologram-to-Phantom Registration. The global time required by the algorithm to compute the transformation matrix $\mathbf{T}_{CT}^{H2}$ and send it back to H2 was 0.26 ± 0.02 s, with a fitness of $92.93 \pm 3.91\%$ which measures the overlapping area between the cleaned PtC and the processed skin model, averaged over the 10 measurements. The median distance with interquartile range between $F_{i,OT}$ and $H_{i,OT}$ for the 4 markers, in the 10 repeated procedures, resulted in $1.25(0.65 - 1.72)$ mm, $1.75(0.98 - 2.42)$ mm, $1.35(0.41 - 1.83)$ mm, for the x, y, z coordinates in the OT space respectively (corresponding to the axial, sagittal and coronal plane of the phantom). The distances referred to each of the 4 markers in the 3 coordinates are reported in Fig. 5A. The average 3D Euclidean distance for all the markers over the 10 procedures resulted in 2.72 ± 0.67 mm. The point-to-point accuracy over the whole model as obtained from Eq. 2 showed a $RMSE$ of 3.19 ± 0.69 mm and 3.61 ± 1.13 mm for the face and ventricle surfaces respectively. Point-to-point distance maps between the surfaces are represented in Fig. 5C.

Ventriculostomy Procedure. Guided by the holographic system, a 42% higher accuracy was shown, on average. Euclidean target distances of blind vs. holographically-guided simulations are reported in Fig. 5B. In the blind procedure, the mean distance to the target was equal to 16.2 ± 3.48 mm, compared to a mean distance of 9.26 ± 3.27 mm for the hologram guided procedure, resulting in a statistically significant improvement on a paired t-test (p=0.001). Figure 5D graphically shows the final tip position in all the procedures done by the neurosurgeons compared to the FoM target in the ventricle.

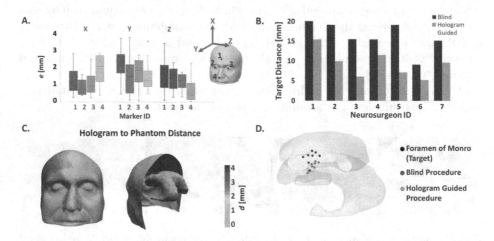

Fig. 5. Experimental results from evaluation tests. A: hologram-to-phantom accuracy evaluated on the 4 markers; B: quantitative targeting accuracy from the tests done by 7 neurosurgeons; C: hologram to phantom accuracy evaluated over the whole model surface; D: visualization of catheter tip positions and target from the EVD simulations.

4 Conclusions

The present study proposed a novel solution for supporting neurosurgeons during EVD procedures. Our system leverages a deep NN for automatic image segmentation and a markerless registration method based on a HMD depth sensor to align internal brain structures with the physical position of the patient. We quantitatively assessed the accuracy of our workflow at each of its steps, proving the added value to ventriculostomy procedures under experimental settings; adopting our technology, neurosurgeons targeting accuracy improved by 42%, on average. Of note, all the described steps require little or no manual intervention, and computations required globally less than 3 min, thus making our system suitable for emergency procedures where timing can be crucial. Our future efforts will focus on including an holographic navigation of the probe using electromagnetic sensors and evaluate the accuracy of adding a further, independent, Azure Kinect (Microsoft, Redmond, WA) depth camera to achieve a more stable and robust point cloud acquisition. Moreover, a phantom reproducing the different resistance of the ventricles with respect to the brain will be adopted to simulate the surgical procedure in a more realistic way.

References

1. AlAzri, A., Mok, K., Chankowsky, J., Mullah, M., Marcoux, J.: Placement accuracy of external ventricular drain when comparing freehand insertion to neuronavigation guidance in severe traumatic brain injury. Acta Neurochir. **159**(8), 1399–1411 (2017). https://doi.org/10.1007/s00701-017-3201-5
2. Arun, K.S., Huang, T.S., Blostein, S.D.: Least-squares fitting of two 3-D point sets. IEEE Trans. Pattern Anal. Mach. Intell. **5**, 698–700 (1987)
3. Azimi, E., et al.: An interactive mixed reality platform for bedside surgical procedures. In: Martel, A.L., et al. (eds.) MICCAI 2020. LNCS, vol. 12263, pp. 65–75. Springer, Cham (2020). https://doi.org/10.1007/978-3-030-59716-0_7
4. Chen, Y., Medioni, G.: Object modelling by registration of multiple range images. Image Vis. Comput. **10**(3), 145–155 (1992)
5. van Doormaal, J.A., Fick, T., Ali, M., Köllen, M., van der Kuijp, V., van Doormaal, T.P.: Fully automatic adaptive meshing based segmentation of the ventricular system for augmented reality visualization and navigation. World Neurosurg. **156**, e9–e24 (2021)
6. Ester, M., et al.: A density-based algorithm for discovering clusters in large spatial databases with noise. In: Proceedings of the Second International Conference on Knowledge Discovery and Data Mining, vol. 96, pp. 226–231 (1996)
7. von Haxthausen, F., Chen, Y., Ernst, F.: Superimposing holograms on real world objects using Hololens 2 and its depth camera. Curr. Dir. Biomed. Eng. **7**(1), 111–115 (2021)
8. Katz, S., Tal, A., Basri, R.: Direct visibility of point sets. ACM Trans. Graph. **26**(3), 24-es (2007). https://doi.org/10.1145/1276377.1276407
9. Li, Y., et al.: A wearable mixed-reality holographic computer for guiding external ventricular drain insertion at the bedside. J. Neurosurg. **131**(5), 1599–1606 (2018)
10. Mahan, M., Spetzler, R.F., Nakaji, P.: Electromagnetic stereotactic navigation for external ventricular drain placement in the intensive care unit. J. Clin. Neurosci. **20**(12), 1718–1722 (2013)
11. Patrick, S., et al.: Comparison of gray values of cone-beam computed tomography with hounsfield units of multislice computed tomography: an in vitro study. Indian J. Dent. Res. **28**(1), 66 (2017)
12. Saitta, S., et al.: A deep learning-based and fully automated pipeline for thoracic aorta geometric analysis and planning for endovascular repair from computed tomography. J. Digit. Imaging **35**(2), 226–239 (2022)
13. Schneider, M., Kunz, C., Pal'a, A., Wirtz, C.R., Mathis-Ullrich, F., Hlaváč, M.: Augmented reality-assisted ventriculostomy. Neurosurg. Focus **50**(1), E16 (2021)
14. Shen, Dinggang, Shen, D., et al. (eds.): MICCAI 2019. LNCS, vol. 11764. Springer, Cham (2019). https://doi.org/10.1007/978-3-030-32239-7
15. Ungureanu, D., et al.: Hololens 2 research mode as a tool for computer vision research. arXiv preprint arXiv:2008.11239 (2020)

16. Wilson, T.J., Stetler, W.R., Al-Holou, W.N., Sullivan, S.E.: Comparison of the accuracy of ventricular catheter placement using freehand placement, ultrasonic guidance, and stereotactic neuronavigation. J. Neurosurg. **119**(1), 66–70 (2013)

17. Zhou, Q.-Y., Park, J., Koltun, V.: Fast global registration. In: Leibe, B., Matas, J., Sebe, N., Welling, M. (eds.) ECCV 2016. LNCS, vol. 9906, pp. 766–782. Springer, Cham (2016). https://doi.org/10.1007/978-3-319-46475-6_47

18. Zhou, Q.Y., Park, J., Koltun, V.: Open3D: a modern library for 3D data processing. arXiv preprint arXiv:1801.09847 (2018)

Deep Regression with Spatial-Frequency Feature Coupling and Image Synthesis for Robot-Assisted Endomicroscopy

Chi Xu[✉], Alfie Roddan, Joseph Davids, Alistair Weld, Haozheng Xu, and Stamatia Giannarou

The Hamlyn Centre for Robotic Surgery, Department of Surgery and Cancer, Imperial College London, London SW7 2AZ, UK
{chi.xu20,a.roddan21,j.davids,a.weld20,haozheng.xu20, stamatia.giannarou}@imperial.ac.uk

Abstract. Probe-based confocal laser endomicroscopy (pCLE) allows *in-situ* visualisation of cellular morphology for intraoperative tissue characterization. Robotic manipulation of the pCLE probe can maintain the probe-tissue contact within micrometre working range to achieve the precision and stability required to capture good quality microscopic information. In this paper, we propose the first approach to automatically regress the distance between a pCLE probe and the tissue surface during robotic tissue scanning. The Spatial-Frequency Feature Coupling network (SFFC-Net) was designed to regress probe-tissue distance by extracting an enhanced data representation based on the fusion of spatial and frequency domain features. Image-level supervision is used in a novel fashion in regression to enable the network to effectively learn the relationship between the sharpness of the pCLE image and its distance from the tissue surface. Consequently, a novel Feedback Training (FT) module has been designed to synthesise unseen images to incorporate feedback into the training process. The first pCLE regression dataset (PRD) was generated which includes *ex-vivo* images with corresponding probe-tissue distance. Our performance evaluation verifies that the proposed network outperforms other state-of-the-art (SOTA) regression networks.

Keywords: Regression · Fourier convolution · Endomicroscopy

1 Introduction

Biophotonics techniques such as pCLE have enabled direct visualisation of tissue at a microscopic level, with recent pilot studies suggesting it may have a role in identifying residual cancer tissue and improving tumour resection rates [1,16,18]. To acquire high quality pCLE data, the probe should be maintained within a working range which depends on the probe type. This is an ergonomically challenging task as this range is of micrometer scale. Recently, robotic tissue scanning has been investigated to control the longitudinal distance between

© The Author(s), under exclusive license to Springer Nature Switzerland AG 2022
L. Wang (Eds.): MICCAI 2022, LNCS 13437, pp. 157–166, 2022.
https://doi.org/10.1007/978-3-031-16449-1_16

the pCLE probe and the tissue [17]. However, this approach provides only an approximation of the probe location with respect to the tissue surface rather than regressing their actual distance.

With rapid development of deep learning applications within the medical imaging field, several regression networks have been proposed to tackle auto-focusing in confocal microscopy. Zhenbo et $al.$ [13] and Tomi et $al.$ [12] both adjusted existing CNN models to predict the optimal distance for in-focus images. Instead of using the spatial domain information, Jiang et $al.$ [8] pre-processed the images to extract a multi-domain representation which is used in a regression CNN model. Zhang et $al.$ [22] utilized Sobel filters of multiple directions to generate gradient images which are used as input to a diversity-learning network, to infer the focus distance. Unlike conventional confocal microscopy techniques, the convergence of the pCLE probe within the working range is more challenging because the degree of improvement in the image clarity decreases as the probe moves towards the optimal scanning position. In contrast, for confocal microscopy the relationship between the focus distance and image quality is more linear. Additionally, there is a larger influence of noise in the pCLE data. These issues present a challenge for deep learning networks to accurately regress the pCLE probe-tissue distance.

In this paper, we propose the first approach to automatically regress the longitudinal distance between a pCLE probe and the tissue surface for robotic tissue scanning. Our method advances existing regression approaches in microscopy by introducing (1) a SFFC-Net which extracts an enhanced pCLE data representation by fusing spatial and frequency domain features; (2) a novel FT module to synthesise pCLE images between fine distance intervals to incorporate image level supervision into the training process. To overcome the lack of available training data, the first PRD dataset has been generated which includes ex-$vivo$ images with corresponding probe-tissue distance. Our performance evaluation verifies that the proposed network outperforms other SOTA regression networks, in terms of accuracy and convergence.

2   Methodology

The proposed framework for pCLE probe-tissue distance estimation is composed of the SFFC-Net and the FT modules, depicted in the grey and blue boxes, respectively in Fig. 1. Given a raw pCLE image $I_{in} \in \mathbb{R}^{1,H,W}$, local normalization is applied as $I_l = \frac{I_{in} - \min(I_{in})}{\max(I_{in}) - \min(I_{in})}$ to increase the image contrast. To increase the contrast of the images along different probe-tissue distances, global normalization is applied as $I_g = \frac{I_{in} - I_{min}}{I_{max} - I_{min}}$, where I_{max} and I_{min} are the maximum and minimum intensity values captured by the pCLE system ($i.e.$ $I_{max} = 8191$ and $I_{min} = -400$ in our system). The normalised images are stacked $[I_l, I_g] \in \mathbb{R}^{2,H,W}$ and passed as input to the neural network.

The SFFC-Net is trained using the L_1 loss to estimate the distance $d_f = f([I_l, I_g], \theta)$ of the probe from the optimal scanning position, where $f(.)$ is the SFFC-Net and θ is the model parameters. During training, the d_f is compared

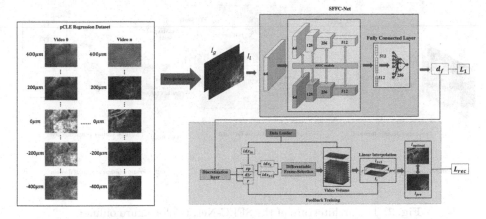

Fig. 1. Our proposed framework for pCLE probe-tissue distance regression. (Color figure online)

with the ground truth distance d_{gt} to infer the new position $(p_{pre} = d_{gt} - d_f)$ of the probe (with respect to the tissue surface) after it has moved according to the above estimation d_f. The FT module $g(.)$ is used to synthesise the pCLE image $I_{pre} = g(d_f)$ at position p_{pre}, as demonstrated in Fig. 1. Then, image I_{pre} is compared to the actual image $I_{optimal}$ at the optimal scanning position to produce additional image-level supervision.

2.1 Spatial-and-Frequency Feature Coupling Network (SFFC-Net)

As demonstrated in Fig. 1, the stacked images $[I_l, I_g]$ firstly enter the shallow feature extraction layer (gray block). Then, the extracted low-level feature maps are analysed in the spatial and frequency domains by multi-scale spatial convolutional (SC) layers (blue blocks) and Fast Fourier Convolution (FFC) [2] layers (light yellow blocks). The SC layers have a local receptive field and detect local features on the input images (*i.e.* texture, edges). The FFC layers have a non-local receptive field and extract global frequency features [2] (*i.e.* image sharpness). The spatial and frequency feature maps are fused at each scale via the SFFC module.

The SFFC-Net has 4 layers to extract spatial and global features. Each layer is composed of the FFC layer, spatial convolutional (SC) layer and SFFC module. Since all the layers have the same architecture, the detailed architecture of the 1st layer only is demonstrated in Fig. 2. In the FFC layer, the input feature maps are converted to the frequency domain via Fast Fourier Transform (FFT) [3] and then, the 1×1 convolutional kernel with batch normalization (BN) and ReLU activator (1×1Conv-BN-ReLU) is applied to analyse each frequency component. Afterwards, the Inverse Fast Fourier Transform (IFFT) converts the extracted frequency features to the spatial domain for further coupling. Each SC layer consists of multiple residual blocks and 1×1 convolution layers (1×1Conv). The residual block is the same as the basic block in ResNet18 [6]. The convolutions

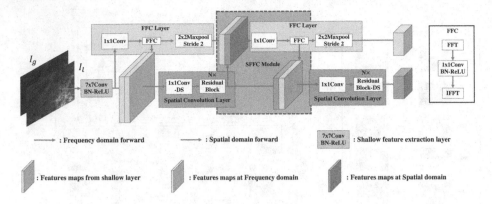

Fig. 2. The architecture of the SFFC-Net. (Color figure online)

with 1×1 kernel are used for channel fusion. The SFFC module takes as input, the output spatial feature map of the residual block from the SC layer and the frequency feature map from the FFC layer. These maps are then concatenated and processed by the 1×1 convolution and the FFC block. The output of the FFC block becomes input to the next FFC layer. In addition, the output of the FFC block is concatenated with the spatial feature map and becomes input to the next SC layer.

The output of the final SC layer and of the FFC layer are converted to feature vectors via global average-pooling. Both feature vectors are stacked across the channel dimension and fed into the fully connected layer (dropout is added to the first dense layer $(P = 0.2)$) to infer the distance d_f of the probe from the tissue surface.

2.2 Feedback Training (FT)

Once the d_f has been estimated by the SFFC-Net, the new position of the probe with respect to the optimal scanning position, after it has moved, is p_{pre}. However, since SFFC-Net regresses continuous values, p_{pre} may fall between positions where pCLE information has been collected. Hence, the aim of the FT module in our framework is to synthesise the pCLE image I_{pre} at position p_{pre} and use this to incorporate feedback into our training process to boost the learning of the network. The main components of FT are the following.

Discretization Layer. This layer is used to decompose d_f into the number of steps $(sp = \left\lfloor \frac{d_f}{S} \right\rfloor)$ and the direction of probe movement $(dir = sign(d_f))$, and the ratio of the interpolation $(r = \frac{d_f - sp \cdot S}{S})$. S is the interval between captured pCLE images. Given a volume V of pCLE video frames and the index (idx_{in}) of the input frame I_{in} in the volume, the sp and dir are used to find the indices of two adjacent frames $idx_t = idx_{in} + dir \cdot sp$ and $idx_{t+1} = idx_{in} + dir \cdot (sp + 1)$.

Differentiable Frame-Selection. After acquiring idx_t and idx_{t+1}, the corresponding frames can be extracted from the video volume. However, the existing index-based frame-selection process is non-differentiable, causing backpropogation failure. To solve this issue, the Differentiable Frame-Selection is proposed in this work. Given the volume range $R = [1, 2, ..., D]$, the one-hot code $(C^{onehot} \in \mathbb{N}^D)$ of each index can be generated in a differentiable manner as:

$$C_n^{onehot}(idx) = \begin{cases} \frac{idx+1}{R_n} & \text{if } R_n = idx + 1 \\ (idx + 1) \cdot 0 & \text{if } R_n \neq idx + 1 \end{cases}, n \in [0, 1, ..., D-1] \quad (1)$$

Then, a frame I_t can be extracted via the sum of the multiplication of C^{onehot} and V across the channel dimension. During this process, the selected frame maintains the gradient of $\frac{1}{idx}$ while the other frames have zero gradient.

$$I_t = \sum_{n=0}^{D-1} V_n \cdot C_n^{onehot}(idx_t) \quad (2)$$

Per-Pixel Linear Interpolation. After extracting two adjacent frames I_t and I_{t+1} from the volume, the predicted frame can be reconstructed using linear interpolation and the ratio of interpolation as follows:

$$I_{pre}(a, b) = \begin{cases} (1 - r) \cdot I_t(a, b) + r \cdot I_{t+1}(a, b) & \text{if } dir = 1 \\ r \cdot I_t(a, b) + (1 - r) \cdot I_{t+1}(a, b) & \text{if } dir = -1 \end{cases} \quad (3)$$

2.3 Loss Function

In order to constrain the network's learning objectives, two different loss functions are combined, namely, L_1 loss and image reconstruction loss L_{rec} as:

$$L = \lambda \cdot L_1 + L_{rec} \quad (4)$$

where, $\lambda = 0.1$ in our work. The L_1 loss for batch size N is defined as $L_1 = \sum_N |d_f - d_{gt}|$. Our reconstruction loss consists of SSIM [5,19] and Mean of Intensity (MoI) losses. The SSIM is sensitive to the sharpness of images so, L_{SSIM} can effectively back-propagate image-level supervision signal for texture-learning. In confocal microscopy, some of the reflected photons are blocked, causing a drop in the intensity of the images. This is observed when the pCLE is out of focus. To reflect this property, the MoI loss enables the network to learn the relationship between the intensity and focusing distance.

$$L_{rec} = \sum_N (\frac{1}{2} \cdot (1 - \text{SSIM}(I_{pre}, I_{optimal})) + |\text{MoI}(I_{pre}) - \text{MoI}(I_{optimal})|) \quad (5)$$

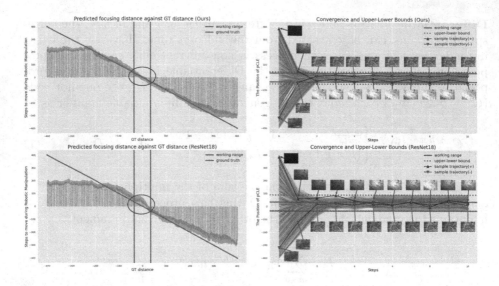

Fig. 3. (Left) Signal direction analysis (Right) Convergence evaluation. Sample pCLE images at different steps of the stabilisation process are shown for two initial probe positions corresponding to 385 μm and −320 μm from the tissue surface. (Color figure online)

3   Experiments and Analysis

PRD Dataset Generation and Model Implementation. Ex-vivo pig brain tissue stained with 0.1% Acriflavin was scanned with a Z 1800 confocal miniprobe (Cellvizio, Mauna Kea Technologies, Paris). The miniprobe was actuated with the Kinesis® K-Cube™ Stepper Motor Controller (Thorlabs, USA) within a distance range of 400 μm and −400 μm from the tissue surface with step size 5 μm. In total, 62 pCLE videos and corresponding pCLE probe positions have been collected from independent samples. The ground truth optimal scanning position was determined first as the position of the pCLE frame with minimum blur, followed by visual verification from an expert neurosurgeon. According to the technical specification of the Z miniprobe, the working range is between −35 μm and 35 μm from the tissue surface. In our experimental work, 50 videos (7539 frames) are used for training and 12 videos (1706 frames) for testing. Examples of images within our dataset are shown in Fig. 1.

All models are implemented with the Pytorch library [11] and executed on two NVIDIA GTX TITAN X with 10GB memory. For training, the optimizer AdamW [10] with weight decay 0.01 is used to train our network with batch size of 16. To make our network generalizable, the cyclical learning rate [15] scheduler is applied (number of epochs = 16, *step size* = 2 epochs, $lr_{base} = 1e^{-6}$ and $lr_{max} = 1e^{-4}$).

Table 1. Comparison with SOTA models on the PRD dataset. The MAE_{1st} is the mean absolute error after the 1st step. The "✗" represents that the trained model cannot converge in the K-step incremental analysis. For the MAE_C and PIR, the standard deviation across the steps are also presented. The unit of errors and width is μm.

Network	MAE_{1st}	MAE_C	W_B	PIR	Acc_{dir}	Params
AlexNet [9]	72.30	62.53 ± 1.76	156.11	39.24% ± 1.43%	91.32%	57M
VGG16 [14]	77.41	56.51 ± 0.75	148.12	44.96% ± 1.21%	93.20%	134M
ResNet18 [6]	78.93	46.08 ± 0.78	121.54	47.18% ± 2.07%	94.02%	11M
WideResNet50 [21]	87.03	57.13 ± **0.73**	130.27	28.10% ± 2.54%	91.91%	67M
DenseNet161 [7]	88.29	84.87 ± 1.79	214.98	35.58% ± 0.77%	87.57%	29M
ViT [4]	99.48	✗	✗	✗	83.59%	85M
Z et al. [22]	152.72	✗	✗	✗	73.09%	**2M**
Z et al. (gray) [22]	84.12	73.90 ± 2.98	205.99	43.30% ± 1.52%	90.27%	**2M**
Ours	**64.29**	**34.90 ± 0.83**	**95.12**	**69.19% ± 0.75%**	**94.49%**	14M

Table 2. Ablation Study.

Network	SFFC	FT	MAE_{1st}	MAE_C	W_B	PIR	Acc_{dir}
ResNet18 [6]			78.93	46.08 ± 0.78	121.54	47.18% ± 2.07%	94.02%
ResNet18 [6]		✔	76.78	47.05 ± **0.40**	101.25	50.87% ± 1.13%	94.07%
SFFC-Net	✔		68.13	44.82 ± 0.72	**75.34**	58.73% ± 1.84%	93.08%
SFFC-Net	✔	✔	**64.29**	**34.90 ± 0.83**	95.12	**69.19% ± 0.75%**	**94.49%**

Convergence and Stability Study. For robotic scanning, the prediction of the network is used to guide the robot to the optimal scanning position incrementally with step $d_r = -d_f$. To evaluate the convergence capabilities and stability of the network for robotic control, two experiments are conducted (1) signal direction analysis to verify that the model's prediction infers the right movement direction; (2) convergence evaluation to show that the model is capable of moving the probe close to the optimal scanning position and stabilising there.

For the signal direction analysis, for each discrete probe-tissue distance d_{gt}, the set $D^{d_{gt}} \in \mathbb{R}^N$ contains the model predictions d_r^i, $i \in [1, N]$ and the N pCLE images that correspond to this position are analysed. The median value $\overline{D^{d_{gt}}}$ and standard error $SE_{D^{d_{gt}}}$ are computed and presented on the bar graphs in Fig. 3 (Left) for our proposed network and ResNet18 [6], respectively. The red solid line is the ground truth step $-d_{gt}$ to be moved by the probe. The accuracy of the predicted signal direction is also computed and visually represented by the difference between the red line and each blue bar in Fig. 3. As shown in the graphs, the accuracy of the prediction increases the closer the probe moves to the optimal scanning position. We attribute this increase to the better image quality when closer to the working range. Better image quality enables the network to extract better and more meaningful features. Within the purple circle in the graph, it becomes apparent that our SFFC-Net has greater accuracy within the working range compared to ResNet18.

The convergence evaluation study of the robotic probe manipulation is done following a K-step incremental analysis with image-feedback steps $k \in [1, K]$.

A pCLE image from the test dataset corresponding to a known probe position d_{gt} is fed into our trained network. At iteration k, the inferred d_r^k is used to update the position of the pCLE probe as $d_{up}^{k+1} = d_{up}^k + d_r^k$, where $d_{up}^1 = d_{gt}$. The pCLE frame within the video volume (V) closer to probe position d_{up}^{k+1} is found using the nearest neighbour method. The selected frame is fed into the trained network, and the process repeats until $k = K$. The convergence of our SFFC-Net is compared to ResNet18 in Fig. 3 (Right) which shows the paths of the updated positions estimated as above for K = 10 steps on all the test data. The green solid lines are the boundaries of the working range. The red dotted lines are the upper and lower bounds of the paths after convergence. This is achieved in models for $k \geq 5$, in the light red regions in Fig. 3 (Right). As shown, our proposed network is more stable than ResNet18 and converges faster. Convergence is also quantitatively evaluated in Table 1 using the width of the upper-lower bound area, W_B (lower being desired). In addition, the percentage of data in the working range (PIR) and the mean absolute error of convergence (MAE_C) is calculated within the light red region.

Comparison and Ablation Study. Due to the lack of distance regression models for pCLE, SOTA deep learning frameworks for general tasks [4,6,7,9, 14,20,21] have been implemented and used for comparison. Furthermore, the diversity-learning network [22], has been included in our comparison as it is the SOTA network for auto-focusing in microscopy. As shown in Table 1, our network outperforms all the other models in the comparison study in $\text{MAE}_{1^{st}}$, MAE_C, W_B, PIR and Acc_{dir}. This verifies that robotic pCLE scanning guided by the proposed network can acquire the most high-quality images, close to the optimal scanning position compared to the other models. The lowest W_B and highest Acc_{dir} demonstrate that our proposed model allows the pCLE probe to remain within the upper-lower bound in the most stable manner. Furthermore, our network's MAE_C outperforms other networks staying within the working range. The lower performance of the other SOTA models is attributed to their use of spatial convolutions and numerical training. Therefore they cannot extract global feature information gained by the novel modules of our network. The gradient-based representation proposed in [22] can not provide accurate predictions for data with sparse texture and significant blur such as our pCLE images. Since [22] did not converge, we used gray scale images as input to their network instead of gradient images, we call this Z *et al.* (gray) in Table 1.

To study the contributions of each proposed component, an ablation study is conducted in Table 2. Since our framework is designed based on ResNet18, this is set as the baseline. This study shows that each contribution is additive to our architecture. Compared with the ResNet18, the SFFC-Net can effectively boost the performance in all performance metrics. For FT, the improvement for ResNet18 is not as significant as SFFC-Net, which illustrates that SFFC-Net can learn more efficient data representations than ResNet18. It is noteworthy that the W_b for the SFFC-Net without the FT is smaller than both contributions together. However, PIR and Acc_{dir} indicate a limited optimal scanning position approach for SFFC-Net without FT whereas SFFC-Net with FT solves this issue. Overall in unison, our contributions significantly improve performance.

4 Conclusion

In this paper, we have introduced the first framework to automatically regress the distance between a pCLE probe and the tissue surface for robotic tissue scanning. Our work demonstrates that our novel SFFC-Net with FT architecture outperforms SOTA models for pCLE distance regression. We provide formula derivations for our contributions and demonstrate an increase in performance for each contribution in an ablation study. Our performance evaluation verifies our network's stability in convergence and overall accuracy on a challenging dataset. Future work includes expanding our PRD dataset on multiple tissue types and testing the network for real time usage.

Acknowledgements. This work was supported by the Royal Society [URF\R\2 01014] , EPSRC [EP/W004798/1] and the NIHR Imperial Biomedical Research Centre.

References

1. Capuano, A., et al.: The probe based confocal laser endomicroscopy (pCLE) in locally advanced gastric cancer: a powerful technique for real-time analysis of vasculature. Front. Oncol. **9**, 513 (2019)
2. Chi, L., Jiang, B., Mu, Y.: Fast fourier convolution. Adv. Neural. Inf. Process. Syst. **33**, 4479–4488 (2020)
3. Cooley, J.W., Tukey, J.W.: An algorithm for the machine calculation of complex fourier series. Math. Comput. **19**(90), 297–301 (1965)
4. Dosovitskiy, A., et al.: An image is worth 16x16 words: transformers for image recognition at scale (2020)
5. Godard, C., Aodha, O.M., Firman, M., Brostow, G.J.: Digging into self-supervised monocular depth prediction (2019)
6. He, K., Zhang, X., Ren, S., Sun, J.: Deep residual learning for image recognition. In: Proceedings of the IEEE Conference on Computer Vision and Pattern Recognition, pp. 770–778 (2016)
7. Huang, G., Liu, Z., Van Der Maaten, L., Weinberger, K.Q.: Densely connected convolutional networks. In: Proceedings of the IEEE Conference on Computer Vision and Pattern Recognition, pp. 4700–4708 (2017)
8. Jiang, S., Liao, J., Bian, Z., Guo, K., Zhang, Y., Zheng, G.: Transform-and multi-domain deep learning for single-frame rapid autofocusing in whole slide imaging. Biomed. Opt. Express **9**(4), 1601–1612 (2018)
9. Krizhevsky, A.: One weird trick for parallelizing convolutional neural networks. arXiv preprint arXiv:1404.5997 (2014)
10. Loshchilov, I., Hutter, F.: Decoupled weight decay regularization. arXiv preprint arXiv:1711.05101 (2017)
11. Paszke, A., et al.: Automatic differentiation in pyTorch (2017)

12. Pitkäaho, T., Manninen, A., Naughton, T.J.: Performance of autofocus capability of deep convolutional neural networks in digital holographic microscopy. In: Digital Holography and Three-Dimensional Imaging, pp. W2A–5. Optical Society of America (2017)
13. Ren, Z., Xu, Z., Lam, E.Y.: Learning-based nonparametric autofocusing for digital holography. Optica **5**(4), 337–344 (2018)
14. Simonyan, K., Zisserman, A.: Very deep convolutional networks for large-scale image recognition. arXiv preprint arXiv:1409.1556 (2014)
15. Smith, L.N.: Cyclical learning rates for training neural networks. In: 2017 IEEE Winter Conference on Applications of Computer Vision (WACV), pp. 464–472. IEEE (2017)
16. Spessotto, P., et al.: Probe-based confocal laser endomicroscopy for in vivo evaluation of the tumor vasculature in gastric and rectal carcinomas. Sci. Rep. **7**(1), 1–9 (2017)
17. Triantafyllou, P., Wisanuvej, P., Giannarou, S., Liu, J., Yang, G.Z.: A framework for sensorless tissue motion tracking in robotic endomicroscopy scanning. In: 2018 IEEE International Conference on Robotics and Automation (ICRA), pp. 2694–2699. IEEE (2018)
18. Wallace, M.B., Fockens, P.: Probe-based confocal laser endomicroscopy. Gastroenterology **136**(5), 1509–1513 (2009)
19. Wang, Z., Bovik, A.C., Sheikh, H.R., Simoncelli, E.P.: Image quality assessment: from error visibility to structural similarity. IEEE Trans. Image Process. **13**(4), 600–612 (2004)
20. Xie, S., Girshick, R., Dollár, P., Tu, Z., He, K.: Aggregated residual transformations for deep neural networks. In: Proceedings of the IEEE Conference on Computer Vision and Pattern Recognition, pp. 1492–1500 (2017)
21. Zagoruyko, S., Komodakis, N.: Wide residual networks. arXiv preprint arXiv:1605.07146 (2016)
22. Zhang, C., Gu, Y., Yang, J., Yang, G.Z.: Diversity-aware label distribution learning for microscopy auto focusing. IEEE Robot. Autom. Lett. **6**(2), 1942–1949 (2021)

Fast Automatic Liver Tumor Radiofrequency Ablation Planning via Learned Physics Model

Felix Meister[1,2](✉), Chloé Audigier[4], Tiziano Passerini[3], Èric Lluch[2], Viorel Mihalef[3], Andreas Maier[1], and Tommaso Mansi[3]

[1] Friedrich-Alexander-Universität Erlangen-Nürnberg, Erlangen, Germany
`felix.meister@fau.de`
[2] Siemens Healthineers, Digital Technology & Innovation, Erlangen, Germany
[3] Siemens Healthineers, Digital Technology & Innovation, Princeton, USA
[4] Advanced Clinical Imaging Technology, Siemens Healthcare, Lausanne, Switzerland

Abstract. Radiofrequency ablation is a minimally-invasive therapy recommended for the treatment of primary and secondary liver cancer in early stages and when resection or transplantation is not feasible. To significantly reduce chances of local recurrences, accurate planning is required, which aims at finding a safe and feasible needle trajectory to an optimal electrode position achieving full coverage of the tumor as well as a safety margin. Computer-assisted algorithms, as an alternative to the time-consuming manual planning performed by the clinicians, commonly neglect the underlying physiology and rely on simplified, spherical or ellipsoidal ablation estimates. To drastically speed up biophysical simulations and enable patient-specific ablation planning, this work investigates the use of non-autoregressive operator learning. The proposed architecture, trained on 1,800 biophysics-based simulations, is able to match the heat distribution computed by a finite-difference solver with a root mean squared error of $0.51 \pm 0.50\,°C$ and the estimated ablation zone with a mean dice score of 0.93 ± 0.05, while being over 100 times faster. When applied to single electrode automatic ablation planning on retrospective clinical data, our method achieves patient-specific results in less than 4 mins and closely matches the finite-difference-based planning, while being at least one order of magnitude faster. Run times are comparable to those of sphere-based planning while accounting for the perfusion of liver tissue and the heat sink effect of large vessels.

Keywords: Radiofrequency ablation planning · Deep learning · Computational modeling · Liver tumor · Patient-specific

Supplementary Information The online version contains supplementary material available at https://doi.org/10.1007/978-3-031-16449-1_17.

1   Introduction

Radiofrequency ablation (RFA) has emerged as a first-line therapy option for early-stage liver tumors and when resection or transplantation is infeasible [4,14]. Using one or more percutaneously inserted electrodes, RFA aims at inducing coagulation necrosis in the tumor by applying a strong electric field that heats surrounding tissue. Treatment success without local recurrences is commonly achieved if the entire tumor as well as a >5 mm safety margin are ablated [7].

To achieve full coverage while minimizing damage to surrounding healthy tissue, the procedure needs to be precisely planned. In particular, the goal is to find an optimal electrode position, i.e. the coordinates of the needle tip, as well as a safe percutaneous pathway that avoids organs at risk [23]. In day-to-day intervention planning, the task is usually performed mentally by the clinician, using only 2D views of pre-operative CT images for reference. To facilitate and speed up the planning process, computer-assisted planning algorithms commonly combine an optimization method to select the electrode position in the liver with a ray-tracing-inspired algorithm to compute feasible needle trajectories [1,3,8, 18]. These methods typically carry out the optimization under the assumption that the induced ablation zone is a sphere or an ellipsoid. For tumors located close to vessels, this assumption is, however, conflicting with reality since large vessels act as a heat sink, thus altering the ablation zone [23]. Patient-specific ablation estimates can be achieved by biophysical computational models. In spite of algorithmic advancements and the use of graphics processing units (GPUs), these methods are, however, not suitable for fast RFA planning due to the high computational burden [2,10].

Over the last couple of years, new research directions based on the success of deep learning have emerged to accelerate computational models. Leveraging the universal function approximation capability of neural networks, several attempts in various domains have been made to replace [19], mimic [16], or interface computational models with data-driven surrogate models [20]. Even though these studies demonstrated that neural networks can learn physical processes with high accuracy, the specific choice of architecture and training strategy has to be tailored to the application.

This work explores the use of operator learning to enable fast, biophysics-based RFA planning by modeling the underlying advection-diffusion-reaction process involved in thermal ablation. Within the operator learning framework, as introduced by DeepONet [9], we propose a specific convolutional neural network that efficiently encodes the voxelized initial and boundary conditions, which are derived from medical images. This method allows non-autoregressive inference, i.e. the temperature distribution at specific points in time can be obtained independently from others, and without sequentially estimating intermediate solutions. Consequently, the method is suitable for massive parallelization.

We train the neural network on a set of simulated RFA scenarios generated by a computational model of RFA. We apply the approach to single electrode planning on retrospective clinical data comprising 14 tumors. We compare the results in terms of compute time and predictive power against the ones obtained by sphere-based and finite-difference-based planning.

2 Methods

2.1 Physics Emulation via Operator Learning

In this work, we aim to learn the solution operator for a advection-diffusion-reaction problem. Specifically, we target a two-compartment model combining the *Pennes* [13] and the *Wulff-Klinger* model [6,21], which has already been successfully applied to RFA simulations [2]. The problem is described by the partial differential equation

$$\rho_{ti}c_{ti}\frac{\partial T}{\partial t} + (1 - \mathbb{1}_V) \cdot \alpha_v \rho_b c_b \boldsymbol{v} \cdot \nabla T = Q + d\nabla^2 T + \mathbb{1}_V \cdot R(T_b - T) \quad (1)$$

where t denotes the time, T the temperature, T_b the blood temperature, R the reaction coefficient, α_v the advection coefficient, d the heat conductivity, and $\boldsymbol{v}$ the blood velocity field inside the liver parenchyma. $\mathbb{1}_V$ is the indicator function evaluating as 1 if a voxel belongs to a vessel and 0 otherwise. c_{ti} and c_b are the heat capacities and ρ_{ti} and ρ_b the densities of the liver tissue and the blood, respectively.

To learn a non-autoregressive physics emulator, this work follows an operator learning approach [9]. Deep operator networks approximate operators $\mathcal{G}{:}i \mapsto s$, mapping an input function i to a solution function s, by means of a branch network $\mathcal{B}$ and a trunk network $\mathcal{T}$. In our case, function i is a 3D image with vectorial voxel values. The solution s is the temperature, function of time, which is expressed by the scalar product

$$s(t) = \mathcal{G}(i)(t) \approx \sum_{k=1}^{p} \mathcal{B}_k(i)\mathcal{T}_k(t) + b_0 \quad (2)$$

between the p-dimensional output latent codes of $\mathcal{B}$ and $\mathcal{T}$, plus a bias term b_0.

The learned operator corresponds to the antiderivative in time $\mathcal{G}(i)(t) = s(t) = T_0 + \int_0^t \frac{\partial T}{\partial y} dy$ evaluated at the voxels of the input image, where T_0 denotes the initial heat distribution. As displayed in Fig. 1, the branch network takes as input a 3D feature map with six features comprising the concatenation of the 3D blood velocity vector field and three 3D binary masks representing the liver, the vessels, and the heat source. The branch network uses a convolutional neural network to efficiently encode the high-dimensional input data. The output is a volume of the same dimension as the input volumes, with each voxel holding a p-dimensional latent code. The trunk network then takes the query time, a single scalar representing the elapsed time t starting at 0 s, as input and processes it by a cascade of fully connected layers. Multiple queries are batched together for maximum efficiency. The output of the trunk network is another p-dimensional latent code, which is applied to the latent code of each voxel via Eq. 2. The resulting volume holds the temperature distribution at time t.

Training and Implementation. To train the proposed method, we first generate a set of reference solutions to Eq. 1 with a finite-difference solver. Since we

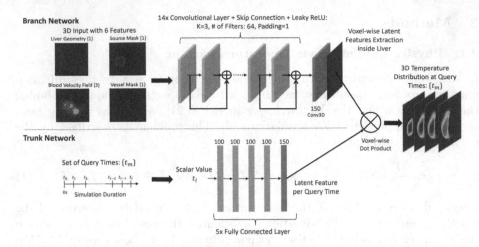

Fig. 1. Visualization of our proposed network architecture. A convolutional branch network encodes the geometry and blood velocity field into a volume of 150-dimensional latent codes. The trunk network computes a single latent code of same dimensionality from the query time. The output is computed according to Eq. 2 for each voxel inside the liver. (Color figure online)

intend to apply the approach to RFA planning where only imaging information is available, we use the nominal parameters for Eq. 1 as in [12]. Furthermore, we follow the approach of [2] and compute a blood velocity field given a specific patient geometry and an image resolution of 1.3 mm. The solution is then downsampled to match the resolution used with the finite-difference method. For simplicity, we further apply a Dirichlet boundary condition to the outside of the liver, which fixes the temperature to the body temperature of 37 °C. For a given geometry, we simulate 200 scenarios by varying the location and radius of a spherical heat source, approximating the adjustable, umbrella-like probe[1]. It is integrated as another Dirichlet boundary condition since the tissue inside the confining radius is expected to be necrotic. A coarse grid resolution of 4 mm is chosen as a good trade-off between runtime efficiency and solution accuracy. Each simulation uses a time step of 0.05 s and covers a 10 min intervention with an initial heating phase of 7 min, in which the heat source is applied.

We build the training set by sampling the finite-difference solutions every 30 s. Our neural network is then fitted to the sequence of M temperature distributions using a mean squared error loss $\mathcal{L}_{\mathrm{MSE}} = \frac{1}{N \cdot M} \sum_{m=1}^{M} \sum_{n=1}^{N} \|T_n^m - \hat{T}_n^m\|^2$ between the predicted temperature $\hat{T}_n^m$ and the ground truth T_n^m for the N voxels inside the liver. We train the network for 100 epochs using Adam as the optimizer with default parameters as in [5], a learning rate of 1×10^{-4}, and a batch size of 5.

For the branch network, we choose 14 convolutional layers with a kernel size of 3, 64 filters per layer, and replication padding of size 1. We add a skip connection to each layer that adds the input of a layer to its output before applying the leaky rectified linear unit (Leaky ReLU) as activation function.

The trunk network comprises four fully connected layers of size 100. A final layer in both the branch and the trunk network maps the preceding features to 150-dimensional latent codes. For maximum performance, the scalar product of Eq. 2 is only applied to the voxels inside the liver. The finite-difference solver and the proposed architecture (see Fig. 1) are implemented using PyTorch [11]. Training and evaluation are performed on a single NVIDIA Tesla V100 graphics card.

2.2   Single Electrode Radiofrequency Ablation Planning

This work relies on a simple planning algorithm that requires segmentations of the body, liver parenchyma, hepatic vessels, tumor, and organs at risk. We dilate the tumor segmentation to add a safety margin of 5mm to form the treatment zone and we define the set of RFA treatment points $\{k\}$ as all voxels within it. Since not all treatment points can be reached by the RFA probe, we select a subset of target points by first extracting a subset of the body surface voxels that are no more than 15 cm (=maximum needle length) away from the tumor and that are not located on the back of the patient. We then compute a signed distance field from a binary mask representing the organs at risk and trace a ray from each possible voxel k inside the treatment zone to each surface skin point. A GPU-accelerated sampling is used to obtain the distance to the critical structures for all points along the ray, excluding all rays with non-positive distance values. The set of target points $\{j\}$ are defined as those reached by valid rays, and the corresponding skin entry point is identified as the one with minimal distance to the target location.

Next, each target location is tested based on a specific ablation zone model. The results are gathered in a binary coverage matrix $\mathbf{C}_{j,k}$ denoting whether a treatment voxel k is part of the ablation zone from electrode target location j. In addition, the volume of overablation $\mathrm{OA}_{vol}(j)$, defined by the sum of voxels within the ablation zone but outside of the treatment zone, is computed for each target location. We seek an electrode target position j minimizing the damage to healthy tissue while achieving 100% coverage of the treatment zone defined by the minimization problem: $\min_j \mathrm{OA}_{vol}(j)$, subject to $\mathbf{C}_{j,k} = 1\ \forall k$, where $\{k\}$ denotes all voxels inside the treatment zone.

3   Experiments and Results

3.1   Data and Pre-processing

For training, evaluation, and planning, retrospective clinical data comprising ten patients with pre- and post-operative CT images were used. There were 14 tumors among the ten patients, which were all treated with RFA. The following clinical RFA protocol was applied: the location and diameter of the ablation electrode was pre-operatively selected according to the size, location, and shape of the tumor. Once a target temperature of 105 °C was reached, the heating was

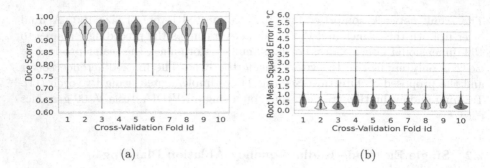

Fig. 2. Illustration of the leave-one-out cross-validation results showing for each fold (a) distribution of dice scores and (b) distribution of root mean squared errors. (red: median; box: 1st and 3rd quartile; whiskers: 5 & 95 percentile) (Color figure online)

maintained for 7 min. For particularly large tumors, the process was repeated with an increased electrode diameter for another 7 min.

A deep learning image-to-image network was applied to obtain segmentations of the body, liver, hepatic and portal veins, and organs at risk (ribs, heart, left and right kidneys, left and right lungs, spleen, and aorta) [22]. The tumor and the post-operative ablation zone were manually annotated by an expert. The post-operative ablation zone was then registered to the pre-operative images.

3.2 Physics Emulator Evaluation

To compensate for the limited number of data sets, we performed a leave-one-out cross-validation over the ten patients. In each of the folds, a deep operator network was trained on a total of 1,800 simulations covering nine patient geometries. The network was then applied to the left-out 200 simulations. For each of the 200 test cases, the temperature distribution predicted every 30 s was compared against the one obtained by the finite-difference solver. We compared the performance of our method in terms of the temperature root mean squared error for all voxels inside the liver and over all time steps. In addition, we used the dice score ($dice = \frac{2|Z_P \cap Z_{GT}|}{|Z_P| + |Z_{GT}|}$) to measure the overlap between the predicted ablation zone Z_P and the ground truth Z_{GT}, which we defined as the isothermal volume comprising all voxels that exceed $\geq 50\,°C$ at any time point in the simulation.

The results as presented in Fig. 2 indicate that the network is able to accurately emulate the RFA simulations. The deep operator network is able to match the ablation zone with a mean dice score of 0.93±0.05 and an average root mean squared error of 0.51±0.50 °C. In addition, the proposed method has a significant runtime advantage. Each simulation can be emulated in less than 0.06 s, which is over 100 times faster than the finite-difference solver. Qualitatively, we observe good agreement in both volume and shape, suggesting that the network is able to account for the heat sink effect of large vessels and the blood velocity field (see Fig. 3). The worst results are typically observed for ablation zones that are particularly large and heavily deformed due to the blood velocity field.

Table 1. Results of the single electrode automatic planning for all 14 tumors, highlighting the highest ablation efficiency score (AE) in the comparison between our method and sphere-based planning. For completeness, we present the finite-difference results. Tumor 07-2 is not ablateable with a single electrode. Time: the compute time; R: the optimal source radius; AR: sphere ablation radius.

Case Id	Tumor ⌀ (mm)	Sphere			Ours			Finite-Difference		
		AR|mm	AE|%	Time|min	R|mm	AE|%	Time|min	R|mm	AE|%	Time|min
01	56.0	34	29.7	1.41	26	**37.9**	3.97	26	38.3	348.81
02	35.8	26	**36.4**	1.45	22	32.1	3.66	26	20.0	194.11
03	22.9	22	29.9	1.24	14	**42.5**	1.37	14	40.9	60.17
04	16.0	18	**30.3**	0.92	18	23.8	1.13	18	23.7	68.53
05-1	24.6	22	26.3	1.53	14	**32.0**	1.85	18	32.1	58.53
05-2	21.4	22	24.5	1.61	18	**37.9**	2.0	18	41.4	47.11
06	44.2	30	**35.3**	1.24	26	31.5	4.18	26	33.0	283.32
07-1	52.0	34	19.1	0.96	30	**25.3**	3.32	30	23.7	260.83
07-2	76.0	-	-	-	-	-	-	-	-	-
07-3	34.4	26	25.0	1.06	18	**33.9**	1.73	18	35.5	78.81
08-1	8.0	14	30.2	1.13	10	**49.5**	1.18	10	52.9	5.89
08-2	14.8	14	**55.3**	1.18	10	34.7	1.31	10	39.1	10.48
09	34.9	26	**36.8**	0.87	18	28.2	1.93	18	26.9	113.17
10	26.0	18	**47.6**	0.99	18	35.4	1.44	18	37.8	51.37

3.3   Application to Automatic Single Needle RFA Planning

To demonstrate the benefits of our fast biophysics model over sphere-based and finite-difference-based planning, we perform single electrode automatic ablation planning for all 14 tumors using the algorithm described in Sect. 2.2 with all three methods. Our trained operator networks are hereby always applied to the unseen geometry of each cross-validation fold.

Specifically, we seek for each tumor an optimal electrode position, i.e. coordinates of the needle tip, and diameter achieving full coverage and minimal damage to surrounding tissue. To this end, we iteratively apply the planning algorithm and increase the source or ablation zone diameter until a solution is found. According to device specifications, an ablation zone of 3-7 cm in diameter can be achieved with a single RFA device[1]. The sphere-based model can directly represent such an ablation zone with a proper choice of the ablation radius AR. We determine a set of equivalent source radii ($R \in \{10, 14, 18, 22, 24, 30\}$ mm) for our biophysical models by simulating, with the finite difference solver, the temperature diffusion in a cubic domain with a central heat source of varying radius resulting in the desired ablation zone diameter.

We report in Table 1 for each tumor and each ablation zone model the optimal radius, the compute time, and the ablation efficiency $AE = \frac{|Z_P \cap Z_T|}{|Z_P|}$ measuring the fraction of the estimated ablation zone Z_P overlapping the treatment

---

[1] e.g. AngioDynamics StarBurst® system.

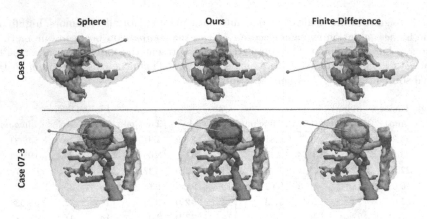

Fig. 3. Illustration of two planning results. Ablation zones and optimal paths are overlayed with the tumor (+safety margin) (white), portal vein (orange), hepatic vein (yellow), and post-operative ablation zone (black wireframe) for the sphere-based model (red), the finite-difference model (green), and the proposed method (blue). Despite marginal differences in the ablation estimates, the biophysical models find similar needle trajectories. The sphere-based approach results instead in significantly different trajectories. The observed difference between tumor and post-operative ablation zone Case 04 may arise from registration errors. (Color figure online)

zone Z_T. In general, ablation efficiency is comparable for sphere-based and both biophysics-based models. Our proposed method finds almost identical optimal source diameter compared to the finite-difference-based approach, with only a small deviation in AE and a significant run time advantage, especially for large tumors that might require multiple optimization iterations. Sphere-based RFA planning was only marginally faster in all tested scenarios.

4   Discussion and Conclusion

Building upon the latest advancements in deep learning, we have presented a data-driven model to simulate RFA interventions faster than real-time using non-autoregressive operator learning. Despite having trained the network on limited amount of data, the method was able to match the ablation zones computed by the finite-difference method on unseen data with an average dice score of 0.93 ± 0.05, while being over 100 times faster. As a reference, an alternative approach to fast approximation of ablation zones [15] achieved a lower median dice score of 0.88 with respect to the finite-element method, while ignoring the effect of heat transport due to the blood perfusion of liver tissue. Another advantage of our proposed method is the ability to estimate the temperature distribution at any time during the ablation procedure, which enables the computation of ablation zones based on more complex necrosis models such as the thermal dose [17] as a potential extension of this work. In addition, future research will aim at increasing the fidelity of the model and specifically at overcoming the challenges posed by the relatively coarse grid resolution, which may lead to disconnected vessels and thus deviations in the estimated ablation zone.

The method was used for single electrode RFA planning in 14 tumors and compared against sphere-based and finite-difference-based approaches. In general, the biophysical models were able to produce viable ablation plans with comparable ablation efficiency to the sphere-based planning, while at the same time integrating information about the heat sink effect of large vessels and the blood velocity field. This is expected to be crucial especially for ablation zones in the proximity of vessels, and is visually demonstrated by the case depicted in Fig. 3 for which sphere-based ablation zone estimation is inconsistent with the post-operative ablation zone. Further research is necessary to assess and quantify the superiority of biophysics-based models, which may require testing optimal candidates in the real world. In addition, understanding the impact of the tumor shape and size, the planning objective, and the modeling assumptions onto the planning results is of great importance and will be the subject of future research.

To the best of our knowledge, the proposed method is the first to achieve the necessary speed-up to enable patient-specific RFA planning subject to the underlying biophysics and with run-times comparable to sphere-based planning, while providing equivalent results as the finite-difference method. This could simplify further validation of the approach also in prospective studies and therefore the collection of larger clinical data sets. Potential extensions of this work include the analysis of multiple electrode ablation scenarios, the inclusion of explicit modeling of cell necrosis, and the efficient estimation of the physical parameters of the liver tissue from temperature measurements.

Disclaimer. This feature is based on research, and is not commercially available. Due to regulatory reasons its future availability cannot be guaranteed.

References

1. Altrogge, I., et al.: Towards optimization of probe placement for radio-frequency ablation. In: Larsen, R., Nielsen, M., Sporring, J. (eds.) MICCAI 2006. LNCS, vol. 4190, pp. 486–493. Springer, Heidelberg (2006). https://doi.org/10.1007/11866565_60
2. Audigier, C., et al.: Efficient lattice Boltzmann solver for patient-specific radiofrequency ablation of hepatic tumors. IEEE Trans. Med. Imaging **34**(7), 1576–1589 (2015)
3. Chaitanya, K., Audigier, C., Balascuta, L.E., Mansi, T.: Automatic planning of liver tumor thermal ablation using deep reinforcement learning. In: Medical Imaging with Deep Learning (2021)
4. Heimbach, J.K., et al.: AASLD guidelines for the treatment of hepatocellular carcinoma. Hepatology **67**(1), 358–380 (2018)
5. Kingma, D.P., Ba, J.: Adam: a method for stochastic optimization. arXiv preprint arXiv:1412.6980 (2014)
6. Klinger, H.: Heat transfer in perfused biological tissue-I: general theory. Bull. Math. Biol. **36**, 403–415 (1974). https://doi.org/10.1007/BF02464617

7. Laimer, G., et al.: Minimal ablative margin (MAM) assessment with image fusion: an independent predictor for local tumor progression in hepatocellular carcinoma after stereotactic radiofrequency ablation. Eur. Radiol. **30**(5), 2463–2472 (2019). https://doi.org/10.1007/s00330-019-06609-7

8. Liang, L., Cool, D., Kakani, N., Wang, G., Ding, H., Fenster, A.: Automatic radiofrequency ablation planning for liver tumors with multiple constraints based on set covering. IEEE Trans. Med. Imaging **39**(5), 1459–1471 (2019)

9. Lu, L., Jin, P., Pang, G., Zhang, Z., Karniadakis, G.E.: Learning nonlinear operators via deepONet based on the universal approximation theorem of operators. Nat. Mach. Intell. **3**(3), 218–229 (2021)

10. Mariappan, P., et al.: GPU-based RFA simulation for minimally invasive cancer treatment of liver tumours. Int. J. Comput. Assist. Radiol. Surg. **12**(1), 59–68 (2016). https://doi.org/10.1007/s11548-016-1469-1

11. Paszke, A., et al.: PyTorch: an imperative style, high-performance deep learning library. In: Wallach, H., Larochelle, H., Beygelzimer, A., d' Alché-Buc, F., Fox, E., Garnett, R. (eds.) Advances in Neural Information Processing Systems 32, pp. 8024–8035. Curran Associates, Inc. (2019)

12. Payne, S., et al.: Image-based multi-scale modelling and validation of radio-frequency ablation in liver tumours. Philos. Trans. A Math. Phys. Eng. Sci. **369**(1954), 4233–4254 (2011)

13. Pennes, H.H.: Analysis of tissue and arterial blood temperatures in the resting human forearm. J. Appl. Physiol. **1**(2), 93–122 (1948)

14. Reig, M., et al.: BCLC strategy for prognosis prediction and treatment recommendation barcelona clinic liver cancer (BCLC) staging system. the 2022 update. J Hepatol. **76**(3), 681–693 (2021)

15. Rieder, C., Kroeger, T., Schumann, C., Hahn, H.K.: GPU-based real-time approximation of the ablation zone for radiofrequency ablation. IEEE Trans. Visual Comput. Graphics **17**(12), 1812–1821 (2011)

16. Sanchez-Gonzalez, A., Godwin, J., Pfaff, T., Ying, R., Leskovec, J., Battaglia, P.: Learning to simulate complex physics with graph networks. In: International Conference on Machine Learning, pp. 8459–8468. PMLR (2020)

17. Sapareto, S.A., Dewey, W.C.: Thermal dose determination in cancer therapy. Int. J. Radiat. Oncol. Biol. Phys. **10**(6), 787–800 (1984)

18. Seitel, A., et al.: Computer-assisted trajectory planning for percutaneous needle insertions. Med. Phys. **38**(6), 3246–3259 (2011)

19. Thuerey, N., Weißenow, K., Prantl, L., Hu, X.: Deep learning methods for Reynolds-averaged Navier-stokes simulations of airfoil flows. AIAA J. **58**(1), 25–36 (2020)

20. Um, K., Brand, R., Fei, Y.R., Holl, P., Thuerey, N.: Solver-in-the-loop: Learning from differentiable physics to interact with iterative PDE-solvers. Adv. Neural. Inf. Process. Syst. **33**, 6111–6122 (2020)

21. Wulff, W.: The energy conservation equation for living tissue. IEEE Trans. Biomed. Eng. **6**, 494–495 (1974)

22. Yang, D., et al.: Automatic liver segmentation using an adversarial image-to-image network. In: Descoteaux, M., Maier-Hein, L., Franz, A., Jannin, P., Collins, D.L., Duchesne, Simon (eds.) MICCAI 2017. LNCS, vol. 10435, pp. 507–515. Springer, Cham (2017). https://doi.org/10.1007/978-3-319-66179-7_58

23. Zhang, R., Wu, S., Wu, W., Gao, H., Zhou, Z.: Computer-assisted needle trajectory planning and mathematical modeling for liver tumor thermal ablation: a review. Math. Biosci. Eng. **16**(5), 4846–4872 (2019)

Multi-task Video Enhancement for Dental Interventions

Efklidis Katsaros[1], Piotr K. Ostrowski[1], Krzysztof Włódarczak[1],
Emilia Lewandowska[1], Jacek Ruminski[1], Damian Siupka-Mróz[4],
Łukasz Lassmann[3,4], Anna Jezierska[1,2], and Daniel Węsierski[1(✉)]

[1] Gdańsk University of Technology, Gdansk, Poland
daniel.wesierski@pg.edu.pl
[2] Systems Research Institute, Polish Academy of Sciences, Warsaw, Poland
[3] Dental Sense Medicover, Gdansk, Poland
[4] Master Level Technologies, Gdansk, Poland

Abstract. A microcamera firmly attached to a dental handpiece allows
dentists to continuously monitor the progress of conservative dental procedures. Video enhancement in video-assisted dental interventions alleviates low-light, noise, blur, and camera handshakes that collectively
degrade visual comfort. To this end, we introduce a novel deep network
for multi-task video enhancement that enables macro-visualization of
dental scenes. In particular, the proposed network jointly leverages video
restoration and temporal alignment in a multi-scale manner for effective
video enhancement. Our experiments on videos of natural teeth in phantom scenes demonstrate that the proposed network achieves state-of-the-
art results in multiple tasks with near real-time processing. We release
Vident-lab at https://doi.org/10.34808/1jby-ay90, the first dataset of
dental videos with multi-task labels to facilitate further research in relevant video processing applications.

Keywords: Multi-task learning · Dental interventions · Video
restoration · Motion estimation

1   Introduction

Computer-aided dental intervention is an emerging field [10,20,33]. In contemporary clinical practice, dentists use various instruments to view the teeth better
for decreased work time and increased quality of conservative dental procedures
[13]. A close and continuous view of the operated tooth enables a more effective

This work was supported in part by The National Centre for Research and Development, Poland, under grant agreement POIR.01.01.01–00–0076/19.

Supplementary Information The online version contains supplementary material
available at https://doi.org/10.1007/978-3-031-16449-1_18.

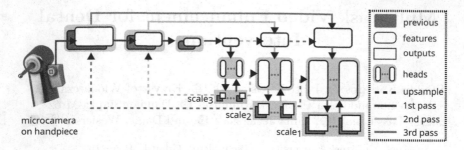

Fig. 1. Multi-task enhancement of videos from dental microcameras. We propose **MOST-Net**, a multi-output, multi-scale, multi-task network that propagates task outputs within and across scales in the encoder and decoder to fuse spatio-temporal information and leverage task interactions.

and safer dental bur maneuver within the tooth to remove caries and limit the risk of exposing pulp tissue to infection. A microcamera in an adapter firmly attached to a dental handpiece near the dental bur allows dentists to inspect the operated tooth during drilling closely and uninterruptedly in a display. However, the necessary miniaturization of vision sensors and optics introduces artifacts. Macro-view translates the slight motion of the bur to its more significant image displacement. The continuous camera shakes increase eye fatigue and blur. Handpiece vibrations, rapid light changes, splashing water and saliva further complicate imaging of intra-oral scenes. This study is the first to address the effectively compromised quality of videos of phantom scenes with an algorithmic solution to integrate cost-effective microcameras into digital dental workflows.

We propose a new multi-task, decoder-focused architecture [25] for video processing and apply it to video enhancement of dental scenes (Fig. 1). The proposed network has multiple heads at each scale level. Provided that task-specific outputs amend themselves to scaling, the network propagates the outputs bottom-up, from the lowest to the highest scale level. It thus enables task synergy by loop-like modeling of task interactions in the encoder and decoder across scales. UberNet [9] and cross-stitch networks [16] are encoder-focused architectures that propagate task outputs across scales in the encoder. Multi-modal distillation in PAD-Net [27] and PAP-Net [29] are decoder-focused networks that fuse outputs of task heads to make the final dense predictions but only at a single scale. MTI-Net [24], which is most similar to our architecture, extends the decoder fusion by propagating task-specific features bottom-up across multiple scales through the encoder. Instead of propagating the *task features* in scale-specific distillation modules across scales to the *encoder*, our network simultaneously propagates *task outputs* to the *encoder* and to the task heads in the *decoder*. Furthermore, the networks make dense task prediction in static images while we extend our network to videos.

We instantiate the proposed model to jointly solve low-, mid-, and high-level video tasks that enhance intra-oral scene footage. In particular, the model formulates color mapping [28], denoising, and deblurring [8,26,30] as a single dense prediction task and leverages, as auxiliary tasks, homography estimation [12] for video stabilization [1] and teeth segmentation [3,32] to re-initialize video stabilization. Task-features interacted in the two-branch decoder in [7] at a single scale for dense motion and blur prediction in dynamic scenes. We are the first to jointly address the tasks of color mapping, denoising, deblurring, motion estimation, and segmentation. We demonstrate that our near real-time network achieves state-of-the-art results in multiple tasks on videos with natural teeth in phantom scenes.

Our contributions are: (i) a novel application of a microcamera in computer-aided dental intervention for continuous tooth macro-visualization during drilling, (ii) a new, asymmetrically annotated dataset of natural teeth in phantom scenes with pairs of frames of compromised and good quality using a beam splitter, (iii) a novel deep network for video processing that propagates task outputs to encoder and decoder across multiple scales to model task interactions, and (iv) demonstration that an instantiated model effectively addresses multi-task video enhancement in our application by matching and surpassing state-of-the-art results of single task networks in near real-time.

2 Proposed Method

Video enhancement tasks are interdependent, e.g. aligning video frames assists deblurring [8,26,30,31] while denoising and deblurring expose image features that facilitate motion estimation [12]. We describe MOST-Net (multi-output, multi-scale, multi-task), a network that models and exploits task interactions across scale levels of the encoder and decoder. We assume the network yields T task outputs at each scale $\{\mathcal{O}_i^s\}_{i,s=1}^{T,S}$, where $s = 1$ denotes the original image resolution. Task outputs are propagated innerscale but also upsampled from the lower scale and propagated to the encoder layers and the task-specific branches of the decoder at higher scales. We require the following task-specific relation:

$$u_i(\mathcal{O}_i^{s+1}) \approx \mathcal{O}_i^s, \qquad (1)$$

where u_i denotes some operator, for instance, the upsampling operator for segmentation or the scaling operator for homography estimation.

Problem Statement. We instantiate MOST-Net to address the video enhancement tasks in dental interventions. In this setting, $T = 3$ and $\mathcal{O}_1$, $\mathcal{O}_2$, $\mathcal{O}_3$ denote the outputs for video restoration, segmentation and homography estimation. A video stream generates observations $\{B_{t-z}\}_{p=0}^P$, where t is the time index and $P > 0$ is a scalar value referring to the number of past frames. The problem is to estimate a clean frame, a binary teeth segmentation mask and approximate the inter-frame motion by a homography matrix, denoted by the triplet $\mathcal{O}_{1,1}^{3,S} = \{R_t^s,$ $M_t^s, H_{t-1 \to t}^s,\}_{s=1}^S$. Let x correspond to pixel location. Given per-pixel blur kernels $k_{x,t}$ of size K, the degraded image at $s = 1$ is generated as:

$$\forall_x \forall_t \ B_{x,t} = \sigma \langle (R_{x,t})^- k_{x,t} \rangle + \eta, \qquad (2)$$

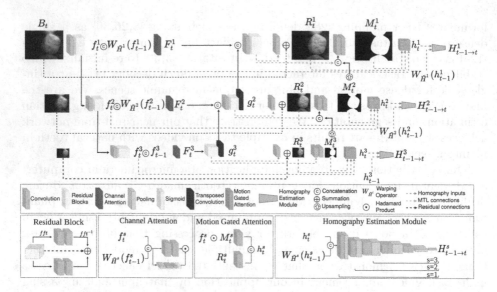

Fig. 2. Our MOST-Net instantiation addresses three tasks for video enhancement: video restoration, teeth segmentation, and homography estimation.

where η and σ denote additive and signal dependent noise, respectively, while $(R_{x,t})^-$ is a window of size K around pixel x in image R_t^1. Next we assume multiple independently moving objects present in the considered scenes, while our task is to estimate only the motion related to the object of interest (i.e. teeth), which is present in the region indicated by non-zero values of mask M:

$$\forall_t\ \forall_x \qquad x_t = H_{t-1\to t}x_{t-1} \quad \text{s.t.} \quad M_t(x) = 1 \qquad (3)$$

Training. In our setting, all tasks share training data in dataset $\mathcal{D} = \{\{B\}_j, \{\mathcal{O}_i^s\}_j\}_{i,s,j=1}^{T,S,N}$, where $\{\mathcal{O}_i^s\}_j$ is a label related to task i at scale s for the j-th training sample $\{B\}_j$, while N denotes number of samples in training data. In the context of deep learning, the optimal set of parameters Θ for some network $\mathcal{F}_\Theta$ is derived by minimizing a penalization criterion:

$$\mathcal{L}(\Theta) = \sum_j^N \sum_i^T \sum_s^S \lambda_i L_i \left(\{\mathcal{O}_i^s\}_j, \{\hat{\mathcal{O}}_i^s(\Theta)\}_j\right), \qquad (4)$$

where λ_i is a scalar weighting value, $\hat{\mathcal{O}}_i^s(\Theta)$ is an estimate of $\mathcal{O}_i^s$ for j-th sample in $\mathcal{D}$, and L_i is a distance measure. In this study, we use the Charbonnier loss [2] as L_1, the binary cross-entropy [17] as L_2 and the Mean Average Corner Error (MACE) [5] as L_3.

Encoders. At each time step, MOST-Net extracts features f_{t-1}^s, f_t^s from two input frames B_{t-1} and B_t independently at three scales. To effectuate the U-shaped [21] downsampling, features are extracted via 3×3 convolutions with strides of $1, 2, 2$ for $s = 1, 2, 3$ followed by ReLU activations and 5 residual blocks [4] at each scale. The residual connections are augmented with an additional branch of convolutions in the Fast Fourier domain, as in [14]. The output channel dimension for features f_t^s is 2^{s+4}. At each scale, features f_t^s and $W_{\tilde{H}_s}(f_{t-1}^s)$ are concatenated and a channel attention mechanism follows [30] to fuse them into F_t^s. MOST-Net uses homography outputs from lower scales to warp encoder features from the previous time step as $W_{\tilde{H}}(f_{t-1}^s)$. Here, $\tilde{H}^s$ is an upscaled version of H^{s+1} for higher scales and the identity matrix for $s = 3$.

Decoders. The attended encoder features F_t^s are passed onto the expanding blocks scale-wisely via the skipping connections. At the lower scale ($s = 3$), the attended features F_t^3 are directly passed on a stack of two residual blocks with 128 output channels. Thereafter, transposed convolutions with strides of 2 are used twice to recover the resolution scale. At higher scales ($s < 3$), features F_t^s are first concatenated with the upsampled decoder features g_t^{s-1} and convolved by 3×3 kernels to halve the number of channels. Subsequently, they are propagated onto two residual blocks with 64 and 32 output channels each. The residual block outputs constitute scale-specific shared backbones. Lightweight task-specific branches follow to estimate the dense outputs. Specifically, one 3×3 convolution estimates R_t^s and two 3×3 convolutions, separated by ReLU, yield M_t^s at each scale. Figure 2 shows MOST-Net enables refinement of lower scale segmentations by upsampling and inputting them at the task-specific branches of higher scales.

At each scale, homography estimation modules estimate 4 offsets, related 1-1 to homographies via the Direct Linear Transformation (DLT) as in [5,12]. The motion gated attention modules multiply features f_t^s with segmentations M_t^s to filter out context irrelevant to the motion of the teeth. The channel dimensionality is then halved by a 3×3 convolution while a second one extracts features from the restored output R_t^s. The concatenation of the two streams forms features h_t^s. At each scale, h_t^s and $W_{\tilde{H}^s}(h_{t-1}^s)$, are employed to predict the offsets with shallow downstream networks. Predicted offsets at lower scales are transformed back to homographies and cascaded bottom-up [12] to refine the higher scale ones. Similarly to [5], we use blocks of 3×3 convolutions coupled with ReLU, batch normalization and max-pooling to reduce the spatial size of the features. Before the regression layer, a 0.2 dropout is applied. For $s = 1$, the convolution output channels are 64, 128, 256, 256 and 256. For s=2,3 the network depth is cropped from the second and third layers onwards respectively.

3   Results and Discussion

Dataset. We describe the generation of *Vident-lab* dataset $\mathcal{D}$ with frames B and labels R, M and H for training, validation, and testing (Table 1). We generate the labels at full resolution as illustrated in Fig. 3. The lower scale labels $\mathcal{O}_i^s$ are obtained from the inverse of Eq. 1, i.e. downsampling for R, M and downscaling for H. The dataset is publicly available at https://doi.org/10.34808/1jby-ay90.

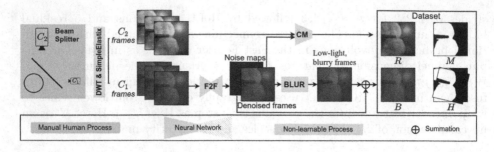

Fig. 3. A flowchart of dataset preparation.

Data Acquisition. A miniaturized camera C_1 has inferior quality to intraoral cameras C_2, which have larger sensors and optics. Our task is to teach C_1 to image the scene equally well as C_2. Both cameras, which are firmly coupled through a 50/50 beam splitter, acquire videos of the same dental scene. Dynamic time warping (DTW) synchronizes the videos and then *SimpleElastix* [15] registers the corresponding 320×416 frames.

Noise, Blur, Colorization: B and R. We use frame-to-frame (F2F) training [6] on static video fragments recorded with camera C_1 and apply the trained image denoiser to obtain *denoised frames* and their *noise maps*. The denoised frames are temporally interpolated [19] 8 times and averaged over a temporal window of 17 frames to synthesize realistic blur. The noise maps are added to the blurry frames to form the *input video frames* B. However, perfect registration of frames between two different modalities C_1 and C_2 is challenging. Instead, we colorize frames of

Data	Train	Val	Test
Videos	300	29	80
Frames	60K	5.6K	15.5K
Segm(h)	300	116	320
Segm(n)	59.7K	5.5K	15.2K

Table 1. Dataset summary ($K = \times 10^3$), (h,n) human- and network-labelled teeth masks.

C_1 as per C_2 to create the ground truth *output video frames* R. Similarly to [28], we learn a color mapping (CM) network to predict parameters of 3D functions used to map colors of dental scenes from C_2 to C_1. In effect, we circumvent local registration errors and obtain exact, pixel-to-pixel spatial correspondence of frames B and R.

Segmentation Masks and Homographies: M, H. We manually annotate one frame R of natural teeth in phantom scenes from each training video and four frames of teeth in each validation and test videos. Following [18] that used a powerful network, an HRNet48 [22] pretrained on ImageNet, is fine-tuned on our annotations to automatically segment the teeth in the remaining frames in all three sets. We compute optical flows between consecutive clean frames with RAFT [23]. Motion fields are cropped with teeth masks M_t to discard other moving objects, such as the dental bur or the suction tube, as we are interested in stabilizing the videos with respect to the teeth. Subsequently, a partial affine homography H is fitted by RANSAC to the segmented motion field.

Setup. We train, validate, and test all methods on our dataset (Table 1). In all MOST-Net training runs, we set $\lambda_1, \lambda_2, \lambda_3$ to 2×10^{-4}, 5×10^{-5} and 1 for balancing tasks in Eq. 4. We train all methods with batch size 16 and use Adam optimizer with learning rate $1e - 4$, decayed to $1e - 6$ with cosine annealing. The training frames are augmented by horizontal and vertical flips with 0.5 probability, random channel perturbations, and color jittering, after [31]. All experiments are performed with PyTorch 1.10 (FP32). The inference speed is reported in frames-per-second (FPS) on GPU NVidia RTX 5000.

Diagnostics. In Fig. 4 we assess the performance gains across scale levels of MOST-Net. To this end, we upsample all outputs at lower scales to original scale and compare them with ground truth. We observe that MOST-Net performance improves via task-output propagation across scales in all measures. We perform an ablation study of MOST-Net (Table 2) by reconfiguring our architecture (Fig. 2) as follows: (i) NS-no segmentation as auxiliary task, (ii) NE-no connection of encoder features f_t^s with Motion Gated Attention module, (iii) NW-no warping of previous encoder features f_{t-1}^s, (iv) NMO-no multi-outputs at scales $s > 1$ so that our network has no task interactions between scales. Ablations show that all network configurations lead to > 0.5dB drops in PSNR and drops in temporal consistency error E(W). Segmentation task and temporal alignment help the video restoration task the most. No multi-task interactions across scales increases MACE error by > 0.6. The NE ablation improves MACE only slightly at the considerable drop in PSNR. We also find that the IoU remains relatively unaffected by the ablations suggesting room for improving task interactions to aid the figure-ground segmentation task. Qualitative results are shown in Fig. 5.

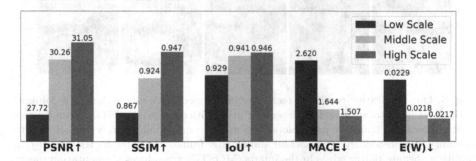

Fig. 4. MOST-Net performance improves with output upscaling.

Table 2. STL and MTL benchmarks (top panel) and MOST-Net (bottom panel). Best results of MOST-Net wrt ESTRNN+MHN+DL are in bold.

Methods	PSNR ↑	SSIM ↑	MACE ↓	IoU ↑	E(W) ↓	#P(M)	FPS
MIMO-UNet [4]	26.66	0.916	-	-	0.0278	5.3	8.4
ESTRNN [30]	30.72	0.943	-	-	0.0229	2.3	68.5
MHN [12]	-	-	1.347	-	-	6.2	89.8
DeepLabv3+(DL) [3]	-	-	-	0.968	-	26.7	108.2
UNet++ [32]	-	-	-	0.969	-	50.0	38.9
ESTRNN+MHN+DL	30.72	0.943	**1.368**	**0.967**	0.0229	35.2	**28.6**
MOST-Net-NS	30.21	0.939	1.426	-	0.0223	9.7	19.0
MOST-Net-NE	30.22	0.941	1.423	0.946	0.0221	9.8	19.2
MOST-Net-NW	30.37	0.943	1.456	0.952	0.0221	9.8	19.3
MOST-Net-NMO	30.48	0.940	2.155	0.946	0.0227	8.5	19.1
MOST-Net	**31.05**	**0.947**	1.507	0.946	**0.0217**	**9.8**	19.3

Input B_{t-1}	*Input* B_t	*Overlaid* B_{t-1}, B_t	*Overlaid* $W_{H_{t-1 \to t}}(R_{t-1}), R_t$	*Restored* R_t, *Mask* M_t

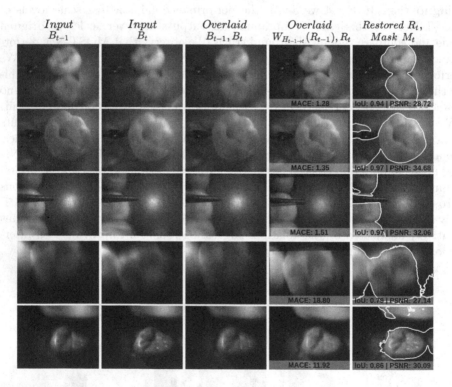

Fig. 5. Our qualitative results of teeth-specific homography estimation (4th column) and full frame restoration and teeth segmentation (5th column). MOST-Net can denoise video frames and translate pale colors (first and second column) into vivid colors (5th column). Simultaneously, it can deblur and register frames wrt to teeth (4th column). In addition, despite blurry edges in the inputs, MOST-Net produces segmentation masks that align well with teeth contours (rows 1-3). Failure cases (bottom panel, 4-5th rows) stem from heavy blur (4th row, and tooth-like independently moving objects (5th row), such as suction devices.

Quantitative Results. We compare MOST-Net with single task state-of-the-art methods for restoration, homography estimation, and binary segmentation in Table 2. MOST-Net outperforms video restoration baseline ESTRNN [30] and image restoration MIMO-UNet [4] in PSNR by > 0.3dB and > 4.3dB, respectively. We posit the low PSNR performance of MIMO-UNet stems from its single frame input that negatively affects its colorization abilities and also leads to high temporal consistency error $E(W)$ [11]. ESTRNN also introduces observable flickering artifacts expressed by higher $E(W)$ than MOST-Net. MACE error in homography estimation is slightly higher for our method than for MHN [12] but MOST-Net has potential for improvement due to its multi-tasking approach. Notably, MHN has a significantly lower error of MACE=0.6 when it is trained and tested on *ground truth* videos, which have vivid colors and no noise and blur. This suggests that video restoration task is necessary to improve homography estimation task. Subsequently, we evaluate our method wrt DeepLabv3+ [3] with ResNet50 encoder and wrt UNet++ [32] for teeth segmentation task using intersection-over-union (IoU) criterion. MOST-Net achieves comparable results with these benchmarks with several times less parameters (#P(M)). Though both methods have several times higher FPS, MOST-Net addresses three tasks instead of a single task and still achieves near real-time efficiency. Finally, we compare our multi-task MOST-Net with single task methods ESTRNN+MHN+DL that are forked, with MHN and DL as heads. ESTRNN restores videos from the training set and MHN and DeepLabv3+ (DL) are trained and tested on the restored frames. The pipeline runs at 28.6 FPS but requires $\times 3.6$ more model parameters than our network. Moreover, MOST-Net achieves higher PSNR, SSIM, and $E(W)$ results than the forked pipeline on the video restoration task, having comparable MACE error and IoU scores while running near real time at 19.3 FPS or **21.3 FPS** (TorchScript-ed).

Conclusions. We proposed MOST-Net, a novel deep network for video processing that models task interactions across scales. MOST-Net jointly addressed the tasks of video restoration, teeth segmentation, and homography-based motion estimation. The study demonstrated the applicability of the network in computer-aided dental intervention on the publicly released *Vident-lab* video dataset of natural teeth in phantom scenes.

References

1. Bradley, A., Klivington, J., Triscari, J., van der Merwe, R.: Cinematic-L1 video stabilization with a log-homography model. In: Proceedings of the IEEE/CVF Winter Conference on Applications of Computer Vision, pp. 1041–1049 (2021)
2. Charbonnier, P., Blanc-Feraud, L., Aubert, G., Barlaud, M.: Two deterministic half-quadratic regularization algorithms for computed imaging. In: Proceedings of 1st International Conference on Image Processing, vol. 2, pp. 168–172. IEEE (1994)
3. Chen, L.C., Zhu, Y., Papandreou, G., Schroff, F., Adam, H.: Encoder-decoder with atrous separable convolution for semantic image segmentation. In: Proceedings of the European Conference on Computer Vision (ECCV), pp. 801–818 (2018)

4. Cho, S.J., Ji, S.W., Hong, J.P., Jung, S.W., Ko, S.J.: Rethinking coarse-to-fine approach in single image deblurring. In: Proceedings of the IEEE/CVF International Conference on Computer Vision, pp. 4641–4650 (2021)
5. DeTone, D., Malisiewicz, T., Rabinovich, A.: Deep image homography estimation. arXiv preprint arXiv:1606.03798 (2016)
6. Ehret, T., Davy, A., Morel, J.M., Facciolo, G., Arias, P.: Model-blind video denoising via frame-to-frame training. In: Proceedings of the IEEE/CVF Conference on Computer Vision and Pattern Recognition, pp. 11369–11378 (2019)
7. Jung, H., Kim, Y., Jang, H., Ha, N., Sohn, K.: Multi-task learning framework for motion estimation and dynamic scene deblurring. IEEE Trans. Image Process. **30**, 8170–8183 (2021)
8. Katsaros, E., Ostrowski, P.K., Węsierski, D., Jezierska, A.: Concurrent video denoising and deblurring for dynamic scenes. IEEE Access **9**, 157437–157446 (2021)
9. Kokkinos, I.: UberNet: Training a universal convolutional neural network for low-, mid-, and high-level vision using diverse datasets and limited memory. In: Proceedings of the IEEE Conference on Computer Vision And Pattern Recognition, pp. 6129–6138 (2017)
10. Kühnisch, J., Meyer, O., Hesenius, M., Hickel, R., Gruhn, V.: Caries detection on intraoral images using artificial intelligence. J. Dent. Res. **101**(2), 158–165 (2021)
11. Lai, W.S., Huang, J.B., Wang, O., Shechtman, E., Yumer, E., Yang, M.H.: Learning blind video temporal consistency. In: Proceedings of the European Conference on Computer Vision (ECCV), pp. 170–185 (2018)
12. Le, H., Liu, F., Zhang, S., Agarwala, A.: Deep homography estimation for dynamic scenes. In: Proceedings of the IEEE/CVF Conference on Computer Vision and Pattern Recognition, pp. 7652–7661 (2020)
13. Low, J.F., Dom, T.N.M., Baharin, S.A.: Magnification in endodontics: a review of its application and acceptance among dental practitioners. Eur. J. Dent. **12**(04), 610–616 (2018)
14. Mao, X., Liu, Y., Shen, W., Li, Q., Wang, Y.: Deep residual fourier transformation for single image deblurring. arXiv preprint arXiv:2111.11745 (2021)
15. Marstal, K., Berendsen, F., Staring, M., Klein, S.: SimpleElastix: a user-friendly, multi-lingual library for medical image registration. In: 2016 IEEE Conference on Computer Vision and Pattern Recognition Workshops (CVPRW), pp. 134–142 (2016)
16. Misra, I., Shrivastava, A., Gupta, A., Hebert, M.: Cross-stitch networks for multi-task learning. In: Proceedings of the IEEE Conference on Computer Vision and Pattern Recognition, pp. 3994–4003 (2016)
17. Murphy, K.P.: Machine learning: a probabilistic perspective. MIT press (2012)
18. Nekrasov, V., Dharmasiri, T., Spek, A., Drummond, T., Shen, C., Reid, I.: Real-time joint semantic segmentation and depth estimation using asymmetric annotations. In: 2019 International Conference on Robotics and Automation (ICRA), pp. 7101–7107. IEEE (2019)
19. Niklaus, S., Mai, L., Liu, F.: Video frame interpolation via adaptive separable convolution. In: Proceedings of the IEEE International Conference on Computer Vision, pp. 261–270 (2017)
20. Rashid, U., et al.: A hybrid mask RCNN-based tool to localize dental cavities from real-time mixed photographic images. PeerJ Comput. Sci. **8**, e888 (2022)
21. Ronneberger, O., Fischer, P., Brox, T.: U-Net: convolutional networks for biomedical image segmentation. In: Navab, N., Hornegger, J., Wells, W.M., Frangi, A.F. (eds.) MICCAI 2015. LNCS, vol. 9351, pp. 234–241. Springer, Cham (2015). https://doi.org/10.1007/978-3-319-24574-4_28

22. Sun, K., Xiao, B., Liu, D., Wang, J.: Deep high-resolution representation learning for human pose estimation. In: Proceedings of the IEEE/CVF Conference on Computer Vision and Pattern Recognition, pp. 5693–5703 (2019)
23. Teed, Z., Deng, J.: RAFT: recurrent all-pairs field transforms for optical flow. In: Vedaldi, A., Bischof, H., Brox, T., Frahm, J.-M. (eds.) ECCV 2020. LNCS, vol. 12347, pp. 402–419. Springer, Cham (2020). https://doi.org/10.1007/978-3-030-58536-5_24
24. Vandenhende, S., Georgoulis, S., Van Gool, L.: MTI-Net: multi-scale task interaction networks for multi-task learning. In: Vedaldi, A., Bischof, H., Brox, T., Frahm, J.-M. (eds.) ECCV 2020. LNCS, vol. 12349, pp. 527–543. Springer, Cham (2020). https://doi.org/10.1007/978-3-030-58548-8_31
25. Vandenhende, S., Georgoulis, S., Van Gansbeke, W., Proesmans, M., Dai, D., Van Gool, L.: Multi-task learning for dense prediction tasks: a survey. IEEE Trans. Pattern Anal. Mach. Intell. **43** (2021)
26. Wang, X., Chan, K.C., Yu, K., Dong, C., Change Loy, C.: EDVR: video restoration with enhanced deformable convolutional networks. In: 2019 IEEE/CVF Conference on Computer Vision and Pattern Recognition Workshops (CVPRW), pp. 1954–1963 (2019)
27. Xu, D., Ouyang, W., Wang, X., Sebe, N.: PAD-Net: multi-tasks guided prediction-and-distillation network for simultaneous depth estimation and scene parsing. In: IEEE/CVF Conference on Computer Vision and Pattern Recognition (CVPR), pp. 675–684 (2018)
28. Zhang, M., et al.: RT-VENet: a convolutional network for real-time video enhancement. In: Proceedings of the 28th ACM International Conference on Multimedia, pp. 4088–4097 (2020)
29. Zhang, Z., Cui, Z., Xu, C., Yan, Y., Sebe, N., Yang, J.: Pattern-affinitive propagation across depth, surface normal and semantic segmentation. In: IEEE/CVF Conference on Computer Vision and Pattern Recognition (CVPR), pp. 4106–4115 (2019)
30. Zhong, Z., Gao, Y., Zheng, Y., Zheng, B.: Efficient spatio-temporal recurrent neural network for video deblurring. In: Vedaldi, A., Bischof, H., Brox, T., Frahm, J.-M. (eds.) ECCV 2020. LNCS, vol. 12351, pp. 191–207. Springer, Cham (2020). https://doi.org/10.1007/978-3-030-58539-6_12
31. Zhou, S., Zhang, J., Pan, J., Xie, H., Zuo, W., Ren, J.: Spatio-temporal filter adaptive network for video deblurring. In: Proceedings of the IEEE International Conference on Computer Vision (2019)
32. Zhou, Z., Rahman Siddiquee, M.M., Tajbakhsh, N., Liang, J.: UNet++: a nested u-net architecture for medical image segmentation. In: Stoyanov, D., et al. (eds.) DLMIA/ML-CDS -2018. LNCS, vol. 11045, pp. 3–11. Springer, Cham (2018). https://doi.org/10.1007/978-3-030-00889-5_1
33. Zhu, G., Piao, Z., Kim, S.C.: Tooth detection and segmentation with mask R-CNN. In: 2020 International Conference on Artificial Intelligence in Information and Communication (ICAIIC), pp. 070–072. IEEE (2020)

Outcome and Disease Prediction

Weighted Concordance Index Loss-Based Multimodal Survival Modeling for Radiation Encephalopathy Assessment in Nasopharyngeal Carcinoma Radiotherapy

Jiansheng Fang[1,2,3], Anwei Li[2], Pu-Yun OuYang[4], Jiajian Li[2], Jingwen Wang[2], Hongbo Liu[2], Fang-Yun Xie[4], and Jiang Liu[3(✉)]

[1] School of Computer Science and Technology, Harbin Institute of Technology, Harbin, China
[2] CVTE Research, Guangzhou, China
[3] Research Institute of Trustworthy Autonomous Systems, Southern University of Science and Technology, Shenzhen, China
liuj@sustech.edu.cn
[4] Department of Radiation Oncology, Sun Yat-sen University Cancer Center, Guangzhou, China

Abstract. Radiation encephalopathy (REP) is the most common complication for nasopharyngeal carcinoma (NPC) radiotherapy. It is highly desirable to assist clinicians in optimizing the NPC radiotherapy regimen to reduce radiotherapy-induced temporal lobe injury (RTLI) according to the probability of REP onset. To the best of our knowledge, it is the first exploration of predicting radiotherapy-induced REP by jointly exploiting image and non-image data in NPC radiotherapy regimen. We cast REP prediction as a survival analysis task and evaluate the predictive accuracy in terms of the concordance index (CI). We design a deep multimodal survival network (MSN) with two feature extractors to learn discriminative features from multimodal data. One feature extractor imposes feature selection on non-image data, and the other learns visual features from images. Because the priorly balanced CI (BCI) loss function directly maximizing the CI is sensitive to uneven sampling per batch. Hence, we propose a novel weighted CI (WCI) loss function to leverage all REP samples effectively by assigning their different weights with a dual average operation. We further introduce a temperature hyper-parameter for our WCI to sharpen the risk difference of sample pairs to help model convergence. We extensively evaluate our WCI on a private dataset to demonstrate its favourability against its counterparts. The experimental

---

J. Fang, A. Li and P.-Y. OuYang—Co-first authors.

Supplementary Information The online version contains supplementary material available at https://doi.org/10.1007/978-3-031-16449-1_19.

L. Wang (Eds.): MICCAI 2022, LNCS 13437, pp. 191–201, 2022.
https://doi.org/10.1007/978-3-031-16449-1_19

results also show multimodal data of NPC radiotherapy can bring more gains for REP risk prediction.

Keywords: Dose-volume histogram · Survival modeling · Multimodal learning · Loss function · Concordance index

1   Introduction

Radiotherapy is the standard radical treatment for nasopharyngeal carcinoma (NPC) and has considerably improved disease control and survival [3]. However, NPC radiotherapy may induce radiation encephalopathy (REP), which makes the brain suffer irreversible damage during the incubation period [22]. REP diagnosis remains challenging in clinical research due to its various clinical symptoms, insidious onset, long incubation period [6]. Currently, conventional magnetic resonance (MR) diagnosis can only discern REP at the irreversible stage [5]. Hence, it is highly desirable to assist clinicians in optimizing the radiotherapy regimen to reduce radiotherapy-induced temporal lobe injury (RTLI) according to the probability of REP onset. In this work, we study how to speculate REP risk in the pre-symptomatic stage by jointly exploiting the diagnosis and treatment data generated in NPC radiotherapy regimen, including computed tomography (CT) images, radiotherapy dose (RD) images, radiotherapy struct (RS) images, dose-volume histogram (DVH), and demographic characteristics.

Recently, the data-driven approach has attracted a wide range of attention in the field of NPC. For example, many works for NPC segmentation have shifted from traditional hand-engineered models [13,14,30] to automatic deep learning models [12,19]. Currently, there are a few studies that use data-driven approaches to predict the onset of REP. Zhao *et al.* predict REP in NPC by analyzing the whole-brain resting-state functional connectivity density (FCD) of pre-symptomatic REP patients using multivariate pattern analysis (MVPA) [29]. Zhang *et al.* explore fractional amplitude of low-frequency fluctuation (fALFF) as an imaging biomarker for predicting or diagnosing REP in patients with NPC by using the support vector machine (SVM) [28]. However, existing studies mainly consider the contribution of hand-engineered features computed from MR images to the REP prediction. The lesion region of REP stemming from NPC radiotherapy usually lie in the bilateral temporal lobe, and its severity is positively related to the dose [24]. Hence, according to the clinical evaluation mechanism, by casting REP risk prediction as a survival analysis task, we aim to build a deep multimodal survival network (MSN) to learn valuable and discriminative features from multiple data types generated during NPC radiotherapy.

The data types of NPC radiotherapy can be grouped into two modalities: image and non-image data. In MSN, we instinctively design two feature extractors to capture the discriminative information from multimodal data. One feature extractor explores the image data by convolution module, and the other mines the one-dimensional data by multi-layer perception (MLP) module. Given the concordance index (CI) [9] is a standard performance metric for survival models, the existing balanced CI (BCI) loss function formulates the learning problem

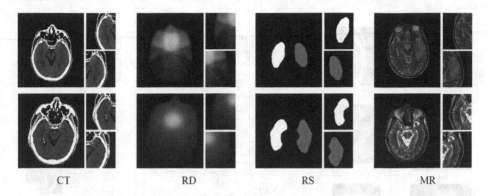

CT   RD   RS   MR

Fig. 1. Schematic of image data in NPC-REP dataset. The upper row is a patient that confirms REP during follow-up, and the lower row is a patient without temporal lobe injury. The bilateral temporal lobe in MR images during follow-up place on column (MR). And by traceback, corresponding CT, RD, RS images generated in NPC radiotherapy lie in columns (CT), (RD), and (RS), respectively.

as directly maximizing the CI [21]. However, BCI exhibits poor predictive accuracy due to its sensitivity to the sample pairing per batch. Hence, we propose a novel weighted CI (WCI) loss function with a temperature hyper-parameter τ to combat the sensitivity to uneven sampling and sharpen the relative risk of sample pairs, thus helping model convergence. We demonstrate our WCI advantage in terms of CI and 3-years area under the curve (AUC).

Our *contributions* are: (a) To advance REP risk prediction to assess the rationality of NPC radiotherapy regimen, we propose a deep multimodal survival network (MSN) equipped with two feature extractors to learn discriminative features from image and non-image data. (b) We propose a novel WCI loss function with a temperature hyper-parameter for our MSN training to effectively leverage REP samples of each batch and help model convergence. (c) We confirm the benefits of our WCI with other loss functions used for survival models by extensively experimenting on a private dataset.

2   Materials and Methods

2.1   NPC-REP Dataset Acquisition

We acquire eligible 4,816 patients from a cancer center. The eligibility of enrolling in the NPC-REP dataset is that diagnosis and treatment data in NPC radiotherapy and follow-up are traceable and recorded. We binarize a label for each patient according to whether the REP confirmation by MR image diagnosis during follow-up. If diagnosing out RTLI, the label of NPC patient is REP, otherwise non-REP. At the same time, we record the time intervals (in Months) between NPC radiotherapy and REP confirmation or the last follow-up of non-REP. Then we trace back the diagnosis and treatment data during NPC radiotherapy for

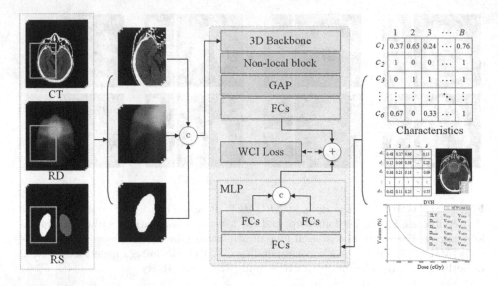

Fig. 2. The architecture of our MSN with two feature extractors. Characteristics $c_{1,2,...6}$ denote age, gender, T stage, N stage, TNM stage and treatment option. $d_{1,2,...18}$ are manually calculated on DVH. FCs includes two fully-connected layers. GAP is a global average pooling layer. B is the batch size, and c indicates a concatenation operator.

each patient. RD images show the radiation dose distribution to the human body during radiation therapy. Clinicians sketch the tumor outline in closed curve coordinate form to yield the RS image. Then we apply the mask of RS images to generate the input ROIs of RD and CT images. As Fig. 1 shows, we crop out the identical regions of interest (ROI) for CT images for diagnosis and RD images according to regions of the bilateral temporal lobe in RS images. Then we feed the three 3D ROIs into the multimodal learning network for extracting visual features. It is confirmed that the imaging dose delivered during NPC radiotherapy is positively related to REP risk [24]. Hence, in addition to considering image data, we introduce 18 features manually calculated from DVH, which relates radiation dose to tissue volume in radiation therapy planning [8]. Moreover, we also observe the affection of demographic information (age, gender), clinic stages (T/N/TNM), and treatment options (radio-chemotherapy and radiotherapy) to REP risk. The detailed statistics of demographic and clinical characteristics for the NPC-REP dataset are shown in Appendix A1.

2.2   Multimodal Survival Modeling

We notice that most NPC patients with REP are diagnosed after radiotherapy for about 37 months. And it is almost impossible to occur REP when the follow-up time is less than two months and more than 73 months. Such a statistics observation about the onset time of REP illustrates that the risk probability of suffering REP is highly related to some factors in NPC radiotherapy. These factors cause the onset of REP in a relatively definite period. At the same time, we can cast REP assessment as a 3-years survival analysis task according to the average REP confirmation time intervals. Thus we build a deep multimodal survival network (MSN) to identify potential factors from the NPC-REP dataset, which consists of a binarized label, time intervals, image data (CT, RD, and RS images), and non-image data (24 features in total). As Fig. 2 shows, one feature extractor learns visual features of 3D ROIs intercepted from CT, RD, and RS images, and the other explores the contributions of non-image data to REP risk. The 24 features of non-image data are directly fed into the MLP module to output the REP risk R_{nv}. In the branch of visual feature extractor, we first apply convolution network as a backbone, followed by non-local attention capturing the global inter-dependencies [23]. Next, we use a global average pooling (GAP) to merge feature maps and two fully-connected layers (FCs) to generate the REP risk R_v. After generating the two risk probabilities for two modality data, we further utilize our WCI loss function to train the network. The combination of the two predicted probabilities is defined as follows:

$$R = w_{nv}R_{nv} + w_v R_v, \tag{1}$$

where $R_{nv}, R_v \in [0, 1]$ and w_{nv} and w_v are their corresponding weights. Both are initially set to 0.5. We can tune the two weights to observe the contribution of different modal data on assessing the REP risk.

2.3   WCI Loss Function

CI is a standard performance metric for survival models that corresponds to the probability that the model correctly orders a randomly chosen pair of patients in terms of event time [4,25]. To calculate this indicator, we first pair all samples in the dataset of sample size n with each other to get $n(n-1)/2$ sample pairs. Then, we filter out evaluation sample pairs from all sample pairs, as follows:

$$\mathbb{E} := \{(i,j)|(O_i = 1 \wedge T_j \geq T_i)\}, \tag{2}$$

where T_i and T_j are the time intervals of ith and j^{th} NPC patients respectively, and $O_i = 1$ denotes that i^{th} NPC patient is observed as REP. $\mathbb{E}$ is the set of sample pairs (i, j) based on the ordering of survival times, $i.e.$, the time intervals of ith NPC patient is shorter than j^{th} NPC patient. Assuming the REP risks predicted by models for sample pairs (i, j) are R_i and R_j, we assert the prediction outcomes are correct if $R_i > R_j$ and $T_i < T_j$. We mark those sample pairs whose predicted results coincide with the actual as a set $\mathbb{E}_t$. The CI value is indicated as the number ratio of $\mathbb{E}_t/\mathbb{E}$. The higher CI explains away the better competence of survival models in predicting REP risks.

Because the assumptions that proportional hazards are invariant over time and log-linear limit the Cox loss function (See Appendix A2) used for survival models, BCI loss function (See Appendix A3) directly formulates the optimizing problem of survival models as maximizing the CI. The Cox depends only on the ranks of the observed survival time rather than on their actual numerical value [21]. Given the CI as a performance measure in survival analysis, BCI can effectively exploit the changes of proportional hazards with survival time. However, BCI is sensitive to uneven sampling, thus yielding poor predictive accuracy. Uneven sampling refers to the sample pairs meeting the constraints of Eq. 2 in each batch failing to equally cover REP samples ($O_i = 1$) which are the key learning objectives. BCI balances CI loss for all REP samples in each batch by adopting a global average operation, not paying more attention to those REP samples with fewer sample pairs.

By referring to the weighted cross-entropy loss function, our WCI enhances the contributions of those REP samples with fewer sample pairs in each batch by a dual average operation, as follows:

$$\mathcal{L}_{wci} = \frac{1}{N_{O=1}} \sum_{i:O_i=1} (\frac{1}{N_{T_j \geq T_i}} \sum_{j:T_j \geq T_i} e^{-(R_i - R_j)/\tau}), \tag{3}$$

where τ is a temperature hyper-parameter [10,26] and initially set as 0.1. $N_{O=1}$ (outer) is the number of REP samples, and $N_{T_j \geq T_i}$ (inner) is the number of sample pairs of ith REP sample. If the relation of predicted risk probabilities is $R_i < R_j$ (wrong prediction), the CI loss is amplified, and if $R_i > R_j$ (correct prediction), the CI loss infinitely approximates to zero. After tuning the risk difference, we use an inner average operation $1/N_{T_j \geq T_i}$ and an outer average operation $1/N_{O=1}$ to prevent fluctuation in the optimization process caused by uneven sampling per batch. The outer average operation is balanced for all REP samples, while the inner average operation assigns different weights for REP samples according to their number of sample pairs.

3   Experiments

3.1   Implementation Details

Dataset. We split the NPC patients in the NPC-REP dataset as a training set with 3,253 patients (264 REP and 2,989 non-REP), a validation set with 575 patients (43 REP and 532 non-REP), and a test set with 988 patients (61 REP and 927 non-REP). Moreover, we view left and right temporal lobes as independent samples during model training and inference.

Loss Functions. We experiment with four loss functions to demonstrate the gains of our WCI, including standard cross-entropy (CE) [27], censored cross-entropy (CCE) [25], BCI [21], and Cox [16]. We adopt the survival time of 3-years (36 months) for training and evaluation in terms of the time intervals in the NPC-REP dataset. For the CE, we define the label as 1 if REP confirmation and as 0 if the last follow-up time of non-REP is beyond 36 months. And the CCE extends CE used for classification models to train survival predictions with right-censored data by discretizing event time into intervals $(0, 36], (36, \infty)$. The BCI maximizes a lower bound on CI, explaining the high CI-scores of proportional

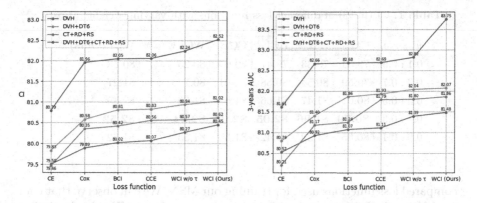

Fig. 3. Comparison of different loss functions on our MSN for the NPC-REP dataset in terms of CI and 3-years AUC (in %). DT6 covers age, gender, T/N/TNM stage, and treatment option, six features in total. (Color figure online)

hazard models observed in practice. The Cox for every sample is a function of all samples in the training data and is usually used for fitting Cox proportional hazard models [2,7].

Evaluation Metrics. CI is the most frequently used evaluation metric of survival models [1,11]. It is defined as the ratio of correctly ordered (concordant) pairs to comparable pairs. Two samples i and j are comparable if the sample with lower observed time T confirmed REP during follow-up, *i.e.*, if $T_j > T_i$ and $O_i = 1$, where $O_i = 1$ is a binary event indicator. A comparable pair $(i, j) \in \mathbb{E}$ is concordant if the estimated risk R by an assessment model is higher for NPC patients with lower REP confirmation time, *i.e.*, $R_i > R_j \wedge T_j > T_i$, otherwise the pair is discordant. We also report the time-dependent AUC [15,20] for all loss functions, apart from CI.

Optimizer. Our MSN model is trained from scratch using the SGD optimizer for all compared loss functions. Specifically, the learning rate is decay scheduled by cosine annealing [18] setting 2e-4 as an initial value and 1e-3 from 5th epochs. We set the batch size as a multiple of 8 for all tasks, *i.e.*, 128 (8 GPUs with 16 input images per GPU). The parameters of networks are optimized in 60 epochs with a weight decay of 3e-5 and a momentum of 0.8.

3.2   Results and Analysis

Multimodal Data Analysis. The intention of acquiring treatment and diagnosis data in NPC radiotherapy launches on the confirmation that the imaging dose delivered during NPC radiotherapy is positively related to REP risk. Hence, we employ our MSN to jointly leverage non-image data (DVH+DT6) and images (CT+RD+RS) to predict REP risk. We report the CI and 3-years AUC in Fig. 3

Table 1. CI (in %) of different loss functions with varying weight ratios.

$w_{nv} : w_v$	CE	Cox	BCI	CCE	WCI w/o τ	WCI (Ours)
0.1 : 0.9	79.54	79.33	79.47	79.65	80.25	80.53
0.3 : 0.7	79.59	80.54	80.58	80.61	80.77	81.27
0.5 : 0.5	80.01	81.47	81.66	81.54	81.66	81.89
0.7 : 0.3	**80.79**	**81.96**	**82.05**	**82.06**	**82.24**	**82.52**
0.9 : 0.1	80.65	81.32	81.42	81.50	81.53	81.62

for compared loss functions used for training our MSN. We can observe that combining DVH and DT6 (orange line) brings more gains than DVH (blue line) alone or CT+RD+RS (green line), and using all three data types (red line) achieves the best predictive accuracy. Such experimental results can demonstrate the contributions of multimodal data of NPC radiotherapy to the improvement of REP prediction and give supportive evidence for the above confirmation. Next, we analyze the contribution proportion of two modality data: images and non-image data. We conduct an ablation study for Eq. 1 to find the best weight ratio for two probabilities predicted by the two branches in our MSN. Table 1 shows the best CI (bold) achieved by setting the proportion of w_{nv} and w_v as 0.7:0.3. The more contributions of non-image data than images coincide with the reports of Fig. 3 in which non-image data (orange line) yields more gains than image data (green line).

Loss Function Analysis. After confirming the utility of multimodal data and analyzing the contributions of different modalities, we further discuss the benefits of our WCI according to the report in Fig. 3. We interpret CI as the probability of correct ranking of REP occurring probability by casting predicting REP risk as a survival analysis task. Hence, CE used for learning classification models can not achieve sound accuracy without considering event times. Compared to CE, Cox can significantly improve the performance by exploiting the ranks of the observed survival time. On the other hand, Cox is beat by BCI due to ignoring the changes of proportional hazards with survival time. Although BCI enjoys the gain of directly maximizing CI, its performance is slightly lower than CCE due to sensitivity to uneven sampling. Identical to BCI, CCE also considers the contributions of the proportional hazard of sample pairs in learning the survival model. Based on the comparable performance of BCI and CCE, WCI without τ can further obtain better performance by encouraging our MSN to effectively learn REP samples on the condition of uneven sampling per batch. By using τ help model convergence, our WCI achieves the highest CI (82.52) and 3-years AUC (83.75) on two modality data (DVH+DT6+CT+RD+RS). Although WCI is designed initially for the CI indicator used for evaluating survival models, the best performance on 3-years AUC can also prove the excellent generalization of our MSN trained by our WCI. We also conduct McNemar tests [17] on CI

Table 2. CI (in %) of different data types with varying τ for our WCI loss function.

Data types	$\tau = 10$	$\tau = 1$	$\tau = 0.1$	$\tau = 0.05$	$\tau = 0.02$
DVH	80.23	<u>80.27</u>	**80.45**	80.23	79.38
DVH+DT6	80.85	<u>80.94</u>	**81.02**	80.65	80.69
CT+RD+RS	80.32	<u>80.57</u>	**80.62**	80.29	79.53
DVH+DT6+CT+RD+RS	80.93	<u>82.24</u>	**82.52**	81.16	80.76

between our WCI and other loss functions, including CE, Cox, BCI, CCE, and WCI w/o τ. All comparisons in a p-value of less than 0.01 can support that our WCI brings significant improvement to CI.

Ablation Studies of τ. By casting REP assessment as a survival analysis task, our WCI exhibits manifest advantage over other loss functions by introducing the dual average operation and temperature hyper-parameter τ. Apart from the dual average operation preventing our MSN from being sensitive to uneven sampling, we also utilize τ to sharpen the risk difference of sample pairs to help model convergence. We conduct an ablation study for τ to observe the heating and cooling effect of tuning the CI loss for our WCI. Table 2 shows that the best value of τ used to train our MSN is 0.1 (bold). And when $\tau = 1.0$, our WCI without imposing the tuning on CI achieves the second-highest CI of 82.24 (underline) over two modality data (DVH+DT6+CT+RD+RS). When $\tau > 1.0$, it occurs the opposite effect. At the same time, the adjustment force should not be too large from the performance of setting $\tau = 0.05$ or $\tau = 0.02$. The ablation studies in Table 2 confirm that τ can help model convergence by sharpening the risk difference. Theoretically, the significant risk difference helps improve the predicted certainty probability, *i.e.*, reduces the entropy of probability distribution. Benefiting from the insensitive to uneven sampling and sharpening risk difference, our WCI exhibits more stability and convergence than BCI by comparing standard deviations of batch loss (See Appendix A4).

4   Conclusions

We present a novel WCI loss function to effectively exploit REP samples per batch by a dual average operation and help model convergence by sharpening risk difference with a temperature hyper-parameter. It is the first exploration of jointly leveraging multimodal data to predict the probability of REP onset to help optimize the NPC radiotherapy regimen. We acquire a private dataset and experiment on it to demonstrate the favorability of our WCI by comparing it to four popular loss functions used for survival models. We also confirm the contributions of multimodal data to REP risk prediction.

References

1. Brentnall, A.R., Cuzick, J.: Use of the concordance index for predictors of censored survival data. Stat. Methods Med. Res. **27**(8), 2359–2373 (2018)
2. Breslow, N.: Covariance analysis of censored survival data. Biometrics **30**(1), 89–99 (1974)
3. Cavanna, A.E., Trimble, M.R.: The precuneus: a review of its functional anatomy and behavioural correlates. Brain **129**(3), 564–583 (2006)
4. Chaudhary, K., Poirion, O.B., Lu, L., Garmire, L.X.: Deep learning-based multi-omics integration robustly predicts survival in liver cancer. Clin. Cancer Res. **24**(6), 1248–1259 (2018)
5. Chen, Q., et al.: Altered properties of brain white matter structural networks in patients with nasopharyngeal carcinoma after radiotherapy. Brain Imaging Behav. **14**(6), 2745–2761 (2020). https://doi.org/10.1007/s11682-019-00224-2
6. Chen, Y.P., Chan, A.T., Le, Q.T., Blanchard, P., Sun, Y., Ma, J.: Nasopharyngeal carcinoma. Lancet **394**(10192), 64–80 (2019)
7. Cox, D.R.: Partial likelihood. Biometrika **62**(2), 269–276 (1975)
8. Drzymala, R., et al.: Dose-volume histograms. Int. J. Radiat. Oncol. Biol. Phys. **21**(1), 71–78 (1991)
9. Harrell, F.E., Jr., Lee, K.L., Mark, D.B.: Multivariable prognostic models: issues in developing models, evaluating assumptions and adequacy, and measuring and reducing errors. Stat. Med. **15**(4), 361–387 (1996)
10. He, K., Fan, H., Wu, Y., Xie, S., Girshick, R.: Momentum contrast for unsupervised visual representation learning. In: Proceedings of the IEEE/CVF Conference on Computer Vision and Pattern Recognition, pp. 9729–9738 (2020)
11. Heller, G., Mo, Q.: Estimating the concordance probability in a survival analysis with a discrete number of risk groups. Lifetime Data Anal. **22**(2), 263–279 (2015). https://doi.org/10.1007/s10985-015-9330-3
12. Huang, J., Zhuo, E., Li, H., Liu, L., Cai, H., Ou, Y.: Achieving accurate segmentation of nasopharyngeal carcinoma in MR images through recurrent attention. In: Shen, D., et al. (eds.) MICCAI 2019. LNCS, vol. 11768, pp. 494–502. Springer, Cham (2019). https://doi.org/10.1007/978-3-030-32254-0_55
13. Huang, K.W., Zhao, Z.Y., Gong, Q., Zha, J., Chen, L., Yang, R.: Nasopharyngeal carcinoma segmentation via HMRF-EM with maximum entropy. In: 2015 37th Annual International Conference of the IEEE Engineering in Medicine and Biology Society (EMBC), pp. 2968–2972. IEEE (2015)
14. Huang, W., Chan, K.L., Zhou, J.: Region-based nasopharyngeal carcinoma lesion segmentation from MRI using clustering-and classification-based methods with learning. J. Digit. Imaging **26**(3), 472–482 (2013)
15. Hung, H., Chiang, C.T.: Estimation methods for time-dependent AUC models with survival data. Can. J. Stat. **38**(1), 8–26 (2010)
16. Katzman, J.L., Shaham, U., Cloninger, A., Bates, J., Jiang, T., Kluger, Y.: Deep-Surv: personalized treatment recommender system using a cox proportional hazards deep neural network. BMC Med. Res. Methodol. **18**(1), 1–12 (2018). https://doi.org/10.1186/s12874-018-0482-1
17. Lachenbruch, P.A.: McNemar Test. Statistics Reference Online, Wiley StatsRef (2014)
18. Loshchilov, I., Hutter, F.: SGDR: stochastic gradient descent with warm restarts. arXiv preprint arXiv:1608.03983 (2016)

19. Men, K., et al.: Deep deconvolutional neural network for target segmentation of nasopharyngeal cancer in planning computed tomography images. Front. Oncol. **7**, 315 (2017)

20. Nuño, M.M., Gillen, D.L.: Censoring-robust time-dependent receiver operating characteristic curve estimators. Stat. Med. **40**(30), 6885–6899 (2021)

21. Steck, H., Krishnapuram, B., Dehing-Oberije, C., Lambin, P., Raykar, V.C.: On ranking in survival analysis: bounds on the concordance index. In: Advances in Neural Information Processing Systems, pp. 1209–1216 (2008)

22. Tang, Y., Zhang, Y., Guo, L., Peng, Y., Luo, Q., Xing, Y.: Relationship between individual radiosensitivity and radiation encephalopathy of nasopharyngeal carcinoma after radiotherapy. Strahlenther. Onkol. **184**(10), 510–514 (2008)

23. Wang, X., Girshick, R., Gupta, A., He, K.: Non-local neural networks. In: Proceedings of the IEEE Conference on Computer Vision and Pattern Recognition, pp. 7794–7803 (2018)

24. Wei, W.I., Sham, J.S.: Nasopharyngeal carcinoma. Lancet **365**(9476), 2041–2054 (2005)

25. Wulczyn, E., et al.: Deep learning-based survival prediction for multiple cancer types using histopathology images. PLoS ONE **15**(6), e0233678 (2020)

26. Yi, X., et al.: Sampling-bias-corrected neural modeling for large corpus item recommendations. In: Proceedings of the 13th ACM Conference on Recommender Systems, pp. 269–277 (2019)

27. Zadeh, S.G., Schmid, M.: Bias in cross-entropy-based training of deep survival networks. IEEE Trans. Pattern Anal. Mach. Intell. **43**(9), 3126–3137 (2020)

28. Zhang, Y-M., et al.: Surface-based Falff: a potential novel biomarker for prediction of radiation encephalopathy in patients with nasopharyngeal carcinoma. Front. Neurosci. **15**, 692575 (2021)

29. Zhao, L.M., et al.: Functional connectivity density for radiation encephalopathy prediction in nasopharyngeal carcinoma. Front. Oncol. **11**, 687127 (2021)

30. Zhou, J., Chan, K.L., Xu, P., Chong, V.F.: Nasopharyngeal carcinoma lesion segmentation from MR images by support vector machine. In: 3rd IEEE International Symposium on Biomedical Imaging: Nano to Macro, pp. 1364–1367. IEEE (2006)

Reducing Positional Variance in Cross-sectional Abdominal CT Slices with Deep Conditional Generative Models

Xin Yu[1]([✉]), Qi Yang[1], Yucheng Tang[2], Riqiang Gao[1], Shunxing Bao[1], Leon Y. Cai[3], Ho Hin Lee[1], Yuankai Huo[1], Ann Zenobia Moore[4], Luigi Ferrucci[4], and Bennett A. Landman[1,2,3]

[1] Computer Science, Vanderbilt University, Nashville, TN, USA
xin.yu@vanderbilt.edu
[2] Electrical and Computer Engineering, Vanderbilt University, Nashville, TN, USA
[3] Biomedical Engineering, Vanderbilt University, Nashville, TN, USA
[4] National Institute on Aging, Baltimore, MD, USA

Abstract. 2D low-dose single-slice abdominal computed tomography (CT) slice enables direct measurements of body composition, which are critical to quantitatively characterizing health relationships on aging. However, longitudinal analysis of body composition changes using 2D abdominal slices is challenging due to positional variance between longitudinal slices acquired in different years. To reduce the positional variance, we extend the conditional generative models to our C-SliceGen that takes an arbitrary axial slice in the abdominal region as the condition and generates a defined vertebral level slice by estimating the structural changes in the latent space. Experiments on 1170 subjects from an in-house dataset and 50 subjects from BTCV MICCAI Challenge 2015 show that our model can generate high quality images in terms of realism and similarity. External experiments on 20 subjects from the Baltimore Longitudinal Study of Aging (BLSA) dataset that contains longitudinal single abdominal slices validate that our method can harmonize the slice positional variance in terms of muscle and visceral fat area. Our approach provides a promising direction of mapping slices from different vertebral levels to a target slice to reduce positional variance for single slice longitudinal analysis. The source code is available at: https://github.com/MASILab/C-SliceGen.

Keywords: Abdominal slice generation · Body composition · Longitudinal data harmonization

1 Introduction

Body composition describes the percentage of fat, muscle and bone in the human body [14] and can be used to characterize different aspects of health and disease,

---

X. Yu and Q. Yang—Equal contribution.

L. Wang (Eds.): MICCAI 2022, LNCS 13437, pp. 202–212, 2022.
https://doi.org/10.1007/978-3-031-16449-1_20

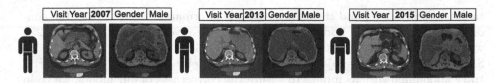

Fig. 1. Longitudinal slices acquired at different axial cross-sectional positions. The figure shows a single abdominal slice being acquired for the same subject in different years. The orange line represent the given CT axial position. The red and green masks represent the visceral fat and muscle, respectively. The area of two masks varies largely due to the slice positional variance. (Color figure online)

including sarcopenia [19], heart disease [6] and diabetes [22]. One widely used method to assess body composition is computed tomography body composition (CTBC) [2]. 2D low-dose axial abdominal single-slice computed tomography (CT) is preferred over 3D CT to reduce unnecessary radiation exposure [14]. However, difficulty in positioning cross-sectional locations leads to challenges acquiring the 2D slices at the same axial position (vertebral level). For instance, in the clinical setting, patients who visit hospitals in the different years can have varied vertebral level abdominal slices being scanned (Fig. 1). This causes significant variation in the organs and tissues captured. Since the specific organs and tissues scanned are highly associated with measures of body composition, increased positional variance in 2D abdominal slices makes body composition analyses difficult. To the best of our knowledge, no method has been proposed to tackle the 2D slice positional variance problem.

We aim to reduce the positional variance by synthesizing slices to a target vertebral level. In other contexts, image registration would be used to correct pose/positioning. However, such an approach cannot address out of plane motion with 2D acquisitions. Recently, deep learning based generative models have shown superior results in generating high-quality and realistic images. The generative model can solve the registration limitation by learning the joint distribution between the given and target slices. Variational autoencoders (VAEs), a class of generative models, encode inputs to an interpretable latent distribution capable of generating new data [13]. Generative adversarial networks (GANs) contain two sub-models: a generator model that aims to generate new data and a discriminator to distinguish between real and generated images. VAEGAN [16] incorporates GAN into a VAE framework to create better synthesized images. In the original VAEs and GANs, the generated images cannot be manipulated. Conditional GAN (cGAN) [18] and conditional VAE (cVAE) [21] tackle this by giving a condition to generate specific data. However, the majority of these conditional methods require the target class, semantic map, or heatmap [5] in the testing phase, which is not applicable in our scenario where no direct target information is available.

We posit that by synthesizing an image at a pre-defined vertebral level with an arbitrary abdominal slice, generated slices will consistently localize to the target vertebral level and the subject-specific information derived from the conditional image such as body habitus will be preserved. Inspired by [5,11,25], we propose the Conditional SliceGen (C-SliceGen) based on VAEGAN. C-SliceGen can generate subject-specific target vertebral level slice given an arbitrary abdominal slice as input. To ensure the correctness of our model, we train and validate our method first on an inhouse 3D volumertric CT dataset and the BTCV MICCAI Challenge 2015 3D CT dataset [15] where target slices are acquired and used as ground truth to compare with generated slices. SSIM, PSNR, and LPIPS are used as evaluation metrics. Our experiments show that our model can capture positional variance in the generated realistic image. Moreover, we apply our method to the Baltimore Longitudinal Study of Aging (BLSA) single-slice dataset [7]. By computing body composition metrics on synthesized slices, we are able to harmonize the longitudinal muscle and visceral fat area fluctuations brought by the slices positional variation.

Our contributions are three-fold: (1) we propose C-SliceGen to successfully capture positional variance in the same subject; (2) the designed generative approach implicitly embeds the target slice without requiring it during the testing phase; and (3) we demonstrate that the proposed method can consistently harmonize the body composition metrics for longitudinal analysis.

2   Method

2.1   Technical Background

VAE. VAEs can be written in probabilistic form as $P(x) = P(z)P(x|z)$, where x denotes the input images and z denotes the latent variables. The models aim to maximize the likelihood $p(x) = \int p(z)p_\theta(x|z)dz$, where $z \sim N(0,1)$ is the prior distribution, $p_\theta(x|z)dz$ is the posterior distribution and θ is the decoder parameters. However, it is intractable to find decoder parameters θ to maximize the log likelihood. Instead, VAEs optimize encoder parameters ϕ, by computing $q_\phi(z|x)$ to estimate $p_\theta(x|z)$ with the assumption that $q_\phi(z|x)$ is a Gaussian distribution whose μ and σ are the outputs of the encoder. VAEs train the encoder and decoder jointly to optimize the Evidence Lower Bound (ELBO),

$$L_{VAE}(\theta, \phi, x, y) = E[\log p_\theta(x|z)] - D_{KL}[q_\phi(z|x)||p_\theta(z)], \tag{1}$$

where $E[\log p_\theta(x|z)]$ represent the reconstruction loss and the KL-divergence encourages the posterior estimate to approximate the prior $p(z)$ distribution. Samples can be generated from the normal distribution $z \sim N(0,1)$ and fed into the decoder to generate new data during testing time. cVAEs add flexibility to the VAEs and can also be trained by optimizing the ELBO.

WGAN-GP. Wasserstein GAN (wGAN) with gradient penalty [10] is an extension of GAN that improves stability when training the model whose loss function can be written as:

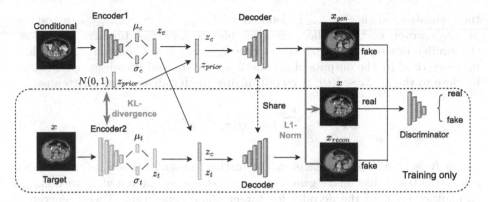

Fig. 2. Overall pipeline. Conditional images, arbitrary slices in the abdominal region, are the input images for both training and test phase. Target images (x) are the ground truth for the reconstruction and generation process that only exist in the training phase. z_c, z_t and z_{prior} are the latent variables derived from conditional images, target images, and the normal Gaussian distribution, respectively. x_{gen} and x_{recon} serve as fake images and target images serve as real images for the discriminator.

$$L_{GAN} = \mathbb{E}_{\tilde{x} \sim \mathbb{P}_g}[D(\tilde{x})] + \mathbb{E}_{x \sim \mathbb{P}_r}[D(x)] + \lambda \mathbb{E}_{\hat{x} \sim \mathbb{P}_{\hat{x}}}[(\| \bigtriangledown_{\hat{x}} D(\hat{x})\|_2 - 1)^2], \quad (2)$$

where $\mathbb{P}_g$ is the model distribution, $\mathbb{P}_r$ is the data distribution and $\mathbb{P}_{\hat{x}}$ is the random sample distribution.

Although the current VAE/GAN based methods have been successful in generating samples, no existing method found can be directly applied to our task.

2.2 C-SliceGen

In our scenario, acquired slice can be in any vertebral level within the abdominal region. The goal is to use these arbitrary slices to synthesize a new slice at a pre-defined target vertebral level. The overall method is shown in Fig. 2. There are two encoders, one decoder and one discriminator. The arbitrary slice is the conditional image for the model, which provides subject-specific information such as organs shape and tissue localization. We assume this information remains interpretable after encoding to latent variables z_c by encoder1.

Training. During training phase, we have the sole target slice for each subject. We select the most similar target slice in terms of organ/tissue structure and appearance for all subjects. The target slice selection method will be covered in the following section. We assume all target slices x to have similar organ/tissue structure and appearance which are preserved in the latent variables z_t, whose distribution can be written as $q_\phi(z_t|x)$. By concatenating the latent variables z_c and z_t, we combine the target slice organ/tissue structure and appearance with given subject-specific information. This combined latent variable encourages the decoder to reconstruct the target slice for the given subject. We regularize this reconstruction step by computing the L1-Norm between the target slice (x) and the reconstructed slice (x_{recon}), denoted as $L_{recon} = \|x - x_{recon}\|$. In testing phase, however, no target slices are available. To solve this problem, we follow the similar practice in VAEs: We assume $q_\phi(z_t|x)$ is a Gaussian distribution parameterized by the outputs of encoder2 μ_t and σ_t, and encourage $q_\phi(z_t|x)$ to be close to the $z_{prior} \sim N(0,1)$ by optimizing the KL-divergence, written as:

$$L_{KL} = \frac{1}{2} \sum_{k=1}^{K} (1 + \log(\sigma_k^2) - \mu_k^2 - \sigma_k^2), \tag{3}$$

where K is latent space dimension. To further constrain this KL term and to mimic the test phase image generation process, we concatenated z_c with z_{prior} as another input of the decoder for target slices generation. These generated slices are denoted as x_{gen}. We compute the L1-Norm between the generated image (x_{gen}) and target image (x), denoted as $L_{gen} = \|x - x_{gen}\|$. The total loss function of the aforementioned steps is written as:

$$L_{cVAE} = L_{recon} + L_{gen} + L_{KL}, \tag{4}$$

However, maximize likelihood function is inherently a difficult problem which can cause blurry generated images. GANs on the other hand increase the image quality in an adversarial manner. Following [16], we combine GAN with our VAE model. The generated image and reconstructed image both serve as fake images, and the target images serve as the real images for the discriminator to perform classification. The decoder serves as the generator. By sharing the same parameters for the generated and reconstructed images, the GAN loss adds another constraint to force them to be similar. The total loss function of our proposed C-SliceGen can be written as:

$$L = L_{cVAE} + \beta L_{GAN}, \tag{5}$$

where β is a weighting factor that determines the adversarial regularization.

Testing. During testing, given a conditional image as input, the encoded latent variable z_c is concatenated with z_{prior} sampled from a normal Gaussian distribution and is fed into the decoder to generate the target slice.

2.3   Target Slice Selection

It is not straightforward to select similar target slices for each subject since body composition and organ structure are subject-sensitive. We adopt two approaches: (1) Select the slices that have the most similar body part regression (BPR) [23] score as the target slices across subject since BPR is efficacious in locating slices. (2) Select a slice from a reference subject as the reference target slice. Registering axial slice of each subject's volume to the reference target slice and identifying the slice with the largest mutual information [4] as the subject target slice.

3   Experiments and Results

3.1   Dataset

The models are trained and validated on a large dataset containing 1170 3D Portal Venous CT volumes from 1170 de-identified subjects under Institutional Review Board (IRB) protocols. Each CT scan is quality checked for normal abdominal anatomy. The evaluations are performed on the MICCAI 2015 Multi-Atlas Abdomen Labeling Challenge dataset which contains 30 and 20 abdominal Portal Venous CT volumes for training and testing, respectively. We further evaluate our method's efficacy on reducing positional variance for longitudinal body composition analysis on 20 subjects from the BLSA non-contrast single slice CT dataset. Each subject has either 2 or 3 visits for the past 15 years.

3.2   Implementation Details and Results

Metrics. We quantitatively evaluate our generative models C-SliceGen with different β (Eq. 5) and the two different target slice selection approaches using three image quality assessment metrics: Structural Similarity Index (SSIM) [26], Peak Signal-to-Noise Ratio (PSNR) [12], and Learned Perceptual Image Patch Similarity (LPIPS) [27].

Training & Testing. All 3D volumes undergo BPR to ensure a similar Field of View (FOV). The 2D axial CT scans have image sizes of 512×512 and are resized to 256×256 before feeding into the models. The data are processed with soft-tissue CT window range $[-125, 275]$ HU and rescaled to $[0.0, 1.0]$ to facilitate training. The proposed methods are implemented using Pytorch. We use Adam optimizer with a learning rate of $1e - 4$ and weight decay of $1e - 4$ to optimize the total loss of the network. We adapt the encoder, decoder and discriminator structures in [9] to fit our input size. Shift, rotation and flip are used for the online data augmentation. 2614 slices from in-house 83 subjects are used for testing. The results are shown in Table 1.

Table 1. Quantitative results on the in-house test set and BTCV test set. Registration: target slice selected using registration, BPR: target slice selected using BPR score alone. β represent the β in Eq. 5.

Method	SSIM ↑	PSNR ↑	LPIPS ↓
the in-house dataset			
$\beta = 0$, Registration	0.636	17.634	0.361
$\beta = 0$, BPR	0.618	16.470	0.381
$\beta = 0.01$, Registration	0.615	17.256	0.209
$\beta = 0.01$, BPR	0.600	16.117	0.226
the BTCV dataset			
$\beta = 0$, Registration	0.603	17.367	0.362
$\beta = 0$, BPR	0.605	17.546	0.376
$\beta = 0.01$, Registration	0.583	16.778	0.208
$\beta = 0.01$, BPR	0.588	16.932	0.211

BTCV Evaluation. Before evaluating on the BTCV test set, the models are fine-tuned on the BTCV training set to minimize the dataset domain gap with the same training settings except that the learning rate is reduced to $1e-5$. The split of train/validation/test is 22/8/20. The quantitative results and qualitative results are shown in Table 1 and Fig. 3, respectively.

BLSA Evaluation. In the BLSA 2D abdominal dataset, each subject only take one axial abdominal CT scan each year. Therefore, there is no ground truth (GT) for target slice generation. Instead of directly comparing the generated images with the GT, we evaluate the model performance on reducing variance in body compositional areas brought by the cross-sectional variance of CT scans in longitudinal data, specifically muscle and visceral fat. We feed the BLSA data into our C-SliceGen model and resize the generated images to the original size of 512×512. Both the real and generated images are fed into a pre-trained UNet [20] for muscle segmentation and a pre-trained Deeplab-v3 [8] to identify visceral fat by inner/outer abdominal wall segmentation. For the inner/outer wall segmentation, we find that the model usually fails to exclude the retroperitoneum in the inner abdominal wall for both real and fake images. Retroperitoneum is

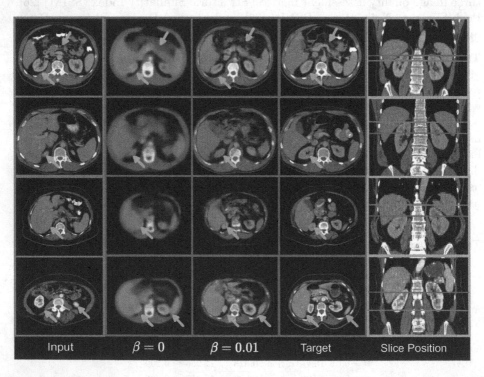

Fig. 3. Qualitative results on the BTCV test set. The image with blue bounding box represent the input slice while the image with red bounding box represents the model results and target slice. In the rightmost column, axial location of the input and target slices are marked with blue line and red line, respectively. (Color figure online)

an anatomical area located behind the abdominal cavity including the left/right kidneys and aorta, often with poorly visualized boundaries. We conduct human assessment on all the results from both the real and generated images to ensure the retroperitoneum is correctly segmented. The adipose tissue of each image is segmented using fuzzy c-means [3,24] in an unsupervised manner. We segment the visceral fat by masking the adipose tissue with the inner abdominal wall. Figure 4 shows the spaghetti plots of the muscle and visceral fat area changes among 2 or 3 visits from 20 subjects before and after harmonization.

4   Discussion and Conclusion

According to the qualitative results shown in Fig. 3, our generated images with $\beta = 0.01$ are realistic and similar to the target slices. Specifically, the results show that our model can generate target slices regardless of whether the conditional slice is at an upper, lower, or similar vertebral level. Comparing the results between $\beta = 0$ and $\beta = 0.01$, the images indicate that the adversarial regularization helps improve image quality significantly. This is consistent with the LPIPS results in Table 1. However, the human qualitative assessment and LPIPS differ from the SSIM and PSNR as shown in Table 1 where SSIM and PSNR have higher scores with $\beta = 0$ on both dataset. This supports that SSIM

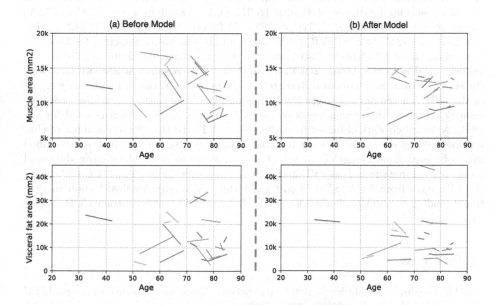

Fig. 4. Spaghetti plot of muscle and visceral fat area longitudinal analysis with 20 subjects from the BLSA dataset. (a) muscle and visceral fat area derived from the original abdominal slices. (b) the corresponding metrics derived from synthesized slices with our model. Each line corresponds to the measurements across visits for one subject. After harmonization, variance in these measures for the population is decreased.

and PSNR score may not fully represent a human perceptional assessment [1,17]. As for the longitudinal data harmonization, according to Fig. 4, before applying our model, both muscle and visceral fat area have large fluctuations. These fluctuations have been reduced after mapping the slices to a similar vertebral level with our model C-SliceGen.

As the first work to use one abdominal slice to generate another slice, our approach currently has several limitations. (1) In most cases, the model is able to identify the position of each organ, but shape and boundary information are not well preserved. (2) It is hard to synthesize heterogeneous soft tissues such as the colon and stomach. (3) There is domain shift when the model trained on Portal Venous phase CT is applied to CT acquired in other phases such as the non-contrast BLSA data.

In this paper, we introduce our C-SliceGen model that conditions on an arbitrary 2D axial abdominal CT slice and generates a subject-specific slice at a target vertebral level. Our model is able to capture organ changes between different vertebral levels and generate realistic and structurally similar images. We further validate our model's performance on harmonizing the body composition measurements fluctuations introduced by positional variance on an external dataset. Our method provides a promising direction for handling imperfect single slice CT abdominal data for longitudinal analysis.

Acknowledgements. This research is supported by NSF CAREER 1452485, 2040462 and the National Institutes of Health (NIH) under award numbers R01EB017230, R01EB006136, R01NS09529, T32EB001628, 5UL1TR002243-04, 1R01MH121620-01, and T32GM007347; by ViSE/VICTR VR3029; and by the National Center for Research Resources, Grant UL1RR024975-01, and is now at the National Center for Advancing Translational Sciences, Grant 2UL1TR000445-06. This research was conducted with the support from the Intramural Research Program of the National Institute on Aging of the NIH. The content is solely the responsibility of the authors and does not necessarily represent the official views of the NIH. The identified datasets used for the analysis described were obtained from the Research Derivative (RD), database of clinical and related data. The inhouse imaging dataset(s) used for the analysis described were obtained from ImageVU, a research repository of medical imaging data and image-related metadata. ImageVU and RD are supported by the VICTR CTSA award (ULTR000445 from NCATS/NIH) and Vanderbilt University Medical Center institutional funding. ImageVU pilot work was also funded by PCORI (contract CDRN-1306-04869).

References

1. Almalioglu, Y., et al.: Endol2h: Deep super-resolution for capsule endoscopy. IEEE Trans. Med. Imaging **39**(12), 4297–4309 (2020)
2. Andreoli, A., Garaci, F., Cafarelli, F.P., Guglielmi, G.: Body composition in clinical practice. Eur. J. Radiol. **85**(8), 1461–1468 (2016)
3. Bezdek, J.C., Ehrlich, R., Full, W.: Fcm: the fuzzy c-means clustering algorithm. Comput. Ggeosci. **10**(2–3), 191–203 (1984)
4. Cover, T.M.: Elements of information theory. John Wiley & Sons (1999)

5. De Bem, R., Ghosh, A., Boukhayma, A., Ajanthan, T., Siddharth, N., Torr, P.: A conditional deep generative model of people in natural images. In: 2019 IEEE Winter Conference on Applications of Computer Vision (WACV), pp. 1449–1458. IEEE (2019)

6. De Schutter, A., Lavie, C.J., Gonzalez, J., Milani, R.V.: Body composition in coronary heart disease: how does body mass index correlate with body fatness? Ochsner J. **11**(3), 220–225 (2011)

7. Ferrucci, L.: The baltimore longitudinal study of aging (blsa): a 50-year-long journey and plans for the future (2008)

8. Florian, L.C., Adam, S.H.: Rethinking atrous convolution for semantic image segmentation. In: Conference on Computer Vision and Pattern Recognition (CVPR). IEEE/CVF (2017)

9. Gao, R., et al.: Lung cancer risk estimation with incomplete data: a joint missing imputation perspective. In: de Bruijne, M., et al. (eds.) MICCAI 2021. LNCS, vol. 12905, pp. 647–656. Springer, Cham (2021). https://doi.org/10.1007/978-3-030-87240-3_62

10. Gulrajani, I., Ahmed, F., Arjovsky, M., Dumoulin, V., Courville, A.C.: Improved training of wasserstein gans. Advances in neural information processing systems 30 (2017)

11. Henderson, P., Lampert, C.H., Bickel, B.: Unsupervised video prediction from a single frame by estimating 3d dynamic scene structure. arXiv preprint arXiv:2106.09051 (2021)

12. Hore, A., Ziou, D.: Image quality metrics: Psnr vs. ssim. In: 2010 20th International Conference on Pattern Recognition, pp. 2366–2369. IEEE (2010)

13. Kingma, D.P., Welling, M.: Auto-encoding variational bayes. arXiv preprint arXiv:1312.6114 (2013)

14. Kuriyan, R.: Body composition techniques. Indian J. Med. Res. **148**(5), 648 (2018)

15. Landman, B., Xu, Z., Igelsias, J., Styner, M., Langerak, T., Klein, A.: Miccai multi-atlas labeling beyond the cranial vault-workshop and challenge. In: Proceedings of MICCAI Multi-Atlas Labeling Beyond Cranial Vault-Workshop Challenge, vol. 5, p. 12 (2015)

16. Larsen, A.B.L., Sønderby, S.K., Larochelle, H., Winther, O.: Autoencoding beyond pixels using a learned similarity metric. In: International Conference on Machine Learning, pp. 1558–1566. PMLR (2016)

17. Ledig, C., et al.: Photo-realistic single image super-resolution using a generative adversarial network. In: Proceedings of the IEEE Conference on Computer Vision and Pattern Recognition, pp. 4681–4690 (2017)

18. Mirza, M., Osindero, S.: Conditional generative adversarial nets. arXiv preprint arXiv:1411.1784 (2014)

19. Ribeiro, S.M., Kehayias, J.J.: Sarcopenia and the analysis of body composition. Adv. Nutr. **5**(3), 260–267 (2014)

20. Ronneberger, O., Fischer, P., Brox, T.: U-Net: convolutional networks for biomedical image segmentation. In: Navab, N., Hornegger, J., Wells, W.M., Frangi, A.F. (eds.) MICCAI 2015. LNCS, vol. 9351, pp. 234–241. Springer, Cham (2015). https://doi.org/10.1007/978-3-319-24574-4_28

21. Sohn, K., Lee, H., Yan, X.: Learning structured output representation using deep conditional generative models. Advances in neural information processing systems 28 (2015)

22. Solanki, J.D., Makwana, A.H., Mehta, H.B., Gokhale, P.A., Shah, C.J.: Body composition in type 2 diabetes: change in quality and not just quantity that matters. Int. J. Preventive Med. **6** (2015)

23. Tang, Y., et al.: Body part regression with self-supervision. IEEE Trans. Med. Imaging **40**(5), 1499–1507 (2021)
24. Tang, Y., et al.: Prediction of type II diabetes onset with computed tomography and electronic medical records. In: Syeda-Mahmood, T., Drechsler, K., Greenspan, H., Madabhushi, A., Karargyris, A., Linguraru, M.G., Oyarzun Laura, C., Shekhar, R., Wesarg, S., González Ballester, M.Á., Erdt, M. (eds.) CLIP/ML-CDS -2020. LNCS, vol. 12445, pp. 13–23. Springer, Cham (2020). https://doi.org/10.1007/978-3-030-60946-7_2
25. Tang, Y., et al.: Pancreas CT segmentation by predictive phenotyping. In: de Bruijne, M., Cattin, P.C., Cotin, S., Padoy, N., Speidel, S., Zheng, Y., Essert, C. (eds.) MICCAI 2021. LNCS, vol. 12901, pp. 25–35. Springer, Cham (2021). https://doi.org/10.1007/978-3-030-87193-2_3
26. Wang, Z., Bovik, A.C., Sheikh, H.R., Simoncelli, E.P.: Image quality assessment: from error visibility to structural similarity. IEEE Trans. Image Process. **13**(4), 600–612 (2004)
27. Zhang, R., Isola, P., Efros, A.A., Shechtman, E., Wang, O.: The unreasonable effectiveness of deep features as a perceptual metric. In: CVPR (2018)

Censor-Aware Semi-supervised Learning for Survival Time Prediction from Medical Images

Renato Hermoza[1]([✉]), Gabriel Maicas[1], Jacinto C. Nascimento[2], and Gustavo Carneiro[1]

[1] Australian Institute for Machine Learning, The University of Adelaide, Adelaide, Australia
renato.hermozaaragones@adelaide.edu.au
[2] Institute for Systems and Robotics, Instituto Superior Tecnico, Lisbon, Portugal

Abstract. Survival time prediction from medical images is important for treatment planning, where accurate estimations can improve healthcare quality. One issue affecting the training of survival models is censored data. Most of the current survival prediction approaches are based on Cox models that can deal with censored data, but their application scope is limited because they output a hazard function instead of a survival time. On the other hand, methods that predict survival time usually ignore censored data, resulting in an under-utilization of the training set. In this work, we propose a new training method that predicts survival time using all censored and uncensored data. We propose to treat censored data as samples with a lower-bound time to death and estimate pseudo labels to semi-supervise a censor-aware survival time regressor. We evaluate our method on pathology and x-ray images from the TCGA-GM and NLST datasets. Our results establish the state-of-the-art survival prediction accuracy on both datasets.

Keywords: Censored data · Noisy labels · Pathological images · Chest x-rays · Semi-supervised learning · Survival time prediction

1 Introduction

Survival time prediction models estimate the time elapsed from the start of a study until an event (e.g., death) occurs with a patient. These models are important because they may influence treatment decisions that affect the health outcomes of patients [3]. Thus, automated models that produce accurate survival time estimations can be beneficial to improve the quality of healthcare.

This work was supported by the Australian Research Council through grants DP180103232 and FT190100525.

Supplementary Information The online version contains supplementary material available at https://doi.org/10.1007/978-3-031-16449-1_21.

L. Wang (Eds.): MICCAI 2022, LNCS 13437, pp. 213–222, 2022.
https://doi.org/10.1007/978-3-031-16449-1_21

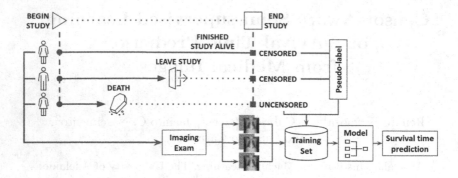

Fig. 1. At the beginning of the study, all patients are scanned for a pathology or chest x-ray image, and then patients are monitored until the end of study. Patients who die during the study represent uncensored data, while patients who do not die or leave before the end of study denote right-censored data. We train our model with uncensored and pseudo-labeled censored data to semi-supervise a censor-aware regressor to predict survival time.

Survival prediction analysis requires the handling of right-censored samples, which represent the cases where the event of interest has not occurred either because the study finished before the event happened, or the patient left the study before its end (Fig. 1). Current techniques to deal with censored data are typically based on Cox proportional hazards models [4] that rank patients in terms of their risks instead of predicting survival time, limiting their usefulness for general clinical applications [2]. The problem is that the survival time prediction from a Cox hazard model requires the estimation of a baseline hazard function, which is challenging to obtain in practice [21]. A similar issue is observed in [17] that treats survival analysis as a ranking problem without directly estimating survival times.

Some methods [1,7,19] directly predict survival time, but they ignore the censored data for training. Other methods [8,20,21] use censored data during training, but in a sub-optimally manner. More specifically, these models use the censoring time as a lower bound for the event of interest. Even though this is better than disregarding the censored cases, such approaches do not consider the potential differences between the censoring time and the hidden survival time. We argue that if the values of these differences can be estimated with pseudo-labeling mechanisms, survival prediction models could be more accurate. However, pseudo labels estimated for the censored data may introduce noisy labels for training, which have been studied for classification [11,12] and segmentation [22], but never for survival prediction in a semi-supervision context.

In this paper, we propose a new training method for deep learning models to predict survival time from medical images using all censored and uncensored training data. The main contributions of this paper are:

– a method that estimates a pseudo label for the survival time of censored data (lower bounded by the annotated censoring time) that is used to semi-supervise a censor-aware survival time regressor; and

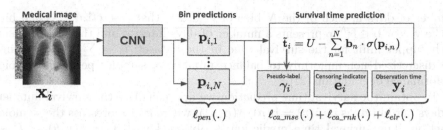

Fig. 2. The proposed model outputs a set of N bin predictions $\{\mathbf{p}_{i,n}\}_{n=1}^{N}$ (each bin representing an amount of time $\mathbf{b}_n$) that are aggregated to produce a survival time prediction for the ith case. This prediction is achieved by taking the maximum survival time U and subtracting it by the activation of each bin $\sigma(p_{i,n})$ times the amount of time in $\mathbf{b}_n$. A set of loss functions are used while training: a censored-aware version of MSE (ℓ_{ca_mse}), a penalization term for bin consistency (ℓ_{pen}), a rank loss (ℓ_{ca_rnk}), and an adapted version of the ELR regularization [12] to survival prediction (ℓ_{elr}).

– two new regularization losses to handle pseudo label noise: a) we adapt the early-learning regularization (ELR) [12] loss from classification to survival prediction; and b) inspired by risk prediction models, we use a censor-aware ranking loss to produce a correct sorting of samples in terms of survival time.

We evaluate our method using the TCGA-GM [13] dataset of pathological images, and the NLST dataset [14,15] of chest x-ray images. Our results show a clear benefit of using pseudo labels to train survival prediction models, achieving the new state-of-the-art (SOTA) survival time prediction results for both datasets. We make our code available at https://github.com/renato145/CASurv.

2   Method

To explain our method, we define the data set as $\mathcal{D} = \{(\mathbf{x}_i, \mathbf{y}_i, \mathbf{e}_i)\}_{i=1}^{|\mathcal{D}|}$, where $\mathbf{x}_i \in \mathcal{X}$ denotes a medical image with $\mathcal{X} \subset \mathbb{R}^{H \times W \times C}$ (H, W, C denote image height, width and number of color channels), $\mathbf{y}_i \in \mathbb{N}$ indicates the observation time defined in days, and $\mathbf{e}_i \in \{0, 1\}$ is the censoring indicator. When $\mathbf{e}_i = 0$ (uncensored observation), $\mathbf{y}_i$ corresponds to a survival time $\mathbf{t}_i$ which indicates the event of death for the individual. In cases where $\mathbf{e}_i = 1$ (censored observation) $\mathbf{t}_i$ is unknown, but is lower bounded by $\mathbf{y}_i$ (i.e., $\mathbf{t}_i > \mathbf{y}_i$).

2.1   Model

The architecture of our model extends the implementation from [7] to work with censored data (Fig. 2). Our survival time prediction model uses a Convolutional Neural Network (CNN) [10] represented by

$$f : \mathcal{X} \times \theta \to \mathcal{P}, \tag{1}$$

which is parameterized by $\theta \in \Theta$ and takes an image $\mathbf{x}$ to output a survival time confidence vector $\mathbf{p} \in \mathcal{P}$ that represents the confidence on regressing to the number of days in each of the N bins of $\mathcal{P} \subset \mathbb{R}^N$ that discretize $\mathbf{y}_i$. Each bin of the vector $\mathbf{b} \in \mathbb{N}^N$ represents a number of days interval and the sum of them is the upper limit for the survival time in the dataset: $U = \sum_{n=1}^{N} \mathbf{b}(n)$. Bins are discretized non-uniformly to balance number of samples per bin to avoid performance degradation [21].

To obtain the survival time prediction we first calculate the survival number of days per bin n, as in $\tilde{\mathbf{p}}(n) = \mathbf{b}(n) \cdot \sigma(\mathbf{p}(n))$, where $\sigma(.)$ represents the sigmoid function. The survival time prediction is obtained with $\tilde{t} = h(\mathbf{x}; \theta) = U - \sum_{n=1}^{N} \tilde{\mathbf{p}}(n)$, where $h(\mathbf{x}; \theta)$ represents the full survival time regression model. Hence, a larger activation $\mathbf{p}(n)$ indicates higher risk, as the predicted $\tilde{t}$ will be smaller. The training of our model minimizes the loss function:

$$\ell(\mathcal{D}, \theta) = \frac{1}{|\mathcal{D}|} \sum_{(\mathbf{x}_i, \mathbf{y}_i, \mathbf{e}_i) \in \mathcal{D}} [\ell_{ca\_mse}(\tilde{\mathbf{t}}_i, \mathbf{y}_i, \mathbf{e}_i, \mathbf{s}_i) + \alpha \cdot \ell_{elr}(\mathbf{p}_i) + \beta \cdot \ell_{pen}(\mathbf{p}_i)$$
$$+ \frac{1}{|\mathcal{D}|} \sum_{j=1}^{|\mathcal{D}|} \delta \cdot \ell_{ca\_rnk}(\tilde{\mathbf{t}}_i, \tilde{\mathbf{t}}_j, \mathbf{y}_i, \mathbf{y}_j, \mathbf{e}_i, \mathbf{e}_j)] \tag{2}$$

where $\ell_{ca\_mse}(.)$ is a censor-aware mean squared error defined in (4), $\mathbf{s}_i$ indicates if sample i is pseudo-labeled, $\ell_{elr}(.)$ is the ELR regularization [12] adapted for survival time prediction defined in (5), $\ell_{pen}(.)$ is the penalization term defined in (6), $\ell_{ca\_rnk}(.)$ is a censor-aware rank loss defined in (7), and α, β and δ control the strength of $\ell_{elr}(.)$, $\ell_{pen}(.)$ and $\ell_{ca\_rnk}(.)$ losses.

Pseudo Labels. We proposed the use of pseudo labels to semi-supervise our censor-aware survival time regressor. For a censored observation i, a pseudo label γ_i is estimated as:

$$\gamma_i = \max(\mathbf{y}_i, \tilde{\mathbf{t}}_i), \tag{3}$$

which takes advantage of the nature of censored data, as being lower bounded by $\mathbf{y}_i$. We estimate pseudo labels at the beginning of each epoch and treat them as uncensored observations by re-labeling $\mathbf{y}_i = \gamma_i$ and setting $\mathbf{s}_i = 1$.

The quality of generated pseudo labels depends on the training procedure stage, where pseudo labels produced during the first epochs are less accurate than the ones at the last epochs. Hence, we control the ratio of censored sample labels to be replaced with pseudo labels at each epoch using a cosine annealing schedule to have no pseudo labels at the beginning of the training and all censored data with their pseudo labels by the end of the training.

Censor-Aware Mean Squared Error (CA-MSE). Using a regular MSE loss is not suitable for survival prediction because the survival time for censored observations is unknown, but lower-bounded by $\mathbf{y}_i$. Therefore, we assume an error exists for censored samples when $\tilde{\mathbf{t}}_i < \mathbf{y}_i$. However, when $\tilde{\mathbf{t}}_i > \mathbf{y}_i$, it will be incorrect to assume an error, as the real survival time is unknown. To mitigate this issue, we introduce the CA-MSE loss:

$$\ell_{ca\_mse}(\tilde{t}_i, y_i, e_i, s_i) = \begin{cases} 0, & \text{if } (e_i = 1) \wedge (\tilde{t}_i > y_i) \\ \tau^{s_i}(\tilde{t}_i - y_i)^2, & \text{otherwise} \end{cases}, \quad (4)$$

where $\tau \in (0, 1)$ reduces the weight for pseudo-labeled data ($s_i = 1$).

Early-Learning Regularization (ELR). To deal with noisy pseudo labels, we modify ELR [12] to work in our survival time prediction setting, as follows:

$$\ell_{elr}(\mathbf{p}_i) = \log\left(1 - (1/N)\left(\sigma(\mathbf{p}_i)^\top \sigma(\mathbf{q}_i)\right)\right), \quad (5)$$

where $\mathbf{q}_i^{(k)} = \psi \mathbf{q}_i^{(k-1)} + (1 - \psi)\mathbf{p}_i^{(k)}$ (with $\mathbf{q}_i \in \mathcal{P}$) is the temporal ensembling momentum [12] of the survival predictions, with k denoting the training epoch, and $\psi \in [0, 1]$. The idea of the loss in (5) is to never stop training for samples where the model prediction coincides with the temporal ensembling momentum (i.e., the 'clean' pseudo-labeled samples), and to never train for the noisy pseudo-labeled samples [12].

Bin Penalization Term. We add the penalization term described in [7] that forces a bin n to be active only when all previous bins (1 to $n - 1$) are active, forcing bins to represent sequential risk levels:

$$\ell_{pen}(\mathbf{p}_i) = \frac{1}{N-1} \sum_{n=1}^{N-1} \max(0, (\sigma(\mathbf{p}_i(n+1)) - \sigma(\mathbf{p}_i(n)))). \quad (6)$$

Censor-Aware Rank Loss. Cox proportional hazards models [4] aim to rank training samples instead of the directly estimating survival time. Following this reasoning, we propose a censor-aware ranking loss to encourage this sample sorting behavior. The rank loss has the following form:

$$\ell_{ca\_rnk}(\tilde{t}_i, \tilde{t}_j, y_i, y_j, e_i, e_j) = \max(0, -\mathcal{G}(y_i, y_j, e_i, e_j) \times (\tilde{t}_i - \tilde{t}_j)) \quad (7)$$

where i and j index all pairs of samples in a training mini-batch, and

$$\mathcal{G}(y_i, y_j, e_i, e_j) = \begin{cases} +1, & \text{if } (y_i > y_j) \wedge e_i = 0 \\ -1, & \text{if } (y_i \leq y_j) \wedge e_i = 0 \\ 0, & \text{otherwise} \end{cases}. \quad (8)$$

In (8), when $\mathcal{G}(.) = +1$ the sample i should be ranked higher than sample j, when $\mathcal{G}(.) = -1$ the sample j should be ranked higher than sample i, and the loss is ignored when $\mathcal{G}(.) = 0$.

3 Experiments

In this section, we describe the datasets and explain the experimental setup and evaluation measures, followed by the presentation of results.

3.1 Data Sets

TCGA-GM Dataset. The Cancer Genome Atlas (TCGA) Lower-Grade Glioma (LGG) and Glioblastoma (GBM) projects: TCGA-GM [13] consists of 1,505 patches extracted from 1,061 whole-slide pathological images. A total of 769 unique patients with gliomas were included in the study with 381 censored and 388 uncensored observations. The labels are given in days and can last from 1 to 6423 days. We use the published train-test split provided with the data set [13].

NLST Dataset. The National Lung Screening Trial (NLST) [14,15] is a randomized multicenter study for early detection of lung cancer of current or former heavy smokers of ages 55 to 74. We only used the chest x-ray images from the dataset and excluded cases where the patient died for causes other than lung cancer. The original study includes 25,681 patients (77,040 images), and after filtering, we had 15,244 patients (47,947 images) with 272 uncensored cases. The labels are given in days and can last from 1 to 2983 days. We patient-wise split the dataset into training (9,133 patients), validation (3,058 patients) and testing (3,053 patients) sets, maintaining the same demographic distribution across sets.

3.2 Experimental Set up

The input image size is adjusted to 512^2 pixels for both datasets and normalized with ImageNet [18] mean and standard deviation. For the model in (1), the space $\mathcal{P}$ has 5 bins for TCGA-GM pathology images and 3 bins for NLST chest x-ray images. The model $f(\mathbf{x}; \theta)$ in (1) is an ImageNet pre-trained ResNet-18 [6]. The training of the model uses Adam [9] optimizer with a momentum of 0.9, weight decay of 0.01 and a mini-batch size of 32 for 40 epochs. To obtain pseudo labels we have $\tau = 0.5$ in (4), the bin penalization term $\beta = 1e6$ in (6), the ELR [12] loss in (5) with $\alpha = 100$ and $\psi = 0.5$, and the rank loss in (7) with $\delta = 1$. We run all experiments using PyTorch [16] v1.8 on a NVIDIA V100 GPU. The total training time is 45 minutes for the TCGA-GM dataset and 240 min for the NLST dataset, and the inference time for a single x-ray image takes 2.8 ms.

To evaluate our method, we use the Mean Absolute Error (MAE) between our predictions and the observed time, where the error is reported patient-wise, aggregating the mean of all image predictions per patient. To account for censored data, we use the following metric proposed by Xiao *et al.* [21].

$$\text{MAE} = \mathbb{E}\left[|\tilde{t} - \mathbf{y}|\mathbb{I}(\tilde{t} < \mathbf{y}) + (1 - e)|\tilde{t} - \mathbf{y}|\mathbb{I}(\tilde{t} \geq \mathbf{y})\right], \qquad (9)$$

where $\mathbb{I}$ is the indicator function. In (9), the censored cases where the model predicts the survival time $\tilde{t} \geq \mathbf{y}$ are not counted as errors. As a result, a higher count of censored records will decrease the average MAE, so we also measure the MAE considering only uncensored records. We also report concordance index (C-index) [5] to measure the correct ranking of the predictions.

3.3   Results

In Table 1, we compare survival time prediction results between our method and SOTA methods [13,21,23] on TCGA-GM, and DeepConvSurv [23] on NLST. Our method sets the new SOTA results in terms of MAE (all samples and only uncensored ones) and C-index for both datasets. The low MAE result for our approach on NLST can be explained by the fact that the vast majority of cases are censored (over 95%), and our method successfully estimates a survival time larger than the censoring time, producing MAE = 0 for many censored cases, according to (9).

Table 1. Comparison of survival time prediction between our approach and SOTA methods on the TCGA-GM and NLST. Evaluation metrics are mean absolute error (MAE) in days, and concordance index (C-index). Best results are highlighted.

Method	MAE (all samples)	MAE (only uncensored)	C-index
TCGA-GM			
DeepConvSurv [23]	439.1	-	0.731
AFT [20]	386.5	-	0.685
SCNNs [13]	424.5	-	0.725
CDOR [21]	321.2	-	0.737
Ours (Pseudo labels + ℓ_{ca_mse} + ℓ_{ca_rnk} + ℓ_{elr})	**286.95**	**365.06**	**0.740**
NLST			
DeepConvSurv [23]	27.98	1334.22	0.690
Ours (Pseudo labels + ℓ_{ca_mse})	**26.28**	**1275.24**	**0.756**

Table 2. Ablation study for our method on the TCGA-GM and NLST datasets. The evaluation metrics are: mean absolute error (MAE) in days and concordance index (C-index). Best results are highlighted in bold.

ℓ_{ca_mse} $+\ell_{pen}$	ℓ_{ca_rnk}	Pseudo labels	ℓ_{elr}	MAE (all samples)	Best MAE (all samples)	Best MAE (only uncensored)	C-index	Best C-index
TCGA-GM								
✓	-	-	-	379.40 ±20.67	356.74	521.91	0.702 ±0.009	0.707
✓	✓	-	-	380.49 ±17.55	361.76	517.41	0.722 ±0.008	0.730
✓	-	✓	-	308.01 ±7.99	302.04	405.48	0.725 ±0.014	**0.740**
✓	-	✓	✓	311.76 ±6.54	304.42	404.32	0.711 ±0.016	0.723
✓	✓	✓	-	298.56 ±6.31	291.53	411.79	0.718 ±0.016	0.736
✓	✓	✓	✓	**294.45 ±6.49**	**286.95**	**365.06**	0.728 ±0.010	0.740
NLST								
✓	-	-	-	27.24 ±0.15	27.10	1283.41	0.729 ±0.004	0.732
✓	✓	-	-	27.68 ±0.33	27.31	1286.27	0.732 ±0.003	0.736
✓	-	✓	-	**26.97 ±0.97**	**26.28**	**1275.24**	0.751 ±0.006	**0.756**
✓	-	✓	✓	29.05 ±0.59	28.37	1397.44	0.703 ±0.002	0.705
✓	✓	✓	-	27.83 ±2.32	26.29	1283.14	0.747 ±0.002	0.749
✓	✓	✓	✓	27.13 ±0.86	26.61	1277.02	**0.752 ±0.002**	0.754

On Table 2, we show the ablation study for the components explained in Sect. 2.1, where we start from a baseline model Base trained with ℓ_{ca_mse} and ℓ_{pen}. Then we show how results change with the use of ℓ_{ca_rnk} (RankLoss), pseudo labels and ℓ_{elr} (ELR). Regarding the MAE results on TCGA-GM and NLST datasets, we can observe a clear benefit of using pseudo labels. For TCGA-GM, the use of pseudo labels improve MAE by more than 20% compared with the Base model with RankLoss. For NLST, the smaller MAE reduction with pseudo labels can be explained by the large proportion of censored cases, which makes the robust estimation of survival time a challenging task. Considering the C-index, pseudo labels show similar benefits for TCGA-GM, with a slightly better result. On NLST, pseudo labels improve the C-index from 0.73 to 0.75 over the Base model with RankLoss. On both datasets, the use of Base model with RankLoss, pseudo labels, and ELR, presents either the best results or a competitive result with best result.

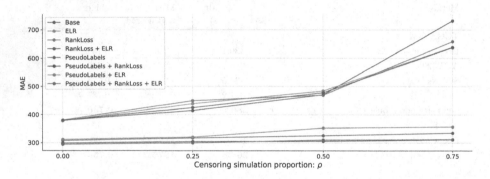

Fig. 3. Performance of our method under different proportions of censored vs uncensored labels on the TCGA-GM dataset. The x-axis shows the ratio of uncensored records transformed to be censored (ρ), and the y-axis shows MAE. Note that pseudo labels combined with RankLoss and ELR show robust results in scenarios with small percentages of uncensored records.

The differences between the results on NLST and TCGA-GM can be attributed to different proportions of censored/uncensored cases in both datasets. We test the hypothesis that such differences have a strong impact in the performance of our algorithm by formulating an experiment using the TCGA-GM dataset, where we simulate different proportion of censored records. Figure 3 shows the result of this simulation, where we show MAE as a function of the proportion of censored data, denoted by ρ. The results show that the PseudoLabels methods (combined with RankLoss and ELR losses) keep the MAE roughly constant, while Base, ELR and RankLoss results deteriorate significantly. The best results are achieved by PseudoLabels + RankLoss and PseudoLabels + RankLoss + ELR, suggesting that pseudo labels and both losses are important for the method to be robust to large proportions of censored data.

4   Discussion and Conclusion

Current techniques for survival prediction discard or sub-optimally use the variability present in censored data. In this paper we proposed a method to exploit such variability on censored data by using pseudo labels to semi-supervise the learning process. Experimental results in Table 1 showed that our proposed method obtains SOTA results for survival time prediction on the TCGA-GM and NLST datasets. In fact, the effect of pseudo labels can be observed in Fig. 3, where we artificially censored samples from the TCGA-GM dataset. It is clear that the use of pseudo labels provide solid robustness to a varying proportion of censored data in the training set. Combining the pseudo labels with a regularisation loss that accounts for noisy pseudo labels and censor-aware rank loss improves this robustness and results on datasets. However, is important to note that the method may not be able to produce good quality pseudo-labels on cases where the dataset contains few uncensored records. We expect our newly proposed method to foster the development of new survival time prediction models that exploit the variability present in censored data.

References

1. Agravat, R.R., Raval, M.S.: Brain tumor segmentation and survival prediction. In: Crimi, A., Bakas, S. (eds.) BrainLes 2019. LNCS, vol. 11992, pp. 338–348. Springer, Cham (2020). https://doi.org/10.1007/978-3-030-46640-4_32
2. Baid, U., et al.: Overall survival prediction in glioblastoma with radiomic features using machine learning. Front. Comput. Neurosci. **14** (2020). https://doi.org/10.3389/fncom.2020.00061, publisher: Frontiers
3. Cheon, S., et al.: The accuracy of clinicians' predictions of survival in advanced cancer: a review. Ann. Palliative Med. **5**(1), 22–29 (2016). https://doi.org/10.3978/j.ISSN:2224-5820.2015.08.04
4. Cox, D.R.: Regression models and life-tables. J. Roy. Stat. Soc. Ser. B (Methodological) **34**(2), 187–220 (1972)
5. Harrell, F.E., Califf, R.M., Pryor, D.B., Lee, K.L., Rosati, R.A.: Evaluating the yield of medical tests. Jama **247**(18), 2543–2546 (1982), publisher: American Medical Association
6. He, K., Zhang, X., Ren, S., Sun, J.: Deep residual learning for image recognition. In: 2016 IEEE Conference on Computer Vision and Pattern Recognition (CVPR), pp. 770–778, June 2016. https://doi.org/10.1109/CVPR.2016.90
7. Hermoza, R., Maicas, G., Nascimento, J.C., Carneiro, G.: Post-hoc overall survival time prediction from brain MRI. In: 2021 IEEE 18th International Symposium on Biomedical Imaging (ISBI), pp. 1476–1480 (2021). https://doi.org/10.1109/ISBI48211.2021.9433877
8. Jing, B., et al.: A deep survival analysis method based on ranking. Artif. Intell. Med. **98**, 1–9 (019). https://doi.org/10.1016/j.artmed.2019.06.001
9. Kingma, D.P., Ba, J.: Adam: a method for stochastic optimization. In: International Conference on Learning Representations (ICLR) (2015)
10. Lecun, Y., Bottou, L., Bengio, Y., Haffner, P.: Gradient-based learning applied to document recognition. Proc. IEEE **86**(11), 2278–2324 (1998). https://doi.org/10.1109/5.726791, conference Name: Proceedings of the IEEE

11. Li, J., Socher, R., Hoi, S.C.: Dividemix: Learning with noisy labels as semi-supervised learning. arXiv preprint arXiv:2002.07394 (2020)
12. Liu, S., Niles-Weed, J., Razavian, N., Fernandez-Granda, C.: Early-learning regularization prevents memorization of noisy labels. arXiv:2007.00151 [cs, stat], October 2020. http://arxiv.org/abs/2007.00151, arXiv: 2007.00151
13. Mobadersany, P., et al.: Predicting cancer outcomes from histology and genomics using convolutional networks. Proc. Natl. Acad. Sci. 115(13), E2970–E2979 (2018). https://doi.org/10.1073/pnas.1717139115. https://www.pnas.org/content/115/13/E2970, publisher: National Academy of Sciences Section: PNAS Plus
14. National Lung Screening Trial Research Team, Aberle, D.R., et al.: Reduced lung-cancer mortality with low-dose computed tomographic screening. New England J. Med. 365(5), 395–409 (2011). https://doi.org/10.1056/NEJMoa1102873. https://pubmed.ncbi.nlm.nih.gov/21714641, edition: 2011/06/29
15. National Lung Screening Trial Research Team, Aberle, D.R., et al.: The national lung screening trial: overview and study design. Radiology 258(1), 243–253 (2011). https://doi.org/10.1148/radiol.10091808. https://pubmed.ncbi.nlm.nih.gov/21045183, edition: 2010/11/02 Publisher: Radiological Society of North America, Inc
16. Paszke, A., et al.: PyTorch: an imperative style, high-performance deep learning library. In: Wallach, H., Larochelle, H., Beygelzimer, A., Alché-Buc, F.d., Fox, E., Garnett, R. (eds.) Advances in Neural Information Processing Systems 32, pp. 8024–8035. Curran Associates, Inc. (2019)
17. Raykar, V.C., Steck, H., Krishnapuram, B., Dehing-Oberije, C., Lambin, P.: On ranking in survival analysis: bounds on the concordance index. In: Proceedings of the 20th International Conference on Neural Information Processing Systems, NIPS 2007, pp. 1209–1216. Curran Associates Inc., Red Hook, December 2007
18. Russakovsky, O., et al.: ImageNet large scale visual recognition challenge. Int. J. Comput. Vision 115(3), 211–252 (2015). https://doi.org/10.1007/s11263-015-0816-y
19. Tang, Z., et al.: Pre-operative overall survival time prediction for glioblastoma patients using deep learning on both imaging phenotype and genotype. In: Shen, D., et al. (eds.) MICCAI 2019. LNCS, vol. 11764, pp. 415–422. Springer, Cham (2019). https://doi.org/10.1007/978-3-030-32239-7_46
20. Wei, L.J.: The accelerated failure time model: a useful alternative to the cox regression model in survival analysis. Stat. Med. 11(14–15), 1871–1879 (1992)
21. Xiao, L., et al.: Censoring-aware deep ordinal regression for survival prediction from pathological images. In: Martel, A.L., et al. (eds.) MICCAI 2020. LNCS, vol. 12265, pp. 449–458. Springer, Cham (2020). https://doi.org/10.1007/978-3-030-59722-1_43
22. Zheng, Z., Yang, Y.: Rectifying pseudo label learning via uncertainty estimation for domain adaptive semantic segmentation. Int. J. Comput. Vis., 1–15 (2021)
23. Zhu, X., Yao, J., Huang, J.: Deep convolutional neural network for survival analysis with pathological images. In: 2016 IEEE International Conference on Bioinformatics and Biomedicine (BIBM), pp. 544–547, December 2016. https://doi.org/10.1109/BIBM.2016.7822579

Prognostic Imaging Biomarker Discovery in Survival Analysis for Idiopathic Pulmonary Fibrosis

An Zhao[1]([envelope]), Ahmed H. Shahin[1,8], Yukun Zhou[1], Eyjolfur Gudmundsson[1], Adam Szmul[1], Nesrin Mogulkoc[2], Frouke van Beek[3], Christopher J. Brereton[4], Hendrik W. van Es[5], Katarina Pontoppidan[4], Recep Savas[6], Timothy Wallis[4], Omer Unat[2], Marcel Veltkamp[3,7], Mark G. Jones[4], Coline H.M. van Moorsel[3], David Barber[8], Joseph Jacob[1], and Daniel C. Alexander[1]

[1] Centre for Medical Image Computing, UCL, London, UK
an.zhao.19@ucl.ac.uk
[2] Department of Respiratory Medicine, Ege University Hospital, Izmir, Turkey
[3] Department of Pulmonology, Interstitial Lung Diseases Center of Excellence,
St. Antonius Hospital, Nieuwegein, Netherlands
[4] NIHR Southampton Biomedical Research Centre and Clinical and Experimental
Sciences, University of Southampton, Southampton, UK
[5] Department of Radiology, St. Antonius Hospital, Nieuwegein, Netherlands
[6] Department of Radiology, Ege University Hospital, Izmir, Turkey
[7] Division of Heart and Lungs, University Medical Center, Utrecht, The Netherlands
[8] Centre for Artificial Intelligence, UCL, London, UK

Abstract. Imaging biomarkers derived from medical images play an important role in diagnosis, prognosis, and therapy response assessment. Developing prognostic imaging biomarkers which can achieve reliable survival prediction is essential for prognostication across various diseases and imaging modalities. In this work, we propose a method for discovering patch-level imaging patterns which we then use to predict mortality risk and identify prognostic biomarkers. Specifically, a contrastive learning model is first trained on patches to learn patch representations, followed by a clustering method to group similar underlying imaging patterns. The entire medical image can be thus represented by a long sequence of patch representations and their cluster assignments. Then a memory-efficient clustering Vision Transformer is proposed to aggregate all the patches to predict mortality risk of patients and identify high-risk patterns. To demonstrate the effectiveness and generalizability of our model, we test the survival prediction performance of our method on two sets of patients with idiopathic pulmonary fibrosis (IPF), a chronic, progressive, and life-threatening interstitial pneumonia of unknown etiology. Moreover, by comparing the high-risk imaging patterns extracted by our model with existing imaging patterns utilised in clinical practice,

---

Supplementary Information The online version contains supplementary material available at https://doi.org/10.1007/978-3-031-16449-1_22.

we can identify a novel biomarker that may help clinicians improve risk stratification of IPF patients.

Keywords: Imaging biomarker discovery · Survival analysis · Contrastive learning · Clustering vision transformer · Idiopathic pulmonary fibrosis

1   Introduction

An imaging biomarker is defined as a characteristic derived from a medical image, that can be used as an indicator of normal biological processes, pathogenic processes, or pharmacologic responses to a therapeutic intervention [2,12]. Clinicians often assess imaging biomarkers through visual assessment of medical images. Though some computer-based methods have been proposed for automated and quantitative measurement of imaging biomarkers [2], these methods often require expert labelling of potential biomarkers as training data. This is not only time-consuming, expensive and susceptible to inter-observer variability, but also restricts computer-based methods to evaluating limited numbers of predefined imaging biomarkers. Such strategies are insufficient to adequately mine the rich information contained within medical images [29].

There remains a long-term unmet need for quantitative and novel imaging biomarker development to inform the prediction of disease outcomes such as mortality. Many deep learning methods have been proposed recently for survival prediction using imaging data without the requirement of manual annotations [13,24]. One major limitation of these end-to-end deep learning survival models is that they often extract high-dimensional biomarkers which are difficult to interpret. Motivated by this, some researchers looking at histopathology images have tried to learn low-level prognostic imaging biomarkers that can underpin image-based survival models [1,30]. These methods often extract patch-level features and aggregate them to predict patient survival. Different strategies then associate the underlying patterns with high mortality risk. Accordingly, learning generalizable representations, aggregating sequences of patch information, and recognizing high-risk patterns become the main challenges for this task.

Recent success in self-supervised learning methods highlights the potential for learning discriminative and generalizable representations from unlabelled data. Contrastive learning, a dominant group of self-supervised learning methods, aims to group similar samples together while separating samples that differ [18]. Transformer neural networks have gained increasing interest from the medical imaging field because of their ability to capture global information [27]. The Vision Transformer (ViT) [8] accepts an image as a sequence of patch embeddings allowing for the fusion of patch information. However, medical images can often be split into thousands of patches, which results in a much longer sequence and a computationally expensive ViT when compared with natural images.

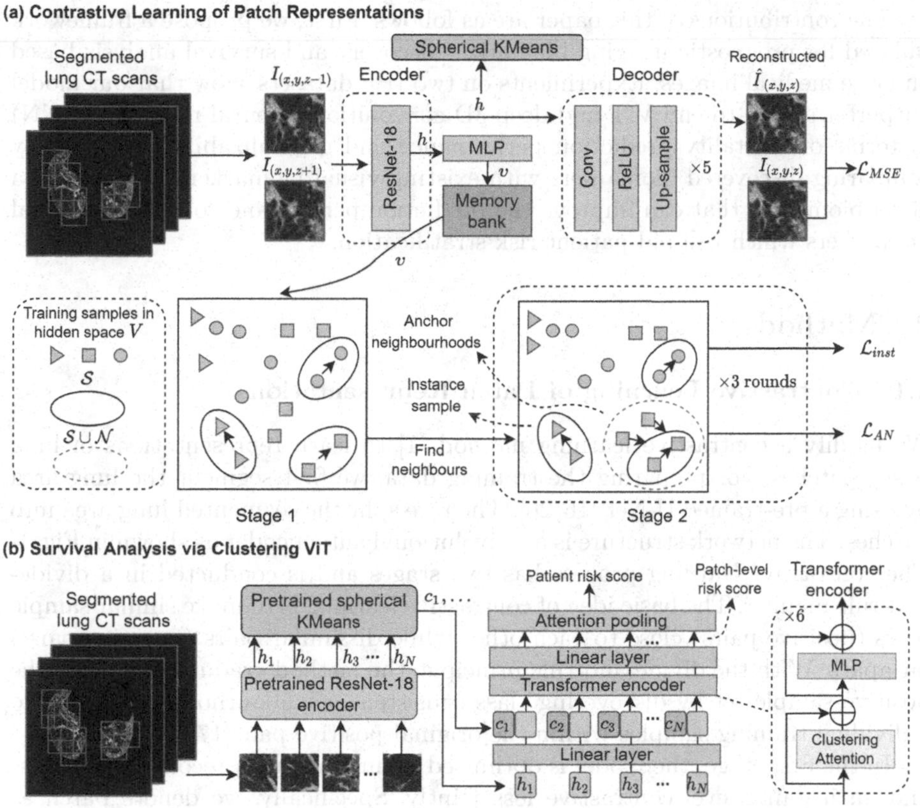

Fig. 1. An overview of the proposed model. (a) We first learn patch representations h by using a modified contrastive learning method [1]. All patch representations are clustered into K clusters by using spherical KMeans. (b) With trained models in (a), CT scans can be represented as a sequence of patch representations and their cluster assignments, which are then fed to a clustering ViT for survival prediction.

Motivated by challenges in prognostic biomarker discovery and survival analysis, we propose a framework that leverages plausible properties of contrastive learning and ViT, and validate its performance in a highly heterogeneous disease, idiopathic pulmonary fibrosis (IPF). IPF is a chronic lung disease of unknown cause, associated with a median survival of between 2.5 to 3.5 years [22]. The generally poor prognosis of IPF belies its highly heterogeneous disease progression between patients. A lack of reliable prognostic biomarkers hampers the ability to accurately predict IPF patient survival [29].

In our study, as shown in Fig. 1, we first learn patch representations via contrastive learning, on computed tomography (CT) imaging of IPF patients, followed by a clustering method to extract underlying patterns in the patches. We then use an efficient clustering ViT to aggregate the extremely long sequence of patch representations for survival prediction and high-risk pattern recognition.

The contributions of this paper are as follows. First, we propose a framework tailored for prognostic imaging biomarker discovery and survival analysis based on large medical images. Experiments on two IPF datasets show that our model outperforms end-to-end ViT and deep 3D convolutional neural networks (CNN) in terms of mortality prediction performance and generalizability. Second, by comparing discovered biomarkers with existing visual biomarkers, we identify a novel biomarker that can improve the prediction performance of existing visual biomarkers which will aid patient risk stratification.

2   Method

2.1   Contrastive Learning of Patch Representations

We modify a contrastive learning method [1] to learn representations of lung tissue patches. For preparing the training data, we first segment the lung area by using a pre-trained U-Net [16, 26]. Then we split the segmented lung area into patches. The network structure is a convolutional autoencoder as shown in Fig. 1. The contrastive learning process has two stages and is conducted in a divide-and-rule manner. The basic idea of contrastive learning is to make similar sample pairs (positive pairs) close to each other while dissimilar pairs (negative pairs) far apart. With the divide-and-rule principle, the method gradually expands the positive sample set by discovering class consistent neighbourhoods anchored to individual training samples within the original positive pair [17].

In the first stage, the model is optimized by minimizing a reconstruction loss and an instance-level contrastive loss jointly. Specifically, we denote patch as $I_{(x,y,z)}$, where (x, y, z) is the location of the patch central voxel in original CT scans. The ResNet-18 [15] encoder maps input patches $I_{(x,y,z-1)}, I_{(x,y,z+1)}$ into the 512-dimensional latent representation h, and the decoder reconstructs the patch $I_{(x,y,z)}$ adjacent to them from h. Mean squared error (MSE) loss $\mathcal{L}_{MSE}$ is used to measure the difference between patch $I_{(x,y,z)}$ and the reconstructed patch. For contrastive learning, we consider patches 50% overlapped with patch i as its similar samples S_i and other patches in the training dataset as dissimilar samples. For calculating the similarity between samples, h is projected to a 128-dimensional variable v by a multi-layer perception (MLP) [5]. The instance-level contrastive loss is defined as $\mathcal{L}_{inst} = -\sum_{i \in B_{inst}} \log(\sum_{j \in S_i} p(j \mid i))$, where $p(j \mid i) = \frac{\exp(v_j^\top v_i / \tau)}{\sum_{k=1}^N \exp(v_k^\top v_i / \tau)}$, B_{inst} is the set of instance samples in a mini-batch.

In the second stage, the method discovers other similar patches (neighbour samples $\mathcal{N}_i$) for a given patch i based on relative entropy to expand the positive sample set. Lower entropy indicates a higher similarity between a patch i and its neighbourhoods, and these patches can be anchored together in the subsequent training process (anchored neighbourhoods, ANs). Higher entropy implies that the given patch is dissimilar with its neighbourhoods and so the pair should remain individuals (instance samples) rather than grouped together. AN-level contrastive loss is defined as $\mathcal{L}_{AN} = -\sum_{i \in B_{AN}} \log(\sum_{j \in S_i \cup \mathcal{N}_i} p(j \mid i))$, where

B_{AN} is the set of ANs in a mini-batch. The total loss for two stages can be defined as $\mathcal{L} = \mathcal{L}_{inst} + \mathbb{1}(\text{stage} = 2) \cdot \mathcal{L}_{AN} + \mathcal{L}_{MSE}$.

The second stage can be performed over multiple rounds with an increasing number of ANs. This will progressively expand the local consistency to find the global class boundaries. A memory bank is used to keep track of the similarity matrix. h is used as patch representation after training the model. All patch representations are clustered into K clusters by using spherical KMeans [33]. These clusters are common patterns found in CT scans.

2.2 Survival Analysis via Clustering ViT

The objective of survival analysis is to estimate the expected duration before an event happens. The Cox proportional hazards model [6] is a widely used survival analysis method. Given the input data x, the hazard function is modeled as $h(t|x) = h_0(t)\exp(f(x))$, where $h_0(t)$ is a base hazard function and $\exp(f(x))$ is a relative risk function learned by models. Deep-learning based Cox methods [9,21] have been proposed to model more complicated nonlinear log-risk function $f(x)$. These methods optimize Cox log partial likelihood $L_{cox} = \prod_{i:E_i=1} \frac{\exp(\hat{f}(x_i))}{\sum_{j \in \mathcal{R}(T_i)} \exp(\hat{f}(x_j))}$, where T_i, E_i, x_i, $\hat{f}(x_i)$ are the event time (or censored), event indicator, baseline data, and estimated log-risk of patient i, and $\mathcal{R}(t)$ is the risk set of patients who are still alive at time t.

Inspired by [32], we propose a clustering ViT for survival analysis and high-risk pattern identification, that can handle long sequences (Fig. 1). For given CT scans, a sequence of patch representations and their cluster assignments are generated using the trained model in Sect. 2.1. We pad all input sequences to be a fixed length N. We first map patch representations to D dimensions with a linear layer, where D is the hidden size, and then pass the sequence into a Transformer encoder with L layers ($L = 6$). Each layer includes a clustering attention layer with 8 heads and a MLP. The original attention layer computes interaction between each pair of input patches, with a $O(N^2)$ complexity. The assumption of the clustering attention is that patches within the same cluster have similar attention maps. Queries within the same cluster can be represented by a prototype, which is the centroid of queries within this cluster. The clustering attention layer only calculates the attention maps between K prototypes and N keys, and then broadcasts it to queries within each cluster. This reduces complexity to $O(NK)$.

The sequential output of the L-layer ViT is defined as $x_L \in \mathbb{R}^{B \times N \times D}$, where B is the batch size. After going through a linear layer g, $g(x_L) \in \mathbb{R}^{B \times N \times 1}$ can be seen as sequences of patch-level risk scores. We then propose attention pooling, a variant of sequence pooling [14], to get the patient-level log-risk $r = \text{softmax}(g(x_L)^\top)g(x_L) \in \mathbb{R}^{B \times 1}$. This pooling method can assign different importance weights across patches with different risks. The network is trained based on average negative partial log-likelihood loss $\mathcal{L}_{neglog} = -\frac{1}{n_{E=1}} \sum_{i:E_i=1} (r_i - \log \sum_{j \in \mathcal{R}(T_i)} \exp(r_j))$, where r is log-risk estimated by the ViT, $n_{E=1}$ is the number of events observed. After training the ViT, we take the mean of risk scores of patches within a cluster as the cluster-level risk $R_k, k \in [1, K]$, which will help us to identify high-risk patterns.

2.3   Novel Prognostic Biomarker Identification

High-risk imaging patterns discovered by our method may overlap the clinically-established patterns. With a CT dataset which includes annotated regions of lung tissue patterns predefined by clinicians (normal lung, ground-glass opacity, emphysema, and fibrosis), we propose an approach to identify novel prognostic biomarkers by disentangling them from the established ones[1] (see Supplementary Fig. 2). The hypotheses are that: 1) The novel biomarker should not have a strong or moderate positive correlation (correlation coefficient > 0.3) with the extents of existing patterns [25]. 2) The novel biomarker should be significantly predictive of mortality (p-value < 0.05) independent of existing biomarkers when inputting both of them into the Cox model. 3) Centroids of representations of novel prognostic patterns should be relatively far from those of existing patterns.

3   Experiment

In this section, we first evaluate mortality risk prediction performance of the proposed model and compare it with CNN-based prediction models. A series of ablation studies are conducted to understand the contribution of each component. We also show representative patches of discovered high-risk patterns.

3.1   Datasets

For training the contrastive learning model, we use a dataset (Dataset 1) that contains 313 CT scans (186 death observed) of IPF patients from the Netherlands and Turkey. 1,547,467 patches are generated by using a 64×64 sliding window with a step size of 32 across the lung area. Dataset 1 is used for training and evaluating the ViT with 5-fold cross-validation. We randomly split the dataset into 5 folds. 1 fold is used as an internal test set, while the remaining 4 folds are randomly split into training and validation sets with a ratio of 4 : 1. To evaluate the generalizability, we introduce an external test dataset (Dataset 2) from University Hospital Southampton, comprising 98 CT scans (48 death observed). For each split, we train the model on the training set in Dataset 1, choose the best model with the lowest loss on the validation set, and test the model on internal and external test sets. For novel prognostic biomarker identification, we use a subset of Dataset 1 and 2 with visual scores in step 1) and 2) of Sect. 2.3. In 253/313 (81%) CTs in Dataset 1, and all CTs in Dataset 2, fibrosis and emphysema extents have been visually scored by radiologists. We also use a publicly available interstitial lung disease dataset with annotated lung tissue patterns [7] for calculating centroids of existing patterns in step 3) of Sect. 2.3.

---

[1] We use the extent of a high-risk pattern as a prognostic biomarker, obtained by measuring the percentage of this pattern within the whole lung.

Table 1. 5-fold cross-validation results on internal and external datasets compared with other survival prediction models. P-value shows significance of better performance of our proposed model ($^*p < 0.05,^\dagger p < 0.01,^\ddagger p < 0.001$).

Methods	# Pars	Internal test set		External test set	
		IPCW C-index	IBS	IPCW C-index	IBS
Ours	21.96M	**0.676 ± 0.033**	**0.165 ± 0.025**	**0.698 ± 0.013**	**0.133 ± 0.003**
K-M method	-	-	0.192±0.028	-	0.144 ± 0.002†
3D ResNet-18	32.98M	0.639 ± 0.032†	0.184±0.028	0.618 ± 0.001‡	0.149 ± 0.006*
3D ResNet-34	63.28M	0.657±0.048	0.180±0.032	0.631 ± 0.032*	0.156±0.028

3.2   Implementation Details

For contrastive learning, we train the model with an Adam optimizer [20,23], a learning rate of 10^{-4}, and batch size of 128. We run the first stage for 1 round and the second stage for 3 rounds with $\tau = 0.05$, which takes about a week. Every round has 25 epochs. The number of clusters K is 64 for spherical KMeans. For clustering ViT, the sequence length N is 15,000 and the batch size is 6. We use Mixup [31] for data augmentation. The hidden size D is 256 and dropout rate is 0.1. We use a pre-trained model [3] for training our ViT. We adopt the Sharpness-Aware Minimization (SAM) algorithm [10] and use Adam as the base optimizer, with a learning rate of 2×10^{-5} and weight decay of 10^{-4}. The ViT is trained for 100 epochs $(3\,h)^2$. We use an inverse-probability-of-censoring weighted version of C-index (IPCW C-index) [28] for assessing the discrimination of model, which quantifies the capability of discriminating patients with different survival times. For measuring the calibration (i.e. the capability of predicting true probabilities), we use the Integrated Brier Score (IBS) [11] which measures the difference between predicted probabilities and observed status by integrating Brier Score across the time span of the test set. The paired t-test is used for testing statistical differences between the proposed method and other methods. More experiments of hyperparameters are provided in Supplementary Table 1.

3.3   Experiment Results

Comparison with Other Survival Prediction Models. We compare our method with end-to-end survival prediction models based on 3D ResNet-18 and 3D ResNet-34 [4]. 3D ResNets are trained with average negative partial log-likelihood loss [6]. Data splitting for each fold is the same as our model to ensure a fair comparison. Kaplan-Meier (K-M) method [19] is used as a baseline of the IBS. As shown in Table 1, our method achieves at least comparable or often significantly better performance in terms of discrimination and calibration with fewer parameters, especially in the external test dataset.

---

^{2} Our method is implemented by Pytorch 1.8. All models were trained on one NVIDIA RTX6000 GPU with 24GB memory. Code is available at https://github. com/anzhao920/PrognosticBiomarkerDiscovery.

Table 2. 5-fold cross-validation results of ablation studies on internal and external datasets. P-values show the significance of better performance of the proposed model ($^*p < 0.05,^\dagger p < 0.01,^\ddagger p < 0.001$).

Methods	Internal test set		External test set	
	IPCW C-index	IBS	IPCW C-index	IBS
Proposed model	**0.676 ± 0.033**	**0.165 ± 0.025**	0.698±0.013	0.133±0.003
w/o contrastive learning	0.633 ± 0.042†	0.183 ± 0.032*	0.561 ± 0.017‡	0.163 ± 0.011†
w/o attention pooling	0.659±0.025	0.177±0.034	**0.699 ± 0.011**	**0.133 ± 0.002**
w/o Mixup	0.660±0.038	0.177±0.034	0.686±0.023	0.139±0.006
w/o SAM	0.666±0.020	0.174±0.030	0.631 ± 0.022†	0.154 ± 0.010*

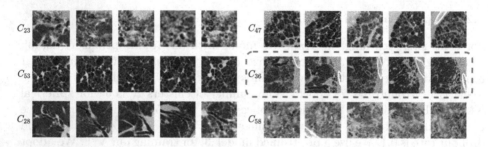

Fig. 2. Representative patches of top-6 high-risk clusters discovered by our model. Cluster 36 (C_{36}) is a novel prognostic pattern.

Ablation Study. To investigate the contribution of each component in the proposed method, we conduct ablation studies as shown in Table 2. The ablation study without contrastive learning in Sect. 2.1 is identical to train a regular ViT [8] end-to-end for survival prediction. We also remove attention pooling (using average pooling instead), Mixup [31] data augmentation and SAM algorithm [10], and compare their performance with the proposed method. Contrastive learning contributes the most to the performance. Using SAM optimizer significantly improves the generalizability, with 6.7% increase of IPCW C-index in the external test set. Mixup and attention pooling provide slightly better performance.

Biomarker Discovery. Based on cluster-level risk scores generated by the ViT, we show representative samples of clusters that have the 6 highest risk scores in Fig. 2. We can observe that these clusters not only focus on imaging patterns but also the location of the patterns. We also identify a novel pattern (C_{36} in Fig. 2) which is morphologically different from existing patterns. C_{36} is centred on the lateral border of the lung by the ribs, in a typical distribution for IPF-related fibrosis. The patches identify the commingling of high and low density extremes. The low density can be a combination of honeycombing, emphysema and traction bronchiectasis. The high density comprises primarily consolidation (not uncommonly representing a radiological pleuroparenchymal fibroelastosis pattern) and reticulation. The mortality prediction performance when using the novel biomarker (extent of C_{36}) and visual scores (fibrosis extent and emphysema extent) is better than using visual scores alone (Supplementary Table 2), with $\approx$ 1% increase of IPCW C-index in both test sets. This suggests the novel biomarker is likely to be complementary to existing biomarkers in predicting mortality risk.

4   Discussion

There are some limitations to this work. First, the homogeneity of some clusters is unsatisfactory (Supplementary Fig. 1a). The challenge is how to set the number of clusters to find a balance between having homogeneous clusters but ones that still have enough patches within the cluster to make them clinically useful. Second, the method of identifying novel biomarkers in Sect. 2.3 is intuitive and more research should be done. Third, this work mainly focuses on axial plane, and other planes need further exploration.

In this work, we propose a framework for prognostic imaging biomarker discovery and survival analysis. Experiments on two IPF datasets demonstrate that the proposed method performs better than its CNN counterparts in terms of discrimination and calibration. The novel biomarker discovered by our method provides additional prognostic information when compared to previously defined biomarkers used in IPF mortality prediction. This method can be potentially extended to broader applications for different diseases and image modalities.

Acknowledgements. AZ is supported by CSC-UCL Joint Research Scholarship. DCA is supported by UK EPSRC grants M020533, R006032, R014019, V034537, Wellcome Trust UNS113739. JJ is supported by Wellcome Trust Clinical Research Career Development Fellowship 209,553/Z/17/Z. DCA and JJ are supported by the NIHR UCLH Biomedical Research Centre, UK.

References

1. Abbet, C., Zlobec, I., Bozorgtabar, B., Thiran, J.-P.: Divide-and-rule: self-supervised learning for survival analysis in colorectal cancer. In: Martel, A.L., et al. (eds.) MICCAI 2020. LNCS, vol. 12265, pp. 480–489. Springer, Cham (2020). https://doi.org/10.1007/978-3-030-59722-1_46
2. Abramson, R.G., et al.: Methods and challenges in quantitative imaging biomarker development. Acad. Radiol. **22**(1), 25–32 (2015)
3. Carion, N., Massa, F., Synnaeve, G., Usunier, N., Kirillov, A., Zagoruyko, S.: End-to-end object detection with transformers. In: Vedaldi, A., Bischof, H., Brox, T., Frahm, J.-M. (eds.) ECCV 2020. LNCS, vol. 12346, pp. 213–229. Springer, Cham (2020). https://doi.org/10.1007/978-3-030-58452-8_13
4. Chen, S., Ma, K., Zheng, Y.: Med3d: transfer learning for 3D medical image analysis. arXiv preprint arXiv:1904.00625 (2019)
5. Chen, T., Kornblith, S., Norouzi, M., Hinton, G.: A simple framework for contrastive learning of visual representations. In: International Conference on Machine Learning, pp. 1597–1607. PMLR (2020)

6. Cox, D.R.: Regression models and life-tables. J. Roy. Stat. Soc.: Ser. B (Methodol.) **34**(2), 187–202 (1972)

7. Depeursinge, A., et al.: Building a reference multimedia database for interstitial lung diseases. Comput. Med. Imaging Graph. **36**(3), 227–238 (2012)

8. Dosovitskiy, A., et al.: An image is worth 16×16 words: transformers for image recognition at scale. arXiv preprint arXiv:2010.11929 (2020)

9. Faraggi, D., Simon, R.: A neural network model for survival data. Stat. Med. **14**(1), 73–82 (1995)

10. Foret, P., Kleiner, A., Mobahi, H., Neyshabur, B.: Sharpness-aware minimization for efficiently improving generalization. arXiv preprint arXiv:2010.01412 (2020)

11. Graf, E., Schmoor, C., Sauerbrei, W., Schumacher, M.: Assessment and comparison of prognostic classification schemes for survival data. Stat. Med. **18**(17–18), 2529–2545 (1999)

12. Group, B.D.W., et al.: Biomarkers and surrogate endpoints: preferred definitions and conceptual framework. Clin. Pharmacol. Therapeutics **69**(3), 89–95 (2001)

13. Haarburger, C., Weitz, P., Rippel, O., Merhof, D.: Image-based survival prediction for lung cancer patients using CNNs. In: 2019 IEEE 16th International Symposium on Biomedical Imaging (ISBI 2019), pp. 1197–1201. IEEE (2019)

14. Hassani, A., Walton, S., Shah, N., Abuduweili, A., Li, J., Shi, H.: Escaping the big data paradigm with compact transformers. arXiv preprint arXiv:2104.05704 (2021)

15. He, K., Zhang, X., Ren, S., Sun, J.: Deep residual learning for image recognition. In: Proceedings of the IEEE Conference on Computer Vision and Pattern Recognition, pp. 770–778 (2016)

16. Hofmanninger, J., Prayer, F., Pan, J., Röhrich, S., Prosch, H., Langs, G.: Automatic lung segmentation in routine imaging is primarily a data diversity problem, not a methodology problem. Europ. Radiol. Exp. **4**(1), 1–13 (2020). https://doi.org/10.1186/s41747-020-00173-2

17. Huang, J., Dong, Q., Gong, S., Zhu, X.: Unsupervised deep learning by neighbourhood discovery. In: International Conference on Machine Learning, pp. 2849–2858. PMLR (2019)

18. Jaiswal, A., Babu, A.R., Zadeh, M.Z., Banerjee, D., Makedon, F.: A survey on contrastive self-supervised learning. Technologies **9**(1), 2 (2021)

19. Kaplan, E.L., Meier, P.: Nonparametric estimation from incomplete observations. J. Am. Stat. Assoc. **53**(282), 457–481 (1958)

20. Kingma, D.P., Ba, J.: Adam: a method for stochastic optimization. arXiv preprint arXiv:1412.6980 (2014)

21. Kvamme, H., Borgan, Ø., Scheel, I.: Time-to-event prediction with neural networks and cox regression. arXiv preprint arXiv:1907.00825 (2019)

22. Ley, B., Collard, H.R., King, T.E., Jr.: Clinical course and prediction of survival in idiopathic pulmonary fibrosis. Am. J. Respir. Crit. Care Med. **183**(4), 431–440 (2011)

23. Loshchilov, I., Hutter, F.: Decoupled weight decay regularization. arXiv preprint arXiv:1711.05101 (2017)

24. Pölsterl, S., Wolf, T.N., Wachinger, C.: Combining 3D image and tabular data via the dynamic affine feature map transform. In: de Bruijne, M., et al. (eds.) MICCAI 2021. LNCS, vol. 12905, pp. 688–698. Springer, Cham (2021). https://doi.org/10.1007/978-3-030-87240-3_66

25. Ratner, B.: The correlation coefficient: its values range between$+ 1/-1$, or do they? J. Target. Meas. Anal. Mark. **17**(2), 139–142 (2009)

26. Ronneberger, O., Fischer, P., Brox, T.: U-Net: convolutional networks for biomedical image segmentation. In: Navab, N., Hornegger, J., Wells, W.M., Frangi, A.F. (eds.) MICCAI 2015. LNCS, vol. 9351, pp. 234–241. Springer, Cham (2015). https://doi.org/10.1007/978-3-319-24574-4_28
27. Shamshad, F., et al.: Transformers in medical imaging: a survey. arXiv preprint arXiv:2201.09873 (2022)
28. Uno, H., Cai, T., Pencina, M.J., D'Agostino, R.B., Wei, L.J.: On the c-statistics for evaluating overall adequacy of risk prediction procedures with censored survival data. Stat. Med. **30**(10), 1105–1117 (2011)
29. Walsh, S.L., Humphries, S.M., Wells, A.U., Brown, K.K.: Imaging research in fibrotic lung disease; applying deep learning to unsolved problems. Lancet Respir. Med. **8**(11), 1144–1153 (2020)
30. Wulczyn, E., et al.: Interpretable survival prediction for colorectal cancer using deep learning. NPJ Digital Med. **4**(1), 1–13 (2021)
31. Zhang, H., Cisse, M., Dauphin, Y.N., Lopez-Paz, D.: mixup: beyond empirical risk minimization. arXiv preprint arXiv:1710.09412 (2017)
32. Zheng, M., et al.: End-to-end object detection with adaptive clustering transformer. arXiv preprint arXiv:2011.09315 (2020)
33. Zhong, S.: Efficient online spherical k-means clustering. In: Proceedings of the 2005 IEEE International Joint Conference on Neural Networks, 2005, vol. 5, pp. 3180–3185. IEEE (2005)

Multi-transSP: Multimodal Transformer for Survival Prediction of Nasopharyngeal Carcinoma Patients

Hanci Zheng[1], Zongying Lin[1], Qizheng Zhou[2], Xingchen Peng[3], Jianghong Xiao[4], Chen Zu[5], Zhengyang Jiao[1], and Yan Wang[1(✉)]

[1] School of Computer Science, Sichuan University, Chengdu, China
wangyanscu@hotmail.com
[2] School of Applied Mathematics, New York State Stony Brook University, Stony Brook, NY, USA
[3] Department of Biotherapy, Cancer Center, West China Hospital, Sichuan University, Chengdu, China
[4] Department of Radiation Oncology, Cancer Center, West China Hospital, Sichuan University, Chengdu, China
[5] Department of Risk Controlling Research, JD.COM, Chengdu, China

Abstract. Nasopharyngeal carcinoma (NPC) is a malignant tumor that often occurs in Southeast Asia and southern China. Since there is a need for a more precise personalized therapy plan that depends on accurate prognosis prediction, it may be helpful to predict patients' overall survival (OS) based on clinical data. However, most of the current deep learning (DL) based methods which use a single modality fail to effectively utilize amount of multimodal data of patients, causing inaccurate survival prediction. In view of this, we propose a Multimodal Transformer for Survival Prediction (Multi-TransSP) of NPC patients that uses tabular data and computed tomography (CT) images jointly. Taking advantage of both convolutional neural network and Transformer, the architecture of our network is comprised of a multimodal CNN-Based Encoder and a Transformer-Based Encoder. Particularly, the CNN-Based Encoder can learn rich information from specific modalities and the Transformer-Based Encoder is able to fuse multimodal feature. Our model automatically gives the final prediction of OS with a concordance index (CI) of 0.6941 on our in-house dataset, and our model significantly outperforms other methods using any single source of data or previous multimodal frameworks. Code is available at https://github.com/gluglurice/Multi-TransSP.

Keywords: Nasopharyngeal carcinoma · Survival prediction · Multimodal · Transformer

1 Introduction

As a malignant cancer, nasopharyngeal carcinoma (NPC) is mainly endemic in Southeast Asia and southern China [1]. Radiotherapy is now the primary treatment for NPC

H. Zheng and Z. Lin—Contribute equally to this work.

L. Wang (Eds.): MICCAI 2022, LNCS 13437, pp. 234–243, 2022.
https://doi.org/10.1007/978-3-031-16449-1_23

patients, but the therapeutic effects of different patients vary a lot due to the highly homogeneous therapy plans for them [2]. Thus, we need a more precise personalized therapy plan which depends on accurate prediction of patients' overall survival (OS) based on clinical data like medical images and tabular records. Here, OS is defined as the duration from the diagnosis to death from any cause or the last follow-up, which is also an aspect of prognosis.

Traditional methods of image analysis for NPC are mainly based on radiomics [3], which requires knowledge of feature engineering. In contrast, deep learning (DL) provides an efficient solution for feature extraction [4–6]. Hence, we turn to DL-based methods for survival prediction. Recently, with the rapid development of DL, a number of methods using convolutional neural network (CNN) have been proposed to predict prognosis. For example, Yang et al. [7] used a weakly-supervised DL-based network to achieve automated T staging of NPC, in which T usually refers to the size and extent of the main tumor. In addition, Liu et al. [8] used the neural network DeepSurv to analyze the pathological microscopic features.

However, despite the impressive progress achieved by those methods, most of them focused solely on the feature found in medical images while neglecting clinical tabular data (i.e., age, BMI, Epstein-Barr virus, etc.), which leads to inefficient utilization of the vast amount of available tabular data as well as inaccurate survival prediction for NPC patients. Meanwhile, Huang et al. [9] proved that uni-modality accesses a worse latent space representation than multi-modality does. Therefore, utilizing multimodal medical data is significant for predicting survival of NPC patients. In recent years, some multimodal approaches have been proposed for prognosis prediction [10–16]. For example, Jing et al. [11] proposed an end-to-end multi-modality deep survival network to predict the risk of NPC tumor progression and showed the best performance compared with the traditional four popular state-of-the-art survival methods. Using MR images and clinical data, Qiang et al. [12] established an advanced 3D convolutional neural network-based prognosis prediction system for locally NPC. Besides, Luis et al. [13] proposed a multimodal deep learning method, MultiSurv, by fusing whole-slide images, clinical data, and multiple physiological data to predict patient risk for 33 types of pancreatic cancers automatically. Based on MR images and clinical variables, Zhang et al. [14] built a model to predict the distant metastasis-free survival of locally advanced NPC. Jointly learning image-tabular representation, Chauhan et al. [15] proposed a novel multimodal deep learning network to improve the performance of pulmonary edema assessment. In addition, Guan et al. [16] combined CT scans and tabular data to improve the prediction of esophageal fistula risks.

Nevertheless, those methods are all based on CNN, which has a local receptive field and cannot catch the long-range information hidden in the multimodal fusion feature. Being able to integrate global feature, Transformer has lately shown comparable performance in natural language processing (NLP) and computer vision (CV) [17], such as image generation [18], objective detection [19], and panoptic segmentation [20]. Besides, benefiting from the cross-modal attention (the self-attention between different modalities) which attends to interactions between multimodal sequences and latently integrates multimodal feature, Transformer has also achieved great success in multimodal data. For instance, Huang et al. [21] utilized Transformer to fuse audio-visual modalities on emotion recognition. Tsai et al. [22] introduced a Multimodal Transformer (MulT) to generically address the inherent data non-alignment and long-range dependencies between elements across modalities in an end-to-end manner with explicitly aligning the data. Hu et al. [23] proposed a Unified Transformer to simultaneously learn the most prominent tasks across different domains.

Consequently, to better extract the image feature and fuse cross-modal information while reducing feature variability and heterogeneity between different modalities, we combine CNN and Transformer, proposing a novel Multimodal Transformer for Survival Prediction (Multi-TransSP). Specifically, the entire model consists of a multimodal CNN-Based Encoder, a Transformer-Based Encoder and a fully connected (FC) layer. The multimodal CNN-Based Encoder processes the data of different modalities separately and learns abundant feature information from each other, after which features of different modalities are fused together. Then, in Transformer-Based Encoder, each Transformer block converts the fusion feature into multimodal sequences and predicts an OS value. At last, all predicted OS values go through an FC layer and the final predicted OS can be obtained.

Our contribution can be summarized as follow: (1) We propose a novel method for predicting the survival of NPC patients, which helps radiotherapists make more precise and personalized clinical therapy plans. (2) A novel end-to-end multimodal network based on Transformer is proposed to jointly learn the global feature correlations among all modalities to predict the survival of NPC patients. *To our knowledge, we are the first to use Transformer for the task of predicting the survival of NPC patients using multimodal information.* (3) Our model jointly learns the CT image data and the tabular data, giving the survival prediction of NPC with a concordance index (CI) of 0.6941 and a mean squared error (MSE) of 0.02498 on our in-house dataset.

2 Methodology

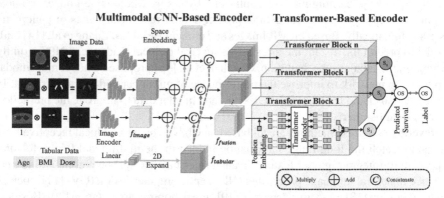

Fig. 1. Illustration of the proposed multimodal prediction network architecture.

The architecture of the proposed network is depicted in Fig. 1, which mainly consists of a multimodal CNN-Based Encoder and a Transformer-Based Encoder. The multimodal CNN-Based Encoder receives image data and the corresponding tabular data as input. Since all slices are processed the same way, we only describe how one slice goes through our model in the following sections. Specifically, the CNN-Based Encoder captures the image feature from a CT slice of a patient, which is multiplied by the corresponding segmentation map indicating where the tumors are in advance. Meanwhile, to maintain the spatial information among the slices, we construct the space embedding and add it to the image feature. As for tabular data, they are represented as high-dimensional feature fusible with other modalities by 2D expansion. Then, different kinds of features from the two modalities are concatenated as the new input to the next Transformer-Based Encoder. The Transformer block absorbs the fusion feature and outputs a predicted survival value. Finally, the predicted survival from all slices is integrated and turns into one survival value through an FC layer. Next, we will describe the two main sub-networks of our proposed multimodal deep network in detail.

2.1 Multimodal CNN-Based Encoder

CT Image Feature Extraction. Before convolution, we first split the 3D CT image into 2D slices along its axis to keep a lower computational consumption. Then the slices are multiplied by corresponding segmentation maps. Next, we use Resnet18 [24] as the backbone to extract image feature from the CT slices.

In the convolution process, we take a 2D CT slice image $I \in \mathbb{R}^{H \times W}$ as input. The image feature is the output from the CT image feature extraction network, represented as $f_{image} = ImageEncoder(I) \in \mathbb{R}^{H_I \times W_I \times C_I}$, where C_I, H_I and W_I are the channel, height and width of the image feature respectively.

Space Embedding. Considering that using only 2D slices will lose part of the spatial location information in 3D CT images and it may be difficult to align the slices of all patients due to the difference among the numbers of slices of every patient, we add each 2D slice with its corresponding space embedding. This embedding, which we denote as $S \in \mathbb{R}^{H_I \times W_I}$, is randomly initialized and learned in the training process, just in case the model aligns different slices of different patients if we simply apply a relative serial number.

Similar to the process of 2D expansion in transforming tabular feature, we expand the spatial feature tensor along the channel dimension to be the same size as the image feature. The expanded spatial position feature is represented as $f_{space} \in \mathbb{R}^{H_I \times W_I \times C_I}$. Then, corresponding voxel in the same spatial position of f_{space} and f_{image} are added up directly and a new image feature including spatial location information is formed. The new image feature can be expressed as $f_{SI} = f_{space} + f_{image}$.

Tabular Feature Construction. Tabular data T has two types, numerical tabular data and categorical tabular data. Specifically, categorical tabular data (e.g., gender, TNM stage) are further numerically processed, and numerical tabular data (e.g., age, BMI) are normalized to the range of [0, 1]. The tabular data concatenated to the image feature right after the space embedding is added to the image feature. Since dimensions of the imaging feature and tabular data are different, we need to align them by either expanding tabular data or reducing the imaging feature. Here, we apply the former method to aggregate image and tabular data in a more detailed dimension.

The two types of tabular data then form a $1 \times D_T$ tabular vector, which is illustrated in Fig. 2. To explore the correlations between the elements in the tabular vector and the slice, we transform the original tabular data linearly and expand the tabular vector to be the same size as the image feature, making them align. More specifically, for each element in the tabular vector, we will copy its value and construct a matrix matching the slice image width W_I, and height H_I. Thus, we obtain the tabular feature denoted as $f_{tabular} \in \mathbb{R}^{H_I \times W_I \times D_T}$.

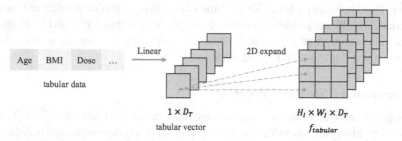

Fig. 2. The process of transforming tabular data to tabular feature.

2.2   Transformer-Based Encoder

In Transformer-Based Encoder, we make f_{SI} and $f_{tabular}$ jointly aligned and fused by concatenating them along the channel dimension, producing the fusion feature denoted as $f_{fusion} \in \mathbb{R}^{H_I \times W_I \times (C_I + D_T)}$. In Transformer block, considering the Transformer encoder expects sequence-style as input, we treat each voxel of the feature along the channel dimension as a "sequence of fusion feature", before which we transform the channel of f_{fusion} to C to fit Transformer encoder. Further, we get multimodal fused sequences by adding position embedding to help Transformer encoder learn the sequence order. The subordinate vectors of the channel dimension C are denoted as $[c_1, c_2, ..., c_L]$, thus generating L sequences, where $L = H_I \times W_I$.

The Transformer encoder receives the L sequences and outputs a vector denoted as $V \in \mathbb{R}^{L*C}$ to form C predicted values. The C values generated from the fusion feature of the i-th slice then assemble to generate an element S_i of the survival vector S through FC. The process can be expressed as:

$$S_i = W_T \times TransformerEncoder\left(f_{fusion}\right) + B_T \tag{1}$$

where W_T and B_T denote the parameters of the FC layer.

As for that survival vector S, we assume each patient has a 3D CT image split to N slices. For each slice, we can get partial information about the survival of that patient, which indicates different spatial and physiological structure information that can determine its weight in predicting the final survival. Therefore, we construct a predicted survival vector S denoted as $S = [s_1, s_2, \dots, s_N]$, which consists of the predicted survival values from each slice of a patient through Transformer block.

Finally, we obtain a predicted survival for the patient by a FC layer and train the entire model by comparing it with the ground truth. At last, the prediction loss function can be expressed as:

$$loss_{MSE} = \frac{1}{N}\sum_{i=1}^{N}\left(y_i - y_i^*\right)^2 \tag{2}$$

where y_i and y_i^* represent the ground truth and the final predicted survival of NPC respectively.

3   Experiment

3.1   Dataset and Preprocessing

Table 1. Description of tabular data.

Characteristics	Number (Proportion)	Characteristics	Number (Proportion)
Staging		Gender (male)	273 (71.09%)
I	1 (0.26%)		
II	58 (15.10%)	PS score (=1)	279 (72.66%)
III	154 (40.10%)		
IVa	122 (31.77%)	EBV before treatment (=1)	178 (46.35%)
IVb	49 (12.76%)	Censored (=1)	376 (97.92%)
		Characteristics	**Average ± standard deviation**
T stage			
T1	138 (35.94%)	Age	47.45 ± 9.94
T2	99 (25.78%)	BMI	23.24 ± 3.11
T3	75 (19.53%)		
T4	72 (18.75%)	Weight loss ratio	-0.02 ± 0.06
		Radiotherapy number	32.94 ± 1.53
N stage		Radiotherapy time (days)	51.14 ± 26.54
N0	25 (6.51%)		
N1	120 (31.25%)	Dosage	71.24 ± 3.52
N2	190 (49.48%)		
N3	49 (12.76%)	OS (month)	52.49 ± 18.44

The experiment is conducted on an in-house dataset, in which there are CT image data, corresponding segmentation map indicating the location of tumors, and tabular data of 384 NPC patients in the period of 2006–2018. Here, we split the dataset to training set and test set with the proportion of 4:1.

For tabular data, which are collected using telephone follow-up, we filled the missing data with mean interpolation and dropped some kinds of data that the contents of almost all patients are the same. The tabular data consist of general, diagnostic, and treatment data, including six categorical attributes and six numerical attributes, which are shown in Table 1 as well. In Table 1, categorical attributes are depicted with a statistical number and its proportion while numerical attributes are depicted with mean ± average deviation. The TNM (tumor node metastasis) staging follows Chinese NPC Staging Standard (2008 version). Since the performance status (PS) score of each patient is either 1 or 0, we only present the statistical data of those whose PS score is 1. "Censored" is not used as one dimension. The label overall survival (OS) takes "month" as unit. Note that for all the tabular data and survival labels, we normalize them to [0, 1] before passing them to the framework, so the unit of MSE loss is also normalized.

And for CT images, we multiplied each CT slice by the corresponding segmentation map, dropped the slices that contain no volume of interest and reconstructed the remaining 3D CT scans by applying a matrix of 332 × 332 pixels at the center, so as to assist the model to focus on the volume of interest and to reduce the computational cost as well. Here, the segmentation maps were manually depicted by experts.

3.2 Experiment Settings

Our network is trained on Pytorch framework and an NVIDIA GeForce GTX 3090 with 24 GB memory. Specifically, we use stochastic gradient descent (SGD) to optimize the network. The proposed network is trained for 300 epochs and the batch size is set to 64. The initial learning rate is set to $1e-3$, and we adjust it with cosine annealing method, where the eta_{min} is set to $1e-6$ and the T_{max} is set to 20.

As for the evaluation metrics, we adopt concordance index (CI) and mean squared error (MSE). Specifically, CI indicates the proportion of patient pairs whose predicted results are consistent with the ground truth in all pairs. In our survival prediction task, for a pair of patients, if the actual survival and predicted survival of one is both longer than the other, we call that pair concordant. CI usually falls in the interval of [0.5, 1], where the higher CI is, the better performance the model will have. As for MSE, it is the mean value of the square sum of the errors between the predicted and the actual value. Contrary to CI, the lower MSE is, the more precise the model is to predict survival.

3.3 Comparison with Other Prognosis Prediction Methods

To demonstrate the superiority of the proposed method in predicting NPC patients' survival, we compare our method with five state-of-the-art prognosis prediction methods: (1) DeepSurv [25]: a Cox proportional hazards deep neural network using clinical and gene expression data for modelling interactions between a patient's covariates and treatment effectiveness, (2) LungNet [26]: a shallow convolutional neural network using only chest radiographs for predicting outcomes of patients with non-small-cell lung cancer, (3) a late fusion technique for fusing multimodal data [27], which was initially designed for skin lesion classification, (4) a joint loss function for multimodal representation learning [15], which was introduced for pulmonary edema assessments from chest radiographs, and (5) MultiSurv [28], a long-term cancer survival prediction using multimodal deep learning. The quantitative comparison results are reported in Table 2, from which we can see that our method attains the best performance. Compared with

[15] which achieves the second-best result, our proposed method still boosts the CI and MSE performance by 0.0471 and 0.00688. Extensive experiments fully demonstrate the efficiency of using multimodal data. Besides, the results also illustrate the excellent performance of our model in fusing cross-modal information and learning the global feature correlations among all modalities. Note that since the forms of data between the comparison methods and our model are different, we replayed them and experimented on our in-house dataset.

Table 2. Quantitative comparison with state-of-the-art methods in terms of MSE and CI

Method		MSE	CI
Tabular only	DeepSurv [25]	0.06224	0.6146
Image only	LungNet [26]	0.03548	0.5089
Tabular + Image	Yap et al. [27]	0.03286	0.6113
	Chauhan et al. [15]	0.03186	0.6470
	MultiSurv [28]	0.04074	0.5739
	Proposed	**0.02498**	**0.6941**

3.4 Ablation Study

Table 3. Ablation study of our method in terms of MSE and CI

Method				MSE	CI
Image	Tabular	Space	Transformer		
	✓		✓	0.02778	0.6563
✓		✓	✓	0.03365	0.6081
✓	✓		✓	0.02716	0.6692
✓	✓	✓		0.03101	0.6299
✓	✓	✓	✓	**0.02498**	**0.6941**

To investigate the contributions of key components of the proposed method, we conduct the ablation study in a leave-one-out way. The experiment arrangement can be concluded as: (1) the proposed model with only tabular data as the input (Tabular + Transformer), (2) the proposed model without tabular data (CT + Space + Transformer), (3) the proposed model without space embedding (CT + Tabular + Transformer), and (4) the proposed model without the transformer block (CT + Tabular + Space). The quantitative results are shown in Table 3, from which we can see that when combining all modalities and Transformer, the performance of the model is the best. Comparing the first three lines with the complete model, we can find that each of the three modalities (i.e., image, tabular and space) improves the performance to certain degree. And we can also find the improvement significant when adding Transformer to this model from the last two lines, revealing its great contribution to the prediction ability of the model.

4 Conclusion

In this paper, we propose a Multimodal Transformer for Survival Prediction (Multi-TransSP) of NPC patients to jointly learn the global feature correlations among all modalities to predict survival. The proposed model effectively combines the local receptive field of CNN and the cross-modal fusion ability of Transformer. Moreover, we convert the tabular data into feature which is fusible with that of other modalities based on 2D expansion, ensuring low-level data to connect with high-dimensional feature. Experiments conducted on our in-house dataset have shown the superiority of our method, compared with other state-of-the-art multimodal approaches in the field of prognosis prediction. Considering the diversity of multimodal data and the scalability of multimodal extraction network, we plan to investigate a more general modal extraction module that can be adapted to more types of modals and more multimodal datasets as well.

Acknowledgement. This work is supported by National Natural Science Foundation of China (NFSC 62071314).

References

1. Hu, L., Li, J., Peng, X., et al.: Semi-supervised NPC segmentation with uncertainty and attention guided consistency. Knowl.-Based Syst. **239**, 108021–108033 (2022)
2. Zhan, B., Xiao, J., Cao, C., et al.: Multi-constraint generative adversarial network for dose prediction in radiotherapy. Med. Image Anal. **77**, 102339–102352 (2022)
3. Lambin, P., Leijenaar, R.T.H., Deist, T.M., et al.: Radiomics: the bridge between medical imaging and personalized medicine. Nat Rev Clin Oncol **14**, 749–762 (2017)
4. Wang, Y., Zhou, L., Yu, B. et al.: 3D auto-context-based locality adaptive multi-modality GANs for PET synthesis. IEEE Trans. Med. Imaging **38**, 1328–1339 (2019)
5. Luo, Y., Zhou, L., Zhan, B., et al.: Adaptive rectification based adversarial network with spectrum constraint for high-quality PET image synthesis. Med. Image Anal. **77**, 102335–102347 (2022)
6. Wang, K., Zhan, B., Zu, C., et al.: Semi-supervised medical image segmentation via a tripled-uncertainty guided mean teacher model with contrastive learning. Med. Image Anal. **79**, 102447–102460 (2022)
7. Yang, Q., Guo, Y., Ou, X., et al.: Automatic T staging using weakly supervised deep learning for nasopharyngeal carcinoma on MR images. J. Magn. Reson. Imaging **52**, 1074–1082 (2020)
8. Liu, K., Xia, W., Qiang, M., et al.: Deep learning pathological microscopic features in endemic nasopharyngeal cancer: prognostic value and protentional role for individual induction chemotherapy. Cancer Med **9**, 1298–1306 (2020)
9. Huang, Y., Zhao, H., Huang, L.: What Makes Multi-modal Learning Better than Single (Provably). arXiv preprint arXiv: 2106.04538 [Cs] (2021)

10. Shi, Y., Zu, C., Hong, M., et al.: ASMFS: adaptive-similarity-based multi-modality feature selection for classification of Alzheimer's disease. Pattern Recogn. **126**, 108566–108580 (2022)
11. Jing, B., Deng, Y., Zhang, T., et al.: Deep learning for risk prediction in patients with nasopharyngeal carcinoma using multi-parametric MRIs. Comput. Methods Programs Biomed. **197**, 105684–105690 (2020)
12. Qiang, M., Li, C., Sun, Y., et al.: A prognostic predictive system based on deep learning for locoregionally advanced nasopharyngeal carcinoma. J. Natl Cancer Inst. **113**, 606–615 (2021)
13. Vale-Silva, L.A., Rohr, K.: Pan-cancer prognosis prediction using multimodal deep learning. In: IEEE 17th International Symposium on Biomedical Imaging, pp. 568–571. IEEE (2020)
14. Zhang, L., Wu, X., Liu, J., et al.: MRI-based deep-learning model for distant metastasis-free survival in locoregionally advanced nasopharyngeal carcinoma. J. Magn. Reson. Imaging **53**, 167–178 (2021)
15. Chauhan, G., et al.: Joint modeling of chest radiographs and radiology reports for pulmonary edema assessment. In: Martel, A.L., et al. (eds.) MICCAI 2020. LNCS, vol. 12262, pp. 529–539. Springer, Cham (2020). https://doi.org/10.1007/978-3-030-59713-9_51
16. Guan, Y., et al.: Predicting esophageal fistula risks using a multimodal self-attention network. In: de Bruijne, M., et al. (eds.) MICCAI 2021. LNCS, vol. 12905, pp. 721–730. Springer, Cham (2021). https://doi.org/10.1007/978-3-030-87240-3_69
17. Lin, T., Wang, Y., Liu, X. et al.: A Survey of Transformers. arXiv preprint arXiv:2106.04554 [cs] (2021)
18. Parmar, N., Vaswani, A., Uszkoreit, J. et al.: Image Transformer. arXiv preprint arXiv:1802.05751v3 [cs] (2018)
19. Carion, N., Massa, F., Synnaeve, G., Usunier, N., Kirillov, A., Zagoruyko, S.: End-to-end object detection with transformers. In: Vedaldi, A., Bischof, H., Brox, T., Frahm, J.-M. (eds.) ECCV 2020. LNCS, vol. 12346, pp. 213–229. Springer, Cham (2020). https://doi.org/10.1007/978-3-030-58452-8_13
20. Wang, H., Zhu, Y., Adam, H. et al.: MaX-DeepLab: end-to-end panoptic segmentation with mask transformers. In IEEE Conference on Computer Vision and Pattern Recognition, pp. 5459–5470. IEEE (2021)
21. Huang, J., Tao, J., Liu, B. et al.: Multimodal transformer fusion for continuous emotion recognition. In: IEEE International Conference on Acoustics, Speech and Signal Processing, pp. 3507–3511. IEEE (2020)
22. Tsai, Y. H., Bai, S., Liang, P. P. et al.: Multimodal transformer for unaligned multimodal language sequences. In: Proceedings of the 57th Annual Meeting of the Association for Computational Linguistics, pp. 6558–6569 (2019)
23. Hu, R., Singh, A.: UniT: multimodal multitask learning with a unified transformer. arXiv preprint arXiv:2102.10772 [cs] (2021)
24. He, K., Zhang, X., Ren, S. et al.: Deep residual learning for image recognition. In: IEEE Conference on Computer Vision and Pattern Recognition, pp. 770–778. IEEE (2016)
25. Katzman, J.L., Shaham, U., Cloninger, A., et al.: DeepSurv: personalized treatment recommender system using a Cox proportional hazards deep neural network. BMC Med. Res. Methodol. **18**, 24–35 (2018)
26. Mukherjee, P., Zhou, M., Lee, E., et al.: A shallow convolutional neural network predicts prognosis of lung cancer patients in multi-institutional CT-Image data. Nat. Mach. Intell. **2**, 274–282 (2020)
27. Yap, J., Yolland, W., Tschandl, P.: Multimodal skin lesion classification using deep learning. Exp. Dermatol. **27**, 1261–1267 (2018)
28. Vale-Silva, L.A., Rohr, K.: Long-term cancer survival prediction using multimodal deep learning. Sci. Rep. **11**, 13505–13516 (2021)

Contrastive Masked Transformers for Forecasting Renal Transplant Function

Leo Milecki[1]([⊠]), Vicky Kalogeiton[2], Sylvain Bodard[3,6], Dany Anglicheau[4,6], Jean-Michel Correas[3,6], Marc-Olivier Timsit[5,6], and Maria Vakalopoulou[1]

[1] MICS, CentraleSupelec, Paris-Saclay University, Inria Saclay,
Gif-sur-Yvette, France
leo.milecki@centralesupelec.fr
[2] LIX, École Polytechnique/CNRS, Institut Polytechnique de Paris,
Palaiseau, France
[3] Department of Adult Radiology, Necker Hospital, LIB, Paris, France
[4] Department of Nephrology and Kidney Transplantation, Necker Hospital,
Paris, France
[5] Department of Urology, HEGP, Necker Hospital, Paris, France
[6] UFR Médecine, Paris-Cité University, Paris, France

Abstract. Renal transplantation appears as the most effective solution for end-stage renal disease. However, it may lead to renal allograft rejection or dysfunction within 15%–27% of patients in the first 5 years post-transplantation. Resulting from a simple blood test, serum creatinine is the primary clinical indicator of kidney function by calculating the Glomerular Filtration Rate. These characteristics motivate the challenging task of predicting serum creatinine early post-transplantation while investigating and exploring its correlation with imaging data. In this paper, we propose a sequential architecture based on transformer encoders to predict the renal function 2-years post-transplantation. Our method uses features generated from Dynamic Contrast-Enhanced Magnetic Resonance Imaging from 4 follow-ups during the first year after the transplant surgery. To deal with missing data, a key mask tensor exploiting the dot product attention mechanism of the transformers is used. Moreover, different contrastive schemes based on cosine similarity distance are proposed to handle the limited amount of available data. Trained on 69 subjects, our best model achieves 96.3% F1 score and 98.9% ROC AUC in the prediction of serum creatinine threshold on a separated test set of 20 subjects. Thus, our experiments highlight the relevance of considering sequential imaging data for this task and therefore in the study of chronic dysfunction mechanisms in renal transplantation, setting the path for future research in this area. Our code is available at https://github.com/leomlck/renal_transplant_imaging.

Keywords: Sequential architectures · Missing data · Contrastive learning · Renal transplant · MRI

Supplementary Information The online version contains supplementary material available at https://doi.org/10.1007/978-3-031-16449-1_24.

1   Introduction

Renal transplantation appears as the most effective solution for end-stage renal disease and highly improves patients' quality of life, mainly by avoiding periodic dialysis [24]. However, a substantial risk of transplant chronic dysfunction or rejection persists and may lead to graft loss or ultimately the patient death [11]. The genesis of such events takes place in heterogeneous causes, complex phenomena, and results from a gradual decrease in kidney function. In clinical practice, the primary indicator of kidney function is based on blood tests and urine sampling (serum creatinine, creatinine clearance). However, when results are irregular, the gold standard method is needle biopsy, an invasive surgical operation. Thus, the need for a non-invasive alternative that could provide valuable information on transplant function post-transplantation through time is crucial.

Medical imaging plays a significant role in renal transplantation. In [21], diverse imaging modalities have been investigated to assess renal transplant functions in several studies. Moreover, in [17] multiple Magnetic Resonance Imaging (MRI) modalities are used for the unsupervised kidney graft segmentation. Beyond the respective limitations of the several imaging modalities, such as the necessity of radiations or the intrinsic trade-off on resolution, to our knowledge, there are no studies focusing on monitoring the evolution of kidney grafts using imaging data. On the other hand, the recent transformer models [26] offer new directions in processing sequential data. Moreover, recent advances in self-supervised learning [25] enable the training of powerful deep learning representations with a limited amount of data. Renal transplantation datasets usually belong to this case, making the use of such methods the way to move forward. Our study is among the first that explore such methods for renal transplantation, solving challenging clinical questions.

In this work, we propose a method to forecast the renal transplant function through the serum creatinine prediction from follow-up exams of Dynamic Contrast-Enhanced (DCE) MRI data post-transplantation. The main contributions of this work are twofold. First, we propose the use of contrastive schemes, generating informative manifolds of DCE MRI exams of patients undergoing renal transplantation. Different self-supervised and weakly-supervised clinical pertinent tasks are explored to generate relevant features using the cosine similarity. Secondly, we introduce a transformer-based architecture for forecasting serum creatinine score, while proposing a tailored method to deal with missing data. In particular, our method is using a key mask tensor that highlights the missing data and does not take them into account for the training of the sequential architecture. Such a design is very robust with respect to the position and number of missing data, while it provides better performance than other popular data imputation strategies. To the best of our knowledge, our study is among the first that propose a novel, robust, and clinically relevant framework for forecasting serum creatinine directly from imaging data.

2   Related Work

Several medical imaging approaches investigated the diagnosis of renal transplant dysfunction. Recent studies focused on detecting specific events such as renal fibrosis [18] or acute rejection [14]. In [22], multi-modal MRI and clinical data are explored to assess renal allograft status at the time of the different exams. Most of those approaches seek to, indirectly through related events or directly through complex automated systems, non-invasively retrieve structural, functional, and molecular information to diagnose chronic kidney disease [1].

When it comes to real clinical settings, missing data is one of the most important issues during data curation. Handling of missing data has been thoroughly studied by data imputation methods, which mainly propose approaches to fill the missing data as a pre-processing step to some downstream task [16]. Beyond simple statistical approaches such as sampling the mean or median of available data, methods can be categorized into two groups: discriminative and generative approaches. The former is mainly developed for structural data (discrete or continuous) with methods such as structured prediction [13]. On the other hand, generative approaches include expectation-maximization algorithms [8] or deep learning models such as Generative Adversarial Imputation Nets (GAIN) [29]. Those latest approaches showed very good performance for medical image tasks, as proposed in [5,28]. However, the training of such models usually is subjective to a big amount of data that are not all the time available [12], especially in a clinical setting.

Considering the use of the transformer models, the attention mechanism showed promising results in missing data imputation for structural [27] and trajectory data [2,9]. In particular, the attention mask was used to investigate the robustness of a vanilla encoder-decoder transformer and a Bidirectional Transformer (BERT) model [7] while missing 1 to 6 point's coordinates out of 32 for forecasting the people trajectories. Among all these methods, our method is the first to handle in an efficient and robust way missing data with high dimensionality, tested on sequences with long time dependencies.

3   Method

In this study, we focus on the prediction of serum creatinine from imaging data and in particular DCE MRI, exploring both anatomical and functional information. An overview of our method is presented in Fig. 1.

3.1   Contrastive Learning for Renal Transplant

In this work, we propose two contrastive learning schemes to explore meaningful data representations: (a) a self-supervised scheme, where we learn meaningful features by solving the proxy task of determining if two MRI volumes belong to the same patient, and (b) a weakly-supervised scheme, where we discriminate samples based on the differences in the value of various clinical variables.

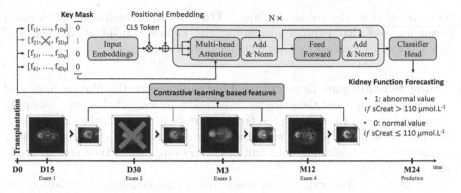

Fig. 1. Overview of the proposed method. Different contrastive schemes are used to represent the different MRIs. These features are used to train a sequential model coupled with a key mask tensor to mark the missing data.

Let us denote $(v_1, v_2) \in (\mathbb{R}^{N_x \times N_y \times N_z})^2$ a pair of MRI regions of interest. Each stream $i = 1, 2$ consists of a ResNet model to extract a latent representation from the MRI volumes, which takes v_i as input and outputs features $z_i \in \mathbb{R}^{D_f}$, with $D_f = 512$ for ResNet18. Then, a feature embedding head associates these features with the underlying task. This is modeled by a linear layer or a Multi-Layer Perceptron (MLP) mapping the features to $(z'_1, z'_2) \in \mathbb{R}^{D_{fe}}$, with $D_{fe} = 256$.

Self-supervised Pre-training. Our first strategy relies on a self-supervised task at the patient level, i.e., we train a model to distinguish if a pair of volumes comes from the same patient or not. $P_j = \{v \in \mathbb{R}^{N_x \times N_y \times N_z} | v$ from patient $j\}$ for $j \in [\![1, N_p]\!]$, where N_p denotes the number of patients, the set of available volumes from MRI series for each exam and patient. Then, our proxy task is to discriminate pairs by knowing if they belong or not to the same patient, i.e., $y = 1$ if $\exists j \ (v_1, v_2) \in (P_j)^2$; else $y = 0$.

Weakly-Supervised Various Clinical Pre-training. Our second strategy discriminates samples based on the difference of certain clinical variable's value, i.e., $y = 1$ if $\|\mathrm{Var}(v_1) - \mathrm{Var}(v_2)\| < \theta$; else $y = 0$, where $Var(\cdot)$ is a clinicobiological variable and θ a clinically relevant threshold. The clinicobiological variables are suggested by nephrology experts to encode clinical priors and information, as they are significantly linked to graft survival [15]. In this paper, we investigate three variables: (1) the transplant incompatibility, (2) the age of the transplant's donor, and (3) the Glomerular Filtration Rate (GFR) value.

Training Loss. From the embedded features (z'_1, z'_2), the optimization is done by the following cosine embedding loss:

$$\mathrm{CosEmbLoss}(z'_1, z'_2, y) = \begin{cases} 1 - \cos(z'_1, z'_2), & \text{if } y = 1, \\ \max(0, \cos(z'_1, z'_2)), & \text{if } y = 0, \end{cases} \quad (1)$$

where cos refers to the cosine similarity. This loss enforces the model to build relevant features that express adequately the kidney transplant imaging and define the way to create strategies to label y each pair.

Training Scheme and Curriculum Learning. Since the dimensionality of our data is very high and the tasks we investigate are very challenging, we apply curriculum learning to facilitate the training process. In particular, for the self-supervised task at the patient level, pairs from the same exam of each patient are enabled in the beginning until half of the training, while they are discarded in the second half.

For the weakly-supervised task based on a clinicobiological variables, the perplexity of the task is determined by the thresholds θ. More specifically, the training labels are adjusted every e_i epochs following the rule: $y = 1$ if $|\text{Var}(v1) - \text{Var}(v2)| < \theta_{i,1}$; $y = 0$ if $|\text{Var}(v_1) - \text{Var}(v_2)| > \theta_{i,2}$; else discard the pair (v_1, v_2), where $\theta_{i,1}$, $\theta_{i,2}$ are set in the image of $Var(\cdot)$ satisfying $\forall i$ (1) $\theta_{i,1} \leq \theta_{i,2}$; (2) $\theta_{i+1,1} \leq \theta_{i,1}$; and (3) $\theta_{i,2} \leq \theta_{i+1,2}$. Our loss enforces the feature pairs to be near or far in the feature embedding space, depending on the label y. The condition (1) enables to form a grey area between the two cases, while the conditions (2) and (3) strengthen the constraint through epochs on the difference of value $Var(\cdot)$ between the two pairs to be correctly arranged.

3.2   Sequential Model Architecture

Our forecasting model takes as input $T = 4$ features $z \in \mathbb{R}^{D_f}$ corresponding to the different follow-ups and relies on a transformer encoder architecture [26]. First, these features are mapped to embeddings of size D_{model} using a linear layer, while a special classification token (CLS) is aggregated in the first position to generate an embedded sequence. Then, the core of the transformer encoder architecture stacks N layers on top of learned positional embeddings added to the embedded sequences. Each layer is first composed of a multi-head self-attention sub-layer, which consists of h heads running in parallel. Each head is based on the scaled dot-product attention. Then, a position-wise fully connected feed-forward sub-layer applies an MLP of hidden dimension D_{model} to each position separately and identically. Finally, to perform the classification task, the CLS token output is fed to a linear layer.

Strategy for Missing Data. Our proposed strategy to deal with missing data is applied to the scaled dot product operation, core of each multi-head self-attention sub-layer. For simplicity, we consider here a sub-layer with one head, $h = 1$. The operation takes as input the query Q, key K and value V, which are linear projections of the embedded sequences, with d_k, d_k and d_v dimensions, respectively and performs $\text{Attention}(Q, K, V) = \text{softmax}(\frac{(QK^t)}{\sqrt{d_k}})V$. In this work, we build a key mask tensor $m_k \in \mathbb{R}^T$ based on the availability of exams for each patient so that zero attention is given to missing data both during the training and inference times, i.e. $\forall t \in [\![1, T]\!]$ $m_k[t] = -\infty$ if exam t is available else 0. Thus, our mask cancels the attention on missing exams by Attention$(Q, K, V) =$ softmax$(\frac{(QK^t)}{\sqrt{d_k}} + M_k)V$ where $M_k = [\![m_k m_k ... m_k]\!] \in \mathbb{R}^{T \times d_k}$. For $h > 1$, keys, values, and queries are linearly projected h times with different, learned linear projections, concatenated, and once again projected after the scaled-dot product.

3.3   Implementation Details

Starting with the contrastive learning, we used data augmentation with horizontal flipping and random affine transformation with a 0.5 probability, as well as random Gaussian blur ($\sigma \in [0, 0.5]$) and random Gaussian noise ($\sigma \in [0, 0.05]$), using TorchIO python library [20]. Having approximately a set of pairs of $\binom{V}{2} = \frac{V(V-1)}{2}$, where V is the number of available volumes, we proposed to fix the training set size to $V_t = 5000$. We decided to fix the number of positive samples, as well as its balance to 25%, and to randomly sample every epoch the remaining from the negative samples.

Concerning the optimization of our models, a 10% dropout has been used for the linear layers of both the contrastive and sequential models. For the contrastive model, the Stochastic Gradient Descent optimizer with a momentum equal to 0.9 was used with a starting learning rate of $1e^{-2}$ following a cosine schedule and preceded by a linear warm-up of 5 epochs. The batch size was set to 20 and the model trained for 60 epochs on 4 NVIDIA Tesla V100 GPU using Pytorch [19]. For the transformer, a binary cross-entropy loss (BCE) was used when binarizing the serum creatinine value using a threshold of $110 \, \mu$ mol.L^{-1}, specified by nephrology experts, as a clinically relevant value to assess normal/abnormal renal transplant function at a specific time point. Adam optimizer was used with a starting learning rate of $1e^{-4}$ following the same learning rate scheduler. The batch size was set to 32 and the model was trained for 30 epochs on 1 NVIDIA Tesla V100 GPU. The architecture's hyperparameters were set by grid search and 10-fold cross-validation, providing $N = 2$, $h = 2$, $D_{model} = 768$.

4   Data

Our study was approved by the Institutional Review Board, which waived the need for patients' consent. The data cohort corresponds to study reference ID-RCB: 2012-A01070-43 and ClinicalTrials.gov identifier: NCT02201537. All the data used in this study were anonymized. Overall, our imaging data are based on DCE MRI series collected from 89 subjects at 4 follow-up exams which took place approximately 15 days (D15), 30 days (D30), 3 months (M3), and 12 months (M12) after the transplant surgery, resulting in respectively 68, 75, 87, and 83 available scans at each follow-up.

The MRI volumes sized $512 \times 512 \times [64-88]$ voxels included spacing ranging in $[0.78-0.94] \times [0.78-0.94] \times [1.9-2.5]$ mm. All volumes were cropped around the transplant using an automatic selection of the region of interest in order to

Table 1. Quantitative evaluation of the proposed method against other methods. sCreat stands for simple statistics from the serum creatinine and Radiomics for predefined radiomics features [10], including shape, intensity, and texture imaging features. We report in format mean(std): Precision (Prec), Recall (Rec), F1 score, and ROC AUC (AUC). Bold indicates the top-performing combination.

Method	Features	Validation				Test			
		Prec	Rec	F1	AUC	Prec	Rec	F1	AUC
LSTM	sCreat	80,5(12,3)	62,9(21,0)	71,1(13,8)	80,4(22,4)	83,3	76,9	80,0	83,5
	Radiomics [10]	86,2(14,9)	73,5(15,5)	77,3(8,2)	80,7(16,0)	90,9	76,9	83,3	84,6
	Imagenet [6]	85,5(15,0)	68,0(17,7)	74,0(12,8)	91,0(10,8)	90,9	76,9	83,3	81,3
	Kinetics [23]	**90,7(9,4)**	74,0(21,5)	78,5(11,3)	**91,4(8,5)**	92,3	92,3	92,3	85,7
	MedicalNet [3]	86,5(13,9)	78,5(21,2)	79,8(13,2)	82,7(18,8)	57,1	61,5	59,3	41,8
	SimCLR [4]	79,8(15,9)	86,5(24,1)	80,9(17,2)	91,8(13,7)	72,2	**100,0**	83,9	64,8
	Proposed GFR	82,8(9,6)	**95,5(9,1)**	**88,3(7,7)**	88,3(13,1)	86,7	**100,0**	92,9	**98,9**
Transformer	sCreat	79,0(28,7)	60,2(31,1)	65,4(29,3)	71,6(24,2)	81,3	**100,0**	89,7	86,8
	Radiomics [10]	81,3(15,7)	66,0(28,6)	69,1(20,2)	65,3(30,5)	90,9	76,9	83,3	91,2
	Imagenet [6]	58,4(22,4)	76,5(34,8)	65,8(27,5)	45,5(21,6)	65,0	**100,0**	78,8	58,2
	Kinetics [23]	53,2(35,8)	66,0(44,8)	58,3(38,9)	64,0(19,7)	65,0	**100,0**	78,8	83,5
	MedicalNet [3]	65,5(27,9)	58,0(33,2)	58,3(28,3)	64,8(19,6)	75,0	46,2	57,1	50,6
	SimCLR [4]	58,9(30,9)	75,5(38,7)	65,6(33,2)	64,8(23,5)	68,4	**100,0**	81,3	72,5
	Proposed GFR	86,3(20,9)	71,5(22,7)	77,4(20,6)	79,7(20,7)	**92,9**	**100,0**	**96,3**	**98,9**

reduce dimensionality while no information about the transplant is discarded. Intensity normalization was executed to each volume independently by applying standard normalization, clipping values to $[-5, 5]$ and rescaling linearly to $[0, 1]$.

As a primary indicator of the kidney function assessment, all patients were subject to blood tests regularly a few days before the transplantation to several years after, to measure the serum creatinine level in $\mu mol.L^{-1}$. The serum creatinine target prediction value is calculated as the mean over an interval of two months before and after the prediction date, 2-year post-transplantation (M24).

5   Experiments and Analysis

To evaluate the performance of our proposed method and compare it with other strategies for the forecasting of serum creatinine, four evaluation metrics are used: recall, precision, F1 score, and the area under the receiver operating characteristic curve (ROC AUC). A testing set of 20 patients is separated from the train set and used to validate the performance of our models. We perform a 10-fold cross-validation (CV) on the train set (69 patients) and report the mean (standard deviation) scores in % for each fold. During CV, the model reaching the minimum loss is saved, and an ensemble approach is used to make the final prediction on the test set from models, which reach more than 50% ROC AUC out of the 10 folds.

We compare our sequential model to an LSTM model, which is a commonly used architecture for sequential data, and which architecture was set using the same approach as our main model, resulting in 2 LSTM cells and a hidden size of 768. Additional sets of feature representations were used to compare the significance of our approach. First simple statistics from the serum creatinine captured from the available blood test results between each follow-up (number of points, mean, median, standard deviation, minimum, maximum) are calculated and used as input to the models. Second, a set of predefined radiomics features [10] are obtained from the segmentation of the kidney transplant following the unsupervised method presented in [17]. Finally, we investigate generating MRI features from SimCLR [4] contrastive scheme, while we report the performance of different transfer-learning approaches, pre-trained on ImageNet [6] by duplicating the weights to 3D, Kinetics [23], and medical image datasets MedicalNet [3].

Quantitative results for all the methods are reported in Table 1. Our proposed approach outperforms the rest of the methods for the test. Both LSTMs and transformers architectures report good performances, with only a few models reporting performance lower than 60% on every metric. Interestingly, our method outperforms the sCreat model which models directly the serum creatinine level. Moreover, our GFR contrastive-based features report the best performance among all the other features for both LSTMs and transformer formulations. The rest of the pre-training performances are summarised in the supplementary materials. Limitations appear as our model seems to misclassify cases where the patient's serum creatinine is stable and close to the used threshold, during the first two years post-transplantation.

5.1 Ablation Study for Missing Data Strategies

The proposed key mask padding approach for handling missing data is specific to the attention mechanism, hence the transformer model. Thus, we investigate 3 other missing data strategies applicable to both the transformer and LSTM model: (1) filling with zeros strategy (None), (2) filling with the nearest available exam (N.A.), and (3) taking the mean for intermediate exams and fill for first and last (M.+N.A.). Results presented in Table 2 are obtained with the best performing imaging features (proposed using the GFR value).

Our proposed approach to handling missing data reports the best precision, recall, and F1 score and the second-best ROC AUC on the test set. Overall, the different strategies report better performance on transformer based architectures than the LSTMs ones indicating the interest in using such models for this task. Moreover, the M.+N.A. strategy reports a lower precision rate for both LSTM and our sequential model, affirming the difficulty to interpolate imaging features. Both None and N.A. strategies appear to report competitive results, lower however from our proposed.

Table 2. Quantitative evaluation of different strategies for missing data. With none we denote the filling with zero strategy, N.A. the filling with the nearest neighbor exam, and with M.+N.A. the filling with the mean and nearest neighbor exam. Bold indicates the top performing combination.

Method	Strategy	Validation				Test			
		Prec	Rec	F1	AUC	Prec.	Rec	F1	AUC
LSTM	None	80,5(11,5)	81,0(14,3)	80,0(9,6)	73,6(16,9)	86,7	**100,0**	92,9	98,9
	N.A.	82,8(9,6)	**95,5(9,1)**	**88,3(7,7)**	**88,3(13,1)**	86,7	**100,0**	92,9	98,9
	M.+N.A.	81,1(10,8)	93,5(10,0)	86,1(6,8)	84,2(11,0)	81,3	**100,0**	89,6	96,7
Transformer	None	86,2(12,9)	78,5(23,2)	79,7(18,2)	71,5(25,3)	92,3	92,3	92,3	98,9
	N.A.	88,8(20,6)	75,5(21,5)	81,3(20,8)	80,5(22,2)	92,3	92,3	92,3	96,7
	M.+N.A.	**90,5(12,3)**	73,5(17,3)	80,0(12,7)	80,0(18,3)	76,5	**100,0**	86,7	**100,0**
	Proposed	86,3(20,9)	71,5(22,7)	77,4(20,6)	79,7(20,7)	**92,9**	**100,0**	**96,3**	98,9

6 Conclusion

This study proposes a novel transformer based architecture tailored to deal with missing data for the challenging task of serum creatinine prediction 2 years post-transplantation using imaging modalities. First, we show the significant use of contrastive learning schemes for this task. Our trained representations outperform common transfer learning and contrastive approaches. Then, a transformer encoder architecture enables to input the sequential features data per follow-up in order to forecast the renal transplant function, including a custom method to handle missing data. Our strategy performs better than other commonly used data imputation techniques. Those promising results encourage the use of medical imaging over time to assist clinical practice for fast and robust monitoring of kidney transplants.

Acknowledgements. This work was performed HPC resources from the "Mésocentre" computing center of CentraleSupélec and École Normale Supérieure Paris-Saclay supported by CNRS and RégionÎ-de-France.

References

1. Alnazer, I., et al.: Recent advances in medical image processing for the evaluation of chronic kidney disease. Med. Image Anal. **69**, 101960 (2021)
2. Becker, S., Hug, R., Huebner, W., Arens, M., Morris, B.T.: MissFormer: (In-)attention-based handling of missing observations for trajectory filtering and prediction. In: Bebis, G., et al. (eds.) ISVC 2021. LNCS, vol. 13017, pp. 521–533. Springer, Cham (2021). https://doi.org/10.1007/978-3-030-90439-5_41
3. Chen, S., Ma, K., Zheng, Y.: Med3D: transfer learning for 3D medical image analysis. arXiv preprint, arXiv:1904.00625 (2019)
4. Chen, T., Kornblith, S., Norouzi, M., Hinton, G.: A simple framework for contrastive learning of visual representations. In: 37th International Conference on Machine Learning (ICML), vol. 119, pp. 1597–1607. PMLR (2020)

5. Dalca, A.V., et al.: Medical image imputation from image collections. IEEE Trans. Med. Imaging **38**, 504–514 (2019)
6. Deng, J., Dong, W., Socher, R., Li, L.J., Li, K., Fei-Fei, L.: Imagenet: a large-scale hierarchical image database. In: Conference on Computer Vision and Pattern Recognition (CVPR), pp. 248–255. IEEE (2009)
7. Devlin, J., Chang, M.W., Lee, K., Toutanova, K.: BERT: Pre-training of deep bidirectional transformers for language understanding. In: North American Chapter of the Association for Computational Linguistics: Human Language Technologies (NAACL), vol. 1, pp. 4171–4186. Association for Computational Linguistics (2019)
8. García-Laencina, P.J., Sancho-Gómez, J.L., Figueiras-Vidal, A.R.: Pattern classification with missing data: a review. Neural Comput. Appl. **19**(2), 263–282 (2010)
9. Giuliari, F., Hasan, I., Cristani, M., Galasso, F.: Transformer networks for trajectory forecasting. In: 25th International Conference on Pattern Recognition (ICPR), pp. 10335–10342. IEEE (2021)
10. van Griethuysen, J.J.M., et al.: Computational radiomics system to decode the radiographic phenotype. Cancer Res. **77**(21), e104–e107 (2017)
11. Hariharan, S., Israni, A.K., Danovitch, G.: Long-term survival after kidney transplantation. N. Engl. J. Med. **385**(8), 729–743 (2021)
12. Kazeminia, S., et al.: Gans for medical image analysis. Artif. Intell. Med. **109**, 101938 (2020)
13. Keshavan, R.H., Montanari, A., Oh, S.: Matrix completion from a few entries. IEEE Trans. Inf. Theory **56**(6), 2980–2998 (2010)
14. Khalifa, F., et al.: Dynamic contrast-enhanced MRI-based early detection of acute renal transplant rejection. IEEE Trans. Med. Imaging **32**(10), 1910–1927 (2013)
15. Loupy, A., et al.: Prediction system for risk of allograft loss in patients receiving kidney transplants: international derivation and validation study. BMJ **366**, l4923 (2019)
16. Mackinnon, A.: The use and reporting of multiple imputation in medical research - a review. J. Intern. Med. **268**(6), 586–593 (2010)
17. Milecki, L., Bodard, S., Correas, J.M., Timsit, M.O., Vakalopoulou, M.: 3D unsupervised kidney graft segmentation based on deep learning and multi-sequence MRI. In: 18th International Symposium on Biomedical Imaging (ISBI), pp. 1781–1785. IEEE (2021)
18. Orlacchio, A., et al.: Kidney transplant: Usefulness of real-time elastography (RTE) in the diagnosis of graft interstitial fibrosis. Ultrasound Med. Biol. **40**(11), 2564–2572 (2014)
19. Paszke, A., et al.: Pytorch: an imperative style, high-performance deep learning library. In: Advances in Neural Information Processing Systems, vol. 32. Curran Associates, Inc. (2019)
20. Pérez-García, F., Sparks, R., Ourselin, S.: Torchio: a python library for efficient loading, preprocessing, augmentation and patch-based sampling of medical images in deep learning. Comput. Methods Programs Biomed. **208**, 106236 (2021)
21. Sharfuddin, A.: Renal relevant radiology: imaging in kidney transplantation. Clin. J. Am. Soc. Nephrol. **9**(2), 416–429 (2014)
22. Shehata, M., et al.: A deep learning-based cad system for renal allograft assessment: diffusion, bold, and clinical biomarkers. In: International Conference on Image Processing (ICIP), pp. 355–359. IEEE (2020)
23. Smaira, L., Carreira, J., Noland, E., Clancy, E., Wu, A., Zisserman, A.: A short note on the kinetics-700-2020 human action dataset. arXiv preprint, arXiv:2010.10864 (2020)

24. Suthanthiran, M., Strom, T.B.: Renal transplantation. N. Engl. J. Med. **331**(6), 365–376 (1994)
25. Taleb, A., et al.: 3D self-supervised methods for medical imaging. In: Advances in Neural Information Processing Systems, vol. 33, pp. 18158–18172. Curran Associates, Inc. (2020)
26. Vaswani, A., et al.: Attention is all you need. In: Advances in Neural Information Processing Systems, vol. 30. Curran Associates, Inc. (2017)
27. Wu, R., Zhang, A., Ilyas, I., Rekatsinas, T.: Attention-based learning for missing data imputation in holoclean. In: Conference on Machine Learning and Systems (MLsys), vol. 2, pp. 307–325 (2020)
28. Xia, Y., et al.: Recovering from missing data in population imaging - cardiac MR image imputation via conditional generative adversarial nets. Med. Image Anal. **67**, 101812 (2021)
29. Yoon, J., Jordon, J., van der Schaar, M.: GAIN: missing data imputation using generative adversarial nets. In: 35th International Conference on Machine Learning (ICML), vol. 80, pp. 5689–5698. PMLR (2018)

Assessing the Performance of Automated Prediction and Ranking of Patient Age from Chest X-rays Against Clinicians

Matthew MacPherson[1]([✉]), Keerthini Muthuswamy[2], Ashik Amlani[2], Charles Hutchinson[1,3], Vicky Goh[4,5], and Giovanni Montana[1,6]

[1] University of Warwick, Coventry CV4 7AL, UK
matthew.macpherson@warwick.ac.uk
[2] Guy's and St Thomas' NHS Foundation Trust, London, UK
[3] University Hospitals Coventry and Warwickshire NHS Trust, Coventry, UK
[4] Department of Radiology, Guy's and St Thomas' NHS Trust, London, UK
[5] School of Biomedical Engineering and Imaging Sciences, King's College London, London, UK
[6] The Alan Turing Institute, London, UK

Abstract. Understanding the internal physiological changes accompanying the aging process is an important aspect of medical image interpretation, with the expected changes acting as a baseline when reporting abnormal findings. Deep learning has recently been demonstrated to allow the accurate estimation of patient age from chest X-rays, and shows potential as a health indicator and mortality predictor. In this paper we present a novel comparative study of the relative performance of radiologists versus state-of-the-art deep learning models on two tasks: (a) patient age estimation from a single chest X-ray, and (b) ranking of two time-separated images of the same patient by age. We train our models with a heterogeneous database of 1.8M chest X-rays with ground truth patient ages and investigate the limitations on model accuracy imposed by limited training data and image resolution, and demonstrate generalisation performance on public data. To explore the large performance gap between the models and humans on these age-prediction tasks compared with other radiological reporting tasks seen in the literature, we incorporate our age prediction model into a conditional Generative Adversarial Network (cGAN) allowing visualisation of the semantic features identified by the prediction model as significant to age prediction, comparing the identified features with those relied on by clinicians.

Keywords: Chest X-rays · Age prediction · GAN · Deep learning

1 Introduction

The aging process is an important part of medical image analysis, since the physiological changes associated with age form a baseline for 'normality' in a

Supplementary Information The online version contains supplementary material available at https://doi.org/10.1007/978-3-031-16449-1_25.

given modality; observations which might be considered in the normal range for an elderly patient can indicate a pathological finding in a younger subject [6]. Understanding the characteristic features of the aging process as manifested in a particular modality is therefore of great importance in accurate and informative clinical reporting, and incorporating this knowledge into machine learning based diagnostic approaches could contribute to greater clinical utility than an 'age blind' model of abnormality presentation. Age prediction from chest X-rays is also known to be a challenging task for radiologists [5], and a greater understanding of the salient features learned in a reliable age prediction model could be of value in improving clinical understanding.

In this paper we leverage a set of 1.8M chest X-rays gathered from a range of clinical settings in six hospitals to train age prediction and ranking models for adult patients. We compare the performance of these models against three radiologists in a novel study testing age prediction and age ranking relative ability to explore clinicians' ability to perceive age-related changes in a single patient over time. Furthermore, to understand the features relied on by the models we use GAN-based age-conditional image generation, allowing progressive re-aging of a synthetic 'patient' and visualisation of age-relevant changes.

Related Work. Deep learning has been used to estimate patient age in several modalities, including neurological MRI scans [2] and paediatric hand X-rays [12]. In the chest X-ray domain, regression models have been applied to the NIH ChestX-ray14 [21] and CheXpert [9] public datasets, and report association of high prediction error with image abnormalities [10,19]. Patient age and gender are predicted from a proprietary dataset with a CNN in [22], with Grad-CAM [20] used to visualise characteristic features. Recent work has shown that over-estimated patient age is predictive of cardiovascular and all-cause mortality rates [8,16]. GANs [4] have previously been used in a variety of medical image processing applications [23]. Controllable image features in GAN images of mammograms, cell histology and brain MRI datasets have been explored recently [1,3,17]. While human face re-aging with GANs has previously been explored [7], to the best of our knowledge GAN-based age-conditional generation of chest X-rays has not been previously demonstrated.

Contributions. The main contributions of this work are: **(i)** We present a state-of-the-art chest X-ray age prediction model trained on a large heterogeneous dataset, with sensitivity analysis to training set size and image resolution, and show generalisation performance on the public dataset NIH ChestX-ray14. **(ii)** We present a study comparing the performance of human radiologists against our model on two tasks: (a) ground truth age prediction from a chest X-ray; (b) ranking two time-separated scans of the same patient in age order. **(iii)** We use GAN-generated synthetic chest X-rays conditioned on patient age to intuitively visualise age-relevant features via simulated age progression.

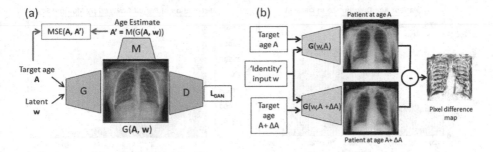

Fig. 1. (a) AC-GAN architecture for age-conditional image generation: G and D are the GAN Generator and Discriminator, and M is the pre-trained age prediction network (frozen during GAN training). (b) Aging feature visualisation via synthetic re-aging.

2 Data and Methods

2.1 Dataset

We train our models using a set of 1.798 million chest X-rays gathered under research partnerships with three Hospital Networks (UK NHS Trusts) comprising six hospitals covering a four million patient population: University Hospitals Coventry and Warwickshire NHS Trust, University Hospitals Birmingham NHS Foundation Trust, and University Hospitals Leicester NHS Trust. The data comprise scans taken from various departments and hardware representing both inpatient and outpatient settings covering the period 2006–2019, along with patient metadata including age, gender, anonymised patient ID, and free text radiologist's report. Images were normalised directly from the DICOM pixel data, and padded to square images with symmetric black borders where required.

After eliminating non-frontal and post-processed images, we obtained a core set of 1,660,060 scans from 955,142 unique patients, comprising 480,147 females and 474,995 males with mean age 61.2 ± 19.0 years. 688,568 patients had a single scan in the set, while 266,574 had multiple scans. The age distribution and longitudinal characteristics are presented in Fig. 2. The images represent an unfiltered, unaligned 'in-the-wild' set of images including both healthy patients and those with abnormalities, and to the best of our knowledge is the largest and most heterogeneous such dataset reported to date. We divide the images into a training set of 1,460,060 images, a validation set of 100k images, and a held-back test set of 100k images sampled from the same age and gender distribution.

2.2 Age Prediction and Ranking Study

We performed a study aiming to address two questions: whether the ability of radiologists to order two images of the same patient by age is superior to their ability to estimate true patient age from a single image, and how their performance on each task compared to data-driven predictive models. To test

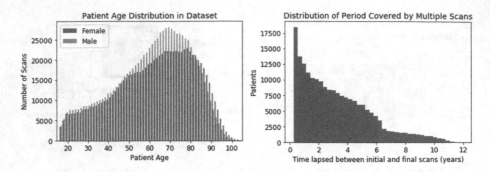

Fig. 2. Characteristics of the dataset. (left) Patient age distribution stratified by gender. (right) Distribution of time separation between first and last images for multiple patient scans (70,493 patients separated by less than 3 months not shown).

this, we developed a web application which presented pairs of X-rays and required the users (working independently) to click on either the image they believed was older or a 'not sure' response, and also give an age estimate for one image (see supplementary material for the website interface). Our test data comprised 200 patients with pairs of images separated by up to 10 years, with 40 pairs from each two year separation bucket. The images were screened as 'normal' by a pre-trained NLP reading of the associated free-text radiologist's reports attached to the image [15], ensuring that only physiological changes resulting from the aging process were present. By presenting pairs of time-separated images, we aimed to establish both the ability to estimate ground truth age (which previous work has indicated is a challenging task for which radiologists are not specifically trained), and over what time period age-related physiological changes become perceptible to an expert human observer with inter-subject differences removed. To the best of our knowledge this aspect has not been previously investigated.

2.3   Age Prediction Models

Given input images x_i with continuous age labels a_i, we aim to train predictive models for patient age formulated as regression and ranking problems. We investigate the accuracy of alternate forms of output prediction model on a convolutional neural network (CNN) backbone: (a) linear regression, taking a single rectified output directly as an age estimate; (b) categorical classification [18], taking an output over nominal integer age labels $c_j, j \in [0, ..., 105]$ and treating the softmax-activated output as a probability distribution with the weighted sum $\sum p_j c_j$ as our estimate; and (c) ordinal regression [13], again outputting over integer age labels but applying sigmoid activation and treating each output as the probability of the age exceeding that label, estimating the age as $\sum p_j$. For pairwise image ranking we train a task-specific model, taking as input two images $\{x_j, x_k\}$ with a binary label $y_{j,k}$ indicating if $a_k > a_j$. The images are passed through a shared CNN and the features concatenated before a fully connected module outputs a binary prediction of the ranking label. By learning simultaneously from a pair of images we expect the model to learn discriminative features directly relevant to fine-grained age discrimination and outperform on the ranking task.

For our CNN backbone we used the feature extraction layers of the EfficientNet-B3 architecture beginning from the ImageNet pre-trained weights. This was followed by the relevant fully connected prediction module with Xavier weight initialisation. An ADAM optimizer with initial learning rate 0.001, adaptive learning rate decay with a 0.5 decay factor and a three epoch patience level was used. We report test set results for the lowest validation set error epoch. All models were trained using an Nvidia DGX1 server with batch sizes 1024 (224^2 resolution), 512 (299^2 resolution), 224 (512^2 resolution) and 52 (1024^2 resolution).

2.4   Age Conditional Image Generation and Manipulation

To visualise the aging features identified by the model, we incorporated our best performing pre-trained age prediction model at 299^2 resolution into a GAN to allow age-conditional synthetic image generation. By comparing images of the same synthetic patient at different ages with nuisance features such as pose held constant, we can visualise the features the age prediction model has identified as relevant to the aging process. To achieve this, we use an auxiliary classifier GAN (AC-GAN) architecture [14], illustrated in Fig. 1, based on StyleGan2 [11]. During training the generated images are passed into both the frozen pre-trained age prediction network M to provide a training gradient for the age targeting accuracy of the generator G, and the discriminator network D to enforce generated image realism. We concatenate the age target A with the noise input $w \in \mathcal{R}^{512}$ into the StyleGan intermediate latent space. The network is trained over a GAN minmax value function with an additional MSE loss term between A and the predicted age of the generated image $G(A, w)$:

$$\min_G \max_D V(D, G) = E_x[\log D(x)] + E_w[\log(D(G(A, w)))] + \lambda(A - M(G(A, w)))^2$$

(1)

Leakage of age-relevant factors between w and A will degrade the age-prediction loss, therefore we expect aging features to be effectively separated from other factors of variation even though we do not explicitly enforce patient identity consistency. We therefore expect modifying A with a fixed w will perform synthetic re-aging of the same synthetic 'patient'. In order to visualise the age-related changes in the 'patient' we generate two images $\tilde{X}_1 = G(A, w)$ and $\tilde{X}_2 = G(A + \delta A, w)$, and generate a difference map of pixel intensity $\tilde{X}_2 - \tilde{X}_1$ to visualise areas of greatest change in the image.

3 Results

3.1 Age Prediction Accuracy

Model Architectures and Resolution Comparison. We aimed to establish the effect of different prediction models on age prediction accuracy, and summarise our results in Table 1 (left) as mean absolute prediction error (MAE) in years. We observe little difference in accuracy between the three modelling approaches at 299^2 resolution with MAEs of 3.33–3.34, leading us to surmise that the performance of the CNN is saturated with this size of dataset and the specific modeling approach is only marginally relevant. We also investigated using DenseNet-169 CNN (14.1M parameters) in place of EfficientNet-B3 (12M parameters), but obtained a slightly lower performance (3.40 vs 3.33 years). We elected to retain the simple regression model 'EfficientNet-LR' for the following work since a more complex approach does not appear to be supported by these results.

Increased image resolution improves model accuracy significantly up to 512^2, with only a minor improvement seen from 3.05 to 2.95 MAE when increasing to 1024^2 (Fig. 3, right). We infer that age-relevant information is not purely limited to the coarse structural features in the image but is also present in intermediate-level features, while fine detail is less relevant. We observe that taking a mean of estimates at each of the four resolution levels for each image reduces MAE further to 2.78 years, which to the best of our knowledge is the best accuracy reported in a heterogeneous non-curated dataset. We note increased error with patient age, possibly attributable to higher levels of clinical abnormalities with age (see supplementary material).

Table 1. Summary MAEs of networks trained. Efficient=EfficientNet-B3, LR = Linear Regression, CL = Classifier, OR = Ordinal Regression.

Network	Resolution	MAE	Network	Resolution	MAE
DenseNet + LR	299 × 299	3.40	Efficient + LR	224 × 224	3.64
Efficient + LR	299 × 299	**3.33**	Efficient + LR	299 × 299	3.33
Efficient + CL	299 × 299	3.34	Efficient + LR	512 × 512	3.05
Efficient + OR	299 × 299	3.34	Efficient + LR	1024 × 1024	**2.95**

Effect of Number of Training Images on Model Accuracy. We investigated the extent to which the large dataset available contributes to prediction accuracy by training the Efficient-LR model at 299^2 resolution on training sets from 10k to 1.597M images. Test set MAE is shown in Fig. 3 (left). MAE vs training set size empirically follows a logarithmic decline (shown as best fit), with diminishing improvements seen up to the full dataset size; we conclude that larger datasets are unlikely to be a major driver of improved performance.

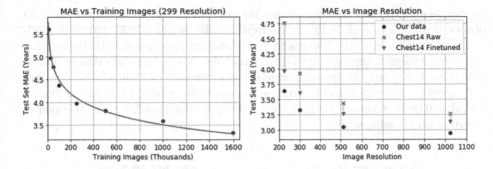

Fig. 3. (left) EfficientNet-LR model test set MAE (299^2 resolution) versus number of images in training data with logarithmic best fit. (right) Test set MAE versus training image resolution for our internal test set and Chest14.

Generalisation Performance on NIH ChestX-ray14. Performance of our pre-trained models on the Chest14 public dataset is shown in Fig. 3 (right). This dataset is provided as jpg files with unknown normalisation; we show results before and after fine-tuning on the Chest14 training data to adapt to the unknown normalisation. After fine-tuning we observe an MAE of 3.10 years at 1024^2 resolution, 5% higher than for our own data; this could be due to information lost in pre-processing and image compression or different population characteristics and warrants further investigation. Direct performance comparisons with the literature are challenging, but we note that for the same dataset [8] reports MAE of 3.78 years, while [10] reports 67.5% within ±4 years vs 87.3% for our model. We conclude that pre-training on our dataset gives a significant performance increase over prior results.

3.2 Clinician vs Algorithm Performance Study

Results from the human vs model study are summarised in Fig. 4. In the age estimation task (left) the regression model achieves an R^2 of 0.94, MAE 3.53 years and mean error (ME) -2.23 ± 3.86 years, whereas the radiologists achieve R^2 of 0.27 with MAE 11.84 years and ME -3.12 ± 14.25 years. The accuracy of the three radiologists varied between MAE 10.86 and 12.69 years with R^2 0.13 to 0.36. On the ranking task (right), both the ranking model and humans see increased accuracy with age separation as expected. The radiologists successfully ranked 48.3% of all pairs and 67.1% of those attempted, vs 82.5% of all pairs for the regression model and 85.5% for the ranking-specific model.

To test our hypothesis that humans are stronger at longitudinal change detection than point age estimation versus a null hypothesis of no benefit, we compare the observed ranking success with that expected if the two ages were estimated independently. Assuming Gaussian errors with the observed standard deviation of 14.25 years and calculating the probability of the sum of the two errors $(\sim \mathcal{N}(0,(2 \times 14.25)^{0.5}))$ exceeding the true age separation of each pair attempted, we derive an expected ranking success rate of 60.1 ± 2.4%. Compared with the observed success rate of 67.1% on attempted pairs, we conclude that the radiologists do benefit from the longitudinal information (p-value 0.002). Performing the same analysis for the models with the observed point estimate standard deviation 3.86 years, we expect a ranking success rate of 79.0 ± 2.7%. We find (as expected) no evidence of the regression model benefiting from paired images (p-value 0.098), whereas we find the ranking model benefits from the paired information (p-value 0.008).

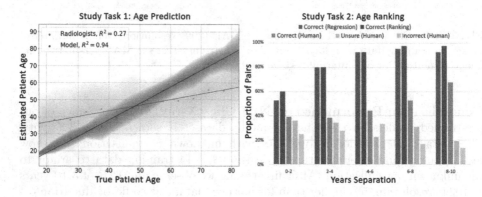

Fig. 4. Age prediction and ranking study results. (Left) Mean age prediction against true age bucketed in five-year intervals, with 2 SD areas shaded. (Right) Image ranking performance bucketed by age separation, regression & ranking models vs humans.

3.3   Aging Feature Visualisation

In Fig. 5 we show two examples of synthetic age interpolations. Each row uses a a constant w 'patient identity' input, showing the same artificial patient at progressively older target ages. The pixel-wise difference map of the 'oldest' image versus the 'youngest' highlights the features changed by the GAN to manipulate the age. We observe a widening of the aortic arch, lowering and widening of the heart, apical shadowing, some narrowing of the rib-cage, and widening of the mediastinum as the major features relevant to age prediction. These features are consistent with the opinion of our consultant radiologists, although they note that decreased transradiancy and increased aortic calcification would be expected which are not evident in the synthetically aged images. Further examples are provided in the supplemental materials.

Target Age: 15   Target Age: 42   Target Age: 68   Target Age: 95   Difference Map

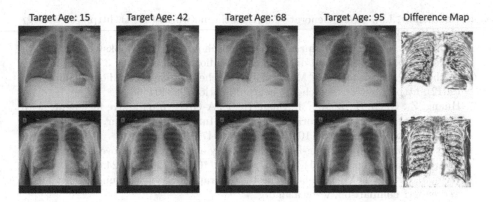

Fig. 5. Synthetic chest x-rays with varying target age and a constant patient 'identity' vector per row. The pixel difference between age 15 and 95 is shown in the final column highlighting the areas changed by the GAN to target the correct age.

4 Conclusion

In this work we present a study comparing the performance of three radiologists on chest X-ray age prediction and ranking tasks, comparing with data-driven models trained on a highly heterogeneous non-curated set of chest X-rays from a variety of clinical settings in six hospitals. We conclude that (a) the radiologists are significantly more accurate at detecting age-related changes in a single patient than at estimating age in single images and (b) the models significantly outperform humans on both tasks. We report a state of the art MAE of 2.78 years on our proprietary test data, and demonstrate the highest reported generalisation performance on the public NIH Chest14 dataset. Our work indicates that accuracy gains are likely to be small from larger datasets, and that the majority of age-relevant information is present at 512^2 resolution in this modality. We further demonstrate a GAN-based 'explainable AI' solution to visualise age-relevant features identified by the model, and compare with those identified by the radiologists based on their clinical experience.

References

1. Blanco, R.F., Rosado, P., Vegas, E., Reverter, F.: Medical image editing in the latent space of generative adversarial networks. Intell.-Based Med. **5**, 100040 (2021). https://doi.org/10.1016/j.ibmed.2021.100040
2. Cole, J.H., et al.: Predicting brain age with deep learning from raw imaging data results in a reliable and heritable biomarker. NeuroImage **163**, 115–124 (2017). https://doi.org/10.1016/j.neuroimage.2017.07.059
3. Fetty, L., et al.: Latent space manipulation for high-resolution medical image synthesis via the stylegan. Zeitschrift fur Medizinische Physik **30**, 305–314 (2020). https://doi.org/10.1016/j.zemedi.2020.05.001

4. Goodfellow, I.J., et al.: Generative adversarial networks (2014). http://arxiv.org/abs/1406.2661

5. Gross, B.H., Gerke, K.F., Shirazi, K.K., Whitehouse, W.M., Bookstein, F.L.: Estimation of patient age based on plain chest radiographs, pp. 141–3 (1985)

6. Hochhegger, B., Zanetti, G., Moreira, J.: The chest and aging: radiological findings (2012). https://www.researchgate.net/publication/233404468

7. Huang, Z., Chen, S., Zhang, J., Shan, H.: PFA-GAN: progressive face aging with generative adversarial network. IEEE Trans. Inf. Forensics Secur. **16**, 2031–2045 (2021). https://doi.org/10.1109/TIFS.2020.3047753

8. Ieki, H., et al.: Deep learning-based chest x-ray age serves as a novel biomarker for 1 cardiovascular aging (2021). https://doi.org/10.1101/2021.03.24.436773

9. Irvin, J., et al.: CheXpert: a large chest radiograph dataset with uncertainty labels and expert comparison. www.aaai.org

10. Karargyris, A., Kashyap, S., Wu, J.T., Sharma, A., Moradi, M., Syeda-Mahmood, T.: Age prediction using a large chest x-ray dataset. SPIE-Int. Soc. Opt. Eng. 66 (2019). https://doi.org/10.1117/12.2512922

11. Karras, T., Aittala, M., Hellsten, J., Laine, S., Lehtinen, J., Aila, T.: Training generative adversarial networks with limited data. In: Proceedings of NeurIPS (2020)

12. Larson, D.B., Chen, M.C., Lungren, M.P., Halabi, S.S., Stence, N.V., Langlotz, C.P.: Performance of a deep-learning neural network model in assessing skeletal maturity on pediatric hand radiographs. Radiology **287**, 313–322 (2018). https://doi.org/10.1148/radiol.2017170236

13. Niu, Z., Zhou, M., Wang, L., Gao, X., Hua, G.: Ordinal regression with multiple output CNN for age estimation. IEEE Comput. Soc. **2016-December**, 4920–4928 (2016). https://doi.org/10.1109/CVPR.2016.532

14. Odena, A., Olah, C., Shlens, J.: Conditional image synthesis with auxiliary classifier GANs (2016). http://arxiv.org/abs/1610.09585

15. Pesce, E., Withey, S.J., Ypsilantis, P.P., Bakewell, R., Goh, V., Montana, G.: Learning to detect chest radiographs containing pulmonary lesions using visual attention networks. Med. Image Anal. **53**, 26–38 (2019). https://doi.org/10.1016/j.media.2018.12.007

16. Raghu, V.K., Weiss, J., Hoffmann, U., Aerts, H.J., Lu, M.T.: Deep learning to estimate biological age from chest radiographs. JACC: Cardiovasc. Imaging **14**(11), 2226–2236 (2021). https://doi.org/10.1016/j.jcmg.2021.01.008, https://www.sciencedirect.com/science/article/pii/S1936878X21000681

17. Ren, Z., Yu, S.X., Whitney, D.: Controllable medical image generation via generative adversarial networks. Electron. Imaging **2021**, 112-1–112-6 (2021). https://doi.org/10.2352/issn.2470-1173.2021.11.hvei-112

18. Rothe, R., Timofte, R., Gool, L.V.: Deep expectation of real and apparent age from a single image without facial landmarks. Int. J. Comput. Vis. **126**, 144–157 (2018). https://doi.org/10.1007/s11263-016-0940-3

19. Sabottke, C.F., Breaux, M.A., Spieler, B.M.: Estimation of age in unidentified patients via chest radiography using convolutional neural network regression (2020). https://doi.org/10.1007/s10140-020-01782-5/Published, https://pytorch.org/

20. Selvaraju, R.R., Cogswell, M., Das, A., Vedantam, R., Parikh, D., Batra, D.: Grad-CAM: visual explanations from deep networks via gradient-based localization. Inst. Electr. Electron. Eng. Inc. **2017-October**, 618–626 (2017). https://doi.org/10.1109/ICCV.2017.74

21. Wang, X., Peng, Y., Lu, L., Lu, Z., Bagheri, M., Summers, R.M.: Chestx-ray8: hospital-scale chest x-ray database and benchmarks on weakly-supervised classification and localization of common thorax diseases. Inst. Electr. Electron. Eng. Inc. **2017-January**, 3462–3471 (2017). https://doi.org/10.1109/CVPR.2017.369
22. Yang, C.Y., et al.: Using deep neural networks for predicting age and sex in healthy adult chest radiographs. J. Clin. Med. **10** (2021). https://doi.org/10.3390/jcm10194431
23. Yi, X., Walia, E., Babyn, P.: Generative adversarial network in medical imaging: a review. Med. Image Anal. **58** (2019). https://doi.org/10.1016/j.media.2019.101552

Transformer Based Multi-task Deep Learning with Intravoxel Incoherent Motion Model Fitting for Microvascular Invasion Prediction of Hepatocellular Carcinoma

Haoyuan Huang[1] ⓘD, Baoer Liu[2], Lijuan Zhang[3], Yikai Xu[2], and Wu Zhou[1]([✉]) ⓘD

[1] Guangzhou University of Chinese Medicine, Guangzhou, China
zhouwu@gzucm.edu.cn
[2] Nanfang Hospital, Guangzhou, China
[3] Chinese Academy of Sciences, Shenzhen, China

Abstract. Prediction of microvascular invasion (MVI) in hepatocellular carcinoma (HCC) has important clinical value for treatment decisions and prognosis. Diffusion-weighted imaging (DWI) intravoxel incoherent motion (IVIM) models have been used to predict MVI in HCC. However, the parameter fitting of the IVIM model based on the typical nonlinear least squares method has a large amount of computation, and its accuracy is disturbed by noise. In addition, the performance of characterizing tumor characteristics based on the feature of IVIM parameter values is limited. In order to overcome the above difficulties, we proposed a novel multi-task deep learning network based on transformer to simultaneously conduct IVIM parameter model fitting and MVI prediction. Specifically, we utilize the transformer's powerful long-distance feature modeling ability to encode deep features of different tasks, and then generalize self-attention to cross-attention to match features that are beneficial to each task. In addition, inspired by the work of Compact Convolutional Transformer (CCT), we design the multi-task learning network based on CCT to enable the transformer to work in the small dataset of medical images. Experimental results of clinical HCC with IVIM data show that the proposed transformer based multi-task learning method is better than the current multi-task learning methods based on attention. Moreover, the performance of MVI prediction and IVIM model fitting based on multi-task learning is better than those of single-task learning methods. Finally, IVIM model fitting facilitates the performance of IVIM to characterize MVI, providing an effective tool for clinical tumor characterization.

Keywords: Transformer · Multi-task learning · Attention mechanism

Supplementary Information The online version contains supplementary material available at https://doi.org/10.1007/978-3-031-16449-1_26.

1 Introduction

Hepatocellular carcinoma (HCC) is the most common primary malignant tumor in the liver, accounting for 75%–85% of all liver cancer cases [15]. At present, microvascular invasion (MVI) has been proved to be a key indicator of HCC recurrence and poor prognosis [4], and Intravoxel incoherent motion (IVIM) diffusion weighted imaging (DWI) has been used to predict MVI in HCC [8]. Usually, the clinical application of IVIM-DWI is mainly based on the quantitative characterization of IVIM parameter values, and the IVIM parameter values include both perfusion and diffusion information and reflect the microstructure information of tissues. It plays an important role in tumor identification, benign and malignant differentiation and efficacy evaluation. However, IVIM parameter values are usually obtained by bi-exponential fitting of multiple b values of each pixel [10]. This parameter fitting is easily affected by noise and has a large amount of calculation [7,9]. Moreover, IVIM parameters are very sensitive to motion and artifacts in DWI, which will significantly reduce the lesion characterization performance of IVIM-DWI.

Recent studies adopt machine learning methods and neural networks to fit IVIM parameter maps, which greatly improve the performance of IVIM parameter estimation. Specifically, an unsupervised learning method is proposed to train the deep neural network (DNN), which greatly reduces the time-consuming of fitting IVIM parameters [1]. Since the previous fitting methods are based on voxels, [18] proposed a self-supervised model based on convolutional neural network (self U-net). On the other hand, recent studies used IVIM-DWI to directly predict the MVI of HCC through deep network [19,21], and achieved promising results. Recent studies have shown that IVIM fitting parameter values are related to MVI information of HCC [11,20,22]. This prompts us to find that IVIM parameter fitting and tumor invasiveness classification based on IVIM-DWI may be two tasks closely related to each other. According to the theory of multi-task learning, when two related tasks share feature representation, their performance will promote each other [2].

Multi-task learning has been widely used in computer vision and medical imaging. At present, most multi-task learning methods use hard parameter sharing to promote each other between tasks, but hard parameter sharing usually requires designers to conduct a large number of comparative experiments to determine which layers should be shared and how many layers should be shared, which greatly increases the cost of training the model [13]. In contrast, soft parameter sharing has better applicability in different multi-task scenarios. Liu et al. proposed the multi-task attention network (MTAN) [12], using task specific attention modules to adaptively extract the features of their respective tasks from a task sharing network. Lyu et al. proposed an attention aware multi-task convolutional neural network (AM-CNN), which uses the channel attention mechanism to suppress the redundant content contained in the representation [13]. However, both methods are based on single head driven attention mechanism, which has limited semantic expression ability and is difficult to consider feature extraction from a more diverse semantic perspective.

In this study, we introduce transformer [17] to solve the above multi-task learning problem, and simultaneously realize IVIM parameter model fitting and MVI prediction. On the one hand, it overcomes the problem of insufficient tumor representation performance of IVIM model fitting parameters. On the other hand, it promotes the tumor representation performance through multi-task learning. More importantly, it improves the soft parameter sharing by introducing transformer to further improve the multi-task learning performance. Inspired by compact convolution transformer (CCT) [5], we design a transformer-based multi-task learning model to be trained from scratch on small data sets. Extensive experiments are conducted to demonstrate the superiority of the proposed method compared with previously reported methods in terms of multi-task learning, IVIM model parameters fitting and MVI prediction.

2 Material and Methods

2.1 Study Population, Image Protocol and IVIM Model

123 HCC from 114 patients from January 2017 to September 2021 were included in this study, and the 3D Region of Interest (ROI) of the lesion was outlined by experienced clinicians. In order to maintain the independence of the experimental samples, we only selected the lesions with the largest area in each patient, so this study finally included only 114 HCC samples. Among them, 71 cases had no MVI (MVI -), and 43 cases were pathologically diagnosed as MVI (MVI +). As part of preoperative imaging, each patient underwent routine IVIM-DWI series examination using 3.0-T MRI system (Achieva, Philips healthcare, the Netherlands). IVIM-DWI sequence is acquired on the axial view (repetition time/echo time $= 1973/57$, 336×336 matrix, 5.5-mm slice thickness, field-of-view $= 400$ mm $\times$ 400 mm $\times$ 198 mm, slice gap $= 0.5$ mm, slices $= 36$, NSA $= 2$). IVIM-DWI uses breath triggered single echo planar imaging (EPI, EPI factor $= 53$). The axial DWI uses the following 9b values (b $= 0$, 10, 20, 40, 80, 200, 400, 600, and 1000 s/mm^2). 3.0-T MRI system is equipped with RF dual transmission source system to ensure the reliability of IVIM. Usually, IVIM uses a bi-exponential function model to describe DWI data and obtain a series of quantitative indicators, including water molecular diffusion coefficient D_t, tissue perfusion (pseudo diffusion coefficient) D_p and tissue perfusion fraction F_p [10]:

$$\frac{S_b}{S_0} = (1 - F_p)e^{-b \times D_t} + F_p \cdot e^{-b \times D_p} \tag{1}$$

where S_b is the diffusion-weighted signal acquired with a specific b-value, S_0 is the signal acquired without diffusion-sensitizing gradient.

2.2 The Proposed Overall Framework

The proposed transformer-based multi-task learning architecture is shown in Fig. 1. The light orange part in the upper half is the backbone network of IVIM

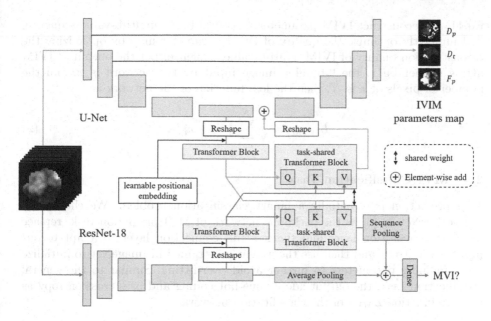

Fig. 1. Proposed network structure. We introduce a compact transformer strategy into the deep layer of convolution network, and extend the self-attention mechanism in multi-head attention to cross attention to realize information sharing between tasks.

parameter fitting task, and the light blue part in the lower half is the backbone network of MVI prediction task. The two backbone networks adopt U-net [14] and ResNet-18 [6], respectively. Since the two tasks use the same IVIM 9-b value sequence as the input, the shallow feature representation may have more identical representations, while the deep feature contains richer task specific information, so we select the deep features of the two tasks for sharing. Specifically, we first select the deep features of the two networks with the same space size and channels, and flatten them in the spatial dimension, and add them with a learnable positional embedding. Then, two independent transformer blocks are used to obtain two sets of task-specific embeddings. Then, both task-specific embeddings are transmitted to the task shared transformer block for cross task information matching to realize information sharing between tasks. Finally, the shared information is integrated into their respective network backbone to enrich the characteristic representation of their respective tasks. The specific operation of combining transformer strategy with convolutional network and the details of task shared transformer block will be described in Sect. 2.5.

2.3   IVIM Model Parameter Fitting Task

For IVIM parameter fitting, we refer to the latest method self U-net [18]. Specifically, it uses U-net [14] as the backbone network to learn the direct mapping from IVIM multi-b-value sequence to IVIM parameter map end-to-end, and then uses

Eq. (1) to reconstruct IVIM parameter map into IVIM multi-b-value sequence, and implicitly optimize the quality of IVIM parameter map by optimizing the reconstruction quality of IVIM multi-b-value sequence. Let the output of IVIM fitting subnet be $\hat{x}$, the 9-b value image input by the network is x, and the number of pixels of x is N, then the loss function of the network is:

$$L_{IVIM} = \frac{1}{N} \sum_{i=1}^{N} | \hat{x}_i - x_i | \tag{2}$$

2.4 MVI Classification Task

MVI prediction is essentially a binary classification problem. We choose the classic ResNet-18 [6] as the backbone network of MVI prediction task, replace the first convolution layer and the last full connection layer to adapt to our input and output, and then use the weight pre-trained in ImageNet to initialize the remaining parameters to further avoid over-fitting. Similar to the general classification task, the output adopts one-hot coding and uses cross entropy as the loss function L_{MVI} of the classification network.

2.5 Transformer-Based Multi-task Learning

We use the method of CCT [5] for reference to combine the transformer with the convolution network. Specifically, CCT proposes a convolutional based patching method. First, the dimension of the feature map ($b \times c \times w \times h$) is flattened as ($b \times c \times n$). Then, it exchanges the last two dimensions to get input embeddings ($b \times n \times c$), where b is the batch size, c and n are the number of channels and pixels respectively, as well as the embedding dimension and sequence length in the transformer block. This method preserves local information and can encode the relationship between patches, unlike the classical Vision Transformer (ViT) [3]. Then, a learnable positional embedding is added to the input embedding and passed to the transformer block. This positional embedding enables the model to understand the positional relationship between the input tokens by adding some additional information to the model.

In addition, CCT proposes a sequential pooling method to eliminate additional [class] tokens. Let X_L be the output of the L-th transformer block and the dimension is ($b \times n \times c$), g is the full connection layer that maps the last dimension c to 1. Sequential pooling is calculated as follows:

$$z = softmax(g(X_L)^T) \times X_L \tag{3}$$

The dimension of output z of sequential pooling is ($b \times 1 \times c$). This sequential pooling allows our network to weigh the sequential embeddings of the latent space generated by the transformer block and better mine the internal associations of the input data. Based on the above operations, the advantages of convolution network and transformer can be combined. For the design of transformer block, we are consistent with ViT and the original transformer [17]. Each transformer block includes a Multi-headed Self Attention (MSA) and a Multi-Layer Perception (MLP) head, and uses Layer Normalization, GELU activation, and dropout.

In order to realize the information sharing among multi tasks, we propose a transformer block based on cross attention, which allows each task to match the information beneficial to itself from its brother tasks by using its own task specific query. Cross attention is a generalization of self attention. Generally, self attention in transformer is calculated as follows [17]:

$$Attention(Q, K, V) = softmax(\frac{QK^T}{\sqrt{d_k}})V \tag{4}$$

Let task 1 and task 2 get the embedding from the $(t-1)$-th task-specific transformer block as e_1^{t-1}, e_2^{t-1}, respectively. We use a task shared transformer block in layer t, which shares a set of embedded mapping weights W^Q, W^K, W^V with all tasks. Output e_1^t, e_2^t of the layer t is calculated as follows:

$$e_1^t = Attention(e_1^{t-1}W^Q, e_2^{t-1}W^K, e_2^{t-1}W^V) \tag{5}$$

$$e_2^t = Attention(e_2^{t-1}W^Q, e_1^{t-1}W^K, e_1^{t-1}W^V) \tag{6}$$

In this cross attention mode, each task can match its own information from the embedding of other tasks according to its own task-specific query, which improves the accuracy of information interaction between tasks.

2.6   Implementation

Since the different sizes of tumors, we first resized the IVIM-DWI of each tumors with 3D ROI to 32×32 in axial-view, and selected three different slices with largest area through axial-view. Therefore, we obtained 342 slices from 114 samples, and each slice corresponds to 9 b-value images, we toke the slice sequence with size 9×32×32 as the input data. Centrally random rotation and centrally scaled was conducted on each slice for data augmentation, which is consistent with the previous study [21]. The model is implemented in the platform of pytorch, and the training and testing are based on GeForceRTX2080Ti. The batch size is set to 64 and epoch is 200. $\beta = (0.9, 0.999)$ is set in the ADAM optimizer. MVI prediction task and IVIM fitting task are optimized at the same time, so the total loss of the model is $\alpha L_{MVI} + \beta L_{IVIM}$, where α, β are separately set to 1.0 and 1e−2 in the experiment. The learning rate of the network is set to 1e−3 and the weight decay is 2e−4. The source code of this study is available at: https://github.com/Ksuriuri/TRMT.

3   Experimental Results

3.1   Evaluation Metrics and Experimental Setup

For MVI classification tasks, accuracy (ACC), sensitivity (SEN), specificity (SPE), and area under the receiver operating characteristic curve (AUC) were

Table 1. Performance comparison of different multi-task learning methods, #P shows the number of network parameters.

#P	Methods	MVI prediction				IVIM fitting	
		ACC (%)	SEN (%)	SPE (%)	AUC (%)	SSIM	RMSE
41.13M	MTAN	76.30 ± 6.09	71.94 ± 10.25	80.54 ± 7.84	80.54 ± 10.23	0.877	0.043
49.67M	AM-CNN	76.94 ± 9.82	**73.94 ± 11.59**	80.47 ± 12.03	82.93 ± 9.81	0.883	0.040
52.42M	Proposed	**79.21 ± 4.41**	73.78 ± 8.93	**85.04 ± 5.07**	**85.51 ± 7.13**	**0.894**	**0.039**

used to evaluate the performance of the proposed method. For the IVIM parameter fitting task, since there is no clinical gold standard for IVIM parameters, we followed the experimental method of [16] to quantitatively evaluate the quality of the generated parameter map. Specifically, we first use Eq. (1) to reconstruct IVIM parameter map into b-value map, and then use structural similarity (SSIM) and root mean square error (RMSE) to evaluate the reconstruction accuracy of b-value map. Because there is a fixed mapping relationship between IVIM parameter map and b-value map, the reconstruction accuracy of b-value map can better reflect the fitting accuracy of IVIM parameter map. In order to obtain more general conclusions on small data sets, all experimental results are based on 4-folded cross validation. Each fold is repeated for 5 times and averaged.

3.2 Performance Comparison of Multi-task Learning Methods

In order to realize the two related multi-task learning networks fairly, we have made some modifications to the implementation of the two methods. For MTAN [12], we use the U-net consistent with this paper as the shared network, and then for the attention subnet predicted by MVI, we only retain the down sampling coding part of the first half, and pass the coded deep feature to the full connection layer for classification. For the implementation of AM-CNN [13], the AM-CNN module is applied to each layer of down sampling of ResNet-18 and U-net networks. Table 1 shows the performance comparison of different multi-task learning methods in MVI classification and IVIM parameter fitting. The proposed transformer based multi-task learning method achieves the best performance in both tasks, which demonstrates that the multi-task learning mechanism based on transformer is more advantageous. Since MTAN only performed local attention mechanism to generate the soft attention mask, while AM-CNN only considered the channel-level attention mechanism. Both of them ignored the global dependencies at the spatial level, while our transformer based model takes it into account.

3.3 Performance Comparison Between the Proposed Method and Single-Task Methods

We compared the performance of the proposed multi-task learning method with the existing single task methods for MVI prediction and IVIM parameter fitting. For MVI prediction, ResNet-18 [6] is the backbone of our MVI prediction

Table 2. Performance comparison between the proposed method and different single task methods, #P shows the number of network parameters.

#P	Methods	MVI prediction				IVIM fitting	
		ACC (%)	SEN (%)	SPE (%)	AUC (%)	SSIM	RMSE
11.20M	ResNet-18	68.76 ± 4.94	61.84 ± 21.49	75.41 ± 8.19	73.56 ± 7.95	-	-
13.57M	ResNet-18 CCT	74.76 ± 5.81	72.70 ± 6.46	78.92 ± 7.80	81.67 ± 9.75	-	-
22.91M	FCSA-net	77.99 ± 7.14	72.11 ± 12.09	83.22 ± 8.23	82.45 ± 9.19	-	-
-	Least-Square	-	-	-	-	0.817	0.058
300	DNN	-	-	-	-	0.821	0.054
37.67M	Self U-net	-	-	-	-	0.862	0.045
52.42M	Proposed	**79.21 ± 4.41**	**73.78 ± 8.93**	**85.04 ± 5.07**	**85.51 ± 7.13**	**0.894**	**0.039**

task. ResNet-18 CCT is equivalent to removing the IVIM fitting task from our proposed model and retaining of transformer (replacing cross attention with self attention). FCSA-Net [21] is a recently proposed module for MVI prediction based on IVIM multi-b-value images, which introduces the spatial and channel attention mechanism simultaneously for MVI prediction. For fair comparison, we use ResNet-18 as the backbone of FCSA-Net, and use the attention module proposed by FCSA-net in each layer. As tabulated in Table 2, it can be seen that our model still achieves the best MVI prediction performance due to the benefits of global dependencies brought about by the transformer based multi-task learning mechanism. For IVIM parameter fitting task, Least-Square is a traditional voxel fitting method [9], DNN is the deep neural network method based on unsupervised learning [1], and self U-net [18] is the backbone of our IVIM fitting task. Our proposed model also achieves the best performance by considering the richer feature representation brought by the transformer based multi-task learning mechanism.

4   Conclusion

In this work, we proposed a new multi-task learning method based on transformer, which realizes MVI prediction task and IVIM parameter fitting task at the same time, and achieves better performance than single task methods. This study shows that IVIM parameter fitting can promote IVIM's characterization of MVI. At the same time, it shows that the proposed multi-task learning based on transformer is better than other multi-task learning models based on attention. On the one hand, this study provides an effective tool for IVIM to characterize clinical tumors, and on the other hand, it provides a new model for multi-task learning.

Acknowledgements. This work is supported by the grant from National Nature Science Foundation of China (No. 81771920).

References

1. Barbieri, S., Gurney-Champion, O.J., Klaassen, R., Thoeny, H.C.: Deep learning how to fit an intravoxel incoherent motion model to diffusion-weighted MRI. Magn. Reson. Med. **83**(1), 312–321 (2020)
2. Baxter, J.: A Bayesian/information theoretic model of learning to learn via multiple task sampling. Mach. Learn. **28**(1), 7–39 (1997)
3. Dosovitskiy, A., Beyer, L., Kolesnikov, A., Weissenborn, D., Houlsby, N.: An image is worth 16x16 words: transformers for image recognition at scale (2020)
4. Erstad, D.J., Tanabe, K.K.: Prognostic and therapeutic implications of microvascular invasion in hepatocellular carcinoma. Ann. Surg. Oncol. **26**(5), 1474–1493 (2019)
5. Hassani, A., Walton, S., Shah, N., Abuduweili, A., Li, J., Shi, H.: Escaping the big data paradigm with compact transformers (2021)
6. He, K., Zhang, X., Ren, S., Sun, J.: Deep residual learning for image recognition. IEEE (2016)
7. Hernando, D., Zhang, Y., Pirasteh, A.: Quantitative diffusion MRI of the abdomen and pelvis. Med. Phys. (2021)
8. Iima, M., Le Bihan, D.: Clinical intravoxel incoherent motion and diffusion MR imaging: past, present, and future. Radiology **278**(1), 13–32 (2016)
9. Lanzarone, E., Mastropietro, A., Scalco, E., Vidiri, A., Rizzo, G.: A novel Bayesian approach with conditional autoregressive specification for intravoxel incoherent motion diffusion-weighted MRI. NMR Biomed. **33**(3), e4201 (2020)
10. Le Bihan, D., Breton, E., Lallemand, D., Aubin, M., Vignaud, J., Laval-Jeantet, M.: Separation of diffusion and perfusion in intravoxel incoherent motion MR imaging. Radiology **168**(2), 497–505 (1988)
11. Li, H., et al.: Preoperative histogram analysis of intravoxel incoherent motion (IVIM) for predicting microvascular invasion in patients with single hepatocellular carcinoma. Eur. J. Radiol. **105**, 65–71 (2018)
12. Liu, S., Johns, E., Davison, A.J.: End-to-end multi-task learning with attention. In: Proceedings of the IEEE/CVF Conference on Computer Vision and Pattern Recognition, pp. 1871–1880 (2019)
13. Lyu, K., Li, Y., Zhang, Z.: Attention-aware multi-task convolutional neural networks. IEEE Trans. Image Process. PP(99), 1–1 (2019)
14. Ronneberger, O., Fischer, P., Brox, T.: U-net: convolutional networks for biomedical image segmentation. In: Navab, N., Hornegger, J., Wells, W.M., Frangi, A.F. (eds.) MICCAI 2015. LNCS, vol. 9351, pp. 234–241. Springer, Cham (2015). https://doi.org/10.1007/978-3-319-24574-4_28
15. Sung, H., et al.: Global cancer statistics 2020: Globocan estimates of incidence and mortality worldwide for 36 cancers in 185 countries. CA Cancer J. Clin. **71**(3), 209–249 (2021)
16. Ulas, C., et al.: Convolutional neural networks for direct inference of pharmacokinetic parameters: application to stroke dynamic contrast-enhanced MRI. Front. Neurol. 1147 (2019)
17. Vaswani, A., et al.: Attention is all you need. arXiv (2017)
18. Vasylechko, S.D., Warfield, S.K., Afacan, O., Kurugol, S.: Self-supervised IVIM DWI parameter estimation with a physics based forward model. Magn. Reson. Med. **87**(2), 904–914 (2022)
19. Wang, A.G., et al.: Prediction of microvascular invasion of hepatocellular carcinoma based on preoperative diffusion-weighted MR using deep learning. Acad. Radiol. (2020)

20. Wei, Y., et al.: IVIM improves preoperative assessment of microvascular invasion in HCC. Eur. Radiol. **29**(10), 5403–5414 (2019)
21. Zeng, Q., Liu, B., Xu, Y., Zhou, W.: An attention-based deep learning model for predicting microvascular invasion of hepatocellular carcinoma using an intra-voxel incoherent motion model of diffusion-weighted magnetic resonance imaging. Phys. Med. Biol. **66**(18), 185019 (2021)
22. Zhao, W., et al.: Preoperative prediction of microvascular invasion of hepatocellular carcinoma with IVIM diffusion-weighted MR imaging and GD-EOB-DTPA-enhanced MR imaging. PLoS ONE **13**(5), e0197488 (2018)

Identifying Phenotypic Concepts Discriminating Molecular Breast Cancer Sub-Types

Christoph Fürböck[1]([✉]), Matthias Perkonigg[1], Thomas Helbich[2],
Katja Pinker[2,3], Valeria Romeo[4], and Georg Langs[1]

[1] Computational Imaging Research Lab,
Department of Biomedical Imaging and Image-guided Therapy,
Medical University of Vienna, Vienna, Austria
christoph.fuerboeck@meduniwien.ac.at
[2] Division of Molecular and Structural Preclinical Imaging,
Department of Biomedical Imaging and Image-guided Therapy,
Medical University of Vienna, Vienna, Austria
[3] Department of Radiology, Breast Imaging Service,
Memorial Sloan Kettering Cancer Center, New York, NY, USA
[4] Department of Advanced Biomedical Sciences, University of Naples Federico II,
Naples, Italy

Abstract. Molecular breast cancer sub-types derived from core-biopsy
are central for individual outcome prediction and treatment decisions.
Determining sub-types by non-invasive imaging procedures would benefit early assessment. Furthermore, identifying phenotypic traits of sub-
types may inform our understanding of disease processes as we become
able to monitor them longitudinally. We propose a model to learn phenotypic appearance concepts of four molecular sub-types of breast cancer. A
deep neural network classification model predicts sub-types from multimodal, multi-parametric imaging data. Intermediate representations of
the visual information are clustered, and clusters are scored based on
testing with concept activation vectors to assess their contribution to
correctly discriminating sub-types. The proposed model can predict sub-
types with competitive accuracy from simultaneous ^{18}F-FDG PET/MRI,
and identifies visual traits in the form of shared and discriminating phenotypic concepts associated with the sub-types.

Keywords: Explainable AI · Breast cancer sub-types · Deep learning

1 Introduction

Molecular sub-types of Breast Cancer (BC) are important for individual outcome prediction, and for the guidance of treatment [5,20,22]. Here, we pro-

This work was supported by the Vienna Science and Technology Fund (WWTF, LS19-
018, LS20-065), the Austrian Research Fund (FWF, P 35189), a CCC Research Grant,
and a European Union's Horizon 2020 research grant (No. 667211).

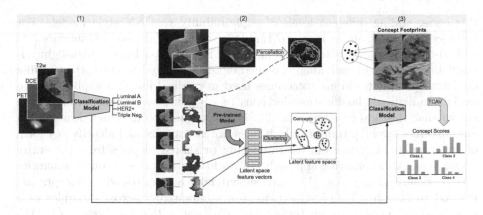

Fig. 1. Overview of method: (1) Breast cancer sub-type classification network; (2) concept calculation via clustering of latent feature representation of image parcels; (3) class-specific importance score for each concept and each class based on TCAV.

pose an interpretable model for the prediction of sub-types from simultaneous multi-parametric Magnetic Resonance Imaging (MRI) and Positron Emission Tomography (PET) data. The model classifies sub-types and identifies phenotypic traits in the form of *concepts* associated with and discriminating between different molecular breast cancer sub-types.

Breast cancer has the highest incidence among cancer types in females worldwide [27] and the highest mortality among the 10 most common cancers in females [27]. This is caused in part by the heterogeneity of breast cancer, as different biological sub-types require different targeted therapy and demonstrate different clinical outcomes [5]. The classification of cancer sub-types is performed through gene expression profiling [25] or immunohistochemistry [18]. Both methods require a tumor biopsy sample, affected by the tissue sampling bias problem as the sample might not be representative of the heterogeneity of the whole tumor lesion. Imaging aims to develop a tool that could non-invasively depict histological and molecular aspects of the entire tumor. For this reason an advanced imaging modality, hybrid ^{18}F-FDG PET/MRI, has been developed which is able to provide morphological, functional and metabolic data at the same time.

Related Work. Machine learning methods to classify breast cancer sub-types from imaging data include radiomics based techniques [12,24] as well as deep learning approaches [10,29]. While they achieve good accuracy (70–82%), none of those approaches offer interpretability of the visual features the models use for classification. Instead of using a prediction model as a black box, it can be beneficial to investigate how the model reaches a decision. Interpretability is crucial for clinical decision making to build trustworthy algorithms [21], and to enable the linking of phenotypic characteristics to underlying molecular processes. While deep learning models are not readily interpretable for humans [21], the field of Explainable AI has emerged in Artificial Intelligence (AI) to tackle this issue, adding different levels of explainability to CNN style models. For instance, *class activation maps (CAM)* [17, 23, 28] or *Integrated Gradients* [26] add ad-hoc explainability to trained models. These approaches can high-light relevant image areas in individual classified examples linked to the corresponding classification result and are sometimes used in medical imaging applications to explain individual classification decisions (e.g. [19]). Nevertheless, studies have shown that this ad-hoc interpretability is not suitable to interpret decisions on an individual level [7]. Importantly, it does not summarize and identify any form of common characteristics of discriminating or shared features from the entire population. An alternative approach is to add explanations through semantics [21]. For natural imaging, [15] have shown how human-friendly concepts are utilized to obtain a high-level explanation summarizing across examples of a specific class. Their method *Testing with Concept Activation Vectors (TCAV)* can be used to score established clinical biomarkers as shown in [2] in a cardiac MRI classification network. The authors of [8] expand the method to automatically detect and score concepts in imaging data. An application of this method in medical imaging is conducted by [13] in a semantic segmentation task for cardiac MRI analysis. In [6] TCAV is used to explain the biomarker prediction of hematoxylin and eosin stained images with both, concepts that have been defined by experts and concepts resulting from unsupervised clustering.

Contribution. In this work, we propose a method to predict the molecular subtypes of breast cancer, and to identify the associated shared and discriminating visual traits from simultaneously acquired MRI and PET data. A deep learning model classifies breast cancer sub-types from multi-modal multi-parametric imaging data. We extract interpretable concepts for different cancer sub-types building upon concept based explanations [8]. Concept expressions are aggregated over an image to form *concept footprints*, and are then used to calculate TCAV scores measuring their impact on the classification decision across the population. The resulting top-scored *phenotypic concepts* offer a novel analysis of the association between molecular cancer sub-types and a set of human-understandable visual traits. The analysis of shared and discriminating visual traits enhances the understanding of the relationships between sub-types.

2   Method

We aim at deriving a set of visual concepts $\mathcal{C} = \langle C_1, \ldots, C_K \rangle$ from a set of segmented example images showing tumors $\{\mathbf{I}_1, \ldots, \mathbf{I}_N\}$, and corresponding class labels $\{y_1, \ldots, y_N\}, y_i \in \mathcal{Y} = \{1, \ldots, 4\}$ indicating the molecular sub-type. Each image contains multiple channels of image information for each voxel derived from registered MRI T2w, MRI DCE, and PET imaging data. For each concept and each class we derive an importance score ϕ measuring the influence of the derived concepts. To this end the proposed method builds upon concept-based explanations [8]. The method has three steps depicted in Fig. 1. (1) First, we train a classification network $f : \mathbf{I}_i \mapsto y_i$ to predict the breast cancer sub-type from the tumor imaging data. (2) Then, concepts $\mathcal{C} = \langle C_1, \ldots, C_K \rangle$ are calculated by clustering the latent space feature representations of tumor image parcels. (3) Finally, class-specific visual traits associated with molecular sub-types are determined by scoring the resulting concepts using the classification network f and the TCAV algorithm.

2.1   Classification Network

The training dataset $\mathcal{D} = \{\langle \mathbf{I}_1, y_1 \rangle, \ldots, \langle \mathbf{I}_N, y_N \rangle\}$ consists of images in the form of tensors $\mathbf{I}_i = [\mathbf{x}_i^1, \ldots, \mathbf{x}_i^M]$ concatenating registered scans from different image modalities, and the label in the form of the molecular sub-type y_i of the lesion. We train a convolutional neural network as a classification model to classify different sub-types $f : \mathbf{I}_i \mapsto y_i$.

2.2   Concept Calculation

We derive concepts from $\mathcal{D}$ in three steps building upon [8] as shown in Fig. 1 (2): **(a) Parcellation.** First, we parcellate each lesion image $\mathbf{I}_i$ segmented with a mask $\mathbf{R}_i$ of the lesion to derive a set of P parcels $\mathcal{P}_i = \{x_i^{(1)}, \ldots, x_i^{(P)}\}$. Different from [8], we use the Felzenszwalb algorithm [4], since it is efficient and produces parcels that can vary significantly in shape and size to capture coinciding regions and the appearance variability of the lesion structure. **(b) Parcel latent feature calculation.** Next, we calculate latent features of the parcels using a pre-trained neural network. The parcels $x_i^{(j)}$ are used as input to an image classification network p pretrained on natural images. The activation $a_l(x_i^{(j)})$ of parcel j of image $\mathbf{I}_i$ of an intermediate layer l of this network is extracted as a latent space feature vector. Those vectors are collected for all parcels of all images in $\mathcal{D}$ to form a set of representations $\mathcal{A}$ in the latent feature space. **(c) Clustering.** Finally, we derive concepts in the form of clusters of $\mathcal{A}$ in the latent feature space. The collected feature vectors in $\mathcal{A}$ are clustered using a clustering algorithm g (k-Means). This way visually similar segments are grouped to the same cluster. These clusters then form the *concepts* $\mathcal{C}$. For each concept, we define a corresponding concept footprint $c_{k,i}$ of concept C_k for the image of an individual subject i as the combination of all parcels from subject i belonging to concept C_k:

$$c_{k,i} = \sum_j \{x_i^{(j)} | g(a_l(x_i^{(j)})) = k\} \tag{1}$$

2.3   Class-Specific Scoring

To obtain an importance score $\phi_{k,t}$ for the k-th concept C_k and a specific target class t, its TCAV score [15] is calculated using the trained classification model f (see Sect. 2.1). In medical imaging the concept footprints are more interpretable and expressive than individual parcels of concepts. To account for the fact that a lesion can contain multiple instances of one concept, such as for instance a certain quality of the border region, in contrast to [8] the concept footprints $c_{k,i}$ are used for the importance score calculation. The TCAV score measures the positive impact of a concept footprint on the prediction of a specific class. With examples of concept footprints as input the score is calculated as the fraction of images with a certain class, whose activation at a given layer is positively influenced by the concept footprint. For the k-th concept C_k, each corresponding concept footprint $c_{k,i}$ and target class t the class-specific score is calculated as $\phi_{k,t} = TCAV(f, c_{k,i}, \{\mathbf{I}_i | y_i = t\})$.

3 Experiments and Results

3.1 Dataset

Participants: The patient population of this IRB-approved study consists of a total number of 154 participants with at least one breast lesion detected by ultrasound and/or mammography who have undergone multiparametric ^{18}F-FDG PET/MRI. Written informed consent was obtained from all participants. All subjects with benign lesions or without histopathological sub-type class, and subjects with missing scans are excluded, leading to a total of 102 subjects with 118 breast cancer lesions included in this work. The number of lesions for each sub-type is: 17 Luminal-A, 65 Luminal-B, 11 HER2+, 25 Triple-Negative. **PET/MRI Acquisition:** A Biograph mMR detector system has been used to perform multiparametric ^{18}F-FDG PET/MRI [3]. After fasting for at least five hours, patients have received ^{18}F-FDG intravenously at a dose of 2.5–3.5 MBq/kg body weight. The uptake time before the PET/MRI has been 60 min. PET is performed simultaneously with a multiparametric MRI protocol which includes three sequences: 1. T2-weighted, 2. Diffusion tensor imaging and 3. Dynamic Contrast Enhanced (DCE) imaging. **Pre-processing:** The three different scans (T2w, DCE, PET) are preprocessed by resampling to have the same pixel-dimension, cropping to a region of interest and rescaling between 0 and 1. For each scan one slice of the scan is selected. This slice is chosen by the amount of pixels in the region of interest (i.e. the lesion) and the slice containing the maximal number is selected. The pre-processed images are concatenated along the channel axis to serve as input for the classification network.

3.2 Experimental Setup

Classification Network: A ResNet-18 [11] with a modified classification head to fit the task of 4-class classification is used as a classification network. The three different input image sequences (T2w, DCE, PET) are concatenated along the channel axis and are treated as the three input channels. Training is performed by using a weighted cross-entropy loss to deal with the class imbalance and an Adam optimizer [16]. Data augmentation is performed by adding Gaussian noise, random flipping and 2D elastic deformations. The learning rate is fixed to $lr = 10^{-4}$ and a weight decay of $\lambda = 0.9$ after a grid search has been performed

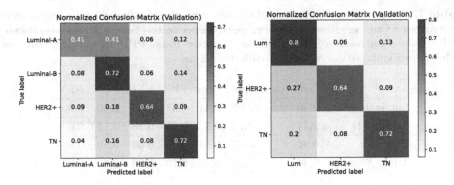

Fig. 2. Normalized confusion matrices for the 4-class (left) and 3-class classification (right) over all splits.

to find the optimal model hyper-parameters. The model is trained and validated using 5-fold cross-validation. To combat overfitting, early stopping is applied by saving the model with the highest validation accuracy for each split.

Concept Calculation: To analyse where the model looks for the prediction and define sub-type specific concept footprints, concept calculation is performed on the T2w images in the train-set. The pre-defined lesion mask includes the lesion segmentation and a surrounding neighborhood. It is obtained by a 2D smoothing convolution and binarization of the mask. The latent space features for concept calculation are extracted from a ResNet-18 [11] pre-trained on ImageNet. The number of clusters used for k-Means is fixed to $K = 10$. This value has been chosen after the application of the elbow method [14] to estimate the optimal cluster number. The best performing split out of the five folds is chosen for concept importance score calculation. For this split the accuracy is 74% with a recall $\geq 60\%$ for each class and the accuracy when grouping Luminal-A and Luminal-B is 85%. TCAV scores are calculated on the validation set using the concept footprints in T2w, DCE, and PET images for the multi-modal classification model f.

3.3 Model Classification Accuracy

The overall accuracy for the classification of the four sub-types over all five splits is 67%. The normalized confusion matrix over all splits is shown in Fig. 2 (left). To facilitate comparison with another state of the art method [29] the classes Luminal-A and Luminal-B are grouped to one class Luminal, yielding a corresponding 3-class overall accuracy of 77% and a normalized confusion matrix shown in Fig. 2 (right). Table 1 provides a more detailed evaluation of the accuracy, specificity, sensitivity, false positive/negative ratios and positive/negative predictive values for each of the four molecular sub-types. The one-vs-rest accuracy for the four sub-types ranges from 74% to 91%, and the clinically important

Table 1. Metrics for the evaluation of the sub-type classification model performance. OvR Accu = One vs Rest Accuracy, Spec = Specificity, Sens = Sensitivity/Recall, FPR = False Positive Rate, FNR = False Negative Rate, PPV = Positive Predictive Value (Precision), NPV = Negative Predictive Value.

	OvR Accu	Spec	Sens	FPR	FNR	PPV	NPV
Luminal-A	0.86	0.93	0.41	0.07	0.59	0.50	0.90
Luminal-B	0.74	0.75	0.72	0.25	0.28	0.78	0.69
HER2+	0.91	0.93	0.64	0.07	0.36	0.50	0.96
Triple-Negative	0.84	0.87	0.72	0.13	0.28	0.60	0.92

discrimination of triple-negative cancer from other sub-types achieves an accuracy of 84%, sensitivity of 0.72, and specificity of 0.87.

3.4 Phenotypic Concepts of Four BC Sub-Types

The concept importance of 10 concepts is plotted in Fig. 3 (a) for all cancer sub-types. The concepts are ordered in an descending order of importance, and for four concepts indicators are added to facilitate the comparison of their position in the ranking. For the classification of different sub-types different concepts are among the top-ranked most important, forming characteristic phenotypic traits of the classes. At the same time several sub-types share top-ranked phenotypic concepts.

To facilitate the interpretation of concepts and their association with sub-types, the class importance for four concepts is visualized in Fig. 3 (b). For each concept the scores for the four sub-types, and example footprints in four BC lesions are depicted. The calculated concepts have a similar distribution across the lesions with different sub-types. Yet the importance of the concepts for the sub-type classification varies significantly. *Concept 1* is the second most important concept for the sub-type Luminal-A. The concept footprints of this concept are distributed in border areas of the lesion with a transition between the lesion and surrounding tissue being defined by a sharp-edged boundary. *Concept 2* is highly specific for the sub-type Luminal-B and has the highest importance score for this sub-type. Its footprints exhibit *dark* spots in the lesion, compared to the surrounding area. The importance scores of *Concept 3* demonstrate a near-linear dependence on the sub-type classes, sorted lowest to highest value: Luminal-A, Luminal-B, HER2+, Triple-Negative. The footprints of *Concept 3* can be interpreted as the main tumor mass which directly translates to the tumor size. Similar to *Concept 1*, the footprints of *Concept 4* focus on the border areas. In contrast to *Concept 1* the transition of the lesion and surrounding tissue is irregular and blurred. This concept is the most important concept for the Triple-Negative sub-type.

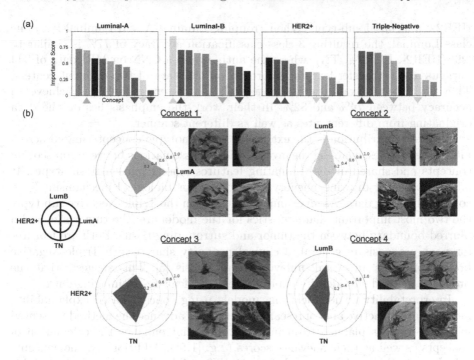

Fig. 3. Concepts occur across imaging data of all breast cancer patients, and are associated with molecular sub-types to different degrees. They discriminate between sub-types (e.g., concept 2 for Luminal-B), or are shared across sub-types (e.g., concept 4 marks all but Luminal-A). (a): TCAV-Scores of concepts (indicated by color) for the molecular sub-types. Four exemplary concepts are marked with triangles indicating high importance ($\triangle$) and low importance ($\triangledown$). (b): Their importance scores plotted for the molecular sub-types in a radial plot and example footprints. (Color figure online)

4   Discussion

We present a multi-modal deep learning model for the classification of breast cancer sub-types Luminal-A, Luminal-B, HER2+ and Triple-Negative. In addition the model is used to identify phenotypic concepts that are shared among or discriminate between different molecular breast cancer sub-types. They constitute appearance traits, offering insight into the composition of lesions captured with multi-modal multi-parametric imaging. By summarizing similar observations across examples, and scoring their contribution to individual classification decisions, they enable the linking of image features to underlying molecular characteristics of tumors.

Evaluation of the classification accuracy is competitive with recent state of the art approaches, while using a smaller dataset as the model takes advantage of the multi-parametric multi-modal data. The 4-class prediction accuracy of our work (67%) is comparable to the value of [10] (70%) in which a CNN processing MRI data of 216 patients predicts the sub-types Luminal-A, Luminal-B,

HER2+ and basal sub-type. When grouping Luminal-A and Luminal-B to one class Luminal, the resulting 3-class classification accuracy of 77% is similar to [29] (HER2-, HER2+, TN), where the authors use a CNN on MRI data of 244 patients and a transfer-learning approach on the data of two different centers. They use different train-test setups as well as different models and achieve an accuracy between 74% and 82%. In their work the emphasis lies on the data originating from different sites as well as different scanners.

The key contribution is the extraction of phenotypic concepts shared across multiple examples, enabling the investigation of associations between top-scoring concepts and shared or discriminating features among molecular sub-types. It offers a basis for drawing phenotypic evidence for biological mechanisms from complex imaging data. As an example, focusing on the Triple-Negative sub-type, the two most important characteristics for the model are: the tumor size and a blurred boundary between the tumor and surrounding tissue. Both are features for poor prognosis in general, a clinical property shared with Triple-Negative breast cancer as an overall unfavorable prognosis [5]. This suggests that the proposed method can identify clinically relevant features in imaging data.

Interpretability of deep learning models using concept-based explainability methods are an active area of research. To further advance the method presented in this work it is planned to include methods that enhance the calculation of concepts as well as their relevance scores (e.g. [1,9]). Additionally, the concepts calculated in this work use parcels from the T2-weighted images exclusively. This limits the information in the clustering process as well as the possible visualisation of phenotypic features. The evaluation of concepts calculated on DCE or PET images or the inclusion of multiple sequences in the concept calculation are therefore a potential future work.

References

1. Chen, Z., Bei, Y., Rudin, C.: Concept whitening for interpretable image recognition. Nat. Mach. Intell. **2**, 1–11 (2020)
2. Clough, J., Oksuz, I., Puyol Anton, E., Ruijsink, B., King, A., Schnabel, J.: Global and local interpretability for cardiac MRI classification (2019). https://arxiv.org/abs/1906.06188
3. Delso, G., et al.: Performance measurements of the siemens MMR integrated whole-body PET/MR scanner. J. Nucl. Med. **52**(12), 1914–1922 (2011)
4. Felzenszwalb, P.F., Huttenlocher, D.P.: Efficient graph-based image segmentation. Int. J. Comput. Vis. **59**(2), 167–181 (2004)
5. Fragomeni, S.M., Sciallis, A., Jeruss, J.S.: Molecular subtypes and local-regional control of breast cancer. Surg. Oncol. Clin. **27**(1), 95–120 (2018)
6. Gamble, P., et al.: Determining breast cancer biomarker status and associated morphological features using deep learning. Commun. Med. **1** (2021)
7. Ghassemi, M., Oakden-Rayner, L., Beam, A.L.: The false hope of current approaches to explainable artificial intelligence in health care. Lancet Digit. Health **3**(11), e745–e750 (2021)
8. Ghorbani, A., Wexler, J., Zou, J., Kim, B.: Towards automatic concept-based explanations. In: Advances in Neural Information Processing Systems, vol. 32, pp. 9277–9286 (2019)

9. Graziani, M., Andrearczyk, V., Marchand-Maillet, S., Müller, H.: Concept attribution: explaining CNN decisions to physicians. Comput. Biol. Med. **123** (2020)
10. Ha, R., et al.: Predicting breast cancer molecular subtype with MRI dataset utilizing convolutional neural network algorithm. J. Digit. Imaging **32**(2), 276–282 (2019)
11. He, K., Zhang, X., Ren, S., Sun, J.: Deep residual learning for image recognition. In: Proceedings of the IEEE Conference on Computer Vision and Pattern Recognition, pp. 770–778 (2016)
12. Huang, Y., et al.: Multi-parametric MRI-based radiomics models for predicting molecular subtype and androgen receptor expression in breast cancer. Front. Oncol. **11** (2021)
13. Janik, A., Dodd, J., Ifrim, G., Sankaran, K., Curran, K.: Interpretability of a deep learning model in the application of cardiac MRI segmentation with an ACDC challenge dataset. In: Proceedings of SPIE - The International Society for Optical Engineering (2021)
14. Ketchen, D.J., Shook, C.L.: The application of cluster analysis in strategic management research: an analysis and critique. Strateg. Manag. J. **17**(6), 441–458 (1996)
15. Kim, B., et al.: Interpretability beyond feature attribution: quantitative testing with concept activation vectors (TCAV). In: International Conference on Machine Learning, pp. 2668–2677. PMLR (2018)
16. Kingma, D.P., Ba, J.: Adam: a method for stochastic optimization. arXiv preprint arXiv:1412.6980 (2014)
17. Lee, J.R., Kim, S., Park, I., Eo, T., Hwang, D.: Relevance-CAM: your model already knows where to look. In: Proceedings of the IEEE/CVF Conference on Computer Vision and Pattern Recognition, pp. 14944–14953 (2021)
18. Nielsen, T.O., et al.: Immunohistochemical and clinical characterization of the basal-like subtype of invasive breast carcinoma. Clin. Cancer Res. **10**(16), 5367–5374 (2004)
19. Pereira, S., Meier, R., Alves, V., Reyes, M., Silva, C.A.: Automatic brain tumor grading from MRI data using convolutional neural networks and quality assessment. In: Stoyanov, D., et al. (eds.) MLCN/DLF/IMIMIC -2018. LNCS, vol. 11038, pp. 106–114. Springer, Cham (2018). https://doi.org/10.1007/978-3-030-02628-8_12
20. Pinker, K., et al.: Improved differentiation of benign and malignant breast tumors with multiparametric 18fluorodeoxyglucose positron emission tomography magnetic resonance imaging: a feasibility study. Clin. Cancer Res. **20**(13), 3540–3549 (2014)
21. Reyes, M., et al.: On the interpretability of artificial intelligence in radiology: challenges and opportunities. Radiol. Artif. Intell. **2**(3), e190043 (2020)
22. Romeo, V., et al.: AI-enhanced simultaneous multiparametric 18F-FDG PET/MRI for accurate breast cancer diagnosis. Eur. J. Nucl. Med. Mol. Imaging, 1–13 (2021)
23. Selvaraju, R.R., Cogswell, M., Das, A., Vedantam, R., Parikh, D., Batra, D.: Gradcam: Visual explanations from deep networks via gradient-based localization. In: Proceedings of the IEEE International Conference on Computer Vision, pp. 618–626 (2017)
24. Son, J., Lee, S.E., Kim, E.K., Kim, S.: Prediction of breast cancer molecular subtypes using radiomics signatures of synthetic mammography from digital breast tomosynthesis. Sci. Rep. **10**(1), 1–11 (2020)

25. Sørlie, T., et al.: Gene expression patterns of breast carcinomas distinguish tumor subclasses with clinical implications. Proc. Natl. Acad. Sci. **98**(19), 10869–10874 (2001)
26. Sundararajan, M., Taly, A., Yan, Q.: Axiomatic attribution for deep networks. In: International Conference on Machine Learning, pp. 3319–3328. PMLR (2017)
27. Sung, H., et al.: Global cancer statistics 2020: Globocan estimates of incidence and mortality worldwide for 36 cancers in 185 countries. CA Cancer J. Clin. **71**(3), 209–249 (2021)
28. Wang, H., et al.: Score-CAM: score-weighted visual explanations for convolutional neural networks. In: Proceedings of the IEEE/CVF Conference on Computer Vision and Pattern Recognition Workshops, pp. 24–25 (2020)
29. Zhang, Y., et al.: Prediction of breast cancer molecular subtypes on DCE-MRI using convolutional neural network with transfer learning between two centers. Eur. Radiol. **31**(4), 2559–2567 (2021)

Fusing Modalities by Multiplexed Graph Neural Networks for Outcome Prediction in Tuberculosis

Niharika S. D'Souza[1(✉)], Hongzhi Wang[1], Andrea Giovannini[2],
Antonio Foncubierta-Rodriguez[2], Kristen L. Beck[1], Orest Boyko[3],
and Tanveer Syeda-Mahmood[1]

[1] IBM Research Almaden, San Jose, CA, USA
niharika.dsouza@ibm.com
[2] IBM Research, Zurich, Switzerland
[3] Department of Radiology, VA Southern Nevada Healthcare System,
North Las Vegas, NV, USA

Abstract. In a complex disease such as tuberculosis, the evidence for
the disease and its evolution may be present in multiple modalities such
as clinical, genomic, or imaging data. Effective patient-tailored outcome
prediction and therapeutic guidance will require fusing evidence from
these modalities. Such multimodal fusion is difficult since the evidence
for the disease may not be uniform across all modalities, not all modality features may be relevant, or not all modalities may be present for all
patients. All these nuances make simple methods of early, late, or intermediate fusion of features inadequate for outcome prediction. In this
paper, we present a novel fusion framework using multiplexed graphs
and derive a new graph neural network for learning from such graphs.
Specifically, the framework allows modalities to be represented through
their targeted encodings, and models their relationship explicitly via multiplexed graphs derived from salient features in a combined latent space.
We present results that show that our proposed method outperforms
state-of-the-art methods of fusing modalities for multi-outcome prediction on a large Tuberculosis (TB) dataset.

Keywords: Multimodal fusion · Graph neural networks · Multiplex
graphs · Imaging data · Genomic data · Clinical data

1 Introduction

Tuberculosis (TB) is one of the most common infectious diseases worldwide [18].
Although the mortality rate caused by TB has declined in recent years, single-
and multi-drug resistance has become a major threat to quick and effective TB
treatment. Studies have revealed that predicting the outcome of a treatment is
a function of many patient-specific factors for which collection of multimodal
data covering clinical, genomic, and imaging information about the patient has

© The Author(s), under exclusive license to Springer Nature Switzerland AG 2022
L. Wang (Eds.): MICCAI 2022, LNCS 13437, pp. 287–297, 2022.
https://doi.org/10.1007/978-3-031-16449-1_28

become essential [17]. However, it is not clear what information is best captured in each modality and how best to combine them. For example, genomic data can reveal the genetic underpinnings of drug resistance and identify genes/mutations conferring drug-resistance [16]. Although imaging data from X-ray or CT can show statistical difference for drug resistance, they alone may be insufficient to differentiate multi-drug resistant TB from drug-sensitive TB [29]. Thus it is important to develop methods that allow simultaneously extraction of relevant information from multiple modalities as well as ways to combine them in an optimal fashion to lead to better outcome prediction.

Recent efforts to study the fusion problem for outcome prediction in TB have focused on a single outcome such as treatment failure or used only a limited number of modalities such as clinical and demographic data [1,20]. In this paper, we take a comprehensive approach by treating the outcome prediction problem as a multiclass classification for multiple possible outcomes through multimodal fusion. We leverage more extensive modalities beyond clinical, imaging, or genomic data, including novel features extracted via advanced analysis of protein domains from genomic sequence as well as deep learning-derived features from CT images as shown in Fig. 2.(a). Specifically, we develop a novel fusion framework using multiplexed graphs to capture the information from modalities and derive a new graph neural network for learning from such graphs. The framework represents modalities through their targeted encodings, and models their relationship via multiplexed graphs derived from projections in a latent space.

Existing approaches often infer matrix or tensor encodings from individual modalities [13] combined with early, late, or intermediate fusion [2,24,26] of the individual representations. Example applications include- CCA for speaker identification [19], autoencoders for video analytics [25], transformers for VQA [10], etc. In contrast, our approach allows for the modalities to retain their individuality while still participating in exploring explicit relationships between the modality features through the multiplexed framework. Specifically, we design our framework to explicitly model relationships within and across modality features via a self-supervised multi-graph construction and design a novel graph neural network for reasoning from these feature dependencies via structured message passing walks. We present results which show that by relaxing the fusing constraints through the multiplex formulation, our method outperforms state-of-the-art methods of multimodal fusion in the context of multi-outcome prediction for TB treatments.

2   A Graph Based Multimodal Fusion Framework

As alluded to earlier, exploring various facets of cross-modal interactions is at the heart of the multimodal fusion problem. To this end, we propose to utilize the representation learning theory of multiplexed graphs to develop a generalized framework for multimodal fusion. A multiplexed graph [3] is a type of multigraph in which the nodes are grouped into multiple planes, each representing an individual edge-type. The information captured within a plane is

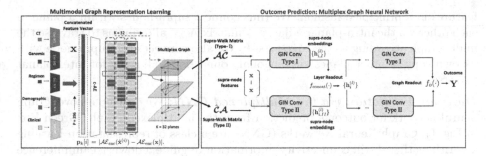

Fig. 1. Graph Based Multimodal Fusion for Outcome Prediction. **Blue Box:** Incoming modality features are concatenated into a feature vector (of size P = 396) and projected into a common latent space (of size K = 32). Salient activations in the latent space are used to form the planes of the multiplexed graph. **Green Box:** The multiplexed GNN uses message passing walks to combine latent concepts for inference. (Color figure online)

multiplexed to other planes through diagonal connections as shown in Fig. 1. Mathematically, we define a multiplexed graph as: $\mathcal{G}_{\text{Mplex}} = (\mathcal{V}_{\text{Mplex}}, \mathcal{E}_{\text{Mplex}})$, where $|\mathcal{V}_{\text{Mplex}}| = |\mathcal{V}| \times K$ and $\mathcal{E}_{\text{Mplex}} = \{(i,j) \in \mathcal{V}_{\text{Mplex}} \times \mathcal{V}_{\text{Mplex}}\}$. There are K distinct types of edges which can link two given nodes. Analogous to ordinary graphs, we have k adjacency matrices $\mathbf{A}_{(k)} \in \mathcal{R}^{P \times P}$, where $P = |\mathcal{V}|$, each summarizing the connectivity information given by the edge-type k. The elements of these matrices are binary $\mathbf{A}_{(k)}[m,n] = 1$ if there is an edge of type k between nodes $m, n \in \mathcal{V}$.

Multimodal Graph Representation Learning: While the multiplexed graph has been used for various modeling purposes in literature [5,6,12,15], we propose to use it for multimodal fusion of imaging, genomic and clinical data for outcome prediction in TB. We adopt the construction shown in the Blue Box in Fig. 1 to produce the multiplexed graph from the individual modality features. First, domain specific autoencoders (d-AE) are used to convert each modality into a compact feature space that can provide good reconstruction using Mean Squared Error (MSE). To capture feature dependencies across modalities, the concatenated features are brought to a common low dimensional subspace through a common autoencoder (c-AE) trained to reconstruct the concatenated features. Each latent dimension of the autoencoder captures an abstract aspect of the multimodal fusion problem, e.g. features projected to be salient in the same latent dimension are likely to form meaningful joint patterns for a specific task, and form a "conceptual" plane of the multiplexed graph. The $|\mathcal{V}_{\text{Mplex}}|$ "supra-nodes" of $\mathcal{G}_{\text{Mplex}}$ are produced by creating copies of features (i.e. nodes) across the planes. The edges between nodes in each plane represent features whose projections in the respective latent dimensions were salient (see Sect. 3.1 for details). Further, each plane is endowed with its own topology and is a proxy for the correlation between features across the corresponding latent dimension. This procedure helps model the interactions between the various modality features in a principled fashion. We thus connect supra-nodes within a plane to each other via the intra-planar adjacency matrix $\mathbf{A}_{(k)}$, allowing us to traverse the multi-graph according to the edge-type k. We also connect each supra-node with its own copy in other planes via diagonal connections, allowing for inter-planar traversal.

Outcome Prediction via the Multiplexed GNN: We develop a novel graph neural network for outcome prediction from the multiplexed graph (Green Box in Fig. 1). Graph Neural Networks (GNN) are a class of representation learning algorithms that distill connectivity information to guide a downstream inference task [21]. A typical GNN schema comprises of two components: (1) a message passing scheme for propagating information across the graph and (2) task-specific supervision to guide the representation learning. For ordinary graphs, the adjacency matrix $\mathbf{A}$ and its matrix powers allows us to keep track of neighborhoods (at arbitrary l hop distance) within the graph during message passing. Conceptually, cascading l GNN layers is analogous to pooling information at each node i from its l-hop neighbors that can be reached by a walk starting at i. The Multiplex GNN is designed to mirrors this behavior.

The *intra-planar adjacency matrix* $\mathcal{A} \in \mathcal{R}^{PK \times PK}$, and the *inter-planar transition control matrix* $\hat{\mathcal{C}} \in \mathcal{R}^{PK \times PK}$ [3] define walks on the multiplex $\mathcal{G}_{\text{Mplex}}$.

$$\mathcal{A} = \bigoplus_k \mathbf{A}_{(k)} \quad ; \quad \hat{\mathcal{C}} = [\mathbf{1}_K \mathbf{1}_K^T] \otimes \mathcal{I}_P \tag{1}$$

where $\bigoplus$ is the direct sum operation, $\otimes$ denotes the Kronecker product, $\mathbf{1}_K$ is the K vector of all ones, and $\mathcal{I}_P$ denotes the identity matrix of size $P \times P$. Thus $\mathcal{A}$ is block-diagonal by construction and captures within plane transitions across supra-nodes. Conversely, $\hat{\mathcal{C}}$ has identity matrices along on off-diagonal blocks. This implicitly restricts across plane transitions to be between supra-nodes which arise from the same multi-graph node (i.e. i and $P(k-1)+i$ for $k \in \{1, \ldots, K\}$). Since supra-nodes across planes can already be reached by combining within and across-planar transitions, this provides comparable representational properties at a reduced complexity ($\mathcal{O}(PK)$) inter-planar edges instead of $\mathcal{O}(P^2K)$).

A walk on $\mathcal{G}_{\text{Mplex}}$ combines within and across planar transitions to reach a supra-node $j \in \mathcal{V}_{\text{Mplex}}$ from a given supra-node $i \in \mathcal{V}_{\text{Mplex}}$. $\mathcal{A}$ and $\hat{\mathcal{C}}$ allow us to define multi-hop transitions on the multiplex in a convenient factorized form. A multiplex walk proceeds according to two types of transitions [3]: (1) A single intra-planar step or (2) A step that includes both an inter-planar step moving from one plane to another (this can be before or after the occurrence of an intra-planar step). To recreate these transitions exhaustively, we have two supra-walk matrices. $\mathcal{A}\hat{\mathcal{C}}$ encodes transitions where *after* an intra-planar step, the walk *can* continue in the same plane or transition to a different plane (Type I). Similarly, using $\hat{\mathcal{C}}\mathcal{A}$, the walk *can* continue in the same plane or transition to a different plane *before* an intra-planar step (Type II).

Message Passing Walks: Let $\mathbf{h}_i^l \in \mathcal{R}^{D^l \times 1}$ denote the (supra)-node representation for (supra)-node i. In matrix form, we can write $\mathbf{H}^{(l)} \in \mathcal{R}^{|\mathcal{V}_{\text{Mplex}}| \times D^l}$, with $\mathbf{H}^{(l)}[i, :] = \mathbf{h}_i^{(l)}$. We then compute this via the following operations:

$$\mathbf{h}_{i,I}^{(l+1)} = \phi_I\Big(\{\mathbf{h}_j^{(l)}, j : [\boldsymbol{\mathcal{A}}\hat{\boldsymbol{\mathcal{C}}}][i,j] = 1\}\Big) \quad ; \quad \mathbf{h}_{i,II}^{(l+1)} = \phi_{II}\Big(\{\mathbf{h}_j^{(l)}, j : [\hat{\boldsymbol{\mathcal{C}}}\boldsymbol{\mathcal{A}}][i,j] = 1\}\Big)$$

$$\mathbf{h}_i^{(l+1)} = f_{\text{concat}}(\mathbf{h}_{i,I}^{(l+1)}, \mathbf{h}_{i,II}^{(l+1)}) \quad ; \quad f_o(\{\mathbf{h}_i^{(L)}\}) = \mathbf{Y} \tag{2}$$

Here, $f_{\text{concat}}(\cdot)$ concatenates the Type I and Type II representations. At the input layer, we have $\mathbf{H}^{(0)} = \mathbf{X} \otimes \mathbf{1}_K$, where $\mathbf{X} \in \mathcal{R}^{|V| \times 1}$ are the node inputs (concatenated modality features). $\{\phi_I(\cdot), \phi_{II}(\cdot)\}$ performs message passing according to the neighborhood relationships given by the supra-walk matrices. Finally, $f_o(\cdot)$ is the graph readout that predicts the outcome $\mathbf{Y}$. The learnable parameters of the Multiplex GNN can be estimated via standard backpropagation.

Implementation Details: We utilize the Graph Isomorphism Network (GIN) [30] with LeakyReLU (neg. slope $= 0.01$) readout for message passing (i.e. $\{\phi_I(\cdot), \phi_{II}(\cdot)\}$). Since the input $\mathbf{x}$ is one dimensional, we have two such layers in cascade with hidden layer width one. $f_o(\cdot)$ is a Multi-Layered Perceptron (MLP) with two hidden layers (size: 100 and 20) and LeakyReLU activation. In each experimental comparison, we chose the model architecture and hyperparameters for our framework (learning rate $= 0.001$ decayed by 0.1 every 20 epochs, weight decay $= 0.001$, number of epochs $= 40$) and baselines using gridsearch and validation set. All frameworks are trained on the Cross Entropy loss between the predicted logits (after a softmax) and the ground truth labels. We utilize the ADAMw optimizer [14]. Models were implemented using the Deep Graph Library (v $= 0.6.2$) in PyTorch (v $= 0.10.1$). We trained all models on a 64 GB CPU RAM, 2.3 GHz 8-Core Intel i9 machine, with 3.5–4 h training time per run (Note: Performing inference via GPUs will likely speed up computation).

3 Experimental Evaluation

3.1 Data and Experimental Setup

We conducted experiments using the Tuberculosis Data Exploration Portal [7]. 3051 patients with five classes of treatment outcomes (Still on treatment, Died, Cured, Completed, or Failure) were used. Five modalities are available (see Fig. 2.(a)). Demographic, clinical, regimen and genomic data are available for each patient, while chest CTs are available for 1015 patients. For clinical and regimen data, information that might be directly related to treatment outcomes,

Table 1. Dataset description of the TB dataset.

Modality	CT	Genomic	Demographic	Clinical	Regimen	Continuous
Native Dimen	2048	4081	29	1726	233	8
Rank	250	300	24	183	112	8
Reduced Dim	128	64	8	128	64	4

such as type of resistance, were removed. For each CT, lung was segmented using multi-atlas segmentation [28]. The pre-trained DenseNet [8] was then applied to extract a feature vector of 1024-dimension for each axial slice intersecting lung. To aggregate the information from all lung intersecting slices, the mean and maximum of each of the 1024 features were used providing a total of 2048 features. For genomic data from the causative organisms *Mycobacterium tuberculosis* (Mtb), 81 single nucleotide polymorphisms (SNPs) in genes known to be related to drug resistance were used. In addition, we retrieved the raw genome sequence from NCBI Sequence Read Archive for 275 patients to describe the biological sequences of the disease-causing pathogen at a finer granularity. The data was processed by the IBM Functional Genomics Platform [23]. Briefly, each Mtb genome underwent an iterative *de novo* assembly process and then processed to yield gene and protein sequences. The protein sequences were then processed using InterProScan [9] to generate the functional domains. Functional domains are sub-sequences located within the protein's amino acid chain. They are responsible for the enzymatic bioactivity of a protein and can more aptly describe the protein's function. 4000 functional features were generated for each patient.

Multiplexed Graph Construction: We note that the regimen and genomic data are categorical features. CT features are continuous. The demographic and clinical data are a mixture of categorical and continuous features. Grouping the continuous demographic and clinical variables together yielded a total of six source modalities (see Table 1). We impute the missing CT and functional genomic features using the mean values from the training set. To reduce the redundancy in each domain, we use d-AEs with fully connected layers, LeakyReLU non-linearities and tied weights trained to reconstruct the raw modality features. The d-AE bottleneck (see Table 1) is chosen via the validation set. The reduced individual modality features are concatenated to form the node feature vector $\mathbf{x}$. To form the multiplexed graph planes, the c-AE projects $\mathbf{x}$ to a 'conceptual' latent space of dimension $K \ll P$ where $P = 128 + 64 + 8 + 128 + 64 + 4 = 396$. We use the c-AE concept space form the planes of the multiplex and explore the correlation between pairs of features. The c-AE architecture mirrors the d-AE, but projects the training examples $\{\mathbf{x}\}$ to $K = 32$ concepts. We infer within plane connectivity along each concept perturbing the features and recording those features giving rise to largest incremental responses. Let $\mathcal{AE}_{\text{enc}}(\cdot) : \mathcal{R}^P \to \mathcal{R}^K$ be the c-AE mapping to the concept space. Let $\hat{\mathbf{x}}^{(i)}$ denote the perturbation of the input by setting $\hat{\mathbf{x}}^{(i)}[j] = \mathbf{x}[j] \; \forall \; j \neq i$ and 0 for $j = i$. Then for concept axis k, the perturbations are $\mathbf{p}_k[i] = |\mathcal{AE}_{\text{enc}}(\hat{\mathbf{x}}^{(i)}) - \mathcal{AE}_{\text{enc}}(\mathbf{x})|$. Thresholding $\mathbf{p}_k \in \mathcal{R}^{P \times 1}$ selects feature nodes with the strongest responses along concept k. To encourage sparsity, we retain the top one percent of salient patterns. We connect all pairs of such feature nodes with edge-type k via a fully connected (complete) subgraph between nodes thus selected (Fig. 1). Across the K concepts, we expect that different sets of features are prominent. The input features $\mathbf{x}$ are one dimensional node embeddings (or the messages at input layer $l = 0$). The latent concepts K, and the feature selection (sparsity) are key quantities that control generalization.

3.2 Baselines

We compared with four multimodal fusion approaches. We also present three ablations, allowing us to probe the inference (Multiplex GNN) and representation learning (Multimodal Graph Construction) separately.

No Fusion: This baseline utilizes a two layered MLP (hidden width: 400 and 20, LeakyReLU activation) on the individual modality features before the d-AE dimensionality reduction. This provides a benchmark for the outcome prediction performance of each modality separately.

Early Fusion: Individual modalities are concatenated before dimensionality reduction and fed through the same MLP architecture as described above.

Intermediate Fusion: In this comparison, we perform intermediate fusion after the d-AE projection by using the concatenated feature x as input to a two layered MLP (hidden width: 150 and 20, LeakyReLU activation). This helps us evaluate the benefit of using graph based fusion via the c-AE latent encoder.

Late Fusion: We utilize the late fusion framework of [27] to combine the predictions from the modalities trained individually in the No Fusion baseline. This framework leverages the uncertainty in the 6 individual classifiers to improve the robustness of outcome prediction. We used the hyperparameters in [27].

Relational GCN on a Multiplexed Graph: This baseline utilizes the multi-graph representation learning (Blue Box of Fig. 1), but replaces the Multiplex GNN feature extraction with the Relational GCN framework of [22]. Essentially, at each GNN layer, the RGCN runs K separate message passing operations on the planes of the multigraph and then aggregates the messages post-hoc. Since the width, depth and graph readout is the same as with the Multiplex GNN, this helps evaluate the expressive power of the walk based message passing in Eq. (2).

Relational GCN w/o Latent Encoder: For this comparison, we utilize the reduced features after the d-AE, but instead create a multi-layered graph with the individual modalities in different planes. Within each plane, nodes are fully connected to each other after which a two layered RGCN [22] model is trained. Effectively, *within modality* feature dependence may still be captured in the planes, but the concept space is not used to infer the *cross-modal* interactions.

GCN on Monoplex Feature Graph: This baseline also incorporates a graph based representation, but does not include the use of latent concepts to model within and cross-modal feature correlations. Essentially, we construct a fully connected graph on x instead of using the (multi-) conceptual c-AE space and train a two layered Graph Convolutional Network [11] for outcome prediction.

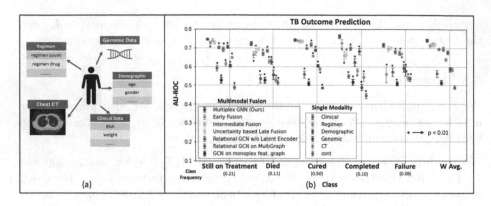

Fig. 2. (a). Multimodal data for Tuberculosis treatment outcome prediction. (b). Outcome prediction performance measured by per-class and weighted average AU-ROC. We display mean performance along with standard errors. * indicates comparisons with the Multiplexed GNN per-class AU-ROC with ($p < 0.01$) according to the DeLong test. Individual class frequencies are listed below the x axis.

3.3   Results

Evaluation Metrics: Since we have unbalanced classes for multi-class classification, we evaluate the performance using AU-ROC (Area Under the Receiver Operating Curve). We also report weighted average AU-ROC as an overall summary. We rely on 10 randomly generated train/validation/test splits of size 2135/305/611 to train the representation learning and GNNs in a fully blind fashion. We use the same splits and training/evaluation procedure for the baselines. For statistical rigour, we indicate significant differences between Multiplex GNN and baseline AU-ROC for each class as quantified by a DeLong [4] test.

Outcome Prediction Performance: Figure 2 illustrates the outcome prediction results. Our framework outperforms common multimodal fusion baselines (Early Fusion, Intermediate Fusion, and Late Fusion), as quantified by the higher mean per-class AU-ROC and weighted average AU-ROC. Our graph based multimodal fusion also provides improved performance over the single modality outcome classifiers. The Relational GCN- Multigraph baseline is an ablation that replaces the Multiplexed GNN with an existing state-of-the art GNN framework. For both techniques, we utilize the same Multi-Graph representation as learned by the c-AE autoencoder latent space. The performance gains within this comparison suggest that the Multiplexed GNN is better suited for reasoning and task-specific knowledge distillation from multigraphs. We conjecture that the added representational power is a direct consequence of our novel multiplex GNN message passing (Eq. (2)) scheme. Along similar lines, the Relational GCN w/o latent encoder and the GCN on the monoplex feature graph baseline comparisons are generic graph based fusion approaches. They allow us to examine the benefit of using the salient activation patterns from the c-AE latent concept space to infer the multi-graph representation. Specifically, the former separates the modality features into plane-specific fully connected graphs within a multi-planar representation. The latter constructs a single fully connected graph on the concatenated modality features. Our framework provides large gains over these baselines. In turn, this highlights the efficacy of our Multimodal graph construction. We surmise that the salient learned conceptual patterns are more successful at uncovering cross modal interactions between features that are explanative of patient outcomes. Overall, these observations highlight key representational aspects of our framework, and demonstrate the efficacy for the TB outcome prediction task. Given the clinical relevance, a promising direction for exploration would be to extend frameworks for explainability in GNNs (for example, via subgraph exploration [31]) to Multiplex GNNs to automatically highlight patterns relevant to downstream prediction.

4 Conclusion

We have introduced a novel Graph Based Multimodal Fusion framework to combine imaging, genomic and clinical data. Our Multimodal Graph Representation Learning projects the individual modality features into abstract concept spaces, wherein complex cross modal dependencies can be mined from the salient patterns. We developed a new multiplexedh Neural Network that can track information flow within the multi-graph via message passing walks. Our GNN formulation provides the necessary flexibility to mine rich representations from multimodal data. Overall, this provides for improved Tuberculosis outcome prediction performance against several state-of-the-art baselines.

References

1. Asad, M., Mahmood, A., Usman, M.: A machine learning-based framework for predicting treatment failure in tuberculosis: a case study of six countries. Tuberculosis **123**, 101944 (2020)
2. Baltrušaitis, T., Ahuja, C., Morency, L.P.: Multimodal machine learning: a survey and taxonomy. IEEE Trans. Pattern Anal. Mach. Intell. **41**(2), 423–443 (2018)
3. Cozzo, E., de Arruda, G.F., Rodrigues, F.A., Moreno, Y.: Multiplex networks (2018). https://doi.org/10.1007/978-3-319-92255-3, http://link.springer.com/10.1007/978-3-319-92255-3
4. DeLong, E.R., DeLong, D.M., Clarke-Pearson, D.L.: Comparing the areas under two or more correlated receiver operating characteristic curves: a nonparametric approach. Biometrics, 837–845 (1988)
5. Domenico, M.D., et al.: Mathematical formulation of multilayer networks. Phys. Rev. X **3**, 041022 (2014). https://doi.org/10.1103/PHYSREVX.3.041022/FIGURES/5/MEDIUM, https://journals.aps.org/prx/abstract/10.1103/PhysRevX.3.041022
6. Ferriani, S., Fonti, F., Corrado, R.: The social and economic bases of network multiplexity: exploring the emergence of multiplex ties. Strateg. Org. **11**, 7–34 (2013). https://doi.org/10.1177/1476127012461576

7. Gabrielian, A., et al.: TB depot (data exploration portal): a multi-domain tuberculosis data analysis resource. PLOS ONE **14**(5), e0217410 (2019). https://doi.org/10.1371/journal.pone.0217410, http://dx.plos.org/10.1371/journal.pone.0217410

8. Huang, G., Liu, Z., Van Der Maaten, L., Weinberger, K.Q.: Densely connected convolutional networks. In: Proceedings of the IEEE Conference on Computer Vision and Pattern Recognition, pp. 4700–4708 (2017)

9. Jones, P., et al.: InterProScan 5: genome-scale protein function classification. Bioinformatics (Oxford, England) **30**(9), 1236–40 (2014). https://doi.org/10.1093/bioinformatics/btu031

10. Kant, Y., et al.: Spatially aware multimodal transformers for TextVQA. In: Vedaldi, A., Bischof, H., Brox, T., Frahm, J.-M. (eds.) ECCV 2020. LNCS, vol. 12354, pp. 715–732. Springer, Cham (2020). https://doi.org/10.1007/978-3-030-58545-7_41

11. Kipf, T.N., Welling, M.: Semi-supervised classification with graph convolutional networks. arXiv preprint arXiv:1609.02907 (2016)

12. Kivelä, M., Arenas, A., Barthelemy, M., Gleeson, J.P., Moreno, Y., Porter, M.A.: Multilayer networks. J. Complex Netw. **2**, 203–271 (2014). https://doi.org/10.1093/COMNET/CNU016, https://academic.oup.com/comnet/article/2/3/203/2841130

13. Lahat, D., Adali, T., Jutten, C.: Multimodal data fusion: an overview of methods, challenges, and prospects. Proc. IEEE **103**(9), 1449–1477 (2015)

14. Loshchilov, I., Hutter, F.: Decoupled weight decay regularization. arXiv preprint arXiv:1711.05101 (2017)

15. Maggioni, M.A., Breschi, S., Panzarasa, P.: Multiplexity, growth mechanisms and structural variety in scientific collaboration networks, **20**, 185–194 (4 2013). https://doi.org/10.1080/13662716.2013.791124, https://www.tandfonline.com/doi/abs/10.1080/13662716.2013.791124

16. Manson, A.L., et al.: Genomic analysis of globally diverse mycobacterium tuberculosis strains provides insights into the emergence and spread of multidrug resistance. Nat. Genet. **49**(3), 395–402 (2017)

17. Muñoz-Sellart, M., Cuevas, L., Tumato, M., Merid, Y., Yassin, M.: Factors associated with poor tuberculosis treatment outcome in the southern region of Ethiopia. Int. J. Tuberc. Lung Dis. **14**(8), 973–979 (2010)

18. World Health Organization: Treatment of Tuberculosis: Guidelines. World Health Organization (2010)

19. Sargin, M.E., Erzin, E., Yemez, Y., Tekalp, A.M.: Multimodal speaker identification using canonical correlation analysis. In: 2006 IEEE International Conference on Acoustics Speech and Signal Processing Proceedings, vol. 1, p. I. IEEE (2006)

20. Sauer, C.M., et al.: Feature selection and prediction of treatment failure in tuberculosis. PLoS ONE **13**(11), e0207491 (2018)

21. Scarselli, F., Gori, M., Tsoi, A.C., Hagenbuchner, M., Monfardini, G.: The graph neural network model. IEEE Trans. Neural Netw. **20**(1), 61–80 (2008)

22. Schlichtkrull, M., Kipf, T.N., Bloem, P., van den Berg, R., Titov, I., Welling, M.: Modeling relational data with graph convolutional networks. In: Gangemi, A., et al. (eds.) ESWC 2018. LNCS, vol. 10843, pp. 593–607. Springer, Cham (2018). https://doi.org/10.1007/978-3-319-93417-4_38

23. Seabolt, E.E., et al.: OMXWare, A Cloud-Based Platform for Studying Microbial Life at Scale (nov 2019), http://arxiv.org/abs/1911.02095

24. Subramanian, V., Do, M.N., Syeda-Mahmood, T.: Multimodal fusion of imaging and genomics for lung cancer recurrence prediction. In: 2020 IEEE 17th International Symposium on Biomedical Imaging (ISBI), pp. 804–808. IEEE (2020)

25. Vu, T.D., Yang, H.J., Nguyen, V.Q., Oh, A.R., Kim, M.S.: Multimodal learning using convolution neural network and sparse autoencoder. In: 2017 IEEE International BigComp, pp. 309–312. IEEE (2017)
26. Wang, H., Subramanian, V., Syeda-Mahmood, T.: Modeling uncertainty in multimodal fusion for lung cancer survival analysis. In: Proceedings - International Symposium on Biomedical Imaging 2021-April, 1169–1172 (2021). https://doi.org/10.1109/ISBI48211.2021.9433823
27. Wang, H., Subramanian, V., Syeda-Mahmood, T.: Modeling uncertainty in multimodal fusion for lung cancer survival analysis. In: 2021 IEEE 18th International Symposium on Biomedical Imaging (ISBI), pp. 1169–1172. IEEE (2021)
28. Wang, H., Yushkevich, P.: Multi-atlas segmentation with joint label fusion and corrective learning-an open source implementation. Front. Neuroinform. 7, 27 (2013)
29. Wáng, Y.X.J., Chung, M.J., Skrahin, A., Rosenthal, A., Gabrielian, A., Tartakovsky, M.: Radiological signs associated with pulmonary multi-drug resistant tuberculosis: an analysis of published evidences. Quant. Imaging Med. Surg. 8(2), 161 (2018)
30. Xu, K., Hu, W., Leskovec, J., Jegelka, S.: How powerful are graph neural networks? arXiv preprint arXiv:1810.00826 (2018)
31. Yuan, H., Yu, H., Wang, J., Li, K., Ji, S.: On explainability of graph neural networks via subgraph explorations. In: International Conference on Machine Learning, pp. 12241–12252. PMLR (2021)

Deep Multimodal Guidance for Medical Image Classification

Mayur Mallya$^{(\boxtimes)}$ and Ghassan Hamarneh

Simon Fraser University, Burnaby, Canada
{mmallya,hamarneh}@sfu.ca

Abstract. Medical imaging is a cornerstone of therapy and diagnosis in modern medicine. However, the choice of imaging modality for a particular theranostic task typically involves trade-offs between the feasibility of using a particular modality (e.g., short wait times, low cost, fast acquisition, reduced radiation/invasiveness) and the expected performance on a clinical task (e.g., diagnostic accuracy, efficacy of treatment planning and guidance). In this work, we aim to apply the knowledge learned from the less feasible but better-performing (*superior*) modality to guide the utilization of the more-feasible yet under-performing (*inferior*) modality and steer it towards improved performance. We focus on the application of deep learning for image-based diagnosis. We develop a light-weight guidance model that leverages the latent representation learned from the superior modality, when training a model that consumes only the inferior modality. We examine the advantages of our method in the context of two clinical applications: multi-task skin lesion classification from clinical and dermoscopic images and brain tumor classification from multi-sequence magnetic resonance imaging (MRI) and histopathology images. For both these scenarios we show a boost in diagnostic performance of the inferior modality without requiring the superior modality. Furthermore, in the case of brain tumor classification, our method outperforms the model trained on the superior modality while producing comparable results to the model that uses both modalities during inference. We make our code and trained models available at: https://github.com/mayurmallya/DeepGuide.

Keywords: Deep learning · Multimodal learning · Classification · Student-teacher learning · Knowledge distillation · Skin lesions · Brain tumors

1 Introduction

Multimodal machine learning aims at analyzing the heterogeneous data in the same way animals perceive the world – by a holistic understanding of the infor-

Supplementary Information The online version contains supplementary material available at https://doi.org/10.1007/978-3-031-16449-1_29.

mation gathered from all the sensory inputs. The complementary and the supplementary nature of this multi-input data helps in better navigating the surroundings than a single sensory signal. The ubiquity of digital sensors coupled with the powerful feature abstraction abilities of deep learning (DL) models has aided the massive interest in this field in the recent years [6,7,12,14,35].

In clinical settings, for clinicians to make informed disease diagnosis and management decisions, a comprehensive assessment of the patient's health would ideally involve the acquisition of complementary biomedical data across multiple different modalities. For instance, the simultaneous acquisition of functional and anatomical imaging data is a common practice in the modern clinical setting [8,29], as the former provides quantitative metabolic information and the latter provides the anatomical spatial context. Similarly, cancer diagnosis and prognosis are increasingly a result of a thorough examination of both genotypic and phenotypic modalities [4,26].

Although the complimentary use of multiple modalities can improve the clinical diagnosis, the acquisition of modalities with stronger performance for a given task, such as diagnosis, may be less feasible due to longer wait times, higher cost of scan, slower acquisition, higher radiation exposure, and/or invasiveness. Hence, generally, a compromise must be struck. For example, higher anatomical or functional resolution imaging may only be possible with a modality that involves invasive surgical procedures or ionizing radiations. Other examples of this trade-off include: anatomical detail provided by computer tomography (CT) versus associated risks of cancer from repeated x-ray exposure [28]; and rich cellular information provided by histology versus expensive, time-consuming invasive biopsy procedure with associated risks of bleeding and infections [21]. Unsurprisingly, in most cases, it is the expensive modality that provides the critical piece of information for diagnosis.

For simplicity, hereinafter, we refer to the over-performing modality with less-feasible acquisition as the *superior* modality, and the more-feasible but underperforming one as the *inferior*. We note that a particular modality may be regarded as inferior in one context and superior in another. For example, magnetic resonance imaging (MRI) is superior to ultrasound for delineating cancerous lesions but inferior to histopathology in deciding cancer grade.

Consequently, it would be advantageous to leverage the inferior modalities in order to alleviate the need for the superior one. However, this is reasonable only when the former can be as informative as the latter. To this end, we propose a novel deep multimodal, student-teacher learning-based framework that leverages existing datasets of paired inferior and superior modalities during the training phase to enhance the diagnosis performance achievable by the inferior modality during the test phase. Our experiments on two disparate multimodal datasets across several classification tasks demonstrate the validity and utility of the proposed method.

2 Related Work

Three sub-fields in particular relate to our work: (i) <u>Multimodal classification:</u> Most of the DL based works on multimodal prediction on paired medical images focus on the classification task that involves the presence of multiple modalities at test time [34]. The primary focus of research being the optimal fusion strategy that aims to answer when and how to efficiently fuse the supposedly heterogeneous and redundant features. *When to fuse?* While the registered multimodal image pairs allow for an input-level data fusion [15,27], a majority of works rely on feature-level fusion not only due to the dimensionality mismatch at the input but also for the flexibility of fusion it offers [9,10,17]. Additionally, some works make use of a decision-level fusion framework that leverages the ensemble learning strategies [18]. *How to fuse?* The most popular fusion strategy to date is the straightforward concatenation of the extracted features [17,24]. However, recent works aim to learn the interactions across multimodal features using strategies like the Kronecker product to model pairwise feature interactions [10] and orthogonalization loss to reduce the redundancies across multimodal features [9]. (ii) Image Translation: One may consider learning image-to-image translation (or style transfer) models to convert the inferior modality to the superior. However, although great success was witnessed in this field [31,33], in the context of multimodal medical imaging, translation is complicated or non-ideal due to the difference in dimensionality (e.g. 2D to 3D) and size (e.g. millions to billions of voxels) between source and target. Additionally, the image translation only optimizes the intermediate task of translation as opposed to the proposed method that also addresses the final classification task. (iii) Student-Teacher (S-T) learning: Also referred to as knowledge distillation (KD), S-T learning aims to transfer the knowledge learned from one model to another, mostly aimed at applications with sparse or no labels, and model compression [32]. Cross-modal distillation, however, aims to leverage the modality specific representation of the teacher to distill the knowledge onto the student model. While most of such applications focus on KD across synchronized visual and audio data [2,3,25], KD methods for cross-modal medical image analysis mainly focus on segmentation [11,16,20]. Recently, Sonsbeek *et al.* [30] proposed a multimodal KD framework for classifying chest x-ray images with the language-based electronic health records as the teacher and X-Ray images as the student network. However, unlike the proposed method, the student network in the prior works only mimics the teacher network, without explicitly incorporating its own learnt classification-specific latent representation.

To summarize, while prior works use multimodal medical images as input during inference to improve the performance, our contribution is that we leverage multimodal data during training in order to enhance inference performance with only unimodal input.

3   Method

Problem Formulation: Given a training dataset $\mathcal{X}$ of N paired images from inferior and superior modalities with corresponding ground truth target labels $\mathcal{Y}$, our goal is to learn a function F that maps novel examples of the inferior type to target labels. Specifically, $\mathcal{X} = \{\mathcal{X}_I, \mathcal{X}_S\}$, with $\mathcal{X}_I = \{x_I^i\}_{i=1}^N$ and $\mathcal{X}_S = \{x_S^i\}_{i=1}^N$,

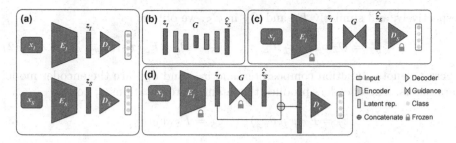

Fig. 1. Overview of the multimodal guidance approach. (a) Two independent modality-specific (inferior $\mathcal{I}$ vs superior $\mathcal{S}$) classifiers are trained, each with encoder E—producing latent representation z—and decoder D. (b) The architecture of the guidance model G. (c) G connects the output of the (frozen) inferior modality encoder $E_\mathcal{I}$ to the input of the (frozen) superior modality decoder $D_\mathcal{S}$. Then G is trained to infer the latent representation of the superior modality from the inferior one. (d) The final model, whose input is the inferior modality alone, uses both the inferior and the estimated superior modality representations to make the final prediction via the trained combined decoder D_c.

is the set of paired inferior and superior images, i.e. $(x_\mathcal{I}^i, x_\mathcal{S}^i)$ is the ith pair. The set of training labels is $\mathcal{Y} = \{y_i\}_{i=1}^N$, where $y_i \in \mathcal{L}$ and $\mathcal{L} = \{l_1, l_2, ..., l_K\}$ is the label space representing the set of all K possible class labels (e.g., disease diagnoses). We represent F using a deep model with parameters θ, i.e., $\hat{y} = F(x_\mathcal{I}; \theta)$, where $\hat{y}$ is the model prediction.

Model Optimization: The proposed method comprises of 3 steps: (i) Train classifiers $C_\mathcal{I}$ and $C_\mathcal{S}$ that each predicts the target label y from $x_\mathcal{I}$ and from $x_\mathcal{S}$, respectively; (ii) train a guidance model G to map the latent representation in $C_\mathcal{I}$ to that of $C_\mathcal{S}$; (iii) construct F that, first, maps $x_\mathcal{I}$ to the latent representation of $C_\mathcal{I}$, then maps that representation using G to estimate the latent representation of $C_\mathcal{S}$ to perform the classification. We now describe these steps in detail.

(i) <u>Classifiers $C_\mathcal{I}$ and $C_\mathcal{S}$ (Fig. 1(a))</u>: Given image pairs $(x_\mathcal{I}^i, x_\mathcal{S}^i)$ and ground-truth labels y^i, we train two independent classification models $C_\mathcal{I}$ and $C_\mathcal{S}$ on the same task. Classifier $C_\mathcal{I}$ is trained to classify images of the inferior modality $x_\mathcal{I}$, whereas $C_\mathcal{S}$ is trained to classify images of the superior modality $x_\mathcal{S}$. Denoting the predictions made by the two networks as $\hat{y}_\mathcal{I}^i$ and $\hat{y}_\mathcal{S}^i$, we have:

$$\hat{y}_\mathcal{I}^i = C_\mathcal{I}(x_\mathcal{I}^i) \qquad \hat{y}_\mathcal{S}^i = C_\mathcal{S}(x_\mathcal{S}^i), \tag{1}$$

where $C_\mathcal{I}$, and similarly $C_\mathcal{S}$, comprise an encoder E, which encodes the high-dimensional input image into a compact low-dimensional latent representation, and a decoder D, which decodes the latent representation by mapping it to one of the labels in $\mathcal{L}$. Denoting the encoder and decoder in $C_\mathcal{I}$ as $E_\mathcal{I}$ and $D_\mathcal{I}$, respectively, and similarly E_S and D_S in C_S, we obtain:

$$\hat{y}_{\mathcal{I}}^i = D_{\mathcal{I}} \circ E_{\mathcal{I}}(x_{\mathcal{I}}^i; \theta_{E_{\mathcal{I}}}) \qquad \hat{y}_{\mathcal{S}}^i = D_{\mathcal{S}} \circ E_{\mathcal{S}}(x_{\mathcal{S}}^i; \theta_{E_{\mathcal{S}}}), \tag{2}$$

where $\circ$ denotes function composition, and $\theta_{E_{\mathcal{I}}}$ and $\theta_{E_{\mathcal{S}}}$ are the encoder model parameters. The encoders produce the latent representations z, i.e.:

$$z_{\mathcal{I}}^i = E_{\mathcal{I}}(x_{\mathcal{I}}^i; \theta_{E_{\mathcal{I}}}) \qquad z_{\mathcal{S}}^i = E_{\mathcal{S}}(x_{\mathcal{S}}^i; \theta_{E_{\mathcal{S}}}). \tag{3}$$

Latent codes $z_{\mathcal{I}}^i$ and $z_{\mathcal{S}}^i$ are inputs to corresponding decoders $D_{\mathcal{I}}$ and $D_{\mathcal{S}}$. Finally, $D_{\mathcal{I}}(z_{\mathcal{I}}^i)$ and $D_{\mathcal{S}}(z_{\mathcal{S}}^i)$ yield predictions $\hat{y}_{\mathcal{I}}^i$ and $\hat{y}_{\mathcal{S}}^i$, respectively.

(ii) <u>Guidance Model G</u> (Fig. 1(b-c)): G is trained to map $z_{\mathcal{I}}^i$ of the inferior image to $z_{\mathcal{S}}^i$ of the paired superior image. Denoting the estimated latent code as $\hat{z}_{\mathcal{S}}^i$ and the parameters of G as θ_G, we obtain:

$$\hat{z}_{\mathcal{S}}^i = G(z_{\mathcal{I}}^i; \theta_G). \tag{4}$$

(iii) <u>Guided Model F</u> (Fig. 1(d)): We incorporate the trained guidance model G into the classification model of the inferior modality $C_{\mathcal{I}}$ so that it is steered to inherit the knowledge captured by the classifier C_S, which was trained on the superior modality but without $C_{\mathcal{I}}$ being exposed to the superior image modality. The model F is thus able, during inference time, to make a prediction based solely on the inferior modality while being steered to generate internal representations that mimic those produced by models trained on superior data. The superior modality encoder can thus be viewed as a teacher distilling its knowledge (the learned latent representation) to benefit the inferior modality student classifier. At inference time, we need not rely only on the guided representations $\hat{z}_{\mathcal{S}}^i$ to make predictions, but rather benefit from the learned representations $z_{\mathcal{I}}$ from the inferior modality as well. Therefore, the two concatenated representations are used to train a common classification decoder $D_c([\hat{z}_{\mathcal{S}}^i \frown z_{\mathcal{I}}^i]; \theta_{D_c})$, with parameters θ_{D_c}, where $\frown$ is the concatenation operator. Thus, our final prediction $\hat{y} = F(x_{\mathcal{I}}; \theta)$ is written as follows, with $\theta = \{\theta_{E_{\mathcal{I}}}, \theta_G, \theta_{D_c}\}$:

$$\hat{y} = D_c\left([G\left(E_{\mathcal{I}}(x_{\mathcal{I}}^i; \theta_{E_{\mathcal{I}}}); \theta_G\right) \frown E_{\mathcal{I}}\left(x_{\mathcal{I}}^i; \theta_{E_{\mathcal{I}}}\right)]; \theta_{D_c}\right). \tag{5}$$

4   Experimental Setup

Datasets: We evaluate our method on two multimodal imaging applications: (i) <u>RadPath 2020</u> [19] is a public dataset that was released as part of the MICCAI 2020 Computational Precision Medicine Radiology-Pathology (CPM RadPath) Challenge for brain tumor classification. The dataset consists of 221 pairs of multi-sequence MRI and digitized histopathology images along with glioma diagnosis labels of the corresponding patients. The labels include glioblastoma ($n = 133$), oligodendroglioma ($n = 34$), and astrocytoma ($n = 54$). The MRI sequences give rise to T1, T2, T1w, and FLAIR 3D images, each of size

$240 \times 240 \times 155$. The histopathology whole slide color images (WSI) are Hematoxylin and Eosin (H&E) stained tissue specimens scanned at $20\times$ or $40\times$ magnifications, with sizes as high as $3 \times 80,000 \times 80,000$. In this scenario, the biopsy-derived WSIs form the superior modality as it provides accurate tumor diagnosis and the MRI is the non-invasive inferior modality. We divide the dataset into 165 training, 28 validation, and 28 testing splits and make 5 of such sets to test the robustness of our methods across different splits.

(ii) Derm7pt [17] is another public dataset consisting of paired skin lesion images of clinical and dermoscopic modalities acquired from the same patients. The dataset includes 1011 pairs of images with their respective diagnosis and 7-point criteria labels for a multi-task ($n = 8$) classification setup. The 7-point criteria [5] comprise of pigment network (PN), blue whitish veil (BWV), vascular structures (VS), pigmentation (PIG), streaks (STR), dots and globules (DaG), and regression structures (RS), each of which contributes to the 7-point score used in inferring melanoma. Both the modalities consist of 2D images of size $3 \times 512 \times 512$. As the dermoscopic images are acquired via a dermatoscope by expert dermatologists and thus reveal more detailed sub-surface intra- and sub-epidermal structures, they are considered the superior modality, while the clinical images may be acquired using inexpensive and ubiquitous cameras and hence regarded as the inferior modality. We adopt the pre-defined train-validation-test splits provided along with the dataset [17] and repeat our training procedure 3 times to test robustness against different random weight initializations.

Implementation Details: We implement the 3D DenseNet based model used by the winners of the RadPath 2019 challenge as our baseline MRI classifier [23] and a recent multiple instance learning-based data efficient learning model, CLAM [22], as our baseline WSI classifier. For our experiments on the Derm7pt dataset, we use the pre-trained clinical and dermoscopic models provided by Kawahara *et al.* [17] as base models. In both sets of experiments, our guidance model uses an autoencoder-like architecture with a bottleneck of 256 and 512 neurons for the RadPath and Derm7pt datasets, respectively. To train the guidance model, we use the mean squared error (MSE) loss between z_S and $\hat{z}_S$ (see Eq. (4)), and a weighted cross-entropy loss in the final combined classification model (see Eq. (5)) to handle the class imbalance. We use PyTorch for all our experiments on the RadPath dataset and TensorFlow for our experiments on Derm7pt dataset, as we build on top of the Keras-based pre-trained models [17]. While the base models for RadPath were trained on a multi-GPU cluster, the rest of the models were trained on an NVIDIA GeForce GTX 1080 Ti GPU. We provide the values of optimal hyperparameters used for different experiments as supplementary table, which reports the batch sizes, optimizer parameters, loss weights, and early stopping parameters.

Evaluation Metrics: Since our experiments involve imbalanced classes, we use balanced accuracy (BA) in conjunction with the micro F1 score. While the macro-averaging of class-wise accuracy values in BA gives equal importance to

Table 1. RadPath results. Radiology MRI sequences (R) are guided using pathology (P). Only the best performing model using ALL MR sequences is used, explaining the '-' in row 1. In row 2, MR is not used, only P, hence '*'. R is the inferior ($\mathcal{I}$) and P is the superior modality ($\mathcal{S}$).

	Method	T1		T2		T1w		FLAIR		ALL	
		BA↑	F1↑	BA↑	F1↑	BA↑	F1↑	BA↑	F1↑	BA↑	F1↑
1	$P+R$	-	-	-	-	-	-	-	-	0.777	0.821
2	P [22]	*	*	*	*	*	*	*	*	0.729	0.792
3	R [23]	0.506	0.578	0.657	0.728	0.641	0.735	0.534	0.621	0.729	0.771
4	$G(R)$	0.443	0.578	0.559	0.692	0.564	0.692	0.419	0.571	0.650	0.764
5	$G(R)+R$	0.587	0.585	0.731	0.757	0.654	0.742	0.648	0.635	0.752	0.799
6	Δ (%)	+16.0	+1.2	+11.2	+3.9	+2.0	+0.9	+21.3	+2.2	+3.1	+3.6

all classes, the micro-averaged F1 score represents the overall correctness of the classifier irrespective of the class performance [13]. Additionally, for the binary task of melanoma inference, we use the AUROC score.

Melanoma Inference: Similar to Kawahara et al. [17], in addition to direct diagnosis of skin lesions, we infer melanoma from the 7-point criteria predictions [5]. Based on these predictions, we compute the 7-point score and use the commonly used thresholds of $t = 1, 3$ for inferring melanoma [5,17]. Finally, we compute the AUROC score to compare the overall performance of the classifiers.

5   Results and Discussion

The results on RadPath and Derm7pt datasets are in Tables 1 and 2, respectively, where we report the classification performance, BA and F1, across multiple tasks, under different input modalities and guidance strategies: when using both superior and inferior modalities as input (referred to as $\mathcal{S} + \mathcal{I}$); superior alone ($\mathcal{S}$); inferior alone without guidance ($\mathcal{I}$); guided inferior ($G(\mathcal{I})$); and guided inferior with inferior ($G(\mathcal{I})+\mathcal{I}$). For RadPath, radiology (R) is inferior and can be either of T1, T2, T1w, FLAIR, or ALL combined, and pathology (P) is superior while for Derm7pt, clinical (C) is inferior and dermoscopic (D) is superior.

$\mathcal{S}$ **outperforms** $\mathcal{I}$ (row 2 vs. 3): The results from the baseline models of both experiments confirm that the superior modality is more accurate for disease diagnosis. Table 1, shows that the classifier P significantly outperforms the individual MRI classifiers while being marginally better than the ALL MRI classifier. Similarly, from Table 2, the classifier D outperforms the classifier C across all the 7-point criteria, diagnosis, and the overall melanoma inference.

$\mathcal{S} + \mathcal{I}$ **outperforms** $\mathcal{S}$ (row 1 vs. 2): When using both modalities for classification, in case of RadPath, we observe the expected improvement over the classification using only superior, affirming the value added by MRI. However,

Table 2. Derm7pt results. The 7-point criteria, diagnosis (DIAG), and melanoma (MEL) are inferred. Clinical (C) is guided by Dermoscopic (D). C is the inferior ($\mathcal{I}$) and D is the superior modality ($\mathcal{S}$).

Method		7-point criteria							DIAG	MEL Inference		
		PN	BWV	VS	PIG	STR	DaG	RS		$t=1$	$t=3$	AUROC
1. $D+C$ [17]	BA↑	0.686	0.772	0.472	0.520	0.611	0.597	0.731	0.484	0.673	0.716	0.788
	F1↑	0.693	0.882	0.816	0.630	0.734	0.617	0.783	0.688	0.591	0.788	
2. D [17]	BA↑	0.666	0.809	0.552	0.573	0.621	0.583	0.719	0.635	0.656	0.702	0.764
	F1↑	0.688	0.859	0.790	0.632	0.716	0.596	0.770	0.716	0.586	0.762	
3. C [17]	BA↑	0.585	0.690	0.513	0.484	0.581	0.530	0.687	0.438	0.663	0.691	0.739
	F1↑	0.584	0.775	0.747	0.571	0.625	0.528	0.714	0.604	0.576	0.716	
4. $G(C)$	BA↑	0.563	0.664	0.395	0.460	0.509	0.503	0.653	0.330	0.675	0.630	0.716
	F1↑	0.588	0.822	0.803	0.611	0.681	0.501	0.763	0.616	0.682	0.744	
5. $G(C)+C$	BA↑	0.591	0.696	0.528	0.502	0.548	0.540	0.695	0.447	0.691	0.704	0.751
	F1↑	0.613	0.810	0.749	0.595	0.630	0.552	0.744	0.618	0.629	0.736	
6. Δ (%)	BA↑	+1.0	+0.8	+2.9	+3.7	(−5.6)	+1.8	+1.1	+2.0	+4.2	+1.8	+1.6
	F1↑	+4.9	+4.5	+0.2	+4.2	+0.8	+4.5	+4.2	+2.3	+9.2	+2.7	

the combined skin classifier does not concretely justify the addition of inferior (as also shown earlier [1,17]). We attribute this to the redundancy of information from the clinical images, which further adds to our motivation of guiding the inferior modality using the superior modality.

$G(\mathcal{I})$ **alone does not outperform** $\mathcal{I}$ (row 3 vs. 4): Our guided model, trained with the solitary goal of learning a mapping from the inferior features to superior using an MSE loss, performs worse than the original inferior model in both the datasets. This is not surprising as the poorer performance of this method suggests an imperfect reconstruction of the superior features, which can be attributed to the small size of the datasets. Consequently, our proposed method does not ignore the importance of the inferior modality, as shown next.

$G(\mathcal{I})+\mathcal{I}$ **outperforms** $\mathcal{I}$ (row 3 vs 5): Our proposed model (row 5), which retains the inferior inputs alongside the reconstructed superior features (using the guidance model), performs better than the baseline method (row 3) – that takes inferior inputs without leveraging the superior features – across all 5 radiology models and 7 of the 8 clinical-skin models. Row 6 in both tables reports the percentage improvement Δ in the performance achieved by the proposed method (row 5) over the baseline inferior classifier (row 3).

$G(\mathcal{I})+\mathcal{I}$ **reaches performance of** $\mathcal{S}+\mathcal{I}$ **for RadPath** (row 1 vs 5): Moreover, our proposed model with ALL sequences (Table 1 row 5) outperforms the superior modality model (row 2) while being comparable to the model that takes both the inferior and superior modalities as inputs during the test time (row 1), essentially alleviating the need for the superior inputs during the inference. However, in the case of Derm7pt, we observe that the improvement in performance over the baseline method does not approach the performance of the superior model, thus underscoring the superiority of information provided by the dermoscopic images over clinical. Additionally, we hypothesize that the performance of $\mathcal{S}+\mathcal{I}$ forms an upper bound on the performance achievable by

$G(\mathcal{I}) + I$ as the guidance $(G(\mathcal{I}))$ aims to mimic the latent representation of the superior modality $(\mathcal{S})$.

6 Conclusion

Motivated by the observation that for a particular clinical task the better-performing medical imaging modalities are typically less feasible to acquire, in this work we proposed a student-teacher method that distills knowledge learned from a better-performing (superior) modality to guide a more-feasible yet under-performing (inferior) modality and steer it towards improved performance. Our evaluation on two multimodal medical imaging based diagnosis tasks (skin and brain cancer diagnosis) demonstrated the ability of our method to boost the classification performance when only the inferior modality is used as input. We even observed (for the brain tumour classification task) that our proposed model, using guided unimodal data, achieved results comparable to a model that uses both superior and inferior multimodal data, i.e. potentially alleviating the need for a more expensive or invasive acquisition. Our future work includes extending our method to handle cross-domain continual-learning and testing on other applications.

Acknowledgements. We thank Weina Jin, Kumar Abhishek, and other members of the Medical Image Analysis Lab, Simon Fraser University, for the helpful discussions and feedback on the work. This project was funded by the Natural Sciences and Engineering Research Council of Canada (NSERC), BC Cancer Foundation-BrainCare BC Fund, and the Mitacs Globalink Graduate Fellowship. Computational resources were provided by Compute Canada (www.computecanada.ca).

References

1. Abhishek, K., Kawahara, J., Hamarneh, G.: Predicting the clinical management of skin lesions using deep learning. Sci. Rep. **11**(1), 1–14 (2021)
2. Afouras, T., Chung, J.S., Zisserman, A.: ASR is all you need: cross-modal distillation for lip reading. In: IEEE ICASSP, pp. 2143–2147 (2020)
3. Albanie, S., Nagrani, A., Vedaldi, A., Zisserman, A.: Emotion recognition in speech using cross-modal transfer in the wild. In: ACM International conference on Multimedia, pp. 292–301 (2018)
4. Aldape, K., Zadeh, G., Mansouri, S., Reifenberger, G., von Deimling, A.: Glioblastoma: pathology, molecular mechanisms and markers. Acta Neuropathol. **129**(6), 829–848 (2015). https://doi.org/10.1007/s00401-015-1432-1
5. Argenziano, G., Fabbrocini, G., Carli, P., De Giorgi, V., Sammarco, E., Delfino, M.: Epiluminescence microscopy for the diagnosis of doubtful melanocytic skin lesions: comparison of the abcd rule of dermatoscopy and a new 7-point checklist based on pattern analysis. Arch. Dermatol. **134**(12), 1563–1570 (1998)
6. Baltrušaitis, T., Ahuja, C., Morency, L.P.: Multimodal machine learning: a survey and taxonomy. IEEE TPAMI **41**(2), 423–443 (2018)

7. Bayoudh, K., Knani, R., Hamdaoui, F., Mtibaa, A.: A survey on deep multimodal learning for computer vision: advances, trends, applications, and datasets. Vis. Comput. **38**, 1–32 (2021). https://doi.org/10.1007/s00371-021-02166-7

8. Beyer, T., et al.: A combined PET/CT scanner for clinical oncology. J. Nucl. Med. **41**(8), 1369–1379 (2000)

9. Braman, N., Gordon, J.W.H., Goossens, E.T., Willis, C., Stumpe, M.C., Venkataraman, J.: Deep orthogonal fusion: multimodal prognostic biomarker discovery integrating radiology, pathology, genomic, and clinical data. In: de Bruijne, M., et al. (eds.) MICCAI 2021. LNCS, vol. 12905, pp. 667–677. Springer, Cham (2021). https://doi.org/10.1007/978-3-030-87240-3_64

10. Chen, R.J., et al.: Pathomic fusion: an integrated framework for fusing histopathology and genomic features for cancer diagnosis and prognosis. In: IEEE TMI (2020)

11. Dou, Q., Liu, Q., Heng, P.A., Glocker, B.: Unpaired multi-modal segmentation via knowledge distillation. IEEE TMI **39**(7), 2415–2425 (2020)

12. Gao, J., Li, P., Chen, Z., Zhang, J.: A survey on deep learning for multimodal data fusion. Neural Comput. **32**(5), 829–864 (2020)

13. Grandini, M., Bagli, E., Visani, G.: Metrics for multi-class classification: an overview. arXiv preprint arXiv:2008.05756 (2020)

14. Guo, W., Wang, J., Wang, S.: Deep multimodal representation learning: a survey. IEEE Access **7**, 63373–63394 (2019)

15. Guo, Z., Li, X., Huang, H., Guo, N., Li, Q.: Deep learning-based image segmentation on multimodal medical imaging. IEEE TRPMS **3**(2), 162–169 (2019)

16. Hu, M., et al.: Knowledge distillation from multi-modal to mono-modal segmentation networks. In: Martel, A.L., et al. (eds.) MICCAI 2020. LNCS, vol. 12261, pp. 772–781. Springer, Cham (2020). https://doi.org/10.1007/978-3-030-59710-8_75

17. Kawahara, J., Daneshvar, S., Argenziano, G., Hamarneh, G.: Seven-point checklist and skin lesion classification using multitask multimodal neural nets. IEEE J. Biomed. Health Inf. **23**(2), 538–546 (2018)

18. Kumar, A., Kim, J., Lyndon, D., Fulham, M., Feng, D.: An ensemble of fine-tuned convolutional neural networks for medical image classification. IEEE J. Biomed. Health Inf. **21**(1), 31–40 (2016)

19. Kurc, T., et al.: Segmentation and classification in digital pathology for glioma research: challenges and deep learning approaches. Front. Neurosci. **14**, 27 (2020)

20. Li, K., Yu, L., Wang, S., Heng, P.A.: Towards cross-modality medical image segmentation with online mutual knowledge distillation. In: AAAI, vol. 34, pp. 775–783 (2020)

21. Loeb, S., et al.: Systematic review of complications of prostate biopsy. Eur. Urol. **64**(6), 876–892 (2013)

22. Lu, M.Y., Williamson, D.F., Chen, T.Y., Chen, R.J., Barbieri, M., Mahmood, F.: Data-efficient and weakly supervised computational pathology on whole-slide images. Nat. Biomed. Eng. **5**(6), 555–570 (2021)

23. Ma, X., Jia, F.: Brain tumor classification with multimodal MR and pathology images. In: Crimi, A., Bakas, S. (eds.) BrainLes 2019. LNCS, vol. 11993, pp. 343–352. Springer, Cham (2020). https://doi.org/10.1007/978-3-030-46643-5_34

24. Mobadersany, P., et al.: Predicting cancer outcomes from histology and genomics using convolutional networks. Proc. Nat. Acad. Sci. **115**(13), E2970–E2979 (2018)

25. Nagrani, A., Albanie, S., Zisserman, A.: Seeing voices and hearing faces: cross-modal biometric matching. In: IEEE CVPR, pp. 8427–8436 (2018)

26. Olar, A., Aldape, K.D.: Using the molecular classification of glioblastoma to inform personalized treatment. J. Pathol. **232**(2), 165–177 (2014)

27. Pei, L., Vidyaratne, L., Rahman, M.M., Iftekharuddin, K.M.: Context aware deep learning for brain tumor segmentation, subtype classification, and survival prediction using radiology images. Sci. Rep. **10**(1), 1–11 (2020)
28. Rehani, M.M., et al.: Patients undergoing recurrent CT scans: assessing the magnitude. Eur. Radiol. **30**(4), 1828–1836 (2020)
29. Shao, Y., et al.: Simultaneous PET and MR imaging. Phys. Med. Biol. **42**(10), 1965 (1997)
30. van Sonsbeek, T., Zhen, X., Worring, M., Shao, L.: Variational knowledge distillation for disease classification in chest X-rays. In: Feragen, A., Sommer, S., Schnabel, J., Nielsen, M. (eds.) IPMI 2021. LNCS, vol. 12729, pp. 334–345. Springer, Cham (2021). https://doi.org/10.1007/978-3-030-78191-0_26
31. Wang, L., Chen, W., Yang, W., Bi, F., Yu, F.R.: A state-of-the-art review on image synthesis with generative adversarial networks. IEEE Access **8**, 63514–63537 (2020)
32. Wang, L., Yoon, K.J.: Knowledge distillation and student-teacher learning for visual intelligence: a review and new outlooks. In: IEEE TPAMI (2021)
33. Wang, T., et al.: A review on medical imaging synthesis using deep learning and its clinical applications. J. Appl. Clin. Med. Phys. **22**(1), 11–36 (2021)
34. Xu, Y.: Deep learning in multimodal medical image analysis. In: Wang, H., et al. (eds.) HIS 2019. LNCS, vol. 11837, pp. 193–200. Springer, Cham (2019). https://doi.org/10.1007/978-3-030-32962-4_18
35. Zhang, Y., Sidibé, D., Morel, O., Mériaudeau, F.: Deep multimodal fusion for semantic image segmentation: a survey. Image Vis. Comput. **105**, 104042 (2021)

Opportunistic Incidence Prediction of Multiple Chronic Diseases from Abdominal CT Imaging Using Multi-task Learning

Louis Blankemeier[1](✉), Isabel Gallegos[1], Juan Manuel Zambrano Chaves[1],
David Maron[1], Alexander Sandhu[1], Fatima Rodriguez[1], Daniel Rubin[1],
Bhavik Patel[2], Marc Willis[1], Robert Boutin[1], and Akshay S. Chaudhari[1]

[1] Stanford University, Stanford, USA
lblankem@stanford.edu
[2] Mayo Clinic, Rochester, USA

Abstract. Opportunistic computed tomography (CT) analysis is a paradigm where CT scans that have already been acquired for routine clinical questions are reanalyzed for disease prognostication, typically aided by machine learning. While such techniques for opportunistic use of abdominal CT scans have been implemented for assessing the risk of a handful of individual disorders, their prognostic power in simultaneously assessing multiple chronic disorders has not yet been evaluated. In this retrospective study of 9,154 patients, we demonstrate that we can effectively assess 5-year incidence of chronic kidney disease (CKD), diabetes mellitus (DM), hypertension (HT), ischemic heart disease (IHD), and osteoporosis (OST) using single already-acquired abdominal CT scans. We demonstrate that a shared multi-planar CT input, consisting of an axial CT slice occurring at the L3 vertebral level, as well as carefully selected sagittal and coronal slices, enables accurate future disease incidence prediction. Furthermore, we demonstrate that casting this shared CT input into a multi-task approach is particularly valuable in the low-label regime. With just 10% of labels for our diseases of interest, we recover nearly 99% of fully supervised AUROC performance, representing an improvement over single-task learning.

Keywords: Multitask learning · Computed tomography · Opportunistic imaging

1 Introduction

Chronic disorders are those that have slow progression, long duration, and are not passed from person to person [2]. For multiple chronic diseases, we aim to predict whether patients that do not have the disease will develop it within 5 years

Supplementary Information The online version contains supplementary material available at https://doi.org/10.1007/978-3-031-16449-1_30.

following a computed tomography (CT) exam. Early detection of such chronic diseases is of great interest since they carry massive financial and societal costs. In 2018, 69% of beneficiaries with 2 or more chronic diseases represented 94% of Medicare fee-for-service expenditures, while 18% of beneficiaries with 6 or more chronic conditions accounted for 54% of Medicare fee-for-service spending [1,24].

Lifestyle and pharmacologic interventions can manage and improve outcomes for many chronic diseases [7,18,21]. However, triggering such interventions requires screening tools that are comprehensive, low-cost, and scalable. Still, the heterogeneity of chronic diseases makes developing such comprehensive screening exams difficult, and existing screening tests can be inequitable, have low compliance and efficacy, and contribute towards rising healthcare costs [21,27].

Recent work has shown that automatic extraction of features from already-acquired abdominal CT scans can be used to screen patients for cardiovascular and osteoporosis risks [9,13,28,31] with higher accuracy than current clinical measures and risk stratification scores. Given that around 20 million abdominal CT scans are performed annually in the US, there is an opportunity to use this imaging for high-value early screening and diagnostics [25,26]. However, prior work that utilizes CT imaging for early diagnostics is mostly limited to single diseases of interest.

Nonetheless, many cardiometabolic disorders share common risk factors. Imaging biomarkers, such as abdominal calcifications, the extent of muscle and fat, and hepatic steatosis are surrogates for well-established risk factors, such as body mass index, blood glucose, and blood pressure. Furthermore, mechanisms of muscle, fat, and bone are highly interconnected, and these tissues are increasingly considered a single unit [22,29].

Motivated by this biological understanding, we show that abdominal CT imaging can be used to simultaneously predict the 5-year incidence of five chronic diseases: chronic kidney disease (CKD), diabetes mellitus (DM), hypertension (HT), ischemic heart disease (IHD), and osteoporosis (OST). We select these diseases due to their (a) high prevalence [3–6,11], (b) close co-morbidity relationships [10,14], (c) costs to the healthcare system, and (d) potential for positive intervention following screening.

Furthermore, leveraging a shared multi-planar CT representation and multitask learning, we investigate a sparse disease label regime. Since we are assessing longitudinal progression, it can be challenging to acquire labels due to the necessary follow-up. This motivates the need for label-efficient approaches.

Our main contributions can be summarized as follows:

1. We predict 5-year incidence of chronic kidney disease (CKD), diabetes (DM), hypertension (HT), ischemic heart disease (IHD), and osteoporosis (OST) using abdominal CT imaging, achieving 0.77, 0.77, 0.73, 0.75, and 0.75 area under the receiver operator characteristic (AUROC) respectively.
2. Going beyond single-plane approaches [9,28,31], we create a multi-planar 2D CT input that reduces the dimensionality of the volumetric 3D data. This method both outperforms 2D single-plane approaches and is simultaneously beneficial for predicting all diseases.

3. Leveraging the comorbidities between CKD, DM, HT, IHD, and OST, we propose a multi-task learning technique for label-scarce scenarios. Our approach recovers nearly 99% of fully-supervised AUROC performance despite using only 10% of associated disease labels, and performs significantly better than single-task approaches. Likewise, using 25% and 100% of associated labels, our multi-task approach improves AUROC performance over single-task learning on average across our diseases of interest.

2   Methods

2.1   Dataset

Following institutional review board approval, we obtained 14,676 CT exams and electronic health records (EHR) for 9,154 patients who presented to our tertiary center emergency department between January 2013 and May 2018 and received at least one abdominal CT exam. See Table 1 in Supplementary Material for subject demographics. From the abdominal CT exams, we selected the series with the most axial images to maximize the superior-inferior field of view.

For each of the 5 chronic diseases (CKD, DM, HT, IHD, and OST), we identified diagnosis criteria based on ICD-10 codes, vital signs, and lab results (see the Data Labeling section in Supplementary Material for defining disease prevalence). We refer to the EHR covering up to and including the day of the CT exam as *baseline* and the EHR covering 5 years after the CT date as *follow-up*. Table 1 shows the number of scans from patients that had each disease at baseline, *"had before"*, scans from patients that did not present with the disease during baseline or follow-up, *"never developed"*, and scans from patients that developed the disease during follow-up, *"developed"*. Scans that did not have 5 years of corresponding follow-up or a diagnosis in the EHR record were not included in any of these classes. As our problem is a binary incidence prediction task, we use only the *"developed"* and *"never developed"* classes for model training.

Table 1. Number of scans in each cohort.

Cohort	CKD	DM	HT	IHD	OST
Had before	1,672	6,377	11,418	2,009	1,761
Never developed	2,146	1,392	438	2,170	2,029
Developed	899	878	713	754	966

2.2   Multi-planar CT Representation

To enable feature sharing across diseases, we created a multi-planar 2D image representation, incorporating axial, coronal, and sagittal planes, such that useful information can be extracted for all diseases, while minimizing the computational resources needed to make predictions. The initial input to our pipeline is a 3D CT volume that is transformed into a coronal maximum intensity projection for locating an axial L3 slice, as described previously in [19] We chose the L3 level due to its predictive power for cardiometabolic diseases [9,28,31] and body composition metrics [8,23]. The aorta and spine are then segmented in the L3 slice using a 2D UNet algorithm. We trained our segmentation algorithm on 300 annotated L3 slices and tested our algorithm on 100 slices. The centroids of the spine and aorta are used to select separate sagittal and coronal slices, respectively (schematic shown in Fig. 1). The extracted axial slice, coronal slice, and sagittal slice (Fig. 1) are used as inputs to a ResNet-18 [17] disease incidence prediction network. We concatenate each of these slices laterally as shown in Supplementary Material Fig. 1. Furthermore, we window/level the Hounsfield Units (HU) of these images in three concatenated channels of 400/50 (soft tissue window), 1800/400 (bone window), and 500/50 (custom window) HU as is done in [31].

To assess the benefit of the multi-planar approach, we used 5-fold cross validation across our full dataset and computed the average area under the receiver operating characteristic curve (AUROC) and standard deviation across all splits.

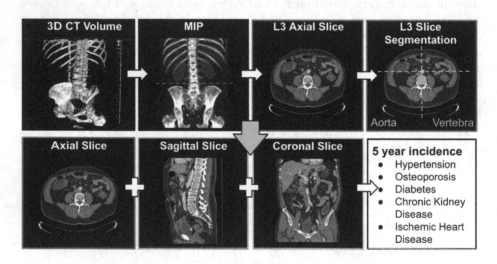

Fig. 1. Pipeline for generating predictions from 3D CT volumes.

2.3   Multi-task Learning

Conventionally, multi-task learning enables parameter sharing, allowing one to satisfy train or inference time budgets. Instead, we assume here that we can train a model for one disease jointly with any secondary disease. The secondary disease serves as an auxiliary task, where its purpose is to boost performance on the primary disease. Following training, we discard predictions about the secondary disease. This is analogous to pretraining where predictions for the pretext task are not utilized, and the pretraining strategy simply aids with downstream task training. Throughout this manuscript, we refer to the secondary disease as the *source disease* and the primary disease as the *target disease*. While any number of source diseases can be used, we chose to study the impact of joint training with a single source disease to mitigate the combinatorial explosion of numerous source and target configurations. This approach allowed us to effectively study task affinity, and for each disease, select the source disease that exhibited the greatest affinity. This approach was inspired by research exploring task groupings that maximize performance across tasks [15,30].

We carried out experiments to assess disease affinities using cross-validation under three target disease labeling regimes - 10%, 25%, and 100% of available labels. For each of these target label regimes, we used all of the available source labels. We divided our dataset into 5 splits, setting one split aside for our test set. For each combination of source and target diseases, we computed cross-validation results on 4 out of the 5 splits. Additionally, we computed the average percent increase of each multi-task run, along each column and row, over single-tasking, indicted by the symbol, $\overline{\%\Delta}$.

To select the source disease that we used for model testing, we chose the configuration that exhibited the greatest benefit over single-tasking during 4-fold cross validation (shaded cells in Table 3). To test the most promising configurations, we ensembled each of the 4 models selected during 4-fold cross validation by averaging the sigmoid outputs. Overall, we used the validation splits to evaluate optimal configurations for both the input images (Table 2) as well as the best source-target pairs for each disease (Table 3). We report the overall results of these ideal configurations in the label-scare 10% and 25% target label regimes in addition to fully-supervised results (Table 4).

2.4 Model Training

Multi-planar experiments were all performed with the same hyperparameters, using a ResNet-18 with a batch size of 8, a learning rate of 1e−5, an Adam [20] optimizer with default parameters, and a binary cross-entropy loss. We chose a ResNet-18 as it is a standard model for computer vision tasks, and its small size makes performing numerous experiments more computationally tractable.

For multi-task experiments, we also used a ResNet-18 model where the output dimensionality was 16. We applied a ReLU activation to the ResNet-18 output, followed by a 16 × 1 fully connected layer for each disease. We used a learning rate of 1e−4, a batchsize of 8, and a binary cross-entropy loss. While our strategy can be paired with any pretraining method, for this paper, we used ImageNet-initialized pretraining. Furthermore, we masked the loss from the *"had before"* class as well as instances where there was not sufficient follow-up to determine membership in the *"never developed"* category. This was relevant when one head in the multi-task model should be masked, while the other should not for a given example. All training was performed using Pytorch 1.10.0 on a single Titan RTX GPU.

3   Results

3.1   Multi-planar CT Representation

We achieved strong performance for the various steps of our pipeline (Fig. 1). Using the L3 selection model from [19] gave us an estimated accuracy of >99% based on visual inspection. Furthermore, on 100 held out axial L3 slices, our 2D UNet segmentation algorithm achieved an aorta dice score of around 0.65 and a mean error of about 0.46 mm in predicted coronal slice location. Likewise, our algorithm achieved a spine dice score of about 0.81 and mean error of approximately 0.92 mm in predicted sagittal slice location.

The results shown in Table 2 represent average AUROC validation set performance across folds using 5-fold cross validation. For model selection, we chose the model from the train epoch where AUROC was maximized. *CMTL* stands for complete multi-task learning where all 5 diseases are trained jointly using the same network. We shaded the cells that exhibited the greatest performance, in terms of t-statistic, for each disease. Excluding multi-planar (*MP*) and *CMTL + MP*, the axial slice performed best for DM, the coronal slice performed best for CKD, HT, and IHD, and the sagittal slice performed best for OST. We hypothesize that the utility of the coronal slice for cardiometabolic disorders is due to the visibility of the aorta and aortic calcifications. Furthermore, we hypothesize that the presence of the vertebral column in the sagittal slice makes this plane useful for osteoporosis predictions. Since there was an improvement for all disease predictions with the multi-planar configuration, we adopted this approach for subsequent experiments. We note that *CMTL + MP* did not perform significantly better than *MP*. This motivates our multi-task learning approach where we first analyze disease affinities on validation data and then select diseases that exhibit greatest affinities for joint training.

Table 2. Assessing the multi-planar configuration using 5-fold cross-validation. Values indicate mean AUROC ± standard deviation across 5 folds. Multi-planar (*MP*) is our 2D CT representation with concatenated axial, coronal, and sagittal slices. Complete multi-task learning (*CMTL*) represents multi-task learning where all 5 disease are trained together and their predictions are provided by a single model.

Inputs	CKD	DM	HT	IHD	OST
Axial	0.77 ± 0.018	0.75 ± 0.008	0.74 ± 0.032	0.76 ± 0.026	0.75 ± 0.021
Coronal	0.77 ± 0.021	0.75 ± 0.018	0.74 ± 0.044	0.77 ± 0.038	0.73 ± 0.037
Sagittal	0.74 ± 0.017	0.75 ± 0.027	0.74 ± 0.011	0.77 ± 0.037	0.76 ± 0.017
MP	0.77 ± 0.020	0.77 ± 0.024	0.76 ± 0.027	0.78 ± 0.033	0.76 ± 0.027
CMTL+MP	0.77 ± 0.020	0.77 ± 0.016	0.75 ± 0.030	0.79 ± 0.031	0.75 ± 0.037

3.2   Multi-task Learning

Table 3 shows the mean AUROC and standard deviation across 4 validation splits. The entries along the diagonals within each label extent give performance of the target task trained by itself. $\overline{\%\Delta}$ indicates the average percent improvement across diseases using multi-task learning over single-task learning. On average across source diseases, validation performance with multi-tasking exceeds or matches single-tasking performance in every label extent regime, except for hypertension with 25% of target labels. Specifically, this benefit is greatest with 10% of target labels. The $\overline{\%\Delta}$ row indicates that OST is most amenable to multi-tasking with 10% of target labels, while the $\overline{\%\Delta}$ column indicates that IHD is the most useful source disease overall within the 100% target label regime. Within the 25% regime, IHD benefits most from multi-tasking.

The significant benefits observed in the low label regimes are likely due to the close comorbidities between diseases, such that the source disease labels serve as a surrogate for the target disease labels. Furthermore, we hypothesize that the source task acts as a regularizer, preventing overfitting on the target task.

Table 3. Multi-task AUROC results ($\pm$ standard deviation) across 4-fold cross-validation. The shaded cells indicate the configurations used on the test set.

Label extent	Source	Target					$\overline{\%\Delta}$
		CKD	DM	HT	IHD	OST	
10%	CKD	0.77 ± 0.012	0.68 ± 0.039	0.74 ± 0.079	0.80 ± 0.040	0.69 ± 0.156	5.04
	DM	0.76 ± 0.014	0.69 ± 0.124	0.70 ± 0.097	0.80 ± 0.047	0.74 ± 0.151	5.44
	HT	0.78 ± 0.024	0.77 ± 0.059	0.69 ± 0.040	0.77 ± 0.072	0.67 ± 0.073	5.49
	IHD	0.78 ± 0.016	0.73 ± 0.111	0.67 ± 0.071	0.76 ± 0.045	0.71 ± 0.134	4.14
	OST	0.78 ± 0.039	0.72 ± 0.094	0.76 ± 0.129	0.79 ± 0.046	0.63 ± 0.085	4.64
	$\overline{\%\Delta}$	0.73	5.08	3.36	3.21	12.37	
25%	CKD	0.72 ± 0.031	0.80 ± 0.023	0.79 ± 0.094	0.77 ± 0.070	0.74 ± 0.045	0.86
	DM	0.71 ± 0.032	0.80 ± 0.035	0.81 ± 0.056	0.78 ± 0.054	0.71 ± 0.057	0.92
	HT	0.69 ± 0.044	0.80 ± 0.013	0.80 ± 0.077	0.77 ± 0.043	0.71 ± 0.010	−0.44
	IHD	0.74 ± 0.055	0.79 ± 0.028	0.81 ± 0.055	0.75 ± 0.066	0.72 ± 0.025	0.46
	OST	0.72 ± 0.043	0.79 ± 0.024	0.80 ± 0.042	0.77 ± 0.068	0.72 ± 0.025	0.25
	$\overline{\%\Delta}$	−0.42	−0.84	0.44	2.93	−0.06	
100%	CKD	0.74 ± 0.006	0.74 ± 0.024	0.76 ± 0.024	0.78 ± 0.027	0.75 ± 0.025	0.79
	DM	0.77 ± 0.021	0.74 ± 0.021	0.75 ± 0.028	0.79 ± 0.025	0.75 ± 0.038	1.26
	HT	0.76 ± 0.012	0.75 ± 0.033	0.74 ± 0.025	0.77 ± 0.039	0.74 ± 0.030	0.50
	IHD	0.78 ± 0.015	0.75 ± 0.024	0.77 ± 0.016	0.78 ± 0.033	0.76 ± 0.026	2.63
	OST	0.76 ± 0.014	0.76 ± 0.015	0.76 ± 0.037	0.78 ± 0.040	0.75 ± 0.039	1.64
	$\overline{\%\Delta}$	3.14	1.13	2.15	0.09	0.33	

3.3   Test Set Analysis

Table 4 provides test set AUROC results using the most significant source/target configurations from the task affinity experiments. The STL, 100% configuration uses 100% of target labels and single-tasking. We refer to this configuration as fully-supervised. Within the multi-task rows (MTL) we provide the source disease, corresponding to each target, in parentheses. $\%\Delta_{S\to M}$ indicates the average percent improvement of using multi-tasking over single-tasking for a common label extent. $\overline{\%}_{FS}$ indicates the average percent of fully-supervised performance exhibited by the other configurations.

Table 4. Test set comparison of single-task learning (STL) with 10% of target labels, multi-task learning (MTL) with 10% of target labels, STL with 25% of labels, MTL with 25% of labels, STL with 100% of labels, and MTL with 100% of labels. $\overline{\%}\Delta_{S\to M}$ represents the average % improvement from multi-task learning vs single-task learning for a given label extent. $\overline{\%}\Delta_{FS}$ represents the average percent deviation from fully-supervised learning (STL with 100% of labels). The diseases provided, labeled by "Source" in the "Config" column, are the selected source diseases used for multi-tasking.

Label extent	Config	Target					$\overline{\%}\Delta_{S\to M}$	$\overline{\%}\Delta_{FS}$
		CKD	DM	HT	IHD	OST		
10%	STL	0.734	0.714	0.662	0.713	0.679		−5.2%
	MTL	0.750	0.732	0.709	0.773	0.712	+5.0%	−0.3%
	Source	IHD	HT	CKD	CKD	DM		
25%	STL	0.728	0.769	0.676	0.708	0.712		−2.7%
	MTL	0.760	0.753	0.676	0.737	0.727	+1.7%	−1.1%
	Source	IHD	CKD	DM	DM	CKD		
100%	STL	0.790	0.766	0.724	0.689	0.731		
	MTL	0.775	0.770	0.739	0.751	0.751	+2.4%	+2.4%
	Source	IHD	OST	IHD	DM	IHD		

4   Discussion and Conclusion

Prior opportunistic CT work focused on targeting single disease outcomes using features extracted from specific CT planes. Extending this work, we showed that we can simultaneously predict incidence of multiple chronic diseases. Furthermore, we demonstrated that we can utilize multi-task learning to improve predictions for label scarce diseases using label rich diseases that we measure to be most helpful for multi-task learning. It is important to note that our proposed multitasking approach is one of several potential solutions [16,32] for label efficient scenarios and future work should investigate whether this is in fact an optimal approach.

We hypothesize that insights from this work can be harnessed to scale predictions to a greater number of diseases. Such scaling has been limited by (1) the difficulty of designing custom CT representations targeting particular diseases and (2) the scarcity of labels with necessary followup for rare or under-diagnosed diseases. Our methods address both challenges.

We note that our prediction model is end-to-end, in the sense that we do not compute intermediate biomarkers. Previous work showed that such an approach can generate stronger predictions at the expense of greater interpretability [31]. We recognize that providing insight about predictions may be important for clinical adoption. As such, future work will explore explanation methods, such as per-tissue saliency [31] and counterfactuals [12].

References

1. Chartbook and charts. https://www.cms.gov/Research-Statistics-Data-and-Systems/Statistics-Trends-and-Reports/Chronic-Conditions/Chartbook_Charts
2. Noncommunicable diseases. https://www.who.int/health-topics/noncommunic able-diseases
3. Chronic kidney disease in the united states, 2021, March 2021. https://www.cdc. gov/kidneydisease/publications-resources/ckd-national-facts.html
4. Facts about hypertension, September 2021. https://www.cdc.gov/bloodpressure/
5. Heart disease facts, February 2022. https://www.cdc.gov/heartdisease/facts.htm
6. National diabetes statistics report, January 2022. https://www.cdc.gov/diabetes/ data/statistics-report/index.html
7. Bangalore, S., Maron, D.J., Hochman, J.S.: Evidence-based management of stable ischemic heart disease. JAMA **314**(18), 1917 (2015). https://doi.org/10.1001/ jama.2015.11219
8. Boutin, R.D., Houston, D.K., Chaudhari, A.S., Willis, M.H., Fausett, C.L., Lenchik, L.: Imaging of sarcopenia. Radiol. Clin. North Am. **60**(4), 575–582 (2022). https://doi.org/10.1016/j.rcl.2022.03.001
9. Boutin, R.D., Lenchik, L.: Value-added opportunistic CT: insights into osteoporosis and sarcopenia. Am. J. Roentgenol. **215**(3), 582–594 (2020). https://doi.org/10. 2214/ajr.20.22874
10. Charlson, M.E., Charlson, R.E., Peterson, J.C., Marinopoulos, S.S., Briggs, W.M., Hollenberg, J.P.: The Charlson comorbidity index is adapted to predict costs of chronic disease in primary care patients. J. Clin. Epidemiol. **61**(12), 1234–1240 (2008). https://doi.org/10.1016/j.jclinepi.2008.01.006
11. Clynes, M.A., Harvey, N.C., Curtis, E.M., Fuggle, N.R., Dennison, E.M., Cooper, C.: The epidemiology of osteoporosis. Br. Med. Bull. (2020). https://doi.org/10. 1093/bmb/ldaa005
12. Cohen, J.P., et al.: Gifsplanation via latent shift: a simple autoencoder approach to progressive exaggeration on chest x-rays. CoRR abs/2102.09475 (2021). https:// arxiv.org/abs/2102.09475
13. Dagan, N., et al.: Automated opportunistic osteoporotic fracture risk assessment using computed tomography scans to aid in FRAX underutilization. Nat. Med. **26**(1), 77–82 (2020). https://doi.org/10.1038/s41591-019-0720-z
14. Rijken, M., Van Kerkhof, M., Dekker, J., Schellevis, F.G.: Comorbidity of chronic diseases: effects of disease pairs on physical and mental functioning. https:// pubmed.ncbi.nlm.nih.gov/15789940/
15. Fifty, C., Amid, E., Zhao, Z., Yu, T., Anil, R., Finn, C.: Efficiently identifying task groupings for multi-task learning. CoRR abs/2109.04617 (2021). https://arxiv.org/ abs/2109.04617
16. Finn, C., Abbeel, P., Levine, S.: Model-agnostic meta-learning for fast adaptation of deep networks. In: Precup, D., Teh, Y.W. (eds.) Proceedings of the 34th International Conference on Machine Learning. Proceedings of Machine Learning Research, vol. 70, pp. 1126–1135. PMLR, 06–11 August 2017. https://proceedings. mlr.press/v70/finn17a.html

17. He, K., Zhang, X., Ren, S., Sun, J.: Deep residual learning for image recognition. CoRR abs/1512.03385 (2015). http://arxiv.org/abs/1512.03385

18. Higgins, T.: Hba1c for screening and diagnosis of diabetes mellitus. Endocrine **43**(2), 266–273 (2012). https://doi.org/10.1007/s12020-012-9768-y

19. Kanavati, F., Islam, S., Arain, Z., Aboagye, E.O., Rockall, A.: Fully-automated deep learning slice-based muscle estimation from CT images for sarcopenia assessment (2020)

20. Kingma, D.P., Ba, J.: Adam: a method for stochastic optimization (2014). https://doi.org/10.48550/ARXIV.1412.6980, https://arxiv.org/abs/1412.6980

21. LaVallee, L.A., Scott, M.A., Hulkower, S.D.: Challenges in the screening and management of osteoporosis. North Carol. Med. J. **77**(6), 416–419 (2016). https://doi.org/10.18043/ncm.77.6.416

22. Liu, C.T., et al.: Visceral adipose tissue is associated with bone microarchitecture in the Framingham osteoporosis study. J. Bone Mineral Res. **32**(1), 143–150 (2017). https://doi.org/10.1002/jbmr.2931, https://asbmr.onlinelibrary.wiley.com/doi/abs/10.1002/jbmr.2931

23. Manzano, W., Lenchik, L., Chaudhari, A.S., Yao, L., Gupta, S., Boutin, R.D.: Sarcopenia in rheumatic disorders: what the radiologist and rheumatologist should know. Skeletal Radiol. **51**(3), 513–524 (2021). https://doi.org/10.1007/s00256-021-03863-z

24. Martin, A.B., Hartman, M., Lassman, D., Catlin, A.: National health care spending in 2019: steady growth for the fourth consecutive year. Health Aff. **40**(1), 14–24 (2021). https://doi.org/10.1377/hlthaff.2020.02022

25. Mettler, F.A., et al.: Patient exposure from radiologic and nuclear medicine procedures in the united states: procedure volume and effective dose for the period 2006–2016 (2020)

26. Oecd. https://stats.oecd.org/index.aspx?queryid=30184

27. Pfohl, S., Marafino, B.J., Coulet, A., Rodriguez, F., Palaniappan, L., Shah, N.H.: Creating fair models of atherosclerotic cardiovascular disease risk. CoRR abs/1809.04663 (2018). http://arxiv.org/abs/1809.04663

28. Pickhardt, P.J., et al.: Automated CT biomarkers for opportunistic prediction of future cardiovascular events and mortality in an asymptomatic screening population: a retrospective cohort study. Lancet Digit. Health **2**(4) (2020). https://doi.org/10.1016/s2589-7500(20)30025-x

29. Reginster, J.Y., Beaudart, C., Buckinx, F., Bruyère, O.: Osteoporosis and sarcopenia. Curr. Opin. Clin. Nutr. Metab. Care **19**(1), 31–36 (2016). https://doi.org/10.1097/mco.0000000000000230

30. Standley, T., Zamir, A.R., Chen, D., Guibas, L.J., Malik, J., Savarese, S.: Which tasks should be learned together in multi-task learning? CoRR abs/1905.07553 (2019). http://arxiv.org/abs/1905.07553

31. Zambrano Chaves, J.M., et al.: Opportunistic assessment of ischemic heart disease risk using abdominopelvic computed tomography and medical record data: a multimodal explainable artificial intelligence approach. medRxiv (2021). https://doi.org/10.1101/2021.01.23.21250197, https://www.medrxiv.org/content/early/2021/01/26/2021.01.23.21250197

32. Zamir, A.R., Sax, A., Shen, W.B., Guibas, L.J., Malik, J., Savarese, S.: Taskonomy: disentangling task transfer learning. CoRR abs/1804.08328 (2018). http://arxiv.org/abs/1804.08328

TMSS: An End-to-End Transformer-Based Multimodal Network for Segmentation and Survival Prediction

Numan Saeed[✉], Ikboljon Sobirov, Roba Al Majzoub, and Mohammad Yaqub

Mohamed bin Zayed University of Artificial Intelligence, Abu Dhabi, UAE
{numan.saeed,ikboljon.sobirov,roba.majzoub,mohammad.yaqub}@mbzuai.ac.ae

Abstract. When oncologists estimate cancer patient survival, they rely on multimodal data. Even though some multimodal deep learning methods have been proposed in the literature, the majority rely on having two or more independent networks that share knowledge at a later stage in the overall model. On the other hand, oncologists do not do this in their analysis but rather fuse the information in their brain from multiple sources such as medical images and patient history. This work proposes a deep learning method that mimics oncologists' analytical behavior when quantifying cancer and estimating patient survival. We propose TMSS, an end-to-end **T**ransformer based **M**ultimodal network for **S**egmentation and **S**urvival predication that leverages the superiority of transformers that lies in their abilities to handle different modalities. The model was trained and validated for segmentation and prognosis tasks on the training dataset from the HEad & NeCK TumOR segmentation and the outcome prediction in PET/CT images challenge (HECKTOR). We show that the proposed prognostic model significantly outperforms state-of-the-art methods with a concordance index of **0.763±0.14** while achieving a comparable dice score of **0.772 ± 0.030** to a standalone segmentation model. The code is publicly available at https://t.ly/V-_W.

Keywords: Survival analysis · Segmentation · Vision transformers · Cancer

1 Introduction

Cancer is one of the most lethal diseases and is among the top leading causes of death as reported by the World Health Organization [28] with almost 10 million deaths in 2020. This fact encourages doctors and medical researchers to strive for more efficient treatments and better care for cancer patients. Head and Neck (H&N) cancer is a collective term used to describe malignant tumors that develop in the mouth, nose, throat, or other head and neck areas. Early prediction and accurate diagnosis of the patient's survival risk (prognosis) can lower the mortality rate to 70% [27] of the H&N cancer patient. Furthermore, an accurate prognosis helps doctors better plan the treatments of patients [19].

© The Author(s), under exclusive license to Springer Nature Switzerland AG 2022
L. Wang (Eds.): MICCAI 2022, LNCS 13437, pp. 319–329, 2022.
https://doi.org/10.1007/978-3-031-16449-1_31

Doctors routinely conduct different types of scans like computed tomography (CT) and positron emission tomography (PET) in clinics and utilize them to extract biomarkers of the tumor area that is used with other information like patients electronic health records (EHR)for treatment plans. However, the manual delineation of the scans is time-consuming and tedious. Automatic prognosis and segmentation can significantly influence the treatment plan by speeding up the process and achieving robust outcomes.

With the advent of the deep learning (DL) field, considerable effort has been put into the automatic prognostic analysis for cancer patients. Brain [24,31], breast [10,23], liver [2,17,30], lung [6,13], rectal [18] and many other cancer types have been previously studied extensively. Long-term survival prediction using 33 different types of cancer was examined in-depth in [26]. Their MultiSurv multimodal network is compromised of several submodules responsible for feature extraction, representation fusion, and prediction. A multimodal deep neural network using gene expression profile, copy-number alteration profile, and clinical data was proposed in [23] for breast cancer prognosis. In [2], an improvement on the prognosis of patients with colorectal cancer liver metastases was studied. The authors proposed an end-to-end autoencoder neural network for this task utilizing radiomics features taken from MRI images. For overall survival prediction of patients with brain cancer, authors in [31] proposed an end-to-end model that extracts features from MRI images, fuses them, and combines outputs of modality-specific submodels to produce the survival prediction.

Using DL to predict a patient's outcome with H&N cancers is understudied. Clinically, H&N squamous cell carcinoma refers to different types of H&N cancers [14], including our topic of interest - oropharynx cancer. Authors in [5] studied H&N squamous cell carcinoma, creating an end-to-end network and arguing that a basic CNN-based model can extract more informative radiomics features from CT scans to predict H&N cancer treatment outcomes. H&N squamous cell carcinoma prognosis and its recurrence using DL were examined in [8]. The authors used CT scans of patients diagnosed with this type of cancer and extracted radiomics features manually using gross tumor volume and planning target volume. They predicted H&N cancer-related death and recurrence of cancer using a DL-driven model. Oropharyngeal squamous cell carcinoma, in particular, was a topic of interest in [9]. PET scans were used to train different popular CNN architectures, such as AlexNet [16], GoogleLeNet [25], and ResNet [12], all of which were pretrained on ImageNet [4], to compare it with the traditional methods that are trained on clinical records. By comparing all four different approaches, they concluded that using PET scans for a diagnostic DL model can predict progression-free survival and treatment outcome.

In [21], authors tackled the prognosis task for oropharyngeal squamous cell carcinoma patients using CT and PET images, along with clinical data. Their proposed solution was ranked the first in the progression-free survival prediction task of MICCAI 2021 HEad and neCK TumOR (HECKTOR) segmentation and outcome prediction challenge [20]. Features from medical images were first extracted using a CNN-based module and concatenated with electronic health records. The outputs were passed through fully connected layers and then to a multi-task logistic regression (MTLR) model [29]. Parallelly, electronic health records were fed to a Cox proportional hazard (CoxPH) model [3] to predict the patients' risk score. Finally, the risk predictions were calculated by taking the average of the outputs from MTLR and CoxPH models. This ensemble model achieved a concordance index (C-index; a common metric to evaluate prognosis accuracy) of 0.72 on the HECKTOR testing set, outperforming other proposed solutions. Although this work utilizes multimodal data, the learning from the medical images and EHR were disjoint, which may lead to less discriminative features learned by the deep learning model and consequently affect the final outcome.

In this paper, we propose a novel end-to-end architecture, TMSS, to predict the segmentation mask of the tumor and the patient's survival risk score by combining CT, PET scans and the EHR of the patient. Standard convolutional neural networks mainly focus on the imaging modality and cannot use other input features. To generally address this concern, we propose a transformer-based encoder that is capable of attending to the available multimodal input data and the interaction between them.

In the current work, our contributions are as follows:

- We propose *TMSS*, a novel end-to-end solution for H&N cancer segmentation and risk prediction. *TMSS* outperforms the SOTA models trained on the same dataset.
- We show that a vision transformer encoder could attend to multimodal data to predict segmentation and disease outcome, where the multimodal data is projected to the same embedding space.
- We propose a combined loss function for segmentation mask and risk score predictions.

2 Proposed Method

In the following section, we describe the build up of the main architecture, depicted in Fig. 1. As can be seen, it comprises four major components, and each is further explained below.

Transformer Encoder. The main advantage of this network is that the encoder itself embeds both the CT/PET and EHR data and encodes positions for them accordingly while extracting dependencies (i.e. **attention**) between the different modalities. The 3D image with dimensions $x \in \mathbb{R}^{H \times W \times D \times C}$ is reshaped into a sequence of flattened 2D patches $x_p \in \mathbb{R}^{n \times (P^3 C)}$, where H, W, and D are the height, width, and depth of the 3D image respectively, C denotes the number of channels, $P \times P \times P$ represents each patch's dimensions, and $n = HWD/P^3$ is the number of patches extracted. These patches are then projected to the embedding dimension h, forming a matrix $I \in \mathbb{R}^{n \times h}$. Simultaneously, EHR data is also projected to a dimension $E \in \mathbb{R}^{1 \times h}$. Both projections of images and EHR

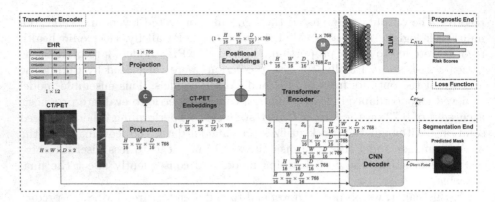

Fig. 1. An illustration of the proposed TMSS architecture and the multimodal training strategy. TMSS linearly projects EHR and multimodal images into a feature vector and feeds it into a Transformer encoder. The CNN decoder is fed with the input images, skip connection outputs at different layers, and the final layer output to perform the segmentation, whereas the prognostic end utilizes the output of the last layer of the encoder to predict the risk score.

are concatenated, forming a matrix $X \in \mathbb{R}^{(n+1) \times h}$. Positional encodings with the same dimension are added to each of the patches and the EHR projection as learnable parameters. The class token is dropped from the ViT [7] as our solution does not address a classification task. The resulting embeddings are fed to a transformer encoder consisting of 12 layers, following the same pipeline as the original ViT, with normalization, multi-head attention, and multi-layer perceptron. The purpose of using self-attention is to learn relations between $n + 1$ number of embeddings, including images and EHR. The self-attention inside the multi-head attention can be written as [7]:

$$Z = softmax\left(\frac{QK^T}{\sqrt{D_q}}\right)V \tag{1}$$

Segmentation End. The segmentation end is a CNN-based decoder, similar to the decoder in [11]. The original images are fed to the decoder along with skip connections passed from ViT layers Z_3, Z_6, Z_9, and Z_{12} (last layer). Only the image latent representations are passed through these skip connections $Z_l \in \mathbb{R}^{(n) \times h}$ and fed to the CNN decoder, where $l \in \{3, 6, 9, 12\}$. Convolution, deconvolution, batch normalization, and Rectified Linear Unit (ReLU) activation are used in the upsampling stage. Please, refer to [22] for more details.

Prognostic End. The prognostic path receives the output of the encoder with dimensions $Z_{12} \in \mathbb{R}^{(n+1) \times h}$, and its mean value is computed, reducing the dimensions down to $Z_{mean} \in \mathbb{R}^{1 \times h}$. This latent vector is then forwarded to two fully connected layers, reducing the dimensions from h to 512 and 128 respectively.

The resulting feature map is then fed to an MTLR model for final risk prediction. The MTLR module divides the future horizon into different time bins, set as a hyperparameter, and for each time bin a logistic regression model is used to predict if an event occurs or not.

Loss Function. Since the network performs two tasks concurrently, a combination of three losses is formulated as the final objective function. The segmentation end is supported by the sum of a dice loss (Eq. 2) and a focal loss (Eq. 3), where N is the sample size, $\hat{p}$ is the model prediction, y is the ground truth, α is the weightage for the trade-off between precision and recall in the focal loss (set to 1), and γ is focusing parameter (empirically set to 2).

$$\mathcal{L}_{Dice} = \frac{2\sum_i^N \hat{p}_i y_i}{\sum_i^N \hat{p}_i^2 + \sum_i^N y_i^2}, \tag{2}$$

$$\mathcal{L}_{Focal} = -\sum_i^N \alpha y_i (1 - \hat{p}_i)^\gamma log(\hat{p}_i) - (1 - y_i)\hat{p}_i^\gamma log(1 - \hat{p}_i), \tag{3}$$

The prognostic end has a negative-log likelihood loss (NLL) as given in Eq. 4. Here, the first line in the NLL loss corresponds to uncensored data, the second line corresponds to censored data and the third line is the normalizing constant, as described in [15]. The product $w_k^T x^{(n)}$ is the model prediction, b_k is the bias term, and y_k is the ground truth.

$$\mathcal{L}_{NLL}(\theta, D) = \sum_{n:\delta^{(n)}=1} \sum_{k=1}^{K-1} (w_k^T x^{(n)} + b_k) y_k^{(n)}$$

$$+ \sum_{n:\delta^{(n)}=0} log \left(\sum_{i=1}^{K-1} \mathbb{1}\{t_i \geq T^{(n)}\} exp \left(\sum_{k=1}^{K-1} \left((w_k^T x^{(n)} + b_k) y_k^{(n)} \right) \right) \right) \tag{4}$$

$$- \sum_{n=1}^N log \left(\sum_{i=1}^K exp \left(\sum_{k=1}^{K-1} w_k^T x^{(n)} + b_k \right) \right),$$

The final loss used for our network training is provided in Eq. 5 as a combination of the three losses. The hyperparameter β, provides weightage to either side of the model paths, and is empirically set to 0.3.

$$\mathcal{L}_{Final} = \beta * (\mathcal{L}_{Dice} + \mathcal{L}_{Focal}) + (1 - \beta) * \mathcal{L}_{NLL} \tag{5}$$

3   Experimental Setup

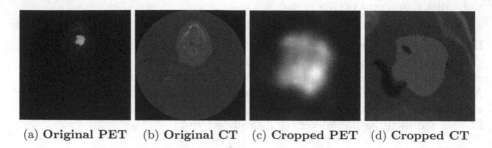

(a) **Original PET**   (b) **Original CT**   (c) **Cropped PET**   (d) **Cropped CT**

Fig. 2. A sample from the imaging dataset. (a) depicts the original PET scan (b) depicts the original CT scan and the imposed ground truth mask (c) shows the 80 × 80 × 48 cropped PET and (d) shows the 80 × 80 × 48 cropped CT with ground truth mask.

3.1   Dataset Description

A multicentric dataset of PET and CT images, their segmentation masks, and electronic health records are available on the HECKTOR challenge platform[1]. The data comes from six different clinical centers; 224 and 101 patient records for training and testing respectively. The testing set ground truths, both for segmentation and prognosis tasks are hidden for competition purposes, thus are not used to validate our method. Therefore, k-fold (where $k = 5$) cross validation was performed on the training set. EHR is comprised of data pertinent to gender, weight, age, tumor stage (N-, M- and T-stage), tobacco and alcohol consumption, chemotherapy experience, human papillomavirus (HPV), TNM edition, and TNM group. Imaging data contains CT, PET, and segmentation masks for tumor; sample slices are illustrated in Fig. 2 respectively.

3.2   Data Preprocessing

Both the CT and PET images are resampled to an isotropic voxel spacing of $1.0\,mm^3$. Their intensity values are then normalized before being fed to the network. The window of HU values of CT images was empirically clipped to (−1024, 1024), after which the images were normalized between (−1, 1). On the other hand, Z-score normalization was used for PET images. Furthermore, the images are cropped down to $80 \times 80 \times 48\,mm^3$ as in [21] for two main purposes; the first is to fairly compare our results to the state-of-the-art in [21], which also used images with these dimensions. The second is that this reduction of image

---

[1] https://www.aicrowd.com/challenges/miccai-2021-hecktor.

dimensions, in turn, speeds up training and inference processes and allows to run multiple experiments.

EHR, being multicentric, is missing some data for tobacco, alcohol consumption, performance status, HPV status, and estimated weight for SUV from most of the centers; therefore, they were dropped. 75% of the total data is censored, assumed to have stopped the follow-up to the hospitals.

3.3 Implementation Details

For our experiments, we used a single NVIDIA RTX A6000 (48 GB). We used PyTorch to implement the network and trained the model for 50 epochs. The batch size was set to 16, the learning rate to 4e–3, and the weight decay to 1e–5. The step decay learning rate strategy was used to reduce the learning rate by a factor of 10 after the 35 epochs.

Table 1. Prognosis performance by different models on the HECKTOR dataset. The reported are the mean and standard deviation for 5-fold cross validation.

	CoxPH	MTLR	Deep MTLR	Ensemble [21]	**Ours**
C-index	0.682 ± 0.06	0.600 ± 0.031	0.692 ± 0.06	0.704 ± 0.07	**0.763 ± 0.14**

The scans are patched into the size of $16 \times 16 \times 16$, and projected to the embedding dimension of 768. The total number of layers used in the encoder was 12, each having 12 attention heads.

The β in the loss function was set to 0.3. All the hyperparameters were chosen empirically, using the framework OPTUNA [1]. The evaluation metrics for the prognosis risk was concordance index (C-index), and for the segmentation was dice similarity coefficient (DSC).

4   Experimental Results

We use the HECKTOR dataset as described above for the diagnosis and prognosis of patients with head and neck cancer. Several experiments were conducted in house, all in 5-fold cross validation using the training dataset from the challenge. All the experiments were trained and cross validated using the same settings.

Table 1 shows the results of all conducted experiments. We started with the commonly used algorithms for survival analysis. CoxPH, MTLR and Deep MTLR were applied as baselines. As is vivid, CoxPH, achieving C-index of 0.68, outperforms the MTLR model by a huge degree of 0.08, yet introducing neural nets to MTLR (i.e. Deep MTLR) boosted the score to 0.692. All three calculate the risk using only the EHR data on account of their architectural nature. An ensemble of CNNs for the images with MTLR, and CoxPH for EHR achieved the highest C-index on the testing set [21] which was also implemented to train and validate using the same fashion as the original work. The ensemble was able to reach C-index of 0.704. Finally, our model, embedding EHR information in the input and using transformers unlike the ensemble, outperforms all the other models, achieving a mean C-index of **0.763**.

We optimized the hyperparameters using only one of the folds and then performed k-fold cross-validation using the entire dataset. However, there is a chance of leakage due to that specific fold; therefore, we redid the testing using a hold-out test set. To rule out the statistical dependence of the model training, we split the dataset randomly into two subsets, train and test, with an 80% and 20% ratio, respectively. The model hyperparameters were optimized using a small subset of the training set and tested using the hold-out set. We got a C-index score of **0.74** on the testing set. The prognosis task score on the hold-out set is slightly lower than the k-fold cross-validation score of 0.76, but it is more reliable and greater than the previous best scores of 0.70.

For segmentation comparison purposes we implement UNETR, a segmentation standalone network using the same settings as [22]. Our model achieved DSC of 0.772 ± 0.03, which was only 0.002 lower than that of UNETR network optimized for segmentation which achieved DSC of 0.774±0.01.

5   Discussion

The traditional approach to automate diagnosis and prognosis of cancerous patients is generally performed in two stages; either a standalone network that extracts tumor radiomics such as tumor volume [15] and feeds it to a prognostic model, or as in SOTA [21] that uses an ensemble of CNNs to extract scans features and concatenate with the EHR, then feeds them to another network for the risk prediction.

However, our approach tackles both problems at once, in an end-to-end network, making it simpler and easier to train. We show how our method outperforms other models by a good margin using vision transformers. Encoding EHR data into the network was newly introduced to mimic the way doctors review patient data. This has effectively boosted the accuracy of prognosis as shown in Table 1. The aforementioned results show the superiority of transformers in handling multimodal data. We hypothesize that the attention embedded in the transformer blocks, along with their ability to accommodate multimodal data, allows them to find relations across the modalities and within their intermediate representations. That can help them better address the tasks at hand. The use of multiple losses boosts the ability of the model to better interpolate within the given data, and hopefully become more robust when subjected to unseen ones. The introduction of the weighting variable β with a value of 0.3 penalizes the model more for prognosis errors coercing it to learn the features better and adjust its weights accordingly for an accurate prognosis.

Although the main goal of our model is prognosis, and not segmentation, we achieve comparable results with UNETR which was optimized for segmentation. This reinforces our hypothesis that both tasks compliment and aid each other for a better performance. It also sheds light on how improving the segmentation task in turn hones the prognosis results and helps the model learn better representations of both images and EHR data.

6 Conclusion and Future Work

We propose an end-to-end multimodal framework for head and neck tumor diagnosis and prognosis in this work. This model takes advantage of the strengths of transformers in dealing with multimodal data and its ability to find long relations within and across modalities for better model performance. We train and validate our model on head and neck CT/PET images with patient EHR and compare our results with the current state-of-the-art methods for prognosis and segmentation. For future work, self-supervised learning and pretraining of the network can be explored. They have proven to help models learn better, especially when the data is limited, as in our case. Additionally, the current network could be applied on similar tasks with different datasets to test the model for generalizability.

References

1. Akiba, T., Sano, S., Yanase, T., Ohta, T., Koyama, M.: Optuna: a next-generation hyperparameter optimization framework. CoRR **abs/1907.10902** (2019). http://arxiv.org/abs/1907.10902
2. Chen, J., Cheung, H.M.C., Milot, L., Martel, A.L.: AMINN: autoencoder-based multiple instance neural network improves outcome prediction in multifocal liver metastases. In: de Bruijne, M., Cattin, P.C., Cotin, S., Padoy, N., Speidel, S., Zheng, Y., Essert, C. (eds.) MICCAI 2021. LNCS, vol. 12905, pp. 752–761. Springer, Cham (2021). https://doi.org/10.1007/978-3-030-87240-3_72
3. Cox, D.R.: Regression models and life-tables. J. Roy. Stat. Soc.: Ser. B (Methodol.) **34**(2), 187–202 (1972)
4. Deng, J., Dong, W., Socher, R., Li, L.J., Li, K., Fei-Fei, L.: Imagenet: a large-scale hierarchical image database. In: 2009 IEEE Conference on Computer Vision and Pattern Recognition, pp. 248–255. IEEE (2009)
5. Diamant, A., Chatterjee, A., Vallières, M., Shenouda, G., Seuntjens, J.: Deep learning in head & neck cancer outcome prediction. Sci. Rep. **9**(1), 1–10 (2019)
6. Doppalapudi, S., Qiu, R.G., Badr, Y.: Lung cancer survival period prediction and understanding: deep learning approaches. Int. J. Med. Inf. **148**, 104371 (2021)
7. Dosovitskiy, A., et al.: An image is worth 16×16 words: transformers for image recognition at scale. arXiv preprint arXiv:2010.11929 (2020)
8. Fh, T., Cyw, C., Eyw, C.: Radiomics AI prediction for head and neck squamous cell carcinoma (hnscc) prognosis and recurrence with target volume approach. BJR—Open **3**, 20200073 (2021)
9. Fujima, N., et al.: Prediction of the local treatment outcome in patients with oropharyngeal squamous cell carcinoma using deep learning analysis of pretreatment fdg-pet images. BMC Cancer **21**(1), 1–13 (2021)
10. Gupta, N., Kaushik, B.N.: Prognosis and prediction of breast cancer using machine learning and ensemble-based training model. Comput. J. (2021)

11. Hatamizadeh, A., et al.: Unetr: transformers for 3D medical image segmentation. In: Proceedings of the IEEE/CVF Winter Conference on Applications of Computer Vision, pp. 574–584 (2022)
12. He, K., Zhang, X., Ren, S., Sun, J.: Deep residual learning for image recognition. In: Proceedings of the IEEE Conference on Computer Vision and Pattern Recognition, pp. 770–778 (2016)
13. Hosny, A., et al.: Deep learning for lung cancer prognostication: a retrospective multi-cohort radiomics study. PLoS Med. 15(11), e1002711 (2018)
14. Johnson, D.E., Burtness, B., Leemans, C.R., Lui, V.W.Y., Bauman, J.E., Grandis, J.R.: Head and neck squamous cell carcinoma. Nat. Rev. Dis. Primers 6(1), 1–22 (2020)
15. Kazmierski, M.: Machine Learning for Prognostic Modeling in Head and Neck Cancer Using Multimodal Data. Ph.D. thesis, University of Toronto (Canada) (2021)
16. Krizhevsky, A., Sutskever, I., Hinton, G.E.: Imagenet classification with deep convolutional neural networks. Adv. Neural Inf. Process. Syst. 25, 1097–1105 (2012)
17. Lee, H., Hong, H., Seong, J., Kim, J.S., Kim, J.: Survival prediction of liver cancer patients from ct images using deep learning and radiomic feature-based regression. In: Medical Imaging 2020: Computer-Aided Diagnosis, vol. 11314, p. 113143L. International Society for Optics and Photonics (2020)
18. Li, H., et al.: Deep convolutional neural networks for imaging data based survival analysis of rectal cancer. In: 2019 IEEE 16th International Symposium on Biomedical Imaging (ISBI 2019), pp. 846–849. IEEE (2019)
19. Mackillop, W.J.: The importance of prognosis in cancer medicine. TNM Online (2003)
20. Oreiller, V., et al.: Head and neck tumor segmentation in pet/ct: the hecktor challenge. Med. Image Anal., 102336 (2021)
21. Saeed, N., Majzoub, R.A., Sobirov, I., Yaqub, M.: An ensemble approach for patient prognosis of head and neck tumor using multimodal data (2022)
22. Sobirov, I., Nazarov, O., Alasmawi, H., Yaqub, M.: Automatic segmentation of head and neck tumor: how powerful transformers are? arXiv preprint arXiv:2201.06251 (2022)
23. Sun, D., Wang, M., Li, A.: A multimodal deep neural network for human breast cancer prognosis prediction by integrating multi-dimensional data. IEEE/ACM Trans. Comput. Biol. Bioinf. 16(3), 841–850 (2018)
24. Sun, L., Zhang, S., Chen, H., Luo, L.: Brain tumor segmentation and survival prediction using multimodal MRI scans with deep learning. Front. Neurosci. 13, 810 (2019)
25. Szegedy, C., et al.: Going deeper with convolutions. In: Proceedings of the IEEE Conference on Computer Vision and Pattern Recognition, pp. 1–9 (2015)
26. Vale-Silva, L.A., Rohr, K.: Long-term cancer survival prediction using multimodal deep learning. Sci. Rep. 11(1), 1–12 (2021)
27. Wang, X., Li, B.b.: Deep learning in head and neck tumor multiomics diagnosis and analysis: review of the literature. Front. Genet. 12, 42 (2021). https://doi.org/10.3389/fgene.2021.624820, https://www.frontiersin.org/article/10.3389/fgene.2021.624820
28. WHO: Cancer. https://www.who.int/news-room/fact-sheets/detail/cancer. Accessed 30 Jan 2022
29. Yu, C.N., Greiner, R., Lin, H.C., Baracos, V.: Learning patient-specific cancer survival distributions as a sequence of dependent regressors. Adv. Neural Inf. Process. Syst. 24, 1845–1853 (2011)

30. Zhen, S.H., et al.: Deep learning for accurate diagnosis of liver tumor based on magnetic resonance imaging and clinical data. Front. Oncol. **10**, 680 (2020)
31. Zhou, T., et al.: M^2Net: Multi-modal multi-channel network for overall survival time prediction of brain tumor patients. In: Martel, A.L., Martel, A.L., et al. (eds.) MICCAI 2020. LNCS, vol. 12262, pp. 221–231. Springer, Cham (2020). https://doi.org/10.1007/978-3-030-59713-9_22

Surgical Data Science

Bayesian Dense Inverse Searching Algorithm for Real-Time Stereo Matching in Minimally Invasive Surgery

Jingwei Song[1,2]($\boxtimes$), Qiuchen Zhu[3], Jianyu Lin[4], and Maani Ghaffari[1]

[1] Michigan Robotics, University of Michigan, Ann Arbor, MI 48109, USA
{jingweso,maanigj}@umich.edu
[2] United Imaging Research Institute of Intelligent Imaging, Beijing 100144, China
[3] School of Electrical and Data Engineering, University of Technology,
Ultimo, NSW 2007, Australia
Qiuchen.Zhu@uts.edu.au
[4] Hamlyn Centre for Robotic Surgery, Imperial College London,
London SW7 2AZ, UK

Abstract. This paper reports a CPU-level real-time stereo matching method for surgical images (10 Hz on 640×480 image with a single core of i5-9400). The proposed method is built on the fast LK algorithm, which estimates the disparity of the stereo images patch-wisely and in a coarse-to-fine manner. We propose a Bayesian framework to evaluate the probability of the optimized patch disparity at different scales. Moreover, we introduce a spatial Gaussian mixed probability distribution to address the pixel-wise probability within the patch. In-vivo and synthetic experiments show that our method can handle ambiguities resulted from the textureless surfaces and the photometric inconsistency caused by the non-Lambertian reflectance. Our Bayesian method correctly balances the probability of the patch for stereo images at different scales. Experiments indicate that the estimated depth has similar accuracy and fewer outliers than the baseline methods in the surgical scenario with real-time performance. The code and data set are available at https://github.com/JingweiSong/BDIS.git.

Keywords: Stereo matching · Bayesian theory · Posterior probability inference

1 Introduction

Real-time 3D intra-operative tissue surface shape recovery from stereo images is important in Computer Assisted Surgery (CAS). The reconstructed depth is a crucial for dense Simultaneous Localization and Mapping (SLAM) [23,24], AR

Supplementary Information The online version contains supplementary material available at https://doi.org/10.1007/978-3-031-16449-1_32.

L. Wang (Eds.): MICCAI 2022, LNCS 13437, pp. 333–344, 2022.
https://doi.org/10.1007/978-3-031-16449-1_32

system [11,28] and diseases diagnosis [13,18]. All stereo matching procedures follow the pinhole camera model [2] and conduct image rectification, undistortion, and disparity estimation. The stereo matching techniques are normally classified into two categories regarding disparity estimation: prior-free and learning-based. Conventional prior-free methods estimate the pixel-wise disparity using the image alignment techniques [8,12,14,20,25]. Based on the left-right image consistency assumption (photo-metric or feature-metric), they either use corner feature registration, dense direct pixel searching, or a combination. Differently, Deep Neural Network (DNN) based techniques directly learn the disparity from the training image pairs [3,16,26,29,30]. Although DNN methods are reported to be efficient, the results may be invalidated with changing parameters such as focal length and baseline or a large texture difference between the training and testing data [1,19]. Moreover, the DNN-based methods heavily depend on the size and quality of the annotated training data, which are not accessible in many CAS scenarios.

In the category of prior-free methods, ELAS [8] is still one of the most widely used stereo matching algorithms due to its robustness and accuracy [23,24,32, 33]. It is also the most popular method in the industry [4,31]. ELAS uses Sobel descriptors to match sparse corners as the supporting points and triangulate the pixel-wise disparity prior. Then, the optimal dense disparity is retrieved with its proposed maximum a-posteriori algorithm. Its two-step process requires around 0.25–1 s on a single modern CPU core. This paper aims for a faster CPU-based stereo matching method.

The Dense Inverse Searching (DIS) [14] shows the potential of dense direct matching without the time-consuming sparse supporting points alignment. By resizing the left and right images to several coarse scales, it adopts and modifies the Lucas-Kanade (LK) optical flow algorithm [17] for fast estimating the pixel-wise optimal disparity. [14] demonstrates that real-time computation is possible with its patch-based coarse-to-fine dense matching, where patch refers to an arbitrary squared image segment. However, DIS is strictly built based on the photometric consistency and surface texture abundance assumptions, which cannot always be satisfied in CAS. The two main challenges are the textureless/dark surfaces and the serious non-Lambertian reflectance. The weak/dark texture, which widely exists in CAS, leads to ambiguous photometric consistency. Meanwhile, non-Lambertian reflectance brings uneven disturbance on the surfaces, and it cannot be eliminated by just enforcing the patch normalization [21].

In this paper, to deal with photometric inconsistency and non-Lambertian reflectance in stereo matching, we propose a Bayesian Dense Inverse Searching (BDIS) to quantify the posterior probability of each optimized patch. A spatial Gaussian Mixture Model (GMM) is further adapted to quantify pixel-wise confidence within the patch. The final pixel-wise disparity is the fusion of multiple local overlapping patches, reducing the impact of those patches suffering from the textureless/dark surfaces or the non-Lambertian reflectance. In extreme cases, it is beneficial to give up the disparity estimation of some patches identified as dubious. In particular, this work has the following contributions:

- A Bayesian approach is developed to quantify the posterior probability of the patch.
- A spatial GMM is introduced to quantify the pixels' confidence within the patch.
- To our knowledge, BDIS is the first single core CPU based stereo matching approach that achieves similar performance to the near real-time method ELAS.

2   Methodology

2.1   Multiscale DIS

Figure 1 shows the DIS (based on fast LK) algorithm for stereo matching proposed by [14]. It is a modified version of the LK algorithm. We use the fast DIS as our base framework. Note that the variational refinement module in [14] is abandoned because it has a small (less than 0.5%) contribution in promoting the accuracy. The modified fast LK based DIS is achieved by minimizing the following objective function:

$$\Delta \mathbf{u} = \mathrm{argmin}_{\Delta \mathbf{u}'} \sum_x \left[I_r\left(\mathbf{x} + \mathbf{u} \right) - I_l(\mathbf{x} + \Delta \mathbf{u}') \right]^2, \tag{1}$$

where $\mathbf{x}$ is the processed location, $\mathbf{u}$ is the estimated disparity in the loop, I_l and I_r are the left image patch and right image, and $\Delta \mathbf{u}$ is the optimal update of $\mathbf{u}$ at one loop. Different from authentic LK, $\Delta \mathbf{u}'$ is moved from the right image to the left image patch. The improvement avoids the expensive re-evaluation of the Hessian on the right image. (1) is traversed on all patches at different scales. The disparity at the fine-scale level is initialized at the optimized coarse scale. The optimal disparity at the location $\mathbf{x}$ is the weighted fusion with all covering patches using inverse residual:

$$\hat{\mathbf{u}}_{\mathbf{x}} = \sum_{k \in \Omega} \frac{1/\max(\|I_l(\mathbf{x} + \mathbf{u}^{(k)}) - I_r(\mathbf{x})\|^2, 1)}{\sum_{k \in \Omega} 1/\max(\|I_l(\mathbf{x} + \mathbf{u}^{(k)}) - I_r(\mathbf{x})\|^2, 1)} \mathbf{u}^{(k)}, \tag{2}$$

where Ω is the set of patches covering the position $\mathbf{x}$, $\mathbf{u}^{(k)}$ is the estimated disparity of the patch k and $\max(\cdot, \cdot)$ selects the maximum value. The pixel-wise disparity $\hat{\mathbf{u}}_{\mathbf{x}}$ is the weighted average of the estimated disparities from all patches, wherein the weight is the inverse residual of brightness.

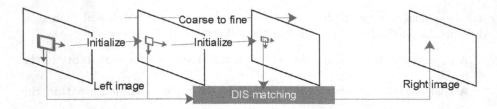

Fig. 1. The framework of the DIS algorithm [14]. It uses 3 scale levels as an example.

2.2   The Bayesian Patch-Wise Posterior Probability

The residual-based weighted average fusion (2) suffers from the ambiguities brought by the textureless/dark surface and non-Lambertian reflectance. The textureless/dark surface leads to ambiguous local minima of the cost function penalizing photometric inconsistency (1) and misleads the algorithm to be over-confident on the estimation. Furthermore, the photometric consistency presumption is seriously violated on the surface affected heavily by the non-Lambertian reflectance. The affine lighting changes formulation in previous large-scale SLAM studies [7] cannot fully tackle the complex and severe non-Lambertian reflectance in CAS. In both situations, the weights retrieved from the photometric residuals (2) are misleading. To overcome the difficulty in defining the confidence of the estimated disparity, we propose a Bayesian model to correctly estimate the confidence in the presence of textureless surface and non-Lambertian reflectance. Since the uncertainty distribution of both the left and right scenes is unclear, it is difficult to conduct the direct inference of the posterior probability in terms of disparity. Thus, we implicitly infer the probability with Bayesian modeling using Conditional Random Fields (CRF) [27]. The posterior probability of the patch-wise disparity $\mathbf{u}^{(k)}$ is

$$p(\mathbf{u}^{(k)}|I_l, I_r) \propto \frac{p(I_r|I_l, \mathbf{u}^{(k)})}{p(I_r, I_l, \mathbf{u}^{(k)})} \propto \frac{p(I_r|I_l, \mathbf{u}^{(k)})}{\Sigma_{\mathbf{u}_i^{(k)} \in \mathcal{P}} p(I_r|I_l, \mathbf{u}_i^{(k)})} \propto \frac{p(I_r|I_l, \mathbf{u}^{(k)})}{\Sigma_{\mathbf{u}_i^{(k)} \in \mathcal{P}'} p(I_r|I_l, \mathbf{u}_i^{(k)})} \mathbf{r},$$
$$(3)$$

where $\mathcal{P}$ is the domain of all possible choice of $\mathbf{u}_i^{(k)}$. To reduce computational load, r is applied as the constant compensation ratio for all patches within the window. $\mathcal{P}$ is reduced to a small window $\mathcal{P}'$ assuming the rest candidates are numerically trivial.

Equation (3) indicates that the posterior probability of the disparity can be obtained by traversing the probability on all possible $\mathbf{u}_i^{(k)}$. And the possible choice of disparity is equal to window size s. Even though the posterior probability suffers from the textureless surface and non-Lambertian reflectance, the illumination consistency probability is proportional to the residuals because the set of neighboring disparities is within one patch, and the impact of the issues is consistent. Thus, we model the illumination consistency probability $p(I_r|I_l, \mathbf{u}_i^{(k)}, \mathbf{x})$

based on the Boltzmann distribution [15] as

$$p(I_r|I_l, \mathbf{u}_i^{(k)}) = \exp\left(-\frac{\|I_l(\mathbf{u}_i^{(k)}) - I_r(\mathbf{u}_i^{(k)})\|_F^2}{2\sigma_r^2 s^2}\right), \tag{4}$$

where $\|\cdot\|_F$ is the Frobenius norm and σ_r is the hyperparameter to describe the variance of the brightness. The relative posterior probability can be obtained with (3) and (4). Generally, the absolute exponential parameter of the Boltzmann distribution denotes the entropy of the state. In our case, such entropy is defined as (4). Image with abundant texture has more entropy loss. Hence, the entropy item is highly related to the photometric inconsistency loss.

Figure 2 shows the relationship between the illumination consistency probability density function and the texture. The response is stronger on the textured surface. The residuals are always small in the textureless surface, no matter how the left and right images are aligned. (2) cannot correctly measure the weights while (4) describes the relative probability of the estimation. Moreover, it tests the local convergence to filter the Saddle point solutions.

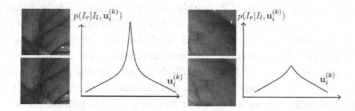

Fig. 2. The probability density function of the textureless region.

2.3 The Prior Spatial Gaussian Probability

In addition to the patch-wise posterior probability of the disparity in the last section, a spatial GMM is adopted to estimate pixel-wise probability within the patch. Considering that medical images are natural images, a multivariate Gaussian distribution is adopted to measure the confidence of the pixel-wise probability using a Gaussian mask. In accordance with the multivariate Gaussian distribution, the center of the patch has higher confidence than the edge pixels since those central pixels preserve more information for inference. Assuming all pixels in the patch are i.i.d, we have

$$p(\mathbf{u}^{(k)}|I_l, I_r, \mathbf{x}) \propto p(\mathbf{u}^{(k)}|I_l, I_r) \exp\left(-\frac{\sum_{\xi^{(k)}(\mathbf{x})}\|\mathbf{x} - \xi^{(k)}(\mathbf{x})\|_F^2}{2\sigma_s^2}\right), \tag{5}$$

where $\xi^{(k)}(\mathbf{x})$ is the set of all pixel positions within the patch k in image coordinate. σ_s is the 2D spatial variance of the probability. Note that (5) is independent of the patch and can therefore be pre-computed before the process. Combining (3), (4) and (5), the final pixel-wise posterior probability distribution can be represented as follows,

$$p(\mathbf{u}^{(k)}|I_l, I_r, \mathbf{x}) \propto \exp\left(-\frac{\sum_{\xi^{(k)}(\mathbf{x})}\|\mathbf{x}-\xi^{(k)}(\mathbf{x})\|_{\mathrm{F}}^2}{2\sigma_s^2}\right) \frac{\exp\left(-\frac{\|I_l(\mathbf{u}^{(k)})-I_r(\mathbf{u}^{(k)})\|_{\mathrm{F}}^2}{2\sigma_r^2 s^2}\right)}{\sum_{\mathbf{u}_i^{(k)}\in\mathcal{P}}\exp\left(-\frac{\|I_l(\mathbf{u}_i^{(k)})-I_r(\mathbf{u}_i^{(k)})\|_{\mathrm{F}}^2}{2\sigma_r^2 s^2}\right)}.$$

Finally, it should be emphasized that (4) and (5) are not the cost functions but probability/weight for each patch or pixel. Costly optimization steps are avoided.

3 Results and Discussion

BDIS was compared with DIS [14], SGBM [12] and ELAS [8] on the in-vivo and the synthetic data sets[1]. The computations were implemented on a commercial desktop (i5-9400) in C++. DNN-based methods PSMNet [5] and GwcNet [10] were also compared for completeness and the computation was conducted on the GTX 1080ti in PyTorch. The public in-vivo stereo videos from [9] were adopted which contains 200 images with size 640×480 and 200 images with size 288×360. All stereo images were rectified, undistorted, calibrated, and vertically aligned with the provided intrinsic and extrinsic parameters. We also provided a synthetic data set generated from an off-the-shelf virtual phantom of a male's digestive system. A virtual handheld colonoscope was placed inside the colon and was manipulated to go through the colon to collect the depth and stereo images. The 3D game engine Unity3D[2] was used to generate the sequential stereo and depth images with a pin-hole camera in size 640×480. The synthetic distortion-free data has accurate intrinsic and extrinsic parameters. Both diffuse lighting (100 frames) and non-Lambertian reflectance (100 frames) were simulated. γ was set to 0.75 for 640×480 and 0.25 for 288×360 data to discard the patch without enough valid pixels. σ_r and σ_s were set to 4; the sampling within one Bayesian window was 5; the disturbance from the convergence was 0.5 and 1 pixel.

3.1 Quantitative Comparisons on the Synthetic Data Set

BDIS was compared quantitatively with the baseline methods ELAS, SGBM, DIS, PSMNet, and GwcNet. The comparison between the prior-based DNN-based method and BDIS is for completeness only. The default setting of PSMNet and GwcNet were strictly followed. The pre-trained networks were adopted and finetuned with the labeled 300 (training) and 50 (validation) synthetic images for training and validation. Both were trained with Adam optimizer in 300 epochs.

Table 1 and Fig. 3 show the comparisons on the synthetic data set, which are unaffected by distortion and inaccurate camera parameters. Considering

---

[1] Readers are encouraged to watch the attached video and test the code.
[2] https://unity.com/.

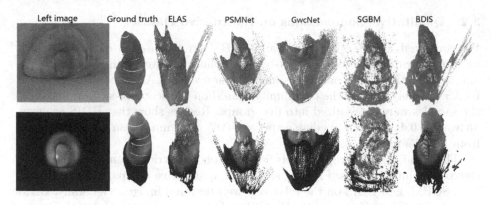

Fig. 3. Sample reconstructions in Diffuse lighting and Lambertian reflectance scenarios.

Table 1. The results on the synthetic data with diffused light and Lambertian reflectance.

		ELAS	SGBM	DIS	BDIS	GwcNet	PSMnet
Diffuse light	Median error	0.178	0.512	0.251	**0.161**	0.542	0.417
	Mean error	**0.220**	1.113	0.753	0.320	0.809	0.641
	Valid pixels (1000)	166.77	103.92	288.41	208.44	100.00	301.42
Non-Lambertian reflectance	Median error	0.198	0.710	0.376	**0.163**	0.271	0.731
	Mean error	**0.235**	1.400	1.051	0.379	0.662	1.027
	Valid pixels (1000)	81.50	74.29	295.92	204.42	106.38	301.46

the median error, BDIS is the best and has 9.55% and 17.68% higher accuracy than ELAS in diffuse lighting and non-Lambertian reflectance. The results indicate that BDIS is more advantageous in the scenario of non-Lambertian reflectance over ELAS, thus more robust in surgical scenarios. Results also show that BDIS cannot handle the edges well. Figure 3 and Table 1 reveal the bad mean error comparison is attributed to the small group of far-out points on the dark regions/edges. The number of valid prediction suggest BDIS produces more predictions but suffers from inaccurate dark region predictions.

Readers may notice the bad performance of DNN, which contradicts the conclusion from [1]. The reason is that the finetuning training process does not yield satisfying model parameters. The synthetic data set for transfer learning and the data used to pre-train the DNN are significantly different in terms of textures. Studies [6,22] indicate that the performance of the convolutional DNN is heavily dependent on the image texture, and efforts were devoted to bridging the domain gap [6,34]. The bad training process indicates its strong dependency on the training data set, which can be avoided using prior-free methods. Further tests will be conducted on labeled in-vivo data set.

3.2   Qualitative Comparisons on the In-vivo Dataset

We compared ELAS, BDIS, DIS, and SGBM on the in-vivo data sets. Since no ground truth is provided, DNN-based methods cannot be implemented. We aim to show that BDIS achieves similar accuracy as ELAS since near real-time ELAS is widely used in the community. Based on the scope-to-surface distance, the samples were categorized into five groups. Results show that BDIS achieves an average 0.4–1.66 mm (median error) and 0.65–2.32 mm (mean error) deviation from ELAS's results.

The invalid/dark/bright pixels lead to photometric inconsistency in the stereo matching process. Figure 4 shows the qualitative comparisons of ELAS, DIS, SGBM, and BDIS on the relatively well-textured images. Generally, BDIS achieves similar performance as ELAS but better matches pixels at the image edge with fewer outliers. DIS and SGBM suffer from the wrong edges. Invalid pixels inevitably exist on the edges of the rectified image after the image undistortion. Thus, in the coarse-level patch disparity estimation, patches with more invalid pixels are more likely to fail in convergence or yield local minima (abnormal depth) due to insufficient information. The dubious predictions, however, substantially influence the prediction and the initialization of the disparity at the finer-scale patch optimization (as in (2)). BDIS solves the problem by quantifying the posterior probability, discarding the patch that does not converge, and lowering the patches' probabilities with invalid pixels. Although the discarded patch does not help yield disparity, other patches compensate for the loss. If one pixel is not covered by any patch, we follow ELAS not to optimize the pixel.

Another noticeable problem is the ambiguous local minima in the cost function, which penalizes the photometric inconsistency. Figure 4 shows BDIS has fewer local minima than DIS and SGBM and is similar to ELAS. Figure 4 (a–b) indicates that the BDIS addresses the patchs' probabilities with textured and alleviates the ambiguous disparity from the textureless surface. Figure 4 (c–e) show that the ambiguities caused by the illumination have been greatly reduced. The quantitative results also provide evidence on its side. It should be emphasized that this work does not enforce any prior smoothness constraint in the optimization process.

We additionally tested BDIS and ELAS on the surfaces with serious non-Lambertian reflectance (Fig. 5). The photometric consistency of this data deteriorates significantly. Figure 5 shows that the center of the soft tissue is exposed to intense lighting while the marginal region is dark. Figure 5 indicates that ELAS suffers from the ambiguity on the marginal dark regions while BDIS can ignore or estimate most dark pixels correctly.

3.3   Processing Rate Comparison

We compared the time consumption of ELAS, DIS, and BDIS on a single core of CPU (i5-9400). BDIS runs 10 Hz on 640 × 480 image and 25 Hz on 360 × 288 image while ELAS achieves 4 Hz and 11 Hz. The two DNN methods GwcNet and PSMnet run 3 Hz and 5 Hz on GTX 1080ti. BDIS consumes double the time of DIS. The majority of the extra time of BDIS is devoted to patch-wise window traversing. Since the sampling window size is 5 in the experiment, 5 more times residual estimations are needed. In general, BDIS achieves similar/better performance over ELAS but runs 2 times faster.

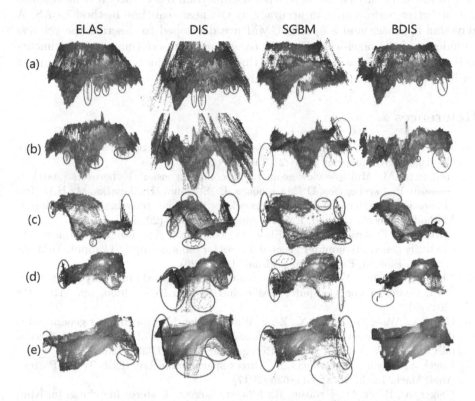

Fig. 4. Sample recovered shapes of the 5 classes. Circles mark the regions with large error.

Fig. 5. The qualitative comparisons on the heavy Lambertian reflectance and dark case.

4   Conclusion

We propose BDIS, the first CPU-level real-time stereo matching approach for CAS. BDIS inherits the fast performance of DIS while being more robust to textureless/dark surface and severe non-Lambertian reflectance. It achieves similar or better performance in accuracy as the near real-time method ELAS. A Bayesian approach and a spatial GMM are developed to describe the relative confidence of the pixel-wise disparity to achieve the performance. Experiments indicate that BDIS has fewer outliers than DIS and achieves a lower amount of outlier predictions than the near real-time ELAS.

References

1. Allan, M., et al.: Stereo correspondence and reconstruction of endoscopic data challenge. arXiv preprint arXiv:2101.01133 (2021)
2. Andrew, A.M.: Multiple view geometry in computer vision. Kybernetes (2001)
3. Brandao, P., Psychogyios, D., Mazomenos, E., Stoyanov, D., Janatka, M.: HAPNet: hierarchically aggregated pyramid network for real-time stereo matching. Comput. Methods Biomech. Biomed. Eng. Imaging Visual. 1–6 (2020)
4. Cartucho, J., Tukra, S., Li, Y.S. Elson, D., Giannarou, S.: VisionBlender: a tool to efficiently generate computer vision datasets for robotic surgery. Comput. Methods Biomech. Biomed. Eng. Imaging Visual. 1–8 (2020)
5. Chang, J.R., Chen, Y.S.: Pyramid stereo matching network. In: Proceedings of the IEEE Conference on Computer Vision and Pattern Recognition, pp. 5410–5418 (2018)
6. Chen, X., Wang, Y., Chen, X., Zeng, W.: S2R-DepthNet: learning a generalizable depth-specific structural representation. In: Proceedings of the IEEE Conference on Computer Vision and Pattern Recognition, pp. 3034–3043 (2021)
7. Engel, J., Koltun, V., Cremers, D.: Direct sparse odometry. IEEE Trans. Pattern Anal. Mach. Intell. **40**(3), 611–625 (2017)
8. Geiger, A., Roser, M., Urtasun, R.: Efficient large-scale stereo matching. In: Kimmel, R., Klette, R., Sugimoto, A. (eds.) ACCV 2010. LNCS, vol. 6492, pp. 25–38. Springer, Heidelberg (2011). https://doi.org/10.1007/978-3-642-19315-6_3
9. Giannarou, S., Visentini-Scarzanella, M., Yang, G.Z.: Probabilistic tracking of affine-invariant anisotropic regions. IEEE Trans. Pattern Anal. Mach. Intell. **35**(1), 130–143 (2013)
10. Guo, X., Yang, K., Yang, W., Wang, X., Li, H.: Group-wise correlation stereo network. In: Proceedings of the IEEE Conference on Computer Vision and Pattern Recognition, pp. 3273–3282 (2019)
11. Haouchine, N., Dequidt, J., Peterlik, I., Kerrien, E., Berger, M.O., Cotin, S.: Image-guided simulation of heterogeneous tissue deformation for augmented reality during hepatic surgery. In: 2013 IEEE International Symposium on Mixed and Augmented Reality (ISMAR), pp. 199–208. IEEE (2013)
12. Hirschmuller, H.: Accurate and efficient stereo processing by semi-global matching and mutual information. In: Proceedings of the IEEE Conference on Computer Vision and Pattern Recognition, vol. 2, pp. 807–814. IEEE (2005)
13. Jia, X., et al.: Automatic polyp recognition in colonoscopy images using deep learning and two-stage pyramidal feature prediction. IEEE Trans. Autom. Sci. Eng. **17**(3), 1570–1584 (2020)

14. Kroeger, T., Timofte, R., Dai, D., Van Gool, L.: Fast optical flow using dense inverse search. In: Leibe, B., Matas, J., Sebe, N., Welling, M. (eds.) ECCV 2016. LNCS, vol. 9908, pp. 471–488. Springer, Cham (2016). https://doi.org/10.1007/978-3-319-46493-0_29
15. Larochelle, H., Bengio, Y.: Classification using discriminative restricted boltzmann machines, pp. 536–543 (2008)
16. Long, Y., et al.: E-DSSR: efficient dynamic surgical scene reconstruction with transformer-based stereoscopic depth perception. arXiv preprint arXiv:2107.00229 (2021)
17. Lucas, B.D., Kanade, T., et al.: An iterative image registration technique with an application to stereo vision. Vancouver, British Columbia (1981)
18. Mahmood, F., Yang, Z., Chen, R., Borders, D., Xu, W., Durr, N.J.: Polyp segmentation and classification using predicted depth from monocular endoscopy. In: Medical Imaging 2019: Computer-Aided Diagnosis, vol. 10950, p. 1095011. International Society for Optics and Photonics (2019)
19. Pratt, P., Bergeles, C., Darzi, A., Yang, G.-Z.: Practical intraoperative stereo camera calibration. In: Golland, P., Hata, N., Barillot, C., Hornegger, J., Howe, R. (eds.) MICCAI 2014. LNCS, vol. 8674, pp. 667–675. Springer, Cham (2014). https://doi.org/10.1007/978-3-319-10470-6_83
20. Rappel, J.K.: Surgical stereo vision systems and methods for microsurgery. US Patent 9,330,477, 3 May 2016
21. Shimasaki, Y., Iwahori, Y., Neog, D.R., Woodham, R.J., Bhuyan, M.: Generating Lambertian image with uniform reflectance for endoscope image. In: IWAIT 2013, pp. 1–6 (2013)
22. Song, J., Patel, M., Girgensohn, A., Kim, C.: Combining deep learning with geometric features for image-based localization in the gastrointestinal tract. Expert Syst. Appl. 115631 (2021)
23. Song, J., Wang, J., Zhao, L., Huang, S., Dissanayake, G.: Dynamic reconstruction of deformable soft-tissue with stereo scope in minimal invasive surgery. IEEE Robot. Autom. Lett. 3(1), 155–162 (2017)
24. Song, J., Wang, J., Zhao, L., Huang, S., Dissanayake, G.: MIS-SLAM: real-time large-scale dense deformable SLAM system in minimal invasive surgery based on heterogeneous computing. IEEE Robot. Autom. Lett. 3(4), 4068–4075 (2018)
25. Stoyanov, D., Scarzanella, M.V., Pratt, P., Yang, G.-Z.: Real-time stereo reconstruction in robotically assisted minimally invasive surgery. In: Jiang, T., Navab, N., Pluim, J.P.W., Viergever, M.A. (eds.) MICCAI 2010. LNCS, vol. 6361, pp. 275–282. Springer, Heidelberg (2010). https://doi.org/10.1007/978-3-642-15705-9_34
26. Turan, M., Almalioglu, Y., Araujo, H., Konukoglu, E., Sitti, M.: Deep endovo: a recurrent convolutional neural network (RCNN) based visual odometry approach for endoscopic capsule robots. Neurocomputing 275, 1861–1870 (2018). https://doi.org/10.1016/j.neucom.2017.10.014, http://www.sciencedirect.com/science/article/pii/S092523121731665X
27. Uzunbas, M.G., Chen, C., Metaxas, D.: An efficient conditional random field approach for automatic and interactive neuron segmentation. Med. Image Anal. 27, 31–44 (2016)
28. Widya, A.R., Monno, Y., Imahori, K., Okutomi, M., Suzuki, S., Gotoda, T., Miki, K.: 3D reconstruction of whole stomach from endoscope video using structure-from-motion. In: 2019 41st Annual International Conference of the IEEE Engineering in Medicine and Biology Society (EMBC), pp. 3900–3904. IEEE (2019)

29. Yang, G., Manela, J., Happold, M., Ramanan, D.: Hierarchical deep stereo matching on high-resolution images. In: Proceedings of the IEEE Conference on Computer Vision and Pattern Recognition, pp. 5515–5524 (2019)
30. Ye, M., Johns, E., Handa, A., Zhang, L., Pratt, P., Yang, G.Z.: Self-supervised Siamese learning on stereo image pairs for depth estimation in robotic surgery. arXiv preprint arXiv:1705.08260 (2017)
31. Zampokas, G., Tsiolis, K., Peleka, G., Mariolis, I., Malasiotis, S., Tzovaras, D.: Real-time 3D reconstruction in minimally invasive surgery with quasi-dense matching. In: 2018 IEEE International Conference on Imaging Systems and Techniques (IST), pp. 1–6. IEEE (2018)
32. Zhan, J., Cartucho, J., Giannarou, S.: Autonomous tissue scanning under free-form motion for intraoperative tissue characterisation. In: Proceedings of the IEEE International Conference on Robotics and Automation, pp. 11147–11154. IEEE (2020)
33. Zhang, L., Ye, M., Giataganas, P., Hughes, M., Yang, G.Z.: Autonomous scanning for endomicroscopic mosaicing and 3D fusion. In: Proceedings of the IEEE International Conference on Robotics and Automation, pp. 3587–3593. IEEE (2017)
34. Zheng, C., Cham, T.J., Cai, J.: T2net: synthetic-to-realistic translation for solving single-image depth estimation tasks. In: Proceedings of the European Conference on Computer Vision, pp. 767–783 (2018)

Conditional Generative Data Augmentation for Clinical Audio Datasets

Matthias Seibold[1,2]([✉]), Armando Hoch[3], Mazda Farshad[3], Nassir Navab[1],
and Philipp Fürnstahl[2,3]

[1] Computer Aided Medical Procedures (CAMP), Technical University of Munich,
85748 Munich, Germany
matthias.seibold@tum.de
[2] Research in Orthopedic Computer Science (ROCS), University Hospital Balgrist,
University of Zurich, 8008 Zurich, Switzerland
[3] Balgrist University Hospital, 8008 Zurich, Switzerland

Abstract. In this work, we propose a novel data augmentation method
for clinical audio datasets based on a conditional Wasserstein Generative
Adversarial Network with Gradient Penalty (cWGAN-GP), operating on
log-mel spectrograms. To validate our method, we created a clinical audio
dataset which was recorded in a real-world operating room during Total
Hip Arthroplasty (THA) procedures and contains typical sounds which
resemble the different phases of the intervention. We demonstrate the
capability of the proposed method to generate realistic class-conditioned
samples from the dataset distribution and show that training with the
generated augmented samples outperforms classical audio augmentation
methods in terms of classification performance. The performance was
evaluated using a ResNet-18 classifier which shows a mean Macro F1-
score improvement of 1.70% in a 5-fold cross validation experiment using
the proposed augmentation method. Because clinical data is often expensive to acquire, the development of realistic and high-quality data augmentation methods is crucial to improve the robustness and generalization capabilities of learning-based algorithms which is especially important for safety-critical medical applications. Therefore, the proposed data
augmentation method is an important step towards improving the data
bottleneck for clinical audio-based machine learning systems.

Keywords: Deep learning · Data augmentation · Acoustic sensing ·
Total hip arthroplasty · Generative adversarial networks

1 Introduction

Acoustic signals are easy and low-cost to acquire, can be captured using air-
borne or contact microphones and show great potentials in medical applications for interventional guidance and support systems. Successful applications

N. Navab and P. Fürnstahl—Equally contributing last authors.

are intra-operative tissue characterization during needle insertion [9] and tissue coagulation [16], the identification of the insertion endpoint in THA procedures [3,20], error prevention in surgical drilling during orthopedic procedures [21] or guidance in orthopedic arthroscopy procedures [23].

Furthermore, acoustic signals have successfully been employed for diagnostic medical applications. Exemplary applications include the assessment of cartilage degeneration by measuring structure borne noise in the human knee during movement [11], the development of a prototype for the detection of implant loosening through an acoustic sensor system [2], a system for monitoring the acoustic emissions of THA implants [19], or the automated analysis of lung sounds captured with a digital stethoscope which allows non-specialists to screen for pulmonary fibrosis [14].

Through recent advances in machine learning research, learning-based methods have replaced and outperformed classical acoustic signal processing-based approaches, as well as classical handcrafted feature-based learning approaches for many acoustic audio signal processing tasks [18]. However, state-of-the-art deep learning methods require large amounts of training data to achieve superior performance and generalize well to unseen data, which are often difficult or infeasible to acquire in a clinical setting. To tackle this issue, the usage of augmentation techniques is a standard approach to increase the diversity and size of training datasets. Hereby, new samples can be synthesized by applying transformations to the existing data, e.g. rotation and cropping for images, replacing words with synonyms for text, and applying noise, pitch shifting, and time stretching to audio samples [28]. Even though these data augmentation methods improve the performance of target applications, they do not necessarily generate realistic samples which is especially crucial in the medical domain where reliability is a key factor. One solution to this problem is for example to exploit the underlying physics for augmentation, e.g. for ultrasound image augmentation [26] which is, however, not applicable for clinical audio data. In the presented work, we will focus on realistic data augmentation of audio datasets for medical applications.

Recently, deep generative models, a family of deep learning models, which are able to synthesize realistic samples from a learned distribution, have been applied for data augmentation of various data modalities outside of the medical domain. For the augmentation of audio data, different generative approaches have been introduced, of which related work to the proposed method is described in the following section. Hu et al. utilized a GAN to synthesize samples of logarithmic Mel-filter bank coefficients (FBANK) from a learned distribution of a speech dataset and subsequently generated soft labels using a pretrained classifier [8]. Madhu et al. trained separate GANs on mel-spectrograms for each class of a dataset to generate augmentation data [12]. Chatziagapi et al. used the Balancing GAN (BAGAN) framework [13] to augment an imbalanced speech dataset [1]. A conditional GAN was employed for data augmentation of speech using FBANK features by Sheng et al. [22] and for respiratory audio signals based on raw waveform augmentation by Jayalakshmy et al. [10].

In this work, we introduce a novel augmentation technique for audio data based on a conditional Wasserstein GAN model with Gradient Penalty (cWGAN-GP) which produces higher-quality samples and is easier to train than standard GANs [5]. The proposed model operates on log-mel spectrograms which have been shown to outperform other feature representations and achieves state-of-the-art performance in audio classification tasks [18]. The proposed model is able to generate realistic and high-quality log-mel spectrograms from the learned dataset distribution. We show that our model can be used for two augmentation strategies, doubling the number of samples and balancing the dataset. While classical audio augmentation techniques might improve the performance of the classifier, they do not generate samples that can be captured in a real environment and might therefore be inconsistent with the real variability of captured real-world acoustic signals. In contrast, the proposed model is able to generate realistic samples from the learned distribution of the original data.

To evaluate the proposed framework on realistic clinical data, we introduce a novel audio dataset containing sounds of surgical actions recorded from five real THA procedures which resemble the different phases of the intervention. We thoroughly evaluate the proposed method on the proposed dataset in terms of classification performance improvement of a ResNet-18 classifier with and without data augmentation using 5-fold cross validation and compare the results with classical audio augmentation techniques.

2 Materials and Method

2.1 Novel Surgical Audio Dataset

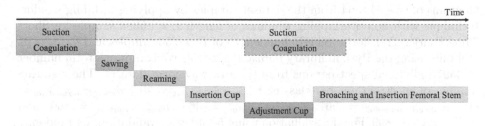

Fig. 1. The classes of the novel clinical dataset resemble the phases of a THA procedure. Occurrences with drawn through lines indicate intensive usage of the respective surgical action, dashed lines correspond to sporadic usage.

Figure 1 illustrates the occurrence of the six classes $C := \{$Suction, Coagulation, Sawing, Reaming, Insertion, Adjustment$\}$ present in the dataset over the course of a THA procedure. Please note that "Insertion Cup" and "Broaching and Insertion Femoral Stem" were joined into a single class ("Insertion") because of the similar acoustic signature generated by hammering onto the metal structure of the insertion tools for the acetabular cup, femoral broach and femoral stem implant, respectively. The "Adjustment" class also contains hammering signals that are, however, performed with a screwdriver-like tool which is used to adjust the orientation of the acetabular cup and generates a slightly different sound. During opening the access to the area of operation in the beginning of the procedure, suction and coagulation is employed intensively, whereas in the rest of the procedure both surgical actions are performed sporadically and on demand (indicated through dashed outlines in Fig. 1). All samples were manually cut from recordings of five THA interventions conducted at our university hospital for which we captured audio with a framerate of 44.1 kHz using a air-borne shotgun microphone (Røde NTG2) pointed towards the area of operation and video captured from the OR light camera (Trumpf TruVidia). The captured video was used to facilitate the labelling process. We labelled the dataset in a way that audio samples do not contain overlapping classes and no staff conversations. An ethical approval has been obtained prior to recording the data in the operating room. The resulting dataset contains 568 recordings with a length of 1 s to 31 s and the following distribution: $n_{raw,Adjustment} = 68$, $n_{raw,Coagulation} = 117$, $n_{raw,Insertion} = 76$, $n_{raw,Reaming} = 64$, $n_{raw,Sawing} = 21$, and $n_{raw,Suction} = 222$. The dataset can be accessed under https://rocs.balgrist.ch/en/open-access/.

2.2 Data Preprocessing and Baseline Augmentations

Log-mel spectrograms are a two-dimensional representation of an audio signal, mapping frequency components of a signal to the ordinate and time to the abscissa. They offer a dense representation of the signal, reduce the dimensionality of the samples, and have been shown to yield superior classification performance for a wide variety of acoustic sensing tasks [18]. We compute log-mel spectrograms of size 64 × 64 from the dataset samples by applying a sliding window technique with non-overlapping windows of length $L = 16380$ samples, a Short Time Fourier Transform (STFT) hop length of $H = 256$ samples and $n_{mels} = 64$ mel bins using the Python library *librosa 0.8.1* [15]. We compute a total number of 3597 individual spectrograms from the raw waveform dataset. The resulting number of spectrograms per-class is: $n_{spec,Adjustment} = 494$, $n_{spec,Coagulation} = 608$, $n_{spec,Insertion} = 967$, $n_{spec,Reaming} = 469$, $n_{spec,Sawing} = 160$, and $n_{spec,Suction} = 899$. For the evaluation using 5-fold cross validation, we randomly split the dataset into five folds on the raw waveform level over all recordings, as the recording conditions are identical.

To compare the proposed augmentation method against classical signal processing augmentation approaches, we implemented the following augmentation strategies which are applied to the raw waveforms directly. We apply Gaussian noise with $\mu = 0$ and $\sigma = 0.01$. We apply Pitch Shifting by 3 semitones upwards. We apply time stretching with a factor of 1.5. Furthermore, we compare our method with *SpecAugment*, a widely used approach for audio augmentation in Automatic Speech Recognition (ASR) tasks which applies random time-warping, frequency- and time-masking to the spectrograms directly [17].

For a fair comparison of all augmentations, we add 100% generated samples for each augmentation strategy, respectively. We normalize the data by computing $X_{norm,mel} = (X_{mel} - \mu)/\sigma$, where (μ) is the mean and (σ) is the standard deviation of the entire dataset.

2.3   Conditional Generative Data Augmentation Method

The architectural details of the proposed GAN are illustrated in Fig. 2. To stabilize the training process and improve the generated sample quality, we apply the Wasserstein loss with Gradient Penalty (GP) as introduced by Gulrajani et al. [5] which enforces a constraint such that the gradients of the discriminator's (critic) output w.r.t the inputs have unit norm. This approach greatly improves the stability of the training and compensates for problems such as mode collapse. We define the critic's loss function as:

$$L_C = \mathbb{E}_{\tilde{x}\sim\mathbb{P}_g}[D(\tilde{x},y)] - \mathbb{E}_{x\sim\mathbb{P}_r}[D(x,y)] + \lambda\,\mathbb{E}_{\hat{x}\sim\mathbb{P}_{\hat{x}}}[(\|\triangle_{\hat{x}}D(\hat{x},y)\|_2 - 1)^2] \quad (1)$$

where $\mathbb{P}_r$ is the real distribution, $\mathbb{P}_g$ is the generated distribution, and $\mathbb{P}_{\hat{x}}$ is the interpolated distribution. The interpolated samples $\hat{x}$ are uniformly sampled along a straight line between real x and generated $\tilde{x}$ samples by computing:

$$\hat{x} = \epsilon x + (1 - \epsilon)\tilde{x} \quad (2)$$

We use the recommended GP weight of $\lambda = 10$, a batch size of 64 and train the discriminator five times for each generator iteration. In order to choose the stopping point for training, we frequently compute the Fréchet Inception Distance (FID) [7] which is calculated from features of a pretrained classifier by:

$$FID = \|\mu_r - \mu_g\|^2 + \text{Tr}(C_r + C_g - 2 * \sqrt{C_r * C_g}) \quad (3)$$

as a measure of the quality for the generated samples and stop the training at epoch 580. Hereby, μ_r and μ_g represent the feature-wise mean of the real and generated spectrograms, C_r and C_g the respective covariance matrices. Because of the structural differences of images and spectrograms, we cannot use an Inception v3 network pretrained on ImageNet to compute the FID. Therefore, we employ a ResNet-18 [6] model pretrained on the proposed dataset, extract the features from the last convolutional layer, and use these features for FID calculation. The proposed model is implemented with *TensorFlow/Keras 2.6* and trained using the Adam optimizer ($LR = 1e - 4$, $\beta_1 = 0.5$, $\beta_2 = 0.9$) in ~6 h on a NVidia RTX 2080 SUPER GPU.

A nonlinear activation function is omitted in the last convolutional layer of the generator because the spectrogram samples are not normalized in the range $[0, 1]$. The mapping layer of the generator employs a dense layer, whereas in the discriminator (critic) we use repeat and reshaping operations for remapping. The generator and discriminator have a total number of 1,526,084 and 4,321,153 parameters, respectively. The implementation, pretrained models, and dataset can be accessed under: https://rocs.balgrist.ch/en/open-access/.

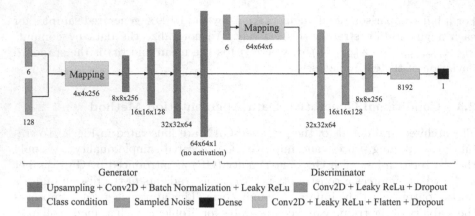

Fig. 2. The architecture of the proposed model including output sizes of each layer. The input for the generator is a noise vector of size 1×128 and a class condition. The generator outputs a spectrogram which is fed to the discriminator together with the class condition. The discriminator (critic) outputs a scalar realness score.

2.4 Classification Model

To evaluate the augmentation performance of our model against classical audio augmentation techniques, we analyze the effect of augmentation on the classification performance in a 5-fold cross validation experiment using a ResNet-18 [6] classifier, an architecture which has been successfully employed for clinical audio classification tasks [20,21]. We train the classifier from scratch for 20 epochs using categorical crossentropy loss, the Adam optimizer ($LR = 1e - 4$, $\beta_1 = 0.9$, $\beta_2 = 0.999$) and a batch size of 32.

3 Results and Evaluation

In Fig. 3, we show a comparison of randomly chosen ground truth and randomly generated samples. The visual quality of the generated samples is comparable to the original data and the model seems to be able to generate samples conditioned on the queried class. By further visual inspection it can be observed that the synthesized samples contain the characteristics of the original dataset, e.g. the hammer strokes are clearly visible for the classes "Adjustment" and "Insertion".

The quantitative evaluation of the classification performance using a ResNet-18 classifier is given in Table 1. We report the mean Macro F1-Score in the format *mean* ± *std*. We compare training without augmentations and classical audio augmentation techniques (adding noise, pitch shifting, time stretching, and SpecAugment [17]) with the proposed method. The cWGAN-GP-based augmentations outperform all classical augmentation strategies when doubling the samples (+1.70%) and show similar performance (+1.07%) as the best performing classical augmentation strategy (Time Stretch) when balancing the dataset.

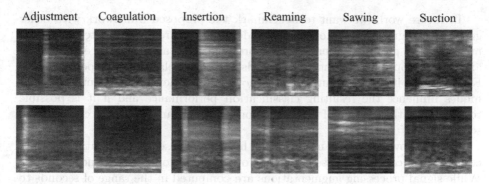

Fig. 3. The top row shows log-mel spectrograms of random samples for each class present in the acquired dataset, the bottom row shows log-mel spectrograms generated by the proposed model for each class, respectively.

Table 1. Results of the proposed model in comparison to classical audio augmentation techniques.

Augmentation technique	Mean macro F1-score	Relative improvement
No augmentation	93.90 ± 2.48%	
Add noise	92.87 ± 0.99%	−1.03%
Pitch shift	94.73 ± 1.28%	+0.83%
Time stretch	95.00 ± 1.49%	+1.10%
SpecAugment [17]	94.23 ± 1.14%	+0.33%
cWGAN-GP (balanced)	94.97 ± 1.71%	+1.07%
cWGAN-GP (doubled)	**95.60 ± 1.26%**	+1.70%

4   Discussion

The proposed augmentation method is an important step towards improving the data limitations by generating synthetic in-distribution augmentation data for clinical applications for which it is often expensive or even impossible to gather large amounts of training data. We showed that our augmentation strategy out-performs classical signal processing approaches and has the capability to balance imbalanced datasets to a certain extent. To balance imbalanced datasets, any arbitrary number of samples can easily be generated for each class with the proposed approach which is not possible using classical signal processing techniques in the same way. However, for the given dataset and configuration, doubling the number of samples using the proposed augmentation method leads to the best final classification results. Furthermore, we show that the proposed method out-performs SpecAugment [17], an established audio augmentation method which applies time-warping, as well as frequency and time masking to the spectrogram data directly.

In future work, we want to benchmark the proposed framework with other generative augmentation models and model architectures, investigate the performance of the proposed approach on more (balanced and imbalanced) datasets and further optimize our model towards improved classification performance. Furthermore, it should be investigated how combinations of augmentation techniques influence the resulting classification performance and if it is possible to maximize the impact of augmentations through an optimized combination scheme.

The improved performance achieved by the proposed augmentation method comes at the cost of increased demands on computational power and resources. While signal processing augmentations are computed in the range of seconds to minutes, our model requires an additional training step which takes ~6 h for the presented dataset and increases with larger datasets.

Because we created the proposed clinical dataset in a way that it resembles the phases and surgical actions executed during a real THA procedure, potential future clinical applications are the prediction of surgical actions from captured audio signals in the operating room which could be used for workflow recognition and surgical phase detection. Therefore, we consider the proposed dataset as an important step towards automated audio-based clinical workflow detection systems, a topic which has only been studied rudimentally so far [24,27]. The proposed approach is designed to work with spectrogram based audio, which can be transformed back to the signal domain, e.g. using the Griffin-Lim algorithm [4] or more recently introduced learning-based transformation approaches, e.g. the work by Takamichi et al. [25]. We reconstructed waveforms from a few generated spectrograms using the Griffin-Lim algorithm and could, despite artifacts being present, recognize acoustic similarities to the original samples for each class, respectively. In future work, the proposed augmentation method could furthermore be transferred to other medical and non-medical grid-like data domains.

5   Conclusion

In the presented work, we introduce a novel data augmentation method for medical audio data and evaluate it on a clinical dataset which was recorded in real-world Total Hip Arthroplasty (THA) surgeries. The proposed dataset contains sound samples of six surgical actions which resemble the different phases of a THA intervention. We show in quantitative evaluations that the proposed method outperforms classical signal and spectrogram processing-based augmentation techniques in terms of Mean Macro F1-Score, evaluated using a ResNet-18 classifier in a 5-fold cross validation experiment. By generating high-quality in-distribution samples for data augmentation, our method has the potential to improve the data bottleneck for acoustic learning-based medical support systems.

Acknowledgment. This work is part of the SURGENT project under the umbrella of Hochschulmedizin Zürich.

References

1. Chatziagapi, A., et al.: Data augmentation using GANs for speech emotion recognition. In: Proceedings of InterSpeech 2019, pp. 171–175 (2019)
2. Ewald, H., Timm, U., Ruther, C., Mittelmeier, W., Bader, R., Kluess, D.: Acoustic sensor system for loosening detection of hip implants. In: 2011 Fifth International Conference on Sensing Technology, pp. 494–497 (2011)
3. Goossens, Q., et al.: Acoustic analysis to monitor implant seating and early detect fractures in cementless THA: an in vivo study. J. Orthop. Res. (2020)
4. Griffin, D., Lim, J.: Signal estimation from modified short-time Fourier transform. IEEE Trans. Acoust. Speech Sig. Process. **32**(2), 236–243 (1984)
5. Gulrajani, I., Ahmed, F., Arjovsky, M., Dumoulin, V., Courville, A.: Improved training of Wasserstein GANs. In: Proceedings of the 31st International Conference on Neural Information Processing Systems, pp. 5769–5779 (2017)
6. He, K., Zhang, X., Ren, S., Sun, J.: Deep residual learning for image recognition. In: 2016 IEEE Conference on Computer Vision and Pattern Recognition (CVPR), pp. 770–778 (2016)
7. Heusel, M., Ramsauer, H., Unterthiner, T., Nessler, B., Hochreiter, S.: GANs trained by a two time-scale update rule converge to a local NASH equilibrium. In: Proceedings of the 31st International Conference on Neural Information Processing Systems, pp. 6629–6640 (2017)
8. Hu, H., Tan, T., Qian, Y.: Generative adversarial networks based data augmentation for noise robust speech recognition. In: 2018 IEEE International Conference on Acoustics, Speech and Signal Processing (ICASSP), pp. 5044–5048 (2018)
9. Illanes, A., et al.: Novel clinical device tracking and tissue event characterization using proximally placed audio signal acquisition and processing. Sci. Rep. **8** (2018)
10. Jayalakshmy, S., Sudha, G.F.: Conditional GAN based augmentation for predictive modeling of respiratory signals. Comput. Biol. Med. **138**, 104930 (2021)
11. Kim, K.S., Seo, J.H., Kang, J.U., Song, C.G.: An enhanced algorithm for knee joint sound classification using feature extraction based on time-frequency analysis. Comput. Methods Programs Biomed. **94**(2), 198–206 (2009)
12. Madhu, A., Kumaraswamy, S.: Data augmentation using generative adversarial network for environmental sound classification. In: 2019 27th European Signal Processing Conference (EUSIPCO) (2019)
13. Mariani, G., Scheidegger, F., Istrate, R., Bekas, C., Malossi, A.C.I.: BaGAN: data augmentation with balancing GAN. arXiv abs/1803.09655 (2018)
14. Marshall, A., Boussakta, S.: Signal analysis of medical acoustic sounds with applications to chest medicine. J. Franklin Inst. **344**(3), 230–242 (2007)
15. McFee, B., et al.: librosa: Audio and music signal analysis in Python. In: 14th Python in Science Conference, pp. 18–25 (2015)
16. Ostler, D., et al.: Acoustic signal analysis of instrument-tissue interaction for minimally invasive interventions. Int. J. Comput. Assist. Radiol. Surg. (2020)
17. Park, D.S., et al.: SpecAugment: a simple data augmentation method for automatic speech recognition. InterSpeech 2019, September 2019
18. Purwins, H., Li, B., Virtanen, T., Schlüter, J., Chang, S.Y., Sainath, T.: Deep learning for audio signal processing. IEEE J. Sel. Top. Sig. Process. **14**, 206–219 (2019)
19. Rodgers, G.W., et al.: Acoustic emission monitoring of total hip arthroplasty implants. IFAC Proc. Vol. **47**(3), 4796–4800 (2014). 19th IFAC World Congress

20. Seibold, M., et al.: Acoustic-based spatio-temporal learning for press-fit evaluation of femoral stem implants. In: International Conference on Medical Image Computing and Computer-Assisted Intervention, pp. 447–456 (2021)

21. Seibold, M., et al.: Real-time acoustic sensing and artificial intelligence for error prevention in orthopedic surgery. Sci. Rep. **11** (2021)

22. Sheng, P., Yang, Z., Hu, H., Tan, T., Qian, Y.: Data augmentation using conditional generative adversarial networks for robust speech recognition. In: 2018 11th International Symposium on Chinese Spoken Language Processing (ISCSLP), pp. 121–125 (2018)

23. Suehn, T., Pandey, A., Friebe, M., Illanes, A., Boese, A., Lohman, C.: Acoustic sensing of tissue-tool interactions - potential applications in arthroscopic surgery. Curr. Direct. Biomed. Eng. **6** (2020)

24. Suzuki, T., Sakurai, Y., Yoshimitsu, K., Nambu, K., Muragaki, Y., Iseki, H.: Intra-operative multichannel audio-visual information recording and automatic surgical phase and incident detection. In: 2010 Annual International Conference of the IEEE Engineering in Medicine and Biology, pp. 1190–1193 (2010)

25. Takamichi, S., Saito, Y., Takamune, N., Kitamura, D., Saruwatari, H.: Phase reconstruction from amplitude spectrograms based on directional-statistics deep neural networks. Sig. Process. **169**, 107368 (2020)

26. Tirindelli, M., Eilers, C., Simson, W., Paschali, M., Azampour, M.F., Navab, N.: Rethinking ultrasound augmentation: a physics-inspired approach. In: Medical Image Computing and Computer Assisted Intervention, pp. 690–700 (2021)

27. Weede, O., et al.: Workflow analysis and surgical phase recognition in minimally invasive surgery. In: 2012 IEEE International Conference on Robotics and Biomimetics (ROBIO), pp. 1080–1074 (2012)

28. Wei, S., Zou, S., Liao, F., Lang, W.: A comparison on data augmentation methods based on deep learning for audio classification. J. Phys: Conf. Ser. **1453**(1), 012085 (2020)

Rethinking Surgical Instrument Segmentation: A Background Image Can Be All You Need

An Wang[1,2], Mobarakol Islam[3], Mengya Xu[4], and Hongliang Ren[1,2,4(✉)]

[1] Department of Electronic Engineering, The Chinese University of Hong Kong,
Shatin, Hong Kong SAR, China
wa09@link.cuhk.edu.hk, hlren@ee.cuhk.edu.hk
[2] Shun Hing Institute of Advanced Engineering,
The CUHK, Shatin, Hong Kong SAR, China
[3] BioMedIA Group, Department of Computing,
Imperial College London, London, UK
m.islam20@imperial.ac.uk
[4] Departmnet of Biomedical Engineering, National University of Singapore,
Singapore, Singapore
mengya@u.nus.edu

Abstract. Data diversity and volume are crucial to the success of training deep learning models, while in the medical imaging field, the difficulty and cost of data collection and annotation are especially huge. Specifically in robotic surgery, data scarcity and imbalance have heavily affected the model accuracy and limited the design and deployment of deep learning-based surgical applications such as surgical instrument segmentation. Considering this, we rethink the surgical instrument segmentation task and propose a one-to-many data generation solution that gets rid of the complicated and expensive process of data collection and annotation from robotic surgery. In our method, we only utilize a single surgical background tissue image and a few open-source instrument images as the seed images and apply multiple augmentations and blending techniques to synthesize amounts of image variations. In addition, we also introduce the chained augmentation mixing during training to further enhance the data diversities. The proposed approach is evaluated on the real datasets of the EndoVis-2018 and EndoVis-2017 surgical scene segmentation. Our empirical analysis suggests that without the high cost of data collection and annotation, we can achieve decent surgical instrument segmentation performance. Moreover, we also observe that our method can deal with novel instrument prediction in the deployment domain. We hope our inspiring results will encourage researchers to emphasize data-centric methods to overcome demanding deep learning limitations besides data shortage, such as class imbalance, domain adaptation, and incremental learning. Our code is available at https://github.com/lofrienger/Single_SurgicalScene_For_Segmentation.

A. Wang and M. Islam—Co-first authors.

L. Wang (Eds.): MICCAI 2022, LNCS 13437, pp. 355–364, 2022.
https://doi.org/10.1007/978-3-031-16449-1_34

1   Introduction

Ever-larger models processing larger volumes of data have propelled the extraordinary performance of deep learning-based image segmentation models in recent decades, but obtaining well-annotated and perfectly-sized data, particularly in the medical imaging field, has always been a great challenge [6]. Various causes, including tremendous human efforts, unavailability of rare disease data, patient privacy concerns, high prices, and data shifts between different medical sites, have made acquiring abundant high-quality medical data a costly endeavor. Besides, dataset imperfection like class imbalance, sparse annotations, noisy annotations and incremental-class in deployment [20] also affects the training and deployment of deep learning models. Moreover, for the recent-developed surgery procedures like the single-port robotic surgery where no dataset of the new instruments is available [5], the segmentation task can hardly be accomplished. In the presence of these barriers, one effective solution to overcome the data scarcity problems is to train with a synthetic dataset instead of a real one.

A few recent studies utilize synthetic data for training and achieve similar and even superior performance than training with real data. For example, in the computer vision community, Tremblay et al. [19] develop an object detection system relying on domain randomization where pose, lighting, and object textures are randomized in a non-realistic manner; Gabriel et al. [7] make use of multiple generative adversarial networks (GANs) to improve data diversity and avoid severe over-fitting compared with a single GAN; Kishore et al. [14] propose imitation training as a synthetic data generation guideline to introduce more underrepresented items and equalize the data distribution to handle corner instances and tackle long-tail problems.

In medical applications, many works have focused on GAN-based data synthesizing [3,10,11,18], while a few works utilize image blending or image composition to generate new samples. For example, mix-blend [8] mixes several synthetic images generated with multiple blending techniques to create new training samples. Nonetheless, one limitation of their work is that they need to manually capture and collect thousands of foreground instrument images and background tissue images, making the data generation process trivial and time-consuming. In addition, E. Colleoni et al. [4] recorded kinematic data as the data source to synthesize a new dataset for the instrument - Large Needle Drivers. In comparison with previous works, our approach only utilizes a single background image and dozens of foreground instrument images as the data source. Without costly data collection and annotation, we show the simplicity and efficacy of our dataset generation framework.

Contributions. In this work, we rethink the surgical instrument segmentation task from a data-centric perspective. Our contributions can be summarized as follows:

- With minimal human effort in data collection and without manual image annotations, we propose a data-efficient framework to generate high-quality synthetic datasets used for surgical instrument segmentation.
- By introducing various augmentation and blending combinations to the foreground and background source images, and training-time chained augmentation mixing, we manage to increase the data diversity and balance the instruments class distribution.
- We evaluate our method on two real datasets. The results suggest that our dataset generation framework is simple yet efficient. It is possible to achieve acceptable surgical instrument segmentation performance, even for novel instruments, by training with synthetic data that only employs a single surgical background image.

2 Proposed Method

2.1 Preliminaries

Data augmentation has become a popular strategy for boosting the size of a training dataset to overcome the data-hungry problem when training the deep learning models. Besides, data augmentation can also be regarded as a regularisation approach for lowering the model generalization error [9]. In other words, it helps boost performance when the model is tested on a distinct unseen dataset during training. Moreover, the class imbalance issue, commonly seen in most surgical datasets, can also be alleviated by generating additional data for the under-represented classes.

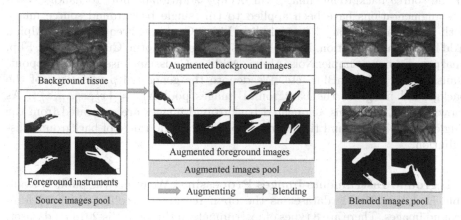

Fig. 1. Demonstration of the proposed dataset generation framework with augmenting and blending. With minimal effort in preparing the source images, our method can produce large amounts of high-quality training samples for the surgical segmentation task.

Blending is a simple yet effective way to create new images simply by image mixing or image composition. It can also be treated as another kind of data augmentation technique that mixes the information contained in different images instead of introducing invariance to one single image. Denote the foreground image and background image as x_f and x_b, we can express the blended image with a blending function Θ as

$$x = \Theta(x_f, x_b) = x_f \oplus x_b \tag{1}$$

where $\oplus$ stands for pixel-wise fusion.

Training-time augmentation can help diversify training samples. By mixing various chained augmentations with the original image, more image variations can be created without deviating too far from the original image, as proposed by AugMix [12]. In addition, intentionally controlling the choices of augmentation operations can also avoid hurting the model due to extremely heavy augmentations. A list of augmentation operations is included in the augmentation chains, such as auto-contrast, equalization, posterization, solarization, etc.

2.2 Synthesizing Surgical Scenes from a Single Background

Background Tissue Image Processing. We collect one background tissue image from the open-source EndoVis-2018 dataset[1] where the surgical scene is the nephrectomy procedures. The critical criterion of this surgical background selection is that the appearance of the instrument should be kept as little as possible. In the binary instrument segmentation task, the background pixels are all assigned with the value 0. Therefore, the appearance of instruments in the source background image will occupy additional effort to handle. Various augmentations have been applied to this single background source image with the imgaug[2] library [13], including LinearContrast, FrequencyNoiseAlpha, AddToHueAndSaturation, Multiply, PerspectiveTransform, Cutout, Affine, Flip, Sharpen, Emboss, SimplexNoiseAlpha, AdditiveGaussianNoise, CoarseDropout, GaussianBlur, MedianBlur, etc. We denote the generated p variations of the background image as the background images pool $X_b^p = \{x_b^1, x_b^2, ..., x_b^p\}$. As shown in Fig. 1, various augmented background images are generated from the single source background tissue image to cover a wide range of background distribution.

Foreground Instruments Images Processing. We utilize the publicly available EndoVis-2018 [1] dataset as the open resource to collect the seed foreground images. There are 8 types of instruments in the EndoVis-2018 [1] dataset, namely Maryland Bipolar Forceps, Fenestrated Bipolar instruments, Prograsp Forceps, Large Needle Driver, Monopolar Curved Scissors, Ultrasound Probe, Clip Applier, and Suction Instrument. We only employ 2 or 3 images for each

---

[1] https://endovissub2018-roboticscenesegmentation.grand-challenge.org/.
[2] https://github.com/aleju/imgaug.

instrument as the source images. We extract the instruments and make their background transparent. The source images are selected with prior human knowledge of the target scenes to ensure their high quality. For example, for some instruments like Monopolar Curved Scissors, the tip states (open or close) are crucial in recognition, and they are not reproducible simply by data augmentation. Therefore, we intentionally select source images for such instruments to make it possible to cover different postures and states. In this way, we aim to increase the in-distribution data diversity to substantially improve generalization to out-of-distribution (OOD) category-viewpoint combinations [15]. Since we get rid of annotation, the instrument masks are applied with the same augmentations as the instruments to maintain the segmentation accuracy. We denote the generated q variations of the foreground images as the foreground image pool $X_f^q = \{x_f^1, x_f^2, ..., x_f^q\}$. Figure 1 shows some new synthetic instruments images. The foreground images pool, together with the background images pool, forms the augmented images pool, which is used for the following blending process.

Blending Images. After obtaining the background image pool X_b^p and the foreground image pool X_f^q, we randomly draw one sample from these two pools and blend them to form a new composited image. Specifically, the foreground image is pasted on the background image with pixel values at the overlapped position taken from the instruments. Furthermore, considering the real surgical scenes, the number of instruments in each image is not fixed. We also paste two instrument images on the background occasionally. Due to this design, we expect the model could better estimate the pixel occupation of the instruments in the whole image. Denoting the blended image as x_s, finally, the blended images pool with t synthetic images can be presented as $X_s^t = \{x_s^1, x_s^2, ..., x_s^t\} = \{\Theta(x_f^i, x_b^j)\}$, where $i = 1, 2, ..., p$ and $j = 1, 2, ..., q$.

In-training Chained Augmentation Mixing. Inspired by AugMix [12], we apply the training-time chained augmentation mixing technique to further make the data more diverse and also improve the generalization and robustness of the model. The number of augmentation operations in each augmentation chain is randomly set as one, two, or three. The parameters in the Beta distribution and the Dirichlet distribution are all set as 1. We create two sets of augmentation collections, namely AugMix-Soft and AugMix-Hard. Specifically, AugMix-Soft includes autocontrast, equalize, posterize and solarize, while AugMix-Hard has additional color, contrast, brightness, and sharpness augmentations. The overall expression of the synthetic training sample after the training-time augmentation mixing with N chains is

$$x_s^{AM} = m \cdot \Theta(x_f, x_b) + (1 - m) \cdot \sum_{i=1}^{N} (w_i \cdot H_i(\Theta(x_f, x_b))) \tag{2}$$

where m is a random convex coefficient sampled from a Beta distribution, w_i is also a random convex coefficient sampled from a Dirichlet distribution controlling the mixing weights of the augmentation chains. Both distribution functions have the same coefficient value of 1. H_i denotes the integrated augmentation operations in the ith augmentation chain.

3   Experiments

3.1   Datasets

Based on effortlessly collected source images and considering the contents in real surgery images, we apply a wide range of augmentation and blending operations to create abundant synthetic images for training. Only one background tissue image is adopted to generate our synthetic datasets. Specifically, for the case of 2 source images per instrument, we first organize the dataset Synthetic-A with 4000 synthetic images, and only one instrument exists in each synthetic image. Then we consider adding up additional 2000 synthetic images to build the dataset Synthetic-B where each image contains 2 distinct instruments. Moreover, we utilize one more source foreground image for each instrument and generate 2000 more synthetic images, among which 80% contain one instrument, and the remaining 20% contain 2 different instruments. This dataset with 8000 samples in total is named Synthetic-C.

To evaluate the quality of the generated surgical scene dataset, we conduct binary segmentation experiments with our synthetic datasets and the real EndoVis-2018 [1] dataset. We also evaluate on EndoVis-2017 [2] dataset to show that the model trained with our synthetic dataset also obtains good generalization ability to handle new domains with unseen instruments like the Vessel Sealer.

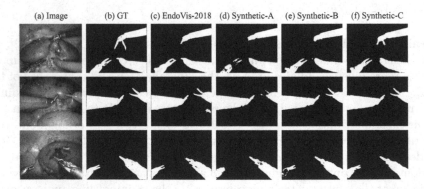

Fig. 2. Qualitative comparison of the binary segmentation results. (b) represents the ground truth. (c), (d), (e), and (f) show the results obtained from models trained with the EndoVis-2018 dataset and our three synthetic datasets.

3.2   Implementation Details

The classic state-of-the-art encoder-decoder network UNet [17] is used as our segmentation model backbone. We adopt a vanilla UNet architecture[3] with Pytorch [16] library and train the model with NVIDIA RTX3090 GPU. The batch size of 64, the learning rate of 0.001, and the Adam optimize are identically used for all experiments. The binary cross-entropy loss is adopted as the loss function. We use the Dice Similarity Coefficient (DSC) to evaluate the segmentation performance. The images are resized to 224×224 to save the training time. Besides, we refer to the implementation[4] of AugMix [12] to apply training-time chained augmentation mixing.

3.3   Results and Discussion

We evaluate the quality and effectiveness of our generated dataset with the EndoVis-2018 [1] and EndoVis-2017 [2] datasets, with the latter one considered as an unseen target domain because it does not contribute to our synthetic dataset generation. The results in Table 1 indicate that our methods can complete the segmentation task with acceptable performance for both datasets. As shown in Fig. 2, the instruments masks predicted by our models only have minimal visual discrepancy from the ground truth. Considering our datasets only depend on a few trivially collected source images and get rid of gathering and annotating hundreds of real data samples, the result is promising and revolutionary for low-cost and efficient surgical instrument segmentation.

Table 1. Overall results of the binary surgical instrument segmentation in DSC (%) with the EndoVis-2018 dataset and our three synthetic datasets. AM is short for the training-time augmentation mixing. Best results of ours are shown in bold.

Train	Test on EndoVis-2018			Test on EndoVis-2017		
	AM-None	AM-Soft	AM-Hard	AM-None	AM-Soft	AM-Hard
EndoVis-2018	81.58	83.15	82.91	83.21	84.06	83.43
Synthetic-A	56.82	66.74	71.03	72.74	72.23	65.13
Synthetic-B	57.37	69.42	72.53	72.65	73.21	72.41
Synthetic-C	**59.28**	**71.48**	**73.51**	**74.37**	**75.69**	**75.16**

3.4   Ablation Studies

To show the efficacy of our training-time chained augmentation mixing, we first conduct experiments with a relevant data augmentation technique - ColorJitter,

---

[3] https://github.com/ternaus/robot-surgery-segmentation.
[4] https://github.com/google-research/augmix.

which randomly changes the brightness, contrast, and saturation of an image. Training with the Synthetic-C dataset, our augmentation strategy outperforms ColorJitter significantly with 5.33% and 4.29% of DSC gain on EndoVis-2018 and EndoVis-2017 datasets.

We then study the effectiveness of training with synthetic data in handling the class-incremental issue in the deployment domain. Compared with EndoVis-2018 [1] dataset, there are two novel instruments in EndoVis-2017 [2], namely the Vessel Sealer and the Grasping Retractor. Following our proposed framework in Fig. 1, we generate 2000 synthetic images for the novel instruments and combine them with EndoVis-2018 [1] for training. As indicated in the highlighted area of Fig. 3(a), the model manages to handle the class-incremental problem to recognize the Vessel Sealer, with only minimal effort of adding synthesized images. The overall performance on the test domain improves significantly, as shown in Fig. 3(b).

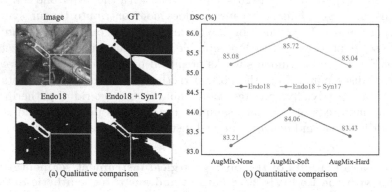

(a) Qualitative comparison   (b) Quantitative comparison

Fig. 3. Qualitative and quantitative results of the class-incremental case. The novel instrument Vessel Sealer is highlighted with the yellow rectangle in (a). The overall performance on EndoVis-2017 [2] gets greatly improved as shown in (b). (Color figure online)

While sufficient well-annotated datasets are not common in practice, a few high-quality data samples are normally feasible to acquire. We further investigate the effect of introducing a small portion of real images when training with synthetic data. We randomly fetch 10% and 20% of the EndoVis-2018 [1] dataset and combine it with our Synthetic-C dataset. The results in Table 2 indicate that only a small amount of real data could provide significant benefits. Compared with training with the real EndoVis-2018 [1] dataset, the models from the synthetic-real joint training scheme can efficiently achieve similar performance regarding adaptation and generalization.

Table 2. Results of synthetic-real joint training. Adding only a small portion of real data can greatly improve the segmentation performance.

Test	Train			
	Synthetic-C	Synthetic-C + 10% Endo18	Synthetic-C + 20% Endo18	Endo18
EndoVis-2018	73.51	80.65	82.45	82.91
EndoVis-2017	75.16	82.10	82.48	83.43

4 Conclusion

In this work, we reevaluate the surgical instrument segmentation and propose a cost-effective data-centric framework for synthetic dataset generation. Extensive experiments on two commonly seen real datasets demonstrate that our high-quality synthetic datasets are capable of surgical instrument segmentation with acceptable performance and generalization ability. Besides, we show that our method can handle domain shift and class incremental problems and greatly improve the performance when only a small amount of real data is available. Future work may be extended to more complicated instrument-wise segmentation and other medical applications. Besides, by considering more prior knowledge in practical surgical scenes, such as cautery smoke and instruments shadow, the quality of the synthetic dataset can be further improved.

Acknowledgements. This work was supported by the Shun Hing Institute of Advanced Engineering (SHIAE project BME-p1-21) at the Chinese University of Hong Kong (CUHK), Hong Kong Research Grants Council (RGC) Collaborative Research Fund (CRF C4026-21GF and CRF C4063-18G), (GRS)#3110167 and Shenzhen-Hong Kong-Macau Technology Research Programme (Type C 202108233000303).

References

1. Allan, M., et al.: 2018 robotic scene segmentation challenge (2020)
2. Allan, M., et al: 2017 robotic instrument segmentation challenge (2019)
3. Cao, B., Zhang, H., Wang, N., Gao, X., Shen, D.: Auto-gan: self-supervised collaborative learning for medical image synthesis. In: Proceedings of the AAAI Conference on Artificial Intelligence, vol. 34, pp. 10486–10493 (2020)
4. Colleoni, E., Edwards, P., Stoyanov, D.: Synthetic and real inputs for tool segmentation in robotic surgery. In: Martel, A.L., et al. (eds.) MICCAI 2020. LNCS, vol. 12263, pp. 700–710. Springer, Cham (2020). https://doi.org/10.1007/978-3-030-59716-0_67
5. Dobbs, R.W., Halgrimson, W.R., Talamini, S., Vigneswaran, H.T., Wilson, J.O., Crivellaro, S.: Single-port robotic surgery: the next generation of minimally invasive urology. World J. Urol. **38**(4), 897–905 (2020)
6. Domingos, P.: A few useful things to know about machine learning. Commun. ACM **55**(10), 78–87 (2012)

7. Eilertsen, G., Tsirikoglou, A., Lundström, C., Unger, J.: Ensembles of gans for synthetic training data generation (2021)
8. Garcia-Peraza-Herrera, L.C., Fidon, L., D'Ettorre, C., Stoyanov, D., Vercauteren, T., Ourselin, S.: Image compositing for segmentation of surgical tools without manual annotations. IEEE Trans. Med. Imaging **40**(5), 1450–1460 (2021)
9. Goodfellow, I., Bengio, Y., Courville, A.: Deep Learning. MIT press, Cambridge (2016)
10. Hamghalam, M., Lei, B., Wang, T.: High tissue contrast MRI synthesis using multi-stage attention-gan for segmentation. In: Proceedings of the AAAI Conference on Artificial Intelligence, vol. 34, pp. 4067–4074 (2020)
11. Han, C., et al.: Synthesizing diverse lung nodules wherever massively: 3d multi-conditional gan-based CT image augmentation for object detection. In: 2019 International Conference on 3D Vision (3DV), pp. 729–737. IEEE (2019)
12. Hendrycks, D., Mu, N., Cubuk, E.D., Zoph, B., Gilmer, J., Lakshminarayanan, B.: Augmix: a simple data processing method to improve robustness and uncertainty. arXiv preprint arXiv:1912.02781 (2019)
13. Jung, A.B., et al.: imgaug. https://github.com/aleju/imgaug. Accessed 01 Feb 2020 (2020)
14. Kishore, A., Choe, T.E., Kwon, J., Park, M., Hao, P., Mittel, A.: Synthetic data generation using imitation training. In: Proceedings of the IEEE/CVF International Conference on Computer Vision, pp. 3078–3086 (2021)
15. Madan, S., et al.: When and how do cnns generalize to out-of-distribution category-viewpoint combinations? arXiv preprint arXiv:2007.08032 (2020)
16. Paszke, A., et al.: Automatic differentiation in pytorch. In: NIPS-W (2017)
17. Ronneberger, O., Fischer, P., Brox, T.: U-Net: convolutional networks for biomedical image segmentation. In: Navab, N., Hornegger, J., Wells, W.M., Frangi, A.F. (eds.) MICCAI 2015. LNCS, vol. 9351, pp. 234–241. Springer, Cham (2015). https://doi.org/10.1007/978-3-319-24574-4_28
18. Shin, H.-C., et al.: Medical image synthesis for data augmentation and anonymization using generative adversarial networks. In: Gooya, A., Goksel, O., Oguz, I., Burgos, N. (eds.) SASHIMI 2018. LNCS, vol. 11037, pp. 1–11. Springer, Cham (2018). https://doi.org/10.1007/978-3-030-00536-8_1
19. Tremblay, J., et al.: Training deep networks with synthetic data: bridging the reality gap by domain randomization. In: Proceedings of the IEEE Conference on Computer Vision and Pattern Recognition Workshops, pp. 969–977 (2018)
20. Xu, M., Islam, M., Lim, C.M., Ren, H.: Class-incremental domain adaptation with smoothing and calibration for surgical report generation. In: de Bruijne, M., Cattin, P.C., Cotin, S., Padoy, N., Speidel, S., Zheng, Y., Essert, C. (eds.) MICCAI 2021. LNCS, vol. 12904, pp. 269–278. Springer, Cham (2021). https://doi.org/10.1007/978-3-030-87202-1_26

Free Lunch for Surgical Video Understanding by Distilling Self-supervisions

Xinpeng Ding[1], Ziwei Liu[2], and Xiaomeng Li[1,3(✉)]

[1] The Hong Kong University of Science and Technology, Kowloon, Hong Kong
eexmli@ust.hk
[2] S-Lab, Nanyang Technological University, Singapore, Singapore
[3] The Hong Kong University of Science and Technology Shenzhen Research Institute,
Kowloon, Hong Kong

Abstract. Self-supervised learning has witnessed great progress in vision and NLP; recently, it also attracted much attention to various medical imaging modalities such as X-ray, CT, and MRI. Existing methods mostly focus on building new pretext self-supervision tasks such as reconstruction, orientation, and masking identification according to the properties of medical images. However, the publicly available self-supervision models are not fully exploited. In this paper, we present a powerful yet efficient self-supervision framework for surgical video understanding. Our key insight is to distill knowledge from *publicly available models trained on large generic datasets* (For example, the released models trained on ImageNet by MoCo v2: https://github.com/facebookresearch/moco) to facilitate the self-supervised learning of surgical videos. To this end, we first introduce a semantic-preserving training scheme to obtain our teacher model, which not only contains semantics from the publicly available models, but also can produce accurate knowledge for surgical data. Besides training with only contrastive learning, we also introduce a distillation objective to transfer the rich learned information from the teacher model to self-supervised learning on surgical data. Extensive experiments on two surgical phase recognition benchmarks show that our framework can significantly improve the performance of existing self-supervised learning methods. Notably, our framework demonstrates a compelling advantage under a low-data regime. Our code is available at https://github.com/xmed-lab/DistillingSelf.

Keywords: Surgical videos · Self-supervised learning · Knowledge distillation

1 Introduction

Generally, training deep neural networks requires a large amount of labeled data to achieve great performance. However, obtaining the annotation for surgical videos is expensive as it requires professional knowledge from surgeons. To

L. Wang (Eds.): MICCAI 2022, LNCS 13437, pp. 365–375, 2022.
https://doi.org/10.1007/978-3-031-16449-1_35

Fig. 1. Left: Existing self-supervised learning approaches aim to design different pretext tasks on a larger dataset to improve the representation ability. **Right:** We proposed a novel distilled contrastive learning to transfer the powerful representation ability from the free available models to the surgical video self-training.

address this problem, self-supervised learning, *i.e.*, training the model on a large dataset (*e.g.*, ImageNet [20]) without annotations and fine-tuning the classifier on the target datasets, has achieved great success for image recognition and video understanding [2,4,10,19]. This paper focuses on designing self-supervised learning methods for surgical video understanding with a downstream task - surgical phase recognition, which aims to predict what phase is occurring for each frame in a video [1,6,9,13–15,24,25,28,29].

Self-supervised learning has been widely applied into various medical images, such as X-ray [30], fundus images [16,17], CT [34] and MRI [30,32]. For example, Zhuang *et al.* [34] developed a rubik's cube playing pretext task to learn 3D representation for CT volumes. Rubik's cube+ [33] introduced more pretext tasks, *i.e.*, cube ordering, cube orientation and masking identification to improve the self-supervised learning performance. Some researchers introduced reconstruction of corrupted images [3,32] or triplet loss [26] for self-supervised learning on nuclei images. Recently, motivated by the great success of contrastive learning in computer vision [2,4,5,10,23], Zhou *et al.* [31] introduced the contrastive loss to 2D radiographs. Taleb *et al.* [22] further improved contrastive predictive coding [18] to a 3D manner for training with 3D medical images. To learn more detailed context, PCRL [30] combined the construction loss with the contrastive one. However, all of these above approaches aim to design different pretext tasks to perform self-supervised learning on medical datasets, as shown in Fig. 1(a).

In this paper, we make a crucial observation that *using the same backbone and self-supervised learning method, the model trained with ImageNet data can yield a comparable performance for surgical phase recognition with that trained with surgical video data (82.3% vs 85.7% Acc.)*; see details in Table 1. This surprising result indicates that the model self-supervised trained with ImageNet data can learn useful semantics that benefit surgical video understanding. There are many publicly available self-supervised models, motivating us to leverage the free knowledge to facilitate the self-supervised learning of the surgical video.

To this end, we propose a novel distilled self-supervised learning method for surgical videos, which distillates the free knowledge from *publicly available self-supervised models (e.g., MoCo v2 trained on the ImageNet)* to improve the self-supervised learning on surgical videos, as shown in Fig. 1(b). We first introduce a semantic-preserving training to train a teacher model, which retains representation ability from ImageNet while producing accurate semantics knowledge for surgical data. Besides training the student model with the only contrastive objective, we boost the self-supervised learning on the surgical data with a distillation objective. The proposed distillation objective forces the similarity matrix of the teacher and the student models to be consistent, which makes the student model to learn extra semantics.

We summarize our key contributions as follows:

- To best of our knowledge, we are the first to investigate the use of self-supervised training on surgical videos. Instead of designing a new pretext task, we provide a new insight that to transfer knowledge from large public dataset improves self-training on surgical data.
- We propose a semantic-preserving training to train a teacher model, which contains representation ability from ImageNet while producing accurate semantic information for the surgical data.
- We propose a distillation objective to enforce the similarity matrix between the teacher model and the student model to be consistent, which makes the student model to learn extra information during self-training.

2   Methodology

This section is divided into three main parts. Firstly, we introduce contrastive learning to perform self-supervised learning on surgical videos in Sect. 2.1. Then, the semantic-preserving training will be presented in Sect. 2.2. Finally, we will describe the distilled self-supervised learning, which transfers useful semantics from publicly available models to the self-training on surgical videos.

2.1   Contrastive Learning on Surgical Videos

Given a frame v sampled from the surgical videos, we apply two different transformations to it, which can be formally as:

$$x^q = t_q(v), \quad x^k = t_k(v), \tag{1}$$

where $t_q(\cdot)$ and $t_k(\cdot)$ are two transformations, sampled from the same transformation distribution. Then, we feed x^q into a query encoder $f_q(\cdot)$, followed by a projection head $f_h(\cdot)$ consists of 2-layer multilayer perceptron (MLP), to obtain the query representation $q = h_q(f_q(x^q))$. Similarly, we can obtain $k_+ = h_k(f_k(x^k))$, where k_+ indicates the positive embedding for q. Similar as [4,10], we obtain a set of encoded samples $B = \{k_i\}_{i=1}^{M}$ that are the keys of a dictionary, where k_i is the i-th key in the dictionary, and M is the size of the dictionary. There is a single key in the dictionary, i.e., k_+, that matches q. In this paper, we use InfoNCE [18], which measures the similarity via dot production, and our contrastive objective is as follows:

$$\mathcal{L}_{con} = -\sum_{i=1}^{N} \log \frac{\exp\left(\text{sim}(q, k_+)/\tau\right)}{\sum_{i=1}^{M} \exp\left(\text{sim}(q, k_i)/\tau\right)}, \tag{2}$$

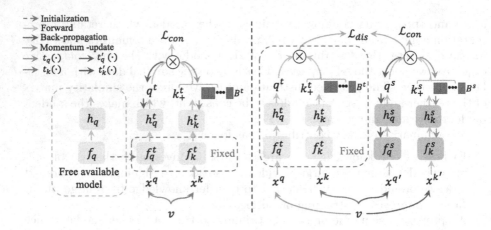

Fig. 2. Left: A semantic-preserving training is adopted to train a teacher model, *i.e.*, f_q^t, f_k^t, h_q^t and h_k^t, from the free available model. **Right:** We introduce a distilled self-supervised learning to distill the knowledge from the teacher model to the student model, *i.e.*, f_q^s, f_k^s, h_q^s and h_k^s.

where $\text{sim}(q, k_+)$ is the cosine similarity between q and k_+, and τ is the temperature that controls the concentration level of the sample distribution [7]. In this paper, the parameters of the query encoder are updated by back-propagation, while the key encoder is momentum-updated encoder [4,10]. Let define the parameters of $f_q(\cdot)$, $h_q(\cdot)$, $f_k(\cdot)$ and $h_k(\cdot)$ as $\theta_q^f(\cdot)$, $\theta_q^h(\cdot)$, $\theta_k^f(\cdot)$ and $\theta_k^h(\cdot)$ respectively. Then, the momentum-updating can be formulated as:

$$\theta_k^f \leftarrow m\theta_k^f + (1-m)\theta_q^f, \tag{3}$$

$$\theta_k^h \leftarrow m\theta_k^h + (1-m)\theta_q^h, \tag{4}$$

where $m \in [0, 1)$ is a momentum coefficient, and is set to 0.999 following [4].

2.2 Semantic-Preserving Training of Free Models

Our motivation is to leverage the semantic information from public free self-supervised model (*i.e.*, a teacher model) trained on large datasets (*e.g.*, ImageNet) to improve the self-supervised learning, (*i.e.*, a student model) on surgical videos. To this end, the teacher model should have two properties. (1) It should retain representation ability learned from ImageNet. Directly self-training on surgical video would make the model forget the learned knowledge from ImageNet. (2) It should produce accurate structured knowledge for surgical videos. Since there is a clear domain gap between ImageNet and surgical videos, transferring the noisy distilled information would degrade the performance.

To achieve these two properties, we propose the semantic-preserving training of free models on surgical video, which is shown in the left of Fig. 2. More specifically, we first initialize the parameters of $f_q(\cdot)$ and $f_k(\cdot)$ from the ImageNet pre-trained model, *i.e.*, using the weights of the free self-supervised model: Moco v2, which can be formulated as $f_q(\cdot) \to f_q^t(\cdot)$ and $f_k(\cdot) \to f_k^t(\cdot)$. During semantic-preserving training, we fixed the parameters of the backbone, $i.e.$, $f_q^t(\cdot)$ and $f_k^t(\cdot)$, to maintain the prior learned semantics. Furthermore, to produce accurate predictions for latter distillation under the large domain gap, we conduct contrastive learning on surgical video to update the parameters of the two-layer MLP, $i.e.$, $h_q^t(\cdot)$ and $h_k^t(\cdot)$, via the contrastive objective. In this way, the trained model would learn to use the prior knowledge from ImageNet to discriminate the surgical frames, and the predictions from it would contain rich semantic information.

2.3   Distilled Self-supervised Learning

Unlike existing self-supervised learning approaches that focus on designing new pretext tasks, we aim to transfer the semantics knowledge from free public self-supervised models into the surgical video self-training. As shown in the right of Fig. 2, besides the standard contrastive learning, we also introduce a distillation objective to transfer the semantics from the teacher model (trained in Sect. 2.2) to the student model, $i.e.$, the model learned by self-supervised on surgical videos. The similarity matrix ($i.e.$, $\mathrm{sim}(q, k_i)$) obtained in Eq. 2 measures the similarities between the query and its keys. The teacher model can generate around 82.3% Acc. for surgical phase recognition, showing that the similarity matrix contains useful semantics information that can benefit to surgical video understanding.

To leverage this information, our idea is to force the similarity matrix of the teacher and student model to be consistent; therefore, the teacher model can provide additional supervision for training the student model, leading to performance improvement. Specifically, we regard the similarity matrix of the teacher model as the soft targets to supervise the self-training of the student. To obtain the soft targets, we apply Softmax with the temperature scale τ to the similarity matrix of the teacher model. We find that the soft targets can be formulated as:

$$p_i^t = \frac{\exp\left(sim(q^t, k_i^t)/\tau\right)}{\sum_{i=j}^M \exp\left(sim(q^t, k_j^t)/\tau\right)}, \tag{5}$$

which is very similar to Eq. 2. Similarly, we can also obtain the similarity matrix of the student model, which can be defined as p_i^s. Finally, the distillation loss is computed by the KL-divergence loss to measure the distribution of p_i^t and p_i^s as follows:

$$\mathcal{L}_{dis} = \sum_{i=1}^M p_i^t \log\left(\frac{p_i^t}{p_i^s}\right), \tag{6}$$

where p_i^s is the similarity matrix of the student model, computed as in Eq. 5. Note that the distillation loss can be regarded as a soft version of the contrastive loss (defined in Eq. 2). This soft version provides much more information per training case than hard targets [12].

Combined with the distillation loss and the contrastive loss, the overall object of the distilled self-supervised can be formulated as:

$$\mathcal{L} = \mathcal{L}_{con} + \lambda\mathcal{L}_{dis}, \tag{7}$$

where λ is the trade-off hyper-parameter.

3   Experiments

3.1   Datasets

Cholec80. The Cholec80 dataset [24] contains 80 cholecystectomy surgeries videos. All videos are recorded as 25 fps, and the resolution of each frame is 1920×1080 or 854×480. We sample the videos into 5 fps for reducing computation cost. Following the same setting in previous works, we split the datasets into 40 videos for training and rest 40 videos for testing [14,28].

MI2CAI16. There are 41 videos in the MI2CAI16 [21] dataset, and the videos in it are recorded at 25 fps. Among them, 27 videos are used for training and 14 videos are used for test, following previous approaches [14,28].

3.2   Implementation Details

We use the ResNet50 [11] as the backbone to extract features for each frame. After that, following [6,24,28], a multi-stage temporal convolution (MS-TCN) [8] is used to extracted temporal relations for frame features and predicts phase results. Following the same evaluation protocols [13–15,24], we employ four commonly-used metrics, i.e., accuracy (AC), precision (PR), recall (RE), and Jaccard (JA) to evaluate the phase prediction accuracy. To evaluate the self-supervised training performance, we self-train the ResNet50 backbone on surgical videos on training set of Cholec80 by different self-supervised learning approaches [2,4,5,27]. Then, we linearly fine-tune the self-trained model on the training set of Cholec80 and MI2CAI16. After that, we extract features for each frame of videos by the fine-tuned model. Finally, based on the fixed extracted features, we use MS-TCN to predict results for surgical phase recognition. We set τ to 0.07 and λ to 5 respectively.

3.3   Comparisons with the State-of-the-Art Methods

To prove the effect of our proposed method, we compare our method with the state-of-the-art self-supervised learning approaches, e.g., MoCo v2 [4], SwAV [2], SimSiam [5] and RegionCL [27]. The results are shown in Table 1. Note that, our method uses MoCo v2 [4] as the self-supervised learning baseline. We find that self-supervised training on ImageNet outperforms the model trained from scratch with a clear margin. This indicates the powerful representation of the ImageNet pre-training, and motivates us to leverage semantics from it to improve the self-supervised training on surgical data. It can be also found that self-supervised

Table 1. Results on Cholec80 dataset and M2CAI16 dataset. "Scratch" indicates training from scratch. "IN-Sup" and "IN-MoCo v2" refer to supervised learning and self-supervised learning on ImageNet, respectively.

Methods	Cholec80				M2CAI16			
	Accuracy	Precision	Recall	Jaccard	Accuracy	Precision	Recall	Jaccard
Scratch	65.0 ± 13.8	67.7 ± 14.3	53.0 ± 21.4	40.7 ± 18.4	59.8 ±12.9	65.0 ± 13.6	56.6 ± 29.7	40.0 ± 23.0
Models Trained on ImageNet								
IN-Sup	83.1 ± 10.5	64.7 ± 11.0	82.3 ± 7.6	77.8 ± 9.8	76.0 ± 6.3	78.8 ± 5.2	73.8 ± 17.6	58.9 ± 15.0
IN-MoCov2	82.3 ± 9.8	64.2 ± 12.0	81.2 ± 6.1	77.7 ± 10.5	75.1 ± 5.8	77.8 ± 6.4	73.5 ± 10.4	58.2 ± 14.2
Models Trained on Surgical Videos								
MoCo v2 [4]	85.7 ± 10.4	72.0 ± 9.1	**85.4 ±6.3**	84.9 ± 5.4	78.4 ± 4.6	80.5 ± 5.1	78.5 ± 12.6	63.5 ± 13.6
SwAV [2]	84.7 ± 2.8	71.6 ± 7.9	84.9 ± 7.6	84.2 ± 6.3	77.3 ± 6.1	79.9 ± 6.3	77.8 ± 11.3	63.1 ± 11.5
SimSiam [5]	83.9 ± 6.7	68.5 ± 5.5	84.5 ± 5.0	82.6 ± 5.7	77.0 ± 5.3	79.1 ± 6.2	77.1 ± 8.2	62.8 ± 9.7
RegionCL [27]	84.8 ± 11.4	69.8 ± 11.1	85.0 ± 6.5	82.8 ± 10.9	78.2 ± 10.4	79.9 ± 8.1	78.0 ± 9.5	63.4 ± 7.4
Ours	**87.3 ± 9.5**	**73.5 ± 9.9**	85.1 ± 7.6	**86.1 ± 6.9**	**81.1 ± 6.5**	**80.7 ± 8.3**	**83.2 ± 6.9**	**67.9 ± 5.1**

Table 2. Ablation study on different transferring methods on Cholec80 dataset.

Method	Accuracy	Precision	Recall	Jaccard
Addition	84.0 ± 10.5	67.3 ± 11.1	82.7 ± 8.7	81.6 ± 8.3
Concatenation	84.2 ± 9.4	67.9 ± 11.7	83.1 ± 8.6	81.0 ± 10.1
Initialization	85.5 ± 10.9	71.7 ± 8.1	83.8 ± 7.2	83.2 ± 8.4
Distillation	**87.3 ± 9.5**	**73.5 ± 9.9**	**85.1 ± 7.6**	**86.1 ± 6.9**

training on surgical videos outperforms that training on ImageNet [20]. Combined with the distillation of semantics from ImageNet, our method can improve the baseline, *i.e.*, MoCo v2, over 1.6% on Cholec80.

3.4 Ablation Studies

Effect of Distillation. Table 2 compares the performance of different transferring methods on Cholec80. Specifically, we denote the feature maps generated from the teacher model and our model as $\mathbf{F}^t$ and $\mathbf{F}^s$ respectively. 'Addition' means the sum of $\mathbf{F}^t$ and $\mathbf{F}^s$, *i.e.*, $\mathbf{F}^t + \mathbf{F}^s$. Similarly, 'Concatenation' indicates the concatenation of $\mathbf{F}^t$ and $\mathbf{F}^s$. 'Initialization' means our model is initialized by the teacher model. 'Distillation' means using the distillation objective defined in Eq. 6. It is clear that the distillation model is the best way to transfer the knowledge from the teacher model.

Effect of the Semantic-Preserving Training of Free Models. We conduct an ablation study on Cholec80 to evaluate the effect of the semantic-preserving training, and report the results in Table 3. It is clear that not training the teacher model, *i.e.*, the first row in Table 3, cannot predict correct knowledge for surgical data, and would degrade the performance, *e.g.*, only achieve 84.7% accuracy. Furthermore, training too much would hurt the semantics learned from ImageNet; See the third and the last rows in Table 3. We can find that our proposed semantic-preserving training (only updating parameters of the head projection

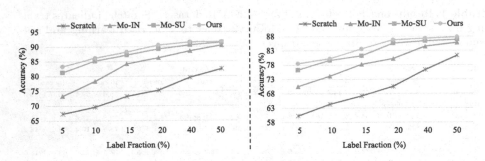

Fig. 3. Accuracy for surgical phase recognition for different methods and varied sizes of label fractions on **Left**: Cholec80 and **Right**: MI2CAI16. "Scratch" indicates training from scratch. "Mo-IN" and "Mo-SU" refer to training from the model pre-trained on ImageNet and surgical data by MoCo v2, respectively.

Table 3. Ablation study on the semantic-preserving training on Cholec80 dataset. 'Backbone' indicates training the backbone of the model, and 'Head' means training the head projection of the model.

Backbone	Head	Accuracy	Precision	Recall	Jaccard
✗	✗	84.7 ± 11.0	67.5 ± 11.4	84.2 ± 8.2	79.6 ± 8.8
✗	✓	**87.3 ± 9.5**	**73.5 ± 9.9**	85.1 ± 7.6	**86.1 ± 6.9**
✓	✗	86.1 ± 11.8	72.4 ± 9.4	**86.1 ± 6.5**	84.8 ± 9.3
✓	✓	85.2 ± 10.9	71.3 ± 10.4	85.1 ± 5.7	85.8 ± 6.5

(*i.e.*, h_q^t and h_k^t) and fixing the backbone (*i.e.*, f_q^t and f_k^t) can achieve the best performance. Since trained with this way, the teacher model would not only maintain the semantics learned from ImageNet, but also can predict accurate knowledge for transferring.

Label-Efficiency Analysis. To evaluate the label-efficiency of our self-training model, we fine-tune the pre-trained model on different fractions of labeled training data. Note that in this ablation study, we train all parameters of the pre-trained model. As shown in Fig. 3, our proposed approach achieves the best label-efficiency performance. Furthermore, fine-tuning with fewer labels, the improvement gain is larger.

4   Conclusion

Unlike existing self-supervised papers that design new pretext tasks, this paper proposes a novel free lunch method for self-supervised learning with surgical videos by distilling knowledge from free available models. Our motivation is that the publicly released model trained on natural images shows comparable performance with the model trained on the surgical data. To leverage the rich semantics from the available models, we propose a semantic-perseving method to train a teacher model that contains prior information and can produce accurate predictions for surgical data. Finally, we introduce a distillation objective to the self-training on surgical data to enable the model to learn extra knowledge from the teacher model. Experimental results indicate that our method can improve the performance of the existing self-supervised learning methods.

Acknowledgement. This work was supported by a research grant from HKUST-BICI Exploratory Fund (HCIC-004) and a research grant from Shenzhen Municipal Central Government Guides Local Science and Technology Development Special Funded Projects (2021Szvup139), and under the RIE2020 Industry Alignment Fund - Industry Collaboration Projects (IAF-ICP) Funding Initiative, as well as cash and in-kind contribution from the industry partner(s).

References

1. Blum, T., Feußner, H., Navab, N.: Modeling and segmentation of surgical workflow from laparoscopic video. In: Jiang, T., Navab, N., Pluim, J.P.W., Viergever, M.A. (eds.) MICCAI 2010. LNCS, vol. 6363, pp. 400–407. Springer, Heidelberg (2010). https://doi.org/10.1007/978-3-642-15711-0_50
2. Caron, M., Misra, I., Mairal, J., Goyal, P., Bojanowski, P., Joulin, A.: Unsupervised learning of visual features by contrasting cluster assignments. Adv. Neural. Inf. Process. Syst. **33**, 9912–9924 (2020)
3. Chen, L., Bentley, P., Mori, K., Misawa, K., Fujiwara, M., Rueckert, D.: Self-supervised learning for medical image analysis using image context restoration. Med. Image Anal. **58**, 101539 (2019)
4. Chen, X., Fan, H., Girshick, R., He, K.: Improved baselines with momentum contrastive learning. arXiv preprint arXiv:2003.04297 (2020)
5. Chen, X., He, K.: Exploring simple siamese representation learning. In: Proceedings of the IEEE/CVF Conference on Computer Vision and Pattern Recognition, pp. 15750–15758 (2021)
6. Ding, X., Li, X.: Exploiting segment-level semantics for online phase recognition from surgical videos. arXiv preprint arXiv:2111.11044 (2021)
7. Ding, X., et al.: Support-set based cross-supervision for video grounding. In: Proceedings of the IEEE/CVF International Conference on Computer Vision, pp. 11573–11582 (2021)
8. Farha, Y.A., Gall, J.: MS-TCN: multi-stage temporal convolutional network for action segmentation. In: Proceedings of the IEEE/CVF Conference on Computer Vision and Pattern Recognition, pp. 3575–3584 (2019)
9. Gao, X., Jin, Y., Long, Y., Dou, Q., Heng, P.A.: Trans-SVNet: accurate phase recognition from surgical videos via hybrid embedding aggregation transformer. arXiv preprint arXiv:2103.09712 (2021)
10. He, K., Fan, H., Wu, Y., Xie, S., Girshick, R.: Momentum contrast for unsupervised visual representation learning. In: Proceedings of the IEEE/CVF Conference on Computer Vision and Pattern Recognition, pp. 9729–9738 (2020)
11. He, K., Zhang, X., Ren, S., Sun, J.: Deep residual learning for image recognition. In: Proceedings of the IEEE Conference on Computer Vision and Pattern Recognition, pp. 770–778 (2016)
12. Hinton, G., Vinyals, O., Dean, J., et al.: Distilling the knowledge in a neural network. arXiv preprint arXiv:1503.02531, vol. 2, no. 7 (2015)

13. Jin, Y., et al.: SV-RCNet: workflow recognition from surgical videos using recurrent convolutional network. IEEE Trans. Med. Imaging **37**(5), 1114–1126 (2017)
14. Jin, Y., et al.: Multi-task recurrent convolutional network with correlation loss for surgical video analysis. Med. Image Anal. **59**, 101572 (2020)
15. Jin, Y., Long, Y., Chen, C., Zhao, Z., Dou, Q., Heng, P.A.: Temporal memory relation network for workflow recognition from surgical video. IEEE Trans. Med. Imaging **40**(7), 1911–1923 (2021)
16. Li, X., et al.: Rotation-oriented collaborative self-supervised learning for retinal disease diagnosis. IEEE Trans. Med. Imaging **40**(9), 2284–2294 (2021)
17. Li, X., Jia, M., Islam, M.T., Yu, L., Xing, L.: Self-supervised feature learning via exploiting multi-modal data for retinal disease diagnosis. IEEE Trans. Med. Imaging **39**(12), 4023–4033 (2020)
18. Van den Oord, A., Li, Y., Vinyals, O.: Representation learning with contrastive predictive coding. arXiv e-prints pp. arXiv-1807 (2018)
19. Pan, T., Song, Y., Yang, T., Jiang, W., Liu, W.: Videomoco: contrastive video representation learning with temporally adversarial examples. In: Proceedings of the IEEE/CVF Conference on Computer Vision and Pattern Recognition, pp. 11205–11214 (2021)
20. Russakovsky, O., et al.: Imagenet large scale visual recognition challenge. Int. J. Comput. Vision **115**(3), 211–252 (2015)
21. Stauder, R., Ostler, D., Kranzfelder, M., Koller, S., Feußner, H., Navab, N.: The TUM LapChole dataset for the M2CAI 2016 workflow challenge. arXiv preprint arXiv:1610.09278 (2016)
22. Taleb, A., et al.: 3D self-supervised methods for medical imaging. Adv. Neural. Inf. Process. Syst. **33**, 18158–18172 (2020)
23. Tian, Y., Krishnan, D., Isola, P.: Contrastive multiview coding. In: Vedaldi, A., Bischof, H., Brox, T., Frahm, J.-M. (eds.) ECCV 2020. LNCS, vol. 12356, pp. 776–794. Springer, Cham (2020). https://doi.org/10.1007/978-3-030-58621-8_45
24. Twinanda, A.P., Shehata, S., Mutter, D., Marescaux, J., De Mathelin, M., Padoy, N.: EndoNet: a deep architecture for recognition tasks on laparoscopic videos. IEEE Trans. Med. Imaging **36**(1), 86–97 (2016)
25. Wang, Z., Ding, X., Zhao, W., Li, X.: Less is more: surgical phase recognition from timestamp supervision. arXiv preprint arXiv:2202.08199 (2022)
26. Xie, X., Chen, J., Li, Y., Shen, L., Ma, K., Zheng, Y.: Instance-aware self-supervised learning for nuclei segmentation. In: Martel, A.L., et al. (eds.) MICCAI 2020. LNCS, vol. 12265, pp. 341–350. Springer, Cham (2020). https://doi.org/10.1007/978-3-030-59722-1_33
27. Xu, Y., Zhang, Q., Zhang, J., Tao, D.: RegionCL: can simple region swapping contribute to contrastive learning? arXiv preprint arXiv:2111.12309 (2021)
28. Yi, F., Jiang, T.: Not end-to-end: explore multi-stage architecture for online surgical phase recognition. arXiv preprint arXiv:2107.04810 (2021)
29. Zappella, L., Béjar, B., Hager, G., Vidal, R.: Surgical gesture classification from video and kinematic data. Med. Image Anal. **17**(7), 732–745 (2013)
30. Zhou, H.Y., Lu, C., Yang, S., Han, X., Yu, Y.: Preservational learning improves self-supervised medical image models by reconstructing diverse contexts. In: Proceedings of the IEEE/CVF International Conference on Computer Vision, pp. 3499–3509 (2021)
31. Zhou, H.-Y., Yu, S., Bian, C., Hu, Y., Ma, K., Zheng, Y.: Comparing to learn: surpassing imagenet pretraining on radiographs by comparing image representations. In: Martel, A.L., et al. (eds.) MICCAI 2020. LNCS, vol. 12261, pp. 398–407. Springer, Cham (2020). https://doi.org/10.1007/978-3-030-59710-8_39

32. Zhou, Z., Sodha, V., Pang, J., Gotway, M.B., Liang, J.: Models genesis. Med. Image Anal. **67**, 101840 (2021)
33. Zhu, J., Li, Y., Hu, Y., Ma, K., Zhou, S.K., Zheng, Y.: Rubik's Cube+: a self-supervised feature learning framework for 3D medical image analysis. Med. Image Anal. **64**, 101746 (2020)
34. Zhuang, X., Li, Y., Hu, Y., Ma, K., Yang, Y., Zheng, Y.: Self-supervised feature learning for 3D medical images by playing a Rubik's cube. In: Shen, D., et al. (eds.) MICCAI 2019. LNCS, vol. 11767, pp. 420–428. Springer, Cham (2019). https://doi.org/10.1007/978-3-030-32251-9_46

Rethinking Surgical Captioning: End-to-End Window-Based MLP Transformer Using Patches

Mengya Xu[1,2,3], Mobarakol Islam[4], and Hongliang Ren[1,2,3(✉)]

[1] Department of Electronic Engineering, The Chinese University of Hong Kong, Shatin, Hong Kong SAR, China
hlren@ee.cuhk.edu.hk
[2] Department of Biomedical Engineering, National University of Singapore, Singapore, Singapore
mengya@u.nus.edu
[3] National University of Singapore (Suzhou) Research Institute (NUSRI), Suzhou, China
[4] BioMedIA Group, Department of Computing, Imperial College London, London, UK
m.islam20@imperial.ac.uk

Abstract. Surgical captioning plays an important role in surgical instruction prediction and report generation. However, the majority of captioning models still rely on the heavy computational object detector or feature extractor to extract regional features. In addition, the detection model requires additional bounding box annotation which is costly and needs skilled annotators. These lead to inference delay and limit the captioning model to deploy in real-time robotic surgery. For this purpose, we design an end-to-end detector and feature extractor-free captioning model by utilizing the patch-based shifted window technique. We propose **Sh**ifted **Win**dow-Based **M**ulti-**L**ayer **P**erceptrons **Tran**sformer **Cap**tioning model (SwinMLP-TranCAP) with faster inference speed and less computation. SwinMLP-TranCAP replaces the multi-head attention module with window-based multi-head MLP. Such deployments primarily focus on image understanding tasks, but very few works investigate the caption generation task. SwinMLP-TranCAP is also extended into a video version for video captioning tasks using 3D patches and windows. Compared with previous detector-based or feature extractor-based models, our models greatly simplify the architecture design while maintaining performance on two surgical datasets. The code is publicly available at https://github.com/XuMengyaAmy/SwinMLP_TranCAP.

M. Xu and M. Islam—Equal technical contribution.

1 Introduction

Automatic surgical captioning is a prerequisite for intra-operative context-aware surgical instruction prediction and post-operative surgical report generation. Despite the impressive performance, most approaches require heavy computational resources for the surgical captioning task, which limits the real-time deployment. The main-stream captioning models contain an expensive pipeline of detection and feature extraction modules before captioning. For example, Meshed-Memory Transformer [4], self-sequence captioning [13] entail bounding box from detection model (Faster R-CNN [12]) to extract object feature using a feature extractor of ResNet-101 [7]. On the other hand, bounding box annotations are used to extract regional features for surgical report generation [17,18]. However, these kinds of approaches arise following issues: (i) require bounding box annotation which is challenging in the medical scene as it requires the use of professional annotators, (ii) region detection operation leads to high computational demand, and (iii) cropping regions may ignore some crucial background information and destroy the spatial correlation among the objects. Thus recent vision-language studies [20] are moving toward the detector-free trend by only utilizing image representations from the feature extractor. Nonetheless, the feature extractor still exists as an intermediate module which unavoidably leads to inadequate training and long inference delay at the prediction stage. These issues restrict the application for real-time deployment, especially in robotic surgery.

To achieve end-to-end captioning framework, ViTCAP model [6] uses the Vision Transformer (ViT) [5] which encodes image patches as grid representations. However, it is very computing intensive even for a reasonably large-sized image because the model has to compute the self-attention for a given patch with all the other patches in the input image. Most recently, the window-based model Swin Transformer [10] outperforms ViT [5] and reduces the computation cost by performing local self-attention between patches within the window. The window is also shifted to achieve information sharing across various spatial locations. However, this kind of approach is yet to explore for captioning tasks.

So far, many Transformer-variants are still based on the common belief that the attention-based token mixer module contributes most to their performance. However, recent studies [15] show that the attention-based module in Transformer can be replaced by spatial multi-layer perceptron (MLPs) and the resulting models still obtain quite good performance. This observation suggests that the general architecture of the transformer, instead of the specific attention-based token mixer module, is more crucial to the success of the model. Based on this hypothesis, [19] replaces the attention-based module with the extremely simple spatial pooling operator and surprisingly finds that the model achieves competitive performance. However, such exploration is mainly concentrated on the image understanding tasks and remains less investigated for the caption generation task.

Our contributions can be summed up as the following points: 1) We design SwinMLP-TranCAP, a detector and feature extractor-free captioning models by utilizing the shifted window-based MLP as the backbone of vision encoder and a Transformer-like decoder; 2) We also further develop a Video SwinMLP-TranCAP by using 3D patch embedding and windowing to achieve the video captioning task; 3) Our method is validated on publicly available two captioning datasets and obtained superior performance in both computational speed and caption generation over conventional approaches; 4) Our findings suggest that the captioning performance mostly relies on transformer type architecture instead of self-attention mechanism.

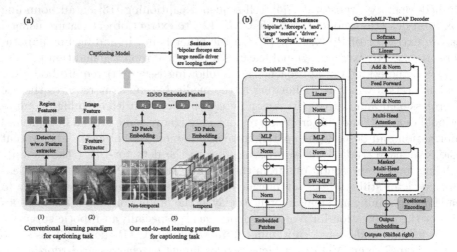

Fig. 1. (a) Comparisons of conventional learning paradigm and our learning paradigm for captioning task and (b) our proposed end-to-end SwinMLP-TranCAP model. (1) Captioning models based on an object detector w/w.o feature extractor to extract region features. (2) To eliminate the detector, the feature extractor can be applied as a compromise to the output image feature. (c) To eliminate the detector and feature extractor, the captioning models can be designed to take the patches as the input representation directly.

2   Methodology

2.1   Preliminaries

Vision Transformer. Vision Transformer (ViT) [5] divides the input image into several non-overlapping patches with a patch size of 16×16. The feature dimension of each patch is $16 \times 16 \times 3 = 768$. We compute the self-attention for a given patch with all the other patches in the input image. This becomes very compute-intensive even for a reasonably large-sized image: $\Omega(MSA) = 4hwC^2 + 2(hw)^2$, where hw indicates the number of patches, C stands for the embedding dimension.

Swin Transformer. The drawback of Multi-Head Self Attention (MSA) in ViT is attention calculation is very compute-heavy. To overcome it, Swin Transformer [10] introduces "window" to perform local self-attention computation. Specifically, a single layer of transformer is replaced by two layers which design Window-based MSA (W-MSA) and Shifted Window-based MSA (SW-MSA) respectively. In the first layer with W-MSA, The input image is divided into several windows and we compute self-attention between patches within that window. The intuition is local neighborhood patches are more important than attending patches that are far away. In the second layer with SW-MSA, the shifted window is designed to achieve information sharing across various spatial locations. The computation cost can be formulated as Ω(W-MSA) = $4hwC^2 + 2M^2hwC$, where M stands for window size and M^2 indicates the number of patches inside the window.

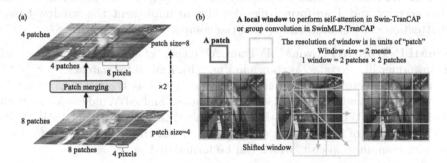

Fig. 2. (a) Patch merging layer in our Swin-TranCAP and SwinMLP-TranCAP. (b) shifted window to perform self-attention in Swin-TranCAP and group convolution in SwinMLP-TranCAP. For space without any pixels, cycle shifting is utilized to copy over the patches on top to the bottom, from left to right, and also diagonally across to make up for the missing patches.

2.2 Our Model

We propose the end-to-end SwinMLP-TranCAP model for captioning tasks, as shown in Fig. 1(b), which takes the embedded patches as the input representation directly. SwinMLP-TranCAP consists of a hierarchical fully multi-layer perceptron (MLP) architecture-based visual encoder and a Transformer-like decoder. Our model creates a new end-to-end learning paradigm for captioning tasks by using patches to get rid of the intermediate modules such as detector and feature extractor (see Fig. 1(a)), reduces the training parameters, and improves the inference speed. In addition, we designed the temporal version of SwinMLP-TranCAP, named Video SwinMLP-TranCAP to implement video captioning.

2D/3D Patch Embedding Layer. The raw image $[3, H, W]$ is partitioned and embedded into N discrete non-overlapping patches $[N, C]$ with the embedding dimension C by using a 2D patch embedding layer which consists of a Conv2d projection layer (See Eq. 1) followed by flattening and transpose operation and LayerNorm. The size of each patch is $[p, p, 3]$ and the number of patches N is $\frac{H}{p} \times \frac{W}{p}$. The raw feature dimension $p \times p \times 3$ of each patch will be projected into an arbitrary embedding dimension C. Thus, the output size from the Conv2d projection layer is $[C, \frac{H}{p}, \frac{W}{p}]$. Next, it is flattened and transposed into embedded patches of $[N, C]$. Eventually, LayerNorm is applied to embedded patches. Similarly, in the 3D patch embedding layer, the video clip $[3, T, H, W]$ is partitioned and embedded into N 3D patches of $[C, \frac{T}{t}, \frac{H}{p}, \frac{W}{p}]$ (t stands for the sequence length of a cubic patch, T stands for the sequence length of a video clip) via a Conv3d projection layer.

$$Conv2d(in = 3, out = C, kernel\_size = p, stride = p) \qquad (1)$$

Swin-TranCAP. We utilize Swin Transformer [10] as the backbone of the vision encoder and Transformer-like decoder to implement the window-based self-attention computation and reduce the computation cost.

SwinMLP-TranCAP. SwinMLP-TranCAP consist of a SwinMLP encoder and a Transformer-like decoder, as shown in Fig. 2(b). Embedded patches $[\frac{H}{p} \times \frac{W}{p}, C]$ are used as the input representation and each patch is treated as a "token". In SwinMLP encoder, we replace the Window/Shifted-Window MSA with Window/Shifted-Window Multi-Head Spatial MLP (W-MLP and SW-MLP) with the number of heads n and window size M via group convolution for less-expensive computation cost, which can be formulated as

$$Conv1d(in = nM^2, out = nM^2, kernel\_size = 1, groups = n) \qquad (2)$$

In two successive Swin MLP blocks, the first block with MLP is applied to image patches within the window independently and the second block with MLP is applied across patches by shifting window. The vision encoder consists of 4 stages and a linear layer. The Swin MLP block in the first stage maintain the feature representation as $[\frac{H}{p} \times \frac{W}{p}, C]$. Each stage in the remaining 3 stages contains a patch merging layer and multiple (2×) Swin MLP Blocks. The patch merging layer (see Fig. 2(a)) concatenates the features of each group of 2×2 neighboring patches and applies a linear layer to the $4C$ dimensional concatenated features. This decreases the amount of "tokens" by $2 \times 2 = 4$, and the output dimension is set to $2C$. Thus the feature representation after patch merging layer is $[\frac{H}{2 \times p} \times \frac{W}{2 \times p}, 2C]$. Afterward, Swin MLP blocks maintain such a shape of feature representation. Swin MLP blocks are created by replacing the standard MSA in a Transformer block with the window-based multi-head MLP module while keeping the other layer the same. The block in the vision encoder consists of a window-based MLP module, a 2-layer MLP with GELU nonlinearity. Before each windowed-based MLP module and each MLP, a LayerNorm layer is applied, and a residual connection is applied after each module. Overall, the number of tokens is reduced and the feature dimension is increased by patch merging layers as the network gets deeper. The output of 4 stages is $[\frac{H}{2^3 \times p} \times \frac{W}{2^3 \times p}, 2^3 \times C]$

(3 means the remaining 3 stages). Eventually, a linear layer is applied to produce the $[\frac{H}{2^3 \times p} \times \frac{W}{2^3 \times p}, 512]$ feature representation to adapt the Transformer-like decoder. The decoder is a stack of 6 identical blocks. Each block consists of two multi-head attention layers (an encoder-decoder cross attention layer and a decoder masked self-attention layers) and a feed-forward network.

Video SwinMLP-TranCAP. We form the video clip $x = \{x_{t-T+1}, ..., x_t\}$ consisting the current frame and the preceding $T-1$ frames. The video clip is partitioned and embedded into $\frac{T}{t} \times \frac{H}{p} \times \frac{W}{p}$ 3D patches and each 3D patch has a C-dimensional feature via a 3D Patch Embedding layer. Each 3D patch of size $[t, p, p]$ is treated as a "token". The major part is the video SwinMLP block which is built by the 3D shifted window-based multi-head MLP module while keeping the other components same. Given a 3D window of $[P, M, M]$, in two consecutive layers, the multi-head MLP module in the first layer employs the regular window partition approach to create non-overlapping 3D windows of $[\frac{T}{t \times P} \times \frac{H}{p \times M} \times \frac{W}{p \times M}]$. In the second layer with multi-head MLP module, the window is shifted along the temporal, height, and width axes by $[\frac{P}{2}, \frac{M}{2}, \frac{M}{2}]$ tokens from the first layer's output.

Our models depart from ViT [5] and Swin Transformer [10] with several points: 1) no extra "class" token; 2) no self-attention blocks in the visual encoder: it is replaced by MLP via group convolutions; 3) Designed for captioning tasks. The window-based visual encoder is paired with a Transformer-like decoder via a linear layer which reduces the dimensional to 512 to adapt to the language decoder; 4) use window size of 14, instead of 7 used in Swin Transformer [10].

3 Experiments

3.1 Dataset Description

DAISI Dataset. The Database for AI Surgical Instruction (DAISI) [14][1] contains 17339 color images of 290 different medical procedures, such as laparoscopic sleeve gastrectomy, laparoscopic ventral hernia repair, tracheostomy, open cricothyroidotomy, inguinal hernia repair, external fetal monitoring, IVC ultrasound, etc. We use the filtered DAISI dataset cleaned by [20] which removes noisy and irrelevant images and text descriptions. We split the DAISI dataset into 13094 images for training, and 1646 images for validation by following [20].

EndoVis 2018 Dataset is from the MICCAI robotic instrument segmentation dataset[2] of endoscopic vision challenge 2018 [1]. The training set consists of 15 robotic nephrectomy procedures acquired by the da Vinci X or Xi system. We use the annotated caption generated by [17] which employs 14 sequences out of the 15 sequences while disregarding the 13th sequence due to the less interaction. The validation set consists of the 1st, 5th, and 16th sequences, and the train set includes the 11 remaining sequences following the work [17].

---

[1] https://engineering.purdue.edu/starproj/_daisi/.

[2] https://endovissub2018-roboticscenesegmentation.grand-challenge.org/.

3.2 Implementation Details

Our models are trained using cross-entropy loss with a batch size of 9 for 100 epochs. We employ the Adam optimizer with an initial learning rate of $3e-4$ and follow the learning rate scheduling strategy with 20000 warm-up steps. Our models are implemented on top of state-of-the-art captioning architecture, Transformer [5]. The vanilla Transformer architecture is realized with the implementation[3] from [5]. All models are developed with Pytorch and trained with NVIDIA RTX3090 GPU. In our models, the embedding dimension $C = 128$, patch size $p = 4$, and window size $M = 14$. In Video SwinMLP-TranCAP, we use the sequence length T of 4 and 3D patch of $[2, 4, 4]$ and window size of $[2, 14, 14]$.

4 Results and Evaluation

Table 1. Comparison of hybrid models and our models. The DAISI dataset does not have object annotation and video captioning annotation. Thus the hybrid of YLv5 w. RN18 and Transformer, and Video SwinMLP-Tran experiments are left blank.

Model			DAISI dataset				EndoVis18 dataset			
Det	FE	Captioning model	B4	MET	SPI	CID	B4	MET	SPI	CID
FasterRCNN [12]	RN18 [7]	Tran [5]	✗				0.363	0.323	0.512	2.017
		Self-Seq [13]					0.295	0.283	0.496	1.801
		AOA [8]					0.377	0.371	0.58	1.811
YLv5x [9]	RN18 [7]	Tran [5]	✗				0.427	0.328	0.577	**3.022**
✗	RN18 [7]	Self-Seq [13]	0.296	0.207	0.330	2.827	0.446	0.353	0.531	2.674
		AOA [8]	0.349	0.246	0.403	3.373	0.427	0.322	0.533	2.903
		Tran [20]	0.454	0.308	**0.479**	**4.283**	0.426	0.335	0.524	2.826
✗	3DRN18	Tran [5]	✗				0.406	0.345	0.586	2.757
✗	✗	Swin-TranCAP	0.346	0.237	0.378	3.280	**0.459**	0.336	0.571	3.002
		SwinMLP-TranCAP	**0.459**	**0.308**	0.478	4.272	0.403	0.313	0.547	2.504
		V-SwinMLP-TranCAP	✗				0.423	**0.378**	**0.619**	2.663

We evaluate the captioning models using the BLEU-4 (B4) [11], METEOR (MET) [3], SPICE (SPI) [2], and CIDEr (CID) [16]. We employ the ResNet18 [7] pre-trained on the ImageNet dataset as the feature extractor (FE), YOLOv5 (YLv5) [9], and FasterRCNN [12] pre-trained on the COCO dataset as the detector (Det). The captioning models includes self-sequence captioning model (Self-Seq) [13], Attention on Attention (AOA) model [8], Transformer model [5]. Self-sequence and AOA originally take the region features extracted from the object detector with feature extractor as input. In our work, we design the hybrid style for them by sending image features extracted by the feature extractor only. 3D ResNet18 is employed to implement the video captioning task. We define the end-to-end captioning models take the image patches directly as input, trained from scratch as **Ours**, including Swin-TranCAP and SwinMLP-TranCAP for image captioning and Video SwinMLP-TranCAP (V-SwinMLP-TranCAP) for video captioning.

Table 2. Proof of "less computation cost" of our approaches with FPS, Num of Parameters, and GFLOPs.

Model			Proof of less-computation cost		
Det	FE	Captioning model	FPS	N_Parameters(M)	GFLOPs
FasterRCNN [12]	RN18 [7]	Tran [5]	8.418	28.32+46.67	251.84+25.88
YLv5x [9]	RN18 [7]	Tran [5]	9.368	97.88+46.67	1412.8+25.88
✗	RN18 [7]	Tran [20]	11.083	11.69+46.67	1.82+25.88
✗	✗	Swin-TranCAP	10.604	165.51	19.59
		SwinMLP-TranCAP	**12.107**	**99.11**	**14.15**

Fig. 3. Qualitative results with our model and hybrid models.

We now briefly describe how to use the detector and feature extractor extract features for captioning model. The spatial features from YOLOv5 w. ResNet are $(X, 512)$. X indicates the number of predicted regions, and X varies from image to image. The first N predicted regions are sent to the captioning model. Zero appending is used to deal with those images where $N > X$ ($N = 6$ in our work). When only using the feature extractor, we take the 2D adaptive average pooling of the final convolutional layer output, which results in $14 \times 14 \times 512$ output for each image. It is reshaped into 196×512 before sending into the captioning model. We find that our models can achieve decent performance even without the need for a detector and feature extractor (see Table 1 and Fig. 3). On the DAISI dataset, our SwinMLP-TranCAP preserves the performance compared with the hybrid Transformer model and outperforms the FC and AOA model including 11% gain in BlEU-4. On EndoVis 2018 Dataset, our Swin-TranCAP model shows 4.7% improvement in SPICE compared with the hybrid Transformer model. Although the performance of SwinMLP-TranCAP drops a bit, it has less computation. In Table 2, less computation cost of our approaches is proven by evaluating FPS, Num of Parameters, and GFLOPs. Our approach achieves better captioning performance/efficiency trade-offs. It is worth highlighting that our purpose is not to provide better results but to remove the object detector and feature extractor from the conventional captioning system training pipeline for more flexible training, less computation cost, and faster

---

[3] https://github.com/ruotianluo/ImageCaptioning.pytorch.

Table 3. Tune patch size p for vanilla transformer.

p	DAISI		EndoVis18	
	B4	CID	B4	CID
4	0.192	1.703	0.339	2.105
8	0.247	2.195	0.364	1.886
16	**0.302**	**2.776**	**0.416**	**3.037**

Table 4. Tune embedding dimension C with patch size p of 4 and window size M of 14, for our models.

Model	Para.		DAISI		EndoVis18	
	C	M	B4	CID	B4	CID
Swin-Tran-S	96	14	0.329	3.109	**0.471**	**3.059**
Swin-Tran-L	128		0.346	3.280	0.459	3.002
SwinMLP-Tran-S	96		0.455	4.188	0.398	2.322
SwinMLP-Tran-L	128		**0.459**	**4.272**	0.403	2.504
Swin-Tran-L	128	7	0.433	4.049	0.389	2.932
SwinMLP-Tran-L	128	7	0.434	4.046	0.403	2.707

inference speed without sacrificing performance. Surprisingly, our approach also obtained a slightly better quantitative performance.

5 Ablation Study

We simply incorporate the patch embedding layer into the vanilla Transformer captioning model and report the performance in Table 3. The embedding dimension C is set to 512. And we also investigate the performance of different configurations for our models (see Table 4). $C = 96$ paired with layer number $(2, 2, 6, 2)$ is **S**mall version and $C = 128$ paired with layer number $(2, 2, 18, 2)$ is **L**arger version. We find that the large version with a window size of 14 performs slightly better.

6 Discussion and Conclusion

We present the end-to-end SwinMLP-TranCAP captioning model, and take the image patches directly, to eliminate the object detector and feature extractor for real-time application. The shifted window and multi-head MLP architecture design make our model less computation. Video SwinMLP-TranCAP is also developed for video captioning tasks. Extensive evaluation of two surgical captioning datasets demonstrates that our models can maintain decent performance without needing these intermediate modules. Replacing the multi-head attention module with multi-head MLP also reveals that the generic transformer architecture is the core design, instead of the attention-based module. Future works will explore whether the attention-based module in a Transformer-like decoder is necessary or not. Different ways to implement patch embedding in the vision encoder are also worth studying.

Acknowledgements. This work was supported by the Shun Hing Institute of Advanced Engineering (SHIAE project BME-p1-21) at the Chinese University of Hong Kong (CUHK), Hong Kong Research Grants Council (RGC) Collaborative Research Fund (CRF C4026-21GF and CRF C4063-18G) and (GRS)#3110167. Key project 2021B1515120035 (B.02.21.00101) of the Regional Joint Fund Project of the Basic and Applied Research Fund of Guangdong Province

References

1. Allan, M., et al.: 2018 robotic scene segmentation challenge. arXiv preprint arXiv:2001.11190 (2020)
2. Anderson, P., Fernando, B., Johnson, M., Gould, S.: SPICE: semantic propositional image caption evaluation. In: Leibe, B., Matas, J., Sebe, N., Welling, M. (eds.) ECCV 2016. LNCS, vol. 9909, pp. 382–398. Springer, Cham (2016). https://doi.org/10.1007/978-3-319-46454-1_24
3. Banerjee, S., Lavie, A.: Meteor: an automatic metric for MT evaluation with improved correlation with human judgments. In: Proceedings of the ACL Workshop on Intrinsic and Extrinsic Evaluation Measures for Machine Translation and/or Summarization, pp. 65–72 (2005)
4. Cornia, M., Stefanini, M., Baraldi, L., Cucchiara, R.: Meshed-memory transformer for image captioning. In: Proceedings of the IEEE/CVF Conference on Computer Vision and Pattern Recognition, pp. 10578–10587 (2020)
5. Dosovitskiy, A., et al.: An image is worth 16x16 words: transformers for image recognition at scale. arXiv preprint arXiv:2010.11929 (2020)
6. Fang, Z., et al.: Injecting semantic concepts into end-to-end image captioning. arXiv preprint arXiv:2112.05230 (2021)
7. He, K., Zhang, X., Ren, S., Sun, J.: Deep residual learning for image recognition. In: Proceedings of the IEEE Conference on Computer Vision and Pattern Recognition, pp. 770–778 (2016)
8. Huang, L., Wang, W., Chen, J., Wei, X.Y.: Attention on attention for image captioning. In: International Conference on Computer Vision (2019)
9. Jocher, G., Chaurasia, A., Stoken, A., Borovec, J., Kwon, Y., et al.: ultralytics/yolov5: v6.1 - TensorRT, TensorFlow Edge TPU and OpenVINO Export and Inference, February 2022. https://doi.org/10.5281/zenodo.6222936
10. Liu, Z., et al.: Swin transformer: Hierarchical vision transformer using shifted windows. In: International Conference on Computer Vision (ICCV) (2021)
11. Papineni, K., Roukos, S., Ward, T., Zhu, W.J.: Bleu: a method for automatic evaluation of machine translation. In: Proceedings of the 40th Annual Meeting of the Association for Computational Linguistics, pp. 311–318 (2002)
12. Ren, S., He, K., Girshick, R., Sun, J.: Faster R-CNN: towards real-time object detection with region proposal networks. In: Advances in Neural Information Processing Systems, vol. 28 (2015)
13. Rennie, S.J., Marcheret, E., Mroueh, Y., Ross, J., Goel, V.: Self-critical sequence training for image captioning. In: Proceedings of the IEEE Conference on Computer Vision and Pattern Recognition, pp. 7008–7024 (2017)
14. Rojas-Muñoz, E., Couperus, K., Wachs, J.: DAISI: database for AI surgical instruction. arXiv preprint arXiv:2004.02809 (2020)
15. Tolstikhin, I.O., et al.: MLP-mixer: an all-MLP architecture for vision. In: Advances in Neural Information Processing Systems, vol. 34 (2021)
16. Vedantam, R., Lawrence Zitnick, C., Parikh, D.: Cider: consensus-based image description evaluation. In: Proceedings of the IEEE Conference on Computer Vision and Pattern Recognition, pp. 4566–4575 (2015)
17. Xu, M., Islam, M., Lim, C.M., Ren, H.: Class-incremental domain adaptation with smoothing and calibration for surgical report generation. In: de Bruijne, M., et al. (eds.) MICCAI 2021. LNCS, vol. 12904, pp. 269–278. Springer, Cham (2021). https://doi.org/10.1007/978-3-030-87202-1_26

18. Xu, M., Islam, M., Lim, C.M., Ren, H.: Learning domain adaptation with model calibration for surgical report generation in robotic surgery. In: 2021 IEEE International Conference on Robotics and Automation (ICRA), pp. 12350–12356. IEEE (2021)
19. Yu, W., et al.: Metaformer is actually what you need for vision. arXiv preprint arXiv:2111.11418 (2021)
20. Zhang, J., Nie, Y., Chang, J., Zhang, J.J.: Surgical instruction generation with transformers. In: de Bruijne, M., et al. (eds.) MICCAI 2021. LNCS, vol. 12904, pp. 290–299. Springer, Cham (2021). https://doi.org/10.1007/978-3-030-87202-1_28

CaRTS: Causality-Driven Robot Tool Segmentation from Vision and Kinematics Data

Hao Ding[1]([✉]), Jintan Zhang[1], Peter Kazanzides[1], Jie Ying Wu[2], and Mathias Unberath[1]

[1] Department of Computer Science, Johns Hopkins University, Baltimore, USA
{hding15,unberath}@jhu.edu
[2] Department of Computer Science, Vanderbilt University, Nashville, USA

Abstract. Vision-based segmentation of the robotic tool during robot-assisted surgery enables downstream applications, such as augmented reality feedback, while allowing for inaccuracies in robot kinematics. With the introduction of deep learning, many methods were presented to solve instrument segmentation directly and solely from images. While these approaches made remarkable progress on benchmark datasets, fundamental challenges pertaining to their robustness remain. We present CaRTS, a causality-driven robot tool segmentation algorithm, that is designed based on a complementary causal model of the robot tool segmentation task. Rather than directly inferring segmentation masks from observed images, CaRTS iteratively aligns tool models with image observations by updating the initially incorrect robot kinematic parameters through forward kinematics and differentiable rendering to optimize image feature similarity end-to-end. We benchmark CaRTS with competing techniques on both synthetic as well as real data from the dVRK, generated in precisely controlled scenarios to allow for counterfactual synthesis. On training-domain test data, CaRTS achieves a Dice score of 93.4 that is preserved well (Dice score of 91.8) when tested on counterfactually altered test data, exhibiting low brightness, smoke, blood, and altered background patterns. This compares favorably to Dice scores of 95.0 and 86.7, respectively, of the SOTA image-based method. Future work will involve accelerating CaRTS to achieve video framerate and estimating the impact occlusion has in practice. Despite these limitations, our results are promising: In addition to achieving high segmentation accuracy, CaRTS provides estimates of the true robot kinematics, which may benefit applications such as force estimation. Code is available at: https://github.com/hding2455/CaRTS.

Keywords: Deep learning · Computer vision · Minimally invasive surgery · Computer assisted surgery · Robustness

---

Supplementary Information The online version contains supplementary material available at https://doi.org/10.1007/978-3-031-16449-1_37.

1   Introduction

With the increasing prevalence of surgical robots, vision-based robot tool segmentation has become an important area of research [7,11,14,15,24,25,36,38, 42,43]. Image-based tool segmentation and tracking is considered important for downstream tasks, such as augmented reality feedback, because it provides tolerance to inaccurate robot kinematics. Deep learning techniques for tool segmentation now achieve respectable performance on benchmark dataset, but unfortunately, this performance is not generally preserved when imaging conditions deteriorate or deviate from the conditions of the training data [18]. Indeed, when encountering conditions of the surgical environment not seen in training, such as smoke and blood as shown in Fig. 1, the performance of contemporary algorithms that seek to infer segmentation masks directly from images deteriorates dramatically. While some of this deterioration can likely be avoided using modern robustness techniques [18], we believe that a lot more can be gained by revisiting how the tool segmentation problem is framed.

In robot tool segmentation, although a rich variety of information is available from the system (e.g. kinematics), contemporary approaches to tool segmentation often neglect the complex causal relationship between the system, the surgical environment, and the segmentation maps. These methods posit a direct causal relationship between the observed images and the segmentation maps, following a causal model of segmentation as shown in Fig. 2a. It follows that the segmentation $\mathbf{S}$ is entirely determined by the image $\mathbf{I}$. Information from robot tools and cameras poses (jointly represented by $\mathbf{T}$), and the environment $\mathbf{E}$ are neither observed nor considered. We refer to this model of determining segmentation maps solely from images as the contemporary model; a causal relation that has previously been suggested for brain tumor segmentation [13].

Under this causal model, there is no confounding bias and one can simply fit $p(\mathbf{S} \mid \mathbf{I})$ which identifies the probability $p(\mathbf{S}(\mathbf{I}))$ for the counterfactual $\mathbf{S}(\mathbf{I})$. However, these models must learn to generalize solely from the image domain. This makes them vulnerable to domain shifts such as those introduced by lighting changes or smoke during surgery [39,44].

We present an alternative causal model of robot tool segmentation, shown in Fig. 2b, and an end-to-end method designed based on this model, which we refer to as CaRTS: causality-driven robot tool segmentation. In this model, image $\mathbf{I}$ has no direct causal effect on the segmentation $\mathbf{S}$. Instead, both of them are directly determined by the robot kinematics and camera poses $\mathbf{T}$, and the environment $\mathbf{E}$. Interactions between tools and environment are represented by direct causal effects between $\mathbf{E}$ and $\mathbf{T}$. In this causal model, we focus on counterfactual $\mathbf{S}(\mathbf{T})$ instead of $\mathbf{S}(\mathbf{I})$.

Since our models of the robot and the environment are imperfect, there are two main sources of errors for algorithms designed based on this causal model - confounding bias and estimation error. The confounding bias arises from the environment. The causal effect of the environment on the segmentation consists of occlusion. Here, we assume occlusion does not affect segmentation. Under this assumption, the causal relationship between the environment and the segmen-

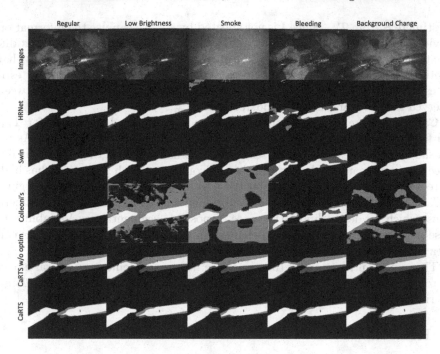

Fig. 1. Comparisons between HRNet, Swin Transformer, Colleoni's method, and CaRTS with/without optimization on images with counterfactual surgical environment. White/black: True positives/negatives; Orange/Red: False positives/negatives. (Color figure online)

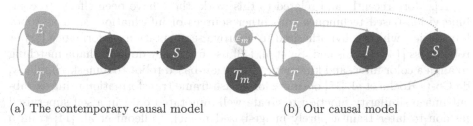

(a) The contemporary causal model   (b) Our causal model

Fig. 2. Illustration of two causal models for robot tool segmentation. Solid lines mean direct causal effect. Light blue represent unobserved factors; dark blue nodes represent observed factors. We note I for images, E for environment, S for segmentation, T for true robot kinematics and camera poses, T_m for measured robot kinematics and camera poses, and ϵ_m for measurement error. (Color figure online)

tation map can be ignored, denoted by the dashed line in Fig. 2b. The second source of error, estimation error, is caused by the fact that surgical robots are designed to be compliant. Therefore, there is significant error in tool pose estimation based on joint positions. This causes us to observe T_m, which contains measurement errors ϵ_m, instead of the ground truth values T. Our goal is then to estimate $\hat{\mathbf{S}}$ from $\mathbf{T_m}$. Since $\mathbf{I}$ is not affected by ϵ_m, we can use the information provided by $\mathbf{I}$ to estimate $\mathbf{T}$ based on $\mathbf{T_m}$. More specifically, CaRTS uses differentiable rendering given robot and camera information to model $p(\mathbf{S}|\mathbf{T})$ that identifies $p(\mathbf{S}(\mathbf{T}))$. To account for measurement error, CaRTS iteratively estimates ϵ_m and infers $\mathbf{T}$ by optimizing a feature-level cosine similarity between observed images $\mathbf{I}$ and images rendered from $\mathbf{T_m}$.

CaRTS has the advantage that the image is solely used to correct for $\mathbf{T_m}$, and feature extraction reduces reliance on specific pixel representations. As we will show, this makes CaRTS less sensitive to domain shifts in the image domain. Since image appearance depends on the environment, which potentially has high degrees of variability, achieving robustness to image domain shift is challenging. In CaRTS, the segmentation map is conditioned on robot kinematics and camera poses, which by design limits the space of possible segmentation. Variability in these parameters is restricted and changes in kinematics and camera poses can often be accounted for by small adjustments.

The main contributions of this paper are: (1) a complementary causal model for robot tool segmentation that is robust to image domain shifts, and (2) an algorithm, CaRTS, based on (1) that generalizes well to new domains.

2　Related Work

Robot Tool Segmentation: Vision-based segmentation is a mature area of research and many deep learning-based methods, including [4–6,17,23,32,33,41], exist. For robot tool segmentation, many techniques follow the same paradigm and solely rely on vision data to achieve scene segmentation, e. g., [7,11,24,25, 36,42]. More recently and related to this work, there have been efforts to combine vision-based techniques with other sources of information, such as robot kinematics, which is hypothesized to improve segmentation performance [38] or robustness [14]. To this end, Su et al. [43] use frequency domain shape matching to align a color filter- and kinematic projection-based robot tool mask. Similarly, da Costa Rocha et al. [15] optimize for a base-frame transformation using a grabcut image similarity function to create well annotated data in a self-supervised fashion to later train a purely image-based model. Colleoni et al. [14] train a post-processing neural network for image-based refinement of a kinematic-based projection mask. While these recent methods use geometric data in addition to images, none of the previous methods enables direct estimation of robot kinematic parameters, including joint angles, base frame and camera transformations, in an end-to-end fashion.

Geometric Information for Robot Vision: Geometric information (e. g. depth, pose) has been explored and previously been used in robot vision. Some methods [9,10,12,20,21,29,45] are proposed to acquire geometric information from vision, other methods [14–16,22,38,43] use geometric information to enhance downstream vision tasks. Our method leverages both directions by optimizing the measured robot and camera parameters from vision and enhancing the segmentation of the image via projection of the robot.

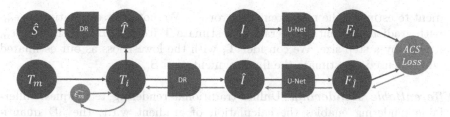

Fig. 3. Illustration of the overall CaRTS architecture. We iteratively optimize ACL Loss w.r.t T_i to estimate robot parameters $\hat{T}$ then generate estimated segmentation $\hat{S}$. solid blue lines in this figure represent information flow and solid orange lines represent gradient flow. DR stands for differentiable rendering. (Color figure online)

Causality in Medical Image Analysis: Recently, causality receives increasing attention in the context of medical image analysis [13]. Reinhold et al. [40] use causal ideas to generate MRI images with a Deep Structural Causal Model [37]. Lenis et al. [3] use the concept of counterfactuals to analyze and interpret medical image classifiers. Lecca [28] provides perspectives on the challenges of machine learning algorithms for causal inference in biological networks. Further, causality-inspired methods [1,30,34,46,47] exist for domain generalization [44]. These techniques focus on feature representation learning directly from images, i. e., the contemporary causal model of segmentation. Different from these methods, our approach frames the robot tool segmentation task using an alternative causal model, and leverages this model to derive a different approach to segmentation that alters the domain shift problem.

3   CaRTS: Causality-Driven Robot Tool Segmentation

We propose the CaRTS architecture as an algorithm designed based on our causal model. CaRTS models $p(S|T)$, which identifies $p(S(T))$ through differentiable rendering, to estimate segmentation maps from robot kinematic parameters. CaRTS estimates the measurement error ϵ_m between true and observed robot kinematics (T and T_m, respectively) by maximizing a cosine similarity between the semantic features of the observed and rendered image extracted by a U-Net.

Overall Architecture: The overall architecture of CaRTS is shown in Fig. 3. We observe T_m and I, the measured kinematic parameters and an image, as input to our model. T_i represents the values of robot and camera parameters at iteration i of the optimization, with T_0 being initialized to T_m.

For each iteration, CaRTS uses differentiable rendering to generate an estimated image $\hat{I}$ from T_i and articulated tool models. Then it uses a pre-trained U-Net to extract features F_I and $F_{\hat{I}}$ from both observed I and rendered image $\hat{I}$, respectively. Attentional cosine similarity loss (ACS Loss) between the feature representations is used to evaluate the similarity of both images. Since the whole pipeline is differentiable, we can use backpropagation to directly calculate the gradient to estimate the measurement error $\epsilon_\mathbf{m}$. We note this estimation as $\hat{\epsilon_\mathbf{m}}$. So with gradient descent, we iteratively estimate $\mathbf{T}$ from $\mathbf{T_i}$ by subtracting $\hat{\epsilon_\mathbf{m}}$ multiplied by a step size. We consider $\mathbf{T_i}$ with the lowest loss as our estimated $\hat{\mathbf{T}}$, which is used to estimate the final segmentation $\hat{\mathbf{S}}$.

Differentiable Rendering: Unlike traditional rendering techniques, differentiable rendering enables the calculation of gradient w.r.t. the 3D quantities which has been found beneficial for related projective spatial alignment tasks [2]. We use the differentiable rendering function (noted as f_{DR}) provided by Pytorch3d [8], implementing Soft Rasterizer [31] and Neural 3D Mesh Renderer [26]. For CaRTS, differentiable rendering enables the calculation of gradient w.r.t. the robot parameters used to calculate the final robot meshes and camera parameters. We calculate forward kinematics (denoted by the function f_{FK}) [19] for the robot to construct meshes $\mathbf{M}$ for rendering. We render image $\hat{\mathbf{I}}$ and segmentation $\hat{\mathbf{S}}$ using one point light centered behind the camera position $\mathbf{L_d}$. This yields:

$$\mathbf{M} = f_{FK}(\mathbf{DH}, \mathbf{M_B}, \mathbf{F_B}, \mathbf{K}) \qquad \hat{\mathbf{I}}, \hat{\mathbf{S}} = f_{DR}(\mathbf{C}, \mathbf{M}, \mathbf{L_d}) \qquad (1)$$

Neural Network Feature Extraction: We train a U-Net [41] on a dataset consisting of two types of images: The originally collected images $\mathbf{I}$ from the training set, and hybrid images denoted as $\mathbf{BG_m} + \hat{\mathbf{I}}$. $\mathbf{BG_m}$ is the pixel-wise average background of all originally collected images in the training set. $+$ means overlaying the average background with the robot tool content of the rendered images $\hat{\mathbf{I}}$. Training on hybrid images ensures that feature extraction yields reasonable features from both acquired and rendered images. The average background replaces 0s in the rendered image to avoid potential numerical issues. All hybrid images and the average background can be obtained "for free" from the training set. We optimize the binary cross-entropy loss between the network output and the ground-truth segmentation for both the original and hybrid images to make the network learn semantic feature representations. At inference time, we extract feature maps from the last feature layer for both $\mathbf{I}$ and $\mathbf{BG_m} + \hat{\mathbf{I}}$. We note the feature extraction function as f_{FE}. This yields:

$$\mathbf{F_I} = f_{FE}(\mathbf{I}) \qquad \mathbf{F_{\hat{I}}} = f_{FE}(\hat{\mathbf{I}} + \mathbf{BG_m}) \qquad (2)$$

Although our feature extractor is trained for segmentation, other training targets beyond segmentation, such as unsupervised representation learning, might be equally applicable and more desirable.

Attentional Cosine Similarity Loss: We design the Attentional Cosine Similarity Loss (ACSLoss) to measure the feature distance between $\mathbf{F_{\hat{I}}}$ and $\mathbf{F_I}$. First, we generate an attention map $\mathbf{Att}$ by dilating the rendered silhouette $\hat{\mathbf{S}}$ of the robot tools. This attention map allows us to focus on the local features around the approximate tool location given by the current kinematics estimate and DR.

Then we calculate cosine similarity along the channel dimension of the two feature maps. The final loss is calculated using Eq. 3. Higher channel-wise cosine similarity suggests better alignment between feature points:

$$\mathbf{Att} = dilate(\hat{\mathbf{S}}) \qquad \mathbf{L} = 1 - \frac{1}{w \times h} \sum_{i=0}^{h-1} \sum_{j=0}^{w-1} sim_{cos}(\mathbf{F_I}, \mathbf{F_{\hat{I}}})_{i,j} \times \mathbf{Att}_{i,j} \quad (3)$$

The complete objective for optimization is shown in euqation 4. While in this paper we fix all but the kinematic parameters and optimize f_L w.r.t the kinematics input $\mathbf{K}$ to achieve tool segmentation, we note that optimization w.r.t. these variables corresponds to possibly desirable robotic applications. For example, if we fix all variables and optimize f_L w.r.t the camera parameters $\mathbf{C}$ and robot base frame $\mathbf{F_B}$, we can perform hand-eye calibration:

$$\arg\min f_L(\mathbf{DH}, \mathbf{M_B}, \mathbf{F_B}, \mathbf{K}, \mathbf{C}, \mathbf{L_d}, \mathbf{I}, \mathbf{BG_m}) \qquad (4)$$

4   Experiments

We use the da Vinci Research Kit (dVRK) [27] as the platform for data collection. The system has a stereo endoscope and two patient side manipulators (PSM). The task is differentiate PSMs from the rest of the endoscope scene through segmentation. We conduct robot tool segmentation experiments on synthetic data generated using AMBF [35] and real-world data collected with the dVRK, respectively. Under this setting, CaRTS perform optimization w.r.t a 6 degree-of-freedom joint space. Implementation details can be found in the released code.

Data Collection: For simulated data, we generate 8 sequences (6 for training, 1 for validation, and 1 for testing), by teleoperating the PSMs in the simulator, recording the image, segmentation and kinematics read from the simulator. Each sequence contains 300 images. Robot base frame and camera information are determined before recording. For test data, we add random offsets to the kinematics to simulate measurement error and create another test set consisting of images that are corrupted with simulated smoke.

For real data, we adopt the data collecting strategy from Colleoni et al. [14], but expand it to allow for counterfactual data generation. We prerecord a kinematics sequence and replay it on different backgrounds. The first replay is on a green screen background to extract ground truth segmentation through color masking. Then, we introduce different tissue backgrounds and set different conditions to counterfactually generate images that differ only in one specific condition - every condition will be referred to as a domain. We choose one domain without corruption for training, validation and test. This regular domain has a specific tissue background, a bright and stable light source, and no smoke or bleeding conditions. Images collected from other domains are only provided for validation and testing. We record 7 sequences for training, 1 for validation, and

regular   simulated smoke

Fig. 4. Examples of the images in the synthetic dataset.

Table 1. Robot tool segmentation results

	R-Reg	R-LB	R-Bl	R-Sm	R-BC	S-Reg	S-SS
Colleoni's	94.9 ± 2.7	87.0 ± 4.5	55.0 ± 5.7	59.7 ± 24.7	75.1 ± 3.6	99.8 ± 0.1	41.4 ± 4.3
HRNet w Aug	95.2 ± 2.7	86.3 ± 3.9	56.3 ± 16.4	77.2 ± 23.6	92.1 ± 4.6	–	–
Swin transformer w Aug	95.0 ± 5.5	93.0 ± 5.5	76.5 ± 9.0	82.4 ± 17.0	94.8 ± 5.3	–	–
CaRTS w/o optim.	89.4 ± 3.9	–	–	–	–	83.4 ± 6.8	–
CaRTS L1	92.7 ± 4.0	89.8 ± 3.9	89.4 ± 3.9	89.8 ± 3.9	89.5 ± 3.9	90.5 ± 10.0	83.7 ± 6.9
CaRTS ACS	93.4 ± 3.0	92.4 ± 3.1	90.8 ± 4.4	91.6 ± 4.7	92.3 ± 4.8	96.6 ± 2.7	96.5 ± 2.4

1 for testing. Each sequence contains 400 images. To ensure replay accuracy, we keep the instrument inserted during the whole data collection procedure.

All segmentation maps are generated under no occlusion condition to avoid confounding issues. Figures 1 and 4 show examples of the real-world and synthetic datasets respectively.

Experiment on Robot Tool Segmentation: In this experiment, we compare the tool's Dice score of CaRTS to the HRNet [4], Swin Transformer [32] trained with simulated smoke augmentation, CaRTS without optimization and method by Colleoni et al. [14] using both kinematics and images.

The results are summarized in Table 1. Each column represents a test domain. "R-" and "S-" represent real-world data and synthetic data, respectively. Reg, LB, Bl, Sm, BC, and SS represent regular, low brightness, bleeding, smoke, background change, and simulated smoke respectively. As the table shows, HRNet and Swin Transformer perform well on the regular domain on both real and synthetic data. However, their performance deteriorates notably when the testing domain changes. Colleoni et al.'s architecture performs similarly well on the regular domain, and compared to HRNet and Swin Transformer and still suffers from performance deterioration in unseen testing domains. CaRTS achieves similarly high performance irrespective of test domain.

Ablation Study: We perform ablation studies on the validation set to explore the impact of the similarity function, iteration numbers, and magnitude of kinematics errors on segmentation performance. Fully-tabled results for the latter two ablation studies are not shown here due to the space limit.

As shown in Table 1, we find that optimizing ACSLoss on features better utilizes the rich information extracted by the U-Net compared to optimizing a pixel-level smooth-L1 loss on predictions, especially on unseen domains. The ablation study on the number of iterations indicates that even though optimization on different test domains performs similarly after a single step. Only some domains continue to improve with larger numbers of iterations (iteration > 30). It takes from 30 to 50 iterations on average for the optimization to converge on different domains. The ablation study on kinematic error magnitude indicates that final segmentation performance depends more on initial mismatch in the image space rather than in the joint space. When initial Dice score is less than a threshold (around 75 in our setting), CaRTS's performance starts dropping.

5   Conclusion

While CaRTS achieved robustness across different domains, its current implementation has some limitations. First, the optimization for **T** typically requires over 30 iterations to converge and is thus not yet suitable for real-time use. Including temporal information may reduce the number of iterations necessary or reduce the rendering time at each iteration. Second, from a theoretical perspective, CaRTS does not allow for occlusions. To address occlusion, future work could evaluate the importance of this theoretical limitation in practice, and explore disentangling environment features from images and use adjustment equation to predict segmentation.

The benchmark results on both synthetic and real dataset generated in precisely controlled scenarios show that, unlike segmentation algorithms based solely on images, CaRTS performs robustly when the test data is counterfactually altered. In addition to high segmentation accuracy and robustness, CaRTS estimates accurate kinematics of the robot, which may find downstream application in force estimation or related tasks.

Acknowledgement:. This research is supported by a collaborative research agreement with the MultiScale Medical Robotics Center at The Chinese University of Hong Kong.

References

1. Ouyang, C., et al.: Causality-inspired single-source domain generalization for medical image segmentation. arxiv:2111.12525 (2021)
2. Gao, C., et al.: Generalizing spatial transformers to projective geometry with applications to 2D/3D registration. arxiv:2003.10987 (2020)
3. Lenis, D., Major, D., Wimmer, M., Berg, A., Sluiter, G., Bühler, Katja: Domain aware medical image classifier interpretation by counterfactual impact analysis. In: Martel, A.L., et al. (eds.) MICCAI 2020. LNCS, vol. 12261, pp. 315–325. Springer, Cham (2020). https://doi.org/10.1007/978-3-030-59710-8_31
4. Wang, J., et al.: Deep high-resolution representation learning for visual recognition. TPAMI (2019)
5. Chen, K., et al.: Hybrid task cascade for instance segmentation. In: Proceedings of Computer Vision and Pattern Recognition Conference, CVPR (2019)

6. Chen, L.-C., Zhu, Y., Papandreou, G., Schroff, F., Adam, H.: Encoder-decoder with Atrous separable convolution for semantic image segmentation. In: Ferrari, V., Hebert, M., Sminchisescu, C., Weiss, Y. (eds.) ECCV 2018. LNCS, vol. 11211, pp. 833–851. Springer, Cham (2018). https://doi.org/10.1007/978-3-030-01234-2_49

7. Peraza-Herrera, L.C.G., et al.: ToolNet: Holistically-nested real-time segmentation of robotic surgical tools. In: 2017 IEEE/RSJ International Conference on Intelligent Robots and Systems (IROS), pp. 5717–5722 (2017)

8. Ravi, N., et al.: Accelerating 3D deep learning with pyTorch3D. arXiv:2007.08501 (2020)

9. Long, Y., et al.: E-DSSR: efficient dynamic surgical scene reconstruction with transformer-based stereoscopic depth perception. In: de Bruijne, M., et al. (eds.) MICCAI 2021. LNCS, vol. 12904, pp. 415–425. Springer, Cham (2021). https://doi.org/10.1007/978-3-030-87202-1_40

10. Li, Z., et al.: Revisiting stereo depth estimation from a sequence-to-sequence perspective with transformers. In: 2021 IEEE/CVF International Conference on Computer Vision (ICCV), pp. 6177–6186 (2021)

11. Zhao, Z., et al.: One to many: adaptive instrument segmentation via meta learning and dynamic online adaptation in robotic surgical video. In: International Conference on Robotics and Automation, ICRA (2021)

12. Allan, M., Ourselin, S., Hawkes, D.J., Kelly, J.D., Stoyanov, D.: 3-D pose estimation of articulated instruments in robotic minimally invasive surgery. IEEE Trans. Med. Imaging **37**(5), 1204–1213 (2018)

13. Castro, D.C., Walker, I., Glocker, B.: Causality matters in medical imaging. Nat. Commun. **11**(1), 3673 (2020)

14. Colleoni, E., Edwards, P., Stoyanov, D.: Synthetic and real inputs for tool segmentation in robotic surgery. In: Martel, A.L., et al. (eds.) MICCAI 2020. LNCS, vol. 12263, pp. 700–710. Springer, Cham (2020). https://doi.org/10.1007/978-3-030-59716-0_67

15. da Costa Rocha, C., Padoy, N., Rosa, B.: Self-supervised surgical tool segmentation using kinematic information. In: 2019 International Conference on Robotics and Automation, ICRA (2019)

16. Couprie, C.: Indoor semantic segmentation using depth information. In: First International Conference on Learning Representations (ICLR), pp. 1–8 (2013)

17. Ding, H., Qiao, S., Yuille, A.L., Shen, W.: Deeply shape-guided cascade for instance segmentation. In: 2021 IEEE/CVF Conference on Computer Vision and Pattern Recognition, CVPR (2021)

18. Drenkow, N., Sani, N., Shpitser, I., Unberath, M.: Robustness in deep learning for computer vision: mind the gap? arxiv:2112.00639 (2021)

19. Fontanelli, G.A., Ficuciello, F., Villani, L., Siciliano, B.: Modelling and identification of the da Vinci research kit robotic arms. In: 2017 IEEE/RSJ International Conference on Intelligent Robots and Systems (IROS), pp. 1464-1469 (2017)

20. Godard, C., Aodha, O.M., Firman, M., Brostow, G.J.: Digging into self-supervised monocular depth estimation. In: 2019 IEEE/CVF International Conference on Computer Vision, ICCV (2019)

21. Guo, X., Yang, K., Yang, W., Wang, X., Li, H.: Group-wise correlation stereo network. In: 2019 IEEE/CVF Conference on Computer Vision and Pattern Recognition, CVPR (2019)

22. Hazirbas, C., Ma, L., Domokos, C., Cremers, D.: FuseNet: Incorporating depth into semantic segmentation via fusion-based CNN architecture. In: Lai, S.-H., Lepetit, V., Nishino, K., Sato, Y. (eds.) ACCV 2016. LNCS, vol. 10111, pp. 213–228. Springer, Cham (2017). https://doi.org/10.1007/978-3-319-54181-5_14

23. He, K., Gkioxari, G., Dollár, P., Girshick, R.B.: Mask R-CNN. In: 2017 IEEE International Conference on Computer Vision (ICCV), pp. 2980-2988 (2017)
24. Islam, M.: Real-time instrument segmentation in robotic surgery using auxiliary supervised deep adversarial learning. In: IEEE Robotics and Automation Letters, vol. 4, no. 2, pp. 2188-2195 (2019)
25. Jin, Y., Cheng, K., Dou, Q., Heng, P.: Incorporating temporal prior from motion flow for instrument segmentation in minimally invasive surgery video. In: Medical Image Computing and Computer Assisted Intervention – MICCAI 2019, 22nd International Conference, pp.440-448 (2019)
26. Kato, H., Ushiku, Y., Harada, T.: Neural 3D mesh renderer. In: 2018 IEEE/CVF Conference on Computer Vision and Pattern Recognition, CVPR (2018)
27. Kazanzides, P., Chen, Z., Deguet, A., Fischer, G.S., Taylor, R.H., DiMaio, S.P.: An open-source research kit for the da Vinci surgical system. In: 2014 IEEE International Conference on Robotics and Automation (ICRA), pp. 6434-6439 (2014)
28. Lecca, P.: Machine learning for causal inference in biological networks: perspectives of this challenge. Front. Bioinform. **1** (2021). https://doi.org/10.3389/fbinf.2021.746712
29. Li, Z.: Temporally consistent online depth estimation in dynamic scenes (2021). https://arxiv.org/abs/2111.09337
30. Liu, C.: Learning causal semantic representation for out-of-distribution prediction. In: Advances in Neural Information Processing Systems 34, NeurIPS (2021)
31. Liu, S., Chen, W., Li, T., Li, H.: Soft rasterizer: a differentiable renderer for image-based 3D reasoning. In: 2019 IEEE/CVF International Conference on Computer Vision (ICCV), pp. 7707-7716 (2019)
32. Liu, Z.: Swin transformer: hierarchical vision transformer using shifted windows. In: International Conference on Computer Vision, ICCV (2021)
33. Long, J., Shelhamer, E., Darrell, T.: Fully convolutional networks for semantic segmentation. In: IEEE/CVF Computer Vision and Pattern Recognition Conference, CVPR (2015)
34. Mitrovic, J., McWilliams, B., Walker, J.C., Buesing, L.H., Blundell, C.: Representation learning via invariant causal mechanisms. In: International Conference on Learning Representations, ICLR (2021)
35. Munawar, A., Wang, Y., Gondokaryono, R., Fischer, G.S.: A real-time dynamic simulator and an associated front-end representation format for simulating complex robots and environments. In: IEEE/RSJ International Conference on Intelligent Robots and Systems, IROS (2019)
36. Pakhomov, D., Premachandran, V., Allan, M., Azizian, M., Navab, N.: Deep residual learning for instrument segmentation in robotic surgery. In: Machine Learning in Medical Imaging, MLMI (2019)
37. Pawlowski, N., de Castro, D.C., Glocker, B.: Deep structural causal models for tractable counterfactual inference. In: Advances in Neural Information Processing Systems, NIPS (2020)
38. Qin, F.: Surgical instrument segmentation for endoscopic vision with data fusion of CNN prediction and kinematic pose. In: 2019 International Conference on Robotics and Automation (ICRA), pp. 9821-9827 (2019)
39. Quionero-Candela, J., Sugiyama, M., Schwaighofer, A., Lawrence, N.D.: Dataset Shift in Machine Learning. The MIT Press (2009)
40. Reinhold, J.C., Carass, A., Prince, J.L.: A structural causal model for MR images of multiple sclerosis. In: de Bruijne, M., et al. (eds.) MICCAI 2021. LNCS, vol. 12905, pp. 782–792. Springer, Cham (2021). https://doi.org/10.1007/978-3-030-87240-3_75

41. Ronneberger, O., Fischer, P., Brox, T.: U-Net: convolutional networks for biomedical image segmentation. In: Navab, N., Hornegger, J., Wells, W.M., Frangi, A.F. (eds.) MICCAI 2015. LNCS, vol. 9351, pp. 234–241. Springer, Cham (2015). https://doi.org/10.1007/978-3-319-24574-4_28
42. Shvets, A.A., Rakhlin, A., Kalinin, A.A., Iglovikov, V.I.: Automatic instrument segmentation in robot-assisted surgery using deep learning. In: 2018 17th IEEE International Conference on Machine Learning and Applications (ICMLA), pp. 624-628 (2018)
43. Su, Y.H., Huang, K., Hannaford, B.: Real-time vision-based surgical tool segmentation with robot kinematics prior. In: 2018 International Symposium on Medical Robotics (ISMR), pp. 1–6. IEEE (2018)
44. Wang, J., Lan, C., Liu, C., Ouyang, Y., Qin, T.: Generalizing to unseen domains: A survey on domain generalization. In: Proceedings of the Thirtieth International Joint Conference on Artificial Intelligence, IJCAI (2021)
45. Ye, M., Zhang, L., Giannarou, S., Yang, G.-Z.: Real-time 3D tracking of articulated tools for robotic surgery. In: Ourselin, S., Joskowicz, L., Sabuncu, M.R., Unal, G., Wells, W. (eds.) MICCAI 2016. LNCS, vol. 9900, pp. 386–394. Springer, Cham (2016). https://doi.org/10.1007/978-3-319-46720-7_45
46. Zhang, C., Zhang, K., Li, Y.: A causal view on robustness of neural networks. In: Advances in Neural Information Processing Systems 33, NeurIPS (2020)
47. Zhang, X., Cui, P., Xu, R., Zhou, L., He, Y., Shen, Z.: Deep stable learning for out-of-distribution generalization. In: 2021 IEEE/CVF Conference on Computer Vision and Pattern Recognition, CVPR (2021)

Instrument-tissue Interaction Quintuple Detection in Surgery Videos

Wenjun Lin[1,2], Yan Hu[2], Luoying Hao[2,3], Dan Zhou[4], Mingming Yang[5], Huazhu Fu[6], Cheekong Chui[1(✉)], and Jiang Liu[2(✉)]

[1] Department of Mechanical Engineering, National University of Singapore, Queenstown, Singapore
mpecck@nus.edu.sg
[2] Research Institute of Trustworthy Autonomous Systems and Department of Computer Science and Engineering, Southern University of Science and Technology, Shenzhen, China
liuj@sustech.edu.cn
[3] School of Computer Science, University of Birmingham, Birmingham, UK
[4] School of Ophthalmology and Optometry, Wenzhou Medical University, Wenzhou, China
[5] Department of Ophthalmology, Shenzhen People's Hospital, Shenzhen, China
[6] Agency for Science, Technology and Research (A*STAR), Queenstown, Singapore

Abstract. Instrument-tissue interaction detection in surgical videos is a fundamental problem for surgical scene understanding which is of great significance to computer-assisted surgery. However, few works focus on this fine-grained surgical activity representation. In this paper, we propose to represent instrument-tissue interaction as ⟨instrument bounding box, tissue bounding box, instrument class, tissue class, action class⟩ quintuples. We present a novel quintuple detection network (QDNet) for the instrument-tissue interaction quintuple detection task in cataract surgery videos. Specifically, a spatiotemporal attention layer (STAL) is proposed to aggregate spatial and temporal information of the regions of interest between adjacent frames. We also propose a graph-based quintuple prediction layer (GQPL) to reason the relationship between instruments and tissues. Our method achieves an mAP of 42.24% on a cataract surgery video dataset, significantly outperforming other methods.

Keywords: Instrument-tissue interaction quintuple detection · Surgical scene understanding · Surgery video

---

W. Lin and Y. Hu—Co-first authors.

Supplementary Information The online version contains supplementary material available at https://doi.org/10.1007/978-3-031-16449-1_38.

L. Wang (Eds.): MICCAI 2022, LNCS 13437, pp. 399–409, 2022.
https://doi.org/10.1007/978-3-031-16449-1_38

1   Introduction

Computer-assisted surgical systems (CAS) assist doctors in performing surgeries based on their understanding of surgical scenes to improve surgical safety and quality [2,3,10]. Surgical scene understanding is an important element of surgeon-robot collaboration, operating room management, and surgical skill evaluation and training systems. Surgical scene understanding explains what happens in the given surgery scenario and a fundamental role of surgical scene understanding is to recognize surgical workflow.

Researchers proposed various algorithms to recognize surgical workflow, which can be divided into three levels based on granularity, including phase, step, and activity [15]. Surgical phase and step are two coarse-grained workflow representations. The main research focus of previous works is phase recognition [12,13,20], which is to recognize the major types of events that occurred during the surgery. Phase recognition increases surgical efficiency by optimizing operating room management such as patient preparation and staff assignment. A surgical phase composes of a series of steps. For step recognition, Hashimoto et al. [8] developed algorithms to recognize steps in laparoscopic sleeve gastrectomy. Coarse-grained workflow representations cannot provide a detailed description of activities in surgical scenes. Fine-grained activity recognition has important implications for surgical training, surgical skill assessment, and the automatic generation of surgical reports. Previous works mainly defined an activity as a sequence of actions [4,14]. This definition of activity overlooks the subject and target of the activity. To gain a fuller understanding of surgical scenes, Nwoye et al. [17] proposed to directly predict ⟨instrument, tissue, action⟩ labels to describe activities in surgical scenes. Some other works focus on inferring instrument-tissue interaction with knowing instrument and target positions and types [11,19,23]. Islam et al. [11] developed a graph network-based method to inference the interaction type. These works focus on category information and lacks the detection of specific locations of instruments and tissues. However, the localization of instruments and tissues has great significance for CAS. The location information is essential to instruct operations of surgical robots and monitor surgical safety, such as alerting surgeons when a dangerous area is touched. A detailed activity description in surgical scenes should be a multivariate combination including the instruments, targets, and the locations and relationships of them. However, current works in surgical scenes tend to address only a part of this combination, with only category information and without location information.

In natural scenes, human-object interaction (HOI) detection [1] with both category and location information to describe human activities is one of the fundamental problems for detailed scene understanding. Most existing works in HOI detection share a similar pipeline [5,21,24], using a pretrained object detection network such as Faster R-CNN [18] or Mask R-CNN [9] to localize all people and objects in the scene first and then pair people and objects to predict the action between them. However, directly applying HOI detection methods to surgical scenes cannot achieve desired results since surgical scenes differ from natural scenes in many ways. Firstly, instruments and tissues are harder to detect. There are a lot of transparent tissues in cataract surgeries and the instruments and tissues are often partially visible due to mutual occlusion, resulting in a large number of missed detections and false detections. Secondly, action prediction is more complicated in surgical scenes. The subject of an action in the HOI task is always human, while a variety of instruments can be the subject of an action type in surgical scenes. It is possible that the types of actions are the same but the instruments used to perform these actions are different in surgical scenes. So even for the same type of actions, there are many different characteristics.

Therefore, we address the above problems from the following aspects. Firstly, comprehensive surgical scene understanding needs multivariate fine-grained activity description to detect instrument-tissue interaction. In this paper, we represent instrument-tissue interaction as ⟨instrument bounding box, tissue bounding box, instrument class, tissue class, action class⟩ quintuples and propose a novel model for quintuple detection. Secondly, in order to better detect instruments and tissues in surgical scenes, we propose to combine temporal and spatial location information to localize instruments and tissues. Finally, considering that relation reasoning between instruments and tissues is vital for quintuple prediction, we propose to learn the relationship based on a graph neural network. Our main contributions are summarized as follows: • We define the instrument-tissue interaction detection as ⟨instrument bounding box, tissue bounding box, instrument class, tissue class, action class⟩ quintuple detection, and present a novel model named quintuple detection network (QDNet) to detect the quintuples in surgical videos. • We propose two network layers, including a spatiotemporal attention layer (STAL) to aggregate spatial and temporal information for the detection of instruments and tissues, and a graph-based quintuple prediction layer (GQPL) for quintuple prediction considering the instrument-tissue relation reasoning. • Our proposed QDNet is validated in a cataract surgery dataset and achieves 42.24% mAP for instrument-tissue interaction detection task.

2 Methodology

The overview of our proposed quintuple detection network (QDNet) is shown in Fig. 1. QDNet follows a two-stage strategy, the instrument and tissue detection stage and the quintuple prediction stage. In the first stage, bounding boxes of instruments and tissues are detected using our modified Faster R-CNN [18] with a spatiotemporal attention layer (STAL). In the second stage, these generated bounding boxes of instruments and tissues are used to extract regions of interest (RoI) features and then these detected instruments and tissues are fed to a graph-based quintuple prediction Layer (GQPL) to predict quintuples.

2.1 Instrument and Tissue Detection Stage

Since instruments are often partially occluded by tissues or other instruments, and tissues in some surgeries are hard to detect like the transparent corneal in cataract surgeries, directly applying Faster R-CNN into our surgery scenes cannot provide desired RoI features. Considering that certain relationships exist between instruments and tissues in a standardized surgery workflow, we propose to aggregate temporal features between frames and spatial features among instruments and tissues to improve detection accuracy.

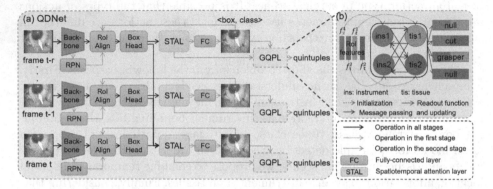

Fig. 1. Illustration of the proposed (a) QDNet for instrument-tissue interaction quintuple detection in the videos. The proposed STAL fuses the spatial and temporal features. (b) A graph-based GQPL for quintuple prediction. A quintuple is represented as ⟨instrument bounding box, tissue bounding box, instrument class, tissue class, action class⟩. Instruments and tissues are paired in GQPL to predict actions between them and these actions and boxes and classes of instrument-tissue pairs form quintuples.

As shown in Fig. 1, a backbone network extracts feature maps for each input frame, and a region proposal network (RPN) generates a set of region proposals based on these feature maps. Then a RoI align network followed by linear box heads extract regional features as visual features $f_v \in \mathbb{R}^c$ (c is the feature dimension) for the generated proposals. To improve the detection accuracy, we propose a novel STAL adopting the instrument-tissue relation between adjacent frames. Visual features and spatial location features of the proposals are sent to our STAL to obtain spatiotemporal attention features. For these attention features of each proposal, fully-connected layers predict its probabilities belonging to a certain instrument or tissue category and refine the bounding box of this proposal via regression. We simply adopt the same losses as the RPN loss $\mathcal{L}_{rpn}$ and the Fast R-CNN loss $\mathcal{L}_{fast\_rcnn}$ in Faster R-CNN [18] for training.

Spatiotemporal Attention Layer (STAL). In a standardized surgery, correspondence exists between instruments and tissues, since instruments can only act on the specified tissues. Such correspondence can be used to assist the detection of instruments and tissues. As an instrument has a close spatial location to the target tissue, spatial location information can provide guidance for feature fusion. Temporal information is also a good aid to detection tasks in videos due to the sequential nature of data. To combine spatial location and temporal information, we design STAL for instrument-tissue interaction quintuple detection.

For current frame t, we regard this frame as the key frame and the first r frames in front of it as reference frames. As shown in Fig. 2, our STAL aggregates N attention features through N spatiotemporal attention modules (STAM) and augments the input's visual features via addition, expressed as:

$$f_v^{k'} = f_v^k + \text{Concat}(f_A^1, f_A^2, ..., f_A^N) \tag{1}$$

where $f_A^n \in \mathbb{R}^{1 \times c/N}$ is the output features from n_{th} STAM, $f_v^k \in \mathbb{R}^{1 \times c}$ is the input visual features of the key frame, $f_v^{k'} \in \mathbb{R}^{1 \times c}$ is the output refined visual features for the key frame via STAL and $\text{Concat}(\cdot)$ is a concatenation function.

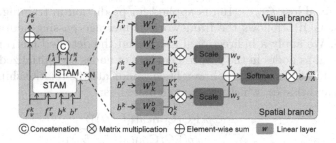

Fig. 2. Spatiotemporal attention layer and spatiotemporal attention module

As illustrated in Fig. 2, our STAM consists of two branches, a visual branch and a spatial branch. The visual branch computes the attention of visual features of RoI from the key frame $f_v^k \in \mathbb{R}^{1 \times c}$ and reference frames $f_v^r \in \mathbb{R}^{r \times c}$ to generate visual weights. RoI features in the key frame are mapped into a query $Q_v^k \in \mathbb{R}^{1 \times c/N}$ via linear layers. RoI features in r reference frames are mapped into a key $K_v^r \in \mathbb{R}^{r \times c/N}$ and a value $V_v^r \in \mathbb{R}^{r \times c/N}$ via linear layers. The spatial branch computes the attention of spatial location information of RoI from the key frame and reference frames to generate spatial location weights. We use binary box maps $b \in \mathbb{R}^{h \times w}$ to encode spatial location information of RoI. Binary box map has $h \times w$ dimension with value one in RoI locations and value zero in other locations. h and w are scaled from the original image shape. Binary box maps of the key frame $b^k \in \mathbb{R}^{1 \times h \times w}$ and reference frames $b^r \in \mathbb{R}^{r \times h \times w}$ are reshaped and mapped into a query $Q_s^k \in \mathbb{R}^{1 \times c/N}$ and a key $K_s^r \in \mathbb{R}^{r \times c/N}$ respectively. The visual similarity weights w_v and the spatial location similarity weights w_s of RoI in the key frame and its reference frames can be calculated by:

$$w_v = \frac{Q_v^k K_v^{rT}}{\sqrt{d_v^r}}, \quad w_s = \frac{Q_s^k K_s^{rT}}{\sqrt{d_s^r}} \tag{2}$$

where T is the transpose function, d_v^r denotes the dimension of K_v^r, and d_s^r denotes the dimension of K_s^r. These two kinds of weights are added to aggregate attention weights. The output attention features of STAM are computed by the multiplication of the mapped RoI features of the reference frames V_v^r and the attention weights. Our STAM can be formulated as $f_A^n = \text{softmax}(w_v + w_s)V_v^r$.

2.2 Quintuple Prediction Stage

In the second stage, as shown in Fig. 1, the backbone network is used to extract feature maps of video frames first. These feature maps and bounding boxes generated in the first stage are passed through RoI align, linear box heads and STAL to generate regional features. Relation reasoning is vital to quintuple prediction, so we propose a graph-based layer, GQPL, to model the relationship of instruments and tissues. Regional features of instruments and tissues are sent into GQPL for feature refinement and quintuple prediction. A binary classification loss, named focal loss $\mathcal{L}_{focal}$ [16], is applied for prediction in training.

Here, our GQPL follows the framework of message passing neural network (MPNN) [6]. We apply a bidirectional graph and the detected instruments and tissues are used as nodes in the graph. Features of nodes are initialized using the input regional features. The graph weights $a_{i,t}$ indicate the relationships between instruments and tissues, which can be calculated as:

$$a_{i,t} = \frac{exp(\text{FC}(\text{Concat}(f_i, f_t)))}{\sum_{n=1}^{N} exp(\text{FC}(\text{Concat}(f_n, f_t)))} \tag{3}$$

where f_i and f_n denotes features of instrument's node, f_t denotes features of tissue's node, and $\text{FC}(\cdot)$ is the fully-connected layer.

To obtain a more discriminative feature representation, we refine node features by bidirectional message passing and updating function:

$$f_i' = \text{Norm}(f_i + \sum_{t=1}^{N_t} a_{i,t}f_t), \quad f_t' = \text{Norm}(f_t + \sum_{i=1}^{N_i} a_{t,i}f_i) \tag{4}$$

where $a_{t,i}$ is the transpose of $a_{i,t}$, N_i and N_t denotes the number of instruments and tissues in the graph, f_i' denotes output features of instrument's node, f_t' denotes output features of tissue's node, and $\text{Norm}(\cdot)$ is the LayerNorm operation.

A fully connected layer is used as the readout function to utilize the refined features of instrument and tissue pairs to predict the action between them, defined as: $action = \text{FC}(\text{Concat}(f_i', f_t'))$. Therefore, we obtain the quintuple by pairing the instrument and tissue detected in the first stage and generating the action between them in the second stage.

3 Experiments

Dataset. To perform the task, we build a cataract surgery video dataset named Cataract Quintuple Dataset based on phacoemulsification, which is a widely used cataract surgery. The dataset consists of 20 videos with frame rate at 1 fps (a total of 13374 frames after removing frames with no instrument-tissue interaction). These video frames are labeled frame by frame with the location of the instruments and tissues as well as the type of instruments, tissues, and actions under the direction of ophthalmologists. Twelve kinds of instruments, 12 kinds of tissues, and 15 kinds of actions form 32 kinds of interaction labels (the details are listed in the supplementary material). The surgical video dataset is randomly split into a training set with 15 videos and a testing set with 5 videos. Some examples of this dataset are presented in the first row of Fig. 3.

Evaluation Metrics. Mean Average Precision (mAP) [7] is used to evaluate the proposed network for quintuple detection on Cataract Quintuple Dataset. A true positive interaction detection should satisfy two conditions: 1. correct instrument-tissue-action class; 2. both instrument and tissue boxes have an Intersection over Union (IoU) higher than 0.5 with corresponding ground truth. To validate the effectiveness of our STAL, mAP is also used to measure instrument and tissue detection performance in the first stage. We denote mAP for the instrument and tissue detection as mAP_{IT} and mAP for the instrument-tissue interaction quintuple detection as mAP_{ITI}.

Table 1. Comparison of methods and ablation studies on Cataract Quintuple Dataset.

Methods	mAP_{IT}	mAP_{ITI}
Faster R-CNN [18]	69.06%	36.89%
iCAN [5]	69.06%	37.29%
Zhang et al. [24]	69.06%	39.01%
QDNet (ours)	**71.30%**	**42.24%**
STALNet with FC	71.30%	40.93%
TALNet with GQPL	70.51%	41.04%
TALNet with FC	70.51%	39.75%
ALNet with GQPL	69.76%	39.99%
ALNet with FC	69.76%	38.76%

Implementation Details. We use ResNet50-FPN [16] as the backbone network. Our STAL aggregates a total of 16 STAM and regards 3 frames in front of the key frame as reference frames. In the first stage, our models are trained for 20 epochs and the initial learning rate is set to 0.001 with 0.1 decayed at 10th and 15th epoch. Detection results of the first stage with scores lower than 0.2 are filtered out and non-maximum suppression with a threshold of 0.5 is applied. Up to 5 instrument boxes and 5 tissue boxes with the highest score are selected to the second stage for quintuple prediction. In the second stage, our models are trained for 20 epochs with the initial learning rate at 0.0001. SGD optimizer with 0.9 momentum and 0.0001 weight decay is used in both stages. Our models are implemented using Pytorch and trained on GeForce GTX 2080 Ti GPUs.

Method Comparison. We build a Faster R-CNN network for instrument and tissue detection in the first stage and a fully-connected layer to predict actions for instrument-tissue pairs as a baseline model. We also reimplement two state-of-the-art HOI detection methods, iCAN [5] and Zhang et al. [24], for instrument-tissue interaction quintuple detection. Quantitative results on the Cataract Quintuple test set are shown in Table 1. Noted that Faster R-CNN is used in the first stage in both iCAN and Zhang et al., so both methods have the same mAP_{IT} for the first stage. Our QDNet outperforms other methods in both instrument and tissue detection stage and quintuple detection stage. Our QDNet exceeds Faster R-CNN baseline by 2.24% at mAP_{IT} by using STAL. The improvement (5.35% higher mAP_{ITI}) is even more obvious for the overall quintuple detection task, since our model also includes relation reasoning by GQPL. Though iCAN and Zhang et al. develop advanced strategies to model relationships between instruments and tissues, our QDNet achieves better results by using spatial and temporal information to assist the detection of instruments and tissues. Results also indicate that directly applying methods in HOI task to surgical scenes cannot achieve satisfactory results. Qualtitative results of Faster R-CNN baseline and our QDNet are presented in Fig. 3. The results show that our QDNet can provide more accurate detection of instruments and tissues. Moreover, when

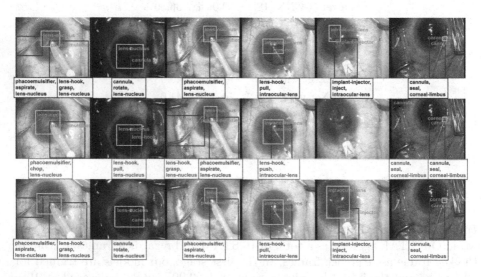

Fig. 3. Example results for instrument-tissue interaction quintuples detection. From top to bottom, we represent the ground-truth, results of Faster R-CNN baseline and our QDNet respectively. The ground-truth or detection bounding boxes of instruments and tissues are marked in blue and light blue. Incorrect quintuple detection results are marked in red and correct detection results are marked in green. (Color figure online)

both the instrument and tissue are correctly detected, our QDNet is able to more accurately determine the action between them.

Ablation Study. Table 1 lists the results of our ablation studies to investigate the effectiveness of key components in our QDNet. STALNet is a Faster R-CNN combine with our proposed STAL for the detection of instruments and tissues as the first stage of our QDNet. TALNet is modified from STALNet by removing the spatial branch in STAL, and ALNet is a Faster R-CNN combined with self-attention modules [22] to compute the attention of proposals within a single frame. FC is a fully-connected layer to predict actions for instrument-tissue pairs in the second stage. ALNet improves over Faster R-CNN baseline by utilizing the self-attention modules, which proves that the relationship between instruments and tissues is helpful in improving detection accuracy. Results of ALNet and TALNet demonstrate that temporal information is important for feature fusion in videos. With the spatial branch in our proposed STAL, our model can achieve better results in instrument-tissue interaction quintuple detection with 0.79% higher mAP_{IT} and 1.18% higher mAP_{ITI}. This shows that spatial information is a vital aid in instrument-tissue interaction quintuple detection. GQPL in QDNet improves the mAP_{ITI} by approximately 1.31%, justifying the need for using GQPL for relation reasoning in instrument-tissue interaction quintuple detection task. Therefore, the two proposed layers are valuable for quintuplet detection.

4   Conclusion

We introduced the task of instrument-tissue interaction detection for surgical scene understanding by detecting ⟨instrument bounding box, tissue bounding box, instrument class, tissue class, action class⟩ quintuples and proposed a quintuple detection network (QDNet) to detect quintuples in surgical videos. Experiments on a cataract surgery video dataset demonstrated that our QDNet is able to aggregate spatial and temporal information and reason relationships between instruments and tissues effectively. Compared with state-of-the-art methods, our model improves the detection accuracy of instruments, tissues and quintuples.

Acknowledgement. This work was supported in part by The National Natural Science Foundation of China(8210072776), Guangdong Provincial Department of Education(2020ZDZX 3043), Guangdong Basic and Applied Basic Research Foundation(2021A1515012195), Shenzhen Natural Science Fund (JCYJ20200109140820699), the Stable Support Plan Program (20200925174052004), and AME Programmatic Fund (A20H4b0141).

References

1. Chao, Y.W., Zhan, W., He, Y., Wang, J., Jia, D.: Hico: A benchmark for recognizing human-object interactions in images. In: IEEE International Conference on Computer Vision (2015)

2. Chen, K., et al.: Application of computer-assisted virtual surgical procedures and three-dimensional printing of patient-specific pre-contoured plates in bicolumnar acetabular fracture fixation. Orthop. Traumatol. Surg. Res. **105**, 877–884 (2019). https://doi.org/10.1016/j.otsr.2019.05.011

3. Chen, Y.W., Hanak, B.W., Yang, T.C., Wilson, T.A., Nagatomo, K.J.: Computer-assisted surgery in medical and dental applications. Expert Rev. Med. Devices **18**(7), 669–696 (2021)

4. DiPietro, R.S., et al.: Recognizing surgical activities with recurrent neural networks. In: MICCAI (2016)

5. Gao, C., Zou, Y., Huang, J.B.: iCAN: instance-centric attention network for human-object interaction detection. In: British Machine Vision Conference (2018)

6. Gilmer, J., Schoenholz, S.S., Riley, P.F., Vinyals, O., Dahl, G.E.: Neural message passing for quantum chemistry. In: International Conference on Machine Learning, pp. 1263–1272. PMLR (2017)

7. Gupta, S., Malik, J.: Visual semantic role labeling. arXiv preprint arXiv:1505.04474 (2015)

8. Hashimoto, D.A., et al.: Computer vision analysis of intraoperative video: automated recognition of operative steps in laparoscopic sleeve gastrectomy. Ann. Surg. **270**(3), 414 (2019)

9. He, K., Gkioxari, G., Dollár, P., Girshick, R.: Mask r-CNN. In: Proceedings of the IEEE International Conference on Computer Vision, pp. 2961–2969 (2017)

10. He, Y., Huang, T., Zhang, Y., An, J., He, L.H.: Application of a computer-assisted surgical navigation system in temporomandibular joint ankylosis surgery: a retrospective study. Int. J. Oral Maxillofac. Surg. **46**, 189–197 (2016). https://doi.org/10.1016/j.ijom.2016.10.006

11. Islam, M., Lalithkumar, S., Ming, L.C., Ren, H.: Learning and reasoning with the graph structure representation in robotic surgery. CoRR arXiv preprint arXiv:2007.03357 (2020)

12. Jin, Y., et al.: SV-RCNet: workflow recognition from surgical videos using recurrent convolutional network. IEEE Trans. Med. Imaging **37**(5), 1114–1126 (2018). https://doi.org/10.1109/TMI.2017.2787657

13. Jin, Y., et al.: Multi-task recurrent convolutional network with correlation loss for surgical video analysis. Med. Image Anal. **59**, 101572 (2020)

14. Khatibi, Toktam, Dezyani, Parastoo: Proposing novel methods for gynecologic surgical action recognition on laparoscopic videos. Multimed. Tools Appl. **79**(41), 30111–30133 (2020). https://doi.org/10.1007/s11042-020-09540-y

15. Lalys, F., Jannin, P.: Surgical process modelling: a review. Int. J. Comput. Assist. Radiol. Surg. **9**(3), 495–511 (2013). https://doi.org/10.1007/s11548-013-0940-5

16. Lin, T.Y., Dollár, P., Girshick, R., He, K., Hariharan, B., Belongie, S.: Feature pyramid networks for object detection. In: Proceedings of the IEEE Conference on Computer Vision and Pattern Recognition, pp. 2117–2125 (2017)

17. Nwoye, C.I., et al.: Recognition of instrument-tissue interactions in endoscopic videos via action triplets. In: Martel, A.L., et al. (eds.) MICCAI 2020. LNCS, vol. 12263, pp. 364–374. Springer, Cham (2020). https://doi.org/10.1007/978-3-030-59716-0_35

18. Ren, S., He, K., Girshick, R., Sun, J.: Faster R-CNN: towards real-time object detection with region proposal networks. In: Advances in Neural Information Processing Systems, vol.28 (2015)

19. Seenivasan, L., Mitheran, S., Islam, M., Ren, H.: Global-reasoned multi-task learning model for surgical scene understanding. IEEE Robot. Autom. Lett. **7**(2), 3858–3865 (2022). https://doi.org/10.1109/LRA.2022.3146544

20. Twinanda, A.P., Shehata, S., Mutter, D., Marescaux, J., De Mathelin, M., Padoy, N.: Endonet: a deep architecture for recognition tasks on laparoscopic videos. IEEE Trans. Med. Imaging **36**(1), 86–97 (2016)
21. Ulutan, O., Iftekhar, A., Manjunath, B.S.: Vsgnet: Spatial attention network for detecting human object interactions using graph convolutions. arXiv preprint arXiv:2003.05541 (2020)
22. Vaswani, A., et al.: Attention is all you need. In: Advances in Neural Information Processing Systems, vol.30 (2017)
23. Xu, M., Islam, M., Ming Lim, C., Ren, H.: Learning domain adaptation with model calibration for surgical report generation in robotic surgery. In: 2021 IEEE International Conference on Robotics and Automation (ICRA), pp. 12350–12356 (2021). https://doi.org/10.1109/ICRA48506.2021.9561569
24. Zhang, F.Z., Campbell, D., Gould, S.: Spatially conditioned graphs for detecting human-object interactions (2020)

Surgical Skill Assessment via Video Semantic Aggregation

Zhenqiang Li[1], Lin Gu[2], Weimin Wang[3]([✉]), Ryosuke Nakamura[4],
and Yoichi Sato[1]

[1] The University of Tokyo, Tokyo, Japan
{lzq,ysato}@iis.u-tokyo.ac.jp
[2] RIKEN, Tokyo, Japan
lin.gu@riken.jp
[3] Dalian University of Technology, Dalian, China
wangweimin@dlut.edu.cn
[4] AIST, Tsukuba, Japan
r.nakamura@aist.go.jp

Abstract. Automated video-based assessment of surgical skills is a promising task in assisting young surgical trainees, especially in poor-resource areas. Existing works often resort to a CNN-LSTM joint framework that models long-term relationships by LSTMs on spatially pooled short-term CNN features. However, this practice would inevitably neglect the difference among semantic concepts such as tools, tissues, and background in the spatial dimension, impeding the subsequent temporal relationship modeling. In this paper, we propose a novel skill assessment framework, **Vi**deo **S**emantic **A**ggregation (ViSA), which discovers different semantic parts and aggregates them across spatiotemporal dimensions. The explicit discovery of semantic parts provides an explanatory visualization that helps understand the neural network's decisions. It also enables us to further incorporate auxiliary information such as the kinematic data to improve representation learning and performance. The experiments on two datasets show the competitiveness of ViSA compared to state-of-the-art methods. Source code is available at: bit.ly/MICCAI2022ViSA.

Keywords: Surgical skill assessment · Surgical video understanding · Video representation learning · Temporal modeling

1 Introduction

Automated assessment of surgical skills is a promising task in computer-aided surgical training [10,26], especially in the resource-poor countries. Surgical skill

Supplementary Information The online version contains supplementary material available at https://doi.org/10.1007/978-3-031-16449-1_39.

assessment involves two major challenges [23]: 1. how to capture the difference between fine-grained atomic actions. 2. how to model the contextual relationships between these actions in the long-term range. Previous works [2,9,13,23,24] counteract above two challenges by stacking convolutional neural networks (CNNs) and LSTMs: CNNs for short-term feature extraction and temporal aggregation networks (e.g., LSTMs) for long-term relationship modeling. For example, Fawaz et al. [1] used 1D CNN to encode kinematic data before the aggregation over time by global average pooling. Wang et al. [23] extract 2D or 3D CNN features from video frames and model their temporal relationship by leveraging an LSTM network. Considerable progress has been made in predicting skill levels [1,2,9,11,19,24,27] or scores [13,23,26] using kinematic data or video frames.

Regarding the video-based methods, most methods [2,9,13,23,24] perform the global pooling over the spatial dimension on CNN features before feeding them to the subsequent network. We argue that this global pooling operation ignores the semantics variance of different features and compresses all the information together in the spatial scale without distinction. As a result, the consequent networks could hardly model the temporal relationship of the local features in different spatial parts separately, e.g., the movements of different tools and the status changes of tissue. This bottleneck is particularly severe for surgical skill assessment because the tracking of tools from a large part of the background is essential to judge the manipulation quality. Similarly, the interactions between tools and tissue across time are also important for the assessment. To make it worse, most of existing methods are end-to-end neural networks, revealing little about what motion or appearance information is captured.

Since surgical videos comprise limited objects with explicit semantic meanings such as tools, tissue, and background, we propose a novel framework, **Vi**deo **S**emantic **A**ggregation (ViSA), for surgical skill assessment by aggregating local features across spatiotemporal dimensions according to these semantic clues. This aggregation allows our method to separate the video parts related to different semantics from the background and further dedicate to tracking the tool or modeling its interaction with tissue.

As shown in Fig. 1, ViSA first aggregates similar local semantic features through clustering and generates the abstract features for each semantic group in the semantic grouping stage. Then we aggregate the features for the same semantic across time and model their temporal contextual relationship via multiple bidirectional LSTMs. The spatially aggregated features can visualize the correspondence between different video parts and semantics in the form of assignment maps, facilitating the transparency of the network. In addition, the explicit semantic group of features allows us to incorporate auxiliary information that further improves the assignment and influences the intermediate features, e.g., using kinematic data to bound features for certain semantics to tools.

Our contribution is threefold: (1) We propose a novel framework, ViSA, to assess skills in surgical videos via explicitly splitting different semantics in video and efficiently aggregating them across spatiotemporal dimensions.

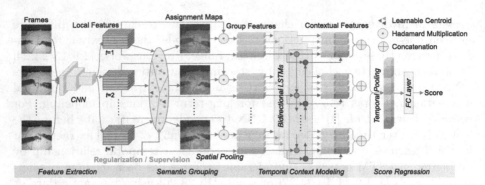

Fig. 1. The proposed framework, **Vi**deo **S**emantic **A**ggregation (ViSA), for surgical skill assessment. It aggregates local CNN features over space and time using semantic clues embedding in the video. We aim to group video parts according to semantics, *e.g.*, tools, tissue and background, and model the temporal relationship for different parts separately via this framework. We also investigate the regularization or additional supervision to the group results for enhancement.

(2) Via aggregating the video representations and regularization, our method can discover different semantic parts such as tools, tissue, and background in videos. This provides explanatory visualization as well as allows integrating auxiliary information like tool kinematics for performance enhancement. (3) The framework achieves competitive performance on two datasets: JIGSAWS [4] and HeiChole [22].

2   Methodology

As shown in Fig. 1, for each video, our framework takes frames from a fixed number of timesteps as input and predicts the final skill assessment score through the following 4 stacked modules: (1) the Feature Extraction Module (FEM) that produces feature maps for each timestep (Sect. 2.1); (2) the Semantic Grouping Module (SGM) that aggregates local features into a specified number of groups based on the embedded semantics (Sect. 2.2); (3) the Temporal Context Modeling Module (TCMM) that models the contextual relationship for the feature series of each group (Sect. 2.3); (4) the Score Prediction Module (SPM) that regresses the final score based on the spatiotemporally aggregated features (Sect. 2.4).

2.1   Feature Extraction Module

Our framework first feeds the input frames to CNNs to collect the intermediate layer responses as the feature maps for the subsequent processing. We stack the feature maps for one video along the temporal dimension and registered them as $X \in \mathbb{R}^{T \times H \times W \times C}$, where T is the number of timesteps, H, W represents the height and width of each feature map, and C is the number of channels. X can also be seen as a group of local features, where each feature is indexed by the temporal and spatial position t, i, j as $X_{tij} \in \mathbb{R}^C$.

2.2 Semantic Grouping Module

This module aggregates the extracted features X into a fixed number of groups, with each representing a specific kind of semantic meaning. Taking inspirations from [5,7], we achieve this by clustering all local CNN features across the entire spatiotemporal range of the video according to K learnable vectors $D = [d_1, \cdots, d_K]^\top \in \mathbb{R}^{K \times C}$ which record the feature centroid of each group (drawn as colored triangles in Fig. 1).

Local Feature Assignment. We softly assign local features to different groups by computing the assignment possibilities $P \in \mathbb{R}^{T \times H \times W \times K}$ subject to $P_{tij}^k = \frac{\exp(-\|(X_{tij}-d_k)/\sigma_k\|_2^2)}{\sum_{k'} \exp(-\|(X_{tij}-d_{k'})/\sigma_{k'}\|_2^2)}$. Component P_{tij}^k represents the possibility of assigning the local CNN feature X_{tij} to the kth semantic group. $\sigma_k \in (0,1)$ is a learnable factor to adjust feature magnitudes and smooth the assignment for each semantic group. Taking the index of the group with the maximum possibility at each position, we obtain the assignment maps $A \in \mathbb{R}^{T \times H \times W}$ with component $A_{tij} = \text{argmax}_{k=1,\ldots,K} P_{tij}^k, \forall t, i, j$. As shown in Fig. 1, the assignment maps visualize the correspondence between video parts and semantic groups.

Group Feature Aggregation. For each timestep, we aggregate the corresponding local features according to the soft-assignment results and generate an abstract representation for each semantic group. We first calculate the normalized residual between the centroid and the averaged local features weighted by assignment possibilities as $z_t^k = \frac{z_t'^k}{\|z_t'^k\|_2}$, where $z_t'^k = \frac{1}{\sigma_k}\left(\sum_{ij} \frac{P_{tij}^k X_{tij}}{\sum_{ij} P_{tij}^k} - d_k\right)$. Then we get one *group feature* g_t^k for capturing the abstract information of the kth semantics at each timestep by transforming z_t^k with a sub-network f_g consisting of 2 linear transformations as $g_t^k = f_g(z_t^k)$.

Group Existence Regularization. To avoid most local features being allocated to one single group, we leverage a regularization term [7] to retain the even distribution among groups at every timestep. Specifically, it regularizes the existence of each group by constraining the max assignment probability of one group, i.e., $\hat{p}_t^k := \max_{i,j} P_{t,i,j}^k$, as close to 1.0 as possible. Instead of tightly constraining $\hat{p}_t^k$ to an exact number, the regularization term restricts the cumulative distribution function of $\hat{p}^k$ to follow a beta distribution. We implement the regularization term as $\mathcal{L}_{exist} = \sum_{i=1}^{T} \left(\log\left(p_i^{\star k} + \epsilon\right) - \log\left(F^{-1}(\frac{2i-1}{2T}; \alpha, \beta) + \epsilon\right)\right)$. $p^{\star k} = \text{sort}([\hat{p}_1^k \cdots \hat{p}_T^k])$ arranges the maximum assignment probabilities of the kth group at all timesteps in ascending order. $\alpha=1$ and $\beta=0.001$ control the shape of the beta distribution. ϵ is a small value for numerical stability. $F^{-1}(\frac{2i-1}{2T}; \alpha, \beta)$ means an inverse cumulative beta distribution function, which returns the ith element of the sequence of T probability values obeying this distribution.

2.3 Temporal Context Modeling Module

For each obtained semantic group feature series, this module aims to keep track of its long-term dependencies and model their contextual relationships independently in a recurrent manner. As shown in Fig. 1, we achieve this by employing K bidirectional LSTMs (BiLSTMs): $\overrightarrow{\boldsymbol{h}}_t^k = \overrightarrow{L}^k(\overrightarrow{\boldsymbol{h}}_{t-1}^k, \boldsymbol{g}_t^k)$, $\overleftarrow{\boldsymbol{h}}_t^k = \overleftarrow{L}^k(\overleftarrow{\boldsymbol{h}}_{t+1}^k, \boldsymbol{g}_t^k)$, where $\overrightarrow{L}^k$ and $\overleftarrow{L}^k$ denote the kth forward and backward LSTMs respectively. The output vectors of every timestep $\overrightarrow{\boldsymbol{h}}_t^k$ and $\overleftarrow{\boldsymbol{h}}_t^k$ from the two directions are further concatenated to form the *contextual feature* $\boldsymbol{c}_t^k = [\overrightarrow{\boldsymbol{h}}_t^{k\top}, \overleftarrow{\boldsymbol{h}}_t^{k\top}]^\top$. To prevent the potential information loss caused by the separated modeling (*e.g.*, the interaction between groups), we also employ another BiLSTM to model the global features $[\boldsymbol{x}_1 \cdots \boldsymbol{x}_T]$ which are obtained by taking spatial average pooling on $\boldsymbol{X}$. $\boldsymbol{c}_t^0$ denotes the generated contextual features by this additional BiLSTM.

2.4 Score Prediction Module

This module concatenates the contextual features $\boldsymbol{c}_t^k$ of different semantic groups at the same timestep into a vector, followed by an average pooling on these vectors across the temporal dimension. The pooled vector is finally regressed to the skill score s by f_s, a fully connected layer. We formulate this module mathematically as follows: $\hat{s} = f_s(\frac{1}{T}\sum_{t=1}^{T}[\boldsymbol{c}_t^{0\top} \cdots \boldsymbol{c}_t^{K\top}]^\top)$. The training loss function is defined as: $\mathcal{L} = (s - \hat{s})^2 + \lambda \mathcal{L}_{exist}$, with s denoting the ground-truth score. The scalar $\lambda = 10$ controls the contribution of the group existence regularization.

2.5 Incorporating Auxiliary Supervision Signals

Our unique representation aggregation strategy also allows enhancing the feature grouping results by incorporating additional supervision signals to assign specific known semantic information to certain groups. Specifically, we propose a Heatmap Prediction Module (HPM), which takes as input the CNN feature maps $\boldsymbol{X}$ and the assignment possibility maps $\boldsymbol{P}^m$ of the mth group. The module predicts the positions of the specified semantics by generating heatmaps $\hat{\boldsymbol{H}} = f_{HPM}(\boldsymbol{P}^m \odot \boldsymbol{X}) \in \mathbb{R}^{T \times H \times W}$, where $\odot$ denotes the Hadamard product, and f_{HPM} is composed of one basic 1×1 conv block followed by a Sigmoid function. Assuming the 2D positions of the specified semantics is known, we generate the position heatmaps $\boldsymbol{H} \in \mathbb{R}^{T \times H \times W}$ from the 2D positions as the supervision signal. The framework integrating HPM is trained by the loss function with an extra position regularization $\mathcal{L}' = (s - \hat{s})^2 + \lambda_1 \mathcal{L}_{exist} + \lambda_2 \mathcal{L}_{pos}$, where $\mathcal{L}_{pos}$ computes the binary cross entropy between H and $\hat{H}$ and $\lambda_1 = 10$, $\lambda_2 = 20$.

3   Experiments

Dataset. We evaluate ViSA framework on 2 datasets for surgical skill assessment: JIGSAWS [4] and HeiChole [22]. JIGSAWS is a widely used dataset consisting of 3 elementary surgical tasks: Knot Tying (KT), Needle Passing (NP), and Suturing (SU). Each task contains more than 30 trials performed on the da Vinci surgical system, which is rated from 6 aspects. Following [13,23,26], the sum score is used as the ground truth. We validate our method on every task by 3 kinds of cross-validation schemes: Leave-one-supertrial-out (LOSO), Leave-one-user-out (LOUO) and 4-Fold [3,14,18]. HeiChole is a challenging dataset containing 24 endoscopic videos in real surgical environments for laparoscopic cholecystectomy. For each video, 2 clips of the phases calot triangle dissection and gallbladder dissection are provided with skill scores from 5 domains. We use the sum score as the ground truth. We train and validate our framework on the 48 video clips by 4-fold cross-validation with a 75/25 partition ratio.

Table 1. Baseline comparison on JIGSAWS. We report the Spearman's Rank Correlations by three cross-validation schemes on every task. **K**: Kinematic data, **V**: video frames, *: extra annotations such as surgical gestures are utilized.

Input	Method	KT			NP			SU			Avg.		
		LOSO	LOUO	4-Fold	LOSO	LOUO	4-Fold	LOSO	LOUO	4-Fold	LOSO	LOUO	4-Fold
K	SMT-DCT-DFT [26]	0.70	0.73	-	0.38	0.23	-	0.64	0.10	-	0.59	0.40	-
	DCT-DFT-ApEn [26]	0.63	0.60	-	0.46	0.25	-	0.75	0.37	-	0.63	0.41	-
V	ResNet-LSTM [23]	0.52	0.36	-	0.84	0.33	-	0.73	0.67	-	0.72	0.59	-
	C3D-LSTM [16]	0.81	0.60	-	0.84	0.78	-	0.69	0.59	-	0.79	0.67	-
	C3D-SVR [16]	0.71	0.33	-	0.75	-0.17	-	0.42	0.37	-	0.65	0.18	-
	USDL [18]	-	-	0.61	-	-	0.63	-	-	0.64	-	-	0.63
	MUSDL [18]	-	-	0.71	-	-	0.69	-	-	0.71	-	-	0.70
	*S3D [25]	0.64	0.14	-	0.57	0.35	-	0.68	0.03	-	-	-	-
	*ResNet-MTL-VF [23]	0.63	0.72	-	0.73	0.48	-	0.79	0.68	-	0.73	0.64	-
	*C3D-MTL-VF [23]	0.89	**0.83**	-	0.75	0.86	-	0.77	0.69	-	0.75	0.68	-
V+K	JR-GCN [14]	-	0.19	0.75	-	0.67	0.51	-	0.35	0.36	-	0.40	0.57
	AIM [3]	-	0.61	0.82	-	0.34	0.65	-	0.45	0.63	-	0.47	0.71
	MultiPath-VTP [13]	-	0.58	0.78	-	0.62	0.76	-	0.45	0.79	-	0.56	0.78
	*MultiPath-VTPE [13]	-	0.59	0.82	-	0.65	0.76	-	0.45	**0.83**	-	0.57	0.80
V	ViSA	**0.92**	0.76	**0.84**	**0.93**	**0.90**	**0.86**	**0.84**	**0.72**	0.79	**0.90**	**0.81**	**0.83**

Evaluation Metric. For JIGSAWS, since previous works [13,23] only included the results on Spearman's Rank Correlation (**Corr**), we report the validation correlations averaged on all folds for one task and compute the average correlation across tasks by Fisher's z-value [15] for baseline comparison. We incorporate the results on Mean Absolute Error (**MAE**) in ablation studies. For HeiChole, we report both the correlation and MAE averaged on all folds. The reported results are averaged on multiple runs with different random seeds.

Implementation Details. We employ R(2+1)D-18 [20] pre-trained on Kinetic-400 dataset [8] as the feature extractor and take the response of the 4th convolutional block as the 3D spatiotemporal feature. Since each 3D feature is extracted from 4 frames, we divide each video into T segments and sample a 4-frame snippet from each. T is set to 32 on JIGSAWS videos and 64 on longer HeiChole videos. Each sampled frame is resized to 160×120 and crop the 112×112 region in the center as input. Hence, the spatial size of the extracted CNN feature maps is 14×14. We also investigate 2D-CNNs feature extractor by using ResNet-18 pre-trained on ImageNet. Since the surgical scenes in our experiments explicitly include 3 parts: represented by tools, tissues, and background, we initialize the number of semantic groups $K = 3$. It should be set according to the scene complexity and fine-tuned based on the experiment results. The framework is implemented by PyTorch. We use SDG optimizer with mean squared error loss. Models are trained in 40 epochs with a batch size of 4. Learning rates are initialized as 3e–5 and decayed by 0.1 times for every 20 epochs.

Table 2. Results on HeiChole. Baseline frameworks are newly constructed and compared with ViSA by MAE and Corr metrics.

Method	MAE	Corr
R2D-18 + FC	1.56	0.32
R2D-18 + LSTM	1.42	0.15
R(2+1)D-18 + FC	1.54	0.29
R(2+1)D-18 + LSTM	1.33	0.31
ViSA (R2D-18)	**1.27**	**0.46**
ViSA (R(2+1)D-18)	**1.27**	**0.46**

Table 3. Ablation studies on JIGSAWS. We train frameworks across videos of three tasks and report Corr and MAEs on LOSO and LOUO settings. K denotes the number of Groups.

FEM	SGM	TCMM	K	LOSO		LOUO	
				MAE	Corr	MAE	Corr
R2D-18	✗	BiLSTMs	-	3.40	0.65	3.53	0.67
R2D-18	✓	BiLSTMs	3	3.27	0.80	3.42	0.74
R(2+1)D-18	✓	LSTMs	3	2.41	0.83	3.23	0.78
R(2+1)D-18	✓	Transformer	3	3.02	0.76	3.27	0.68
R(2+1)D-18	✗	BiLSTMs	-	2.90	0.73	3.30	0.72
R(2+1)D-18	✓	BiLSTMs	2	2.34	0.84	3.07	0.72
R(2+1)D-18	✓	BiLSTMs	4	2.32	0.85	**2.68**	**0.79**
R(2+1)D-18	✓	BiLSTMs	3	**2.24**	**0.86**	2.86	0.76

JIGSAWS   HeiChole

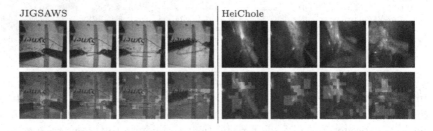

Fig. 2. Visualization of the assignment maps that display the correspondence between video parts and different semantic groups.

3.1   Baseline Comparison

Table 1 shows the quantitative baseline comparison results of ViSA on JIGSAWS. Baseline results are taken from previous papers. We find that ViSA outperforms many competitive methods on most tasks and cross-validation schemes. ViSA has a CNN-RNN framework akin to C3D-MTL-VF [23] but surpasses it with obvious margins on most LOSO and LOUO metrics. On 4-Fold, ViSA achieves nearly equivalent performance as MultiPath-VTPE [13] which needs extra input sources such as frame-wise surgical gestures information and tool movement paths. For HeiChole, since no baseline is available, we newly constructed four basic frameworks by combining 2D or 3D CNN feature extractors with temporal aggregation modules of fully connected layers (FC) or LSTMs, which share similar CNN architectures and pre-trained parameters as ViSA. Table 2 shows the out-performance of ViSA against the four baselines on both metrics.

3.2   Assignment Visualization

Figure 2 visualizes the assignment maps generated by ViSA on two datasets. The three semantic groups generally correspond to the tools, the manipulated tissues, and the background regions as expected. Notably, ViSA gets this assignment result without taking any supervision of semantics, which further indicates its effectiveness in modeling surgical actions over spatiotemporal dimensions.

3.3   Ablation Study

We also conduct the ablation analysis on three key components and one hyperparameter of ViSA: Feature Extraction Module (**FEM**), Semantic Grouping Module (**SGM**), Temporal Context Modeling Module (**TCMM**), and the number of semantic groups (**K**). Referring to [23], we train and validate the ablative frameworks across videos of 3 JIGSAWS tasks by forming them together, in order for stable results on more samples. Results in Table 3 indicate that: (1) leveraging SGM boosts the performance on either R(2+1)D-18 or R2D-18; (2) BiLSTMs perform better than LSTMs and Transformer [21] in modeling temporal context; (3) increasing K from 2 to 3 causes more improvements than raising it from 3 to 4, which indicates separating semantics into 3 groups fits the JIGSAWS dataset. Transformer consists of two layers of LayerNorm+Multi-Head-Attention+MLP. Considering the data-hungry nature of Transformer [6], we attribute its unremarkable performance to the small amount of training data.

Figure 3 presents the visual explanation results generated by the post-hoc interpretation method Grad-CAM [12,17] which localizes the input regions used by networks for decision making. Compared to the network without using the semantic grouping module (SGM), the full framework employs more task-related parts for predictions (*e.g.*, regions about robotic tools) and discards many unrelated regions. We attribute the improvement to explicitly discovering different video parts and modeling their spatiotemporal relationship in our framework.

3.4   Improved Performance with Supervision

ViSA also supports the explicit supervision on the grouping process that allocates specific semantics to certain expected groups. On JIGSAWS, the kinematic

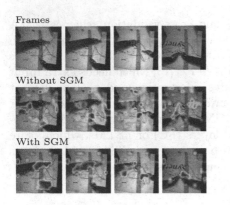

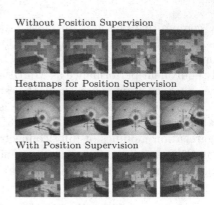

Fig. 3. Visual explanations generated by Grad-CAM. SGM facilitates the concentration on the tools and discarding the unrelated background regions.

Fig. 4. Improved semantic grouping results after using the tool position supervision. Green regions turn to focus on tool clips after using supervision. (Color figure online)

data recording the positions of two robotic clips in 3D space is available. We first approximate the two clips' positions on the 2D image plane by projecting their 3D kinematic positions with estimated transformation matrix. Then we generate the heatmaps $H \in \mathbb{R}^{T \times H \times W}$ from the 2D positions and train the framework as described in Sect. 2.5. Figure 4 illustrates one example of the generated position heatmaps. On the Suturing task, we achieve the improvement in average validation correlations as $0.84 \rightarrow 0.87$, $0.72 \rightarrow 0.80$ and $0.80 \rightarrow 0.86$ on LOSO, LOUO, and 4-Fold schemes respectively. In Fig. 4, we show one imperfect assignment maps generated by the model without supervision and its corresponding maps after using the position supervision. Although the position supervision is not fully precise, it still benefits the network to discover the tools' features. Hence, in Fig. 4, the green regions become less noisy and are guided to the regions of tool clips and tool-tissue interactions after using the supervision.

4   Conclusion

In this paper, a novel framework called ViSA is proposed to predict the skill score from surgical videos by discovering and aggregating different semantic parts across spatiotemporal dimensions. The framework can achieve competitive performance on two datasets: JIGSAWS and HeiChole as well as support explicit supervision on the feature semantic grouping for performance improvement.

Acknowledgment. This work is supported by JST AIP Acceleration Research Grant Number JPMJCR20U1, JSPS KAKENHI Grant Number JP20H04205, JST ACT-X Grant Number JPMJAX190D, JST Moonshot R&D Grant Number JPMJMS2011, Fundamental Research Funds for the Central Universities under Grant DUT21RC(3)028 and a project commissioned by NEDO.

References

1. Ismail Fawaz, H., Forestier, G., Weber, J., Idoumghar, L., Muller, P.-A.: Evaluating Surgical skills from kinematic data using convolutional neural networks. In: Frangi, A.F., Schnabel, J.A., Davatzikos, C., Alberola-López, C., Fichtinger, G. (eds.) MICCAI 2018. LNCS, vol. 11073, pp. 214–221. Springer, Cham (2018). https://doi.org/10.1007/978-3-030-00937-3_25

2. Funke, I., Mees, S.T., Weitz, J., Speidel, S.: Video-based surgical skill assessment using 3d convolutional neural networks. Int. J. Comput. Assist. Radiol. Surg. **14**(7), 1217–1225 (2019)

3. Gao, J., et al.: An asymmetric modeling for action assessment. In: Vedaldi, A., Bischof, H., Brox, T., Frahm, J.-M. (eds.) ECCV 2020. LNCS, vol. 12375, pp. 222–238. Springer, Cham (2020). https://doi.org/10.1007/978-3-030-58577-8_14

4. Gao, Y., et al.: Jhu-isi gesture and skill assessment working set (jigsaws): a surgical activity dataset for human motion modeling. In: MICCAI workshop: M2cai, vol. 3, p. 3 (2014)

5. Girdhar, R., Ramanan, D., Gupta, A., Sivic, J., Russell, B.: Actionvlad: learning spatio-temporal aggregation for action classification. In: Proceedings of the IEEE Conference on Computer Vision and Pattern Recognition, pp. 971–980 (2017)

6. Hassani, A., Walton, S., Shah, N., Abuduweili, A., Li, J., Shi, H.: Escaping the big data paradigm with compact transformers. arXiv preprint arXiv:2104.05704 (2021)

7. Huang, Z., Li, Y.: Interpretable and accurate fine-grained recognition via region grouping. In: Proceedings of the IEEE Conference on Computer Vision and Pattern Recognition, pp. 8662–8672 (2020)

8. Kay, W., et al.: The kinetics human action video dataset. arXiv preprint arXiv:1705.06950 (2017)

9. Kelly, J.D., Petersen, A., Lendvay, T.S., Kowalewski, T.M.: Bidirectional long short-term memory for surgical skill classification of temporally segmented tasks. Int. J. Comput. Assist. Radiol. Surg. **15**(12), 2079–2088 (2020). https://doi.org/10.1007/s11548-020-02269-x

10. Lavanchy, J.L., et al.: Automation of surgical skill assessment using a three-stage machine learning algorithm. Sci. Rep. **11**(1), 1–9 (2021)

11. Li, Z., Huang, Y., Cai, M., Sato, Y.: Manipulation-skill assessment from videos with spatial attention network. In: Proceedings of the IEEE/CVF International Conference on Computer Vision Workshops. pp. 0–0 (2019)

12. Li, Z., Wang, W., Li, Z., Huang, Y., Sato, Y.: Spatio-temporal perturbations for video attribution. IEEE Trans. Circuits Syst. Video Technol. **32**(4), 2043–2056 (2021)

13. Liu, D., Li, Q., Jiang, T., Wang, Y., Miao, R., Shan, F., Li, Z.: Towards unified surgical skill assessment. In: Proceedings of the IEEE/CVF Conference on Computer Vision and Pattern Recognition, pp. 9522–9531 (2021)

14. Pan, J.H., Gao, J., Zheng, W.S.: Action assessment by joint relation graphs. In: Proceedings of the IEEE/CVF International Conference on Computer Vision. pp. 6331–6340 (2019)

15. Parmar, P., Morris, B.: Action quality assessment across multiple actions. In: IEEE Winter Conference on Applications of Computer Vision. pp. 1468–1476 (2019)

16. Parmar, P., Tran Morris, B.: Learning to score olympic events. In: Proceedings of the IEEE Conference on Computer Vision and Pattern Recognition Workshops, pp. 20–28 (2017)

17. Selvaraju, R.R., Cogswell, M., Das, A., Vedantam, R., Parikh, D., Batra, D.: Grad-cam: visual explanations from deep networks via gradient-based localization. In: Proceedings of the IEEE/CVF International Conference on Computer Vision (2017)
18. Tang, Y., Ni, Z., Zhou, J., Zhang, D., Lu, J., Wu, Y., Zhou, J.: Uncertainty-aware score distribution learning for action quality assessment. In: Proceedings of the IEEE Conference on Computer Vision and Pattern Recognition (2020)
19. Tao, L., Elhamifar, E., Khudanpur, S., Hager, G.D., Vidal, R.: Sparse hidden Markov models for surgical gesture classification and skill evaluation. In: Abolmaesumi, P., Joskowicz, L., Navab, N., Jannin, P. (eds.) IPCAI 2012. LNCS, vol. 7330, pp. 167–177. Springer, Heidelberg (2012). https://doi.org/10.1007/978-3-642-30618-1_17
20. Tran, D., Wang, H., Torresani, L., Ray, J., LeCun, Y., Paluri, M.: A closer look at spatiotemporal convolutions for action recognition. In: Proceedings of the IEEE Conference on Computer Vision and Pattern Recognition, pp. 6450–6459 (2018)
21. Vaswani, A., et al.: Attention is all you need. Advances in Neural Information Processing Systems 30 (2017)
22. Wagner, M., et al.: Comparative validation of machine learning algorithms for surgical workflow and skill analysis with the heichole benchmark. arXiv preprint arXiv:2109.14956 (2021)
23. Wang, T., Wang, Y., Li, M.: Towards accurate and interpretable surgical skill assessment: a video-based method incorporating recognized surgical gestures and skill levels. In: Martel, A.L., Abolmaesumi, P., Stoyanov, D., Mateus, D., Zuluaga, M.A., Zhou, S.K., Racoceanu, D., Joskowicz, L. (eds.) MICCAI 2020. LNCS, vol. 12263, pp. 668–678. Springer, Cham (2020). https://doi.org/10.1007/978-3-030-59716-0_64
24. Wang, Z., Majewicz Fey, A.: Deep learning with convolutional neural network for objective skill evaluation in robot-assisted surgery. Int. J. Comput. Assist. Radiol. Surg. 13(12), 1959–1970 (2018). https://doi.org/10.1007/s11548-018-1860-1
25. Xiang, X., Tian, Y., Reiter, A., Hager, G.D., Tran, T.D.: S3d: stacking segmental p3d for action quality assessment. In: IEEE International Conference on Image Processing, pp. 928–932. IEEE (2018)
26. Zia, A., Essa, I.: Automated surgical skill assessment in RMIS training. Int. J. Comput. Assist. Radiol. Surg. 13(5), 731–739 (2018)
27. Zia, A., Sharma, Y., Bettadapura, V., Sarin, E.L., Essa, I.: Video and accelerometer-based motion analysis for automated surgical skills assessment. Int. J. Comput. Assist. Radiol. Surg. 13(3), 443–455 (2018). https://doi.org/10.1007/s11548-018-1704-z

Nonlinear Regression of Remaining Surgical Duration via Bayesian LSTM-Based Deep Negative Correlation Learning

Junyang Wu, Rong Tao, and Guoyan Zheng[✉]

Institute of Medical Robotics, School of Biomedical Engineering, Shanghai Jiao Tong University, No. 800, Dongchuan Road, Shanghai 200240, China
guoyan.zheng@sjtu.edu.cn

Abstract. In this paper, we address the problem of estimating remaining surgical duration (RSD) from surgical video frames. We propose a Bayesian long short-term memory (LSTM) network-based Deep Negative Correlation Learning approach called BD-Net for accurate regression of RSD prediction as well as estimating prediction uncertainty. Our method aims to extract discriminative visual features from surgical video frames and model the temporal dependencies among frames to improve the RSD prediction accuracy. To this end, we propose to ensemble a group of Bayesian LSTMs on top of a backbone network by the way of deep negative correlation learning (DNCL). More specifically, we deeply learn a pool of decorrelated Bayesian regressors with sound generalization capabilities through managing their intrinsic diversities. BD-Net is simple and efficient. After training, it can produce both RSD prediction and uncertainty estimation in a single inference run. We demonstrate the efficacy of BD-Net on a public video dataset containing 101 cataract surgeries. The experimental results show that the proposed BD-Net achieves better results than the state-of-the-art (SOTA) methods. A reference implementation of our method can be found at: https://github.com/jywu511/BD-Net.

Keywords: Remaining surgical duration · Bayesian LSTM · Deep negative correlation learning · Nonlinear regression

This study was partially supported by Shanghai Municipal S&T Commission via Project 20511105205 and by the Natural Science Foundation of China via project U20A20199.

Supplementary Information The online version contains supplementary material available at https://doi.org/10.1007/978-3-031-16449-1_40.

L. Wang (Eds.): MICCAI 2022, LNCS 13437, pp. 421–430, 2022.
https://doi.org/10.1007/978-3-031-16449-1_40

1   Introduction

The past few years have witnessed remarkable progress in developing context-aware systems to monitor surgical process [3], schedule surgeons [2], plan and schedule operating rooms [4]. Particularly, automatic estimation of remaining surgical duration (RSD) has become a key component when developing the context-aware systems. This is because surgery is a complex service, requiring human resources with different types of medical expertise, tools, and equipment, which involve high investment and operation costs. Therefore, data on the duration of surgery is required in order to plan and schedule operating rooms, which is an important activity that contributes to effectiveness of surgery and improves the quality of service, by providing timely treatment to patients and avoiding waiting time for patients to receive the treatment.

There exist previous attempts to develop surgery duration prediction model using predictors such as patient, surgeon and clinical characteristics [6,7]. However, accurate prediction of RSD is a challenging task due to multiple factors, such as complex patient conditions, surgeon's experience, and the variety of intra-operative circumstances [21]. To meet the challenges, lots of studies have been dedicated to extracting discriminative visual features from video frames and modeling the temporal dependencies among frames to improve the RSD prediction accuracy. For example, Aksamentov et al. [1] proposed a pipeline consisting of a convolutional neural network (CNN) and a long short-term memory (LSTM) network to predict RSD via regression from the video information available up to the current time. The method was evaluated on surgical videos of cholecystectomy surgeries, by pretraining its CNN using phase annotations. Twinanda et al. [20] introduced RSDNet for RSD prediction using unlabeled surgical videos of cholecystectomy surgeries. Aiming to predict RSD for cataract surgery, Marafioti et al. [16] proposed to explicitly incorporate information from observed surgical phases, the operating surgeon's experience and the elapsed time at any given time point to infer RSD prediction.

Despite these efforts, however, it is still challenging for existing methods to fully solve the problem and there are great potentials to improve the RSD prediction performance for the following reasons. First, most of previously used RSD prediction approaches are based on single deep models. As studied in [5], a single model may be lacking due to the statistical, computational and representational limitations. There exist methods for ensemble CNNs [13,19] but have not been utilized for RSD prediction. Additionally, existing methods for ensemple CNNs [13,19] typically trained multiple CNNs, which usually led to much larger computational complexity and hardware consumption. Second, due to the inherent ambiguity of future events, predicting RSD is challenging and benefits from estimating uncertainty scores alongside model predictions. Bayesian deep learning through Monte-Carlo dropout provides a framework for estimating uncertainties in deep neural networks [8,9]. However, in order to estimate uncertainty at the inference stage, the posterior distribution is approximated by Monte Carlo sampling of multiple predictions with dropout [17], which is not desired when developing context-aware computer-assisted intervention systems.

In this paper, to tackle the challenges, we propose a Bayesian LSTM-based Deep Negative Correlation Learning approach for accurate regression of RSD prediction as well as estimating prediction uncertainty. Concretely, our method ensembles multiple Bayesian LSTMs via deep negative correlation learning (DNCL) [22]. Negative correlation learning (NCL) [14] has been shown, both theoretically and empirically, to work well for regression-based problems. NCL controls the *bias-variance-covariance* trade-off systematically, aiming for a regression ensemble where each base model is both "accurate and diversified". Following a "divide and conquer" strategy, previous work [22] utilized group convolution on top of a backbone network to learn a group of regressors. Each regressor is jointly optimized with the backbone network by an amended cost function which penalizes correlation with others to make better trade-offs among the *bias-variance-covariance* in the ensemble. Different from theirs [22], here we propose to use Bayesian LSTMs on top of a backbone network to learn a group of regressors. Each Bayesian LSTM is initialized with different distributions using dropout variational inference and is trained via DNCL strategy. We call our method BD-Net where "B" means Bayesian and "D" means decorrelated. BD-Net based RSD prediction framework has following advantages:

- BD-Net generates multiple RSD predictions with a single network. By managing diversities among predictions, it has a sound generalization capability and a better *bias-variance-covariance* trade-off.
- BD-Net is simple and efficient. After training, it can produce both RSD prediction and uncertainty estimation in a single inference run.
- BD-Net achieves better results than the state-of-the-art (SOTA) methods when applied to predicting RSD for cataract surgery.

2   Methods

We are aiming to predict RSD for cataract surgery. Following the strategy proposed in [16], we explicitly incorporate information from observed surgical phases, the operating surgeon's experience and the elapsed time at any given time point into our model. Conceretely, the elapsed time, which is available at both training and inference time, is incorporated as an additional channel to the input video frames. We then construct a multi-task learning problem, aiming for jointly estimating both the surgeon's experience and the surgical phase together with the RSD.

2.1   Network Architecture

Figure 1-(a) shows the overall network architecture of our BD-Net. It consists of a feature extraction backbone CNN and a prediction head incorporating multiple Bayesian LSTMs (referred as B-LSTM) trained with DNCL. The input to BD-Net is $\hat{x}_t = [x_t, 1\frac{t}{T_{max}}] \in \mathbb{R}^{H \times W \times (3+1)}$, where x_t is the video frame at time t, and $\frac{t}{T_{max}}$ is the normalized elapsed time ranged [0,1]; H, W denote the height

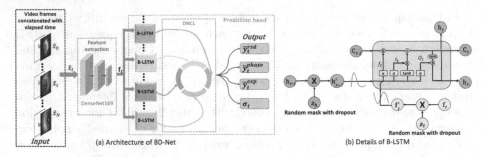

Fig. 1. The proposed BD-Net. (a) Overview of the network architecture of BD-Net. Input is a concatenation of the video frame at time t and the elapsed time of the surgery, which is fed to the backbone network to extract feature $\mathbf{f}_t$. $\mathbf{f}_t$ is then taken as the input to all B-LSTMs. The output of each B-LSTM is finally passed through four independent fully connected layers to predict surgeon's experience, surgical phase, RSD and the submodel's observation noise parameter. (b) A schematic illustration of the k-th B-LSTM where $\mathbf{z}_\mathbf{f}$ and $\mathbf{z}_\mathbf{h}$ are random masks with dropout probability p_k. $\mathbf{f}_t$ and $\mathbf{h}_{t-1}$ are multiplied by $\mathbf{z}_\mathbf{f}$ and $\mathbf{z}_\mathbf{h}$, respectively, before being fed to LSTM.

and width of input picture; $\mathbf{1}$ is an all-one tensor of size $H \times W \times 1$; and "3+1" represents three channels of RGB plus the appended time channel. We set T_{max} to the maximum video length of 20 min [16].

For each input $\hat{x}_t$, its feature $\mathbf{f}_t$ is extracted by the backbone CNN. Then the feature is taken as the input to the prediction head, which contains K B-LSTMs. In Fig. 1-(a), different colors indicate that each B-LSTM as a submodel obeys a different statistical distribution. The video descriptor at time t generated by the k-th B-LSTM is finally processed with four independent fully connected layers to estimate the surgeon's experience $\widehat{\mathbf{y}}_{k,t}^{\exp}$, the surgical phase $\widehat{\mathbf{y}}_{k,t}^{phase}$, the RSD $\widehat{\mathbf{y}}_{k,t}^{rsd}$ and the submodel's observation noise parameter $\sigma_{k,t}$. The probabilities $\widehat{\mathbf{y}}_{k,t}^{\exp}$ and $\widehat{\mathbf{y}}_{k,t}^{phase}$ are obtained with softmax layers.

Bayesian LSTM. Gal and Ghahramani [8,9] demonstrated that Bayesian CNN offers better robustness to over-fitting on small data than traditional approaches. Given the training data, Bayesian CNN requires to find the posterior distribution over the convolutional weights of the network. In general, this posterior distribution is not analytically tractable. Gal and Ghahramani [8,9] suggested to use variational dropout to tackle this problem for neural networks. To this end, in order to encourage each submodel in the DNCL to be diverse and accurate, and at the same time to capture the uncertainty to tackle the challenges of complex surgery circumstances, we apply Bayesian deep learning to our LSTM components. As shown in Fig. 1-(b), we multiplied the input to each LSTM component by random masks with dropout to realize Bayesian LSTM [10]. Different LSTM components can have different dropout probabilities to encourage diversity.

At training stage, all submodels are ensembled via DNCL where each submodel is optimized separately as described below. At testing stage, for each estimation in $\{\widehat{\mathbf{y}}_{k,t}^{\exp}, \widehat{\mathbf{y}}_{k,t}^{phase}, \widehat{\mathbf{y}}_{k,t}^{rsd}\}$, we average outputs of all submodels to obtain the final results $\{\widetilde{\mathbf{y}}_t^{\exp}, \widetilde{\mathbf{y}}_t^{phase}, \widetilde{\mathbf{y}}_t^{rsd}\}$.

2.2   Training Objectives

The ground truth labels of each input video sequence $\{\mathbf{x}_t\}$ include the remaining surgical duration y_t^{rsd} per frame, the surgical phase label y_t^{phase} per frame, and the surgeon's experience label $y^{\exp}$ per sequence. The index t is the elapsed time of the sequence.

In DNCL [14,22], each submodel is trained with its own objective function. Specifically, given the labeled data and for the k-th submodel, we need to minimize the cross-entropies between the frame-level predictions of surgical phase $\widehat{\mathbf{y}}_{k,t}^{phase}$ as well as surgeon's experience $\widehat{\mathbf{y}}_{k,t}^{\exp}$, and the associated ground truth labels:

$$L_{k,t}^{phase} = \mathbf{CE}(\widehat{\mathbf{y}}_{k,t}^{phase}, y_t^{phase}), \quad L_{k,t}^{\exp} = \mathbf{CE}(\widehat{\mathbf{y}}_{k,t}^{\exp}, y^{\exp}) \tag{1}$$

For RSD prediction, to capture uncertainty in each submodel, we follow [12] to minimize following objective function:

$$L_{k,t}^{rsd} = \frac{1}{2\sigma_{k,t}^2}||\widehat{\mathbf{y}}_{k,t}^{rsd} - y_t^{rsd}||^2 + \frac{1}{2}\log\sigma_{k,t}^2 \tag{2}$$

where $\widehat{\mathbf{y}}_{k,t}^{rsd}$ is the prediction; y_t^{rsd} is the ground truth; $\sigma_{k,t}$ is the submodel's observation noise parameters, which can be interpreted as follows: when the error $||\widehat{\mathbf{y}}_{k,t}^{rsd} - y_t^{rsd}||^2$ is large, $\sigma_{k,t}^2$ tends to be large. In contrast, when the error is small, $\sigma_{k,t}$ cannot be too large due to the penalty $\frac{1}{2}\log\sigma_{k,t}^2$, i.e., $\sigma_{k,t}$ can be used to indicate the uncertainty about the submodel's estimation.

In DNCL [14,22], all the individual submodels in the ensemble are trained simultaneously and interactively through the correlation penalty term in their individual error functions. The correlation penalty term appended to the training objective functions of each submodel is defined as:

$$L_{k,t}^{rsd-DNCL} = -(\widehat{\mathbf{y}}_{k,t}^{rsd} - \frac{1}{K}\sum_{k=1}^{K}\widehat{\mathbf{y}}_{k,t}^{rsd})^2 \tag{3}$$

where K is the number of submodels in the ensemble.

Combing all above together, we have the overall training objective function for the k-th submodel at the t-th video frame:

$$L_{k,t}^{Overall} = L_{k,t}^{rsd} + \lambda \cdot L_{k,t}^{rsd-DNCL} + \gamma \cdot (L_{k,t}^{phase} + L_{k,t}^{\exp}) \tag{4}$$

where λ and γ are hyperparameters controlling the relatively weights of different losses. For example, setting $\gamma = 0$ means that we do not incorporate information about surgical phase and surgeon's experience for RSD prediction while setting $\lambda = 0$ means that each submodel is trained separately and independently without any interaction with other submodels.

2.3   Uncertainty Estimation at Inference Stage

BD-Net allows for training K submodels with different dropout probabilities. After training, at inference stage, we can estimate our model's prediction as well as uncertainty by drawing one time parameter samples for each submodel and then assembling the results from each submodel. The model's outputs are obtained by averaging outputs of all submodels. This is different from previous work [17] where they draw multiple times of parameter samples in order to approximate the predictive expectation. We follow the method introduced by Kendall and Gal [12] to calculate our model's uncertainty:

$$Var_t^{rsd} \approx \underbrace{\frac{1}{K}\sum_{k=1}^{K}\sigma_{k,t}^2}_{aleatoric} + \underbrace{(\frac{1}{K}\sum_{k=1}^{K}(\widehat{\mathbf{y}}_{k,t}^{rsd})^2 - (\frac{1}{K}\sum_{k=1}^{K}\widehat{\mathbf{y}}_{k,t}^{rsd})^2)}_{epistemic} \tag{5}$$

where the first term is the aleatoric uncertainty and the second term is the epistemic uncertainty.

2.4   Implementation Details

We use DenseNet-169 [11] as the feature extraction backbone network. Each input video frame is reshaped and cropped to size of 224×224. For each image frame, the backbone network produces a feature vector of 1664-dimension, which is taken as the input to the prediction head. We implement each B-LSTM in the prediction head as an one-layer LSTM of 128 cells. We empirically set $K = 5$, $\lambda = 0.001$ and $\gamma = 0.3$. The dropout probabilities for the 5 submodels are set as 0.0, 0.2, 0.4, 0.4, 0.6, respectively. We implement BD-Net in Python using Pytorch framework. We adopt a two-stage strategy to train BD-Net. Specifically, at the first stage, we train the backbone CNN for 3 epochs with a learning rate of 1e−4. We then freeze the weights of the backbone CNN. At the second stage, we train all B-LSTMs to minimize $\{L_{k,t}^{Overall}\}$ for 40 epochs on full video sequences with a learning rate of 8e−4.

3   Experiments

Experimental Setup. We evaluated our method on the cataract-101 dataset [18], which is publicly available. The cataract-101 dataset contains 101 videos ranging from 5 min to 20 min. The resolution is 720×540 pixels acquired at 25fps. Limited by memory, we downsampled the videos to 2.5 fps. The dataset was randomly split into 64 videos for training, 17 videos for validation and 20 videos for testing.

Although BD-Net can predict surgical phase as well as surgeon's experience in addition to RSD, we only evaluated the RSD prediction accuracy in order to do a fair comparison of the performance of BD-Net with other SOTA methods. To quantify the accuracy of RSD prediction, we adopted the mean absolute error (MAE) per video, which is defined as the average of the MAE over all T frames contained in the video: $MAE - ALL = \frac{1}{T}\sum_{t=1}^{T}|\widehat{y}_t^{rsd} - y_t^{rsd}|$. Similarly, we also calculated MAE averaged over the last two ($MAE - 2$) and five ($MAE - 5$) minutes.

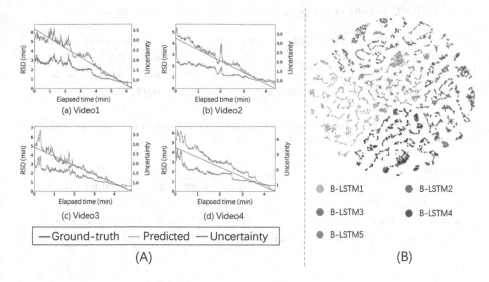

Fig. 2. Qualitative results of the proposed BD-Net for RSD prediction. (A) Four examples of our method's outputs. For each plot, we show the ground truth RSD, the predicted RSD, and the estimated uncertainty. (B) Visualization of the output features from each B-LSTM using t-SNE [15].

We compared the proposed BD-Net with three SOTA RSD estimation methods, i.e., TimeLSTM [1], RSDNet [20] and CataNet [16]. The first two methods were originally developed for RSD prediction of cholecystectomy surgery and the third method was specifically developed for RSD prediction of cataract surgery. All three methods combined CNN with LSTM.

Results of Comparison with the SOTA Methods. Table 1 shows the results when the proposed BD-Net was compared with the SOTA methods. From this table, one can see the best results are acquired by the proposed BD-Net. Specifically, the average $MAE - ALL$, $MAE - 5$, and $MAE - 2$ achieved by the proposed BD-Net were 0.94 min, 0.53 min and 0.27 min, respectively. In contrast, the average $MAE - ALL$, $MAE - 5$, and $MAE - 2$ achieved by the second-best method (CataNet [16]) were 0.99 min, 0.64 min and 0.35 min, respectively. We conducted a paired t-test to compare the results achieved by the proposed BD-Net with the CataNet [16]. The rightest column of Table 1 lists the p-value. From the results, one can see that the differences on $MAE - 5$ and $MAE - 2$ achieved by these two methods are statistically significant ($p < 0.05$) while the difference on $MAE - All$ is not, indicating that it is still challenging to estimate the RSD at the beginning of the operation, as there is little information available for prediction RSD.

Table 1. Results of comparison with the SOTA methods (Unit: min).

	TimeLSTM [1]	RSDNet [20]	CataNet [16]	BD-Net (Ours)	p-value
$MAE - 5$	1.47 ± 0.78	1.37 ± 0.83	0.64 ± 0.56	$\mathbf{0.53 \pm 0.22}$	**0.011**
$MAE - 2$	1.22 ± 0.32	1.23 ± 0.53	0.35 ± 0.20	$\mathbf{0.27 \pm 0.13}$	**0.022**
$MAE - ALL$	1.66 ± 0.79	1.59 ± 0.69	0.99 ± 0.65	$\mathbf{0.94 \pm 0.65}$	0.116

Table 2. Ablation study of the influence of different components on the performance of the proposed BD-Net (Unit: min). w/o: without

Models	$MAE - 5$	$MAE - 2$	$MAE - ALL$
BD-Net	$\mathbf{0.53 \pm 0.22}$	$\mathbf{0.27 \pm 0.13}$	$\mathbf{0.94 \pm 0.65}$
BD-Net w/o DNCL	0.61 ± 0.29	0.30 ± 0.15	0.95 ± 0.60
BD-Net w/o Bayesian learning	0.58 ± 0.27	0.36 ± 0.17	0.95 ± 0.70
BD-Net w/o phase & experience branches	0.61 ± 0.23	0.30 ± 0.12	1.00 ± 0.65

Figure 2 shows qualitative results of the proposed BD-Net. We first show four examples of our method's outputs at Fig. 2-(A). It is clear that the larger the prediction error is, the bigger the uncertainty. At Fig. 2-(B), we also visualize the output features from each B-LSTM using t-SNE [15]. One can see that there is little overlap among output features from B-LSTMs, indicating that the proposed BD-Net learns diversified submodels.

Ablation Study Results. The effectiveness of BD-Net depends on three components, i.e., a) DNCL; b) Bayesian learning; and c) incorporating information on surgical phase and surgeon's experience at training stage. To investigate the influence of each component on the RSD prediction accuracy, we conducted an ablation study using the same setup as described above. Each time, one of the three components was removed from the proposed approach. The results are shown in Table 2. When any one of the three components was removed from the proposed approach, the average $MAE - ALL$, $MAE - 5$, and $MAE - 2$ all increased. It is worth to note that when the branches incorporating information on surgical phase and surgeon's experience were removed, the obtained average $MAE - 5$ and $MAE - 2$ were larger than the proposed BD-Net but smaller than the results achieved by CataNet [16]. Considering the logistic efforts on organizing such information, this is a clear advantage.

4　Conclusion and Future Work

In this paper, we proposed a Bayesian LSTM network-based Deep Negative Correlation Learning approach called BD-Net for accurate regression of RSD prediction as well as estimating prediction uncertainty. We conducted comprehensive experiments on a publicly available video dataset containing 101 cataract surgeries. The experimental results demonstrated the efficacy of the proposed method. In future work, we will evaluate the performance of our method on estimating RSD of other types of surgery such as cholecystectomy surgery, and explore our method on more challenging tasks such as image super resolution.

References

1. Aksamentov, I., Twinanda, A.P., Mutter, D., Marescaux, J., Padoy, N.: Deep Neural Networks Predict Remaining Surgery Duration from Cholecystectomy Videos. In: Descoteaux, M., Maier-Hein, L., Franz, A., Jannin, P., Collins, D.L., Duchesne, S. (eds.) MICCAI 2017. LNCS, vol. 10434, pp. 586–593. Springer, Cham (2017). https://doi.org/10.1007/978-3-319-66185-8_66
2. Bhatia, B., Oates, T., Xiao, Y., Hu, P.: Real-time identification of operating room state from video. In: AAAI, vol. 2, pp. 1761–1766 (2007)
3. Bricon-Souf, N., Newman, C.R.: Context awareness in health care: a review. Int. J. Med. Inform. **76**(1), 2–12 (2007)
4. Devi, S.P., Rao, K.S., Sangeetha, S.S.: Prediction of surgery times and scheduling of operation theaters in optholmology department. J. Med. Syst. **36**(2), 415–430 (2012)
5. Dietterich, T.G.: Ensemble methods in machine learning. In: Kittler, J., Roli, F. (eds.) MCS 2000. LNCS, vol. 1857, pp. 1–15. Springer, Heidelberg (2000). https://doi.org/10.1007/3-540-45014-9_1
6. Edelman, E.R., Van Kuijk, S.M., Hamaekers, A.E., De Korte, M.J., Van Merode, G.G., Buhre, W.F.: Improving the prediction of total surgical procedure time using linear regression modeling. Front. Med. **4**, 85 (2017)
7. Eijkemans, M.J., Van Houdenhoven, M., Nguyen, T., Boersma, E., Steyerberg, E.W., Kazemier, G.: Predicting the unpredictable: a new prediction model for operating room times using individual characteristics and the surgeon's estimate. J. Am. Soc. Anesthesiol. **112**(1), 41–49 (2010)
8. Gal, Y., Ghahramani, Z.: Bayesian convolutional neural networks with Bernoulli approximate variational inference. arXiv preprint arXiv:1506.02158 (2016)
9. Gal, Y., Ghahramani, Z.: Dropout as a Bayesian approximation: representing model uncertainty in deep learning. In: International Conference on Machine Learning, pp. 1050–1059. PMLR (2016)
10. Gal, Y., Ghahramani, Z.: A theoretically grounded application of dropout in recurrent neural networks. Adv. Neural Inf. Process. Syst. **29** (2016)
11. Huang, G., Liu, Z., Van Der Maaten, L., Weinberger, K.Q.: Densely connected convolutional networks. In: Proceedings of the IEEE Conference on Computer Vision and Pattern Recognition, pp. 4700–4708 (2017)
12. Kendall, A., Gal, Y.: What uncertainties do we need in Bayesian deep learning for computer vision? Adv. Neural Inf. Process. Syst. **30** (2017)
13. Kumar, A., Kim, J., Lyndon, D., Fulham, M., Feng, D.: An ensemble of fine-tuned convolutional neural networks for medical image classification. IEEE J. Biomed. Health Inform. **21**(1), 31–40 (2016)
14. Liu, Y., Yao, X., Higuchi, T.: Evolutionary ensembles with negative correlation learning. IEEE Trans. Evol. Comput. **4**(4), 380–387 (2000)
15. Van der Maaten, L., Hinton, G.: Visualizing data using t-SNE. J. Mach. Learn. Res. **9**(11) (2008)
16. Marafioti, A., et al.: CataNet: predicting remaining cataract surgery duration. In: de Bruijne, M., et al. (eds.) MICCAI 2021. LNCS, vol. 12904, pp. 426–435. Springer, Cham (2021). https://doi.org/10.1007/978-3-030-87202-1_41

17. Rivoir, D., et al.: Rethinking anticipation tasks: uncertainty-aware anticipation of sparse surgical instrument usage for context-aware assistance. In: Martel, A.L., et al. (eds.) MICCAI 2020. LNCS, vol. 12263, pp. 752–762. Springer, Cham (2020). https://doi.org/10.1007/978-3-030-59716-0_72

18. Schoeffmann, K., Taschwer, M., Sarny, S., Münzer, B., Primus, M.J., Putzgruber, D.: Cataract-101: video dataset of 101 cataract surgeries. In: Proceedings of the 9th ACM Multimedia Systems Conference, pp. 421–425 (2018)

19. Sert, E., Ertekin, S., Halici, U.: Ensemble of convolutional neural networks for classification of breast microcalcification from mammograms. In: 2017 39th Annual International Conference of the IEEE Engineering in Medicine and Biology Society (EMBC), pp. 689–692. IEEE (2017)

20. Twinanda, A.P., Yengera, G., Mutter, D., Marescaux, J., Padoy, N.: RSDNet: learning to predict remaining surgery duration from laparoscopic videos without manual annotations. IEEE Trans. Med. Imaging **38**(4), 1069–1078 (2019)

21. Yuniartha, D.R., Masruroh, N.A., Herliansyah, M.K.: An evaluation of a simple model for predicting surgery duration using a set of surgical procedure parameters. Inform. Med. Unlocked **25**, 100633 (2021)

22. Zhang, L., et al.: Nonlinear regression via deep negative correlation learning. IEEE Trans. Pattern Anal. Mach. Intell. **43**(3), 982–998 (2021)

Neural Rendering for Stereo 3D Reconstruction of Deformable Tissues in Robotic Surgery

Yuehao Wang[1], Yonghao Long[1], Siu Hin Fan[2], and Qi Dou[1(✉)]

[1] Department of Computer Science and Engineering,
The Chinese University of Hong Kong, Shatin, Hong Kong
qidou@cuhk.edu.hk

[2] Department of Biomedical Engineering, The Chinese University of Hong Kong,
Shatin, Hong Kong

Abstract. Reconstruction of the soft tissues in robotic surgery from endoscopic stereo videos is important for many applications such as intra-operative navigation and image-guided robotic surgery automation. Previous works on this task mainly rely on SLAM-based approaches, which struggle to handle complex surgical scenes. Inspired by recent progress in neural rendering, we present a novel framework for deformable tissue reconstruction from binocular captures in robotic surgery under the single-viewpoint setting. Our framework adopts dynamic neural radiance fields to represent deformable surgical scenes in MLPs and optimize shapes and deformations in a learning-based manner. In addition to non-rigid deformations, tool occlusion and poor 3D clues from a single viewpoint are also particular challenges in soft tissue reconstruction. To overcome these difficulties, we present a series of strategies of tool mask-guided ray casting, stereo depth-cueing ray marching and stereo depth-supervised optimization. With experiments on DaVinci robotic surgery videos, our method significantly outperforms the current state-of-the-art reconstruction method for handling various complex non-rigid deformations. To our best knowledge, this is the first work leveraging neural rendering for surgical scene 3D reconstruction with remarkable potential demonstrated. Code is available at: https://github.com/med-air/EndoNeRF.

Keywords: 3D reconstruction · Neural rendering · Robotic surgery

1 Introduction

Surgical scene reconstruction from endoscope stereo video is an important but difficult task in robotic minimally invasive surgery. It is a prerequisite for many downstream clinical applications, including intra-operative navigation and augmented reality, surgical environment simulation, immersive education, and robotic surgery automation [2,12,20,25]. Despite much recent

L. Wang (Eds.): MICCAI 2022, LNCS 13437, pp. 431–441, 2022.
https://doi.org/10.1007/978-3-031-16449-1_41

progress [10,22,28–30,33], several key challenges still remain unsolved. First, surgical scenes are deformable with significant topology changes, requiring dynamic reconstruction to capture a high degree of non-rigidity. Second, endoscopic videos show sparse viewpoints due to constrained camera movement in confined space, resulting in limited 3D clues of soft tissues. Third, the surgical instruments always occlude part of the soft tissues, which affects the completeness of surgical scene reconstruction.

Previous works [1,13] explored the effectiveness of surgical scene reconstruction via depth estimation. Since most of the endoscopes are equipped with stereo cameras, depth can be estimated from binocular vision. Follow-up SLAM-based methods [23,31,32] fuse depth maps in 3D space to reconstruct surgical scenes under more complex settings. Nevertheless, these methods either hypothesize scenes as static or surgical tools not present, limiting their practical use in real scenarios. Recent work SuPer [8] and E-DSSR [11] present frameworks consisting of tool masking, stereo depth estimation and SurfelWarp [4] to perform single-view 3D reconstruction of deformable tissues. However, all these methods track deformation based on a sparse warp field [16], which degenerates when deformations are significantly beyond the scope of non-topological changes.

As an emerging technology, neural rendering [6,26,27] is recently developed to break through the limited performance of traditional 3D reconstruction by leveraging differentiable rendering and neural networks. In particular, neural radiance fields (NeRF) [15], a popular pioneering work of neural rendering, proposes to use *neural implicit field* for continuous scene representations and achieves great success in producing high-quality view synthesis and 3D reconstruction on diverse scenarios [14,15,17]. Meanwhile, recent variants of NeRF [18,19,21] targeting dynamic scenes have managed to track deformations through various neural representations on non-rigid objects.

In this paper, we endeavor to reconstruct highly deformable surgical scenes captured from single-viewpoint stereo endoscopes. We embark on adapting the emerging neural rendering framework to the regime of deformable surgical scene reconstruction. We summarize our contributions as follows: 1) To accommodate a wide range of geometry and deformation representations on soft tissues, we leverage neural implicit fields to represent dynamic surgical scenes. 2) To address the particular tool occlusion problem in surgical scenes, we design a new mask-guided ray casting strategy for resolving tool occlusion. 3) We incorporate a depth-cueing ray marching and depth-supervised optimization scheme, using stereo prior to enable neural implicit field reconstruction for single-viewpoint input. To the best of our knowledge, this is the first work introducing cutting-edge neural rendering to surgical scene reconstruction. We evaluate our method on 6 typical in-vivo surgical scenes of robotic prostatectomy. Compared with previous methods, our results exhibit great performance gain, both quantitatively and qualitatively, on 3D reconstruction and deformation tracking of surgical scenes.

2   Method

2.1   Overview of the Neural Rendering-Based Framework

Given a single-viewpoint stereo video of a dynamic surgical scene, we aim to reconstruct 3D structures and textures of surgical scenes without occlusion of surgical instruments. We denote $\{(\mathbf{I}_i^l, \mathbf{I}_i^r)\}_{i=1}^T$ as a sequence of input stereo video frames, where T is the total number of frames and $(\mathbf{I}_i^l, \mathbf{I}_i^r)$ is the pair of left and right images at the i-th frame. The video duration is normalized to $[0, 1]$. Thus, time of the i-th frame is i/T. We also extract binary tool masks $\{\mathbf{M}_i\}_{i=1}^T$ for the left views to identify the region of surgical instruments. To utilize stereo clues, we estimate coarse depth maps $\{\mathbf{D}_i\}_{i=1}^T$ for the left views from the binocular captures. We follow the modeling in D-NeRF [21] and represent deformable surgical scenes as a canonical neural radiance field along with a time-dependent neural displacement field (cf. Sect. 2.2). In our pipeline, each training iteration consists of the following six stages: i) randomly pick a frame for training, ii) run tool-guided ray casting (cf. Sect. 2.3) to shoot camera rays into the scene, iii) sample points along each camera ray via depth-cueing ray marching (cf. Sect. 2.4), iv) send sampled points to networks to obtain color and space occupancy of each point, v) evaluate volume rendering integral on sampled points to produce rendering results, vi) optimize the rendering loss plus depth loss to reconstruct shapes, colors and deformations of the surgical scene (cf. Sect. 2.5). The overview of key components in our approach is illustrated in Fig. 1. We will describe the detailed methods in the following subsections.

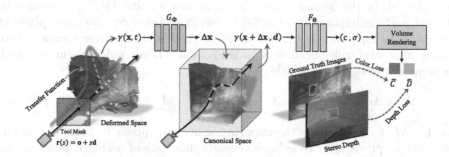

Fig. 1. Illustration of our proposed novel approach of neural rendering for stereo 3D reconstruction of deformable tissues in robotic surgery.

2.2   Deformable Surgical Scene Representations

We represent a surgical scene as a canonical radiance field and a time-dependent displacement field. Accordingly, each frame of the surgical scene can be regarded as a deformation of the canonical field. The canonical field, denoted as $F_\Theta(\mathbf{x}, \mathbf{d})$, is an 8-layer MLP with network parameter Θ, mapping coordinates $\mathbf{x} \in \mathbb{R}^3$ and unit view-in directions $\mathbf{d} \in \mathbb{R}^3$ to RGB colors $c(\mathbf{x}, \mathbf{d}) \in \mathbb{R}^3$ and space occupancy $\sigma(\mathbf{x}) \in \mathbb{R}$. The time-dependent displacement field $G_\Phi(\mathbf{x}, t)$ is encoded in another 8-layer MLP with network parameters Φ and maps input space-time coordinates $(\mathbf{x}, t)$ into displacement between the point $\mathbf{x}$ at time t and the corresponding point in the canonical field. For any time t, the color and occupancy at $\mathbf{x}$ can be retrieved as $F_\Theta(\mathbf{x} + G_\Phi(\mathbf{x}, t), \mathbf{d})$. Compared with other dynamic modeling approaches [18,19], a displacement field is sufficient to explicitly and physically express all tissue deformations. To capture high-frequency details, we use positional encoding $\gamma(\cdot)$ to map the input coordinates and time into Fourier features [24] before feeding them to the networks.

2.3   Tool Mask-Guided Ray Casting

With scene representations, we further leverage the differentiable volume rendering used in NeRF to yield renderings for supervision. The differentiable volume rendering begins with shooting a batch of camera rays into the surgical scene from a fixed viewpoint at an arbitrary time t. Every ray is formulated as $\mathbf{r}(s) = \mathbf{o} + s\mathbf{d}$, where $\mathbf{o}$ is a fixed origin of the ray, $\mathbf{d}$ is the pointing direction of the ray and s is the ray parameter. In the original NeRF, rays are shot towards a batch of randomly selected pixels on the entire image plane. However, there are many pixels of surgical tools on the captured images, while our goal is to reconstruct underlying tissues. Thus, training on these tool pixels is unexpected. Our main idea for solving this issue is to bypass those rays traveling through tool pixels over the training stage. We utilize binary tool masks $\{M_i\}_{i=1}^T$, where 0 stands for tissue pixels and 1 stands for tool pixels, to inform which rays should be neglected. In this regard, we create importance maps $\{\mathcal{V}\}_{i=1}^T$ according to M_i and perform importance sampling to avoid shooting rays for those pixels of surgical tools. Equation (1) exhibits the construction of importance maps, where $\otimes$ is element-wise multiplication, $\|\cdot\|_F$ is Frobenius norm and $\mathbf{1}$ is a matrix with the same shape as M_i while filled with ones:

$$\mathcal{V}_i = \Lambda \otimes (\mathbf{1} - M_i), \quad \Lambda = \left(1 + \sum\nolimits_{j=1}^T M_j \Big/ \Big\| \sum\nolimits_{j=1}^T M_j \Big\|_F \right). \quad (1)$$

The $\mathbf{1} - M_i$ term initializes the importance of tissue pixels to 1 and the importance of tool pixels to 0. To balance the sampling rate of occluded pixels across frames, the scaling term Λ specifies higher importance scaling for those tissue areas with higher occlusion frequencies. Normalizing each importance map as $\widehat{\mathcal{V}}_i = \mathcal{V}_i / \|\mathcal{V}_i\|_F$ will yield a probability mass function over the image plane. During our ray casting stage for the i-th frame, we sample pixels from the distribution $\widehat{\mathcal{V}}_i$ using inverse transform sampling and cast rays towards these sampled pixels. In this way, the probability of shooting rays for tool pixels is guaranteed to be zero as the importance of tool pixels is constantly zero.

2.4   Stereo Depth-Cueing Ray Marching

After shooting camera rays over tool occlusion, we proceed ray marching to sample points in the space. Specifically, we discretize each camera ray $\mathbf{r}(s)$ into batch of points $\{x_j | x_j = r(s_j)\}_{j=1}^{m}$ by sampling a sequence of ray steps $s_1 \leq s_2 \leq \cdots \leq s_m$. The original NeRF proposes hierarchical stratified sampling to obtain $\{s_j\}_{j=1}^{m}$. However, this sampling strategy hardly exploits accurate 3D structures when NeRF models are trained on single-view input. Drawing inspiration from early work in iso-surface rendering [7], we create Gaussian transfer functions with stereo depth to guide point sampling near tissue surfaces. For the i-th frame, the transfer function for a ray $r(s)$ shooting towards pixel (u, v) is formulated as:

$$\delta(s; u, v, i) = \exp\left(-(s - D_i[u, v])^2 / 2\xi^2\right). \tag{2}$$

The transfer function $\delta(s; u, v, i)$ depicts an impulse distribution that continuously allocates sampling weights for every location on $r(s)$. The impulse is centered at $D_i[u, v]$, i.e., the depth at the (u, v) pixel. The width of the impulse is controlled by the hyperparameter ξ, which is set to a small value to mimic Dirac delta impulse. In our ray marching, $s_1 \leq s_2, \leq \cdots \leq s_m$ are drawn from the normalized impulse distribution $\frac{1}{\xi\sqrt{2\pi}}\delta(s; u, v, i)$. By this means, sampled points are concentrated around tissue surfaces, imposing stereo prior in rendering.

2.5   Optimization for Deformable Radiance Fields

Once we obtain the sampled points in the space, the emitted color $\widehat{C}$ and optical depth $\widehat{D}$ of a camera ray $r(s)$ can be evaluated by volume rendering [5] as:

$$\widehat{C}(r(s)) = \sum_{j=1}^{m-1} w_j c(x_j, d), \quad \widehat{D}(r(s)) = \sum_{j=1}^{m-1} w_j s_j,$$
$$w_j = (1 - \exp(-\sigma(x_j)\Delta s_j)) \exp\left(-\sum_{k=1}^{j-1} \sigma(x_k)\Delta s_k\right), \quad \Delta s_j = s_{j+1} - s_j. \tag{3}$$

To reconstruct the canonical and displacement fields from single-view captures, we optimize the network parameters Θ and Φ by jointly supervising the rendered color and optical depth [3]. Specifically, the loss function for training the networks is defined as:

$$\mathcal{L}(r(s)) = \left\|\widehat{C}(r(s)) - I_i[u, v]\right\|_2^2 + \lambda \left|\widehat{D}(r(s)) - D_i[u, v]\right|, \tag{4}$$

where (u, v) is the location of the pixel that $r(s)$ shoots towards, λ is a hyperparameter weighting the depth loss.

Last but not least, we conduct statistical depth refinement to handle corrupt stereo depth caused by fuzzy pixels and specular highlights on the images of surgical scenes. Direct supervision on the estimated depth will overfit corrupt depth in the end, leading to abrupt artifacts in reconstruction results (Fig. 3). Our preliminary findings reveal that our model at the early training stage would produce smoother results both in color and depth since the underfitting model tends to average learned colors and occupancy. Thus, minority corrupt depth is smoothed by majority normal depth. Based on this observation, we propose to patch the corrupt depth with the output from underfitting radiance fields. Denoting $\widehat{D}_i^K$ as the underfitting output depth maps for the i-th frame after

K iterations of training, we firstly find residual maps through $\epsilon_i = |\widehat{\boldsymbol{D}}_i^K - \boldsymbol{D}_i|$, then we compute a probabilistic distribution over the residual maps. After that, we set a small number $\alpha \in [0, 1]$ and locate those pixels with the last α-quantile residuals. Since those located pixels statistically correspond to large residuals, we can identify them as occurrences of corrupt depth. Finally, we replace those identified corrupt depth pixels with smoother depth pixels in $\widehat{\boldsymbol{D}}_i^K$. After this refinement procedure, the radiance fields are optimized on the patched depth maps in the subsequent training iterations, alleviating corrupt depth fitting.

3   Experiments

3.1   Dataset and Evaluation Metrics

We evaluate our proposed method on typical robotic surgery stereo videos from 6 cases of our in-house DaVinci robotic prostatectomy data. We totally extracted 6 clips with a total of 807 frames. Each clip lasts for $4 \sim 8$ s with 15 fps. Each case is captured from stereo cameras at a single viewpoint and encompasses challenging scenes with non-rigid deformation and tool occlusion. Among the selected 6 cases, 2 cases contain traction on thin structures such as fascia, 2 cases contain significant pushing and pulling of tissue, and 2 cases contain tissue cutting, which altogether present the typical soft tissue situations in robotic surgery. For comparison, we take the most recent state-of-the-art surgical scene reconstruction method of E-DSSR [11] as a strong comparison. For qualitative evaluation, we exhibit our reconstructed point clouds and compare textural and geometric details obtained by different methods. We also conduct an ablation study on our depth-related modules through qualitative comparison. Due to clinical regulation in practice, it is impossible to collect ground truth depth for numerical evaluation on 3D structures. Following the evaluation method in [11] and wide literature in neural rendering, we alternatively use photometric errors, including PSNR, SSIM and LPIPS, as evaluation metrics for quantitative comparisons.

3.2   Implementation Details

In our implementation, we empirically set the width of the transfer function $\xi = 1$, the weight of depth loss $\lambda = 1$, depth refinement iteration $K = 4000$ and $\alpha = 0.1$. Other training hyper-parameters follow the settings in the state-of-the-art D-NeRF [21]. We calibrate the endoscope in advance to acquire its intrinsics. In all of our experiments, tool masks are obtained by manually labeling and coarse stereo depth maps are generated by STTR-light [9] pretrained on Scene Flow. We optimize each model over $100K$ iterations on a single case. To recover explicit geometry from implicit fields, we render optimized radiance fields to RGBD maps, smooth rendered depth maps via bilateral filtering, and back-project RGBD into point clouds based on the endoscope intrinsics.

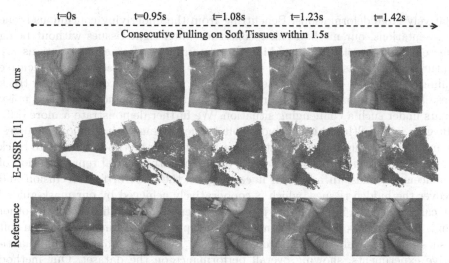

(a) Results on the case "pulling tissues", where soft tissues are drastically pulled within 2s. We exhibit 5 reconstruction results of our method and E-DSSR over time.

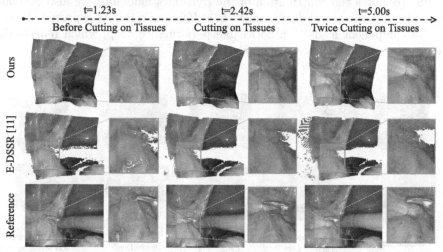

(b) Results on the case "cutting tissues twice". We show 3 frames corresponding to no deformation, cutting once and cutting twice, respectively. The close-ups of the cutting areas display reconstructed tissue details before and after cutting.

Fig. 2. Qualitative comparisons of 2 cases, demonstrating reconstruction of soft tissues with large deformations and topology changes.

3.3 Qualitative and Quantitative Results

For qualitative evaluation, Fig. 2 illustrates the reconstruction results of our approach and the comparison method, along with a reference to the original video. In the test case of Fig. 2(a), the tissues are pulled by surgical instruments, yielding relatively large deformations. Benefitting from the underlying continuous scene representations, our method can reconstruct water-tight tissues without being affected by the tool occlusion. More importantly, per-frame deformations are captured continuously, achieving stable results over the episode of consecutive pulling. In contrast, the current state-of-the-art method [11] could not fully track these large deformations and its reconstruction results include holes and noisy points under such a challenging situation. We further demonstrate a more difficult case in Fig. 2(b) which includes soft tissue cutting with topology changes. From the reconstruction results, it is observed that our method manages to track the detailed cutting procedures, owing to the powerful neural representation of displacement fields. In addition, it can bypass the issue of tool occlusion and recover the hidden tissues, which is cooperatively achieved by our mask-guided ray casting and the interpolation property of neural implicit fields. On the other hand, the comparison method is not able to capture these small changes on soft tissues nor patch all the tool-occluded areas. Table 1 summarizes our quantitative experiments, showing overall performance on the dataset. Our method dramatically outperforms E-DSSR by ↑ 16.433 PSNR, ↑ 0.295 SSIM and ↓ 0.342 LPIPS. To assess the contribution of the dynamics modeling, we also evaluate our model without neural displacement field (Ours w/o D). As expected, removing this component leads to a noticeable performance drop, which reflects the effectiveness of the displacement modeling.

| Complete Model | w/o Depth-Supervised Loss | w/o Depth Refinement | Corrupt Stereo Depth | w/o Depth-Cueing Ray Marching |

Fig. 3. Ablation study on our depth-related modules, i.e., depth-supervised loss, depth refinement and depth-cueing ray marching.

We present a qualitative ablation study on our depth-related modules in Fig. 3. Without depth-supervision loss, we observe that the pipeline is not capable of learning correct geometry from single-viewpoint input. Moreover, when depth refinement is disabled, abrupt artifacts occur on the reconstruction results due to corruption in stereo depth estimation. Our depth-cueing ray marching can further diminish artifacts on 3D structures, especially for boundary points.

Table 1. Quantitative evaluation on photometric errors of the dynamic reconstruction on metrics of PSNR, SSIM and LPIPS.

Methods	PSNR ↑	SSIM ↑	LPIPS ↓
E-DSSR [11]	13.398 ± 1.387	0.630 ± 0.057	0.423 ± 0.047
Ours w/o D	24.088 ± 2.567	0.849 ± 0.023	0.230 ± 0.023
Ours	**29.831 ± 2.208**	**0.925 ± 0.020**	**0.081 ± 0.022**

4   Conclusion

This paper presents a novel neural rendering-based framework for dynamic surgical scene reconstruction from single-viewpoint binocular images, as well as addressing complex tissue deformation and tool occlusion. We adopt the cutting-edge dynamic neural radiance field method to represent surgical scenes. In addition, we propose mask-guided ray casting to handle tool occlusion and impose stereo depth prior upon the single-viewpoint situation. Our approach has achieved superior performance on various scenarios in robotic surgery data such as large elastic deformations and tissue cutting. We hope the emerging NeRF-based 3D reconstruction techniques could inspire new pathways for robotic surgery scene understanding, and empower various down-stream clinical-oriented tasks.

Acknowledgements. This work was supported in part by CUHK Shun Hing Institute of Advanced Engineering (project MMT-p5-20), in part by Shenzhen-HK Collaborative Development Zone, and in part by Multi-Scale Medical Robotics Centre InnoHK.

References

1. Brandao, P., Psychogyios, D., Mazomenos, E., Stoyanov, D., Janatka, M.: Hapnet: hierarchically aggregated pyramid network for real-time stereo matching. Comput. Methods Biomech. Biomed. Eng. Imaging Vis. **9**(3), 219–224 (2021)
2. Chen, L., Tang, W., John, N.W., Wan, T.R., Zhang, J.J.: Slam-based dense surface reconstruction in monocular minimally invasive surgery and its application to augmented reality. Comput. Methods Programs Biomed. **158**, 135–146 (2018)
3. Deng, K., Liu, A., Zhu, J.Y., Ramanan, D.: Depth-supervised nerf: fewer views and faster training for free. arXiv preprint arXiv:2107.02791 (2021)
4. Gao, W., Tedrake, R.: Surfelwarp: efficient non-volumetric single view dynamic reconstruction. In: Robotics: Science and Systems XIV (2019)
5. Kajiya, J.T., Von Herzen, B.P.: Ray tracing volume densities. ACM SIGGRAPH Comput. Graph. **18**(3), 165–174 (1984)
6. Kato, H., Ushiku, Y., Harada, T.: Neural 3D mesh renderer. In: CVPR, pp. 3907–3916 (2018)
7. Kniss, J., Ikits, M., Lefohn, A., Hansen, C., Praun, E., et al.: Gaussian transfer functions for multi-field volume visualization. In: IEEE Visualization, 2003, VIS 2003, pp. 497–504. IEEE (2003)

8. Li, Y., et al.: Super: a surgical perception framework for endoscopic tissue manipulation with surgical robotics. IEEE Rob. Autom. Lett. **5**(2), 2294–2301 (2020)

9. Li, Z., et al.: Revisiting stereo depth estimation from a sequence-to-sequence perspective with transformers. In: ICCV, pp. 6197–6206 (2021)

10. Liu, X., et al.: Reconstructing sinus anatomy from endoscopic video – towards a radiation-free approach for quantitative longitudinal assessment. In: Martel, A.L., et al. (eds.) MICCAI 2020. LNCS, vol. 12263, pp. 3–13. Springer, Cham (2020). https://doi.org/10.1007/978-3-030-59716-0_1

11. Long, Y., et al.: E-DSSR: efficient dynamic surgical scene reconstruction with transformer-based stereoscopic depth perception. In: de Bruijne, M., et al. (eds.) MICCAI 2021. LNCS, vol. 12904, pp. 415–425. Springer, Cham (2021). https://doi.org/10.1007/978-3-030-87202-1_40

12. Lu, J., Jayakumari, A., Richter, F., Li, Y., Yip, M.C.: Super deep: a surgical perception framework for robotic tissue manipulation using deep learning for feature extraction. In: ICRA, pp. 4783–4789. IEEE (2021)

13. Luo, H., Wang, C., Duan, X., Liu, H., Wang, P., Hu, Q., Jia, F.: Unsupervised learning of depth estimation from imperfect rectified stereo laparoscopic images. Comput. Biol. Med. **140**, 105109 (2022)

14. Martin-Brualla, R., Radwan, N., Sajjadi, M.S., Barron, J.T., Dosovitskiy, A., Duckworth, D.: Nerf in the wild: neural radiance fields for unconstrained photo collections. In: CVPR, pp. 7210–7219 (2021)

15. Mildenhall, B., Srinivasan, P.P., Tancik, M., Barron, J.T., Ramamoorthi, R., Ng, R.: NeRF: representing scenes as neural radiance fields for view synthesis. In: Vedaldi, A., Bischof, H., Brox, T., Frahm, J.-M. (eds.) ECCV 2020. LNCS, vol. 12346, pp. 405–421. Springer, Cham (2020). https://doi.org/10.1007/978-3-030-58452-8_24

16. Newcombe, R.A., Fox, D., Seitz, S.M.: Dynamicfusion: reconstruction and tracking of non-rigid scenes in real-time. In: CVPR, pp. 343–352 (2015)

17. Niemeyer, M., Geiger, A.: Giraffe: representing scenes as compositional generative neural feature fields. In: CVPR, pp. 11453–11464 (2021)

18. Park, K., et al.: Nerfies: deformable neural radiance fields. In: CVPR, pp. 5865–5874 (2021)

19. Park, K., et al.: Hypernerf: a higher-dimensional representation for topologically varying neural radiance fields. ACM Trans. Graph. (TOG) **40**(6), 1–12 (2021)

20. Penza, V., De Momi, E., Enayati, N., Chupin, T., Ortiz, J., Mattos, L.S.: envisors: enhanced vision system for robotic surgery. a user-defined safety volume tracking to minimize the risk of intraoperative bleeding. Front. Rob AI **4**, 15 (2017)

21. Pumarola, A., Corona, E., Pons-Moll, G., Moreno-Noguer, F.: D-nerf: neural radiance fields for dynamic scenes. In: CVPR, pp. 10318–10327 (2021)

22. Recasens, D., Lamarca, J., Fácil, J.M., Montiel, J., Civera, J.: Endo-depth-and-motion: reconstruction and tracking in endoscopic videos using depth networks and photometric constraints. IEEE Rob. Autom. Lett. **6**(4), 7225–7232 (2021)

23. Song, J., Wang, J., Zhao, L., Huang, S., Dissanayake, G.: Dynamic reconstruction of deformable soft-tissue with stereo scope in minimal invasive surgery. IEEE Rob. Autom. Lett. **3**(1), 155–162 (2017)

24. Tancik, M., et al.: Fourier features let networks learn high frequency functions in low dimensional domains. NeurIPS **33**, 7537–7547 (2020)

25. Tang, R., et al.: Augmented reality technology for preoperative planning and intraoperative navigation during hepatobiliary surgery: a review of current methods. Hepatobiliary Panc. Dis. Int. **17**(2), 101–112 (2018)

26. Tewari, A., et al.: Advances in neural rendering. In: ACM SIGGRAPH 2021 Courses, pp. 1–320 (2021)
27. Tewari, A., al.: State of the art on neural rendering. In: Computer Graphics Forum, vol. 39, pp. 701–727. Wiley Online Library (2020)
28. Tukra, S., Marcus, H.J., Giannarou, S.: See-through vision with unsupervised scene occlusion reconstruction. IEEE Trans. Pattern Anal. Mach. Intell. 01, 1–1 (2021)
29. Wei, G., Yang, H., Shi, W., Jiang, Z., Chen, T., Wang, Y.: Laparoscopic scene reconstruction based on multiscale feature patch tracking method. In: 2021 International Conference on Electronic Information Engineering and Computer Science (EIECS), pp. 588–592. IEEE (2021)
30. Wei, R., et al.: Stereo dense scene reconstruction and accurate laparoscope localization for learning-based navigation in robot-assisted surgery. arXiv preprint arXiv:2110.03912 (2021)
31. Zhou, H., Jagadeesan, J.: Real-time dense reconstruction of tissue surface from stereo optical video. IEEE Trans. Med. Imaging 39(2), 400–412 (2019)
32. Zhou, H., Jayender, J.: EMDQ-SLAM: real-time high-resolution reconstruction of soft tissue surface from stereo laparoscopy videos. In: de Bruijne, M., et al. (eds.) MICCAI 2021. LNCS, vol. 12904, pp. 331–340. Springer, Cham (2021). https://doi.org/10.1007/978-3-030-87202-1_32
33. Zhou, H., Jayender, J.: Real-time nonrigid mosaicking of laparoscopy images. IEEE Trans. Med. Imaging 40(6), 1726–1736 (2021)

Towards Holistic Surgical Scene Understanding

Natalia Valderrama[1]([⊠]), Paola Ruiz Puentes[1], Isabela Hernández[1],
Nicolás Ayobi[1], Mathilde Verlyck[1], Jessica Santander[2], Juan Caicedo[2],
Nicolás Fernández[3,4], and Pablo Arbeláez[1]([⊠])

[1] Center for Research and Formation in Artificial Intelligence,
Universidad de los Andes, Bogotá, Colombia
`{nf.valderrama,p.arbelaez}@uniandes.edu.co`
[2] Fundación Santafé de Bogotá, Bogotá, Colombia
[3] Seattle Children's Hospital, Seattle, USA
[4] University of Washington, Seattle, USA

Abstract. Most benchmarks for studying surgical interventions focus on a specific challenge instead of leveraging the intrinsic complementarity among different tasks. In this work, we present a new experimental framework towards holistic surgical scene understanding. First, we introduce the Phase, Step, Instrument, and Atomic Visual Action recognition (PSI-AVA) Dataset. PSI-AVA includes annotations for both long-term (Phase and Step recognition) and short-term reasoning (Instrument detection and novel Atomic Action recognition) in robot-assisted radical prostatectomy videos. Second, we present Transformers for Action, Phase, Instrument, and steps Recognition (TAPIR) as a strong baseline for surgical scene understanding. TAPIR leverages our dataset's multi-level annotations as it benefits from the learned representation on the instrument detection task to improve its classification capacity. Our experimental results in both PSI-AVA and other publicly available databases demonstrate the adequacy of our framework to spur future research on holistic surgical scene understanding.

Keywords: Holistic surgical scene understanding · Robot-assisted surgical endoscopy · Dataset · Vision transformers

1 Introduction

Surgeries can be generally understood as compositions of interactions between instruments and anatomical landmarks, occurring inside sequences of deliberate activities. In this respect, surgical workflow analysis aims to understand

---

P.R. Puentes and I. Hernández–Equal contribution.

Supplementary Information The online version contains supplementary material available at https://doi.org/10.1007/978-3-031-16449-1_42.

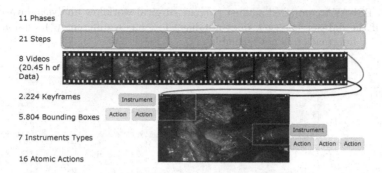

Fig. 1. PSI-AVA Dataset enables a holistic analysis of surgical videos through annotations for both long-term (Phase and Step Recognition) and short-term reasoning tasks (Instrument Detection and Atomic Action Recognition).

the complexity of procedures through data analysis. This field of research is essential for providing context-aware intraoperative assistance [23], streamlining surgeon training, and as a tool for procedure planning and retrospective analysis [20]. Furthermore, holistic surgical scene understanding is essential to develop autonomous robotic assistants in the surgical domain.

This work addresses the comprehension of surgical interventions through a systematical deconstruction of procedures into multi-level activities performed along robot-assisted radical prostatectomies. We decompose the procedure into short-term activities performed by the instruments and surgery-specific long-term processes, termed phases and steps. We refer to the completion of a surgical objective as a step, whereas phases are primary surgical events, and are composed of multiple steps. Further, each step contains certain instruments that perform a set of actions to achieve punctual goals. Stressing on their intrinsic complementarity, the spatio-temporal relationship between multi-level activities and instruments is essential for a holistic understanding of the scene dynamics. Most of the currently available datasets for surgical understanding do not provide enough information to study whole surgical scenes, as they focus on independent tasks of a specific surgery [1,4,28,29]. These task-level restrictions lead to approaches with a constrained understanding of surgical workflow. Only a fraction of surgical datasets enable multi-level learning [17,24,25,30], despite the proven success of multi-level learning in complex human activity recognition [22].

The first main contribution of this work is a novel benchmark for Phase, Step, Instrument, and Atomic Visual Action (PSI-AVA) recognition tasks for holistic surgical understanding (Fig. 1), which enables joint multi-task and multi-level learning for robot-assisted radical prostatectomy surgeries. Inspired by [15], we propose a surgical Atomic Action Recognition task where actions are not limited to a specific surgery nor dependent of interacted objects, and cannot be broken down into smaller activities.

Table 1. PSI-AVA compared against current surgical workflow analysis datasets. We indicate if the dataset has (✓), does not have (−), has a different (*) version of the attribute, or information is not available (?).

Dataset	Multi-level understanding	Phase recognition	Step recognition	Instrument recognition	Instrument detection	Action task	Spatial annotations	Video hours	Publicly available
LSRAO [28]	−	−	−	−	−	✓	−	?	−
JIGSAWS [1]	−	−	−	−	−	✓*	*	2.62	✓
TUM LapChole [29]	−	✓	−	−	−	−	−	23.24	✓
Cholec80 [30]	✓	✓	−	✓	−	−	−	51.25	✓
CaDIS [14]	−	✓	−	✓	−	−	−	9.1	✓
HeiCo [24]	✓	✓	−	−	✓	−	✓	96.12	✓
CholecT40,CholecT50 [25, 26]	−	−	−	✓	−	✓	−	?	−
MISAW [17]	✓	✓	✓	−*	−	✓	−	1.52	✓
ESAD [4]	−	−	−	−	−	✓	✓	9.33	✓
AVOS [13]	−	−	−	✓	✓	✓	✓	47	−
PSI-AVA (Ours)	✓	✓	✓	✓	✓	✓	✓	20.45	✓

Table 1 compares the main datasets for surgical workflow analysis. Long-term surgical phase recognition is the most studied task and is frequently considered with instrument recognition. Finer action tasks fit in the context of gesture recognition through verb labels [1, 4, 13, 17, 25, 26], but are specific to one surgery type, thus motivating the added atomization of actions in our dataset. Current spatial annotations include instance-wise bounding boxes [4, 13], segmentation masks [2, 3, 24] and kinematic data [1]. These annotations can be used to support coarser classification-based tasks. Lastly, multi-level learning is duly approached in [24], enabling the concurrent study of tasks in the outermost levels of surgical timescale analysis. Our dataset extends this analysis to intermediate scales, solely using the more widespread video format while providing a solid experimental framework for each of the proposed tasks.

Recently, Vision Transformers have achieved outstanding performance in both natural image and video related tasks [5, 8, 9, 33], stressing on the benefit of analyzing complex scenes through attention mechanisms. Specifically, transformers can exploit general temporal information in sequential data, providing convenient architectures for surgical video analysis, and overcoming the narrow receptive fields of traditional CNNs. Recent methods leverage transformers' attention modules and achieve outstanding results in phase recognition [6, 7, 11], interaction recognition [26], action segmentation [31] and tools detection, and segmentation [19, 32]. Additionally, [26] proposes a weakly supervised method for instrument-action-organ triplets recognition. Nonetheless, no previous research has employed transformers for surgical atomic action recognition.

As a second main contribution towards holistic surgical scene understanding, we introduce the Transformers for Action, Phase, Instrument, and steps Recognition (TAPIR) method. Figure 2 shows our general framework that consists of two main stages: a video feature extractor followed by a task-specific classification head. Our method benefits from box-specific features extracted by an instrument detector to perform the more fine-grained recognition task. TAPIR performance on the PSI-AVA dataset highlights the benefits of multi-level annotations and its relevance to holistic understanding on surgical scenes.

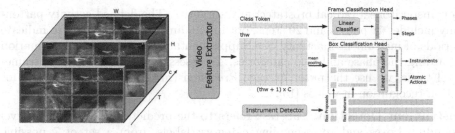

Fig. 2. TAPIR. Our approach leverages the global, temporal information extracted from a surgical video sequence with localized appearance cues. The association of complementary information sources enables the multi-level reasoning over dynamic surgical scenes. Best viewed in color. (Color figure online)

Our main contributions can be summarized as follows: (1) We introduce and validate the PSI-AVA dataset with annotations for phase and step recognition, instrument detection, and the novel task of atomic action recognition in surgical scenes. (2) We propose TAPIR, a transformer-based method that leverages the multi-level annotations of PSI-AVA dataset and establishes a baseline for future work in our holistic benchmark.

To ensure our results' reproducibility and promote further research on holistic surgical understanding, we will make publicly available the PSI-AVA dataset, annotations, and evaluation code, as well as pre-trained models and source code for TAPIR under the MIT license at https://github.com/BCV-Uniandes/TAPIR.

2 Data and Ground-Truth Annotations

The endoscopic videos were collected at Fundación Santafé de Bogotá during eight radical prostatectomy surgeries, performed with the Da Vinci SI3000 Surgical System. Data collection and post-processing protocols were done under approval of the ethics committees of all institutions directly involved in this work. Regarding the experimental framework, we formulate a two-fold cross-validation setup. Each fold contains data related to four procedures and a similar distribution of phase, step, instrument, and atomic action categories. Each surgery is uniquely assigned to one of the folds. Our dataset contains annotations for four tasks that enable a comprehensive evaluation of surgical workflow at multiple scales. Next, we describe each of the tasks, their annotation strategies, and evaluation methodology, followed by a report of dataset statistics.

Phase and Step Recognition. These tasks share the common objective of identifying surgical events, but differ in the temporal scale at which these events occur. Annotations for these frame-level tasks were done by two urologist experts in transperitoneal radical prostatectomy surgeries. The defined activity partonomy includes 11 phases and 20 steps, both including an *Idle* category to indicate periods of instrument inactivity. See Supplementary Fig. 1 for more information regarding the phase and step categories. To assess performance in these frame-level tasks, we use the mean Average Precision (mAP), a standard metric for action recognition in videos [15].

Instrument Detection. This task refers to the prediction of instrument-level bounding boxes and corresponding category labels, from a set of 7 possible classes. A hybrid strategy was carried out to annotate all bounding boxes in our dataset's keyframes. A domain-specific Faster R-CNN model pretrained on the task of instrument detection on the EndoVis 2017 [2] and 2018 [3] datasets is used to generate initial bounding box estimates. Next, crowd-source annotators verify and refine the bounding box predictions. Annotators are instructed to verify that all bounding boxes enclose complete instruments, and to select one category label per bounding box. The evaluation of this task uses the mAP metric at an Intersection-over-Union threshold of 0.5 (mAP@0.5 IoU).

Atomic Action Recognition. Once the spatial location of instruments is defined, the aim of this novel task is to describe their localized activity by fine action categories. Similarly to [15], our atomic action vocabulary is based on two principles: generality and atomicity. The annotations for this task associate the previously defined bounding boxes to a maximum of three atomic actions per instrument, from a set of 16 classes. To evaluate this task, we employ the standard metric for action detection in video: mAP@0.5 IoU.

Dataset Statistics. After data collection, an anonymization and curation process was carried out, removing video segments with non-surgical content and potential patient identifiers. As a result, our dataset contains approximately 20.45 h of the surgical procedure performed by three expert surgeons. Given that phase and step annotations are defined over fixed time intervals, their recognition tasks benefit from dense, frame-wise annotations at a rate of one keyframe per second. As a result, 73,618 keyframes are annotated with one phase-step pair. Due to the fine-grained nature of instruments and atomic actions annotations, we sample frames every 35 s of video, to obtain 2238 keyframe candidates. These keyframes are annotated with 5804 instrument instances, with corresponding instrument category and atomic action(s). Supplementary Fig. 2 depicts the distribution of instances within the different categories per task. We highlight the imbalanced class distribution in the proposed tasks inherent in surgical-based problems. The less frequent classes in our dataset correspond to sporadic and case-specific surgical events.

3   Method

3.1   TAPIR

We propose TAPIR: a Transformer model for Action, Phase, Instrument, and step Recognition in robot-assisted radical prostatectomy surgery videos. Figure 2 depicts an overview of this baseline method. We build upon the Multiscale Vision Transformer (MViT) model developed for video and image recognition tasks [9]. The connection between multiscale feature hierarchies with a transformer model, provides rich features to model the complex temporal cues of the surgical scene around each keyframe in PSI-AVA. The extraction of these features constitutes the first stage of TAPIR, followed by a frame classification head for phase and step recognition. Supplementary Fig. 3 provides a detailed architecture of the Video Feature Extractor stage of our method. To account for instrument detection and atomic action recognition tasks, we employ the Deformable DETR [33] method to obtain bounding box estimates and box-specific features, which are combined with the keyframe features by an additional classification head to predict instrument category or atomic action(s). The following subsections provide further details on the methods we build upon.

Instrument Detector. Deformable DETR [33] is a transformer-based detector that implements multi-scale deformable attention modules to reduce the complexity issues of transformer architectures. This model builds upon DETR [5], and comprises a transformer encoder-decoder module that predicts a set of object detections based on image-level features extracted by a CNN backbone. Furthermore, this method applies an effective iterative bounding box refinement mechanism, to enhance the precision of the bounding boxes coordinates.

Video Feature Extractor. MViT [9] is a transformer-based architecture for video and image recognition that leverages multiscale feature hierarchies. This method is based on stages with channel resolution rescaling processes. The spatial resolution decreases from one stage to another while the channel dimension expands. A Multi-Head Pooling Attention (MHPA) pools query, key, and value tensors to reduce the spatiotemporal resolution. This rescaling strategy produces a multiscale pyramid of features inside the network, effectively connecting the principles of transformers with multi-scale feature hierarchies, and allows capturing low-level features at the first stages and high-level visual information at deeper stages.

Classification Heads. Depending on the granularity of the task, we use a frame classification head or a box classification head. The former receives the features from the Video Feature Extractor. As described in [9], MViT takes the embedding of the token class and performs a linear classification into the phase or step category. The box classification head receives the features extracted from the video sequence and the box-specific features from the Instrument Detector. After an average pooling of the spatiotemporal video features, we concatenate this high-level frame information with the specific box features. Lastly, a linear layer maps this multi-level information to an instrument or atomic action class.

Implementation Details. We pretrained Deformable DETR [33] with a ResNet-50 [16] backbone on the EndoVis 2017 [2] and EndoVis 2018 dataset [3] (as defined in [12]). We train our instrument detection with the iterative bounding box refinement module in PSI-AVA for 100 epochs in a single Quadro RTX 8000 GPU, and a batch size of 6. As in [9,10], we select the bounding

Table 2. Performance on PSI-AVA dataset. We report the number of parameters, inference cost in FLOPs, and mean Average Precision (mAP) on our two-fold cross-validation setup along with the standard deviation.

Method	FLOPs (G)	Param (M)	Long-term Tasks mAP (%)		Short-term Tasks mAP@0.5IoU (%)	
			Phases	Steps	Instruments	Atomic actions
SlowFast [10]	81.24	33.7	46.80 ± 2.07	33.86 ± 1.93	80.29 ± 0.92	18.57 ± 0.62
TAPIR-VST	65.76	87.7	51.63 ± 0.89	41.00 ± 3.34	**81.14 ± 1.51**	22.49 ± 0.34
TAPIR (Ours)	70.80	36.3	**56.55 ± 2.31**	**45.56 ± 0.004**	80.82 ± 1.75	**23.56 ± 0.39**

boxes with a confidence threshold of 0.75. As Video Feature Extractor, we adapt and optimize MViT-B pretrained on Kinetics-400 [18] for the long-term recognition tasks. For the instrument detection and atomic action recognition tasks, we optimized an MViT model pretrained for the step recognition task. For video analysis, we select a 16-frame sequence centered on an annotated keyframe with a sampling rate of 1. We train for 30 epochs, with a 5 epoch warm-up period and a batch size of 6. We employ a binary cross-entropy loss for the atomic action recognition task and a cross-entropy loss for the remaining tasks, all with an SGD optimizer, a base learning rate of 0.0125, and a cosine scheduler with a weight decay of $1e^{-4}$.

4   Experiments

4.1   Video Feature Extractor Method Comparison

Deep Convolutional Neural Networks-Based (DCNN) Method. Similar to the methodology described in Sect. 3, we replace the Instrument Detector and the Video Feature Extractor based on transformers with two DCNN methods. We use Faster R-CNN as detector [27], which we pretrain on the EndoVis 2017 dataset [2] and the training set of the 2018 dataset [3] (as described in [12]). We train on PSI-AVA for 40 epochs, with a batch size of 16 in 1 GPU, a base learning rate of 0.01, and use a SGD optimizer. Finally, we keep the predictions with a confidence score higher than 0.75. Additionally, as video feature extractor we use SlowFast [10] pretrained on CaDIS [14] and likewise to TAPIR we select clips of 16 consecutive frames centered on an annotated keyframe. We set the essential SlowFast parameters α and β to 8 and $\frac{1}{8}$, respectively. We train for 30 epochs, with a 5 epoch warm-up period and a batch size of 45. We employ SGD optimizer, a base learning rate of 0.0125, and a cosine scheduler with a weight decay of $1e^{-4}$. For instrument detection and atomic action recognition tasks we perform a learnable weighted sum of the instrument detector features and SlowFast temporal features, and we apply a linear classification layer to obtain their corresponding predictions.

TAPIR-VST. To complement our study, we compare the use of MViT as Video Feature Extractor with another state-of-the-art transformer method: Video Swin Transformer [21] (VST). We follow the methodology and training curriculum described in Sect. 3, replacing MViT in TAPIR for Video Swin Transformer with Swin-B as a backbone, pretrained on ImageNet-22K. We keep Deformable DETR as the Instrument Detector.

4.2 Results and Discussion

Table 2 shows a comparison study between transformer-based and DCNN-based methods on the PSI-AVA dataset. TAPIR and TAPIR-VST achieve a significant improvement in all tasks over the DCNN-based method. However, TAPIR overcomes TAPIR-VST in the three recognition tasks with a lower computational cost. Our best method outperforms SlowFast [10] by 11.7%, and 9.75% in mAP for steps and phases recognition tasks, respectively. Furthermore, TAPIR improves 4.99% mAP@0.5IoU in our challenging atomic action recognition task, stressing on the relevance of understanding the whole scene to accurately identify the actions performed by each instrument. Qualitative comparisons of SlowFast and TAPIR in all tasks are shown in Supplementary Fig. 4. Overall, these results demonstrate the superior capacity of transformer-based methods to understand the surgical workflow in video and a higher recognition capacity of atomic actions. Also, indicate that TAPIR represents a strong baseline for future research on surgical workflow understanding.

Table 3. Comparison of PSI-AVA with publicly available datasets. We report the number of parameters, inference cost in FLOPs, and mAP performance with corresponding standard deviations.

Method	FLOPs (G)	Param. (M)	Phase recognition task mAP (%)		Step recognition task mAP (%)	
			MISAW [17]	PSI-AVA	MISAW [17]	PSI-AVA
SlowFast [10]	81.24	33.7	94.18	46.80 ± 2.07	76.51	33.86 ± 1.93
TAPIR (Ours)	70.80	36.3	**94.24**	**56.55 ± 2.31**	**79.18**	**45.56 ± 0.004**
Method	FLOPs (G)	Param. (M)	Instruments detection task mAP@0.5IoU (%)			
			Endovis 2017 [2]		PSI-AVA	
Faster R-CNN [27]	180	42	47.51 ± 6.35		80.12 ± 2.32	
Def. DETR [33]	173	40.6	**61.71 ± 3.79**		**81.19 ± 0.87**	

Table 4. TAPIR results for detection-based tasks on PSI-AVA dataset. We report mAP performance on our cross-validation setup with standard deviation.

Video feature extractor	Box features in classification head	mAP@0.5IoU (%)	
		Instrument detection	Action recognition
TAPIR (Ours)	W/o features	13.96 ± 0.33	9.69 ± 0.07
	Faster R-CNN	80.40 ± 0.63	21.18 ± 0.35
	Deformable DETR	**80.82 ± 1.75**	**23.56 ± 0.39**

PSI-AVA Validation. To validate PSI-AVA's benchmark, we study the behavior of the proposed baselines in our dataset and other public databases whose task definition is equivalent to ours. We compare our image-level recognition tasks with their counterparts in the MISAW [17] dataset, and compare our instrument detection task with EndoVis 2017 [2] dataset. Table 3 summarizes the results of the benchmark validation. The consistency in the relative order of the two families of methods across datasets provides empirical support for the adequacy of PSI-AVA as a unified testbed for holistic surgical understanding. Furthermore, the superiority of TAPIR and its instrument detector across datasets further support our method as a powerful baseline for multi-level surgical workflow analysis.

Multi-level Annotations Allow Better Understanding. In this set of experiments, we study the complementarity between the task of detection and classification, specifically instrument detection and atomic action recognition. Such complementarity is inherent to the concept of surgical scene understanding and supports the benefits of having multi-task annotations in PSI-AVA. Table 4 shows that TAPIR effectively exploits the complementary, multi-task, and multi-level information. Without instrument-specific features, TAPIR's head classifies from the high-level temporal representation and loses the spatially localized information. Adding instrument-specific features improves significantly the performance of both instrument detection and atomic action recognition tasks, which entails that the box classification head is capable of leveraging localized and temporal information. Consequently, higher performance in instrument detection, related to more instrument-specific features, reflects in a more accurate recognition by TAPIR (Table 4).

5 Conclusions

In this work, we present a novel benchmark for holistic, multi-level surgical understanding based on a systematical deconstruction of robot-assisted radical prostatectomies. We introduce and validate PSI-AVA, a novel dataset that will allow the study of intrinsic relationships between phases, steps and instrument dynamics, for which we introduce the atomic action definition and vocabulary. Furthermore, we propose TAPIR as a strong transformer-based baseline method tailored for multi-task learning. We hope our framework will serve as a motivation for future work to steer towards holistic surgery scene understanding.

Acknowledgements. The authors would like to thank Laura Bravo-Sánchez, Cristina González and Gabriela Monroy for their contributions along the development of this research work. Natalia Valderrama, Isabela Hernández and Nicolás Ayobi acknowledge the support of the 2021 and 2022 UniAndes-DeepMind Scholarships.

References

1. Ahmidi, N., et al.: A dataset and benchmarks for segmentation and recognition of gestures in robotic surgery. IEEE Trans. Biomed. Eng. **64**(9), 2025–2041 (2017)

2. Allan, M., et al.: 2017 robotic instrument segmentation challenge. arXiv preprint arXiv:1902.06426 (2019)
3. Allan, M., et al.: 2018 robotic scene segmentation challenge. arXiv preprint arXiv:2001.11190 (2020)
4. Bawa, V.S., Singh, G., KapingA, F., Skarga-Bandurova, I., Oleari, E., et al.: The SARAS endoscopic surgeon action detection (ESAD) dataset: challenges and methods. arXiv preprint arXiv:2104.03178 (2021)
5. Carion, N., Massa, F., Synnaeve, G., Usunier, N., Kirillov, A., Zagoruyko, S.: End-to-End object detection with transformers. In: Vedaldi, A., Bischof, H., Brox, T., Frahm, J.-M. (eds.) ECCV 2020. LNCS, vol. 12346, pp. 213–229. Springer, Cham (2020). https://doi.org/10.1007/978-3-030-58452-8_13
6. Czempiel, T., Paschali, M., Ostler, D., Kim, S.T., Busam, B., Navab, N.: OperA: attention-regularized transformers for surgical phase recognition. In: de Bruijne, M., et al. (eds.) MICCAI 2021. LNCS, vol. 12904, pp. 604–614. Springer, Cham (2021). https://doi.org/10.1007/978-3-030-87202-1_58
7. Ding, X., Li, X.: Exploiting segment-level semantics for online phase recognition from surgical videos. arXiv preprint arXiv:2111.11044 (2021)
8. Dosovitskiy, A., et al.: An image is worth 16×16 words: transformers for image recognition at scale. arXiv preprint arXiv:2010.11929 (2020)
9. Fan, H., et al.: Multiscale vision transformers. In: Proceedings of the IEEE/CVF International Conference on Computer Vision, pp. 6824–6835 (2021)
10. Feichtenhofer, C., Fan, H., Malik, J., He, K.: SlowFast networks for video recognition. In: Proceedings of the IEEE/CVF International Conference on Computer Vision, pp. 6202–6211 (2019)
11. Gao, X., Jin, Y., Long, Y., Dou, Q., Heng, P.-A.: Trans-SVNet: accurate phase recognition from surgical videos via hybrid embedding aggregation transformer. In: de Bruijne, M., et al. (eds.) MICCAI 2021. LNCS, vol. 12904, pp. 593–603. Springer, Cham (2021). https://doi.org/10.1007/978-3-030-87202-1_57
12. González, C., Bravo-Sánchez, L., Arbelaez, P.: ISINet: an instance-based approach for surgical instrument segmentation. In: Martel, A.L., et al. (eds.) MICCAI 2020. LNCS, vol. 12263, pp. 595–605. Springer, Cham (2020). https://doi.org/10.1007/978-3-030-59716-0_57
13. Goodman, E.D., et al.: A real-time spatiotemporal AI model analyzes skill in open surgical videos. arXiv preprint arXiv:2112.07219 (2021)
14. Grammatikopoulou, et al.: Cadis: Cataract dataset for image segmentation. arXiv preprint arXiv:1906.11586 (2019)
15. Gu, C., et al.: AVA: a video dataset of spatio-temporally localized atomic visual actions. In: Proceedings of the IEEE Conference on Computer Vision and Pattern Recognition, pp. 6047–6056 (2018)
16. He, K., Zhang, X., Ren, S., Sun, J.: Deep residual learning for image recognition. In: Proceedings of the IEEE Conference on Computer Vision and Pattern Recognition, pp. 770–778 (2016)
17. Huaulmé, A., et al.: Micro-surgical anastomose workflow recognition challenge report. Comput. Methods Programs Biomed. 212, 106452 (2021)
18. Kay, W., et al.: The kinetics human action video dataset. arXiv preprint arXiv:1705.06950 (2017)
19. Kondo, S.: LapFormer: surgical tool detection in laparoscopic surgical video using transformer architecture. Comput. Meth. Biomech. Biomed. Eng. Imaging Visual. 9(3), 302–307 (2021)
20. Lalys, F., Jannin, P.: Surgical process modelling: a review. Int. J. Comput. Assist. Radiol. Surg. 9(3), 495–511 (2013). https://doi.org/10.1007/s11548-013-0940-5

21. Liu, Z., et al.: Video swin transformer. In: Proceedings of the IEEE/CVF Conference on Computer Vision and Pattern Recognition, pp. 3202–3211 (2022)

22. Luo, Z., et al.: MOMA: multi-object multi-actor activity parsing. Adv. Neural Inf. Process. Syst. **34**, 17939–17955 (2021)

23. Maier-Hein, L., Vedula, S.S., Speidel, S., et al.: Surgical data science for next-generation interventions. Nat. Biomed. Eng. **1**(9), 691–696 (2017)

24. Maier-Hein, L., Wagner, M., Ross, T., et al.: Heidelberg colorectal data set for surgical data science in the sensor operating room. Scient. Data **8**(1), 1–11 (2021)

25. Nwoye, C.I., et al.: Recognition of instrument-tissue interactions in endoscopic videos via action triplets. In: Martel, A.L., et al. (eds.) MICCAI 2020. LNCS, vol. 12263, pp. 364–374. Springer, Cham (2020). https://doi.org/10.1007/978-3-030-59716-0_35

26. Nwoye, C.I., et al.: Rendezvous: attention mechanisms for the recognition of surgical action triplets in endoscopic videos. arXiv preprint arXiv:2109.03223 (2021)

27. Ren, S., He, K., Girshick, R., Sun, J.: Faster R-CNN: towards real-time object detection with region proposal networks. arXiv preprint arXiv:1506.01497 (2015)

28. Sharghi, A., Haugerud, H., Oh, D., Mohareri, O.: Automatic operating room surgical activity recognition for robot-assisted surgery. In: Martel, A.L., et al. (eds.) MICCAI 2020. LNCS, vol. 12263, pp. 385–395. Springer, Cham (2020). https://doi.org/10.1007/978-3-030-59716-0_37

29. Stauder, R., Ostler, D., et al.: The tum lapchole dataset for the m2cai 2016 workflow challenge. arXiv preprint arXiv:1610.09278 (2016)

30. Twinanda, A.P., Shehata, S., Mutter, D., Marescaux, J., De Mathelin, M., Padoy, N.: EndoNet: a deep architecture for recognition tasks on laparoscopic videos. IEEE Trans. Med. Imaging **36**(1), 86–97 (2016)

31. Zhang, B., et al.: Towards accurate surgical workflow recognition with convolutional networks and transformers. Comput. Meth. Biomech. Biomed. Eng. Imaging Visual. **10**(4), 1–8 (2021)

32. Zhao, Z., Jin, Y., Heng, P.A.: TraSeTR: track-to-segment transformer with contrastive query for instance-level instrument segmentation in robotic surgery. arXiv preprint arXiv:2202.08453 (2022)

33. Zhu, X., Su, W., Lu, L., Li, B., Wang, X., Dai, J.: Deformable DETR: deformable transformers for end-to-end object detection. arXiv preprint arXiv:2010.04159 (2020)

Multi-modal Unsupervised Pre-training for Surgical Operating Room Workflow Analysis

Muhammad Abdullah Jamal$^{(\boxtimes)}$ and Omid Mohareri

Intuitive Surgical Inc., Sunnyvale, CA, USA
abdullah.jamal@intusurg.com

Abstract. Data-driven approaches to assist operating room (OR) workflow analysis depend on large curated datasets that are time consuming and expensive to collect. On the other hand, we see a recent paradigm shift from supervised learning to self-supervised and/or unsupervised learning approaches that can learn representations from unlabeled datasets. In this paper, we leverage the unlabeled data captured in robotic surgery ORs and propose a novel way to fuse the multi-modal data for a single video frame or image. Instead of producing different augmentations (or "views") of the same image or video frame which is a common practice in self-supervised learning, we treat the multi-modal data as different views to train the model in an unsupervised manner via clustering. We compared our method with other state of the art methods and results show the superior performance of our approach on surgical video activity recognition and semantic segmentation.

Keywords: OR workflow analysis · Surgical activity recognition · Semantic segmentation · Self-supervised learning · Unsupervised learning

1 Introduction

Robotic Surgery has allowed surgeons to perform complex surgeries with more precision and with potential benefits such as shorter hospitalization, fast recoveries, less blood lost etc. However, it might not be easy to adopt due to cost, training, integration with the existing systems, and OR workflow complexities [6] that can eventually lead to human errors in OR.

Recently, data-driven based workflow analysis has been proposed to help identify and mitigate these errors. Sharghi et al. [29] proposes a new dataset that includes 400 full length videos captured from several surgical cases. They used 4 Time of Flight (ToF) cameras to generate a multi-view dataset. Moreover, they also propose a framework to automatically detect surgical activities

---

Supplementary Information The online version contains supplementary material available at https://doi.org/10.1007/978-3-031-16449-1_43.

inside OR. Li et al. [21] proposes a new dataset and framework for semantic segmentation in OR. Schmidt et al. [28] leverages multi-view information and proposes a new architecture for surgical activity detection. However, all these approaches require expert people to manually annotate the data which is expensive and time consuming. Multi-model data/sensing can provide richer information about the scene that can benefit higher performing visual perception tasks like action recognition or object detection. Cameras such as RGB-D, Time of Flight (ToF) have been used in OR to capture the depth information, 3D point clouds, and intensity maps etc. For example [31] uses RGB-D cameras to capture data for 2D/3D pose estimation in hybrid OR. To the best of our knowledge, very little or no work has been done to leverage multi-modalities for surgical OR understanding under unsupervised setting. In this paper, we are focusing on leveraging unlabeled multi-modal image and video data collected in the OR. We propose an unsupervised representation learning approach based on clustering that can help alleviate the annotation time in data-driven approaches. Recently, unsupervised or self-supervised methods have been proposed which have significantly lessen the performance gap with the supervised learning [8,15,22,23]. These methods mostly rely on contrastive loss and set of transformations. They generate different augmentations or views of an image and then directly compare the features using contrastive loss so that they can push away representations from different images and pull together the representations from different views of the same image. But this may not be a scalable approach as it could require computation of all pairwise comparisons on large datasets, and often need larger memory banks [15] or larger batch size [8]. On the other hand, clustering-based approaches group the semantically similar features instead of individual images. We are also interested in semantically similar group of features that can provide better representations especially using multi-modal data for image and video understanding tasks in surgical operating rooms. To address the privacy concerns, and preserve the anonymity of people in OR, we limit ourselves to intensity and the depth maps captured from the Time of Flight (ToF) cameras.

To this end, we propose a novel way to fuse the intensity and the depth map of a single video frame or an image to learn representations. Inspired by [4], we learn the representations in an unsupervised manner via clustering by considering the intensity and the depth map as two different 'views' instead of producing different augmentations ('views') of the same video frame or image. The features of the two views are mapped to the set of learnable prototypes to compute cluster assignments or codes which are subsequently predicted from the features. If the features capture the same information, then it should be possible to predict the code of one view from the feature of the other view. We validate the efficacy of our approach by evaluating on surgical video activity recognition and semantic segmentation in OR. In particular, we achieve superior performance as compared to self-supervised approaches designed especially for video action recognition under various data regime. We also show that our approach achieve better results on semantic segmentation task under low-data regime as compared to clustering based self-supervised approaches, namely Deep Cluster [3], and SELA [36].

2 Related Work

Our work is closely related to data-driven approaches for OR workflow analysis briefly reviewed in Sect. 1. In this section, we will further review some data-driven approaches for OR workflow analysis, and some self-supervised approaches.

Data-Driven Approaches for OR Workflow. [17,18] use multi-view RGB-D data for clinician detection and human pose estimation in OR. [30] introduced a multi-scale super-resolution architecture for human pose estimation using low-resolution depth images. [16] compared several SOTA face detectors using MVOR dataset, and then propose a self-supervised approach which learns from unlabeled clinical data.

Self-supervised Learning. Self-supervised approaches learn representations using unlabeled data by defining a pre-text task which provides the supervised signal. Earlier approaches use reconstruction loss with the autoencoders as pre-text task to learn unsupervised features [26,32]. Masked-patch based prediction models [1,14] which use autoencoders have also been proposed in the similar context. Recently, self-supervised learning paradigm has shifted towards the instance discrimination based contrastive learning [12,23]. It considers every image in the data as its own class and learns the representations by pulling together features of different augmentations of the same image. We have also discussed few contrastive learning based approaches in Sect. 1. Besides, there are several methods that learn visual features by grouping samples via clustering [3,36]. Our work is closely related to the clustering-based approaches. [3] learns representations using k-means assignments which are used as pseudo-labels. [36] uses optimal transport to solve the pseudo-labels assignment problem. [4] contrasts cluster assignments instead of features which enforces the consistency between different augmentations of the image. Finally, several pretext tasks have also been proposed for video domain. It includes pace prediction [33,35], frame and clip order prediction [20,34], and contrastive prediction [11,25].

3 Method

Our goal is to learn representations using unlabeled OR data via clustering by fusing intensity and the depth maps captured using Time of flight (ToF) cameras. Previous clustering-based unsupervised approaches [3,36] work in offline manner where they first cluster the features of the whole dataset and then compute the cluster assignments for different views. This is not feasible in large-scale setting as it requires to compute features of the entire dataset multiple times.

Inspired by recent clustering-based approach called SwAV [4] which works in an online manner by computing cluster assignments using features within a batch, we also limit our selves to an online learning and computes the cluster assignments within a batch using the representations from our multi-modal data. But, unlike SwAV, we don't produce different augmentations (views) of the same image, we treat the intensity and the depth map as two different views of the same video frame or image.

3.1   Fusion of Intensity and Depth Maps

Given an intensity $\mathbf{x}_1$ and the depth map $\mathbf{x}_2$ of a single image or video frame, we first learn the representation through encoder f which is parameterized by θ as $\mathbf{z}_1 = f_\theta(\mathbf{x}_1)$ and $\mathbf{z}_2 = f_\theta(\mathbf{x}_2)$ respectively. Then, we compute the codes $\mathbf{q}_1$ and $\mathbf{q}_2$ by mapping these representations to the set of K learnable prototypes $\{c_1, ..., c_K\} \in \mathcal{C}$. Finally, codes and representations are then used in the following loss function. Figure 1 illustrates the main idea.

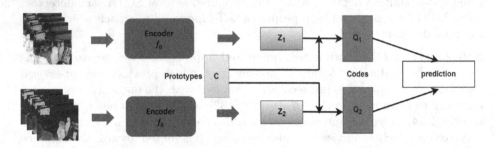

Fig. 1. Our approach takes an intensity and the depth map, and then it extracts features $\mathbf{z}_1$ and $\mathbf{z}_s$ from the encoder f_θ. Next, it computes the codes $\mathbf{q}_1$ and $\mathbf{q}_2$ by mapping these features to the set of K learnable prototypes $\mathcal{C}$. Finally, it predicts the code of one sample from the representation of the other sample.

$$\mathcal{L}(\mathbf{z}_1, \mathbf{z}_2) = l(\mathbf{z}_1, \mathbf{q}_2) + l(\mathbf{z}_2, \mathbf{q}_1) \tag{1}$$

where $l(\mathbf{z}, \mathbf{q})$ represents the cross-entropy loss between the probability obtained by softmax on the dot product between $\mathbf{z}$ and $\mathcal{C}$ and the code $\mathbf{q}$ which is given as:

$$l(\mathbf{z}_1, \mathbf{q}_2) = -\sum_k \mathbf{q}_2^{(k)} \log \mathbf{p}_1^{(k)}, \text{where } \mathbf{p}_1^{(k)} = \frac{\exp(\mathbf{z}_1 \cdot c_k / \tau)}{\sum_{k'} \exp(\mathbf{z}_1 \cdot c_{k'} / \tau)} \tag{2}$$

where τ is a temperature hyperparameter. The intuition here is that if the representations $\mathbf{z}_1$ and $\mathbf{z}_2$ share the same information, then it should be possible to predict the code from the other representation. The total loss is calculated over all the possible pair of intensity and depth maps, which is then minimized to update the encoder f_θ and prototype $\mathcal{C}$ which is implemented as a linear layer in the model. It is also straight forward to apply our fusion approach to learn representations for unlabeled video. Instead of using a single intensity and depth pair, we can sample a video clip, where each frame in the clip can be splitted into intensity and depth map. We will empirically show the superiority of our approach in the surgical video activity recognition in the Sect. 4.

3.2   Estimating Code q

The codes $\mathbf{q}_1$ and $\mathbf{q}_2$ are computed within a batch by mapping the features $\mathbf{z}_1$, and $\mathbf{z}_2$ to the prototypes $\mathcal{C}$ in an online setting. Following [4], we ensure that all the instances in the batch should be equally partitioned into different clusters, thus avoiding the collapse of assigning them to a single prototype. Given feature vectors $\mathbf{Z}$ whose columns are $\mathbf{z}_1, ..., \mathbf{z}_B$ which are mapped to prototypes $\mathcal{C}$. We are interested in optimizing this mapping or codes $\mathbf{Q} = \mathbf{q}_1,..., \mathbf{q}_B$. We optimize using Optimal Transport Solver [24] as:

$$\max_{\mathbf{Q} \in \mathcal{Q}} \mathrm{Tr}(\mathbf{Q}^T \mathbf{C}^T \mathbf{Z}) + \epsilon \mathcal{H}(\mathbf{Q}) \tag{3}$$

where $\mathcal{H}(\mathbf{Q})$ corresponds to the entropy and ϵ is the hyperparameter that avoids the collapsing. Similar to [4], we restricted ourselves to mini-batch settings, and adapt their solution for transportation polytope as:

$$\mathcal{Q} = \left\{ \mathbf{Q} \in \mathcal{R}^{K \times B} \mid \mathbf{Q} \mathbb{1}_B = \frac{1}{\mathbf{K}} \mathbb{1}_\mathbf{K}, \mathbf{Q}^T \mathbb{1}_\mathbf{K} = \frac{1}{\mathbf{B}} \mathbb{1}_\mathbf{K} \right\} \tag{4}$$

where $\mathbb{1}_\mathbf{K}$ corresponds to the vector of ones with dimension $\mathbf{K}$. The soft assignments $\mathbf{Q}^*$ found for Eq. 3 are estimated using iterative Sinkhorn-Knopp algorithm [9]. It can be written as:

$$\mathbf{Q}^* = \mathrm{Diag}(\lambda) \exp \left(\frac{\mathbf{C}^T \mathbf{Z}}{\epsilon} \right) \mathrm{Diag}(\mu) \tag{5}$$

where λ and μ correspond to renormalization vectors. We also use a queue to store the features from the previous iteration because the batch size is usually smaller as compared to the number of prototypes.

4   Experiments

The goal of our pre-training approach is to provide better initialization for downstream tasks such as surgical activity recognition, semantic segmentation etc., and it is applicable to both video and image domain. We evaluated our learned representation on two different tasks i.e., surgical video activity recognition and semantic segmentation under low-data regime. We use mean average precision (mAP) and mean intersection over union (IOU) as evaluation metrics.

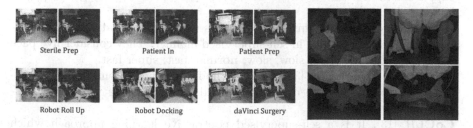

Fig. 2. a). OR Activity Dataset [28, 29] with the intensity on the left side and the depth map on the right side for each activity. b). Semantic Segmentation Dataset [21].

4.1   Datasets

We evaluated our approach on the following two datasets.

Surgical Video Activity Recognition. This dataset has been proposed by [28,29] that consists of 400 full-length videos captured from 103 surgical cases performed by the da Vinci Xi surgical system. The videos are captured from four Time of Flight (ToF) cameras placed on two vision carts inside the operating room. The videos are annotated by a trained user with 10 clinical activities. The intensity and the depth maps are extracted from the raw data from ToF camera. The dataset is splitted into 70% training set and 30% test set. During dataset preparation, we also make sure that all 4 videos belonging to the same case are either in train or test set. Please refer to left panel of the Fig. 2 for example activities.

Semantic Segmentation. [21] has proposed densely annotated dataset for semantic segmentation task. It consists of 7980 images captured by four different Time of Flight (ToF) camera attached in the OR. We first split the dataset into training set (80%) and testing set (20%). From the training set, we create four different subsets of training data by varying the percentage of labels. The right panel of Fig. 2 shows few overlay images from the dataset.

4.2   Surgical Activity Recognition

We use I3D [5] as a backbone architecture for this experiment, but our approach can be applied to any of the other SOTA models like TimeSFormer [2], SlowFast [10] etc. For all methods, we first train the I3D model on video clips similar to [29], and then extract features for full videos from I3D to train Bi-GRU to detect surgical activities. Please refer to supplementary material for implementation details.

Competing Methods. We compare our approach to the following competing baselines.

- **Baseline.** This is the baseline that either trains the I3D model on video clips from the scratch or trains the model pre-trained on ImageNet [27] + Kinetics-400 [19].
- **Pace Prediction** [33]. This method learns video representations by predicting pace of the video clips in self-supervised manner. They consider five pace candidates that are super slow, slow, normal, fast, super fast.
- **Clip Order Prediction** [34]. This method learns spatiotemporal representations by predicting the order of shuffled clips from the video in a self-supervised manner.
- **CoCLR** [13]. It is a self-supervised contrastive learning approach which learns visual-only features using multi-modal information like RGB stream and optical flow.

Table 1. Mean average precision (mAP) % for Bi-GRU under different data regime for surgical activity recognition.

Methods	Pre-train	5% Labeled	10% Labeled	20% Labeled	50% Labeled	100% Labeled
Baseline	None	36.31	55.77	75.70	86.09	90.71
Baseline	ImageNet + Kinetics-400	37.00	55.86	76.50	87.58	90.45
Pace Prediction [33]	Dataset [29]	39.34	63.97	82.69	91.63	91.86
Clip Order Prediction [34]	Dataset [29]	38.03	62.65	82.44	89.34	91.76
CoCLR [13]	Dataset [29]	-	64.87	83.74	-	-
Ours	Dataset [29]	**40.27**	**67.52**	**85.20**	**91.13**	**92.40**

Table 2. Mean average precision (mAP) % for Bi-GRU under different data regime for surgical activity recognition when I3D model is frozen during the training.

Methods	Pre-train	5% Labeled	10% Labeled	20% Labeled	50% Labeled	100% Labeled
Baseline	ImageNet + Kinetics-400	35.56	40.96	50.93	53.32	53.55
Pace Prediction [33]	Dataset [29]	36.88	44.03	54.09	72.51	80.44
Clip Order Prediction [34]	Dataset [29]	36.33	41.98	50.71	55.67	69.00
Ours	Dataset [29]	**37.63**	**43.43**	**57.19**	**73.27**	**80.47**

Results. Table 1 shows the mean average precision of Bi-GRU with I3D backbone under different data regime. It is clear that our approach outperforms the competing ones in each data regime. In general, the advantages of our approach over existing ones become more significant as the number of labeled data decreases. We outperform the pace prediction and clip order prediction by +3.5% and +2.5% on 10% and 20% labeled data respectively. Moreover, when we have 100% labeled data available in training set, our approach still outperforms the baselines despite the fact that our main goal is to have better pre-trained model for low-data regime.

Table 2 shows the mean average precision of Bi-GRU with I3D backbone under different data regime. In this setting, I3D backbone is frozen during training. The purpose of this experiment is to check the quality of the representations learned by different approaches. We can draw the same observation that our approach outperforms the competing ones in each data regime.

Influence of Number of Prototypes K. We study the impact of the number of prototypes K. In Table 3, we report the mAP % on low-data regime for surgical activity recognition by varying the number of prototypes. We don't see significant impact on the performance when using 5% labeled data. For 10% labelled data, we see a slight drop in the performance for $K = 1000$. Our work shares the same spirit of [4] which also observe no significant impact for ImageNet by varying the number of K.

Table 3. We report the mean average precision (mAP) % on low-data regime for surgical activity recognition by varying the number of **K**.

Labeled Data	Number of prototypes		
	1000	2000	3000
5%	40.39	40.48	40.27
10%	66.84	67.40	67.52

4.3 Semantic Segmentation

We use deeplab-v2 [7] with ResNet-50 as a backbone architecture for this experiment. We only train the ResNet-50 during pre-training step for all the competing methods and our approach. Please refer to supplementary material for implementation details.

Competing Methods. We compare our approach to the following clustering-based competing baselines.

- **Baseline.** This is the baseline that trains the deeplabv2 from the scratch.
- **SELA** [36]. This method trains a model to learn representations based on clustering by maximizing the information between the input data samples and labels. The labels are obtained using self-labeling method that cluster the samples into K distinct classes.
- **Deep Cluster** [3]. It is a clustering based approach to learn visual features. It employs k-means for cluster assignments, and subsequently uses these assignments as supervision to train the model.

Table 4. Mean intersection over union (mIoU) for deeplab-v2 with ResNet-50 backbone under low-data regime for semantic segmentation.

Methods	Pre-train	2% Labeled	5% Labeled	10% Labeled	15% Labeled
Baseline	None	0.452±0.008	0.483±0.007	0.500±0.012	0.521±0.010
SELA [36]	Dataset [28, 29]	0.464±0.009	0.499±0.010	0.518±0.013	0.532±0.006
DeepCluster [3]	Dataset [28, 29]	0.481±0.006	0.498±0.011	0.520±0.004	0.535±0.008
SwAV [4]	Dataset [28, 29]	0.484±0.006	0.502±0.006	0.522±0.004	-
Ours	Dataset [28, 29]	**0.494±0.014**	**0.516±0.005**	**0.538±0.012**	**0.553±0.010**

Results. Table 4 shows the mean intersection over union (mIoU) under low-data regime. For evaluation, we regenerate the subsets and train the model five times to report mean and standard deviation. It is clear that our approach outperforms the competing approaches on low-data regime. This shows that our approach is task-agnostic, and can be used as pre-training step if provided with multi-modal data.

4.4 Remarks

While SwAV [4] relies on producing difference augmentations (views) of the same image, we propose a multi-modal fusion approach in which different modalities (intensity and depth in our case) are treated as two different views of the same video frame. Moreover, unlike SwAV [4], we show the effectiveness of our approach on video domain. Finally, Table 4 shows that our approach still outperforms the SwAV baseline for semantic segmentation.

5 Conclusion

In this paper, we propose an unsupervised pretraining approach for video and image analysis task for surgical OR that can enable workflow analysis. Our novel approach combine the intensity and the depth map of a single video frame or image captured from the surgical OR to learn unsupervised representations. While the recent self-supervised or unsupervised learning methods require different augmentations ('views') of a single image, our method considers the intensity and the depth map as two different views. While we demonstrate the effectiveness of our approach on surgical video activity recognition and semantic segmentation in low-data regime, it can also be extended to other tasks where similar multi-modal data is available such as video and image analysis task in laparoscopic surgery. Furthermore, it can be used in pre-training stage for other downstream tasks such as 2D/3D pose estimation, person detection and tracking etc.

References

1. Bao, H., Dong, L., Wei, F.: Beit: BERT pre-training of image transformers. CoRR abs/2106.08254 (2021)
2. Bertasius, G., Wang, H., Torresani, L.: Is space-time attention all you need for video understanding? CoRR abs/2102.05095 (2021)
3. Caron, M., Bojanowski, P., Joulin, A., Douze, M.: Deep clustering for unsupervised learning of visual features. CoRR abs/1807.05520 (2018)
4. Caron, M., Misra, I., Mairal, J., Goyal, P., Bojanowski, P., Joulin, A.: Unsupervised learning of visual features by contrasting cluster assignments. CoRR abs/2006.09882 (2020)
5. Carreira, J., Zisserman, A.: Quo vadis, action recognition? a new model and the kinetics dataset. CoRR abs/1705.07750 (2017)
6. Catchpole, K., et al.: Safety, efficiency and learning curves in robotic surgery: a human factors analysis. Surg. Endosc. **30**(9), 3749–3761 (2015). https://doi.org/10.1007/s00464-015-4671-2
7. Chen, L., Papandreou, G., Kokkinos, I., Murphy, K., Yuille, A.L.: Deeplab: Semantic image segmentation with deep convolutional nets, atrous convolution, and fully connected crfs. CoRR abs/1606.00915 (2016)
8. Chen, T., Kornblith, S., Norouzi, M., Hinton, G.E.: A simple framework for contrastive learning of visual representations. CoRR abs/2002.05709 (2020)
9. Cuturi, M.: Sinkhorn distances: Lightspeed computation of optimal transport. In: Burges, C.J.C., Bottou, L., Welling, M., Ghahramani, Z., Weinberger, K.Q. (eds.) Advances in Neural Information Processing Systems (2013)

10. Feichtenhofer, C., Fan, H., Malik, J., He, K.: Slowfast networks for video recognition. CoRR abs/1812.03982 (2018)
11. Feichtenhofer, C., Fan, H., Xiong, B., Girshick, R.B., He, K.: A large-scale study on unsupervised spatiotemporal representation learning. CoRR abs/2104.14558 (2021)
12. Grill, J., et al.: Bootstrap your own latent: a new approach to self-supervised learning. CoRR abs/2006.07733 (2020)
13. Han, T., Xie, W., Zisserman, A.: Self-supervised co-training for video representation learning. CoRR abs/2010.09709 (2020)
14. He, K., Chen, X., Xie, S., Li, Y., Dollár, P., Girshick, R.B.: Masked autoencoders are scalable vision learners. CoRR abs/2111.06377 (2021)
15. He, K., Fan, H., Wu, Y., Xie, S., Girshick, R.B.: Momentum contrast for unsupervised visual representation learning. CoRR abs/1911.05722 (2019)
16. Issenhuth, T., Srivastav, V., Gangi, A., Padoy, N.: Face detection in the operating room: Comparison of state-of-the-art methods and a self-supervised approach. CoRR abs/1811.12296 (2018)
17. Kadkhodamohammadi, A., Gangi, A., de Mathelin, M., Padoy, N.: 3d pictorial structures on RGB-D data for articulated human detection in operating rooms. CoRR abs/1602.03468 (2016)
18. Kadkhodamohammadi, A., Gangi, A., de Mathelin, M., Padoy, N.: A multi-view RGB-D approach for human pose estimation in operating rooms. CoRR abs/1701.07372 (2017)
19. Kay, W., et al.: The kinetics human action video dataset. CoRR abs/1705.06950 (2017)
20. Lee, H., Huang, J., Singh, M., Yang, M.: Unsupervised representation learning by sorting sequences. CoRR abs/1708.01246 (2017)
21. Li, Z., Shaban, A., Simard, J., Rabindran, D., DiMaio, S.P., Mohareri, O.: A robotic 3d perception system for operating room environment awareness. CoRR abs/2003.09487 (2020)
22. Misra, I., van der Maaten, L.: Self-supervised learning of pretext-invariant representations. CoRR abs/1912.01991 (2019)
23. van den Oord, A., Li, Y., Vinyals, O.: Representation learning with contrastive predictive coding. CoRR abs/1807.03748 (2018)
24. Peyré, G., Cuturi, M.: Computational optimal transport (2020)
25. Qian, R., et al.: Spatiotemporal contrastive video representation learning. CoRR abs/2008.03800 (2020)
26. Ranzato, M., Huang, F.J., Boureau, Y.L., LeCun, Y.: Unsupervised learning of invariant feature hierarchies with applications to object recognition. In: CVPR (2007)
27. Russakovsky, O., et al.: Imagenet large scale visual recognition challenge. CoRR abs/1409.0575 (2014)
28. Schmidt, A., Sharghi, A., Haugerud, H., Oh, D., Mohareri, O.: Multi-view surgical video action detection via mixed global view attention. In: de Bruijne, M., et al. (eds.) MICCAI 2021. LNCS, vol. 12904, pp. 626–635. Springer, Cham (2021). https://doi.org/10.1007/978-3-030-87202-1_60
29. Sharghi, A., Haugerud, H., Oh, D., Mohareri, O.: Automatic operating room surgical activity recognition for robot-assisted surgery. In: Martel, A.L., et al. (eds.) MICCAI 2020. LNCS, vol. 12263, pp. 385–395. Springer, Cham (2020). https://doi.org/10.1007/978-3-030-59716-0_37
30. Srivastav, V., Gangi, A., Padoy, N.: Human pose estimation on privacy-preserving low-resolution depth images. CoRR abs/2007.08340 (2020)

31. Srivastav, V., Issenhuth, T., Kadkhodamohammadi, A., de Mathelin, M., Gangi, A., Padoy, N.: MVOR: a multi-view RGB-D operating room dataset for 2d and 3d human pose estimation. CoRR abs/1808.08180 (2018)
32. Vincent, P., Larochelle, H., Bengio, Y., Manzagol, P.A.: Extracting and composing robust features with denoising autoencoders. In: ICML 2008 (2008)
33. Wang, J., Jiao, J., Liu, Y.: Self-supervised video representation learning by pace prediction. CoRR abs/2008.05861 (2020)
34. Xu, D., Xiao, J., Zhao, Z., Shao, J., Xie, D., Zhuang, Y.: Self-supervised spatiotemporal learning via video clip order prediction. In: Proceedings of the IEEE/CVF Conference on Computer Vision and Pattern Recognition (CVPR) (2019)
35. Yao, Y., Liu, C., Luo, D., Zhou, Y., Ye, Q.: Video playback rate perception for self-supervisedspatio-temporal representation learning. CoRR abs/2006.11476 (2020)
36. Asano, Y.M., Rupprecht, C., Vedald, A.: Self-labelling via simultaneous clustering and representation learning. In: International Conference on Learning Representations (2020)

Deep Laparoscopic Stereo Matching
with Transformers

Xuelian Cheng[1], Yiran Zhong[2,3], Mehrtash Harandi[1,4], Tom Drummond[5],
Zhiyong Wang[6], and Zongyuan Ge[1,7,8(✉)]

[1] Faculty of Engineering, Monash University, Melbourne, Australia
[2] SenseTime Research, Shanghai, China
[3] Shanghai AI Laboratory, Shanghai, China
[4] Data61, CSIRO, Sydney, Australia
[5] University of Melbourne, Melbourne, Australia
[6] The University of Sydney, Sydney, Australia
[7] eResearch Centre, Monash University, Melbourne, Australia
zongyuan.ge@monash.edu
[8] Monash-Airdoc Research Centre, Melbourne, Australia
https://mmai.group

Abstract. The self-attention mechanism, successfully employed with
the transformer structure is shown promise in many computer vision
tasks including image recognition, and object detection. Despite the
surge, the use of the transformer for the problem of stereo matching
remains relatively unexplored. In this paper, we comprehensively investigate the use of the transformer for the problem of stereo matching,
especially for laparoscopic videos, and propose a new hybrid deep stereo
matching framework (HybridStereoNet) that combines the best of the
CNN and the transformer in a unified design. To be specific, we investigate several ways to introduce transformers to volumetric stereo matching pipelines by analyzing the loss landscape of the designs and in-
domain/cross-domain accuracy. Our analysis suggests that employing
transformers for feature representation learning, while using CNNs for
cost aggregation will lead to faster convergence, higher accuracy and
better generalization than other options. Our extensive experiments on
Sceneflow, SCARED2019 and dVPN datasets demonstrate the superior
performance of our HybridStereoNet.

Keywords: Stereo matching · Transformer · Laparoscopic video

1 Introduction

3D information and stereo vision are important for robotic-assisted minimally
invasive surgeries (MIS) [2,17]. Given the success of modern deep learning

---

X. Cheng and Y. Zhong—Equal contribution.

Supplementary Information The online version contains supplementary material
available at https://doi.org/10.1007/978-3-031-16449-1_44.

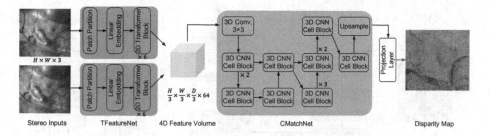

Fig. 1. The overall pipeline of the HybridStereoNet network.

systems [4,5,34,36,37] on natural stereo pairs, a promising next challenge is surgical stereo vision, *e.g.*, laparoscopic and endoscopic images. It has received substantial prior interest as its promise for many medical down-streaming tasks such as surgical robot navigation [21], 3D registration [2,17], augmented reality (AR) [20,33] and virtual reality (VR) [6].

In recent years, we have witnessed a substantial progress of deep stereo matching in natural images such as KITTI 2015, Middlebury and ETH3D. Several weaknesses of conventional stereo matching algorithms (*e.g.*, handling occlusion [23], and textureless areas [19]) have been largely alleviated through deep convolutional networks [5,14,26,27,35] and large training data. However, recovering dense depth maps for laparoscopic stereo videos is still a non-trivial task. First, the textureless problem in laparoscopic stereo images is much severe than natural images. Greater demands were placed on the stereo matching algorithms to handle large textureless areas. Second, there are only few laparoscopic stereo datasets for training a stereo network due the hardness of retrieving the ground truth. Lack of large-scale training data requires the network to be either an effective learner (*i.e.*, to be able to learn stereo matching from few samples) or a quick adapter that can adjust to the new scene with few samples. Also, there is a large domain gap between natural images and medical images. In order to be a quick adapter, the network needs superior generalisation ability to mitigate the gap.

The transformer has become the horsepower of neural architectures recently, due to its generalization ability [15,25,28]. Transformers have been successfully applied in natural language processing [22] and high-level computer vision tasks, *e.g.*, image classification, semantic segmentation and object detection. Yet, its application to low level vision is yet to be proven, *e.g.*, the performance of STTR [13] is far behind convolution-based methods in stereo matching. In this paper, we investigate the use of transformers for deep stereo matching in laparoscopic stereo videos. We will show that by using transformers to extract features, while employing convolutions for aggregation of matching cost, a deep model with higher domain specific and cross domain performances can be achieved.

Following the volumetric deep stereo matching pipeline in LEAStereo [5], our method consists of a feature net, a 4D cost volume, a matching net and a projection layer. The feature net and the matching net are the only two

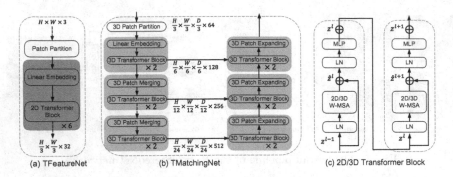

Fig. 2. The architecture of our transformer-based TFeatureNet and TMatchNet.

modules that contain trainable parameters. We substitute these modules with our designed transformer-based structure and compare the accuracy, generalization ability, and the loss landscapes to analyze the behavior of the transformer for the stereo matching task. The insights obtained from those analyses have led us to propose a new hybrid architecture for laparoscopic stereo videos that achieves better performance than both convolution-based methods and pure transformer-based methods. Specifically, we find that the transformers tend to find flatter local minima while the CNNs can find a lower one but sharper. By using transformers as the feature extractor, and CNNs for cost aggregation, the network can find a local minima both lower and flatter than pure CNN-based methods, which leads to better generalization ability. Thanks to the global field of view of transformers, our HybridStereoNet has better textureless handling capability as well.

2 Method

In this section, we first illustrate our hybrid stereo matching architecture and then provide a detailed analysis of transformers in stereo matching networks. Our design is inspired by LEAStereo which is an state-of-the-art model for the natural stereo matching task. For the sake of discussion, we denote the feature net in LEAStereo [5] as CFeatureNet, and the matching net as CMatchNet.

2.1 The HybridStereoNet

Our proposed HybridStereoNet architecture is shown in Fig. 1. We adapt the standard volumetric stereo matching pipeline that consists of a feature net to extract features from input stereo images, a 4D feature volume that is constructed by concatenating features from stereo image pairs through epipolar lines [32], a matching net that regularizes the 4D volume to generate a 3D cost volume, and a projection layer to project the 3D cost volume to a 2D disparity

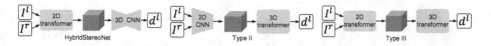

Fig. 3. The overall pipeline of variant networks. We follow the feature extraction - 4D feature volume construction - dense matching pipeline for deep stereo matching. The variants change the FeatureNet and Matching Net with either transformer or CNNs. The gray box represents 4D cost volume.

map. In this pipeline, only the feature net and the matching net contain trainable parameters. We replace both networks with transformer-based structures to make them suitable for laparoscopic stereo images.

We show our transformer-based feature net (TFeatureNet) in Fig. 2(a) and matching net (TMatchNet) in Fig. 2(b). For a fair comparison with the convolutional structure of the LEAStereo [5], we use the same number of layers L for our TFeatureNet and TMatchNet as in the LEAStereo, *i.e.*, $L^F = 6$ for the TFeatureNet and $L^M = 12$ for the TMatchNet. The 4D feature volume is also built in 1/3 resolution.

TFeatureNet is a Siamese network with shared weights to extract features from input stereo pairs of size $H \times W$. We adapt the same patching technique as in ViT [7] and split the image into N non-overlapping patches before feeding them to a vision transformer. We set the patch size to 3×3 to obtain $\frac{H}{3} \times \frac{W}{3}$ tokens and thus ensure the cost volume built in 1/3 resolution. A linear embedding layer is applied on the raw-valued features and project them to a C dimensional space. We set the embedding feature channel to 32. We empirically select the swin transformer block [15] as our 2D transformer block.

TMatchNet is a U-shaped encoder-decoder network with 3D transformers. The encoder is a three-stage down-sampling architecture with stride of 2. In contrast, the decoder is a three-stage up-sampling architecture as shown in Fig 2(b). Similar to the TFeatureNet, we use a 3D Patch Partitioning layer to split the 4D volume $H' \times W' \times D' \times C$ into N non-overlapping 3D patches and feed them into a 3D transformer. Unlike previous 3D transformers [16] that keep the dimension D unchanged when down-sampling the spatial dimensions H, W, we change D accordingly with the stride to enforce the correct geometrical constraints.

We compare the functionality of transformers in feature extraction and cost aggregation. As we will show soon, we find that transformers are good for feature representation learning while convolutions are good for cost aggregation. Therefore, in our HybridStereoNet, we use our TFeatureNet as the feature extractor and CMatchNet for the cost aggregation. We provide a detailed comparison and analysis in the following section.

2.2 Analyzing Transformer in Laparoscopic Stereo

To integrate the transformer in the volumetric stereo pipeline, we have three options as shown in Fig. 3. Type I (HybridStereoNet): TFeatureNet with CMatchNet; Type II: CFeatureNet with TMatchNet; and Type III: TFeatureNet

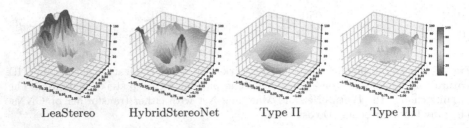

LeaStereo   HybridStereoNet   Type II   Type III

Fig. 4. Loss Landscape Visualization on SCARED2019 dataset. 3D surfaces of the gradient variance from HybridStereoNet and its variants on SceneFlow. The two axes mean two random directions with filter-wise normalization. The height of the surface indicates the value of the gradient variance.

with TMatchNet. In this section, we investigate these options in terms of loss landscapes, projected learning trajectories, in-domain/cross-domain accuracy, and texture-less handling capabilities in laparoscopic stereo.

Loss Landscape. We visualize the loss landscape on the SCARED2019 dataset for types described above along with LEAStereo using the random direction approach [12] in Fig. 4. The plotting details can be found in the supplemental material. The visualization suggests that Type II and Type III tend to find a flatter loss landscape but with a higher local minima. Compared with LEAStereo, the landscape of the HybridStereoNet shares a similar local minima but is flatter, which leads to better cross-domain performance [3,11].

Projected Learning Trajectory. We also compare the convergence curve of each type with the projected learning trajectory [12]. Figure 5 shows the learning trajectories along the contours of loss surfaces. Let θ_i denote model parameters at epoch i. The final parameters of the model after n epochs of training are shown by θ_n. Given n training epochs, we can apply PCA [24] to the matrix $M = [\theta_0 - \theta_n; \cdots; \theta_{n-1} - \theta_n]$, and then select the two most explanatory directions. This enables us to visualize the optimizer trajectory (blue dots in Fig. 5) and loss surfaces along PCA directions. On each axis, we measure the amount of variation in the descent path captured by that PCA direction. Note that the loss landscape

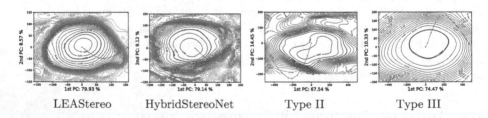

LEAStereo   HybridStereoNet   Type II   Type III

Fig. 5. Projected learning trajectories use normalized PCA directions for our variants. Please zoom in for better visualization. (Color figure online)

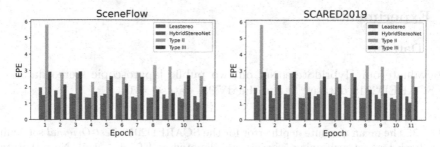

Fig. 6. Accuracy comparison for our proposed variants. (a) In-domain results on SceneFlow dataset. (b) Cross domain results on SCARED2019 dataset.

Table 1. Memory footprint for the proposed variant networks.

Method	LeaStereo	Type II	Type III	HybridStereoNet
Params [M]	1.81	9.54	9.62	1.89
Flops [M]	3713.53	4144.85	811.68	380.01
Runtime [s]	0.30	0.48	0.50	0.32

dynamically changes during training and we only present the "local" landscape around the final solution.

As we can see from the Fig. 5, the Type II variant misses the local minima; Type III directly heads to a local minima after circling around the loss landscape in its first few epochs. It prohibits the network from finding a lower local minima. The LEAStereo and our HybridStereoNet both find some lower local minimas but the HybridStereoNet uses a sharper descending pathway on the loss landscape and therefore leads to a lower local minima. We use the SGD optimizer and the same training setting with LEAStereo [5] for side-by-side comparison.

Accuracy. In Fig. 6, we compare the in-domain and cross-domain accuracy of all types for the first several epochs. All of these variants are trained on a large nature scene synthetic stereo dataset called SceneFlow [18] with over 30k stereo pairs. For in-domain performance, we plot the validation end point error (EPE) on the SceneFlow dataset in Fig. 6(a). The HybridStereoNet consistently achieves low error rates than all the other variants. For the cross-domain performance, we directly test the trained models on the SCARED2019 laparoscopic stereo dataset and plot the EPE in Fig. 6(b). Again, the HybridStereoNet achieves lower cross-domain error rates in most epochs.

Memory Footprint. We provide details of four variants regarding the running time, and memory footprint in Table 1, which were tested on a Quadro GV100 with the input size 504 × 840. We keep relatively similar number of learnable parameters to make a fair comparison.

3   Experiments

3.1   Datasets

We evaluate our HybridStereoNet on two public laparoscopic stereo datasets: the SCARED2019 dataset [1] and the dVPN dataset [31].

Table 2. The mean absolute depth error for the SCARED2019 *Test-Original* set (unit: mm). Each test set containing 5 keyframes, denoted as $kf_n, n \in [1,5]$. Note that our method and STTR are not fine-tuned on the target dataset. The lower the better.

	Method	Test Set 1						Test Set 2					
		kf_1	kf_2	kf_3	kf_4	kf_5	Avg.	kf_1	kf_2	kf_3	kf_4	kf_5	Avg.
Supervised	Lalith Sharan [1]	30.63	46.51	45.79	38.99	53.23	43.03	35.46	50.09	25.24	62.37	70.45	48.72
	Xiaohong Li [1]	34.42	20.66	17.84	27.92	13.00	22.77	24.58	16.80	29.92	11.37	19.93	20.52
	Huoling Luo [1]	29.68	16.36	13.71	22.42	15.43	19.52	20.83	11.27	35.74	8.26	14.97	18.21
	Zhu Zhanshi [1]	14.64	7.77	7.03	7.36	11.22	9.60	14.41	12.55	16.30	27.87	34.86	21.20
	Wenyao Xia [1]	**5.70**	7.18	6.98	8.66	5.13	6.73	13.80	6.85	13.10	5.70	7.73	9.44
	Trevor Zeffiro [1]	7.91	2.97	**1.71**	2.52	2.91	3.60	5.39	1.67	4.34	3.18	2.79	3.47
	Congcong Wang [1]	6.30	2.15	3.41	3.86	4.80	4.10	6.57	2.56	6.72	4.34	1.19	4.28
	J.C. Rosenthal [1]	8.25	3.36	2.21	2.03	1.33	3.44	8.26	2.29	7.04	2.22	0.42	4.05
	D.P. 1 [1]	7.73	2.07	1.94	2.63	**0.62**	3.00	4.85	**0.65**	**1.62**	**0.77**	0.41	**1.67**
	D.P. 2 [1]	7.41	**2.03**	1.92	2.75	0.65	2.95	4.78	1.19	3.34	1.82	**0.36**	2.30
	STTR [13]	9.24	4.42	2.67	**2.03**	2.36	4.14	7.42	7.40	3.95	7.83	2.93	5.91
	HybridStereoNet	7.96	2.31	2.23	3.03	1.01	3.31	**4.57**	1.39	3.06	2.21	0.52	2.35

SCARED2019 is released during the Endovis challenge at MICCAI 2019, including 7 training subsets and 2 test subsets captured by a da Vinci Xi surgical robot. The original dataset only provides the raw video data, the depth data of each key frame and corresponding camera intrinsic parameters. We perform additional dataset curation to make it suitable for stereo matching. After curation, the SCARED2019 contains 17206 stereo pairs for training and 5907 pairs for testing. We use the official code to assess the mean absolute depth error on all the subsequent frames provided by [1], named *Test-Original*.

However, as pointed by STTR [13], the depth of following frames are interpolated by forwarding kinematics information of the point cloud. This would lead to the synchronization issues and kinematics offsets, resulting in inaccurate depth values for subsequent frames. Following STTR [13], we further collect the first frame of each video and build our *Test-19* set, which subset consists of 19 images of resolution 1080×1024 with the maximum disparity of 263 pixels. The left and right 100 pixels were cropped due to invalidity after rectification. We further provide a complete *one-key evaluation toolbox* for disparity evaluation.

dVPN is provided by Hamlyn Centre Laparoscopic, with 34320 pairs of rectified stereo images for training and 14382 pairs for testing. There is no ground truth depth for these frames. To compare the performance of our model with other methods, we use the image warping accuracy [32] as our evaluation metrics, *i.e.*, Structural Similarity Index Measure (SSIM) [29], and Peak-Signal-to-Noise

Ratio (PSNR) [9]. Note that for a fair comparison, we exclude self-supervised methods in our comparison as they directly optimize the disparity with image warping losses [32].

3.2 Implementation

We implemented all the architectures in Pytorch. A random crop with size 336×336 is the only data argumentation technique used in this work. We use the SGD optimizer with momentum 0.9, cosine learning rate that decays from 0.025 to 0.001, and weight decay 0.0003. Our pretrained models on SceneFlow are conduced on two Quadro GV100 GPUs. Due to the limitation of public laparoscopic data and the ground truth, we train the proposed variant models on a synthetic dataset, SceneFlow [18], which has per-pixel ground truth disparities. It contains 35,454 training and 4,370 testing rectified image pairs with a typical resolution of 540×960. We use the "finalpass" version as it is more realistic (Fig. 7).

Table 3. Quantitative results on the *Test-19* set (evaluated on all pixels). We compare our method with various state-of-the-art methods, by bad pixel ratio disparity errors.

Methods	EPE [px] ↓	RMSE [px] ↓	bad 2.0 [%] ↓	bad 3.0 [%] ↓	bad 5.0 [%] ↓
STTR [13]	6.1869	20.4903	8.4266	8.0428	7.5234
LEAStereo [5]	1.5224	**4.1135**	4.5251	3.6580	2.1338
HybridStereoNet	**1.4096**	4.1336	**4.1859**	**3.4061**	**2.0125**

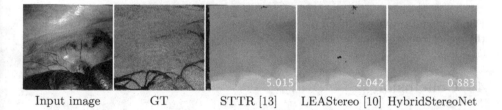

| Input image | GT | STTR [13] | LEAStereo [10] | HybridStereoNet |

Fig. 7. Qualitative results with bad 3.0 value on the *Test-19* set. Our model predicts dense fine-grained details even for occlusion areas.

3.3 Results

SCARED2019. We summarized the evaluation results on *Test-Original* in Table 2, including methods reported in the challenge summary paper [1]. We also provided unsupervised methods from [10] in the supplementary material. Note that our model never seen the training set. As shown in Table 3, our results show an improvement compared with the state-of-the-art Pure CNN method LEAStereo and a transformer-based method STTR [13] on our reorganized *Test-19* set. Please refer to supplementary material for more results on non-occluded areas.

dVPN. As shown in Table 4, our results are better than other competitors. Noting that DSSR [17] opts for the same structure with STTR [13]. Several unsupervised methods, *e.g.,* SADepth [10], are not included in this table as they are trained with reconstruction losses which will lead to a high value of the evaluated SSIM metric. However, for the sake of completeness, we provide their results in supplementary material (Fig. 8).

Table 4. Evaluation on the *dVPN* test set (↑ means higher is better). We directly report results of ELAS and SPS from [10]. Results of E-DSSR and DSSR are from [17].

Method	Training	Mean SSIM ↑	Mean PSNR ↑
ELAS [8]	No training	47.3	–
SPS [30]	No training	54.7	–
E-DSSR [17]	No training	41.97±7.32	13.09 ± 2.14
DSSR [17]	No training	42.41 ± 7.12	12.85 ± 2.03
LEAStereo [5]	No training	55.67	15.25
HybridStereoNet	No training	**56.98**	**15.45**

Left   Right   Predict Disp   Reconstructed Left

Fig. 8. Quantitative results on the dVPN dataset. The invalid areas on the left side of the reconstructed image are the occluded areas.

4   Conclusion

In this paper, we extensively investigated the effect of transformers for laparoscopic stereo matching in terms of loss landscapes, projected learning trajectories, in-domain/cross-domain accuracy, and proposed a new hybrid stereo matching framework. We empirically found that for laparoscopic stereo matching, using transformers to learn feature presentations and CNNs to aggregate matching costs can lead to faster convergence, higher accuracy and better generalization. Our proposed HybridStereoNet surpasses state-of-the-art methods on SCARED2019 and dVPN datasets.

References

1. Allan, M., et al.: Stereo correspondence and reconstruction of endoscopic data challenge. arXiv preprint arXiv:2101.01133 (2021)

2. Cartucho, J., Tukra, S., Li, Y., S. Elson, D., Giannarou, S.: Visionblender: a tool to efficiently generate computer vision datasets for robotic surgery. CMBBE Imaging Vis. **9**(4), 331–338 (2021)

3. Chaudhari, P., et al.: Entropy-SGD: biasing gradient descent into wide valleys. J. Stat. Mech. Theory Exp. **2019**(12), 124018 (2019)

4. Cheng, X., Zhong, Y., Dai, Y., Ji, P., Li, H.: Noise-aware unsupervised deep lidar-stereo fusion. In: CVPR (2019)

5. Cheng, X., et al.: Hierarchical neural architecture search for deep stereo matching. In: NeurIPS, vol. 33 (2020)

6. Chong, N., et al.: Virtual reality application for laparoscope in clinical surgery based on siamese network and census transformation. In: Su, R., Zhang, Y.-D., Liu, H. (eds.) MICAD 2021. LNEE, vol. 784, pp. 59–70. Springer, Singapore (2022). https://doi.org/10.1007/978-981-16-3880-0_7

7. Dosovitskiy, A., et al.: An image is worth 16x16 words: transformers for image recognition at scale. arXiv preprint arXiv:2010.11929 (2020)

8. Geiger, A., Roser, M., Urtasun, R.: Efficient large-scale stereo matching. In: Kimmel, R., Klette, R., Sugimoto, A. (eds.) ACCV 2010. LNCS, vol. 6492, pp. 25–38. Springer, Heidelberg (2011). https://doi.org/10.1007/978-3-642-19315-6_3

9. Hore, A., Ziou, D.: Image quality metrics: Psnr vs. ssim. In: 2010 20th International Conference on Pattern Recognition, pp. 2366–2369. IEEE (2010)

10. Huang, B., et al.: Self-supervised generative adversarial network for depth estimation in laparoscopic images. In: de Bruijne, M., et al. (eds.) MICCAI 2021. LNCS, vol. 12904, pp. 227–237. Springer, Cham (2021). https://doi.org/10.1007/978-3-030-87202-1_22

11. Keskar, N.S., Mudigere, D., Nocedal, J., Smelyanskiy, M., Tang, P.T.P.: On large-batch training for deep learning: Generalization gap and sharp minima. ICLR (2017)

12. Li, H., Xu, Z., Taylor, G., Studer, C., Goldstein, T.: Visualizing the loss landscape of neural nets. NeurIPS **31** (2018)

13. Li, Z., et al.: Revisiting stereo depth estimation from a sequence-to-sequence perspective with transformers. In: ICCV, pp. 6197–6206, October 2021

14. Lipson, L., Teed, Z., Deng, J.: RAFT-Stereo: Multilevel Recurrent Field Transforms for Stereo Matching. arXiv preprint arXiv:2109.07547 (2021)

15. Liu, Z., et al.: Swin transformer: Hierarchical vision transformer using shifted windows. In: ICCV, pp. 10012–10022 (2021)

16. Liu, Z., et al.: Video swin transformer. arXiv preprint arXiv:2106.13230 (2021)

17. Long, Y., et al.: E-DSSR: efficient dynamic surgical scene reconstruction with transformer-based stereoscopic depth perception. In: de Bruijne, M., et al. (eds.) MICCAI 2021. LNCS, vol. 12904, pp. 415–425. Springer, Cham (2021). https://doi.org/10.1007/978-3-030-87202-1_40

18. Mayer, N., et al.: A large dataset to train convolutional networks for disparity, optical flow, and scene flow estimation. In: CVPR, pp. 4040–4048 (2016)

19. Menze, M., Geiger, A.: Object scene flow for autonomous vehicles. In: CVPR (2015)

20. Nicolau, S., Soler, L., Mutter, D., Marescaux, J.: Augmented reality in laparoscopic surgical oncology. Surg. Oncol. **20**(3), 189–201 (2011)

21. Overley, S.C., Cho, S.K., Mehta, A.I., Arnold, P.M.: Navigation and robotics in spinal surgery: where are we now? Neurosurgery **80**(3S), S86–S99 (2017)

22. Qin, Z., et al.: Cosformer: Rethinking softmax in attention. In: ICLR (2022)

23. Scharstein, D., et al.: High-resolution stereo datasets with subpixel-accurate ground truth. In: Jiang, X., Hornegger, J., Koch, R. (eds.) GCPR 2014. LNCS, vol. 8753, pp. 31–42. Springer, Cham (2014). https://doi.org/10.1007/978-3-319-11752-2_3

24. Schölkopf, B., Smola, A., Müller, K.R.: Nonlinear component analysis as a kernel eigenvalue problem. Neural Comput. **10**(5), 1299–1319 (1998)

25. Sun, W., Qin, Z., Deng, H., Wang, J., Zhang, Y., Zhang, K., Barnes, N., Birchfield, S., Kong, L., Zhong, Y.: Vicinity vision transformer. In: arxiv. p. 2206.10552 (2022)

26. Wang, J., et al.: Deep two-view structure-from-motion revisited. In: CVPR, pp. 8953–8962, June 2021

27. Wang, J., Zhong, Y., Dai, Y., Zhang, K., Ji, P., Li, H.: Displacement-invariant matching cost learning for accurate optical flow estimation. In: NeurIPS (2020)

28. Wang, W., et al.: Pyramid vision transformer: a versatile backbone for dense prediction without convolutions. In: ICCV, pp. 568–578 (2021)

29. Wang, Z., Bovik, A.C., Sheikh, H.R., Simoncelli, E.P.: Image quality assessment: from error visibility to structural similarity. TIP **13**(4), 600–612 (2004)

30. Yamaguchi, K., McAllester, D., Urtasun, R.: Efficient joint segmentation, occlusion labeling, stereo and flow estimation. In: Fleet, D., Pajdla, T., Schiele, B., Tuytelaars, T. (eds.) ECCV 2014. LNCS, vol. 8693, pp. 756–771. Springer, Cham (2014). https://doi.org/10.1007/978-3-319-10602-1_49

31. Ye, M., Johns, E., Handa, A., Zhang, L., Pratt, P., Yang, G.Z.: Self-supervised siamese learning on stereo image pairs for depth estimation in robotic surgery. arXiv preprint arXiv:1705.08260 (2017)

32. Zhong, Y., Dai, Y., Li, H.: Self-supervised learning for stereo matching with self-improving ability (2017)

33. Zhong, Y., Dai, Y., Li, H.: 3d geometry-aware semantic labeling of outdoor street scenes. In: ICPR (2018)

34. Zhong, Y., Dai, Y., Li, H.: Stereo computation for a single mixture image. In: Ferrari, V., Hebert, M., Sminchisescu, C., Weiss, Y. (eds.) ECCV 2018. LNCS, vol. 11213, pp. 441–456. Springer, Cham (2018). https://doi.org/10.1007/978-3-030-01240-3_27

35. Zhong, Y., Ji, P., Wang, J., Dai, Y., Li, H.: Unsupervised deep epipolar flow for stationary or dynamic scenes. In: CVPR (2019)

36. Zhong, Y., Li, H., Dai, Y.: Open-world stereo video matching with deep RNN. In: Ferrari, V., Hebert, M., Sminchisescu, C., Weiss, Y. (eds.) ECCV 2018. LNCS, vol. 11206, pp. 104–119. Springer, Cham (2018). https://doi.org/10.1007/978-3-030-01216-8_7

37. Zhong, Y., et al.: Displacement-invariant cost computation for stereo matching. In: IJCV, March 2022

4D-OR: Semantic Scene Graphs for OR Domain Modeling

Ege Özsoy[1(✉)], Evin Pınar Örnek[1], Ulrich Eck[1], Tobias Czempiel[1],
Federico Tombari[1,2], and Nassir Navab[1,3]

[1] Computer Aided Medical Procedures, Technische Universität München, Garching,
Germany
{ege.oezsoy,evin.oernek}@tum.de
[2] Google, Mountain View, USA
[3] Computer Aided Medical Procedures, Johns Hopkins University, Baltimore, USA

Abstract. Surgical procedures are conducted in highly complex operating rooms (OR), comprising different actors, devices, and interactions. To date, only medically trained human experts are capable of understanding all the links and interactions in such a demanding environment. This paper aims to bring the community one step closer to automated, holistic and semantic understanding and modeling of OR domain. Towards this goal, for the first time, we propose using semantic scene graphs (SSG) to describe and summarize the surgical scene. The nodes of the scene graphs represent different actors and objects in the room, such as medical staff, patients, and medical equipment, whereas edges are the relationships between them. To validate the possibilities of the proposed representation, we create the first publicly available 4D surgical SSG dataset, 4D-OR, containing ten simulated total knee replacement surgeries recorded with six RGB-D sensors in a realistic OR simulation center. 4D-OR includes 6734 frames and is richly annotated with SSGs, human and object poses, and clinical roles. We propose an end-to-end neural network-based SSG generation pipeline, with a rate of success of 0.75 macro F1, indeed being able to infer semantic reasoning in the OR. We further demonstrate the representation power of our scene graphs by using it for the problem of clinical role prediction, where we achieve 0.85 macro F1. The code and dataset are publicly available at github.com/egeozsoy/4D-OR.

Keywords: Semantic scene graph · 4D-OR dataset · 3D surgical scene understanding

---

E. Özsoy, E. P. Örnek—Both authors share first authorship.

Supplementary Information The online version contains supplementary material available at https://doi.org/10.1007/978-3-031-16449-1_45.

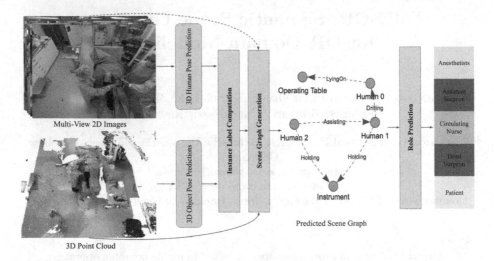

Fig. 1. An overview of our scene graph generation pipeline. We predict 3D human poses from images and object bounding boxes from point clouds and assign an instance label to every point. The scene graph generation then uses the fused point cloud, instance labels and images to predict the relations between the nodes, resulting in a semantically rich representation.

1   Introduction

An automatic and holistic understanding of all the processes in the operating room (OR) is a mandatory building block for next-generation computer-assisted interventions [11,14,15,17]. ORs are highly variant, unpredictable, and irregular environments with different entities and diverse interactions, making modeling surgical procedures fundamentally challenging. The main research directions in surgical data science (SDS) focus on the analysis of specific tasks such as surgical phase recognition, tool recognition, or human pose estimation [2,4,7,13,19,25]. Even though some methods combine different tasks such as surgical phase and tool detection, most previous works operate on a perceptual level, omitting the parallel nature of interactions. To fully understand the processes in an OR, we have to create models capable of disentangling different actors, objects, and their relationships, regarding the OR as one interwoven system rather than multiple individual lines of action. Such models would allow digital systems, *e.g.*, *medical robots, imaging systems*, or *user interfaces*, to automatically adapt their state, function and data visualization, creating an optimized working environment for the surgical staff and improving the final outcome. In the computer vision field, Johnson et al. [10] introduced scene graphs as an abstraction layer for images, where nodes represent objects or humans in the scene and edges describe their relationships. Since then, scene graphs have been employed for a variety of tasks such as image editing, caption generation or scene retrieval [1,6,8–10,22,28]. The analysis of a scene with scene graphs allows combining different perceptual tasks in one model with holistic knowledge about the entire scene. Most computer vision and scene graph generation methods are designed for everyday-task benchmark datasets [3,5,12,18,23], where content and interactions are simpler compared to the high complexity in modern OR. Despite the flexible application possibilities, scene graphs have not been used to model the OR specific 3D dynamic environment with complex semantic interactions between different entities.

The learning process of a scene graph generation model for the OR requires a dataset containing annotations for multiple tasks that summarize the events in the scene. Sharghi et al. [24] create a dataset capturing different robot-assisted interventions with two Time-of-Flight video cameras. Their research focuses on the phase recognition of the OR scene on an in-house dataset. Srivastav et al. [25] created and published the MVOR dataset, which contains synchronized multi-view frames from real surgeries with human pose annotations. This dataset aimed to promote human pose recognition in the operating room and does not provide semantic labels to describe a surgical scene. Currently, no suitable publicly available dataset in the SDS domain exists to supervise the training of a semantic scene graph generation method for the OR.

As we believe in the potential impact of semantic scene graphs for the SDS community, we propose a novel, end-to-end, neural network-based method to generate SSGs for the OR. Our network builds an SSG that is structured, generalizable, and lightweight, summarizing the entire scene with humans, objects, and complex interactions. To this end, we introduce a new 4D dynamic operating room dataset (4D-OR) of simulated knee surgeries, annotated with human and object poses, SSG labels, and clinical roles. 4D-OR is the first publicly available 4D (3D+time) operating room dataset promoting new research directions in SDS. Finally, we demonstrate the capability of our SSG representation by using them for clinical role prediction.

2 Methodology

2.1 Semantic Scene Graphs

Scene graphs are defined by a set of tuples $G = (N, E)$, with $N = \{n_1, ..., n_n\}$ a set of nodes and $E \subseteq N \times R \times N$ a set of edges with relationships $R = \{r_1, ..., r_M\}$ [10]. A 3D scene graph considers the entire 3D representation of an environment. For the OR, the medical equipment and humans are the nodes of our graph, e.g., *anesthesia machine* or *operating table*. The edges describe the interactions between nodes, such as: human *cutting* the patient.

2.2 4D-OR Dataset

To enable and evaluate the modeling of the complex interactions in an OR through SSGs, we propose a novel 4D operating room dataset. 4D-OR consists of ten simulated total knee replacement surgeries captured at a medical simulation center by closely following a typical surgery workflow under the consultancy

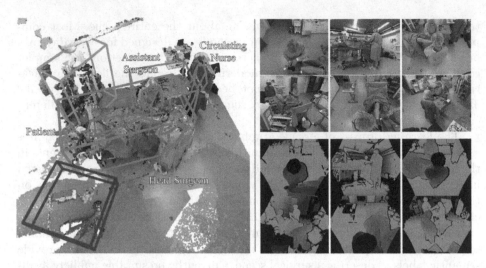

Fig. 2. A sample 4D-OR scene with multiview RGB-D frames and fused 3D point cloud with detected 3D object bounding boxes, human poses and clinical roles.

of medical experts. Our workflow includes major workflow steps such as bone resections, trial implants, and cement preparation. For the type of intervention, we decided on total knee replacement, a representative orthopedic surgery, which involves different steps and interactions. 4D-OR includes a total of 6734 scenes, recorded by six calibrated RGB-D Kinect sensors[1] mounted to the ceiling of the OR, with one frame-per-second, providing synchronized RGB and depth images. 4D-OR is the first publicly available and semantically annotated dataset. We provide fused point cloud sequences of entire scenes, automatically annotated human 6D poses and 3D bounding boxes for OR objects. Furthermore, we provide SSG annotations for each step of the surgery together with the clinical roles of all the humans in the scenes, *e.g.*, *nurse, head surgeon, anesthesiologist*. More details are provided in the supplementary material.

2.3   Scene Graph Generation

Scene graph generation aims at estimating the objects and their semantic relationships from a visual input, *e.g.*, image or point cloud. In this work, we propose an SSG generation pipeline, visualized in Fig. 1. Our method detects the humans and objects, extracts their visual descriptors, and builds the scene graph by predicting the pair-wise relations. For human and object pose estimation, we use the state-of-the-art methods, VoxelPose [27] and Group-Free [16]. For scene graph generation, we develop an instance label computation algorithm, which assigns every point in the point cloud to an object instance label using the predicted poses. Additionally, we introduce a virtual node called "instrument", to model

---

[1] https://azure.microsoft.com/en-us/services/kinect-dk/.

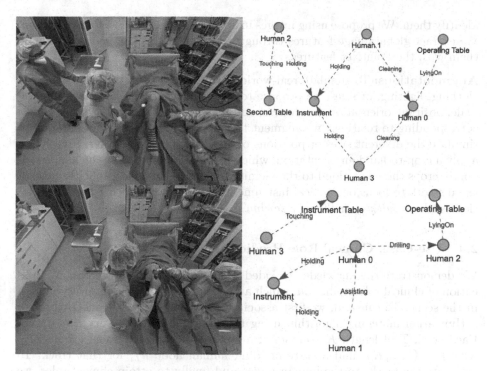

Fig. 3. Scene graph generation results on two sample scenes. Visualization from one camera viewpoint, and the scene graph generated from the 3D point cloud.

interactions of humans with small/invisible medical instruments. Finally, we use a modified version of 3DSSG [28] for generating the scene graphs.

3DSSG is a neural network-based approach to predict node relations. As input, it takes a point cloud and the corresponding instance labels for the entire scene. Then, two PointNet [20] based neural networks are used to encode the point clouds into a feature space. The first one, ObjPointNet uses the point clouds extracted at object level. The input to the second one, RelPointNet, is the union of the object pairs' point clouds. Afterward, a graph convolutional network is employed to enhance the features of nodes and edges taking into account the objects and their pair-wise relationships. Finally, the updated representations are processed by multi-layer perceptrons to predict the object and relation classes, trained using the cross-entropy loss. For our scene graph generation method we applied the following domain specific modifications to 3DSSG:

Use of Images: The OR environment typically contains many objects of various sizes. Especially smaller objects are not always captured well enough by a point cloud *e.g., scissors, lancet,* etc. Further, reflective or transparent items are often omitted due to hardware limitations of depth sensors, even though they can be critical for the correct identification of many OR relations. As these objects are not appropriately represented in point clouds, vanilla 3DSSG often fails to identify them. We propose using images in addition to point clouds to fill this gap. We extract global image features through EfficientNet-B5 [26] and concatenate them with the PointNet features.

Augmentations: To simulate real-world variations, such as different shades of clothing, lighting, or sizes, we propose to augment the point clouds by random scale, position, orientation, brightness, and hue during training. For point clouds corresponding to relations, we augment the points of both objects separately to simulate the different sizes or positions of the objects to each other. Finally, we apply a crop-to-hand augmentation, which for relations involving the hands randomly crops the point cloud to the vicinity of the hands. This implicitly teaches the network to focus on medical instruments when differentiating between relations such as *cutting, drilling,* or *sawing.*

2.4   Use-Case: Clinical Role Prediction

We demonstrate the knowledge encoded in the SSGs in the downstream application of clinical role prediction, which aims to predict the role for every human in the scene. To this end, we first associate every human to a track T through a Hungarian matching algorithm using the detected pose at every timestamp. Each track T of length K consists of a subset of generated scene graphs G_{Ti} with $i = \{1, ..., K\}$, and a corresponding human node n_{Ti} for that track. To associate the tracks to the human nodes and finally to attain clinical roles, we follow these two steps:

Track Role Scoring. For each track T, we first compute a role probability score representing the likelihood for each role. For this, we employ a the state-of-the-art self-attention graph neural network Graphormer [29]. We use all the scene graphs within the track G_T, and rename nodes n_{Ti} as "TARGET" in corresponding graph G_{Ti}, so that the network understands which node embedding to associate the role with. Then, we calculate the mean "TARGET" node embedding and predict the role scores with a linear layer, trained with cross-entropy loss. In addition to this method, we also provide a non-learning baseline for comparison, that is a heuristic-based method which only considers the frequency of relations with respect to each human node. As an example, for every *sawing* relation, we increase the score for *head surgeon*, while for every *lying on* relation, increase the score for *patient*.

Unique Role Assignment. After the track role scoring, we infer the clinical roles of human nodes through solving a matching problem. For every human node in the graph, we retrieve the role probabilities for each track, and bijectively match the roles to the nodes according to their probabilities. Our algorithm guarantees that each human node is assigned to a distinct role.

Table 1. Quantitative results for scene graph generation on test split with precision, recall, and F1-score. "Avg" stands for macro average, which is the unweighted average over all classes. We use images, augmentations, linear loss weighting, and PointNet++.

Relation	Assist	Cement	Clean	CloseTo	Cut	Drill	Hammer	Hold	LyingOn	Operate	Prepare	Saw	Suture	Touch	Avg
Prec	0.42	0.78	0.53	0.97	0.49	0.87	0.71	0.55	1.00	0.55	0.62	0.69	0.60	0.41	0.68
Rec	0.93	0.78	0.63	0.89	0.49	1.00	0.89	0.95	0.99	0.99	0.91	0.91	1.00	0.69	0.87
F1	0.58	0.78	0.57	0.93	0.49	0.93	0.79	0.70	0.99	0.71	0.74	0.79	0.75	0.51	0.75

3   Experiments

Model Training: We split the 4D-OR dataset into six train, two validation and two test takes. We configure VoxelPose to use 14 joints, and train it for 20 epochs with a weighted loss for patient poses. We train the 3D object detection network Group-Free for 180 epochs. In our scene graph generation method, we use PointNet++ [21] for feature extraction, with a balancing loss to deal with rare relations, with 3e–5 lr, 4000 points for object and 8000 for relations. Our method is implemented in PyTorch and trained on a single GPU, achieves a 2.2 FPS inference run-time.

Evaluation Metrics: We evaluate human poses with Percentage of Correct Parts (PCP3D), and object poses with average precision (AP) at an intersection over union (IoU) threshold. Further, the scene graph relations and role predictions are evaluated with precision, recall, and F1-score for each class separately and a macro averaged over all classes. Macro averages are unweighted, thus they are sample sizes agnostic, evaluating the performance on all relationships equally. This is crucial in our setting, as rare relation types such as *cutting* or *drilling* are important for scene understanding. For all metrics, higher means better.

4   Results and Discussion

Human and Object Pose Prediction. In the task of human pose recognition, we reach a PCP3D of 71.23 in the test split. In the task of object pose recognition, we attain an AP value of 0.9893 for IoU@25 and 0.9345 for IoU@50. We further confirmed the reliability of our methods with qualitative results which are visualized in Fig. 2 through the detected human and object poses.

Scene Graph Generation. In Table 1, we present our scene graph generation results in terms of relation prediction from an unlabeled point cloud. We consider a relation *"correct"* if both of the entities are present in the scene and the relation between them is predicted correctly. Overall, we achieve the best result with 0.75 macro F1 using images with point clouds and proposed augmentation strategies. In Fig. 3, we visualize two scene graph generation samples qualitatively verifying that our method can indeed understand and reason over semantics in the complex OR and generate correct scene graphs. For scenes with high visual similarities

Table 2. Clinical role prediction results on test takes, comparing a non-learned heuristic-based method and the neural network based Graphormer for track scoring.

Role	Heuristic-Based			Graphormer		
	Prec	Rec	F1	Prec	Rec	F1
Patient	0.99	0.98	0.99	0.99	0.98	0.99
Head Surgeon	0.93	1.00	0.96	0.96	0.99	0.97
Assistant Surgeon	0.71	0.72	0.71	0.98	0.98	0.98
Circulating Nurse	0.61	0.59	0.60	0.87	0.78	0.82
Anaesthetist	0.60	0.32	0.41	0.53	0.48	0.51
Macro Avg	0.77	0.72	0.74	**0.87**	**0.84**	**0.85**

Table 3. Ablation study on SSG generation using 3D point clouds with different configurations.

exp #	image	augment	GT	F1
(a)	✗	✗	✗	0.65
(b)	✓	✗	✗	0.66
(c)	✗	✓	✗	0.70
(d)	✓	✓	✗	0.76
(e)	✓	✓	✓	0.78

but different tools (*e.g. drilling, sawing*) our model sometimes fails to predict the correct relation when the instruments are occluded.

Clinical Role Prediction. We present the results on test takes when using heuristics or Graphormer [29] in Table 2. Example role predictions can be seen in Fig. 2. Overall, we achieve near-perfect performance for patient and head surgeon, and good performance for assistant surgeon and circulating nurse. Only the anaesthetist, who is often at least partially occluded, is hard to predict correctly. Unsurprisingly, we observe lower scores with the non-learning heuristic-based score assignment method. While heuristic-based method has the advantage of being transparent, the Graphormer is easier to adapt to new roles or surgeries, as it does not need any tweaking of heuristics.

Ablation Studies. We conduct ablation studies in Table 3 to show the impact of our two model contributions, using images in addition to point clouds, and augmentations. We also show the impact of relying on ground truth human and object pose annotations instead of predictions. We see that using images (c-d) and augmentations (a-c) significantly improve F1 results, performing the best when both are applied (a-d), confirming the benefits of our method. Further, using ground truth instead of predictions (d-e) does not alter the result much, showing that our method can rely on off-the-shelf pose prediction methods.

Clinical Translation. Our 4D-OR dataset is the first fully semantically and geometrically annotated 3D and temporal OR dataset. We provide simulations of complex clinical procedures as a pioneering step toward further research in SSG. We closely followed the clinical workflow and introduced different sets of variability to create the first dataset representing a subset of the real-world OR challenges. The translation of our results to a real clinical setting involves additional challenges in terms of data privacy concerns for the acquisition, storage, and usage of hospital data. Nonetheless, in the view of the benefits, we believe that the community will overcome these limitations.

5   Conclusion

We propose a semantic scene graph generation method capable of modeling a challenging OR environment. We introduce 4D-OR, a novel scene graph dataset consisting of total knee replacement surgeries to supervise our training. Finally, we demonstrate utilizing generated scene graphs for the exemplary downstream task of clinical role prediction. We believe our work brings the community one step closer to achieving a holistic understanding of the OR.

Acknowledgements. The authors have been partially supported by the German Federal Ministry of Education and Research (BMBF), under Grant 16SV8088, by a Google unrestricted gift, by Stryker and J&J Robotics & Digital Solutions. We thank INM and Frieder Pankratz in helping setting up the acquisition environment, and Tianyu Song, Lennart Bastian and Chantal Pellegrini for their help in acquiring 4D-OR.

References

1. Armeni, I., et al.: 3d scene graph: a structure for unified semantics, 3d space, and camera. In: Proceedings of the IEEE/CVF International Conference on Computer Vision, pp. 5664–5673 (2019)
2. Bodenstedt, S., et al.: Active learning using deep bayesian networks for surgical workflow analysis. Int. J. Comput. Assist. Radiol. Surg. **14**, 1079–1987 (2019)
3. Caesar, H., et al.: nuscenes: a multimodal dataset for autonomous driving. In: Proceedings of the IEEE/CVF Conference on Computer Vision and Pattern Recognition, pp. 11621–11631 (2020)
4. Czempiel, T., et al.: TeCNO: surgical phase recognition with multi-stage temporal convolutional networks. In: Martel, A.L., et al. (eds.) MICCAI 2020. LNCS, vol. 12263, pp. 343–352. Springer, Cham (2020). https://doi.org/10.1007/978-3-030-59716-0_33
5. Dai, A., Chang, A.X., Savva, M., Halber, M., Funkhouser, T., Nießner, M.: Scannet: richly-annotated 3d reconstructions of indoor scenes. In: Proceedings of the IEEE Conference on Computer Vision and Pattern Recognition, pp. 5828–5839 (2017)
6. Dhamo, H., et al.: Semantic image manipulation using scene graphs. In: CVPR (2020)
7. Garrow, C.R., et al.: Machine learning for surgical phase recognition: A systematic review. Annals of surgery, November 2020

8. Ji, J., Krishna, R., Fei-Fei, L., Niebles, J.C.: Action genome: actions as compositions of spatio-temporal scene graphs. In: Proceedings of the IEEE/CVF Conference on Computer Vision and Pattern Recognition, pp. 10236–10247 (2020)

9. Johnson, J., Gupta, A., Fei-Fei, L.: Image generation from scene graphs. In: Proceedings of the IEEE Conference on Computer Vision and Pattern Recognition, pp. 1219–1228 (2018)

10. Johnson, J., et al.: Image retrieval using scene graphs. In: 2015 IEEE Conference on Computer Vision and Pattern Recognition (CVPR), pp. 3668–3678 (2015)

11. Kennedy-Metz, L.R., et al.: Computer vision in the operating room: opportunities and caveats. IEEE Trans. Med. Robot. Bionics, 1 (2020). https://doi.org/10.1109/TMRB.2020.3040002

12. Krishna, R., et al.: Visual genome: Connecting language and vision using crowdsourced dense image annotations. Int. J. Comput. Vision **123**(1), 32–73 (2017)

13. Laina, I., et al.: Concurrent segmentation and localization for tracking of surgical instruments. In: Descoteaux, M., Maier-Hein, L., Franz, A., Jannin, P., Collins, D.L., Duchesne, S. (eds.) MICCAI 2017. LNCS, vol. 10434, pp. 664–672. Springer, Cham (2017). https://doi.org/10.1007/978-3-319-66185-8_75

14. Lalys, F., Jannin, P.: Surgical process modelling: a review. Int. J. Comput. Assist. Radiol. Surg. **9**(3), 495–511 (2013). https://doi.org/10.1007/s11548-013-0940-5

15. Li, Z., Shaban, A., Simard, J., Rabindran, D., DiMaio, S.P., Mohareri, O.: A robotic 3d perception system for operating room environment awareness. CoRR abs/2003.09487 (2020). https://arxiv.org/abs/2003.09487

16. Liu, Z., Zhang, Z., Cao, Y., Hu, H., Tong, X.: Group-free 3d object detection via transformers. In: Proceedings of the IEEE/CVF International Conference on Computer Vision, pp. 2949–2958 (2021)

17. Maier-Hein, L., et al.: Surgical data science for next-generation interventions. Nature Biomed. Eng. **1**(9), 691–696 (2017)

18. Nathan Silberman, Derek Hoiem, P.K., Fergus, R.: Indoor segmentation and support inference from rgbd images. In: ECCV (2012)

19. Nwoye, C.I., Mutter, D., Marescaux, J., Padoy, N.: Weakly supervised convolutional lstm approach for tool tracking in laparoscopic videos. Int. J. Comput. Assist. Radiol. Surg. **14**(6), 1059–1067 (2019)

20. Qi, C.R., Su, H., Mo, K., Guibas, L.J.: Pointnet: Deep learning on point sets for 3d classification and segmentation. In: Proceedings of the IEEE Conference on Computer Vision and Pattern Recognition, pp. 652–660 (2017)

21. Qi, C.R., Yi, L., Su, H., Guibas, L.J.: Pointnet++: deep hierarchical feature learning on point sets in a metric space. Advances in neural information processing systems 30 (2017)

22. Rosinol, A., Gupta, A., Abate, M., Shi, J., Carlone, L.: 3d dynamic scene graphs: actionable spatial perception with places, objects, and humans. arXiv preprint arXiv:2002.06289 (2020)

23. Russakovsky, O., et al.: Imagenet large scale visual recognition challenge. Int. J. Comput. Vision **115**(3), 211–252 (2015)

24. Sharghi, A., Haugerud, H., Oh, D., Mohareri, O.: Automatic operating room surgical activity recognition for robot-assisted surgery. In: Martel, A.L., Abolmaesumi, P., Stoyanov, D., Mateus, D., Zuluaga, M.A., Zhou, S.K., Racoceanu, D., Joskowicz, L. (eds.) MICCAI 2020. LNCS, vol. 12263, pp. 385–395. Springer, Cham (2020). https://doi.org/10.1007/978-3-030-59716-0_37

25. Srivastav, V., Issenhuth, T., Abdolrahim, K., de Mathelin, M., Gangi, A., Padoy, N.: Mvor: a multi-view rgb-d operating room dataset for 2d and 3d human pose estimation (2018)

26. Tan, M., Le, Q.: Efficientnet: rethinking model scaling for convolutional neural networks. In: International Conference on Machine Learning, pp. 6105–6114. PMLR (2019)
27. Tu, H., Wang, C., Zeng, W.: VoxelPose: towards multi-camera 3D human pose estimation in wild environment. In: Vedaldi, A., Bischof, H., Brox, T., Frahm, J.-M. (eds.) ECCV 2020. LNCS, vol. 12346, pp. 197–212. Springer, Cham (2020). https://doi.org/10.1007/978-3-030-58452-8_12
28. Wald, J., Dhamo, H., Navab, N., Tombari, F.: Learning 3D semantic scene graphs from 3D indoor reconstructions. In: Proceedings of the IEEE/CVF Conference on Computer Vision and Pattern Recognition, pp. 3961–3970 (2020)
29. Ying, C., et al.: Do transformers really perform badly for graph representation? Advances in Neural Information Processing Systems 34 (2021)

AutoLaparo: A New Dataset of Integrated Multi-tasks for Image-guided Surgical Automation in Laparoscopic Hysterectomy

Ziyi Wang[1,2], Bo Lu[1,2,3], Yonghao Long[4], Fangxun Zhong[1,2],
Tak-Hong Cheung[5], Qi Dou[2,4], and Yunhui Liu[1,2(✉)]

[1] Department of Mechanical and Automation Engineering,
The Chinese University of Hong Kong, Hong Kong, China
`yhliu@mae.cuhk.edu.hk`
[2] T Stone Robotics Institute, The Chinese University of Hong Kong, Hong Kong,
China
[3] Robotics and Microsystems Center, School of Mechanical and Electric Engineering,
Soochow University, Suzhou, China
[4] Department of Computer Science and Engineering,
The Chinese University of Hong Kong, Hong Kong, China
[5] Department of Obstetrics and Gynaecology, Prince of Wales Hospital,
The Chinese University of Hong Kong, Hong Kong, China

Abstract. Computer-assisted minimally invasive surgery has great potential in benefiting modern operating theatres. The video data streamed from the endoscope provides rich information to support context-awareness for next-generation intelligent surgical systems. To achieve accurate perception and automatic manipulation during the procedure, learning based technique is a promising way, which enables advanced image analysis and scene understanding in recent years. However, learning such models highly relies on large-scale, high-quality, and multi-task labelled data. This is currently a bottleneck for the topic, as available public dataset is still extremely limited in the field of CAI. In this paper, we present and release the first integrated dataset (named AutoLaparo) with multiple image-based perception tasks to facilitate learning-based automation in hysterectomy surgery. Our AutoLaparo dataset is developed based on full-length videos of entire hysterectomy procedures. Specifically, three different yet highly correlated tasks are formulated in the dataset, including surgical workflow recognition, laparoscope motion prediction, and instrument and key anatomy segmentation. In addition, we provide experimental results with state-of-the-art models as reference benchmarks for further model developments and evaluations on this dataset. The dataset is available at https://autolaparo.github.io.

Keywords: Surgical video dataset · Image-guided robotic surgical automation · Laparoscopic hysterectomy

---

Supplementary Information The online version contains supplementary material available at https://doi.org/10.1007/978-3-031-16449-1_46.

1   Introduction

New technologies in robotics and computer-assisted intervention (CAI) are widely developed for surgeries, to release the burden of surgeons on manipulating instruments, identifying anatomical structures, and operating in confined spaces [4,24,28]. The advancements of these technologies will thereby facilitate the development of semi- and fully-automatic robotic systems which can understand surgical situations and even make proper decisions in certain tasks. Nowadays, surgeries conducted in a minimally invasive way, e.g. laparoscopy, have become popular [30], and videos recorded through the laparoscope are valuable for image-based studies [23,29]. To enhance surgical scene understanding for image-guided automation, one promising solution is to rely on learning-based methods.

As deep learning methods are data-driven, a large amount of surgical data is required to train and obtain reliable models for task executions. Especially in surgical field, due to various types of surgical procedures, corresponding image and video data vary in surgical scenes, surgical workflows, and the instruments used [14,23]. For laparoscopic minimally invasive surgery (MIS), some datasets have been established and released to improve the learning-based algorithm for different surgical tasks, such as action recognition [13], workflow recognition [26], instrument detection and segmentation [27]. However, these datasets are not collected from real-surgery nor large enough to develop applicable and robust models in practice. Recently, several research teams have worked on developing dataset at large scales [1,3,31], but most are only designed and annotated for one certain task. In terms of clinical applicability, data from different modalities are needed to better understand the whole scenario, make proper decisions, as well as enrich perception with multi-task learning strategy [9,16]. Besides, there are few datasets designed for automation tasks in surgical application, among which the automatic laparoscopic field-of-view (FoV) control is a popular topic as it can liberate the assistant from such tedious manipulations with the help from surgical robots [5]. Therefore, surgical tasks with multiple modalities of data should be formulated and proposed for image-guided surgical automation.

For application, we target laparoscopic hysterectomy, a gynaecologic surgery commonly performed for patients diagnosed with adenomyosis, uterine fibroids or cancer [10,25]. By now, some obstacles remain in the development of learning-based approaches for surgical automation in laparoscopic hysterectomy as public datasets for this procedure are only sparsely available. Although several datasets have been presented for endometriosis diagnosis [19], action recognition [20], and anatomical structures and tool segmentation [33], they are small in scale of cases or annotations and insufficient for learning-based model development.

In this paper, we present AutoLaparo, the first large-scale dataset with integrated multi-tasks towards image-guided surgical automation in laparoscopic hysterectomy. Three tasks along with the corresponding data and annotations are designed: Task 1 surgical workflow recognition, where 21 videos of complete procedures of laparoscopic hysterectomy with a total duration of 1388 min are collected and annotated into 7 phases, proposed as the first dataset dedicating

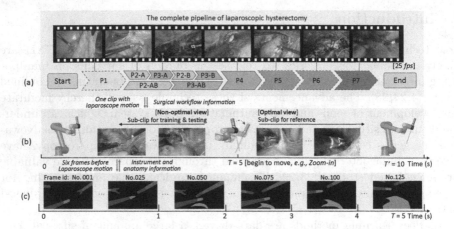

Fig. 1. Overview of AutoLaparo, the proposed integrated dataset of laparoscopic hysterectomy. (a) Order information of the procedure with sample frames of each phase (P1-P7). (b) One clip that contains the *"Zoom-in"* motion of laparoscope. (c) Six frames are sampled from the above clip and pixel-wise annotated for segmentation.

workflow analysis in this surgery; Task 2 laparoscope motion prediction, where 300 clips are selected and annotated with 7 types of typical laparoscope motion towards image-guided automatic FoV control; and Task 3 instrument and key anatomy segmentation, where 1800 frames are sampled from the above clips and annotated with 5936 pixel-wise masks of 9 types, providing rich information for scene understanding. It is worth noting that these tasks, data and annotations are highly correlated to support multi-task and multi-modality learning for advanced surgical perception towards vision-based automation. In addition, benchmarks of state-of-the-art methods for each task are presented as reference to facilitate further model development and evaluation on this dataset.

2   Dataset Design and Multi-tasks Formulation

Towards image-guided surgical automation, we propose AutoLaparo, integrating multi-tasks to facilitate advanced surgical scene understanding in laparoscopic hysterectomy, as illustrated in Fig. 1. The dataset consists of three tasks that are essential components for automatic surgical perception and manipulation. Besides, a three-tier annotation process is carefully designed to support learning-based model development of three tasks with uni- and multi-modality of data. Details of the task and dataset design are presented in Sects. 2.2 and 2.3.

2.1   Dataset Collection

In this work, we collect 21 videos of laparoscopic hysterectomy from Prince of Wales Hospital, Hong Kong, and then conduct pre-processing and annotation to build the dataset. These 21 cases were performed from October to December 2018, using the Olympus imaging platform. The usage of these surgical video data is approved by The Joint CUHK-NTEC Clinical Research Ethics Committee and the data are protected without disclosure of any personal information.

The 21 videos are recorded at 25 fps with a standard resolution of 1920 × 1080 pixels. The duration of videos ranges from 27 to 112 min due to the varying difficulties of the surgeries. It should be noticed that the laparoscope is inevitably taken out from the abdomen when changing instruments or cleaning contamination on the lens, and the video clips during these periods are invalid for visual-based analysis and removed from our dataset to ensure a compact connection. After pre-processing, the average duration is 66 min and the total duration is 1388 min, reaching a large-scale dataset with high-quality.

2.2   Task Formulation

Three tasks are designed in the proposed AutoLaparo dataset:

Task 1: Surgical Workflow Recognition. This task focuses on the workflow analysis of laparoscopic hysterectomy procedure, which is fundamental to the scene understanding in surgery that helps to recognize current surgical phase and provide high-level information to the other two tasks [31].

Task 2: Laparoscope Motion Prediction. In MIS, surgeons need to operate within a proper FoV of the laparoscope, which in conventional surgery is held by an assistant. Recently, some kinds of surgical robots have been designed for laparoscope motion control through human-machine interaction such as foot pedals, eye tracking [11], etc., thereby liberating the assistant from tedious work but distracting surgeons from the surgical manipulation. Therefore, state-of-the-art studies explored learning laparoscope motion patterns through image and video feedback [21], where appropriate datasets need to be developed for this vision-based automatic FoV control approach.

Task 3: Instrument and Key Anatomy Segmentation. The segmentation of instrument and anatomy structure plays an important role in the realization of surgical automation, and the results of detection and segmentation can also serve high-level tasks, such as instrument tracking, pose estimation [2], etc.

2.3   Dataset Annotation and Statistics

Three sub-datasets are designed for the aforementioned three tasks and they are also highly-correlated by applying a three-tier annotation process. The annotation is performed by a senior gynaecologist with more than thirty years of clinical experience and a specialist with three years of experience in hysterectomy. The annotation results are proofread and cross-checked to ensure accuracy and consistency, and then stored in the dataset.

Workflow Annotation at Video-Level. Based on the clinical experience and domain knowledge [6], the 21 videos of laparoscopic hysterectomy are manually

Table 1. Type and number of the clips with laparoscope motion annotation.

Motion Type	Up	Down	Left	Right	Zoom-in	Zoom-out	Static	All
Train set	14	25	18	12	30	17	54	170
Validation set	3	4	10	3	15	12	10	57
Test set	5	16	9	5	9	15	14	73
All Num	22	45	37	20	54	44	78	300

Table 2. Total number of presences in clips and frames, average number of pixels per frame, and total number of annotations of each class in the segmentation dataset.

Type		Presence in Clips	Presence in Frames	Average Pixels	Annotation number			
					All	Train set	Val. set	Test set
Anatomy	Uterus	164	941	253968	1016	663	214	139
Instrument	1-m	121	575	57042	577	371	107	99
	1-s	102	439	92545	442	291	89	62
	2-m	261	1476	113569	1481	842	283	356
	2-s	228	1071	137380	1074	630	206	238
	3-m	109	518	59341	538	287	92	159
	3-s	85	365	94417	373	189	81	103
	4-m	39	200	7042	202	104	50	48
	4-s	40	233	179904	233	124	55	54
All		300	1800	2073600	5936	3501	1127	1258

annotated into 7 phases: *Preparation, Dividing Ligament and Peritoneum, Dividing Uterine Vessels and Ligament, Transecting the Vagina, Specimen Removal, Suturing,* and *Washing.* Figure 1(a) represents the order information and sample frames of each phase (P1-P7). Specifically, since Phase 2 and 3 are performed symmetrically on the left and right sides of the uterus (denoted by A and B), the sequence of these two phases is related to the surgeon's operating habits. Some surgeons prefer to perform in a stage-by-stage sequence, while others prefer to handle the tissues on one side before proceeding to the other side.

Motion Annotation at Clip-Level. To further promote the automatic FoV control, we propose a sub-dataset containing selected clips that corresponds the laparoscope visual feedback to its motion mode. In surgical videos, the manual laparoscope movements are usually momentary motions, lasting for < 1 s, so that we denote the time T as the beginning time of the motion. The main idea of this task is to predict the laparoscope motion at time T using the visual information of the clip before that time, i.e., 0-T (defined as "non-optimal view"). For the clip after T (defined as the "optimal view"), the video information will be provided as reference to present the scene after this motion.

Regarding the statistics, 300 clips are carefully selected from Phase 2–4 of the aforementioned 21 videos and each clip lasts for 10 s. These clips contain

(a) Instrument-1 (b) Instrument-2 (c) Instrument-3 (d) Instrument-4 (e) Key anatomy

Fig. 2. Illustrations of four instruments and the key anatomy (uterus) annotated in the AutoLaparo dataset. The background is blurred to highlight the object.

typical laparoscope motions and the motion time T is set as the fifth second as illustrated in Fig. 1(b). Seven types of motion modes are defined, including one static mode and six non-static mode: *Up, Down, Left, Right, Zoom-in*, and *Zoom-out*, and the number of video clips in each category are listed in Table 1.

Segmentation Annotation at Frame-Level. Based on the selected clips above, we further develop a sub-dataset with frame-level pixel-wise segmentation annotations of the surgical instrument and key anatomy. It provides ground truth for Task 3 and also supports Task 1 and 2 as an additional modality for advanced scene understanding. Following the configuration of motion annotation, where the first five seconds of each clip are used as visual information to infer the motion at the fifth second, six frames are sampled at 1 fps, that are, frames at time $t=0,1,2,3,4,5$ (i.e., frame id 1,25,50,75,100,125), as illustrated in Fig. 1.

There are four types of instruments and one key anatomy appearing in these frames: Instrument-1 *Grasping forceps*, Instrument-2 *LigaSure*, Instrument-3 *Dissecting and grasping forceps*, Instrument-4 *Electric hook*, and *Uterus*, as shown in Fig. 2. The annotations are performed with an open annotation tool called "LabelMe" [32] available online, following the settings of *instrument part segmentation* [3], i.e., the shaft and manipulator of each instrument are annotated separately, as illustrated in Fig. 1(c). Finally, we reach a large-scale sub-dataset for segmentation containing 1800 annotated frames with 9 types and 5936 annotations. Dataset split and statistics are presented in Table 2, where '1-m' denotes the manipulation of Instrument-1 and '1-s' denotes its shaft. It can also be observed that the Instrument-2, the main instrument in the surgery, appears most frequently, while the uterus appears in more than half of the frames.

2.4 Dataset Analysis and Insights

AutoLaparo is designed towards image-guided automation by achieving a comprehensive understanding of the entire surgical scene with three integrated tasks and data. In specific, tool usage information is beneficial for recognizing phase and vice versa, so that these two kinds of annotation help to exploit the complementary information for Task 1 and 3. Besides, the workflow and segmentation results provide rich information about the surgical scene that can help predict the laparoscope motion in Task 2 for further automatic FoV control. In this way, the impact of each modality and the added value of combining

Table 3. Experimental results (%) of four state-of-the-art models on Task 1.

Method	Validation set				Test set			
	AC	PR	RE	JA	AC	PR	RE	JA
SV-RCNet	77.22	72.28	65.73	51.64	75.62	64.02	59.70	47.15
TMRNet	80.96	79.11	67.36	54.95	78.20	66.02	61.47	49.59
TeCNO	81.42	66.64	68.76	55.24	77.27	66.92	64.60	50.67
Trans-SVNet	82.01	66.22	68.48	55.56	78.29	64.21	62.11	50.65

several modalities for advanced perception can be explored with our integrated dataset.

3 Experiments and Benchmarking Methods

In this section, a set of experiments are performed for each task on AutoLaparo and benchmarks are provided for further model evaluation on this dataset.

3.1 Workflow Recognition

State-of-the-Art Methods. Four models are evaluated on workflow recognition task, that are, SV-RCNet [17], TMRNet [18], TeCNO [8], and Trans-SVNet [12]. SV-RCNet integrates visual and temporal dependencies in an end-to-end architecture. TMRNet relates multi-scale temporal patterns with a long-range memory bank and a non-local bank operator. TeCNO exploits temporal modelling with higher temporal resolution and large receptive field. Trans-SVNet attempts to use Transformer to fuse spatial and temporal embeddings.

Evaluation Metrics and Results. In AutoLaparo, 21 videos are divided into three subsets for training, validation, and test, containing 10, 4, and 7 videos, respectively. The performances are comprehensively evaluated with four commonly-used metrics: accuracy (AC), precision (PR), recall (RE), and Jaccard (JA). AC is defined as the percentage of correctly classified frames in the whole video, and PR, RE and JA are calculated by: $PR = \frac{|GT \cap P|}{|P|}$, $RE = \frac{|GT \cap P|}{|GT|}$, $JA = \frac{|GT \cap P|}{|GT \cup P|}$, where GT and P denotes the ground truth and prediction set, respectively.

Experimental results are listed in Table 3. All approaches present satisfactory results with accuracies above 75% on both validation and test sets. The overall results present the potential to be applied to online automatic surgical workflow recognition and also used as guidance for the following tasks.

Table 4. Experimental results (%) of two proposed methods on Task 2.

Method	Validation set	Test set
Resnet-50	28.07	26.03
Resnet-50+LSTM	29.82	27.40

Table 5. Experimental results (%) of three state-of-the-art models on Task 3.

Method	Backbone	box AP	mask AP
Mask R-CNN	ResNet-50	52.10	51.70
YOLACT	ResNet-50	48.16	54.09
YolactEdge	ResNet-50	47.04	52.58

3.2 Laparoscope Motion Prediction

State-of-the-Art Methods. In this task, we propose two methods to predict the laparoscope motion with uni-modality of visual input, i.e., the 5-second sub-clip before the motion time T. The clips are downsampled from 25 fps to 3 fps to capture rich information from sequential frames, and in this way, 16 frames in each sub-clip (including the first and last frames) are input to the models.

In the first method, a ResNet-50 structure is developed to extract visual features of each frame, which are then averaged to represent each clip and fed into a fully-connected (FC) layer for motion mode prediction. For the second method named ResNet-50+LSTM, a long short-term memory (LSTM) module is seamlessly integrated following the ResNet-50 backbone to exploit the long-range temporal information inherent in consecutive frames and output spatial-temporal embeddings. Then the motion prediction results are generated from an FC layer connected to the LSTM module.

Evaluation Metrics and Results. The metric used to assess the performance of the prediction results is the accuracy, which is defined as the percentage of the clips correctly classified into the ground truth in the entire dataset.

Experimental results are presented in Table 4. The accuracies of these two models are 28.07% and 29.82% on validation set, and 26.03% and 27.40% on test set. It can be seen that the second model achieves a higher accuracy value since it jointly learns both spatial representations and temporal motion patterns. The overall results need to be improved and advanced methods combining multi-modality data could be proposed for this important but challenging task.

3.3 Instrument and Key Anatomy Segmentation

State-of-the-Art Methods. Three models for object detection and instance segmentation are evaluated to give benchmarks: Mask R-CNN [15], a representative two-stage approach known as a competitive baseline model; YOLACT [7], the first algorithm that realizes real-time instance segmentation, reaching a considerable speed of 30 fps with satisfactory results; and YolactEdge [22] that yields a speed-up over existing real-time methods and also competitive results.

Evaluation Metrics and Results. The experimental results are reported using the standard COCO metrics where the average precision (AP) is calculated by averaging the intersection over union (IoU) with 10 thresholds of 0.5: 0.05: 0.95 for both bounding boxes (box AP) and segmentation masks (mask AP).

For fair comparison of baseline models, We conduct experiments with a same backbone ResNet-50 and the results on the test set are presented in Table 5. The box AP are 52.10%, 48.16%, and 47.04% for the three methods respectively. For segmentation, Mask R-CNN gives mask AP of 51.70%, YOLACT of 54.09%, and YolactEdge of 52.58%, showing promising potential for practical applications.

4   Conclusion and Future Work

In this paper, we propose AutoLaparo, the first integrated dataset to facilitate visual perception and image-guided automation in laparoscopic hysterectomy. The dataset consists of raw videos and annotations towards multi-task learning and three tasks are defined: surgical workflow recognition, laparoscope motion prediction, and instrument and key anatomy segmentation, which are not independent but also highly correlated for advanced scene understanding. Besides, experiments are performed on each task and benchmarks are presented as reference for the future model development and evaluation on this dataset.

The remaining challenges for this dataset are how to develop models to improve the performances of each task with uni-modality data and achieve multi-task learning by combining various modalities. By making this dataset publicly available, we and the whole community will together dedicate to addressing these issues in the future based on our integrated data and advanced learning strategies. In addition, we plan to extend this dataset by expanding the amount of images and videos and defining more tasks towards vision-based applications.

Acknowledgement. This work is supported in part by Shenzhen Portion of Shenzhen-Hong Kong Science and Technology Innovation Cooperation Zone under HZQB-KCZYB-20200089, in part of the HK RGC under T42-409/18-R and 14202918, in part by the Multi-Scale Medical Robotics Centre, InnoHK, and in part by the VC Fund 4930745 of the CUHK T Stone Robotics Institute.

References

1. Allan, M., et al.: 2018 robotic scene segmentation challenge. arXiv preprint arXiv:2001.11190 (2020)
2. Allan, M., Ourselin, S., Hawkes, D.J., Kelly, J.D., Stoyanov, D.: 3-d pose estimation of articulated instruments in robotic minimally invasive surgery. IEEE Trans. Med. Imaging **37**(5), 1204–1213 (2018)
3. Allan, M., et al.: 2017 robotic instrument segmentation challenge. arXiv preprint arXiv:1902.06426 (2019)
4. Barbash, G.I.: New technology and health care costs-the case of robot-assisted surgery. N. Engl. J. Med. **363**(8), 701 (2010)
5. Bihlmaier, A., Woern, H.: Automated endoscopic camera guidance: a knowledge-based system towards robot assisted surgery. In: ISR/Robotik 2014; 41st International Symposium on Robotics, pp. 1–6. VDE (2014)
6. Blikkendaal, M.D., et al.: Surgical flow disturbances in dedicated minimally invasive surgery suites: an observational study to assess its supposed superiority over conventional suites. Surg. Endosc. **31**(1), 288–298 (2016). https://doi.org/10.1007/s00464-016-4971-1
7. Bolya, D., Zhou, C., Xiao, F., Lee, Y.J.: Yolact: real-time instance segmentation. In: Proceedings of the IEEE/CVF International Conference on Computer Vision, pp. 9157–9166 (2019)
8. Czempiel, T., et al.: TeCNO: surgical phase recognition with multi-stage temporal convolutional networks. In: Martel, A.L., Abolmaesumi, P., Stoyanov, D., Mateus, D., Zuluaga, M.A., Zhou, S.K., Racoceanu, D., Joskowicz, L. (eds.) MICCAI 2020. LNCS, vol. 12263, pp. 343–352. Springer, Cham (2020). https://doi.org/10.1007/978-3-030-59716-0_33
9. Dergachyova, O., Bouget, D., Huaulmé, A., Morandi, X., Jannin, P.: Automatic data-driven real-time segmentation and recognition of surgical workflow. Int. J. Comput. Assist. Radiol. Surg. **11**(6), 1081–1089 (2016). https://doi.org/10.1007/s11548-016-1371-x
10. Farquhar, C.M., Steiner, C.A.: Hysterectomy rates in the united states 1990–1997. Obstetrics Gynecology **99**(2), 229–234 (2002)
11. Fujii, K., Gras, G., Salerno, A., Yang, G.Z.: Gaze gesture based human robot interaction for laparoscopic surgery. Med. Image Anal. **44**, 196–214 (2018)
12. Gao, X., Jin, Y., Long, Y., Dou, Q., Heng, P.A.: Trans-svnet: accurate phase recognition from surgical videos via hybrid embedding aggregation transformer. In: International Conference on Medical Image Computing and Computer-Assisted Intervention. pp. 593–603. Springer (2021)
13. Gao, Y., Vedula, S.S., Reiley, C.E., Ahmidi, N., Varadarajan, B., Lin, H.C., Tao, L., Zappella, L., Béjar, B., Yuh, D.D., et al.: Jhu-isi gesture and skill assessment working set (jigsaws): A surgical activity dataset for human motion modeling. In: MICCAI workshop: M2cai. vol. 3, p. 3 (2014)
14. Grammatikopoulou, M., Flouty, E., Kadkhodamohammadi, A., Quellec, G., Chow, A., Nehme, J., Luengo, I., Stoyanov, D.: Cadis: Cataract dataset for surgical rgb-image segmentation. Med. Image Anal. **71**, 102053 (2021)
15. He, K., Gkioxari, G., Dollár, P., Girshick, R.: Mask R-CNN. In: Proceedings of the IEEE international conference on computer vision. pp. 2961–2969 (2017)
16. Huaulmé, A., et al.: Peg transfer workflow recognition challenge report: Does multi-modal data improve recognition? arXiv preprint arXiv:2202.05821 (2022)

17. Jin, Y., et al.: Sv-rcnet: workflow recognition from surgical videos using recurrent convolutional network. IEEE Trans. Med. Imaging **37**(5), 1114–1126 (2017)
18. Jin, Y., Long, Y., Chen, C., Zhao, Z., Dou, Q., Heng, P.A.: Temporal memory relation network for workflow recognition from surgical video. IEEE Trans. Med. Imaging (2021)
19. Leibetseder, A., Kletz, S., Schoeffmann, K., Keckstein, S., Keckstein, J.: GLENDA: gynecologic laparoscopy endometriosis dataset. In: Ro, Y.M., et al. (eds.) MMM 2020. LNCS, vol. 11962, pp. 439–450. Springer, Cham (2020). https://doi.org/10.1007/978-3-030-37734-2_36
20. Leibetseder, A., et al.: Lapgyn4: a dataset for 4 automatic content analysis problems in the domain of laparoscopic gynecology. In: Proceedings of the 9th ACM Multimedia Systems Conference, pp. 357–362 (2018)
21. Li, B., Lu, B., Wang, Z., Zhong, B., Dou, Q., Liu, Y.: Learning laparoscope actions via video features for proactive robotic field-of-view control. IEEE Robotics and Automation Letters (2022)
22. Liu, H., Soto, R.A.R., Xiao, F., Lee, Y.J.: Yolactedge: Real-time instance segmentation on the edge. arXiv preprint arXiv:2012.12259 (2020)
23. Maier-Hein, L., et al.: Surgical data science-from concepts toward clinical translation. Med. Image Anal. **76**, 102306 (2022)
24. Maier-Hein, L., et al.: Surgical data science for next-generation interventions. Nature Biomed. Eng. **1**(9), 691–696 (2017)
25. Merrill, R.M.: Hysterectomy surveillance in the united states, 1997 through 2005. Med. Sci. Monitor **14**(1), CR24–CR31 (2008)
26. Nakawala, H., Bianchi, R., Pescatori, L.E., De Cobelli, O., Ferrigno, G., De Momi, E.: "deep-onto" network for surgical workflow and context recognition. Int. J. Comput. Assisted Radiol. Surg. **14**(4), 685–696 (2019)
27. Sarikaya, D., Corso, J.J., Guru, K.A.: Detection and localization of robotic tools in robot-assisted surgery videos using deep neural networks for region proposal and detection. IEEE Trans. Med. Imaging **36**(7), 1542–1549 (2017)
28. Taylor, R.H., Kazanzides, P.: Medical robotics and computer-integrated interventional medicine. In: Biomedical Information Technology, pp. 393–416. Elsevier (2008)
29. Topol, E.J.: High-performance medicine: the convergence of human and artificial intelligence. Nat. Med. **25**(1), 44–56 (2019)
30. Tsui, C., Klein, R., Garabrant, M.: Minimally invasive surgery: national trends in adoption and future directions for hospital strategy. Surg. Endosc. **27**(7), 2253–2257 (2013)
31. Twinanda, A.P., Shehata, S., Mutter, D., Marescaux, J., De Mathelin, M., Padoy, N.: Endonet: a deep architecture for recognition tasks on laparoscopic videos. IEEE Trans. Med. Imaging **36**(1), 86–97 (2016)
32. Wada, K.: labelme: Image Polygonal Annotation with Python (2016). https://github.com/wkentaro/labelme
33. Zadeh, S.M., et al.: Surgai: deep learning for computerized laparoscopic image understanding in gynaecology. Surg. Endosc. **34**(12), 5377–5383 (2020)

Retrieval of Surgical Phase Transitions Using Reinforcement Learning

Yitong Zhang[1]([✉]), Sophia Bano[1], Ann-Sophie Page[2], Jan Deprest[2], Danail Stoyanov[1], and Francisco Vasconcelos[1]

[1] Wellcome/EPSRC Centre for Interventional and Surgical Sciences(WEISS) and Department of Computer Science, University College London, London, UK
ucabyyz@ucl.ac.uk
[2] Department of Development and Regeneration, University Hospital Leuven, Leuven, Belgium

Abstract. In minimally invasive surgery, surgical workflow segmentation from video analysis is a well studied topic. The conventional approach defines it as a multi-class classification problem, where individual video frames are attributed a surgical phase label. We introduce a novel reinforcement learning formulation for offline phase transition retrieval. Instead of attempting to classify every video frame, we identify the timestamp of each phase transition. By construction, our model does not produce spurious and noisy phase transitions, but contiguous phase blocks. We investigate two different configurations of this model. The first does not require processing all frames in a video (only $< 60\%$ and $< 20\%$ of frames in 2 different applications), while producing results slightly under the state-of-the-art accuracy. The second configuration processes all video frames, and outperforms the state-of-the art at a comparable computational cost. We compare our method against the recent top-performing frame-based approaches TeCNO and Trans-SVNet on the public dataset Cholec80 and also on an in-house dataset of laparoscopic sacrocolpopexy. We perform both a frame-based (accuracy, precision, recall and F1-score) and an event-based (event ratio) evaluation of our algorithms.

Keywords: Surgical workflow segmentation · Machine learning · Laparoscopic sacrocolpopexy · Reinforcement learning

1 Introduction

Surgical workflow analysis is an important component to standardise the timeline of a procedure. This is useful for quantifying surgical skills [7], training progression [11,18], and can also provide contextual support for further computer analysis both offline for auditing and online for surgeon assistance and

---

Supplementary Information The online version contains supplementary material available at https://doi.org/10.1007/978-3-031-16449-1_47.

L. Wang (Eds.): MICCAI 2022, LNCS 13437, pp. 497–506, 2022.
https://doi.org/10.1007/978-3-031-16449-1_47

automation [5,17,19]. In the context of laparoscopy, where the main input is video, the current approaches for automated workflow analysis focus on frame-level multi-label classification. The majority of the state-of-the-art models can be decomposed into two components: feature extractor and feature classifier. The feature extractor normally is a Convolutional Neural Network (CNN) backbone converting images or batches of images (clips) into feature vectors. Most of the features extracted at this stage are spatial features or fine-level temporal features depending on the type of the input. Considering that long-term information in surgical video sequences aids the classification process, the following feature classifier predicts phases based on a temporally ordered sequence of extracted features. Following from natural language processing (NLP) and computer vision techniques, the architecture behind this feature classifier has evolved from Long Short-Term Memory (EndoNet, SVRCNet) [10,20] , Temporal Convolution Network (TeCNO) [3], to Transformer (OperA, TransSV) [4,6] in workflow analysis. Although these techniques have improved over the years, the main problem formulation remains unchanged in that phase labels are assigned to individual units of frames or clips.

These conventional models achieve now excellent performance on the popular Cholec80 benchmark [20], namely on frame-based evaluation metrics (accuracy, precision, recall, f1-score). However, small but frequent errors still occur throughout the classification of large videos, causing a high number of erroneous phase transitions, which make it very challenging to pinpoint exactly where one phase ends and another starts. To address this problem, we propose a novel methodology for surgical workflow segmentation. Rather than classifying individual frames in sequential order, we attempt to locate the phase transitions directly. Additionally, we employ reinforcement learning as a solution to this problem, since it has shown good capability in similar retrieval tasks [14,16]. Our contributions can be summarised as follows:

- We propose a novel formulation for surgical workflow segmentation based on phase transition retrieval. This strictly enforces that surgical phases are continuous temporal intervals, and immune to frame-level noise.
- We propose Transition Retrieval Network (TRN) that actively searches for phase transitions using multi-agent reinforcement learning. We describe a range of TRN configurations that provide different trade-offs between accuracy and amount of video processed.
- We validate our method both on the public benchmark Cholec80 and on an in-house dataset of laparoscopic sacropolpopexy, where we demonstrate a single phase detection application.

2   Methods

We consider the task of segmenting the temporal phases of a surgical procedure from recorded video frames. The main feature of our proposed formulation can be visualised in Fig. 1. While previous work attempts to classify every frame of a video according to a surgical phase label, we attempt to predict

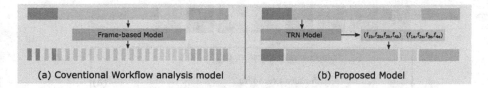

(a) Coventional Workflow analysis model   (b) Proposed Model

Fig. 1. Comparison of network architecture between (a) conventional model and (b) our proposed model with potential error illustration. The conventional model assigns labels for each individual frames and our proposed model predicts frame indices for the starts and end position of phases.

the frame index of phase transitions. More specifically, for a surgical procedure with N different phases, our goal is to predict the frame indices where each phase starts $\{f_{1b}, f_{2b}...f_{Nb}\}$, and where each phase ends $\{f_{1e}, f_{2e}...f_{Ne}\}$. If we can assume that surgical phases are continuous intervals, as it is often the case, then our approach enforces this by design. This is unlike previous frame-based approaches where spurious transitions are unavoidable with noisy predictions. To solve this problem we propose the Transition Retrieval Network (TRN), which we described next.

2.1 Transition Retrieval Network (TRN)

Figure 2 shows the architecture of our TRN model. It has three main modules: an averaged ResNet feature extractor, a multi-agent network for transition retrieval, and a Gaussian composition operator to generate the final workflow segmentation result.

Averaged ResNet Feature Extractor. We first train a standard ResNet50 encoder (outputs 2048 dimension vector) [8] with supervised labels, in the same way as frame-based models. For a video clip of length K, features are averaged into a single vector. We use this to temporally down-sample the video through feature extraction. In this work we consider $K = 16$.

DQN Transition Retrieval. We first discuss the segmentation of a single phase n. We treat it as a reinforcement learning problem with 2 discrete agents W_b and W_e, each being a Deep Q-Learning Network (DQN). These agents iteratively move a pair of search windows centered at frames f_{nb} and f_{ne}, with length L. We enforced a temporal constraint where $f_{nb} \leq f_{ne}$ is always true. The state of the agents s_k is represented by the $2L$ features within the search window, obtained with the averaged ResNet extractor. Based on their state, the agents generate actions $a_{kb} = W_b(s_k)$, and $a_{ke} = W_e(s_k)$, which move the search windows either one clip to the left or to the right within the entire video. During network training, we set a $+1$ reward for actions that move the search window center towards the groundtruth transition, and -1 otherwise. Therefore, we learn

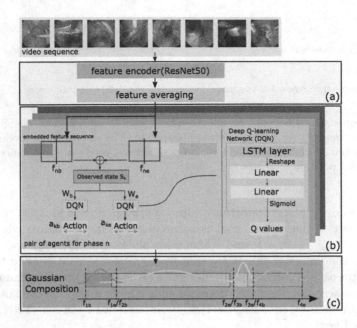

Fig. 2. TRN architecture with (a) averaged ResNet feature extractor, (b) multi-agent network for transition retrieval and (c) Gaussian composition operator

direction cues from image features inside the search windows. As our input to DQN is a sequence of feature vectors, a 3-layer LSTM [9] of dimension 2048 is introduced to DQN architecture for encoding the temporal features into action decision process. The LSTMs are followed by 2 fully connected layer (fc1 and fc2). Fc1 maps input search window of dimension $20L$ to 50 and fc2 maps temporal features of dimension 50 to the final 2 Q-values of 'Right' and 'Left'. We implemented the standard DQN training framework for our network. [15] At inference time, we let the agents explore the video until they converge to a fixed position (i.e. cycling between left and right actions). Two important characteristics of this solution should be highlighted: 1) we do not need to extract clip features from the entire video, just enough for the agent to reach the desired transition; 2) the agents need to be initialised at a certain position in the video, which we discuss later.

Agent Initialization Configurations: We propose two different approaches to initialise the agents: fixed initialization (FI) and, ResNet modified initialization (RMI). FI initializes the search windows based on the statistical distribution (frame index average with relative position in video) of each phase transition on the entire training data. With FI, TRN can make predictions without viewing the entire video and save computation time. On the other hand, RMI initialises the search windows based on the averaged-feature ResNet-50 predictions by averaging the indices of all possible transitions to generate an estimation. In this way, we are very likely to have more accurate initialization positions to FI configuration and yield better performance.

Merging Different Phases with Gaussian Composition: So far, we have only explained how our DQN transition retrieval model segments a single phase. To generalise this, we start by running an independently trained DQN transition retrieval model for each phase. If we take the raw estimations of these phase transitions, we inevitably create overlapping phases, or time intervals with no phase allocated, due to errors in estimation. To address this, we perform a Gaussian composition of the predicted phases. For each predicted pair of transitions f_{nb}, f_{ne}, we draw a Gaussian distribution centred at $\frac{f_{nb}+f_{ne}}{2}$, with standard deviation $\frac{|f_{nb}-f_{ne}|}{4}$. For each video clip, the final multi-class prediction corresponds to the phase with maximum distribution value.

2.2   Training Details

The DQN model is trained in a multi-agent mode where W_b, W_e for a single phase are trained together. The input for individual DQNs in each agent shares a public state concatenated from the content of both search windows, allowing the agents to be able to aware information of others. The procedures of training the DQN are showing in pseudo code in Algorithm 1. For one episode, videos are trained one by one and the maximum number of steps an agent can explore in a video is 200 without early stopping. For every steps the agents made, movement information (s_k, s_{k+1}, a_k, r_k) are stored in its replay memories, and sampled with a batch size of 128 in computing Huber loss. [15] This loss is optimized with gradient descent algorithm, where α is the learning rate and $\nabla_{W_k}\mathcal{L}_k$ is the gradient of loss in the direction of the network parameters.

Algorithm 1. The procedures of training DQN

Initialize parameters of agents W_b and W_e as W_{0b} and W_{0e}
Initialize individual replay memories for agents W_b and W_e
for $episode \leftarrow 0$ to $episode_{MAX}$ **do**
    Initialize search window positions (FI or RMI)
    for $video \leftarrow 0$ to $range(videos)$ **do**
        for $k \leftarrow 0$ to 200 **do**
            $s_k \leftarrow$ read ResNet features in search window
            $a_{kb} \leftarrow W_{kb}(s_k)$ and $a_{ke} \leftarrow W_{ke}(s_k)$
            $s_{k+1} \leftarrow$ update search window position by (a_{kb}, a_{ke}) , read new features
            $r_{kb}, r_{ke} \leftarrow$ compare s_k and s_{k+1} with reward function
            Save $(s_k, s_{k+1}, a_k, r_{kb})$ and $(s_k, s_{k+1}, a_k, r_{ke})$ into agent memory
            Compute loss $(\mathcal{L}_{kb}, \mathcal{L}_{ke})$ from random 128 samples from each memory
            Optimize W_{kb}: $W_{k+1b} \leftarrow W_{kb} + \alpha\nabla_{W_{kb}}\mathcal{L}_{kb}$
            Optimize W_{ke}: $W_{k+1e} \leftarrow W_{ke} + \alpha\nabla_{W_{ke}}\mathcal{L}_{ke}$
        end for
    end for
end for

---

3 Experiment Setup and Dataset Description

The proposed network is implemented in PyTorch using a single Tesla V100-DGXS-32GB GPU of an NVIDIA DGX station. For the ResNet-50 part, PyTorch default ImageNet pretrained parameters are loaded for transfer learning and fine tuned for 100 epochs on Cholec80 and Sacrocolpopexy respectively. The videos are subsampled to 2.4 fps, centre cropped, and resized into resolution 224 * 224 to match the input requirement of ResNet-50. These ResNet parameters are shared for RMI, search windows, TecNO and TransSV in this research We train both ResNet-50 and DQN with Adam [12] at a learning rate of 3e-4. For ResNet-50, we use a batch size of 100, where phases are sampled with equal probability. For DQN, the batch size is 128 sampled from a memory of size 10000 for each agent.

We tested the performance of TRN model on two datasets. Cholec80 is a publicly available benchamark that contains 80 videos of cholecystectomy surgeries [20] divided into 7 phases. We use 40 videos for training, 20 for validation and 20 for testing. We also provide results on an in-house dataset of laparoscopic sacrocolpopexy [1] containing 38 videos. It contains up to 8 phases (but only 5 in most cases), however, here we consider the simplified binary segmentation of the phases related to suturing a mesh implant (2 contiguous phases), given that suturing time is one of the most important indicators of the learning curve [2] in this procedure. We performed a 2-fold cross-validation with 20 videos for training, 8 for validation, and 10 for testing. For Sacrocolpopexy, we train our averaged ResNet extractor considering all phases, but train a single DQN transition retrieval for a suturing phase. We also do not require to apply Gaussian composition since we're interested in a single phase classification.

Evaluation Metrics: We utilise the commonly utilised frame-based metrics for surgical workflow: macro-averaged (per phase) precision and recall, F1-score calculated through this precision and recall, and micro-averaged accuracy. Additionally, we also provide event-based metrics that look at accuracy of phase transitions. An event is defined as block of consecutive and equal phase labels, with a start time and a stop time. We define event ratio as $\frac{E_{gt}}{E_{det}}$ where E_{gt} is the number of ground truth events, and E_{det} is the number of detected events by each method. We define a second ratio based on the Ward metric [21] which allocates events into sub-categories as deletion(D), insertion(I'), merge(M, M'), fragmentation(F, F'), Fragmented and Merged(FM, FM') and Correct(C) events. Here, we denote the Ward event ratio as $(\frac{C}{E_{gt}})$. For both of these ratios, values closer to 1 indicate better performance. Finally, whenever fixed initialisation (FI) is used, we also provide a coverage rate, indicating the average proportion of the videos that was processed to perform the segmentation. Lower values indicate fewer features need to be extracted and thus lower computation time.

4 Results and Discussion

We first provide an ablation of different configurations of our TRN model in Table 1, for Cholec80. It includes two search window sizes (21 and 41 clips) and

Table 1. TRN ablation in the Cholec80 dataset (F1-scores). The values per-phase are computed before Gaussian Composition, while the overall F1-score is for the complete TRN method.

Window size	Phase 1	Phase 2	Phase 3	Phase 4	Phase 5	Phase 6	Phase 7	Overall F1-score
TRN21 FI	0.854	0.917	0.513	0.903	0.687	0.549	0.83	0.782
TRN41 FI	0.828	0.943	0.636	0.922	0.558	0.694	0.85	0.808
TRN21 RMI	0.852	0.942	0.619	0.939	0.727	0.747	0.837	0.830
TRN41 RMI	0.828	0.940	0.678	0.945	0.753	0.738	0.861	0.846

Table 2. Evaluation metric results summary of ResNet-50, our implementation of TeCNO and Trans-SV, and ablative selected TRN result on Cholec80 and Sacrocolpopexy. The computatinla cost is in average second to process a single video

Dataset	Method	Accuracy	Precision	Recall	F1-Score	Event ratio	Ward event ratio	Coverage rate (%)	Computational cost (s)
Cholec80	ResNet-50	79.7±7.5	73.5±8.4	78.5±8.9	0.756	0.120	0.375	full	96.6
	TeCNO	88.3±6.5	78.6±9.9	76.7±12.5	0.774	0.381	0.691	full	99.6
	Trans-SVNet	89.1±5.7	81.7±6.5	79.1±12.6	0.800	0.316	0.566	full	99.6
	TRN21 FI	85.3±9.6	78.1±11.1	78.9±13.5	0.782	1	0.934	57.6	60.6
	TRN41 FI	87.8±8.1	80.3±9.1	81.7±12.4	0.808	1	0.956	59.1	64.9
	TRN41 RMI	90.1±5.7	84.5±5.9	85.1±8.2	0.846	1	0.985	full	105.5
Sacrocol--popexy	ResNet-50	92.5±3.8	94.9±2.8	84.5±8.4	0.892	0.029	0.016	full	493.7
	TeCNO	98.1±1.7	97.7±1.9	97.5±3.0	0.976	0.136	0.438	full	493.8
	Trans-SVNet	97.8±2.2	96.5±4.5	98.0±3.5	0.971	0.536	0.813	full	493.9
	TRN21 FI	89.8±6.2	88.6±11.7	85.3±11.1	0.860	0.971	0.875	14.6	78.1
	TRN81 FI	90.7±6.1	88.6±11.5	88.5±11.1	0.875	0.941	0.860	18.3	104.0

two initialisations (FI, RMI). The observations are straightforward. Larger windows induce generally better f1-scores, and RMI outperforms FI. This means that heavier configurations, requiring more computations, lead to better accuracies. Particular choice of a TRN configuration would depend on a trade-off analysis between computational efficiency and frame-level accuracy.

Table 2 shows a comparison between TRN and state-of-the-art frame-based methods on both Cholec80 and Sacrocolpopexy. The utilised baselines are TeCNO [3], Trans-SVNet [6], which we implemented and trained ourselves. Instead of simple ResNet50, we use the same feature averaging process as the TRN for consistency. Also for consistency, we disabled causal convolution in TCN (it is a provided flag in their code) that allowing Trans-SV and TCN to be trained in off-line mode .

For Cholec80, our full-coverage model (TRN41 RMI) surpasses the best baseline (Trans-SVNet) in all frame-based metrics, while having significantly better even-based metrics (event ratio, Ward event ratio). This can be explained by TRN's immunity to frame-level noisy predictions, which can be visualised on a sample test video in Fig. 3a. Remaining visualisations for all test data are provided in supplementary material.

Still for Cholec80, our partial-coverage models (TRN21/41 FI) have frame-based metrics below the state-of-the-art baselines, however, they have the advantage of performing segmentation by only processing below 60% of the video samples. The trade-off between coverage and accuracy can be observed.

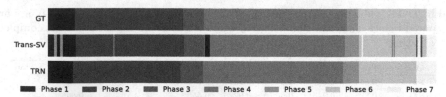

(a) An example of video77 from Cholec80 processed by Trans-SV and TRN41 RMI

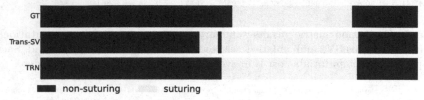

(b) An example video from Sacrocolpopexy processed by Trans-SV and TRN81 FI

Fig. 3. Color-coded ribbon illustration for two complete surgical videos from (a) Cholec80 and (b) Sacrocolpopexy processed by Trans-SV and TRN models.

Additionally, TRN21/41 FI also have substantially better event-based metrics than frame-based methods due to its formulation. An operation may not have a complete set of phases. For the missing phases in Cholec80 test videos, it has little impact as shown in supplementary document. The RL agent makes begin and end labels converge towards the same/consecutive timestamps. Sometimes there are still residual frames erroneously predicted as the missing phase. These errors are counted in reported statistics resulted in imperfect event ratios.

For sacrocolpopexy, we display a case where our partial-coverage models (TRN21/81 FI) are at their best in terms of computational efficiency. These are very long procedures and we are interested in only the suturing phases as an example of clinical interest. [13] Therefore, a huge proportion of the video can be ignored for a full segmentation. Our models slightly under perform all baselines in frame-based metrics, but achieve this result by only looking at under 20% of the videos on average.

5   Conclusion

In this work we propose a new formulation for surgical workflow analysis based on phase transition retrieval (instead of frame-based classification), and a new solution to this problem based on multi-agent reinforcement learning (TRN). This poses a number of advantages when compared to the conventional frame-based methods. Firstly, we avoid any frame-level noise in predictions, strictly enforcing phases to be continuous blocks. This can be useful in practice if, for example, we are interested in time-stamping phase transitions, or in detecting unusual surgical workflows (phases occur in a non-standard order), both of which are challenging to obtain from noisy frame-based classifications. In addition, our models with partial coverage (TRN21/41/81 FI) are able to significantly reduce the number of frames necessary to produce a complete segmentation result.

There are, however, some limitations. First, there may be scenarios where phases occur with an unknown number of repetitions, which would render our formulation unsuitable. Our TRN method is not suitable for real-time application, since it requires navigating the video in arbitrary temporal order. TRN may have scalability issues, since we need to train a different agent for each phase, which may be impractical if a very large number of phases is considered. This could potentially be alleviated by expanding the multi-agent framework to handle multiple phase transitions simultaneously. TRN is also sensitive to agent initialisation, and while we propose 2 working strategies (FI, RMI), they can potentially be further optimised.

Acknowledgements. This research was supported by the Wellcome/EPSRC Centre for Interventional and Surgical Sciences (WEISS) [203145/Z/16/Z]; the Engineering and Physical Sciences Research Council (EPSRC) [EP/P027938/1, EP/R004080/1, EP/P012841/1]; the Royal Academy of Engineering Chair in Emerging Technologies Scheme, and Horizon 2020 FET Open (863146).

References

1. Claerhout, F., Roovers, J.P., Lewi, P., Verguts, J., De Ridder, D., Deprest, J.: Implementation of laparoscopic sacrocolpopexy-a single centre's experience. Int. Urogynecol. J. **20**(9), 1119–1125 (2009)
2. Claerhout, F., Verguts, J., Werbrouck, E., Veldman, J., Lewi, P., Deprest, J.: Analysis of the learning process for laparoscopic sacrocolpopexy: identification of challenging steps. Int. Urogynecol. J. **25**(9), 1185–1191 (2014). https://doi.org/10.1007/s00192-014-2412-z
3. Czempiel, T., et al.: TeCNO: surgical phase recognition with multi-stage temporal convolutional networks. In: Martel, A.L., et al. (eds.) MICCAI 2020. LNCS, vol. 12263, pp. 343–352. Springer, Cham (2020). https://doi.org/10.1007/978-3-030-59716-0_33
4. Czempiel, T., Paschali, M., Ostler, D., Kim, S.T., Busam, B., Navab, N.: Opera: attention-regularized transformers for surgical phase recognition. arXiv preprint arXiv:2103.03873 (2021)
5. DiPietro, R., et al.: Recognizing surgical activities with recurrent neural networks. In: Ourselin, S., Joskowicz, L., Sabuncu, M.R., Unal, G., Wells, W. (eds.) MICCAI 2016. LNCS, vol. 9900, pp. 551–558. Springer, Cham (2016). https://doi.org/10.1007/978-3-319-46720-7_64
6. Gao, X., Jin, Y., Long, Y., Dou, Q., Heng, P.A.: Trans-SVNet: accurate phase recognition from surgical videos via hybrid embedding aggregation transformer. arXiv preprint arXiv:2103.09712 (2021)
7. Goodman, E.D., et al.: A real-time spatiotemporal AI model analyzes skill in open surgical videos. arXiv preprint arXiv:2112.07219 (2021)
8. He, K., Zhang, X., Ren, S., Sun, J.: Deep residual learning for image recognition. In: Proceedings of the IEEE Conference on Computer Vision and Pattern Recognition, pp. 770–778 (2016)

9. Hochreiter, S., Schmidhuber, J.: Long short-term memory. Neural comput. **9**(8), 1735–1780 (12 1997)

10. Jin, Y., et al.: SV-RCNet: workflow recognition from surgical videos using recurrent convolutional network. IEEE Trans. Med. Imaging **37**(5), 1114–1126 (2018)

11. Kawka, M., Gall, T.M., Fang, C., Liu, R., Jiao, L.R.: Intraoperative video analysis and machine learning models will change the future of surgical training. Intell. Surg. **1** (2021)

12. Kingma, D.P., Ba, J.: Adam: a method for stochastic optimization. arXiv preprint arXiv:1412.6980 (2014)

13. Lamblin, G., Chene, G., Warembourg, S., Jacquot, F., Moret, S., Golfier, F.: Glue mesh fixation in laparoscopic sacrocolpopexy: results at 3 years' follow-up. Int. Urogynecol. J. **33**(9), 2533–2541 (2021)

14. Lu, Y., Li, Y., Velipasalar, S.: Efficient human activity classification from egocentric videos incorporating actor-critic reinforcement learning. In: 2019 IEEE International Conference on Image Processing (ICIP), pp. 564–568 (2019). https://doi.org/10.1109/ICIP.2019.8803823

15. Mnih, V., et al.: Human-level control through deep reinforcement learning. Nature **518**(7540), 529–533 (2015)

16. Nikpour, B., Armanfard, N.: Joint selection using deep reinforcement learning for skeleton-based activity recognition. In: 2021 IEEE International Conference on Systems, Man, and Cybernetics (SMC), pp. 1056–1061 (2021). https://doi.org/10.1109/SMC52423.2021.9659047

17. Park, J., Park, C.H.: Recognition and prediction of surgical actions based on online robotic tool detection. IEEE Robot. Autom. Lett. **6**(2), 2365–2372 (2021). https://doi.org/10.1109/LRA.2021.3060410

18. Rojas-Muñoz, E., Couperus, K., Wachs, J.: DAISI: database for AI surgical instruction. arXiv preprint arXiv:2004.02809 (2020)

19. Sarikaya, D., Jannin, P.: Towards generalizable surgical activity recognition using spatial temporal graph convolutional networks. arXiv preprint arXiv:2001.03728 (2020)

20. Twinanda, A.P., Shehata, S., Mutter, D., Marescaux, J., De Mathelin, M., Padoy, N.: EndoNet: a deep architecture for recognition tasks on laparoscopic videos. IEEE Trans. Med. Imaging **36**(1), 86–97 (2016)

21. Ward, J.A., Lukowicz, P., Gellersen, H.W.: Performance metrics for activity recognition. ACM Trans. Intell. Syst. Technol. (TIST) **2**(1), 1–23 (2011)

SGT: Scene Graph-Guided Transformer for Surgical Report Generation

Chen Lin[1,2], Shuai Zheng[1,2], Zhizhe Liu[1,2], Youru Li[1,2], Zhenfeng Zhu[1,2(✉)], and Yao Zhao[1,2]

[1] Institute of Information Science, Beijing Jiaotong University, Beijing, China
zhfzhu@bjtu.edu.cn
[2] Beijing Key Laboratory of Advanced Information Science and Network Technology, Beijing, China

Abstract. The robotic surgical report reflects the operations during surgery and relates to the subsequent treatment. Therefore, it is especially important to generate accurate surgical reports. Given that there are numerous interactions between instruments and tissue in the surgical scene, we propose a $\underline{S}$cene $\underline{G}$raph-guided $\underline{T}$ransformer (SGT) to solve the issue of surgical report generation. The model is based on the structure of transformer to understand the complex interactions between tissue and the instruments from both global and local perspectives. On the one hand, we propose a relation driven attention to facilitate the comprehensive description of the interaction in a generated report via sampling of numerous interactive relationships to form a diverse and representative augmented memory. On the other hand, to characterize the specific interactions in each surgical image, a simple yet ingenious approach is proposed for homogenizing the input heterogeneous scene graph, which plays an effective role in modeling the local interactions by injecting the graph-induced attention into the encoder. The dataset from clinical nephrectomy is utilized for performance evaluation and the experimental results show that our SGT model can significantly improve the quality of the generated surgical medical report, far exceeding the other state-of-the-art methods. The code is public available at: https://github.com/ccccchenllll/SGT_master.

Keywords: Surgical report generation · Transformer · Scene graph

1 Introduction

Deep learning has been widely used in computer-aided diagnosis (CAD) for the past few years [15,29,30]. Thereinto, computer assisted surgery (CAS) expands the concept of general surgery, which uses computer technology for surgical planning and to guide or perform surgical interventions. With the advent of CAS,

---

Supplementary Information The online version contains supplementary material available at https://doi.org/10.1007/978-3-031-16449-1_48.

general surgery has made great strides in minimally invasive approaches. For example, in the field of urology, surgical robots perform pyeloplasty or nephrectomy for laparoscopic surgery. The surgical reports are required to record the surgical procedure performed by the microsurgical robot. Automatic generation of surgical reports frees surgeons and nurses from the tedious task of report recording, allowing them to focus more on patients' conditions and providing a detailed reference for post-operative interventions.

Surgical report generation, also called image caption, involves the understanding of surgical scenes and the corresponding text generation. In natural scenes, image caption algorithms have achieved good performance on MS-COCO [14], flicker30k [28] and Visual Genome [11], which have evolved from earlier approaches based on template stuffing [1,27] and description retrieval [6,17] to current deep learning-based generative approaches [22,25]. However, in the biomedical field, studies on image caption are relatively rare. Existing studies on medical image caption more focus on radiological images, such as chest X-rays [9,23]. However, with the spread of microsurgical robots, surgical reports generation is supposed to receive more attention. Hence, in our work, we focus on understanding surgical scene and generating accurate descriptions.

Many report generation methods follow the architecture of CNN-LSTM. Despite their widespread adoption, LSTM-based models are still affected by their sequential nature and limited representational power. To tackle this shortcoming, a fully-attentive paradigm named Transformer has been proposed [20] and has made great success in machine translation tasks [5]. Similarly, The Transformer-based model has also been applied to the report generation task. Xiong *et al.* [24] proposed hierarchical neural network architecture - Reinforced Transformer to generate coherent informative medical imaging report. Hou *et al.* [7] developed a Transformer-based CNN-Encoder to RNN-Decoder architecture for generating chest radiograph reports. Considering the excellent performance of Transformer in terms of report generation, it is also adopted as main architecture in our work.

Different from the above report generation tasks, the various instruments and complex interactive relationships in the surgical scene make it difficult to generate surgical reports. To generate accurate reports, it is necessary to understand the interactive relations in the surgical scene. However, the previous works mentioned above lack considerations of modeling the inherent interactive relations between objects. To address this issue, we propose a scene graph-guided transformer model. Unlike the previous transformer-based model, we exploit the inputs and interactive relationships in both global and local ways. In summary, the following contributions can be highlighted:

- We propose a novel surgical report generation model via scene graph-guided transformer (SGT), in which the visual interactive relationships between tissues and instruments are well exploited from both global and local ways.
- To reinforce the description of interactions in the generated report, a global relation driven attention is proposed. It uses the sampled interactive relationships with diversity as augmented memory, instead of the traditional way of utilizing the inputs directly.

- To characterize the interactive relationships in each specific surgical image, we also propose a simple yet ingenious approach to homogenize the given heterogeneous scene graph, by which a graph-induced attention is injected into the encoder to encode the above local interactions.

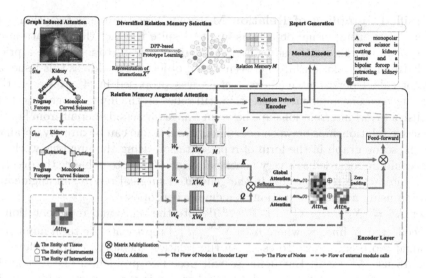

Fig. 1. The framework of the proposed SGT for surgical report generation.

2 Methodology

2.1 Overview of the Proposed Framework

For each surgical image I, a pre-built scene graph $\mathcal{G}_{he}(\mathcal{V}_{he}, \mathcal{E}_{he}, X_{he}^v, X_{he}^r)$ is assumed to be available by [8], where $\mathcal{V}_{he} = \{v_{he}^i\}_{i=1,\ldots,|\mathcal{V}_{he}|}$ and $\mathcal{E}_{he} = \{e_{he}(i,j)\}_{i,j\in\mathcal{V}_{he}}$ are the sets of nodes and edges, respectively. $X_{he}^v \in \mathbb{R}^{|\mathcal{V}_{he}|\times d}$ and $X_{he}^r \in \mathbb{R}^{|\mathcal{E}_{he}|\times d}$ denote the associated representations of graph nodes and the interactive relationships between nodes, respectively. Here, the nodes refer to the visual objects extracted from image I. In particular, for a real surgical scene, $\mathcal{V}_{he}$ can be classified into two types of nodes according to their intention roles: tissue node t and instrument node o, i.e., we have $v_{he}^i \in \{t, o\}$ for $i = 1, \ldots, |\mathcal{V}_{he}|$. Here, it is worth noting that the graph $\mathcal{G}_{he}$ can be considered a heterogeneous graph, because the links between nodes are also depicted in a representation space, which is different from the general homogeneous graph.

The overall framework of the proposed **S**cene **G**raph-guided **T**ransformer for surgical report generation, also named by SGT, is shown in Fig. 1. It is divided into two main modules: the *relation driven encoder* is responsible for encoding the input heterogeneous scene graph, and the *meshed decoder* tends to read from each encoder layer to generate report words by words. Specifically, for the encoder, it receives the injections of the relation memory augmented attention and graph induced attention, to guide the representation learning of visual scene from both the global and local perspectives.

2.2   Relation Driven Attention

Diversified Sampling of Relation Memory. For the task of caption generation, when using visual objects as input, despite the fact that self-attention can encode pairwise relationships between regions, it fails to model the a priori knowledge of the relationships between visual objects. To address this issue, an operator called persistent memory vectors was proposed in [4]. However, this persistent memory augmented attention still relies on the input image objects to establish a priori interrelationships between image objects. Different from [4], the interactive relationships between instruments and tissue can be directly obtained from the scene graph in the form of representation using the previous work [8], which could be regarded as a very beneficial prior. Hence, to enhance the description of interactive relationships in the generated surgical report, an augmented attention using a prior interaction as memory is proposed.

Let $X^r = [X_{he}^{r,1}, \ldots, X_{he}^{r,N}] \in \mathbb{R}^{n_r \times d}$ denote the relational representation of all of collected N images, where $n_r = \sum_{p=1}^{N} |\mathcal{E}_{he}^p|$ represents the total number of relational representations. To establish the relation memory, a straightforward way is to use X^r for it. But doing so will pose two problems. The first is the high computational complexity, and the other one is the over-smoothing of the learned attention due to the excessive redundancy of these relational representations, which will lead inevitably to the loss of focus. For this reason, a sampling scheme with diversity was considered. In particular, the determinant point process (DPP) [16] is used to sample X^r to obtain a rich diversified prototype subset of interaction representation, so that the sampled subset could cover as much of the representation space of the interactive relationships as possible.

Given Z as the metric matrix of X^r, DPP is capble of selecting a diversified subset $X_S^r \subseteq X^r$, whose items are indexed by $S \subseteq L = \{1, \ldots, n_r\}$, by maximizing the following sampling probability $P_Z(X_S^r)$ of X_S^r:

$$P_Z(X_S^r) = \frac{det(Z_S)}{det(Z + I)} \tag{1}$$

where $\sum_{S \subseteq L} det(Z_S) = det(Z + I)$, I is the identity matrix, $Z_S \equiv [Z_{ij}]_{i,j \in S}$, and $det(\cdot)$ denotes the determinant of a matrix. As given by Eq. 1, it can be known that any subset of L corresponds to a probability, which will result in a large search range for the prototype index subset. In order to eliminate the uncertainty of the sample set capacity in the standard DPP, Kulesza [12] proposed a variant of standard DPP with fixed subset size $|S| = k$, to realize the controllability of the sampling process. Through k-DPP, the most appropriate k interaction relations can be selected as the prototypes to serve for the relation memory $M = X_S^r \in \mathbb{R}^{k \times d}$.

Relation Memory Augmented Attention. To enhance the intervention of interactive relationships in the encoder, the relation memory M are concatenated with the transformed input X to obtain the augmented key $K = [XW_k; M] \in \mathbb{R}^{(m+k) \times d}$ and value $V = [XW_v; M] \in \mathbb{R}^{(m+k) \times d}$, respectively, where $W_k \in \mathbb{R}^{d \times d}$ and $W_v \in \mathbb{R}^{d \times d}$ are the corresponding projection matrix, $X = [X_{he}^v; X_{he}^r] \in \mathbb{R}^{m \times d}$ with $m = |\mathcal{V}_{he}| + |\mathcal{E}_{he}|$ is the input node representation of the homogeneous graph $\mathcal{G}_{ho}$ to be described in the following section, and $[\cdot; \cdot]$ indicates the concatenation of two matrices. Furthermore, the relation memory augmented attention can be defined as:

$$Attn_m(\boldsymbol{Q}, \boldsymbol{K}) = softmax\left(\frac{\boldsymbol{K}\boldsymbol{Q}^T}{\sqrt{d}}\right) = \begin{bmatrix} Attn_m(2) \\ Attn_m(1) \end{bmatrix} \quad (2)$$

where $\boldsymbol{Q} = XW_q \in \mathbb{R}^{m \times d}$, and $\sqrt{d}$ denotes a scaling factor relevant to d. Clearly, the relation memory augmented attention $Attn_m$ consists of two attentional block matrices $Attn_m(1)$ and $Attn_m(2)$, in which $Attn_m(2)$ is obtained by calculating the pairwise similarity according to the input X itself, and $Attn_m(1)$ carries the information from the diversified relation memory prototypes M that is globally sampled in the representation space of interactive relationships.

2.3 Graph Induced Attention

The relation memory augmented attention mentioned above establishes a global perception of various interactions by the entities, i.e., nodes in the scene graph. However, such global perception is still inadequate for the portrayal of the detailed interactions among entities in a given specific scene graph. In fact, it is also crucial to seek the local perception of a particular visual scene, as a complement to global perception, when generating surgical reports.

Homogenization of Heterogeneous Graph. For the pre-established heterogeneous scene graph $\mathcal{G}_{he}$, it reflects the unique interaction between the visual objects of the associated image. However, due to its heterogeneity, direct application of the existing graph structure could achieve the local perception to a certain extent, while it is difficult to make such use of various interactions in graphs in such a way.

To address this issue, a simple yet ingenious way is to homogenize the heterogeneous link '$t \xleftrightarrow{r} o$' to the form of '$t \longleftrightarrow r \longleftrightarrow o$', where r denotes the interaction between the tissue node t and instrument node o. In this way, the interaction information hidden in the links in the heterogeneous graph can also be represented as nodes in a re-build homogeneous graph. Specifically, the homogenization of heterogeneous graph can be illustrated as $\mathcal{G}_{he}(\mathcal{V}_{he}, \mathcal{E}_{he}, X_{he}^v, X_{he}^r) \rightarrow \mathcal{G}_{ho}(\mathcal{V}_{ho}, \mathcal{E}_{ho}, X_{ho}^v)$, where $\mathcal{V}_{ho} = \{\mathcal{V}_{he} \cup \mathcal{V}_r\}$ with $v_{ho}^i \in \{t, o, r\}$, $i = 1, \ldots, |\mathcal{V}_{ho}|$, and $\mathcal{E}_{ho} = \{e_{ho}(i, j)\}_{i,j \in \mathcal{V}_{ho}}$ are the sets of nodes and edges of homogeneous graph $\mathcal{G}_{ho}$, respectively. $\mathcal{V}_r$ is the set of nodes with each node representing a specific interactive relationship and $|\mathcal{V}_r| = |\mathcal{E}_{he}|$.

$X_{ho}^v = [X_{he}^v; X_{he}^r] \in \mathbb{R}^{m \times d}$, i.e., X mentioned above, denotes the associated representation of each node with $m = |\mathcal{V}_{ho}| = |\mathcal{V}_{he}| + |\mathcal{E}_{he}|$ being the number of nodes.

Particularly, without loss of generality, taking the visual scene in graph induced attention module of Fig. 1 as an example, there are two instruments, *prograsp forceps* and *monopolar curved scissors*, a *kidney* tissue, and two *retracting* and *cutting* interactions in $\mathcal{G}_{he}$. By the homogenization of $\mathcal{G}_{he}$, the two interactions including *rectracting* and *cutting* in the new converted homogeneous graph $\mathcal{G}_{ho}$ will be treated as nodes.

Attention Based on Homogeneous Graph. Given the obtained heterogeneous graph $\mathcal{G}_{ho}$, its graph structure can be represented using the adjacency matrix A_{ho} with $A_{ho}(i, j) = 1$, if $e_{ho}(i, j) \in \mathcal{E}_{ho}$, and $A_{ho}(i, j) = 0$, otherwise. As known to all, the essence of self-attention is to look for the intrinsic correlations between inputs. Considering the adjacency matrix A_{ho} describes the connections between various entities including the tissue, instruments, and interactions, which just felicitously reflect the correlation mentioned above, we can define the graph induced attention as $Attn_g = D^{-\frac{1}{2}} A_{ho} D^{-\frac{1}{2}}$, where D is a diagonal matrix with $D_{ii} = \sum_j A_{ho}(i, j)$.

Furthermore, we will have a fused attention $Attn$ via a linear combination of the relation memory driven attention $Attn_m$ with the graph induced attention $Attn_g$:

$$Attn = \begin{bmatrix} Attn(1) \\ Attn(2) \end{bmatrix} = \left[(1 - \gamma) \cdot \begin{bmatrix} Attn_m(1) \\ Attn_m(2) \end{bmatrix} + \gamma \cdot \begin{bmatrix} 0 \\ Attn_g \end{bmatrix} \right] \qquad (3)$$

where γ is a trade-off coefficient. Obviously, we can find from Eq. 3 that $Attn$ is also composed of two attentional block metrices $Attn(1)$ and $Attn(2)$. As far as $Attn(1)$ is concerned, it explicitly carries the global interactive information contained in relation memory via global sampling based on DPP. Different from $Attn(1)$, $Attn(2)$ can be seen as a local perception of an input surgical image to some extent, since it exploits effectively the information of the specific scene graph itself from two different views, respectively, i.e., the node representation in $Attn_m(2)$ and the graph structure. Finally, the node embeddings can be calculated as $H = Attn^T \cdot V \in \mathbb{R}^{m \times d}$.

2.4   Caption Generation

To generate the word in the report sequentially, the decoder takes both the words generated in the previous stage and the output H by the encoder as input. Just like in M^2T [4], we use the same backbone structure, i.e., meshed decoder, for caption generation. Specifically, the encoder encodes X to H via the proposed dual attention, and then the decoder reads from the output encoder and finally performs a probabilistic calculation to determine the output of the next word.

3   Experiment Results and Analysis

3.1   Dataset

The dataset comes from Robotic Instrument Segmentation Sub-Challenge of 2018 MICCAI the Endoscopic Vision Challenge [2], which consists of 14 nephrectomy record sequences. The surgical report of the frames in each sequence were annotated by an experienced surgeon in robotic surgery [26] and the scene graph of each frame is generated by [8]. Specifically, there are a total of 9 objects in the scene graphs of the dataset, including 1 tissue and 8 instruments. Besides, a total of 11 interactions exists among these surgical instruments and tissues, such as manipulating, grasping, etc. For the surgical report generation task, 11 sequences including 1124 frames with surgery report are selected as the training set, and the other 3 sequences including 394 frames with surgery report as the test set. For fairness, most interactions are presented in both the training and test sets.

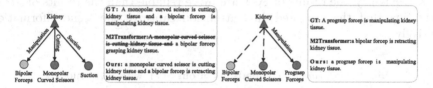

Fig. 2. Examples of the generated reports by our SGT, M^2T, and the corresponding ground-truths. $\rightarrow$ denotes interactions existing between two nodes. $\dashrightarrow$ represents there are no interactions. The words in red and the red ones with strikethrough indicate the generated incorrectly and the ground truth but are not generated, respectively. (Color figure online)

3.2   Experimental Settings

Metrics. To quantitatively verify the effectiveness of our proposed method, the evaluation metrics used in the image caption task are applied for the surgical report generation: BLEU [19], METEOR [3], ROUGE [13], and CIDEr [21].

Implementation Details. Following the preprocessing in [26], we change all the words to lowercase in each surgical report, the punctuation is removed as well. Thus, there are 45 words in the vocabulary. In our model, we set the number of heads to 8, the number of the selected prototypes k to 48, the value of γ to 0.3, and the dimensionality of the node feature d is set to 512. We train the model using cross-entropy loss and fine-tune the sequence generation using reinforcement learning following [4]. The model uses Adam [10] as the optimizer with a batch size of 10, the beam size equals to 5, and is implemented in Pytorch.

3.3 Experimental Results

Performance Comparison. To evaluate the performance of the proposed SGT, we choose to compare with several state-of-the-art transformer-based models: M^2T [4], X-LAN [18], and CIDA [26]. As shown in Table 1, our SGT significantly outperforms all the other methods in terms of all evaluation metrics. Particularly, the relative improvement achieved by SGT grows larger as the standard of BLEU becomes more stringent, indicating that the generated reports of SGT are more approximate to the real reports provided by the doctor compared to the others. Besides, to alleviate the overfitting in the case of small test set, we further conduct the experiment of 5-fold cross validation on random division of the original dataset. The results of ours (BLEU-1:0.7566, CIDEr:5.1139) are slightly lower than the original. Nevertheless, our method still outperformed other methods, the CIDEr score of ours is around 105% than M^2T. Owing to space constraints, the results of 5-fold cross validation are presented in the supplementary material. For a more straightforward comparison of SGT with other methods, some cases of the reports generated by SGT and M^2T are shown in Fig. 2. Obviously, the results of SGT are more reliable than the results of M^2T, illustrating that the proposed dual attention captures the scene information effectively.

Table 1. Performance comparisons of our SGT with other models.

Method	BLEU-1	BLEU-2	BLEU-3	BLEU-4	METEOR	ROUGE	CIDEr
M^2T [4]	0.5881	0.5371	0.4875	0.4445	0.4691	0.6919	2.8240
X-LAN [18]	0.5733	0.5053	0.4413	0.3885	0.3484	0.5642	2.0599
CIDA [26]	0.6246	0.5624	0.5117	0.4720	0.3800	0.6294	2.8548
SGT(Ours)	**0.8030**	**0.7665**	**0.7300**	**0.6997**	**0.5359**	**0.8312**	**5.8044**
Improv	28.56%	36.29%	42.66%	48.24%	14.24%	20.13%	103.32%

Table 2. Ablation study of SGT.

Method		BLEU-1	BLEU-2	BLEU-3	BLEU-4	METEOR	ROUGE	CIDEr
M	$Attn_g$							
×	×	0.7232	0.6799	0.6410	0.6104	0.5088	0.7805	5.1892
✓	×	0.7429	0.7027	0.6640	0.6343	0.5139	0.7890	5.2265
×	✓	0.7776	0.7373	0.6998	0.6716	0.5009	0.7890	5.2578
✓	✓	**0.8030**	**0.7665**	**0.7300**	**0.6997**	**0.5359**	**0.8312**	**5.8044**

Ablation Study. To fully validate the effectiveness of our proposed relation memory prototype M and graph-induced attention $Attn_g$, we conduct an ablation study to compare different variants of SGT. As shown in Table 2, M and $Attn_g$ respectively already bring an improvement in terms of the base model. It's obvious that the proposed M and $Attn_g$ bring significant performance gains. In summary, we can observe that M and $Attn_g$ are designed reasonably and the performance degrades to some extent when removing any of them. In addition, we also perform 5-fold cross validation on ablation study. The experimental results are slightly lower than the original results. But it can still be seen the effectiveness of M and $Attn_g$. The results of the 5-fold cross validation are shown in the supplementary material.

Hyper-parameter Sensitivity Analysis. We then evaluate the role of the tradeoff γ and the number of the selected prototypes. Figure 3 intuitively shows the change trend of γ. Notably, when γ is set to 0.3, SGT achieves the best performance on all metrics. Furthermore, we report the performance of our approach when using a varying number of the selected prototypes k. As shown in Table 3, the best results in terms of all the metrics is achieved with k set to 48.

Table 3. Sensitivity analysis of k.

Method		BLEU-1	BLEU-2	BLEU-3	BLEU-4	METEOR	ROUGE	CIDEr
DPP	$k = 6$	0.7123	0.6636	0.6255	0.5975	0.4801	0.7468	4.8093
	$k{=}12$	0.6884	0.6397	0.5957	0.5634	0.4855	0.7555	4.7155
	$k{=}24$	0.7982	0.7594	0.7247	0.6979	0.5148	0.8113	5.6977
	$k{=}48$	**0.8030**	**0.7665**	**0.7300**	**0.6997**	**0.5359**	**0.8312**	**5.8044**
	$k{=}96$	0.7776	0.7381	0.7026	0.6750	0.4951	0.7948	5.6844

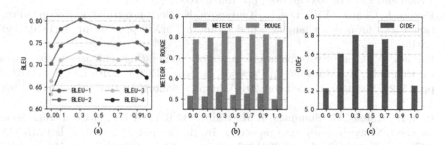

Fig. 3. Impact of the tradeoff γ on BLEU, METEOR, ROUGE and CIDEr.

4   Conclusion

In this work, we mainly focus on how to generate precise surgical reports and propose a novel scene graph-guided Transformer (SGT) model. It takes full advantage of the interactive relations between tissue and instruments from both global and local perspectives. As for the global relation driven attention, the globally sampled representative relation prototypes are utilized as augmented relation memory, thus enhancing the description of the interactions in the generated surgical report. Additionally, a graph-induced attention is proposed to characterize from a local aspect the specific interactions in each surgical image. The experiments on a clinical nephrectomy dataset demonstrate the effectiveness of our model.

Acknowledgement. This work was supported in part by Science and Technology Innovation 2030 – New Generation Artificial Intelligence Major Project under Grant 2018AAA0102100, National Natural Science Foundation of China under Grant No. 61976018, Beijing Natural Science Foundation under Grant No. 7222313.

References

1. Aker, A., Gaizauskas, R.: Generating image descriptions using dependency relational patterns. In: Proceedings of the 48th Annual Meeting of the Association for Computational Linguistics, pp. 1250–1258 (2010)
2. Allan, M., et al.: 2018 robotic scene segmentation challenge. arXiv preprint arXiv:2001.11190 (2020)
3. Banerjee, S., Lavie, A.: Meteor: an automatic metric for MT evaluation with improved correlation with human judgments. In: Proceedings of the ACL Workshop on Intrinsic and Extrinsic Evaluation Measures for Machine Translation and/or Summarization, pp. 65–72 (2005)
4. Cornia, M., Stefanini, M., Baraldi, L., Cucchiara, R.: Meshed-memory transformer for image captioning. In: Proceedings of the IEEE/CVF Conference on Computer Vision and Pattern Recognition, pp. 10578–10587 (2020)
5. Devlin, J., Chang, M.W., Lee, K., Toutanova, K.: Bert: pre-training of deep bidirectional transformers for language understanding. arXiv preprint arXiv:1810.04805 (2018)
6. Farhadi, A., et al.: Every picture tells a story: generating sentences from images. In: Daniilidis, K., Maragos, P., Paragios, N. (eds.) ECCV 2010. LNCS, vol. 6314, pp. 15–29. Springer, Heidelberg (2010). https://doi.org/10.1007/978-3-642-15561-1_2
7. Hou, B., Kaissis, G., Summers, R.M., Kainz, B.: RATCHET: medical transformer for chest X-ray diagnosis and reporting. In: de Bruijne, M., et al. (eds.) MICCAI 2021. LNCS, vol. 12907, pp. 293–303. Springer, Cham (2021). https://doi.org/10.1007/978-3-030-87234-2_28
8. Islam, M., Seenivasan, L., Ming, L.C., Ren, H.: Learning and reasoning with the graph structure representation in robotic surgery. In: Martel, A.L., et al. (eds.) MICCAI 2020. LNCS, vol. 12263, pp. 627–636. Springer, Cham (2020). https://doi.org/10.1007/978-3-030-59716-0_60

9. Jing, B., Xie, P., Xing, E.: On the automatic generation of medical imaging reports. arXiv preprint arXiv:1711.08195 (2017)
10. Kingma, D.P., Ba, J.: Adam: a method for stochastic optimization. arXiv preprint arXiv:1412.6980 (2014)
11. Krishna, R., et al.: Visual genome: connecting language and vision using crowd-sourced dense image annotations. Int. J. Comput. Vision **123**(1), 32–73 (2017)
12. Kulesza, A., Taskar, B.: k-DPPs: fixed-size determinantal point processes. In: Proceedings of the 28th International Conference on Machine Learning (ICML) (2011)
13. Lin, C.Y.: Rouge: a package for automatic evaluation of summaries. In: Proceedings of the Workshop on Text Summarization Branches Out (WAS 2004), pp. 74–81 (2004)
14. Lin, T.-Y., et al.: Microsoft COCO: common objects in context. In: Fleet, D., et al. (eds.) ECCV 2014. LNCS, vol. 8693, pp. 740–755. Springer, Cham (2014). https://doi.org/10.1007/978-3-319-10602-1_48
15. Liu, Z., Zhu, Z., Zheng, S., Liu, Y., Zhou, J., Zhao, Y.: Margin preserving self-paced contrastive learning towards domain adaptation for medical image segmentation. IEEE J. Biomed. Health Inform. **26**(2), 638–647 (2022). https://doi.org/10.1109/JBHI.2022.3140853
16. Macchi, O.: The coincidence approach to stochastic point processes. Adv. Appl. Probab. **7**(1), 83–122 (1975)
17. Pan, J.Y., Yang, H.J., Duygulu, P., Faloutsos, C.: Automatic image captioning. In: 2004 IEEE International Conference on Multimedia and Expo (ICME)(IEEE Cat. No. 04TH8763), vol. 3, pp. 1987–1990. IEEE (2004)
18. Pan, Y., Yao, T., Li, Y., Mei, T.: X-linear attention networks for image captioning. In: Proceedings of the IEEE/CVF Conference on Computer Vision and Pattern Recognition, pp. 10971–10980 (2020)
19. Papineni, K., Roukos, S., Ward, T., Zhu, W.J.: BLEU: a method for automatic evaluation of machine translation. In: Proceedings of the 40th annual meeting of the Association for Computational Linguistics, pp. 311–318 (2002)
20. Vaswani, A., et al.: Attention is all you need. Advances in Neural Information Processing Systems 30 (2017)
21. Vedantam, R., Lawrence Zitnick, C., Parikh, D.: CIDEr: consensus-based image description evaluation. In: Proceedings of the IEEE Conference on Computer Vision and Pattern Recognition, pp. 4566–4575 (2015)
22. Vinyals, O., Toshev, A., Bengio, S., Erhan, D.: Show and tell: a neural image caption generator. In: Proceedings of the IEEE Conference on Computer Vision and Pattern Recognition, pp. 3156–3164 (2015)
23. Wang, X., Peng, Y., Lu, L., Lu, Z., Summers, R.M.: TieNet: text-image embedding network for common thorax disease classification and reporting in chest X-rays. In: Proceedings of the IEEE Conference on Computer Vision and Pattern Recognition, pp. 9049–9058 (2018)
24. Xiong, Y., Du, B., Yan, P.: Reinforced transformer for medical image captioning. In: Suk, H.I., Liu, M., Yan, P., Lian, C. (eds.) MLMI 2019. LNCS, vol. 11861, pp. 673–680. Springer, Cham (2019). https://doi.org/10.1007/978-3-030-32692-0_77
25. Xu, K., et al.: Show, attend and tell: neural image caption generation with visual attention. In: International Conference on Machine Learning, pp. 2048–2057. PMLR (2015)
26. Xu, M., Islam, M., Lim, C.M., Ren, H.: Class-incremental domain adaptation with smoothing and calibration for surgical report generation. In: de Bruijne, M., et al. (eds.) MICCAI 2021. LNCS, vol. 12904, pp. 269–278. Springer, Cham (2021). https://doi.org/10.1007/978-3-030-87202-1_26

27. Yao, B.Z., Yang, X., Lin, L., Lee, M.W., Zhu, S.C.: I2T: image parsing to text description. Proc. IEEE **98**(8), 1485–1508 (2010)
28. Young, P., Lai, A., Hodosh, M., Hockenmaier, J.: From image descriptions to visual denotations: new similarity metrics for semantic inference over event descriptions. Trans. Assoc. Comput. Linguist. **2**, 67–78 (2014)
29. Zhang, W., et al.: Deep learning based torsional nystagmus detection for dizziness and vertigo diagnosis. Biomed. Signal Process. Control **68**, 102616 (2021)
30. Zheng, S., et al.: Multi-modal graph learning for disease prediction. IEEE Trans. Med. Imaging **41**(9), 2207–2216 (2022). https://doi.org/10.1109/TMI.2022.3159264

CLTS-GAN:
Color-Lighting-Texture-Specular Reflection Augmentation for Colonoscopy

Shawn Mathew[1], Saad Nadeem[2(✉)], and Arie Kaufman[1]

[1] Department of Computer Science, Stony Brook University, New York, USA
{shawmathew,ari}@cs.stonybrook.edu
[2] Department of Medical Physics, Memorial Sloan Kettering Cancer Center,
New York, USA
nadeems@mskcc.org

Abstract. Automated analysis of optical colonoscopy (OC) video frames (to assist endoscopists during OC) is challenging due to variations in color, lighting, texture, and specular reflections. Previous methods either remove some of these variations via preprocessing (making pipelines cumbersome) or add diverse training data with annotations (but expensive and time-consuming). We present CLTS-GAN, a new deep learning model that gives fine control over color, lighting, texture, and specular reflection synthesis for OC video frames. We show that adding these colonoscopy-specific augmentations to the training data can improve state-of-the-art polyp detection/segmentation methods as well as drive next generation of OC simulators for training medical students. The code and pre-trained models for CLTS-GAN are available on Computational Endoscopy Platform GitHub (https://github.com/nadeemlab/CEP).

Keywords: Colonoscopy · Augmentation · Polyp detection

1 Introduction

Colorectal cancer is the fourth deadliest cancer. Polyps, anomalous protrusions on the colon wall, are precursors of colon cancer and are often screened and removed using optical colonoscopy (OC). During OC, variations in color, texture, lighting, specular reflections, and fluid motion make polyp detection by a gastroenterologist or an automated method challenging. Previous methods deal with these variations either by removing specular reflections [13,14], removing

---

S. Mathew and S. Nadeem—Equal contribution.

Supplementary Information The online version contains supplementary material available at https://doi.org/10.1007/978-3-031-16449-1_49.

color/texture [15], and correcting lighting [23] in the preprocessing steps (making pipelines cumbersome) or by adding more diverse training data with expert annotations (but expensive and time-consuming). If the automated methods can be made invariant to color, lighting, texture, and specular reflections without adding any preprocessing overhead or additional annotations, then these methods can act as effective second readers to gastroenterologists, improving the overall polyp detection accuracy and potentially reducing the procedure time (end-to-end colon wall inspection from rectum to cecum and back).

We present a new deep learning model, CLTS-GAN, that provides fine-grained control over creation of colonoscopy-specific color, lighting, texture, and specular reflection augmentations. Specifically, we use a one-to-many image-to-image translation model with Adaptive Instance Normalization (AdaIn) and noise input (StyleGAN [12]) to create these augmentations. Color and lighting augmentations are performed by injecting 1D vectors (sampled from a uniform distribution) using AdaIn, while texture and specular reflection augmentations are incorporated by directly adding 2D matrices (sampled from a uniform distribution) to the latent features. The color and lighting vectors can be extracted from one OC image and used to modify the color and lighting of another OC image. We show that these colonoscopy-specific augmentations to the training data can improve accuracy of the state-of-the-art deep learning polyp detection methods as well as drive next generation OC simulators for teaching medical students [7]. The contributions of this work are as follows:

1. CLTS-GAN, an unsupervised one-to-many image-to-image translation model
2. A novel texture loss to encourage a larger variety in texture and specular generation for OC images
3. A method for augmenting colonoscopy frames that produces state-of-the-art results for polyp detection
4. Latent space analysis to make CLTS-GAN more interpretable for generating color, lighting, texture, and specular reflection

2   Related Works

The image-to-image translation task aims to translate an image from one domain to another. Certain applications have access to ground truth information providing supervision for models like pix2pix [11]. Zhu et al. developed CycleGAN, an image-to-image translation model without needing ground truth correspondence [5]. This is done using a cycle consistency loss that drives other unsupervised domain translation models. Examples include MUNIT [9] and Augmented CycleGAN [1] which additionally incorporated noise to learn a many-to-many domain translation. This many-to-many mapping lacks control over specific image attributes. XDCycleGAN [17] and FoldIt [16] model one-to-many image-to-image translation, however their networks functionally learn a one-to-one mapping.

Generating realistic OC from CT scans has been used for OC simulators. VRCaps uses a rendering approach to simulate a camera inside organs captured in CT scans [10]. For the colon, a simple texture is mapped on a mesh where OC artifacts (e.g., specular reflections, fish-eye lense distortion) are added. However, it cannot produce complex textures and colors normally found in OC. OfGAN uses image-to-image translation with optical flow to transform colon simulator images to OC [22]. It uses synthetic colonoscopy frames embedded with texture and specular reflection, which improve the realism of generated images. The texture and specular mapping in the synthetic frames, however, restrict additional texture and specular generation. Rivoir et al. use neural textures to create realistic and temporally consistent textures [19]. They require a full 3D mesh to embed the neural textures making it difficult to augment annotated real videos.

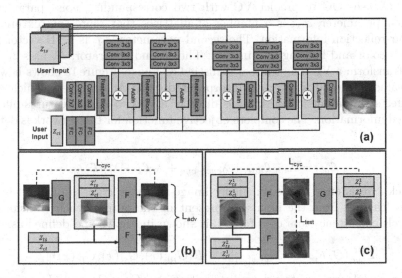

Fig. 1. (a) shows user specified noise being used in F. z_{ts} is a set of 2D matrices that goes through convolutional layers and is added to latent features throughout the network. z_{cl} is a 1D vector that goes through fully connected layers and is distributed to AdaIn layers. Both z_{ts} and z_{cls} are sampled from a uniform distribution and can be sampled until the user is satisfied with the result. (b) depicts the forward cycle where OC passes through G, predicting its noise vectors and VC. These are then passed into F to reconstruct the image. F produces another OC image using different noise vectors where $\mathcal{L}_{adv}$ is applied. (c) depicts the backwards cycle where a VC image with different Z_{ts} is passed into F. The resulting two OC images have $\mathcal{L}_{text}$ applied. One OC image is used for reconstruction via G where $\mathcal{L}_{cyc}$ is applied.

3   Data

10 OC videos and 10 abdominal CT scans for virtual colonoscopy (VC) were obtained at Stony Brook University Hospital. The OC videos were rescaled to 256×256 and cropped to remove borders. Since the colon is deformable and CT scans capture a single time point, there is no ground truth correspondence

between OC and VC. The VC data uses triangulated meshes from abdominal CT scans similar to [18]. Flythroughs were generated using Blender with two lights on both sides of the camera to replicate a colonoscope. Additionally, the inverse square fall-off property was applied to accurately simulate lighting conditions in OC. A total of 3000 VC and OC frames were extracted. 1500 were used for training while 900 and 600 were used for validation and testing.

4   Methods

CLTS-GAN is composed of two generators and three discriminators. One generator, G, uses OC to predict VC with two corresponding noise parameters. The first parameter, z_{ts}, is a number of matrices that represent texture and specular reflection information. The second parameter, z_{cl}, is a 1D vector that contains color and lighting information. The second generator, F, uses z_{ts} and z_{cl} to transform a VC image into a realistic OC image. Figure 1a shows how the noise values are used in F. z_{cl} is incorporated using AdaIn layers, which globally affects the latent features. z_{ts} is directly added to latent features offering localized information. The complete objective function for the network is defined as:

$$\mathcal{L}_{obj} = \lambda_{adv}\mathcal{L}_{adv} + \lambda_{cyc}\mathcal{L}_{cyc} + \lambda_t\mathcal{L}_t + \lambda_{idt}\mathcal{L}_{idt} \tag{1}$$

Cycle consistency is used in many image-to-image translation models and ensures features from the input are present in the output when transformed. The cycle consistency loss used for OC is shown in Fig. 1b and defined as:

$$\mathcal{L}_{cyc}^{OC}(G, F, A) = \mathbb{E}_{x \sim p(A)}\|x - F(G_{im}(x), G_{cl}(x), G_{ts}(x))\|_1 \tag{2}$$

where $x \sim p(A)$ represents a data distribution and G_{im}, G_{cl} and G_{ts} represents G's output. Since G has additional outputs, the cycle consistency loss should incorporate these extra vectors as seen in Fig. 1c.

$$\mathcal{L}_{cyc}^{VC}(G, F, A, Z) = \mathbb{E}_{x \sim p(A), z \sim p(Z)}\|x - G_{im}(F(x, z_{cl}, z_{ts}))\|_1 + \\ \|z_{cl} - G_{cl}(F(x, z_{cl}, z_{ts}))\|_1 + \tag{3} \\ \|z_{ts} - G_{ts}(F(x, z_{cl}, z_{ts}))\|_1$$

The cycle consistency component of the objective loss function is defined as:

$$\mathcal{L}_{cyc} = \mathcal{L}_{cyc}^{OC}(G, F, OC) + \mathcal{L}_{cyc}^{VC}(G, F, VC, Z) \tag{4}$$

Each generator has a discriminator, D, which adds an adversarial loss so the output resembles the output domain. The adversarial loss for each GAN is:

$$\mathcal{L}_{GAN}(G, D, A, B) = \mathbb{E}_{y \sim p(B)}\big[\log(D(y))\big] + \mathbb{E}_{x \sim p(A)}\big[\log(1 - D(G(x)))\big], \tag{5}$$

G has noise vectors in its output so an additional discriminator is required. Rather than distinguishing noise values, a discriminator is applied to recreated images since our concern lies with the imaging rather than the noise. The discriminator compares images produced by random noise vectors and vectors produced by F. This adversarial loss is shown in Fig. 1b and is defined as:

$$\mathcal{L}_{GAN}^{rec}(G, F, D, A) = \mathbb{E}_{x \sim p(A)}\big[\log(D(F(G_{im}(x), G_{cl}(x), G_{ts}(x))))\big] +$$
$$\mathbb{E}_{x \sim p(A), z \sim p(Z)}\big[\log(1 - D(F(G_{im}(x), z_{cl}, z_{ts})))\big], \tag{6}$$

The adversarial portion of the objective loss is as follows:

$$\mathcal{L}_{adv} = \mathcal{L}_{GAN}(G, D_G, OC, VC) + \mathcal{L}_{GAN}(F, D_F, VC, OC) +$$
$$\mathcal{L}_{GAN}^{rec}(G, F, D_{rec}, OC) \tag{7}$$

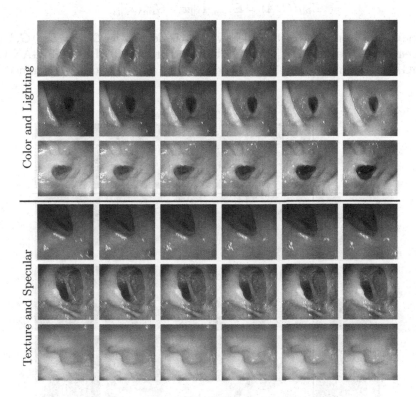

Fig. 2. To understand how z_{cl} and z_{ts} affect the output, z_{cl} and z_{ts} are individually linearly interpolated. The top half shows interpolation between z_{cl} values, while z_{ts} is fixed. The colon-specific color and lighting gradually changes with z_{cl}. The bottom half shows z_{cl} fixed, while z_{ts} is interpolated. The specular reflection shapes and texture gradually change. The last row also shows fecal matter changing between images.

During training, F may ignore z_{ts}. To encourage using noise input, $\mathcal{L}_t$ is added to penalize the network when different noise inputs have similar results. The function penalizing the network is defined as:

$$\mathcal{L}_{text}(I_1, I_2) = \begin{cases} \alpha - \|I_1 - I_2\|_1 & \text{if } \alpha > \|I_1 - I_2\|_1 \\ 0 & \text{else} \end{cases}$$

where I is an image and α they differ. F is applied to two different images, and the OC images are compared using L_t as seen in Fig. 1c and defined as:

$$\mathcal{L}_t = \mathbb{E}_{x \sim p(VC), z \sim p(Z)} \mathcal{L}_{text}(F(x, z_{cl}, z_{ts}^1), F(x, z_{cl}, z_{ts}^2))$$

Lastly, an identity loss is added for stability. An image should be unchanged if the input is from the output domain. It is only applied to G to encourage texture and specular reflection generation. The identity loss is defined as:

$$\mathcal{L}_{idt}(G, A) = \mathbb{E}_{x \sim p(A)} \|x - G_{im}(x)\|_1 \tag{8}$$

The identity portion of the objective loss is defined as $\mathcal{L}_{idt} = \mathcal{L}_{idt}(G, VC)$. The generators are ResNets [8] with 9 blocks that use 23 MB. CLTS-GAN uses PatchGAN discriminators [11], each using 3MB. The network was trained for

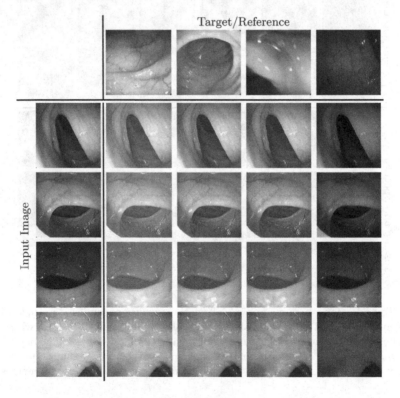

Fig. 3. Showing the z_{cl} vector being extracted from various reference images (top most) and applied to target images (left most) to transfer its colon-specific color and lighting.

200 epochs on an Nvidia RTX 6000 GPU with the following parameters: $\lambda_{adv} = 1, \lambda_T = 10, \lambda_{text} = 20, \lambda_{idt} = 1$, and $\alpha = .1$. Inference time is .04 s.

CLTS-GAN controls the output using z_{ts} and z_{cl}. For VC, if two z_{cl} values are selected with a fixed z_{ts} they can be linearly interpolated and passed into F creating gradual changes in the colon-specific color and lighting as seen in Fig. 2. The strength of the specular reflections change with z_{cl} since the lighting is being altered. Similarly, z_{ts} can be linearly interpolated to provide gradual changes in texture and specular reflection as well as fecal matter. Here the shape of the specular reflections and texture fade in and out. Since changes in z_{ts} and z_{cl} do not lead to sporadic changes, they can be used in more meaningful ways.

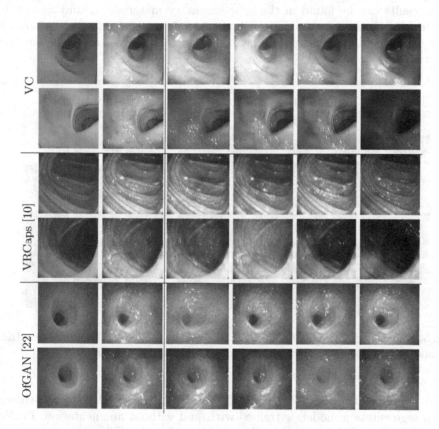

Fig. 4. Depicting our model using various z_{cl} and z_{ts} values to generate realistic OC images. The left most image is the input image for CLTS-GAN followed by the output OC images. We show results on VC, VRCaps [10] data, and OfGAN [22] synthetic input. Additional results can be found in Fig. 1 of the supplementary material.

Figure 3 shows the transfer of colon-specific color and lighting information from one OC image to another. G extracts the z_{cl} vector from the reference and the VC and z_{ts} from the target. When these values are input to F it transfers the color and lighting from the reference to the target. z_{ts} remains fixed since it is intended for generating realistic textures and specular for VC instead of altering geometry dependant texture and specular of OC.

5   Results and Discussion

Figure 4 shows qualitative results for CLTS-GAN's realistic OC generation using VC images and data from VRCaps [10] and OfGAN [22]. The input was passed to F with z_{ts} and z_{cl} randomly sampled from a uniform distribution to show a large variety in colon-specific color, lighting, texture and specular reflection. More results can be found in the supplementary material. z_{ts} and z_{cl} can be individually changed to control the texture and specular reflection separately from the color and lighting as shown in Figs. 2 and 3 of the supplementary material.

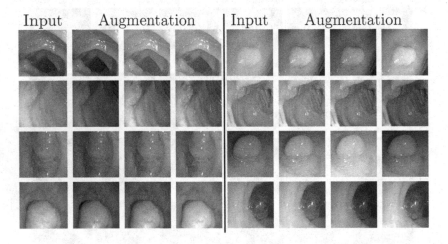

Fig. 5. Augmented data from CVC Clinic DB [2]. The images go through G to extract VC and z_{ts}. z_{cl} is sampled from a uniform distribution and passed into F.

To show quantitative evaluation of CLTS-GAN, PraNet [6], a state-of-the-art polyp segmentation model, is trained with and without augmentation. PraNet uses CVC Clinic DB [2] and HyperKvasir [4] for training. The images were augmented with colon-specific color and lighting, while polyp specific textures and speculars were preserved. Random z_{cl} values are applied to training images by extracting the VC and z_{ts} using G and passing the three values to F. Examples are shown in Fig. 5. PraNet was trained having each image augmented 0, 1, and 3 times. When there was no augmentation or one augmentation the network was trained for 20 epochs. To avoid overfitting on the shapes of the polyps, the network was trained for 10 epochs when augmented 3 times. Testing results are shown in Table 1. Data augmentation from CLTS-GAN improves the DICE, IoU, and MAE scores for various testing datasets. For the CVC-T dataset, using only one augmentation appeared to have marginal improvement over using 3.

Table 1. PraNet results with and without dataset augmentation. Colon-specific color and lighting augmentation was applied to avoid altering polyp specific textures. Results for 1 and 3 additional images are shown in the second and third rows. Both show improvement over PraNet without augmentation. PraNet with 1 augmentation is better for CVC-T which indicates the network may have overfit on the shapes of polyps.

	CVC-Colon DB [3]			ETIS [20]			CVC-T [21]		
	Dice↑	IoU↑	MAE↓	Dice↑	IoU↑	MAE↓	Dice↑	IoU↑	MAE↓
PraNet w/out Aug	0.712	0.640	0.043	0.628	0.567	0.031	0.871	0.797	0.10
PraNet w/ 1 Aug	0.750	0.671	0.037	0.704	0.626	**0.019**	**0.893**	**0.824**	**0.007**
PraNet w/ 3 Aug	**0.781**	**0.697**	**0.030**	**0.710**	**0.639**	0.027	0.884	0.815	0.010

In this work we present CLTS-GAN, a one-to-many image-to-image translation model for dataset augmentation and OC synthesis with control over color, lighting, texture, and specular reflections. z_{ts} and z_{cl} control these attributes, but can be further disentangled. High intensity specular reflections can be extracted with a loss and stored in a separate parameter. CLTS-GAN does not contain temporal components. Adding multiple frames as input can get the network to use the texture and specular information in a temporally consistent manner. Moreover, in the future, we will also explore the utility of CLTS-GAN augmentations in depth inference [15,17] and folds detection [16]. We hypothesize that the full gamut of color-lighting-texture-specular augmentations can be used in these scenarios to improve performance.

Acknowledgements. This project was supported by MSK Cancer Center Support Grant/Core Grant (P30 CA008748) and NSF grants CNS1650499, OAC1919752, and ICER1940302.

References

1. Almahairi, A., Rajeswar, S., Sordoni, A., Bachman, P., Courville, A.: Augmented CycleGAN: learning many-to-many mappings from unpaired data. arXiv preprint arXiv:1802.10151 (2018)
2. Bernal, J., Sánchez, F.J., Fernández-Esparrach, G., Gil, D., Rodríguez, C., Vilariño, F.: WM-DOVA maps for accurate polyp highlighting in colonoscopy: validation vs. saliency maps from physicians. Comput. Med. Imaging Graph. **43**, 99–111 (2015)
3. Bernal, J., Sánchez, J., Vilarino, F.: Towards automatic polyp detection with a polyp appearance model. Pattern Recogn. **45**(9), 3166–3182 (2012)
4. Borgli, H., et al.: Hyper-Kvasir: a comprehensive multi-class image and video dataset for gastrointestinal endoscopy (2019). https://doi.org/10.31219/osf.io/mkzcq

5. Chu, C., Zhmoginov, A., Sandler, M.: CycleGAN, a master of steganography. arXiv preprint arXiv:1712.02950 (2017)

6. Fan, D.-P., et al.: PraNet: parallel reverse attention network for polyp segmentation. In: Martel, A.L., et al. (eds.) MICCAI 2020. LNCS, vol. 12266, pp. 263–273. Springer, Cham (2020). https://doi.org/10.1007/978-3-030-59725-2_26

7. Fazlollahi, A.M., et al.: Effect of artificial intelligence tutoring vs expert instruction on learning simulated surgical skills among medical students: a randomized clinical trial. JAMA Netw. Open **5**(2), e2149008–e2149008 (2022)

8. He, K., Zhang, X., Ren, S., Sun, J.: Deep residual learning for image recognition. In: Proceedings of the IEEE Conference on Computer Vision and Pattern Recognition, pp. 770–778 (2016)

9. Huang, X., Liu, M.Y., Belongie, S., Kautz, J.: Multimodal unsupervised image-to-image translation. In: Proceedings of the European Conference on Computer Vision (ECCV), pp. 172–189 (2018)

10. İncetan, K., et al.: VR-Caps: a virtual environment for capsule endoscopy. Med. Image Anal. **70**, 101990 (2021)

11. Isola, P., Zhu, J.Y., Zhou, T., Efros, A.A.: Image-to-image translation with conditional adversarial networks. In: Proceedings of the IEEE Conference on Computer Vision and Pattern Recognition, pp. 1125–1134 (2017)

12. Karras, T., Laine, S., Aila, T.: A style-based generator architecture for generative adversarial networks. In: Proceedings of the IEEE/CVF Conference on Computer Vision and Pattern Recognition, pp. 4401–4410 (2019)

13. Ma, R., Wang, R., Pizer, S., Rosenman, J., McGill, S.K., Frahm, Jan-Michael.: Real-time 3D reconstruction of colonoscopic surfaces for determining missing regions. In: Shen, D., et al. (eds.) MICCAI 2019. LNCS, vol. 11768, pp. 573–582. Springer, Cham (2019). https://doi.org/10.1007/978-3-030-32254-0_64

14. Ma, R., et al.: RNNSLAM: reconstructing the 3D colon to visualize missing regions during a colonoscopy. Med. Image Anal. **72**, 102100 (2021)

15. Mahmood, F., Chen, R., Durr, N.J.: Unsupervised reverse domain adaptation for synthetic medical images via adversarial training. IEEE Trans. Med. Imaging **37**(12), 2572–2581 (2018)

16. Mathew, S., Nadeem, S., Kaufman, A.: FoldIt: haustral folds detection and segmentation in colonoscopy videos. In: International Conference on Medical Image Computing and Computer-Assisted Intervention, pp. 221–230 (2021)

17. Mathew, S., Nadeem, S., Kumari, S., Kaufman, A.: Augmenting colonoscopy using extended and directional cycleGAN for lossy image translation. In: Proceedings of the IEEE/CVF Conference on Computer Vision and Pattern Recognition, pp. 4696–4705 (2020)

18. Nadeem, S., Kaufman, A.: Computer-aided detection of polyps in optical colonoscopy images. SPIE Med. Imaging **9785**, 978525 (2016)

19. Rivoir, D., et al.: Long-term temporally consistent unpaired video translation from simulated surgical 3D data. In: Proceedings of the IEEE/CVF International Conference on Computer Vision, pp. 3343–3353 (2021)

20. Silva, J., Histace, A., Romain, O., Dray, X., Granado, B.: Toward embedded detection of polyps in wce images for early diagnosis of colorectal cancer. Int. J. Comput. Assist. Radiol. Surg. **9**(2), 283–293 (2014)

21. Vázquez, D., et al.: A benchmark for endoluminal scene segmentation of colonoscopy images. J. Healthc. Eng. **2017** (2017)

22. Xu, J., et al.: OfGAN: realistic rendition of synthetic colonoscopy videos. In: Martel, A.L., et al. (eds.) MICCAI 2020. LNCS, vol. 12263, pp. 732–741. Springer, Cham (2020). https://doi.org/10.1007/978-3-030-59716-0_70

23. Zhang, Y., Wang, S., Ma, R., McGill, S.K., Rosenman, J.G., Pizer, Stephen M..: Lighting enhancement aids reconstruction of colonoscopic surfaces. In: Feragen, A., Sommer, S., Schnabel, J., Nielsen, M. (eds.) IPMI 2021. LNCS, vol. 12729, pp. 559–570. Springer, Cham (2021). https://doi.org/10.1007/978-3-030-78191-0_43

Adaptation of Surgical Activity Recognition Models Across Operating Rooms

Ali Mottaghi[1,2(✉)], Aidean Sharghi[1], Serena Yeung[2], and Omid Mohareri[1]

[1] Intuitive Surgical Inc., Sunnyvale, CA, USA
[2] Stanford University, Stanford, CA, USA
mottaghi@stanford.edu

Abstract. Automatic surgical activity recognition enables more intelligent surgical devices and a more efficient workflow. Integration of such technology in new operating rooms has the potential to improve care delivery to patients and decrease costs. Recent works have achieved a promising performance on surgical activity recognition; however, the lack of generalizability of these models is one of the critical barriers to the wide-scale adoption of this technology. In this work, we study the generalizability of surgical activity recognition models across operating rooms. We propose a new domain adaptation method to improve the performance of the surgical activity recognition model in a new operating room for which we only have unlabeled videos. Our approach generates pseudo labels for unlabeled video clips that it is confident about and trains the model on the augmented version of the clips. We extend our method to a semi-supervised domain adaptation setting where a small portion of the target domain is also labeled. In our experiments, our proposed method consistently outperforms the baselines on a dataset of more than 480 long surgical videos collected from two operating rooms.

Keywords: Surgical activity recognition · Semi-supervised domain adaptation · Surgical workflow analysis

1 Introduction

Surgical workflow analysis is the task of understanding and describing a surgical process based on videos captured during the procedure. Automatic surgical activity recognition in operating rooms provides information to enhance the efficiency of surgeons and OR staff, assess the surgical team's skills, and anticipate failures [17]. Video activity recognition models have been previously utilized for this task; however, one of the main drawbacks of previous works is the lack of generalizability of the trained machine learning models. Models trained on

---

Supplementary Information The online version contains supplementary material available at https://doi.org/10.1007/978-3-031-16449-1_51.

videos from one operating room perform poorly in a new operating room with a distinct environment [13,14]. This prevents the widespread adoption of these approaches for scalable surgical workflow analysis.

This paper explores approaches for adapting a surgical activity recognition model from one operating room to a new one. We consider two cases where we have access to only unlabeled videos on the target operating room and the case where a small portion of target operating room videos is annotated. Our new approach generates pseudo labels for unlabeled video clips on which the model is confident about the predictions. It then utilizes an augmented version of pseudo-annotated video clips and originally annotated video clips for training a more generalizable activity recognition model. Unlike most previous works in the literature that study semi-supervised learning and domain adaptation separately, we propose a unified solution for our semi-supervised domain adaptation problem. This allows our method to exploit labeled and unlabeled data on the source and the target and achieve more generalizability on the target domain.

Our work is the first to study semi-supervised domain adaptation in untrimmed video action recognition. Our setting of untrimmed video action recognition (1) requires handling video inputs, which limits the possible batch size and ability to estimate prediction score distributions as needed by prior image classification works [3]; and (2) requires handling highly imbalanced data, which causes model collapse when using the pseudo labeling strategies in prior image classification works. Our method addresses the first challenge by introducing the use of prediction queues to maintain better estimates of score distributions and the second challenge through pretraining and sampling strategies of both video clips and pseudo labels to prevent model collapse. Additionally, we use a two-step training approach as described in Sect. 3.2.

A dataset of more than 480 full-length surgical videos captured from two operating rooms is used to evaluate our method versus the baselines. As we show in our experiments, our new method outperforms existing domain adaptation approaches on the task of surgical action recognition across operating rooms since it better utilizes both annotated and unannotated videos from the new operating room.

2   Related Works

Surgical Activity Recognition also known as surgical phase recognition, is the task of finding the start and end time of each surgical action given a video of a surgical case. [16,19,20]. This task has been studied for both laparoscopic videos [4,7,19,20] and videos captured in the operating room [13,14]. These videos are often hours long, containing several activities with highly variable lengths. Unlike trimmed video action recognition, which is studied extensively in computer vision literature, untrimmed video action recognition does not assume that the action of interest takes nearly the entire video duration. Despite the relative success of surgical activity recognition models, previous works [13,14] have shown that these models suffer from the domain shift as a model trained in one OR performs significantly worse in a new OR. In this work, we address this problem by proposing a new method for training a more generalizable model with minimal additional annotations from the new OR.

Unsupervised Domain Adaptation (UDA) is the problem of generalizing the model trained on the source domain to the target domain where we only have unlabeled data. The goal is to achieve the highest performance on the target domain without any labeled examples on target. Most of the works in UDA literature focus on reducing the discrepancy of representations between source and target domains. For example, [9] and [10] use the maximum mean discrepancy to align the final representations while [8,15] suggest matching the distribution of intermediate features in deep networks. Maximum classifier discrepancy (MCD) [12] has been proven more successful where two task classifiers are trained to maximize the discrepancy on the target sample. Then the features generator is trained to minimize this discrepancy.

Semi-Supervised Domain Adaptation (SSDA) setting has access to a labeled set on the target domain, unlike UDA. The goal is still achieving the highest performance on the target domain, however, with the use of both labeled and unlabeled data on source and target. SSDA is less explored than DA, and most of the works have focused on image classification tasks [1,11,18]. For example, in Minimax Entropy (MME) [11] approach, adaptation is pursued by alternately maximizing the conditional entropy of unlabeled target data with respect to the classifier and minimizing it with respect to the feature encoder. Recently, [3] suggested AdaMatch, where the authors extend FixMatch to the SSDA setting. Our work borrows some ideas from this work, but unlike AdaMatch, we focus on video action recognition. We also proposed a new method for addressing the long-tailed distribution of data on the target domain, which is very common in surgical activity recognition.

3   Method

3.1   Unsupervised Domain Adaptation (UDA)

Notation. Let $V_s = \{v_s^{(1)}, \ldots, v_s^{(n_s)}\}$ be the set of videos in the source domain and $V_t = \{v_t^{(1)}, \ldots, v_t^{(n_t)}\}$ be videos on target where n_s and n_t are the number of source and target videos. As is common in UDA, we assume source data is labeled while the data from the target is unlabeled. Denote the set of surgical activities as $C = \{c_1, \ldots, c_K\}$ where K is the number of surgical activities of interest. For each source video $v_s^{(i)}$, we are given a set of timestamps $s_s^{(i)} = \{t_{c_1}, \ldots, t_{c_k}\}$ where t_{c_j} is defined as the start and end time for activity c_j.

Sampling Video Clips. We start with sampling short video clips from long and variable length videos. Let $x_s \in \mathbb{R}^{H \times W \times 3 \times M}$ denote a sample clip from source videos and x_t be a sample clip from the target. All sampled clips have a fixed height H, width W, and number of frames M. We sample clips uniformly from surgical activities for labeled videos to ensure each clip only contains one

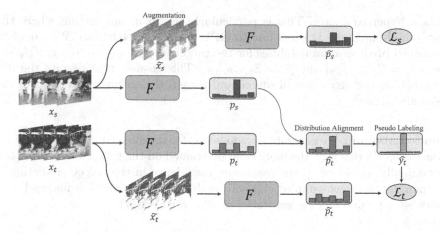

Fig. 1. Our proposed domain adaptation method. First, the labeled video clip from the source x_s and the unlabeled video clip form the target x_t are augmented to create $\tilde{x}_s$ and $\tilde{x}_t$ respectively. Then all four video clips are fed into the clip-based activity recognition model F to generate probability distributions p_s, $\tilde{p}_s$, p_t, and $\tilde{p}_t$. For the source x_s, we already have the label, so we compute the $\mathcal{L}_s$. For the target, we generate and sample a pseudo label as described in Sect. 3.1 before computing $\mathcal{L}_t$. Note that the loss is computed only for the augmented version of the video clips to improve generalizability.

activity. Let $y_s \in \mathcal{C}$ be the corresponding label for x_s. We don't have any labels for target clip x_{TU}.

Augmentations. In our method, we feed both the sampled video clip x and an augmented version $\tilde{x}$ to our model. Augmentation is performed both frame-by-frame and temporally. For frame-wise augmentation, we use RandAugment [6], and we change the playback speed for temporal augmentation. More formally, with probability p, the video clip is up-sampled or down-sampled by a factor α. The activity recognition model F takes an input video clip x and outputs logits $z \in \mathbb{R}^K$ for each surgical activity. For augmented clip from source we have $\tilde{z}_s = F(\tilde{x}_s)$ and $\tilde{p}_s = softmax(\tilde{z}_s)$ where $\tilde{p}_s$ is a vector of predicted probabilities. Similarly, we define z_s, z_t, $\tilde{z}_t$, p_s, p_t, and $\tilde{p}_t$.

Distribution Alignment. Before generating pseudo labels for unlabeled clips from target, we align the distributions of predictions on source and target domains by aligning their expected values. Unlike AdaMatch [3] that uses only one batch to estimate the source and target distributions, we estimate them by maintaining queues of previous predictions $\mathcal{P}_s = \{p_s^{(0)}, p_s^{(1)}, \dots\}$, $\mathcal{P}_t = \{p_t^{(0)}, p_t^{(1)}, \dots\}$. $\mathcal{P}_s$ and $\mathcal{P}_t$ are defined on the fly as queues with the current mini-batch enqueued and the oldest mini-batch dequeued. The size of the queues can be much larger than mini-batch size and can be independently and flexibly set as a hyperparameter. This is particularly useful in our setting where the typical mini-batch size is small due to GPU memory restrictions. We calculate the aligned prediction probabilities for the target by $\hat{p}_t = normalize(p_t \times \bar{p}_s/\bar{p}_t)$ where $\bar{p}_s = \mathbb{E}_{p_s \sim \mathcal{P}_s}[p_s]$ and $\bar{p}_t = \mathbb{E}_{p_t \sim \mathcal{P}_t}[p_t]$. This ensures that despite the difference in input distribution of source and target, the output predictions of the model are aligned.

Pseudo Labeling. We generate pseudo labels for the most confident predictions of the model. In this way, the model can be trained on the target video clips that are originally unlabeled. If the maximum confidence in the target prediction is less than τ, the target sampled clip will be discarded, i.e. $mask = \max(\hat{p}_t) > \tau$, otherwise, a pseudo label is generated as $\hat{y}_t = \arg\max(\hat{p}_t)$.

Sampling Pseudo Labels. In practice, the generated pseudo labels have an imbalanced distribution as the model is usually more confident about the dominant or easier classes. If we train the model on all generated pseudo labels, the classifier could predict the most prevalent class too often or exhibit other failure modes. We only use a subset of pseudo labels based on their distribution to mitigate this problem. We maintain another queue $\mathcal{Y}_t = \{\hat{y}_t^{(0)}, \hat{y}_t^{(1)}, \dots\}$ to estimate the distribution of generated pseudo labels. Let Q be the vector of frequencies of pseudo labels in $\mathcal{Y}_t$ where Q_i is the number of repeats for class i. We sample generated pseudo labels with the probability proportional to the inverse of frequency of that class, i.e. $sample \sim bernoulli(\min(Q)/Q_{\hat{y}_t})$. This adds more representation to the hard or infrequent classes on target.

Loss Functions. The overview of our domain adaptation method is depicted in Fig. 1. To sum up the discussion on UDA, we define loss function $\mathcal{L}$:

$$\mathcal{L}_s = \mathbb{E}_{x_s \sim X_s, y_s \sim Y_s}[H(y_s, \tilde{z}_s)]$$
$$\mathcal{L}_t = \mathbb{E}_{x_s \sim X_s, x_t \sim X_t, y_s \sim Y_s}[H(\text{stopgrad}(\hat{y}_t), \tilde{z}_t) \cdot mask \cdot sample]$$
$$\mathcal{L} = \mathcal{L}_s + \lambda \mathcal{L}_t$$

where H is the categorical cross-entropy loss function and stopgrad stops the gradients from back-propagating. λ is a hyperparameter that weighs the domain adaptation loss. We use a similar scheduler as [3] to increase this weight during the training. Note that the augmented video clips are utilized during the back-propagation, and the original video clips are used only to generate pseudo labels.

3.2   Training on Untrimmed Surgical Videos

Clip-based activity recognition models cannot model long-term dependencies in long untrimmed videos. However, surgical videos captured in operating rooms usually have a natural flow of activities during a full-length surgery. We train a temporal model on top of the clip-based activity recognition model to capture

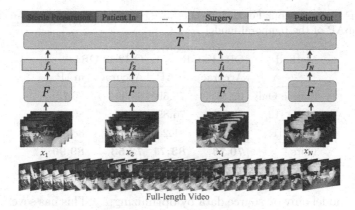

Fig. 2. The original operating room videos are usually long, so we split them into short video clips $(x_1, x_2, \ldots, x_N)$ and feed them into a trained clip-based activity recognition model F. Next, the extracted features $(f_1, f_2, \ldots, f_N)$ are used to train a temporal model T, which predicts the surgical activity of each frame in the original video given the temporal context.

these long-term dependencies. As depicted in Fig. 2, first, we extract features from all videos by splitting long videos into clips with M frames and feeding them into the trained clip-based model. Features are extracted from the last hidden layer of model F. Let $f \in \mathbb{R}^{D \times N}$ be the features extracted from video $v \in \mathbb{R}^{H \times W \times 3 \times N}$ and $l \in \mathbb{R}^{K \times N}$ be the corresponding ground-truth label, where N is the number of frames in the video and D is the latent dimension of the clip-based model. We train the temporal model by minimizing $\mathcal{L}_{temporal} = H(l, \hat{l})$ where $\hat{l} = T(f)$ is the prediction and H is the binary cross-entropy loss.

3.3   Extension to Semi-supervised Domain Adaptation (SSDA)

Our domain adaption method can be extended to SSDA setting. In SSDA, we assume the source videos $\mathcal{V}_s$ are fully labeled, while only a part of target videos $\mathcal{V}_{tl}$ are labeled; the rest are denoted by $\mathcal{V}_{tu}$ are unlabeled. Therefore, in addition to sampling video clips x_s and x_{tu} as described in Sect. 3.1, we also sample video clips x_{tl} from target labeled videos. This adds an additional term $\mathcal{L}_{tu}$ to the loss function:

$$\mathcal{L}_{tl} = \mathbb{E}_{x_{tl} \sim X_{tl}, y_{tl} \sim Y_{tl}} [H(y_{tl}, \tilde{z}_{tl})]$$

The total loss function is $\mathcal{L} = \mathcal{L}_s + \mathcal{L}_{tl} + \lambda \mathcal{L}_{tu}$ where $\mathcal{L}_s$ and $\mathcal{L}_{tu}$ are defined as before. Our distribution alignment and pseudo labeling strategies are similar to UDA setting.

3.4   The Importance of Pretraining

We pretrain our activity recognition model to boost its performance in domain adaptation training. More specifically, in the beginning, we train the activity

Table 1. UDA experiments. For each method, we report the accuracy of the clip-based model and mAP of the temporal model.

Method	OR1 to OR2		OR2 to OR1	
	Accuracy	mAP	Accuracy	mAP
Source Only	62.13	76.39	66.99	86.72
MCD [12]	63.29	76.89	65.17	86.07
MME [11]	67.87	81.20	68.02	88.96
Ours	**70.76**	**83.71**	**73.53**	**89.96**

recognition model only on source data by optimizing $\mathcal{L}_s$. This has several advantages: 1) Predicted probabilities are more meaningful since the beginning of the training. As a result, generated pseudo labels will be more accurate and reliable. 2) Distributions are estimated more accurately; therefore, we can better align distributions and sample more representative pseudo labels. 3) If we train the temporal model on source data, we can also use its predictions on unlabeled targets for sampling video clips. Videos are usually long, containing multiple activities with highly variable lengths. Therefore, we sample a more uniform set of video clips covering all surgical activities with an approximate segmentation of surgical activities. We achieve this by first sampling uniformly from surgical activities and then sampling short video clips from each.

4   Experiments

Dataset. We use a dataset of 484 full-length surgery videos captured from two robotic ORs equipped with Time-of-Flight sensors. The dataset covers 28 types of procedures performed by 16 surgeons/teams using daVinci Xi surgical system. 274 videos are captured in the first OR, and the remaining 210 videos are from the second OR with a distinct layout and type of procedures and teams. These videos are, on average, 2 h long and are individually annotated with ten clinically significant activity classes such as sterile preparation, patient roll-in, etc. Our classes are highly imbalanced; patient preparation class contains ten times more frames than robot docking class.

Model Architectures and Hyperparameters. We use a TimeSformer [2] model with proposed hyperparameter as our clip-based activity recognition model which operates on $224 \times 224 \times 3 \times 16$ video clips. Clips are augmented with RandAugment [6] with the magnitude of 9 and standard deviation of 0.5, as well as temporal augmentation with a factor of 2 and probability of 0.5. Given the size of our dataset, we use a GRU [5] as our temporal model. We generate pseudo labels with $\tau = 0.9$ and keep a queue of 1000 previous predictions and pseudo labels for distribution alignment and sampling. We set $\lambda = 1$ in our experiments. The sensitivity analysis of hyperparameters and further implementation details of our approach are available in the supplementary material.

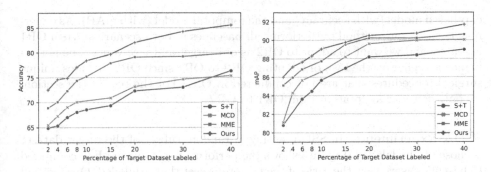

Fig. 3. Performance of different methods in the SSDA setting. The left figure shows the accuracy of the clip-based activity recognition model, and the right figure shows the corresponding mAP of the temporal model trained on extracted features. S+T denotes training a supervised model on the source plus the labeled part of the target dataset.

Table 2. Ablation study on two key components of our model.

Case	Distribution alignment	Sampling pseudo labels	Accuracy	mAP
Only pseudo labeling			55.20	71.98
AdaMatch [3]	✓		68.01	82.13
Semi-supervised learning		✓	62.04	80.78
Ours	✓	✓	**70.76**	**83.71**

Baselines. We compare the performance of our domain adaptation strategy to three baselines: 1)*Supervised* method serves as a fundamental baseline where only labeled data is used during the training. 2)*Maximum Classifier Discrepancy (MCD)* [12] is a well-known domain adaption method that has been applied to various computer vision tasks. 3)*Minimax Entropy (MME)* [11] is specially tailored to the SSDA settings where adversarial training is used to align features and estimate prototypes. Our method borrows some components from AdaMatch [3], and we study the effect of each component in our ablation studies.

Evaluation. The performance of each method is evaluated based on the accuracy of the clip-based activity recognition model and the mean average precision (mAP) of the temporal model on the target domain. The accuracy of the activity recognition model evaluated on class-balanced sampled video clips acts as a proxy for the informativeness of the extracted features and is more aligned with most prior works focused on training a classifier. In all experiments, we start from a pre-trained model as discussed in Sect. 3.4.

UDA Experiments. Table 1 shows the performance of our method compared to the baselines in UDA setting. We measure the performance of both the clip-based model (with accuracy) and the temporal model (with mAP). As experiments show, our method outperforms all baselines in both scenarios: when OR1 is source, and we want to adapt to OR2 as our target and vice versa. The performance is higher on the task of adapting OR2 to OR1 since OR2 includes unique surgical procedures that are not conducted in OR1. For the rest of the section, we only consider adaptation from OR1 to OR2.

SSDA Experiments. In SSDA setting, a random subset of the target dataset is chosen to be labeled. Figure 3 shows the performance of our method compared to baselines as we vary the ratio of the target dataset that is labeled. Our method outperforms all of the baselines on our surgical activity recognition task.

Ablation Study. We perform an ablation study to better understand the importance of each component in our method. In Table 2, we analyze the effect of distribution alignment and sampling pseudo labels as the key components in our method. As we discussed in Sect. 3.1, the classifier could predict the dominant class on target without these components. Although AdaMatch [3] uses distribution alignment to generate pseudo labels, they are usually unbalanced. We show that in datasets with a long-tail distribution, we also need to sample them to ensure more representative and balanced pseudo labels for training. We conduct our study in the UDA setting with OR1 used as source. Please see supplementary materials for more details.

5   Conclusion

In this paper, we studied the generalizability of surgical workflow analysis models as these models are known to suffer from a performance drop when deployed in a new environment. We proposed a new method for domain adaption, which relies on generating pseudo labels for the unlabeled videos from the target domain and training the model using the most confident ones. We showed that our method trains a more generalizable model and boosts the performance in both UDA and SSDA settings. Furthermore, our method is model-agnostics, as a result, it could be applied to other tasks using a suitable clip-based activity recognition and temporal model. For example, one of the areas is surgical activity recognition in endoscopic videos. We hope that our approach will inspire future work in medical machine learning to develop models that are more generalizable.

References

1. Ao, S., Li, X., Ling, C.: Fast generalized distillation for semi-supervised domain adaptation. In: Proceedings of the AAAI Conference on Artificial Intelligence, vol. 31 (2017)
2. Bertasius, G., Wang, H., Torresani, L.: Is space-time attention all you need for video understanding. arXiv preprint arXiv:2102.05095 (2021)

3. Berthelot, D., Roelofs, R., Sohn, K., Carlini, N., Kurakin, A.: Adamatch: A unified approach to semi-supervised learning and domain adaptation. arXiv preprint arXiv:2106.04732 (2021)
4. Chen, W., Feng, J., Lu, J., Zhou, J.: Endo3D: online workflow analysis for endoscopic surgeries based on 3D CNN and LSTM. In: Stoyanov, D., et al. (eds.) CARE/CLIP/OR 2.0/ISIC -2018. LNCS, vol. 11041, pp. 97–107. Springer, Cham (2018). https://doi.org/10.1007/978-3-030-01201-4_12
5. Chung, J., Gulcehre, C., Cho, K., Bengio, Y.: Empirical evaluation of gated recurrent neural networks on sequence modeling. arXiv preprint arXiv:1412.3555 (2014)
6. Cubuk, E.D., Zoph, B., Shlens, J., Le, Q.V.: RandAugment: practical automated data augmentation with a reduced search space. In: Proceedings of the IEEE/CVF Conference on Computer Vision and Pattern Recognition Workshops, pp. 702–703 (2020)
7. Funke, I., Jenke, A., Mees, S.T., Weitz, J., Speidel, S., Bodenstedt, S.: Temporal coherence-based self-supervised learning for laparoscopic workflow analysis. In: Stoyanov, D., et al. (eds.) CARE/CLIP/OR 2.0/ISIC -2018. LNCS, vol. 11041, pp. 85–93. Springer, Cham (2018). https://doi.org/10.1007/978-3-030-01201-4_11
8. Ganin, Y., et al.: Domain-adversarial training of neural networks. J. Mach. Learn. Res. 17(1), 2030–2096 (2016)
9. Long, M., Cao, Y., Wang, J., Jordan, M.: Learning transferable features with deep adaptation networks. In: International Conference on Machine Learning, pp. 97–105. PMLR (2015)
10. Long, M., Zhu, H., Wang, J., Jordan, M.I.: Unsupervised domain adaptation with residual transfer networks. In: Advances in Neural Information Processing Systems 29 (2016)
11. Saito, K., Kim, D., Sclaroff, S., Darrell, T., Saenko, K.: Semi-supervised domain adaptation via minimax entropy. In: Proceedings of the IEEE/CVF International Conference on Computer Vision, pp. 8050–8058 (2019)
12. Saito, K., Watanabe, K., Ushiku, Y., Harada, T.: Maximum classifier discrepancy for unsupervised domain adaptation. In: Proceedings of the IEEE Conference on Computer Vision and Pattern Recognition, pp. 3723–3732 (2018)
13. Schmidt, A., Sharghi, A., Haugerud, H., Oh, D., Mohareri, O.: Multi-view surgical video action detection via mixed global view attention. In: de Bruijne, Marleen, de Bruijne, M., et al. (eds.) MICCAI 2021. LNCS, vol. 12904, pp. 626–635. Springer, Cham (2021). https://doi.org/10.1007/978-3-030-87202-1_60
14. Sharghi, A., Haugerud, H., Oh, D., Mohareri, O.: Automatic operating room surgical activity recognition for robot-assisted surgery. In: Martel, A.L., et al. (eds.) MICCAI 2020. LNCS, vol. 12263, pp. 385–395. Springer, Cham (2020). https://doi.org/10.1007/978-3-030-59716-0_37
15. Sun, B., Saenko, K.: Deep CORAL: correlation alignment for deep domain adaptation. In: Hua, G., Jégou, H. (eds.) ECCV 2016. LNCS, vol. 9915, pp. 443–450. Springer, Cham (2016). https://doi.org/10.1007/978-3-319-49409-8_35
16. Tran, D.T., Sakurai, R., Yamazoe, H., Lee, J.H.: Phase segmentation methods for an automatic surgical workflow analysis. Int. J. Biomed. Imaging 2017, 1–17 (2017)
17. Vercauteren, T., Unberath, M., Padoy, N., Navab, N.: CAI4CAI: the rise of contextual artificial intelligence in computer-assisted interventions. Proc. IEEE 108(1), 198–214 (2019)
18. Yao, T., Pan, Y., Ngo, C.W., Li, H., Mei, T.: Semi-supervised domain adaptation with subspace learning for visual recognition. In: Proceedings of the IEEE Conference on Computer Vision and Pattern Recognition, pp. 2142–2150 (2015)

19. Yengera, G., Mutter, D., Marescaux, J., Padoy, N.: Less is more: surgical phase recognition with less annotations through self-supervised pre-training of CNN-LSTM networks. arXiv preprint arXiv:1805.08569 (2018)
20. Zia, A., Hung, A., Essa, I., Jarc, A.: Surgical activity recognition in robot-assisted radical prostatectomy using deep learning. In: Frangi, A.F., Schnabel, J.A., Davatzikos, C., Alberola-López, C., Fichtinger, G. (eds.) MICCAI 2018. LNCS, vol. 11073, pp. 273–280. Springer, Cham (2018). https://doi.org/10.1007/978-3-030-00937-3_32

Video-Based Surgical Skills Assessment Using Long Term Tool Tracking

Mona Fathollahi[1], Mohammad Hasan Sarhan[2(✉)], Ramon Pena[3],
Lela DiMonte[1], Anshu Gupta[3], Aishani Ataliwala[4], and Jocelyn Barker[1]

[1] Johnson & Johnson, Santa Clara, CA, USA
mfatholl@its.jnj.com
[2] Johnson & Johnson Medical GmbH, Hamburg, Germany
msarhan@its.jnj.com
[3] Johnson & Johnson, Raritan, NJ, USA
[4] Johnson & Johnson, Seattle, WA, USA

Abstract. Mastering the technical skills required to perform surgery is
an extremely challenging task. Video-based assessment allows surgeons to
receive feedback on their technical skills to facilitate learning and development. Currently, this feedback comes primarily from manual video
review, which is time-intensive and limits the feasibility of tracking a
surgeon's progress over many cases. In this work, we introduce a motion-
based approach to automatically assess surgical skills from surgical case
video feed. The proposed pipeline first tracks surgical tools reliably to
create motion trajectories and then uses those trajectories to predict surgeon technical skill levels. The tracking algorithm employs a simple yet
effective re-identification module that improves ID switch compared to
other state-of-the-art methods. This is critical for creating reliable tool
trajectories when instruments regularly move on- and off-screen or are
periodically obscured. The motion-based classification model employs a
state-of-the-art self-attention transformer network to capture short- and
long-term motion patterns that are essential for skill evaluation. The
proposed method is evaluated on an in-vivo (Cholec80) dataset where
an expert-rated GOALS skill assessment of the Calot Triangle Dissection
is used as a quantitative skill measure. We compare transformer-based
skill assessment with traditional machine learning approaches using the
proposed and state-of-the-art tracking. Our result suggests that using
motion trajectories from reliable tracking methods is beneficial for assessing surgeon skills based solely on video streams.

Keywords: Video based assessment · Surgical skill assessment · Tool
tracking

M. Fathollahi, M. H. Sarhan—Equal contribution.

1   Background and Introduction

Tracking a surgeon's instrument movements throughout a minimally invasive procedure (MIP) is a critical step towards automating the measurement of a surgeon's technical skills. Tool trajectory-based metrics have been shown to correlate with surgeon experience [6], learning curve progression [21], and patient outcome measures [9]. Typically, it is only feasible to calculate these metrics for training simulators or robot-assisted MIP procedures directly from the robot kinematic output. Since most MIP cases worldwide are performed laparoscopically rather than robotically, a broadly applicable solution is needed for instrument tracking. Using computer vision techniques to generate tool positional data would enable calculation of the motion metrics based on surgical case video alone and not rely on data outputs of a specific surgical device.

A Surgeon's technical skills are typically evaluated during video review by experts, using rating scales that assign numerical values to specific characteristics exhibited by the surgeon's movements. The Global Operative Assessment of Laparoscopic Skills (GOALS) scale uses a five-point Likert scale to evaluate a surgeon's depth perception, bimanual dexterity, efficiency, and tissue handling [24]. GOALS scores are sometimes used as a measure for which to compare other methods of surgeon technical skill classification [3]. Instead of relying on manual review, many studies have looked at methods to automate the evaluation of technical skills based on a surgeon's movement patterns [16]. Metrics of interest such as path length, velocity, acceleration, turning angle, curvature, and tortuosity can be calculated and used to distinguish surgeon skill level [5]. Motion smoothness metrics such as jerk have been shown to indicate a lack of mastery of a task, indicating a surgeon's progression along the learning curve [1].

Tool trajectory-based metrics have been calculated using robotic system kinematic output [20,27,30]. Estimating instrument trajectories from the laparoscope video feed using image-based tracking can unlock the ability to further explore automated assessment of technical skills, as no extra sensors or robotic system access are needed. Metrics are then calculated directly from the image-based tracking results [7,11,14,15,19]. However, there are limitations to these approaches that we try to address in this work. For example, a surgical tool detection model is trained in [11] to estimate instrument usage timelines and tool trajectory maps per video. But skill assessment step is not automatic, and each visualization is qualitatively evaluated by a human to estimate surgeon skill. Third-party software in semi-automatic mode is used in [7] to track tools. This is not a feasible solution for processing thousands of videos. The duration of each video in their dataset is 15 s and the dataset is not publicly released. Some approaches register 3D tool motion by employing extra sensors or a special lab setting. For example, [19] use two cameras in an orthogonal configuration in addition to marking the distal end of surgical tools with colored tapes to make them recognizable in video feeds.

Image-based surgical instrument tracking is an attractive solution to the temporal localization problem but requires robustness towards variations in the surgical scene which may lead to missed detections or track identity switches.

Approaches achieve state-of-the-art performance on public tracking benchmarks such as MOTChallenges [17] and OTB datasets [28]. However, when it comes to real-world or long videos, most of them have several shortcomings because they are designed for evaluating short-term tracking methods. For example, the longest sequence in the MOT17 and OTB datasets is only 30 and 129 s respectively. To alleviate these problems and obtain suitable motion features for skill assessment, we propose a video-based framework for extracting motion tracks from surgical videos reliably and using them in surgeon skill assessment. The framework utilizes a novel tracking algorithm that is suitable for long sequence tracking with a minimal amount of identity switches. This is achieved with a tailored cost criterion that takes semantic, spatial, and class similarity into consideration when assigning detections to tracklets. The reliable motion features are utilized in assessing surgeon efficiency in laparoscopic cholecystectomy segment of calot triangle dissection. We compare the tracking method with a state-of-the-art tracking method and validate the motion-based skill assessment using feature extraction and learning-based transformer approaches.

2 Materials and Methods

2.1 Dataset Description

An open-source dataset, Cholec80 [23], was used to perform this analysis. The Cholec80 dataset consists of laparoscopic cholecystectomy surgical case videos performed by 13 surgeons, and are labeled with phase annotations. The Calot Triangle Dissection phase was chosen to be evaluated as it showcases a surgeon's fine dissection skills. To account for the variance of the duration of this phase in different videos, we only consider the last three minutes of Calot Triangle Dissection for all our experiments. To evaluate our tracking algorithm, 15 videos of cholec80 dataset have been annotated at half the original temporal resolution. The rest of the videos were used for training the detection and reIdentification models. In addition, to evaluate our Skill assessment model, two expert surgeons annotate the Calot Triangle Dissection segment of all 80 videos of Cholec80 dataset using the GOALS assessment tool from 1 to 5. The GOALS efficiency category in particular was chosen to be representative of surgeon movement patterns that relate to technical skill progression. Therefore, efficiency scores from the two experts were averaged for each case. We decided to binarize the averaged score due to the limited size of the dataset. With 5 classes, the individual class representation was too small for learning. The cutoff point (3.5) is selected as it shows the best agreement between the annotators. Using this cutoff threshold, 29 cases belong to the low-performing group and the other 51 in the high-performing group. The performance may degrade on different thresholds as the agreement between annotators is lower.

2.2 Tracking Algorithm

Our proposed tracking algorithm belongs to the Tracking-by-detection category to solve the problem of tracking multiple objects. Tracking-by-detection algorithms consist of two steps: (i) objects in the frame are detected independent of other frames. (ii) Tracks are created by linking the corresponding detections across time. A corst or similarity score is calculated for every pair of new detections and active tracks to match new data to tracks through data association. When a detection can not be matched with active tracks, either due to the emergence of a new object or a low similarity score, a new track is initialized. Regarding the tracking, we have two main contributions: 1) We propose a new cost function and 2) a new policy in track recovery.

The proposed tracking algorithm is summarized in Algorithm 1. At each frame, the detection algorithm detects all tools in the frame. We utilize yolov5 [12] that is trained on a subset of CholecT50 [10]. We have annotated the dataset in-house with the bounding box location and class of each surgical instrument present in the scene. This resulted in more than 87k annotated frames and 133k annotated bounding boxes. We use a pre-trained model on the COCO dataset and freeze the model backbone (first 10 layers) while training. Each detection is passed through a re-Identification network to extract appearance features. Furthermore, for each active track, the locations of the corresponding tool are predicted using a Kalman filter. This information is used to construct a cost matrix that is fed to the Hungarian algorithm to assign detections to tracks. The next step is "track recovery" which is mainly matching unassigned detections in the previous step to some inactive tracks if it is possible to do so.

Algorithm 1: Tracking Aglorithm

---

1 **foreach** *frame*(i) **do**
2   *detected_tools* ← detect tools in frame(i);
3   **foreach** $d \in detected\_tools$ **do**
4 | *appearance_features*[d] ← extract appearance features(d);
5   **foreach** $t \in active\_tracks$ **do**
6   **foreach** $d \in detected\_tools$ **do**
7 | | $cost\_matrix[t, d] \leftarrow cost((t, d))$;
8   *assignments* ← run hungarian assignment(*cost_matrix*);
9   **foreach** $d \in unassigned\_detections$ **do**
10   **if** *d is matchable to an inactive track* **then**
11 | recover inacive track;
12   **else**
13 | initialize a new track using d;
14   update/inactive tracks(*assignments*);

Re-Identification Network. The goal of the reidentification network is to describe each image from a tool track with a feature embedding such that the feature distance between any two crops in a single track (a) or (b) is small, while the distance of a pair that is made out of track (a) and (b) is large comparatively.

To accomplish this goal, we train a TriNet [8] architecture to represent each tool bounding box with a 128-dimensional vector.

Cost Function Definition. To be able to associate detected objects with the tracks, first a cost matrix between each active track, t, and new detection d is constructed. Next, the cost matrix is minimized using the Hungarian algorithm [13]. The cost function is defined as Eq. 1 which is a combination of appearance feature, detection, track ID matching, and spatial distance, where t is a track, d is a detection and $D(.,.)$ is the distance function.

$$cost(t,d) = D_{feat}(t,d) + M.\mathbb{1}_{D_{spatial}(t,d)>\lambda_{sp}} + M.\mathbb{1}_{d.DetClass \neq t.classID} \qquad (1)$$

In this equation, the first term represents the dissimilarity between the appearance re-ID feature of track t and detection d. To achieve faster tracking and adapt to the appearance changes of a track across frames, we employed short-term re-identification [2] where only the N recent frames of a track are used to calculate the distance to a new detection in the learned embedding space. The second term is the spatial distance between the center of detection, d, and the predicted Kalman position of the track, t. We assume that surgical tools have moved only slightly between frames, which is usually ensured by high frame rates, 25 fps in our case. If this distance is greater than the λ_{sp} threshold, then the corresponding element in the cost matrix is given a very high number to prohibit this assignment. The last term is referred to 'classID' match. This term adds a high bias, $M = 1000$, to the cost if the predicted class label of detection does not match the class label of the object presented in the track.

Track Recovery. In the data association step, any detection that can not be assigned to an active track is fed into the Track Recovery submodule. In this submodule, the cost of assigning an unassigned detection d to any inactive track t is calculated as follows:

$$cost(t,d) = D_{feat}(t,d).M.\mathbb{1}_{D_{spatial}(t,d)>\lambda_{sp} \wedge d.DetClass \neq t.classID} \qquad (2)$$

This step helps to re-identify the tool that becomes occluded or moves out of the screen. The intuition behind this cost function is that, if the tool emerges from an occluded organ, the detection algorithm usually fails to correctly identify its classID, while their spatial distance can be reliably used to match them.

2.3 Feature Based Skill Assessment

Trajectories from the tracking algorithm can be decomposed into coordinates in the 2D pixel space of the video feed. These trajectories and their variations over time can then be used in the calculation of motion metrics that describe the path of the tool. In our feature-based skill assessment, metrics from the literature [4,18] including distance (path length), velocity, acceleration, jerk,

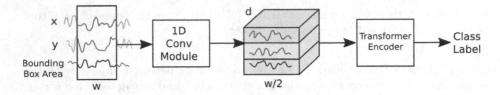

Fig. 1. Transformer-based model to learn surgical skill.

curvature, tortuosity, turning angle, and motion ratio are calculated between each time step. Motion ratio was simply defined as the ratio of time the tool is in motion to the total time of the task. These features were used in a random forest model to classify surgeons into the high and low efficiency classes.

2.4 Learning Based Skill Assessment

We compare feature-based skill assessment classification with a learning-based approach. The motion features (i.e. location and bounding box area information) from the tracking method are used as inputs to the learning model and the model weights are learned by optimizing a cross-entropy loss function to classify surgeon skills using a predefined skill criterion. A transformer model [25] was used due to prior excellent performance across a wide range of time-series applications [26]. Both short and long-range relationships of tool location are needed to get an insight quality of the surgeon's motion which is enabled by the self-attention mechanism. Figure 1 shows our network. 3D tracking time series, x, y and bounding box area, is first fed through a convolutional module with Batch-Normalization and Relu after each convolution: $Conv1D(d = 128, k = 11, s = 1) + Conv1D(d = 128, k = 3, s = 2)$ where kernel size, k, is the width in number of time-steps, d is num of output channel and s is stride. The rationale behind this module is to learn some atomic motion from tool trajectory in addition to reducing the temporal resolution of the input. The second module is a trans-former encoder with $num\_heads = 7, num\_layers = 2, dim\_feedforward = 56$. The output of this model is the binary prediction of the efficiency of the surgeon which serves as a proxy for the surgeon's skill. To our knowledge, this is the first work that uses a transformer-based framework to classify surgeon skills from dominant surgical tools in a video sequence. Moreover, we build a multi-channel 1-dimensional convolution model for skill assessment based on inception-v3 [22]. The model shares the same pre-inception convolution blocks and the first inception block of [22] all changed to 1D operations rather than 2D.

3 Results

3.1 Tracking Model Performance

This section compares a state-of-the-art tracking method (ByteTrack [29]) with the proposed tracking (*cf.* Sect. 2.2). ByteTrack is a simple yet effective tracking method that utilizes high confidence and low confidence detection to create and update the tracks. Both trackers use the same detector for proposing bounding boxes and are evaluated on 15 video segments of Cholec80 dataset that were annotated at half the original temporal resolution. The results of the trackers are shown in Table 1. While the proposed re-identification module has less multi-object tracking accuracy (3% relative decline), it greatly reduces the identity switches (141% relative improvement) which is an important factor in tracking surgical instruments. It is worth noting the higher number of false negatives in the re-identification module is due to not using the lower confidence bounding boxes in the tracking which mainly come from the assistant tools as they appear partially in the scene. This should not hinder the skill assessment performance as we use the longest paths motion data for assessing surgeon skills.

Table 1. Tracking results of ByteTrack and our proposed tracker.

Method	IDs ↓	MOTA ↑	FP ↓	FN ↓	Precision ↑	Recall ↑	#Objects
ByteTrack	210	89.2%	2214	2615	95.2%	94.4%	46479
Proposed tracking	87	86.6%	783	5367	98.1%	88.5%	46479

3.2 Skill Assessment on Cholec80

Data Preprocessing and Augmentation: The Cholec80 data contains sequences of real in-vivo surgeries meaning. The number of present tools is not guaranteed as tools may leave and come back to the scene, or the camera might be out of the body for a few frames creating missing data points in the motion sequences. We focus on the tool that is present the most in the frames of the last 3 min of the Calot's triangle dissection step assuming this is the main tool in the scene that is performing the dissection. We create a temporal mask for motion features of the main tool that is true in the time stamps where the tool is present in the scene and false otherwise to utilize the presence information of the main tool. Lastly, we use a fixed window of motion features for each experiment and randomly change this window in different training iterations, enabling models that are invariant to the length of the task and more robust towards variations of unseen data. In all experiments, we use the same dataset with 5-fold cross-validation. To augment the training dataset, we randomly choose a different temporal window of the length of 700 (28s) out of 3 min duration of the target surgical phase.

Feature-based Skill Assessment: Results from the feature-based skill assessment classification models indicate very low reliability between motion metrics and performance scores. As indicated in Table 2, all the model assessment metrics, i.e. precision, recall, accuracy, and kappa were generally below acceptable standards. This difference from prior literature is likely due to the challenges of working with 2D computer-vision-based tracks compared to 3D robot kinematic sensor data.

Learning-Based Skill Assessment: Learning skill directly from motion allows automatic extraction of features that are more suitable for the task. We run a Bayesian hyper-parameter search of learning-based models and show the results of the best-performing models. The data is normalized before it is fed to the model and random oversampling is employed to mitigate class imbalance problems. The learning-based methods outperformed manual feature extraction methods which suggests that the learning process is able to extract features that are more meaningful for the efficiency calculation.

Table 2. Skill assessment results on Cholec80 dataset.

Method	Tracking	Efficiency				
		Precision	Recall	Acuuracy	Kappa	p-value
Feature-based	Proposed	0.68	0.52	0.65	0.30	0.0090320
1D Convolution	Proposed	0.83	0.75	0.74	0.45	0.0382200
Transformer	ByteTrack	0.73	0.73	0.69	0.36	0.0483700
Transformer	Proposed	**0.88**	**0.84**	**0.83**	**0.63**	**0.0001962**

4   Discussion

Automated surgical skill assessment has the potential to improve surgical practice by providing scalable feedback to surgeons as they develop their practice. Current manual methods are limited by the subjectivity of the metrics resulting in the need for multiple evaluators to obtain reproducible results. Here our method demonstrated the ability to agree with the consensus of our two annotators to a greater degree than the individual annotators agreed with each other (Kappa 0.63 model vs 0.41 annotators). This indicates the classification created by the model is comparable to human-level performance.

A key contributor is the ability of our model to track objects over long temporal periods. Our model showed a large reduction in the number of ID switches compared to the state of the art, meaning objects could be continuously tracked for much longer periods than with prior methods. This ability to consistently track the tools resulted in an improved ability to classify surgeon performance.

Another novel contribution of this work is the ability to classify skills directly from the tool tracking information. Prior work has shown the ability to use feature-based approaches to distinguish surgeons of differing skill levels [30]. However, these approaches have generally been performed in lab tasks or using kinematic data where tools cannot be obfuscated and the scale of the motion is standardized. In procedure video data, tools may frequently be not visible to the camera and the scale of the motion is highly dependent on where and how the camera is moved and placed during the operation. In these more realistic scenarios, we found a feature-based approach to be much less successful at classifying surgeon skills than directly classifying based on the tracked motion.

One limitation of this work is that to standardize our inputs we only assessed the end of the Dissection of Triangle of Calot surgical step. Future studies can determine how well this method will generalize to other dissection steps or even other types of surgical actions such as suturing or stapling. Additionally, we would like to explore this method's ability to assess other qualities of surgeon action such as bimanual dexterity and depth perception.

Acknowledgements. We would like to thank Derek Peyton for orchestrating annotation collection for this project.

References

1. Azari, D.P., et al.: Modeling surgical technical skill using expert assessment for automated computer rating. Ann. Surg. **269**, 574–581 (2019)
2. Bergmann, P., Meinhardt, T., Leal-Taixe, L.: Tracking without bells and whistles. In: Proceedings of the IEEE/CVF International Conference on Computer Vision, pp. 941–951 (2019)
3. Dubin, A.K., Julian, D., Tanaka, A., Mattingly, P., Smith, R.: A model for predicting the gears score from virtual reality surgical simulator metrics. Surg. Endosc. **32**, 3576–3581 (2018)
4. Estrada, S., O'Malley, M.K., Duran, C., Schulz, D., Bismuth, J.: On the development of objective metrics for surgical skills evaluation based on tool motion. In: 2014 IEEE International Conference on Systems, Man, and Cybernetics (SMC), pp. 3144–3149. IEEE (2014)
5. Fard, M.J., Ameri, S., Ellis, R.D., Chinnam, R.B., Pandya, A.K., Klein, M.D.: Automated robot-assisted surgical skill evaluation: predictive analytics approach. Int. J. Med. Robot. Comput. Assist. Surg. **14**, e1850 (2018)
6. Fard, M.J., Ameri, S., Chinnam, R.B., Ellis, R.D.: Soft boundary approach for unsupervised gesture segmentation in robotic-assisted surgery. IEEE Robot. Autom. Lett. **2**, 171–178 (2017)
7. Ganni, S., Botden, S.M., Chmarra, M., Li, M., Goossens, R.H., Jakimowicz, J.J.: Validation of motion tracking software for evaluation of surgical performance in laparoscopic cholecystectomy. J. Med. Syst. **44**(3), 1–5 (2020)
8. Hermans, A., Beyer, L., Leibe, B.: In defense of the triplet loss for person re-identification. arXiv preprint arXiv:1703.07737 (2017). https://github.com/VisualComputingInstitute/triplet-reid
9. Hung, A.J., Chen, J., Gill, I.S.: Automated performance metrics and machine learning algorithms tomeasure surgeon performance and anticipate clinical outcomes in robotic surgery. JAMA Surg. **153**(8), 770–771 (2018)
10. Innocent, N.C., et al.: Rendezvous: attention mechanisms for the recognition of surgical action triplets in endoscopic videos. arXiv e-prints pp. arXiv-2109 (2021)
11. Jin, A., et al.: Tool detection and operative skill assessment in surgical videos using region-based convolutional neural networks. In: 2018 IEEE Winter Conference on Applications of Computer Vision (WACV), pp. 691–699 (2018)
12. Jocher, G., et al.: ultralytics/yolov5: v3.1 - bug fixes and performance improvements, October 2020
13. Kuhn, H.W.: The Hungarian method for the assignment problem. Naval Res. Logistics Q. **2**(1–2), 83–97 (1955)

14. Law, H., Ghani, K., Deng, J.: Surgeon technical skill assessment using computer vision based analysis (2017)
15. Lee, D., Yu, H.W., Kwon, H., Kong, H.J., Lee, K.E., Kim, H.C.: Evaluation of surgical skills during robotic surgery by deep learning-based multiple surgical instrument tracking in training and actual operations. J. Clin. Med. **9**, 1964 (2020)
16. Levin, M., Mckechnie, T., Khalid, S., Grantcharov, T.P., Goldenberg, M.: Automated methods of technical skill assessment in surgery: a systematic review. J. Surg. Educ. **76**(6), 1629–1639 (2019)
17. Milan, A., Leal-Taixé, L., Reid, I., Roth, S., Schindler, K.: MOT16: a benchmark for multi-object tracking. arXiv:1603.00831 [cs], March 2016
18. Oropesa, I., et al.: Relevance of motion-related assessment metrics in laparoscopic surgery. Surg. Innov. **20**(3), 299–312 (2013)
19. Pérez-Escamirosa, F., et al.: Construct validity of a video-tracking system based on orthogonal cameras approach for objective assessment of laparoscopic skills. Int. J. Comput. Assist. Radiol. Surg. **11**(12), 2283–2293 (2016)
20. Rivas-Blanco, I., et al: A surgical dataset from the da vinci research kit for task automation and recognition, pp. 1–6 (2021)
21. Shafiei, S.B., Guru, K.A., Esfahani, E.T.: Using two-third power law for segmentation of hand movement in robotic assisted surgery. In: vol. 5C–2015. American Society of Mechanical Engineers (ASME) (2015)
22. Szegedy, C., Vanhoucke, V., Ioffe, S., Shlens, J., Wojna, Z.: Rethinking the inception architecture for computer vision. In: Proceedings of the IEEE Conference on Computer Vision and Pattern Recognition, pp. 2818–2826 (2016)
23. Twinanda, A.P., Shehata, S., Mutter, D., Marescaux, J., De Mathelin, M., Padoy, N.: Endonet: a deep architecture for recognition tasks on laparoscopic videos. IEEE Trans. Med. Imaging **36**(1), 86–97 (2016)
24. Vassiliou, M.C., et al.: A global assessment tool for evaluation of intraoperative laparoscopic skills. Am. J. Surg. **190**, 107–113 (2005)
25. Vaswani, A., et al.: Attention is all you need. Adv. Neural Inf. Process. Syst. **30** (2017)
26. Wen, Q., et al.: Transformers in time series: a survey. arXiv preprint arXiv:2202.07125 (2022)
27. Witte, B.D., Barnouin, C., Moreau, R., Lelevé, A., Martin, X., Collet, C., Hoyek, N.: A haptic laparoscopic trainer based on affine velocity analysis: engineering and preliminary results. BMC Surg. **21**, 1–10 (2021)
28. Wu, Y., Lim, J., Yang, M.H.: Online object tracking: a benchmark. In: Proceedings of the IEEE Conference on Computer Vision and Pattern Recognition, pp. 2411–2418 (2013)
29. Zhang, Y., et al.: Bytetrack: multi-object tracking by associating every detection box. arXiv preprint arXiv:2110.06864 (2021)
30. Zia, A., Essa, I.: Automated surgical skill assessment in RMIS training. Int. J. Comput. Assist. Radiol. Surg. **13**(5), 731–739 (2018). https://doi.org/10.1007/s11548-018-1735-5

Surgical Scene Segmentation Using Semantic Image Synthesis with a Virtual Surgery Environment

Jihun Yoon[1], SeulGi Hong[1], Seungbum Hong[1], Jiwon Lee[1], Soyeon Shin[1], Bokyung Park[1], Nakjun Sung[1], Hayeong Yu[1], Sungjae Kim[1], SungHyun Park[2], Woo Jin Hyung[1,2], and Min-Kook Choi[1(✉)]

[1] Hutom, Seoul, Republic of Korea
mkchoi@hutom.io
[2] Department of Surgery, Yonsei University College of Medicine, Seoul, Republic of Korea

Abstract. The previous image synthesis research for surgical vision had limited results for real-world applications with simple simulators, including only a few organs and surgical tools and outdated segmentation models to evaluate the quality of the image. Furthermore, none of the research released complete datasets to the public enabling the open research. Therefore, we release a new dataset to encourage further study and provide novel methods with extensive experiments for surgical scene segmentation using semantic image synthesis with a more complex virtual surgery environment. First, we created three cross-validation sets of real image data considering demographic and clinical information from 40 cases of real surgical videos of gastrectomy with the da Vinci Surgical System (dVSS). Second, we created a virtual surgery environment in the Unity engine with five organs from real patient CT data and 22 the da Vinci surgical instruments from actual measurements. Third, We converted this environment photo-realistically with representative semantic image synthesis models, SEAN and SPADE. Lastly, we evaluated it with various state-of-the-art instance and semantic segmentation models. We succeeded in highly improving our segmentation models with the help of synthetic training data. More methods, statistics, and visualizations on https://sisvse.github.io/.

Keywords: Surgical instrument localization · Class imbalance · Domain randomization · Synthetic data · Semantic image snythesis

---

J. Yoon and S. Hong—Co-first authors.

Supplementary Information The online version contains supplementary material available at https://doi.org/10.1007/978-3-031-16449-1_53.

L. Wang (Eds.): MICCAI 2022, LNCS 13437, pp. 551–561, 2022.
https://doi.org/10.1007/978-3-031-16449-1_53

1   Introduction

Recent advances in deep neural network architectures and learning large-scale datasets have significantly improved model performance. Because of that, many large-scale datasets have been published for computer vision applications such as natural scene recognition [1] or autonomous driving [2], remote sensing [3], computer-assisted design [4], medical imaging [5], surgical vision [10], etc.

Unlike other computer vision applications having abundant datasets, surgical vision research always has had a problem of public data shortages. Fortunately, several surgical vision datasets were published [6–10] and encouraged the research. However, building those large-scale surgical vision datasets cost more than common datasets due to the rareness of surgery videos and the difficulties of annotations. Because of that, several researchers conducted image synthesis research for surgical vision [11–13] to decrease data generation costs.

However, the previous works had limited results for real-world applications with simple simulators, including only a few organs and surgical tools and outdated segmentation models to evaluate the quality of the image. In addition, none of the research provided real image data, including annotations, which is necessary for the open research. Therefore, we release a new dataset to encourage further study and provide novel methods with extensive experiments for surgical scene segmentation using semantic image synthesis with a more complex virtual surgery environment.

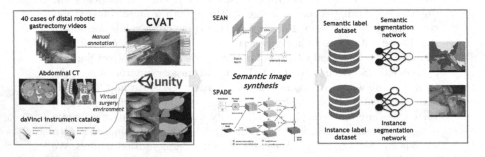

Fig. 1. The schematic diagram of surgical scene segmentation using semantic image synthesis with a virtual surgery environment. SPADE [19] and SEAN [20] translate synthetic data generated from a virtual surgery environment photorealistically and are used for training surgical scene segmentation models.

We compare state-of-the-art instance [14–16] and semantic [17,18] segmentation models trained with real data and synthetic data. First, we created three cross-validation datasets considering demographic and clinical information from 40 cases of real surgical videos of distal gastrectomy for gastric cancer with the da Vinci Surgical System (dVSS). We annotated six organs and 14 surgical instruments divided into 24 instrument parts according to head, wrist, and body structures commonly appearing during the surgery. We also introduced a class-balanced frame sampling to suppress many redundant classes and to reduce unnecessary data labeling costs. Second, we created a virtual surgery environment in the Unity engine with 3D models of five organs from real patient CT data and 11 surgical instruments from actual measurements of the da Vinci surgical instruments. The environment provides cost-free pixel-level annotations enabling segmentation. Third, we used state-of-the-art semantic image synthesis models [19,20] to convert our 3D simulator photo-realistically. Lastly, we created ten combinations of real and synthetic data for each cross-validation set to train/evaluate segmentation models. The analysis showed interesting results: synthetic data improved low-performance classes and was very effective for Mask AP improvement while improving the segmentation models overall. Finally, we release the first large-scale semantic/instance segmentation dataset, including both real and synthetic data that can be used for visual object recognition and image-to-image translation research for gastrectomy with the dVSS. However, our method is not only limited to gastrectomy but can be generalized into all minimally invasive surgeries such as laparoscopic and robotic surgery.

Table 1. Previous semantic image synthesis research for surgical vision. (1) This table shows limitations of previous works for real-world applications and the dataset contribution. (2) Segm. stands for segmentation.

Research	How many classes (#) to translate?	Recognition models (#)	Real image data	Do they provide real image data with annotations?
[11]	Liver class only (1)	Semantic segm. (1)	Re-annotate Chole80	No
[12]	Liver class only (1)	Semantic segm. (1)	Re-annotate Chole80	No
[13]	Laparoscopic tools (5)	Semantic segm. (1)	Own laparoscopic data	No
Our work	Laparoscopic/Robotic tools (22) + Organs (5)	Instance segm. (3) + Semantic segm. (2)	Own robotic gastrectomy data	Yes

The contribution of our work is summarized as follows:

- We release the first large-scale instance and semantic segmentation dataset, including both real and synthetic data that can be used for visual object recognition and image-to-image translation research for gastrectomy with the dVSS.
- We systematically analyzed surgical scene segmentation using semantic image synthesis with state-of-the-art models with ten combinations of real and synthetic data.
- We found interesting results that synthetic data improved low-performance classes and was very effective for Mask AP improvement while improving the segmentation models overall.

2   Background

Surgical Vision Datasets for Minimally Invasive Surgery. The Cholec80 dataset [7] is one of the widely used datasets providing surgical phases and a presence of the surgical instruments for 80 cases of laparoscopic cholecystectomy. However, because of its limitations, several researchers expanded the original dataset into new datasets, including bounding boxes [6,10] and semantic segmentation masks [9,11]. Although [9] provides semantic segmentation masks for Cholec80, it does not provide a performance analysis of any segmentation models. The recently released Heidelberg dataset [8] provides surgical phases and segmentation masks of surgical instruments for laparoscopic colectomy and multi-modal sensor information from medical devices in the operating room. Unlike the datasets mentioned above, [10] provides new 40 cases of gastrectomy data with synthetic data and the Cholec80 for bounding boxes. However, the research utilized synthetic data itself with domain randomization [34] rather than image-to-image translation.

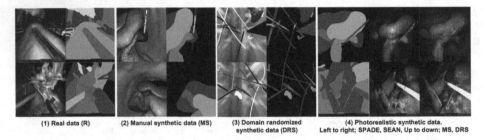

(1) Real data (R)   (2) Manual synthetic data (MS)   (3) Domain randomized
                                                          synthetic data (DRS)   (4) Photorealistic synthetic data.
                                                                                Left to right; SPADE, SEAN, Up to down; MS, DRS

Fig. 2. Examples of real, manual synthetic, and domain randomized synthetic data. SPADE(MS) and SEAN(MS) means manual synthetic data converted photo-realistically by SPADE [19] and SEAN [20]

Semantic Image Synthesis. Semantic image synthesis is an image synthesis task that controls a semantic layout input style with a Generative adversarial network (GAN) [21], converting a semantic segmentation mask to a photorealistic image. Representative state-of-the-art semantic image synthesis models are SPADE [19], SEAN [20]. SPADE overcame a shortcoming of the previous image synthesis models [22,23], normalization layers tend to wash information contained in the input semantic masks by proposing the spatially-adaptive normalization. After that, SEAN proposed the region-adaptive normalization to enable precise control of styles for photo-realistic image synthesis based on the SPADE structure.

Semantic image synthesis has also been applied to generate training data for surgical vision. [24] proposed the DavinciGAN, introducing background consistency loss using a self-attention mechanism without ground truth mask data to supplement insufficient training data for the surgical instrument recognition. [11] proposed an extension of MUNIT [25], incorporating an additional multi-scale structural similarity loss to preserve image content during the translation process to generate realistically looking synthetic images from a simple laparoscopy simulation. [12] proposed a combination of unpaired image translation with neural rendering to transfer simulated to more photo-realistic surgical abdominal scenes. However, both methods were only used for liver segmentation. [13] used CycleGAN [26] to generate realistic synthetic data for five surgical instruments.

Image Segmentation. Image segmentation is largely divided into semantic segmentation [17,18,27,28], instance segmentation [14,15,29]. Semantic segmentation classifies each pixel according to a category, clustering parts of an image that belong to the same object class. Many semantic segmentation architectures follow the CNN encoder-decoder structure and succeed in achieving high-performance [17,18,27,28]. Instance segmentation combines object detection with semantic segmentation, detecting objects with bounding boxes and finding binary masks for each object in the bounding boxes. Unlike semantic segmentation, instance segmentation differentiates the same class objects into different instances. Many instance segmentation architectures follow a joint multi-task structure [14,15,29] between object detection and binary segmentation. Recently, a transformer-based backbone network [16,30] significantly improved the performance of both segmentation tasks compared to conventional CNN-based backbone networks.

3   Data Generation

We collected 40 cases of real surgical videos of distal gastrectomy for gastric cancer with the da Vinci Surgical System (dVSS), approved by an institutional review board at the medical institution. In order to evaluate generalization performance, we created three cross-validation datasets considering demographic and clinical variations such as gender, age, BMI, operation time, and patient bleeding. Each cross-validation set consists of 30 cases for train/validation and 10 cases for test data. Appendix Table 1 shows the overall statistics and demographic and clinical information details.

Categories. We list five organs (Gallbladder, Liver, Pancreas, Spleen, and Stomach) and 13 surgical instruments that commonly appear from surgeries (Hamonic Ace; HA, Stapler, Cadiere Forceps; CF, Maryland Bipolar Forceps; MBF, Medium-large Clip Applier; MCA, Small Sclip Applier; SCA, Curved Atraumatic Graspers; CAG, Suction, Drain Tube; DT, Endotip, Needle, Specimenbag, Gauze). We classify some rare organs and instruments as "other tissues" and "other instruments" classes. The surgical instruments consist of robotic and laparoscopic instruments and auxiliary tools mainly used for robotic subtotal gastrectomy. In addition, we divide some surgical instruments according to their head, H, wrist; W, and body; B structures, which leads to 24 classes for instruments in total.

Class Balanced Frame Sampling. It is widely known that neural networks struggle to learn from a class-imbalanced dataset, which also happens in surgical vision [31]. The problem can be alleviated by loss functions or data augmentation techniques; however, a data sampling method in the data generation step can also easily release it. We introduce a *class balanced frame sampling*, considering a class distribution while creating the dataset. We select major scenes(frames) to learn while keeping a class distribution. We select the same number of frames for each instrument and organ in each surgery video. This approach also reduces unnecessary data labeling costs by suppressing many redundant frames, which originate from the temporal redundant characteristics of the video.

Virtual Surgery Environment and Synthetic Data. Abdominal computed tomography (CT) DICOM data of a patient and actual measurements of each surgical instrument are used to build a virtual surgery environment. We aim to generate meaningful synthetic data from a sample patient. We annotated five organs listed for real data and reconstructed 3D models by using VTK [33]. In addition, we precisely measured the actual size of each instrument commonly used for laparoscopic and robotic surgery with dVSS. We built 3D models with commercial software such as 3DMax, Zbrush, and Substance Painter. After that, we integrated 3D organ and instrument models into the unity environment for virtual surgery. A user can control a camera and two surgical instruments like actual robotic surgery through a keyboard and mouse in this environment. To reproduce the same camera viewpoint as dVSS, we set the exact parameters of an endoscope used in the surgery. While the user simulates a surgery, a snapshot function projects a 3D scene into a 2D image. According to the projected 2D image, the environment automatically generates corresponding segmentation masks.

Qualified Annotations. Seven annotators trained for surgical tools and organs annotated six organs and 14 surgical instruments divided into 24 instruments according to head, wrist, and body structures with a web-based computer vision annotation tool (CVAT) [32]. We call this *real data (R)*. After that, three medical professionals with clinical experience inspected the annotations to ensure their quality. Appendix Table 2 shows the class distribution and the number of frames for each cross-validation set. The three medical professionals also manually simulated virtual surgeries to generate virtual surgical scenes. We call this *manual synthetic data (MS)*. On the other hand, we also use an automatic data generation method called domain randomization, a technique to put objects randomly in a scene to cover the variability in real-world data [34]. We call this *domain randomized synthetic data (DRS)*. Appendix Table 2 shows the number of classes for synthetic data.

4   Experimental Results

We trained and evaluated the state-of-the-art instance and semantic segmentation models with ten combinations of real and synthetic data for each cross-validation set. All combinations of data are listed in Table 2. Manually simulated synthetic data or domain randomized synthetic data translated photo-realistically with SEAN or SPADE models are denoted as *Model(Data)*, e.g., SEAN(MS). We used MMSegmentation [35] and MMDetection [36] libraries for training and testing the models. Moreover, official implementations of SEAN and SPADE are used for semantic image synthesis. In addition, we *fixed a random seed* during all training and testing not to see performance benefits from randomness. Appendix Table 3 shows hyper-parameters of the models.

Table 2. Overall performance of instance and semantic segmentation for Test1. The best performance for each model compared to a baseline (trained with real data) is highlighted in **bold**. Other cross-validation results can be found on the website.

Data set	R		R+MS		R+DRS		R+MS+DRS	
Model	Box AP	Mask AP	Box AP	Mask AP	Box AP	Mask AP	Box AP	Mask AP
CMR [14]	51.2	51.0	51.9	53.0	51.5	51.7	52.3	53.3
HTC [15]	53.9	55.0	53.8	55.8	53.9	55.8	**54.3**	56.8
Data set			R+SPADE(MS)		R+SPADE(DRS)		R+SPADE(MS+DRS)	
Model			Box AP	Mask AP	Box AP	Mask AP	Box AP	Mask AP
CMR [14]			52.1	53.1	51.1	51.4	**52.5**	**53.6**
HTC [15]			**54.3**	57.0	53.7	55.4	54.1	57.1
Data set			R+SEAN(MS)		R+SEAN(DRS)		R+SEAN(MS+DRS)	
Model			Box AP	Mask AP	Box AP	Mask AP	Box AP	Mask AP
CMR [14]			51.7	52.8	51.7	51.8	51.9	53.4
HTC [15]			**54.3**	**57.2**	53.7	55.5	**54.3**	**57.2**

Data set	R			R+MS			R+DRS			R+MS+DRS		
Model	mIoU	mAcc	aAcc	mIoU	mAcc	aAcc	mIoU	mAcc	aAcc	mIoU	mAcc	aAcc
DLV3+ [17]	74.68	82.99	87.72	74.66	83.17	87.85	74.66	83.19	87.77	74.56	82.85	87.66
UperNet [18]	75.33	83.83	87.97	75.32	84.01	87.94	**75.62**	**84.28**	**88.03**	75.60	84.20	87.86
Data set				R+SPADE(MS)			R+SPADE(DRS)			R+SPADE(MS+DRS)		
Model				mIoU	mAcc	aAcc	mIoU	mAcc	aAcc	mIoU	mAcc	aAcc
DLV3+ [17]				75.57	**84.20**	88.02	74.66	83.17	87.77	74.74	83.35	87.80
UperNet [18]				74.44	83.31	87.54	74.37	83.43	87.67	74.15	83.10	87.64
Data set				R+SEAN(MS)			R+SEAN(DRS)			R+SEAN(MS+DRS)		
Model				mIoU	mAcc	aAcc	mIoU	mAcc	aAcc	mIoU	mAcc	aAcc
DLV3+ [17]				**75.58**	83.81	**88.09**	74.93	83.32	87.71	74.14	82.8	87.63
UperNet [18]				74.72	83.55	87.73	74.31	83.41	87.52	74.55	83.46	87.69

Instance Segmentation. We evaluated Cascade Mask R-CNN [14], HTC [15] with ResNet101, and HTC with Dual-swin-L [16] by measuring, mean average precisions for bounding box (box AP) and segmentation mask (mask AP). We follow the metrics from MS-COCO benchmark [1]. Table 2 shows Cascade Mask R-CNN achieves the best performance, mask AP; 53.6(+2.6), box AP; 52.5(+1.3) with R+SPADE(MS+DRS) dataset, and HTC, mask AP; 57.2(+2.2), box AP; 54.3(+0.4), with R+SEAN(MS) and R+SEAN(MS+DRS). HTC achieves higher

Table 3. Top-8 class-wise performance of instance and semantic segmentation for Test1. Each class performance is sorted by decreasing order in a synthetic data set highlighted in **bold**. Relative performance (AP/IoU) is defined in *Class-wise performance* section.

CMR [14]	Mask AP	Relative Mask AP	Box AP	Relative Box AP	HTC [15]	Mask AP	Relative Mask AP	Box AP	Relative Box AP
Data set	R	*R+SPADE (MS+DRS)*	R	*R+SPADE (MS+DRS)*	Data set	R	*R+SEAN (MS)*	R	*R+SEAN (MS)*
CAG_H	17.00	11.18%	17.50	**29.14%**	Specimenbag	40.80	**94.12%**	49.40	−6.68%
Needle	16.70	19.76%	20.40	**28.43%**	Needle	11.60	**34.48%**	24.60	−6.91%
OT	41.50	17.83%	28.10	**26.69%**	CAG_B	29.40	**20.41%**	35.80	−1.40%
Gauze	62.50	5.92%	26.10	**9.96%**	Liver	28.60	**16.08%**	19.30	7.77%
Specimenbag	53.90	2.60%	39.80	**9.05%**	OT	34.50	**15.36%**	32.60	4.91%
Stapler_B	60.30	5.97%	53.40	**8.80%**	DT	31.50	**13.33%**	27.50	−10.18%
HA_H	35.50	1.13%	32.80	**7.32%**	Gauze	50.90	**11.00%**	30.00	2.33%
Stapler_H	62.20	8.04%	63.00	**7.30%**	CAG_H	14.80	**10.14%**	19.90	27.14%

DLV3+ [17]	mIoU	Relative mIoU	UperNet [18]	mIoU	Relative mIoU
Data set	R	*R+SEAN (MS)*	Data set	R	*R+DRS*
Spleen	31.29	**14.54%**	Spleen	42.00	**13.95%**
Needle	42.91	**8.81%**	CAG_H	39.92	**6.21%**
Gallbladder	47.89	**7.87%**	CF_B	78.40	**2.05%**
MBF_B	70.69	**6.79%**	CAG_B	68.29	**1.65%**
SCA_B	83.04	**5.52%**	Specimenbag	87.28	**1.56%**
CF_B	76.62	**4.32%**	MBF_W	75.57	**1.42%**
Suction	73.62	**2.68%**	Suction	75.20	**1.40%**
CF_W	77.35	**2.44%**	OI	78.12	**1.32%**

Mask AP than Cascade Mask R-CNN because it is an advanced Cascade Mask R-CNN leveraging *mask information*. From these results, we assume *synthetic data is effective for mask AP* when a model utilizes semantic mask information. Furthermore, we also trained real data for the same number of iterations as the synthetic data to see whether performance improvement comes from learning more iterations when synthetic data is added. However, Appendix Fig. 1 shows longer training of the real data underperforms than the synthetic data.

Semantic Segmentation. We evaluated DeepLabV3+ [17] with ResNeSt [37] and UperNet [18] with Swin Transformer [30] by measuring mean intersection over union (mIoU), mean accuracy (mAcc), and overall accuracy on all images (aAcc), which are the standard evaluation metrics for semantic segmentation. Table 2 shows DeepLabV3+ achieves the best performance, mIoU 75.58(+0.9), with R+SEAN(MS), and UperNet, mIoU 75.62(+0.29) with R+DRS. From these results, we assume synthetic data is no longer helpful when the models already achieve high performance with the real data. Training the models with smaller datasets will remain for future work.

Class-wise Performance. To take a closer look at performance improvement, we calculate *relative performance of synthetic data* to real data as $(\text{Metric}_{Syn} - \text{Metric}_{Real}) \div \text{Metric}_{Real} \times 100$. Table 3 shows top-8 class-wise performance for the models. The relative performances show us that there is a tendency for lower performance classes to improve more than others overall. Another interesting point is that in the case of HTC, Specimenbag (+94.12%) and Gauze (+11.00%), which are not included in the synthetic data, also improve significantly. Semantic segmentation models also show the same results (highest 14.54% for DLV3+ and 13.95% for UperNet). The class-wise performance analysis also found that overall performance improvement was less significant because some high-performant classes lost performances. However, there is more good than harm because although some classes lose performances, those are still very high for applications. We will study further to improve low-performant classes with the synthetic dataset while not losing high-performant classes.

Semantic Image Synthesis. We applied two representative semantic image synthesis methods, SPADE [19], and SEAN [20], to minimize the semantic gap between real data and synthetic data. We trained each model with real train/valid data for each cross-validation set and made inferences of photo-realistic images from synthetic masks. We also follow the standard quantitative evaluation method from [19,20] to evaluate translation performance. We trained the synthesis models on real train/valid data for the test 1 cross-validation set and made inferences of photo-realistic images from the test masks of the set. After that, we trained DeepLabV3+ on the same train/valid for the test 1 set and evaluated with the real test 1 image set (76.68 mIoU) and the synthesized test 1 image set (66.8 mIoU for SEAN, 62.66 mIoU for SPADE). Figure 2 shows examples of translated both synthetic data. More details of the methods can be found on the website[1].

5   Conclusion

We release the first large-scale instance and semantic segmentation dataset, including both real and synthetic data that can be used for visual object recognition and image-to-image translation research for gastrectomy with the dVSS. We systematically analyzed surgical scene segmentation using semantic image synthesis with state-of-the-art models with ten combinations of real and synthetic data. There are two interesting results. One is *synthetic data improves low-performant classes.* Looking at the relative performances from Table 3, we can see that our method significantly improves low-performant classes while improving the segmentation models overall. Another is *synthetic data is more effective for mask AP for instance segmentation.* However, our experiments also showed a limitation that synthetic data is no longer helpful when the models already achieve high performance with the real data. Our work also has values for a data generation method for surgical scene segmentation using semantic image synthesis with a complex virtual surgery environment. In addition, the method is not only limited to gastrectomy but can be generalized into all minimally invasive surgeries such as laparoscopic and robotic surgery.

Acknowledgement. This work was supported by the Korea Medical Device Development Fund grant funded by the Korea government (the Ministry of Science and ICT, the Ministry of Trade, Industry and Energy, the Ministry of Health & Welfare, the Ministry of Food and Drug Safety) (Project Number: 202012A02-02)

---

[1] https://sisvse.github.io/.

References

1. Lin, T.-Y., et al.: Microsoft COCO: common objects in context. In: Fleet, D., Pajdla, T., Schiele, B., Tuytelaars, T. (eds.) ECCV 2014. LNCS, vol. 8693, pp. 740–755. Springer, Cham (2014). https://doi.org/10.1007/978-3-319-10602-1_48
2. Cordts, M., et al.: The cityscapes dataset for semantic urban scene understanding. In: Proceedings of CVPR (2016)
3. Bondi, E., et al.: BIRDSAI: a dataset for detection and tracking in aerial thermal infrared videos. In: Proceedings of WACV (2019)
4. Koch, S., et al.: ABC: a big CAD model dataset for geometric deep learning. In: Proceedings of CVPR (2019)
5. Yang, T., et al.: IntrA: 3D intracranial aneurysm dataset for deep learning. In: Proceedings of CVPR (2020)
6. Jin, A., et al.: Tool detection and operative skill assessment in surgical videos using region-based convolutional neural networks. In: Proceedings of WACV (2018)
7. Twinanda, A.P., et al.: EndoNet: a deep architecture for recognition tasks on laparoscopic videos. IEEE Trans. Med. Imaging 36, 86–97 (2017)
8. Maier-Hein, L., et al.: Heidelberg colorectal data set for surgical data science in the sensor operating room. Sci. Data 8, 2025–2041 (2020)
9. Hong, W.Y., et al.: CholecSeg8k: a semantic segmentation dataset for laparoscopic cholecystectomy based on Cholec80. In: Proceedings of IPCAI (2018)
10. Yoon, J., et al.: hSDB-instrument: instrument localization database for laparoscopic and robotic surgeries. In: Proceedings of MICCAI (2021)
11. Pfeiffer, M., et al.: Generating large labeled data sets for laparoscopic image processing tasks using unpaired image-to-image translation. In: Proceedings of MICCAI (2019)
12. Rivoir, D., et al.: Long-term temporally consistent unpaired video translation from simulated surgical 3D data. ArXiv. abs/2103.17204 (2021)
13. Ozawa, T., et al.: Synthetic laparoscopic video generation for machine learning-based surgical instrument segmentation from real laparoscopic video and virtual surgical instruments. Comput. Methods Biomech. Biomed. Eng. Imaging Vis. 9, 225–232 (2021)
14. Cai, Z., Vasconcelos, N.: Cascade R-CNN: high quality object detection and instance segmentation. IEEE Trans. TPAMI 43, 1483–1498 (2019)
15. Chen, K., et al.: Hybrid task cascade for instance segmentation. In: Proceedings of CVPR (2019)
16. Liang, T., et al.: CBNetV2: a composite backbone network architecture for object detection. ArXiv:2107.00420. (2021)
17. Chen, L.C., et al.: Encoder-decoder with atrous separable convolution for semantic image segmentation. In: Proceedings of ECCV (2018)
18. Xiao, T., et al.: Unified perceptual parsing for scene understanding. In: Proceedings of ECCV (2018)
19. Park, T., et al.: Semantic image synthesis with spatially-adaptive normalization. In: Proceedings of CVPR (2019)
20. Zhu, P., et al.: SEAN: image synthesis with semantic region-adaptive normalization. In: Proceedings of CVPR (2020)
21. Goodfellow, I., et al.: Generative adversarial nets. In: NIPS (2014)
22. Isola, P., et al.: Image-to-image translation with conditional adversarial networks. In: 2017 IEEE Conference On Computer Vision And Pattern Recognition (CVPR), pp. 5967–5976 (2017)

23. Wang, T.C., et al.: High-resolution image synthesis and semantic manipulation with conditional GANs. In: 2018 IEEE/CVF Conference On Computer Vision and Pattern Recognition, pp. 8798–8807 (2018)

24. Lee, K., Choi, M., Jung, H.: DavinciGAN: unpaired surgical instrument translation for data augmentation. In: Proceedings of MIDL (2019)

25. Huang, X., et al.: Multimodal unsupervised image-to-image translation. ArXiv. abs/1804.04732 (2018)

26. Zhu, J.Y., et al.: Unpaired image-to-image translation using cycle-consistent adversarial networks. In: 2017 IEEE International Conference On Computer Vision (ICCV), pp. 2242–2251 (2017)

27. Wang, J., et al.: Deep high-resolution representation learning for visual recognition. IEEE Trans. TPAMI **43**, 3349–3364 (2019)

28. Yuan, Y., Chen, X., Wang, J.: Object-contextual representations for semantic segmentation. In: Proceedings of ECCV (2020)

29. Vu, T., Kang, H., Yoo, C.: SCNet: training inference sample consistency for instance segmentation. In: Proceedings of AAAI (2021)

30. Liu, Z., et al.: Swin transformer: hierarchical vision transformer using shifted windows. In: Proceedings of ICCV (2021)

31. Yoon, J., et al.: Semi-supervised learning for instrument detection with a class imbalanced dataset. In: Proceedings of MICCAIW (2020)

32. Computer Vision Annotation Tool (CVAT). https://github.com/opencv/cvat

33. Schroeder, W., Martin, K., Lorensen, B.: The Visualization Toolkit. Kitware (2006)

34. Tremblay, J., et al.: Training deep networks with synthetic data: bridging the reality gap by domain randomization. In: Proceedings of CVPRW (2018)

35. Contributors, M.: MMSegmentation: OpenMMLab semantic segmentation toolbox and benchmark (2020). https://github.com/open-mmlab/mmsegmentation

36. Chen, K., et al.: MMDetection: open MMLab detection toolbox and benchmark (2019). ArXiv:1906.07155

37. Zhang, H., et al.: ResNeSt: split-attention networks (2020). ArXiv:2004.08955

Surgical Planning and Simulation

Deep Learning-Based Facial Appearance Simulation Driven by Surgically Planned Craniomaxillofacial Bony Movement

Xi Fang[1], Daeseung Kim[2], Xuanang Xu[1], Tianshu Kuang[2], Hannah H. Deng[2],
Joshua C. Barber[2], Nathan Lampen[1], Jaime Gateno[2],
Michael A. K. Liebschner[3], James J. Xia[2(✉)], and Pingkun Yan[1(✉)]

[1] Department of Biomedical Engineering and Center for Biotechnology and
Interdisciplinary Studies, Rensselaer Polytechnic Institute, Troy, NY 12180, USA
yanp2@rpi.edu
[2] Department of Oral and Maxillofacial Surgery, Houston Methodist Research
Institute, Houston, TX 77030, USA
JXia@houstonmethodist.org
[3] Department of Neurosurgery, Baylor College of Medicine,
Houston, TX 77030, USA

Abstract. Simulating facial appearance change following bony movement is a critical step in orthognathic surgical planning for patients with jaw deformities. Conventional biomechanics-based methods such as the finite-element method (FEM) are labor intensive and computationally inefficient. Deep learning-based approaches can be promising alternatives due to their high computational efficiency and strong modeling capability. However, the existing deep learning-based method ignores the physical correspondence between facial soft tissue and bony segments and thus is significantly less accurate compared to FEM. In this work, we propose an Attentive Correspondence assisted Movement Transformation network (ACMT-Net) to estimate the facial appearance by transforming the bony movement to facial soft tissue through a point-to-point attentive correspondence matrix. Experimental results on patients with jaw deformity show that our proposed method can achieve comparable facial change prediction accuracy compared with the state-of-the-art FEM-based approach with significantly improved computational efficiency.

Keywords: Deep learning · Surgical planning · Simulation · Correspondence assisted movement transformation · Cross point-set attention

---

X. Fang and D. Kim—Contributed equally to this paper.

Supplementary Information The online version contains supplementary material available at https://doi.org/10.1007/978-3-031-16449-1_54.

1   Introduction

Orthognathic surgery is a bony surgical procedure (called "osteotomy") to correct jaw deformities. During orthognathic surgery, the maxilla and the mandible are osteotomized into multiple segments, which are then individually moved to a desired (normalized) position. While orthognathic surgery does not directly operate on facial soft tissue, the facial appearance automatically changes following the bony movement [13]. Surgeons now can accurately plan the bony movement using computer-aided surgical simulation (CASS) technology in their daily practice [17]. However, accurate and efficient prediction of the facial appearance change following bony movement is still a challenging task due to the complicated nonlinear relationship between facial soft tissues and underlying bones [3].

Finite-element method (FEM) is currently acknowledged as the most physically-relevant and accurate method for facial change prediction. However, despite the efforts to accelerate FEM [1,2], FEM is still time-consuming and labor-intensive because it requires heavy computation and manual mesh modeling to achieve clinically acceptable accuracy [3]. In addition, surgical planning for orthognathic surgery often requires multiple times of revisions to achieve ideal surgical outcomes, therefore preventing facial change prediction using FEM from being adopted in daily clinical setting [5].

Deep learning-based approaches have been recently proposed to automate and accelerate the surgical simulation. Li et al. [6] proposed a spatial transformer network based on the PointNet [11] to predict tooth displacement for malocclusion treatment planning. Xiao et al. [18] developed a self-supervised deep learning framework to estimate normalized facial bony models to guide orthognathic surgical planning. However, these studies are not applicable to facial change simulation because they only allows single point set as input whereas facial change simulations require two point sets, i.e., bony and facial surface points. Especially, in-depth modeling of the correlation between bony and facial surfaces is the key factor for accurate facial change prediction using deep learning technology. Ma et al. [9] proposed a facial appearance change simulation network, FC-Net, that embedded the bony-facial relationship into facial appearance simulation. FC-Net takes both bony and facial point sets as input to jointly infer the facial change following the bony movement. However, instead of explicitly establishing spatial correspondence between bony and facial point sets, the movement vectors of all bony segments in FC-Net are represented by a single global feature vector. Such a global feature vector ignores local spatial correspondence and may lead to compromised accuracy in facial change simulation.

In this study, we hypothesize that establishing point-to-point correspondence between the bony and facial point sets can accurately transfer the bony movement to the facial points and in turn significantly improve the postoperative facial change prediction. To test our hypothesis, we propose an Attentive Correspondence assisted Movement Transformation network (ACMT-Net) that equipped with a novel cross point-set attention (CPSA) module to explicitly model the spatial correspondence between facial soft tissue and bony

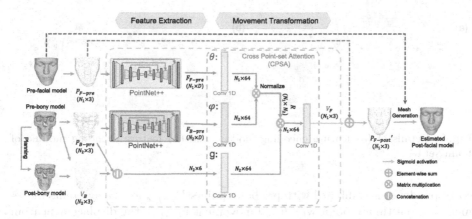

Fig. 1. Scheme of the proposed Attentive Correspondence assisted Movement Transformation network (ACMT-Net) for facial change simulation.

segments by computing a point-to-point attentive correspondence matrix between the two point sets. Specifically, we first utilize a pair of PointNet++ networks [12] to extract the features from the input bony and facial point sets, respectively. Then, the extracted features are fed to the CPSA module to estimate the point-to-point correspondence between each bony-facial point pair. Finally, the estimated attention matrix is used to transfer the bony movement to the preoperative facial surface to simulate postoperative facial change.

The contributions of our work are two-fold. 1) From the technical perspective, an ACMT-Net with a novel CPSA module is developed to estimate the change of one point set driven by the movement of another point set. The network leverages the local movement vector information by explicitly establishing the spatial correspondence between two point sets 2) From the clinical perspective, the proposed ACMT-Net can achieve a comparable accuracy of the state-of-the-art FEM simulation method, while substantially reducing computational time during the surgical planning.

2 Method

ACMT-Net predicts postoperative facial model (three-dimensional (3D) surface) based on preoperative facial and bony model and planned postoperative bony model. ACMT-Net is composed of two major components: 1) point-wise feature extraction and 2) point-wise facial movement prediction (Fig. 1). In the first component, pre-facial point set P_{F-pre}, pre-bony point set P_{B-pre}, and post-bony point set P_{B-post} are subsampled from the pre- and post- facial/bony models for computational efficiency. Then the pre-facial/bony point sets (P_{F-pre}, P_{B-pre}) are fed into a pair of PointNet++ networks to extract semantical and topological features F_{F-pre} and F_{B-pre}, respectively. In the second component, F_{F-pre} and F_{B-pre} are fed into the CPSA module to estimate the point-to-point correspondence between facial and bony points. Sequentially, the estimated point-to-point

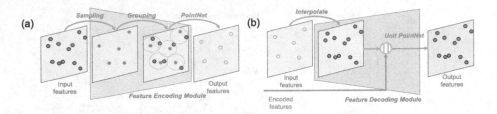

Fig. 2. Details of (a) feature encoding module and (b) feature decoding module of PointNet++.

correspondence is combined with pre bony points P_{B-pre} and bony points movement V_B to estimate point-wise facial movement V_F (i.e., the displacement from the pre-facial points P_{F-pre} to the predicted post-facial points $P_{F-post'}$). Finally, the estimated point-wise facial movement is added to the pre-facial model to generate the predicted post-facial model. Both are described below in detail.

2.1   Point-Wise Feature Extraction

Point-wise facial/bony features are extracted from preoperative facial/bony point sets using PointNet++ networks that are modified by removing the classification layer. The extracted features contain topological and semantic information so that they can be further used to calculate spatial correspondence between facial and bony points. Each modified PointNet++ network is composed of a group of feature encoding and decoding modules as shown in Fig. 2. In each feature encoding module, a subset of points is firstly down-sampled from the input points using iterative farthest point sampling (FPS). Then each sampled point and its neighboring points (with a searching radius of 0.1) are grouped together and fed into a PointNet to abstract the feature into higher level representations. The PointNet [11], composed of a shared MLP layer and a max-pooling layer, captures high-level representations from sets of local points. In each feature decoding module, the point-wise features are interpolated to the same number of points in corresponding encoding module. The features are concatenated with corresponding features from the encoding module, then a unit PointNet layer extracts the point-wise features. Unit PointNet layer, which is similar to PointNet, extracts each point's feature vector without sampling and grouping.

2.2   Attentive Correspondence Assisted Movement Transformation

Attention mechanism can capture the semantical dependencies between two feature vectors at different positions (e.g., space, time) [15,16]. Unlike most previous approaches that aim to enhance the feature representation of a position in a sequence (e.g., sentence, image patch) by attending to all positions, our CPSA module learns point-to-point correspondence between positions of two sequences (i.e. point sets) for movement transformation.

In this step, F_{F-pre}, F_{B-pre} and the concatenated pre-bony points and point-wise bony movement vectors $[P_{B-pre}, V_B]$ are fed into a CPSA module to predict the facial movement vectors V_F. Firstly, two 1D convolutional layers, θ and φ , are used to map the extracted features of facial and bony point sets to the same embedding space. Then a dot product similarity is computed between the facial feature embedding and bony feature embedding, which represents the relationship, i.e., affinity, between each pair of bony and facial points.

$$f(P_{F-pre}, \ P_{B-pre}) \ = \ \theta(F_{F-pre})^T \varphi(F_{B-pre}) \tag{1}$$

The correlation matrix $f(P_{F-pre}, P_{B-pre})$ is normalized by the number of bony points sampled from P_{B-pre}. The normalized relation matrix R models the attentive point-to-point correspondence between bony points and facial points. The movement of facial points is estimated based on the movement of bony points by exploiting the correspondence between bony and facial points. Due to the non-linear relationship between facial change and bony movement, we use a 1D convolutional layer, g, to compute the representation of point-wise bony movement vectors based on $[P_{B-pre}, V_B]$,

$$F_{V_B} = g([P_{B-pre}, V_B]) \tag{2}$$

Then R is utilized to estimate the facial movement features F_{V_F} by transforming the bony movement features F_{V_B}. Specifically, the estimated facial movement feature $F_{V_F}^i$ of the i-th facial point is a normalized summary of all bony movement features $\{F_{V_B}^j\}_{\forall j}$ weighted by the corresponding affinities with it:

$$F_{V_F}^i \ = \ \frac{1}{N_2} \sum_{\forall j} f(P_{F-pre}, \ P_{B-pre})^i F_{V_B}^j, 1 \leq i \leq N_1, \tag{3}$$

where N_1 and N_2 denote the number of facial/bony points of P_{F-pre} and P_{B-pre}, respectively. Sequentially, another 1D convolutional layers reduce the dimension of the facial movement feature to 3. After a sigmoid activation function, the predicted facial movement vector V_F is constrained in $[-1,1]$. Finally, V_F is added to P_{F-pre} to predict the post-facial points P'_{F-post}.

To train the network, we adopt a hybrid loss function, $Loss = L_{shape} + \alpha L_{density} + \beta L_{LPT}$, to compute the difference between P'_{F-post} and P_{F-post}. The loss includes a shape loss L_{shape} [19] to minimize the distance between prediction and target shape, a point density loss $L_{density}$ [19] to measure the similarity between prediction and target shape, and a local-point-transform (LPT) loss L_{LPT} [9] to constraint relative movements between one point and its neighbors.

Since our ACMT-Net processed the point data in a normalized coordinate system, all post-facial points are scaled to the physical coordinate system to generate post-operative faces according to the scale factor of the input points. The movement vectors of all vertices in the pre-facial model are estimated by interpolating the movement vector of 4096 facial points based on the structure information. Predicted post-facial mesh grid is reconstructed by adding the predicted movement vectors to the vertices of pre-facial model.

3   Experiments and Results

3.1   Dataset

The performance of ACMT-Net was evaluated using 40 sets of patient CT data using 5-fold cross validation. The CT scans were randomly selected from our digital archive of patients who had undergone double-jaw orthognathic surgery (IRB# Pro00008890). The resolution of the CT images used in this study was $0.49 \text{ mm} \times 0.49 \text{ mm} \times 1.25 \text{ mm}$. The segmentation of facial soft tissue and the bones were completed automatically using deep learning-based SkullEngine segmentation tool [7], and the surface models were reconstructed using Marching Cube [8]. In order to retrospectively recreate the surgical plan that could "achieve" the actual postoperative outcomes, the postoperative facial and bony surface models were registered to the preoperative ones based on surgically unaltered bony volumes, i.e., cranium, and used as a roadmap [9]. Virtual osteotomies were then reperformed on preoperative bones. Subsequently, the movement of the bony segments, i.e., the surgical plan to achieve the actual postoperative outcomes, were retrospectively established by manually registering each bony segment to the postoperative roadmap. Finally, the facial appearance change was predicted on the preoperative facial model based on the surgical plan as described below in Sect. 3.2. The actual postoperative face served as the ground truth in the evaluation. For efficient training, 4,096 points were down-sampled from each original pre-bony and facial models, respectively.

3.2   Implementation and Evaluation Methods

We implemented three methods to predict postoperative facial appearance based on the bony movement. They were finite element model with realistic lip sliding effect (FEM-RLSE) [4], FC-Net [6], and our ACMT-Net. While the implementation details of FEM-RLSE and FC-Net can be found in [4,6], the implementation detail of ACMT-Net is described below.

During the point-wise pre-facial and pre-bony feature extractions, both PointNet++ networks were composed of 4 feature-encoding modules, followed by 4 feature-decoding modules. The output feature dimensions for each module were $\{128, 256, 512, 1024, 512, 256, 128, 128\}$, respectively. The input point number of the first feature-encoding module was 4,096 and the output point numbers of the extracted point-wise features from each module were $\{1024, 512, 256, 64, 256, 512, 1024, 4096\}$, respectively. In the CPSA module, the dimension of feature embedding was set to 64. We trained the ACMT-Net for 500 epochs with a batch size of 2 using Adam optimizer [4], after which the trained model was used for evaluation. The learning rate was initialized as 1e–3 and decayed to 1e–4 at discrete intervals during training. And α and β were empirically set to be 0.3 and 5, respectively. The models were trained on an NVIDIA DGX-1 deep learning server equipped with eight V100 GPUs.

The prediction accuracy of ACMT-Net was evaluated quantitatively and qualitatively by comparing it to FEM-RLSE and FC-Net. The quantitative

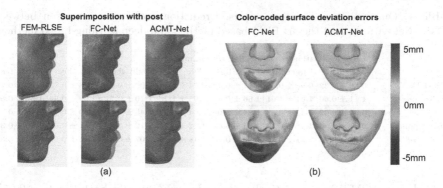

Fig. 3. Examples of the simulation results. (a) comparison of the simulated facial models (blue) with ground truth (red) (b) color-coded error map of simulated facial models compared with ground truth. (Color figure online)

evaluation was to assess the accuracy using the average surface deviation error between the predicted and the actual postoperative facial models [3]. In addition, using the facial landmarks, we divided the entire face into six regions, including nose, upper lip, lower lip, chin, right cheek and left cheek [3], and calculated the average surface deviation error individually. Repeated measures Analysis of Variances (ANOVA) was performed to detect whether there was a statistically significant difference among the FEM-RLSE, FC-Net and our ACMT-Net. If significant, post-hoc tests were used to detect the pairwise difference. A qualitative evaluation of the upper and lower lips, the most clinically critical region, was carried out by comparing the three predicted lips to the actual postoperative lips. One experienced oral surgeon performed the qualitative analysis twice using 2 points Likert scale (1: clinically acceptable; 2: clinically unacceptable). The second analysis was performed a week after the first one to minimize the memory effect. The final decision was made based on the two analysis results. Wilcoxon signed-rank test was performed to detect statistically significant difference. Finally, the time spent on data preparation and computation was recorded.

3.3 Results

The results of quantitative evaluation are shown in Table 1. The results of repeated measures ANOVA showed there was a statistically significant difference among the three methods ($p < 0.05$), but no significant difference was found among six different regions ($p > 0.05$). The results of post-hoc tests showed that our ACMT-Net statistically significantly outperformed the FC-Net ($p < 0.05$) in all regions. In addition, there was no statistically significance in nose, right cheek, left cheek and entire face ($p > 0.05$) between our ACMT-Net and the state-of-the-art FEM-RLSE method [3]. The results of the qualitative evaluation showed that among the 40 patients, the simulations of the upper/lower lips were clinically acceptable in 38/38 patients using FEM-RLSE, 32/30 using ACMT-Net, and 24/21 using FC-Net, respectively. The results of Wilcoxon signed-rank tests

Table 1. Quantitative evaluation results. Prediction accuracy comparison between ACMT-Net with state-of-the-art FEM-based method and deep learning-based method.

Method	Surface deviation error (mean ± std) in millimeter						
	Nose	Upper lip	Lower lip	Chin	Right cheek	Left cheek	Entire face
FEM-RLSE [3]	0.80 ± 0.32	0.96 ± 0.58	0.97 ± 0.44	0.74 ± 0.54	1.09 ± 0.72	1.08 ± 0.73	0.97 ± 0.48
FC-Net [9]	1.11 ± 0.60	1.61 ± 1.01	1.66 ± 0.82	1.95 ± 1.26	1.70 ± 0.77	1.59 ± 0.94	1.56 ± 0.58
No correspondence	0.90 ± 0.32	1.41 ± 0.89	1.84 ± 1.05	2.64 ± 1.71	1.45 ± 0.73	1.59 ± 0.73	1.48 ± 0.51
Closest point	0.81 ± 0.30	1.25 ± 0.59	1.34 ± 0.66	1.32 ± 0.75	1.12 ± 0.63	1.11 ± 0.61	1.10 ± 0.38
ACMT-Net	0.74 ± 0.27	1.21 ± 0.76	1.33 ± 0.72	1.18 ± 0.79	1.07 ± 0.56	1.06 ± 0.62	1.04 ± 0.39

showed that ACMT-Net statistically outperform FC-Net in both lips ($p < 0.05$). In addition, there was no statistically difference between ACMT-Net and FEM-RLSE in the upper lip ($p > 0.05$). Figure 3 shows two randomly selected patients. It clearly demonstrated that the proposed method was able to successfully predict the facial change following bony movement. Finally it took 70-80 s for our ACMT-Net to complete one simulation of facial appearance change. In contrast, it took 30 min for FEM-RLSE to completed one simulation.

3.4 Ablation Studies

We also conducted ablation studies to demonstrate the importance of the spatial correspondence between F_{F-pre} and F_{B-pre} in facial change simulation, and the effectiveness of our CPSA module in modeling such correspondence. Specifically, we considered two ablation models, denoted as 'No correspondence' and 'Closest point' in Table 1, respectively. For 'No correspondence', we solely used the pre-facial point set P_{F-pre} to infer the relation matrix R (by changing the output dimension of θ from $N_1 \times 64$ to $N_1 \times N_2$, shown in S1). For "Closest point", we determined the point-to-point correspondence by finding the nearest neighbor point from the pre-facial point set P_{F-pre} to the pre-bony point set P_{B-pre}. If the j-th bony point was the closest point to the i-th facial point, $R(i,j)$ was assigned to 1. Otherwise $R(i,j)$ was assigned to 0. As shown in Table 1, compared to the baseline model 'No correspondence', the two models with spatial correspondence ('Closest point' and our 'CPSA') lowered the error from 1.48mm to 1.10mm and 1.04mm, respectively. It suggested that 1) the spatial correspondence was crucial in modeling the physical interaction between facial and bony points; and 2) our design of the CPSA module was more effective than 'Closest point' in simulating such correspondence.

4 Discussions and Conclusions

The prediction accuracy evaluation using clinical data proved that ACMT-Net significantly outperformed FC-Net both quantitatively and qualitatively. The results also demonstrated that our ACMT-Net achieved a comparable quantitative accuracy compared with state-of-the-art FEM-RLSE while significantly reducing the prediction time in orthognathic surgical planning. The total prediction time is further reduced because our ACMT-Net does not require time-consuming FE mesh generation, thus significantly increases the potentials of using it as a prediction tool in daily clinical practice. One limitation of our proposed ACMT-Net is the accuracy. In FEM-RLSE, 95% (38/40) of the simulations using FEM-RLSE are clinically acceptable, whereas only about 75% (32 and 30 of 40) of the simulations using our ACMT-Net are clinically acceptable. This may be because that we do not include the anatomical details and the explicit physical interaction during the training due to the efficiency consideration. In the future, we will improve the simulation accuracy by adaptively sampling the points to include more anatomical details in clinically critical regions and integrate incremental learning strategy into our network. In particular, our work can be easily extended to other surgical applications that require correspondence mapping between different shapes/point sets [10,14].

In conclusion, a deep learning-based framework, ACMT-Net with CPSA module, is successfully implemented to accurately and efficiently predict facial change following bony movements. The developed network utilizes the spatial correspondence between two associated point sets to transform the movement of one point set to another. Ablation studies also proves the importance and effectiveness of the CPSA module in modeling the spatial correspondence between the associated point sets for movement transformation.

Acknowledgements. This work was partially supported by NIH under awards R01 DE022676, R01 DE027251 and R01 DE021863.

References

1. Faure, F., et al.: SOFA: a multi-model framework for interactive physical simulation. In: Payan, Y. (eds.) SMTEB 2012, vol. 11, pp. 283–321. Springer, Heidelberg (2012). https://doi.org/10.1007/8415_2012_125
2. Johnsen, S.F., et al.: NiftySim: a GPU-based nonlinear finite element package for simulation of soft tissue biomechanics. Int. J. Comput. Assist. Radiol. Surg. **10**(7), 1077–1095 (2015)
3. Kim, D., et al.: A novel incremental simulation of facial changes following orthognathic surgery using fem with realistic lip sliding effect. Med. Image Anal. **72**, 102095 (2021)
4. Kingma, D.P., Ba, J.: Adam: a method for stochastic optimization. arXiv preprint arXiv:1412.6980 (2014)
5. Lampen, N., et al.: Deep learning for biomechanical modeling of facial tissue deformation in orthognathic surgical planning. Int. J. Comput. Assist. Radiol. Surg. **17**(5), 945–952 (2022). https://doi.org/10.1007/s11548-022-02596-1
6. Li, X., et al.: Malocclusion treatment planning via pointNet based spatial transformation network. In: Martel, A.L., et al. (eds.) MICCAI 2020. LNCS, vol. 12263, pp. 105–114. Springer, Cham (2020). https://doi.org/10.1007/978-3-030-59716-0_11

7. Liu, Q., et al.: SkullEngine: a multi-stage CNN framework for collaborative CBCT image segmentation and landmark detection. In: Lian, C., Cao, X., Rekik, I., Xu, X., Yan, P. (eds.) MLMI 2021. LNCS, vol. 12966, pp. 606–614. Springer, Cham (2021). https://doi.org/10.1007/978-3-030-87589-3_62

8. Lorensen, W.E., Cline, H.E.: Marching cubes: a high resolution 3D surface construction algorithm. ACM SIGGRAPH Comput. Graph. **21**(4), 163–169 (1987). https://doi.org/10.1145/37402.37422

9. Ma, L., et al.: Deep simulation of facial appearance changes following craniomaxillofacial bony movements in orthognathic surgical planning. In: de Bruijne, M., et al. (eds.) MICCAI 2021. LNCS, vol. 12904, pp. 459–468. Springer, Cham (2021). https://doi.org/10.1007/978-3-030-87202-1_44

10. Pfeiffer, M., et al.: Non-rigid volume to surface registration using a data-driven biomechanical model. In: Martel, A.L., et al. (eds.) MICCAI 2020. LNCS, vol. 12264, pp. 724–734. Springer, Cham (2020). https://doi.org/10.1007/978-3-030-59719-1_70

11. Qi, C.R., Su, H., Mo, K., Guibas, L.J.: PointNet: deep learning on point sets for 3D classification and segmentation. In: Proceedings of the IEEE Conference on Computer Vision and Pattern Recognition, pp. 652–660 (2017)

12. Qi, C.R., Yi, L., Su, H., Guibas, L.J.: PointNet++: deep hierarchical feature learning on point sets in a metric space. In: Advances in Neural Information Processing Systems 30 (2017)

13. Shafi, M., Ayoub, A., Ju, X., Khambay, B.: The accuracy of three-dimensional prediction planning for the surgical correction of facial deformities using Maxilim. Int. J. Oral Maxillofac. Surg. **42**(7), 801–806 (2013)

14. Tagliabue, E., et al.: Intra-operative update of boundary conditions for patient-specific surgical simulation. In: de Bruijne, M., et al. (eds.) MICCAI 2021. LNCS, vol. 12904, pp. 373–382. Springer, Cham (2021). https://doi.org/10.1007/978-3-030-87202-1_36

15. Vaswani, A., et al.: Attention is all you need. In: Advances in Neural Information Processing Systems 30 (2017)

16. Wang, X., Girshick, R., Gupta, A., He, K.: Non-local neural networks. In: Proceedings of the IEEE Conference on Computer Vision and Pattern Recognition, pp. 7794–7803 (2018)

17. Xia, J.J., Gateno, J., Teichgraeber, J.F.: New clinical protocol to evaluate craniomaxillofacial deformity and plan surgical correction. J. Oral Maxillofac. Surg. **67**(10), 2093–2106 (2009)

18. Xiao, D., et al.: A self-supervised deep framework for reference bony shape estimation in orthognathic surgical planning. In: de Bruijne, M., et al. (eds.) MICCAI 2021. LNCS, vol. 12904, pp. 469–477. Springer, Cham (2021). https://doi.org/10.1007/978-3-030-87202-1_45

19. Yin, K., Huang, H., Cohen-Or, D., Zhang, H.: P2P-NET: bidirectional point displacement net for shape transform. ACM Trans. Graph. (TOG) **37**(4), 1–13 (2018). https://doi.org/10.1145/3197517.3201288

Deep Learning-Based Head and Neck Radiotherapy Planning Dose Prediction via Beam-Wise Dose Decomposition

Bin Wang[1,2], Lin Teng [3,1], Lanzhuju Mei[1], Zhiming Cui[1,4], Xuanang Xu[5], Qianjin Feng[3(✉)], and Dinggang Shen[1,6(✉)]

[1] School of Biomedical Engineering, ShanghaiTech University, Shanghai, China
`dgshen@shanghaitech.edu.cn`
[2] School of Information Science and Technology, ShanghaiTech University, Shanghai, China
[3] School of Biomedical Engineering, Southern Medical University, Guangzhou, China
`fengqj99@smu.edu.cn`
[4] Department of Computer Science, The University of Hong Kong, Hong Kong, China
[5] Department of Biomedical Engineering, Rensselaer Polytechnic Institute, Troy, USA
[6] Shanghai United Imaging Intelligence Co., Ltd., Shanghai, China[3,1]

Abstract. Accurate dose map prediction is key to external radiotherapy. Previous methods have achieved promising results; however, most of these methods learn the dose map as a black box without considering the beam-shaped radiation for treatment delivery in clinical practice. The accuracy is usually limited, especially on beam paths. To address this problem, this paper describes a novel "disassembling-then-assembling" strategy to consider the dose prediction task from the nature of radiotherapy. Specifically, a global-to-beam network is designed to first predict dose values of the whole image space and then utilize the proposed innovative beam masks to decompose the dose map into multiple beam-based sub-fractions in a beam-wise manner. This can disassemble the difficult task to a few easy-to-learn tasks. Furthermore, to better capture the dose distribution in region-of-interest (ROI), we introduce two novel value-based and criteria-based dose volume histogram (DVH) losses to supervise the framework. Experimental results on the public OpenKBP challenge dataset show that our method outperforms the state-of-the-art methods, especially on beam paths, creating a trustable and interpretable AI solution for radiotherapy treatment planning. Our code is available at https://github.com/ukaukaaaa/BeamDosePrediction.

Keywords: Head and neck cancer · Radiation therapy · Dose prediction · Beam mask · Dose volume histogram

---

B. Wang and L. Teng—Equal Contribution

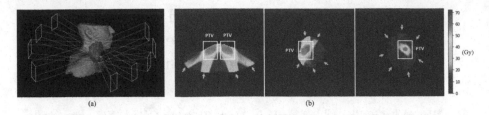

Fig. 1. (a) Demonstration of external radiotherapy with beam-shaped radiation. (b) Different slices of dose map with clear beam paths.

1   Introduction

External radiotherapy is a mainstream therapy widely used for head and neck cancer treatment. Its efficacy highly relies on high-quality treatment plans, in which a dose volume is elaborately designed to deliver a prescribed dose of radiation to the tumor while minimizing the irradiation received by organs-at-risk (OARs). In clinical workflow, this procedure is often accomplished by physicians manually adjusting numerous planning parameters and weights in a trial-and-error manner [2,4,7], which is not only time-consuming but also requires a great level of expertise. Hence, it is greatly demanded to develop an automatic method to predict accurate dose map for cancer treatment planning.

In recent years, due to the explosive development of machine learning techniques [11,13], many deep learning-based methods have been proposed to handle this challenging task. The prior efforts can be generally categorized into three groups. The first group of methods focuses on designing variant neural network architectures. For example, Liu et al. [6] have designed a cascaded 3D U-Net model, incorporating global and local anatomical features. The second group uses novel loss functions. Ngyuen et al. [10] have demonstrated that more accurate dose map can be generated when DVH loss is included. The last line methods propose to exploit additional prior knowledge and integrate into the network learning, including the gradient information [12] and distance information [15].

In external radiotherapy, treatment is achieved by delivering the radiation from several different directions (Fig. 1). Each direction of radiation will result in a beam-shaped dose volume. Due to the beam-wise delivery manner, the resulting dose volume often exhibits sharp edges near the beam boundary. The dose intensities inside the beam regions are much higher than those outside the beam regions. However, such critical prior knowledge is hardly considered in the previous methods, which often leads to some unsatisfying dose distributions on the beam paths and finally affects the prediction performance. This is because the input CT images do not contain any knowledge of the radiation beams. Thus, it is difficult for the deep network to infer the dose distribution accurately without extra prior knowledge of the beam shapes.

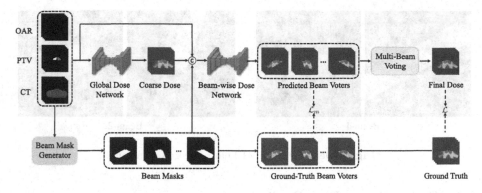

Fig. 2. An overview of our proposed method framework, including (a) global coarse dose prediction and (b) beam-wise dose prediction based on decomposition and multi-beam voting mechanism.

In this paper, to tackle the aforementioned challenges, we present a novel beam mask definition and take it as a prior knowledge for deep network training. To fully utilize this knowledge, a global-to-beam network embedded with multiple innovative strategies is used to conduct the dose prediction task from global to local, denoted as "disassembling-then-assembling" strategy. Specifically, in the first stage, we adopt Global Dose Network to coarsely predict the dose map, and simultaneously generate beam masks according to pre-defined angles and the planning target volume (PTV) masks. Then, in the second stage, we fine-tune the dose map through three novel strategies described as follows.

The main contributions of this work are three-fold. 1) The coarse dose map is guided by multiple beam masks from different angles, and decomposed into several sub-fractions of dose map (i.e., disassembling). Each sub-fraction is responsible for the prediction of corresponding beam path. 2) A multi-beam voting mechanism is proposed to reconstruct the final dose map, in which each voxel value is only determined by the sub-fractions containing that voxel (i.e., assembling). 3) We introduce a value-based DVH loss and a criteria-based DVH loss to focus on ROI regions more accurately and efficiently. To validate our proposed method, extensive experiments have been conducted and show that our method outperforms the state-of-art approaches both qualitatively and quantitatively.

2 Method

Although previous methods have achieved promising performance, we still find many regions with inaccurately-predicted dose values, especially along the beam paths. To solve this problem, we introduce a novel beam mask generator and a global-to-beam network by first predicting dose values of the whole image space and then decomposing the dose map into multiple beam voters by utilizing the proposed beam masks to conduct beam-wise dose prediction. The framework of our proposed method is illustrated in Fig. 2 with details described in the following sub-sections.

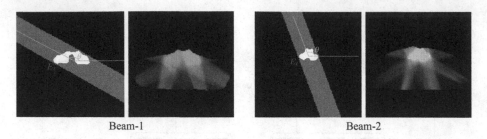

Beam-1   Beam-2

Fig. 3. Demonstration of beam mask generation.

2.1 Global Coarse Dose Prediction

In our framework, the input includes CT image, PTV, and OAR masks. Specifically, CT provides anatomical information. PTV indicates cancer region information (requiring high dose radiation), and OAR masks refer to normal tissues. As shown in Fig. 2, we first employ Global Dose Network to coarsely predict dose values of the whole image space. More importantly, before Beam-wised Dose Network at second-staged, we introduce additional prior knowledge (i.e., beam masks) that is helpful for dose map refinement.

2.2 Beam-Wise Dose Prediction

The output of the first stage is a coarse dose map, in which the basic shape of the PTV region and dose values are predicted roughly, especially for the results on the beam paths. To solve this problem, we generate beam masks according to the pre-defined angles and PTV masks. Then, Beam-wise Dose Network at the second stage refines the dose map in the manner of decomposition and voting strategies. The critical components in beam-wise dose prediction model are elaborated as below.

Beam Mask Generator. Introducing the beam information as a prior knowledge can provide a more targeted manner for predicting dose values on the beam paths. However, the beam information is not provided in most datasets. Thus, the main problem is how to acquire and represent such important information. To address this challenge, we propose a beam mask generator (Fig. 3). The beam path is mainly related to the location of PTV and the angle of the beam. Since the PTV location is different on each slice, we build the beam mask slice-by-slice to make it more suitable for the beam path, which can be defined as:

$$B(t) = F(\theta, E_t), \tag{1}$$

where θ refers to the angle and E_t includes the coordinates of PTV edges on the t-th slice. The detailed process of $F(\cdot)$ is to first calculate the slope of the beam and then utilize the points on edge boundary to find the intercept, resulting in two lines at the end. The region between these two lines represents the respective beam mask.

Decomposition of Dose Map. In each slice of dose map, there are usually more than one beam paths, whose angles and dose value distributions are different, causing the difficulty to jointly fine-tune all beam path regions. Different from the existing dose prediction techniques, our proposed method predicts multiple sub-fractions of the dose map, named beam voter. Each beam voter is responsible for the dose prediction on one beam mask. By decomposing the dose prediction of a whole image space into a set of beam-based prediction tasks, it is relatively easier for the refinement network to learn features of each beam-mask region. In the training process, we use a multi-beam MAE loss to supervise the prediction on each beam-mask region independently, which can be defined as:

$$\mathcal{L}_m = \frac{1}{N_i} \sum_{i=1}^{N_i} |P_i - G_i|, \tag{2}$$

where P_i and G_i refer to the prediction and the ground truth of the i-th beam voter. N_i denotes the maximum number of the beam voters.

Multi-beam Voting. The output of the refinement network in the second stage is multiple beam voters, each of which represents one sub-region of the final dose map. To merge these beam voters, we propose a Multi-Beam Voting mechanism, in which the dose value of each voxel is finally predicted and voted by the beams passing this voxel. Note that if it is only passed by one single beam, its dose value will be set by the same location on the corresponding beam voter. On the other hand, if a voxel is passed by multiple beams, all beam voters are responsible for this voxel to generate the final dose value by an average voting operation. After voting, we use MAE loss $\mathcal{L}_r$ to supervise the final prediction.

2.3 Training Objective

Value-Based DVH Loss. As a commonly used metric in radiotherapy, DVH has been considered as an important tool to enhance the dose prediction quality in ROI. Although many existing methods have proposed DVH loss [10] to calculate the volume difference between prediction and ground-truth DVH curves, it requires repeated image-based computations since a processing of the whole 3D image is needed for each dose value threshold. If we reduce the computational consumption by increasing the threshold interval, it would lose the accuracy to fit the ground-truth DVH curve. Hence, we proposed a value-based DVH loss to balance the efficiency and accuracy, which is defined as follows:

$$\mathcal{L}_{vDVH} = \frac{\sum_{s=1} \sum_{n=1}^{N_s} \left| R(\hat{Y} \cdot M_s)_n - R(Y \cdot M_s)_n \right|}{\sum_{s=1} N_s}, \tag{3}$$

where M_s denotes the ROI masks. $R(\cdot)$ is a sort operation. $\hat{Y}$ and Y refer to the prediction and ground-truth dose maps, respectively.

Specifically, we first use the ROI mask to extract the dose values in ROI region and obtain the DVH curve by conducting the sort operation. Then, we compute the difference between the prediction and the ground-truth DVH to supervise the network. In this way, the volume information can be represented by the rank of sorted dose values, with no hyper-parameter set in this loss function. Note that only one round computation is needed on the whole image, and the processing is directly employed on the dose values, which is more efficient and accurate.

Criteria-Based DVH Loss. In clinical treatment planning, the quality of the planning dose is evaluated by checking a set of critical points on the DVH curve, such as $C_1^{ptv}, C_{95}^{ptv}, C_{99}^{ptv}, C_{0.1cc}^{oar}$, and C_{mean}^{oar}, which indicate the dose values received by the top-ranked 1%, 95%, 99% volume of PTV, the dose value received by the top-ranked 0.1cc volume of OAR, and the mean dose received by OAR, respectively. We represent these criteria by calculating the sorted dose values ranked at the 99th, 5th, and 1st percentile inside the PTVs, and the maximum and mean of sorted dose values inside the OARs, respectively. Then, we define criteria-based DVH loss as

$$\mathcal{L}_{cDVH} = D_1^{ptv} + D_{95}^{ptv} + D_{99}^{ptv} + D_{0.1cc}^{oar} + D_{mean}^{oar}, \qquad (4)$$

where $D_1^{ptv}, D_{95}^{ptv}, D_{99}^{ptv}, D_{0.1cc}^{oar}, D_{mean}^{oar}$ are the difference between prediction and the ground truth calculated at the corresponding criteria.

Finally, the total loss function will be elaborated as below:

$$\mathcal{L} = \mathcal{L}_r + \alpha \mathcal{L}_{vDVH} + \beta \mathcal{L}_{cDVH} \qquad (5)$$

where α and β are hyper-parameters to balance the two DVH loss terms.

3   Experiment

We evaluate our proposed method on the public OpenKBP dataset [1] from 2020 AAPM Grand Challenge, which includes 340 head and neck cancer patients (200 for training, 40 for validation, and 100 for testing). For each patient, paired CT scan, OAR mask, PTV mask, possible dose mask, and ground-truth dose map are provided with the same size of $128 \times 128 \times 128$. The model performance is tracked with two metrics used in the Challenge above, i.e., Dose score and DVH score. Moreover, we introduce the DVH curve as another tool to evaluate the accuracy of dose prediction in ROI.

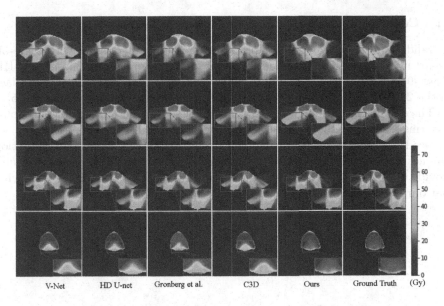

V-Net   HD U-net   Gronberg et al.   C3D   Ours   Ground Truth   (Gy)

Fig. 4. Visualization of our proposed method and SOTA methods. An obvious improvement on beam paths can be seen in the red enlarged boxes. (Color figure online)

For the first stage of coarse prediction, random rotation ranged from -20° to 20° and random flip along Z-axis are performed to augment the data. Note that the PTV masks and OAR masks are merged in one mask, where the different mask was assigned with a different label. The beam masks are generated based on Eq. (1), where θ is set according to the pre-defined angles in the dataset [1]. Both Global Dose Net and Beam-wise Dose Net have a U-Net structure. To supervise the networks, the Adam optimizer with an initial learning rate of 1e−4 is used during the training process. And each epoch takes about two minutes on a GPU of NVIDIA Tesla M40 24 GB.

Table 1. Quantitative comparison with state-of-the-art methods (*: no released code; †: p-value < 0.05).

Method	Dose score [Gy]	DVH score [Gy]
V-Net [8]	$2.922 \pm 1.166^{\dagger}$	$1.545 \pm 1.178^{\dagger}$
Xu et al. [14]	2.753^{*}	1.559^{*}
Zimmermann et al. [16]	$2.620 \pm 1.100^{*}$	$1.520 \pm 1.060^{*}$
HD U-net [9]	$2.592 \pm 1.048^{\dagger}$	$1.643 \pm 1.123^{\dagger}$
Gronberg et al. [3]	$2.563 \pm 1.143^{\dagger}$	$1.704 \pm 1.096^{\dagger}$
C3D [6]	$2.429 \pm 1.031^{\dagger}$	$1.478 \pm 1.182^{\dagger}$
Lin et al. [5]	2.357^{*}	1.465^{*}
Ours	$\mathbf{2.276 \pm 1.013}$	$\mathbf{1.257 \pm 1.163}$

3.1   Comparison with State-of-the-Art Methods

To validate the advantage of our proposed method, we compare with state-of-the-art methods, including V-Net [8], Xu et al. [14], Zimmermann et al. [16], HD U-net [9], Gronberg et al. [3] and C3D [6], which won the first and second place on the 2020 AAPM challenge learderboard [1], respectively, and Lin et al. [5].

The quantitative results are shown in Table 1. Our method significantly outperforms existing methods, in terms of dose score and DVH score. Moreover, we can see from the visualization results of dose maps in Fig. 4 that our prediction matches better with the ground truth, especially on the beam paths. We also provide the DVH curves of the prediction and the ground truth in Fig. 5 showing that our DVH curves are closer to the ground-truth DVH curves.

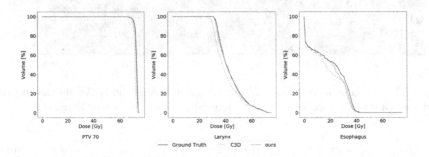

Fig. 5. Visualization of DVH curves by our method and SOTA methods, including DVH curves of PTV70, Larynx, and Esophagus.

3.2   Ablation Study

In this study, we take a cascaded 3D U-Net as the Baseline model and evaluate the effectiveness of two key components of our proposed method: (1) beam-wise dose prediction (BDP) including the decomposition of dose map and multi-beam voting scheme, and (2) valued-based DVH loss (L_{vDVH}) and criteria-based DVH loss (L_{cDVH}). The quantitative results are presented in Table 2. It can be seen that BDP improves the prediction accuracy (46.1% improvements in terms of dose score), indicating that the introduction of the beam masks facilitates the network to learn more features from beam paths region. Then, adding L_{vDVH} shows 10.8% improvement of DVH score and also makes prediction of dose map more efficient, i.e., reducing the computational time from 6 min an epoch by previous DVH loss [10] to 2 min an epoch. Additionally, due to the fact that L_{cDVH} incorporates key criteria of treatment plannings in clinics, it helps promote dose score by 12.2% and dvh score by 20.2%, making our prediction the best result in this dose prediction task.

Table 2. Ablation study of our method, evaluated with dose score and DVH score.

Baseline	BDP	L_{vDVH}	L_{cDVH}	Dose score [Gy]	DVH score [Gy]
✓				2.862 ± 1.049	1.586 ± 1.146
✓	✓			2.401 ± 1.033	1.567 ± 1.179
✓	✓	✓		2.398 ± 1.011	1.459 ± 1.212
✓	✓	✓	✓	$\mathbf{2.276 \pm 1.014}$	$\mathbf{1.257 \pm 1.163}$

4 Conclusion

We describe a novel "disassembling-then-assembling" strategy and propose a global-to-beam framework to accurately conduct the dose prediction task. The proposed method first learns the whole image space of the dose map and then decomposes it into beam-based sub-fractions by proposed beam masks. Moreover, we get the final dose map by utilizing multi-beam voting strategy. Besides, we propose two novel value-based and criteria-based DVH loss to focus on ROI region efficiently. Experimental results demonstrate our method can more precisely predict dose map compared with state-of-the-art methods. The predicted dose is very close to the physically deliverable one and thus can be used as a good starting point in treatment planning, substantially reducing the time and inter-observer variations in clinical workflow. We aim to leverage this method to create trustable and interpretable AI solution for radiotherapy.

References

1. Babier, A., et al.: OpenKBP: the open-access knowledge-based planning grand challenge and dataset. Med. Phys. **48**(9), 5549–5561 (2021)
2. Dias, J., Rocha, H., Ferreira, B., do Carmo Lopes, M.: Simulated annealing applied to IMRT beam angle optimization: a computational study. Physica Medica **31**(7), 747–756 (2015)
3. Gronberg, M.P., Gay, S.S., Netherton, T.J., Rhee, D.J., Court, L.E., Cardenas, C.E.: Dose prediction for head and neck radiotherapy using a three-dimensional dense dilated u-net architecture. Med. Phys. **48**(9), 5567–5573 (2021)
4. Kearney, V., Chan, J.W., Valdes, G., Solberg, T.D., Yom, S.S.: The application of artificial intelligence in the IMRT planning process for head and neck cancer. Oral Oncol. **87**, 111–116 (2018)
5. Lin, Y., Liu, Y., Liu, J., Liu, G., Ma, K., Zheng, Y.: LE-NAS: learning-based ensemble with NAS for dose prediction. arXiv preprint arXiv:2106.06733 (2021)
6. Liu, S., Zhang, J., Li, T., Yan, H., Liu, J.: A cascade 3D U-Net for dose prediction in radiotherapy. Med. Phys. **48**(9), 5574–5582 (2021)
7. Men, C., Romeijn, H.E., Taşkın, Z.C., Dempsey, J.F.: An exact approach to direct aperture optimization in IMRT treatment planning. Phys. Med. Biol. **52**(24), 7333 (2007)
8. Milletari, F., Navab, N., Ahmadi, S.A.: V-Net: fully convolutional neural networks for volumetric medical image segmentation. In: 2016 Fourth International Conference on 3D Vision (3DV), pp. 565–571. IEEE (2016)

9. Nguyen, D., et al.: 3D radiotherapy dose prediction on head and neck cancer patients with a hierarchically densely connected u-net deep learning architecture. Phys. Med. Biol. **64**(6), 065020 (2019)
10. Nguyen, D., et al.: Incorporating human and learned domain knowledge into training deep neural networks: a differentiable dose-volume histogram and adversarial inspired framework for generating pareto optimal dose distributions in radiation therapy. Med. Phys. **47**(3), 837–849 (2020)
11. Pan, Y., Liu, M., Lian, C., Zhou, T., Xia, Y., Shen, D.: Synthesizing missing PET from MRI with cycle-consistent generative adversarial networks for Alzheimer's disease diagnosis. In: Frangi, A.F., Schnabel, J.A., Davatzikos, C., Alberola-López, C., Fichtinger, G. (eds.) MICCAI 2018. LNCS, vol. 11072, pp. 455–463. Springer, Cham (2018). https://doi.org/10.1007/978-3-030-00931-1_52
12. Tan, S., et al.: Incorporating isodose lines and gradient information via multi-task learning for dose prediction in radiotherapy. In: de Bruijne, M., et al. (eds.) MICCAI 2021. LNCS, vol. 12907, pp. 753–763. Springer, Cham (2021). https://doi.org/10.1007/978-3-030-87234-2_71
13. Xiang, L., et al.: Deep embedding convolutional neural network for synthesizing CT image from t1-weighted MR image. Med. Image Anal. **47**, 31–44 (2018)
14. Xu, X., et al.: Prediction of optimal dosimetry for intensity-modulated radiotherapy with a cascaded auto-content deep learning model. Int. J. Radiat. Oncol. Biol. Phys. **111**(3), e113 (2021)
15. Zhang, J., Liu, S., Yan, H., Li, T., Mao, R., Liu, J.: Predicting voxel-level dose distributions for esophageal radiotherapy using densely connected network with dilated convolutions. Phys. Med. Biol. **65**(20), 205013 (2020)
16. Zimmermann, L., Faustmann, E., Ramsl, C., Georg, D., Heilemann, G.: Dose prediction for radiation therapy using feature-based losses and one cycle learning. Med. Phys. **48**(9), 5562–5566 (2021)

Ideal Midsagittal Plane Detection Using Deep Hough Plane Network for Brain Surgical Planning

Chenchen Qin[1], Wenxue Zhou[1,3], Jianbo Chang[2], Yihao Chen[2], Dasheng Wu[1], Yixun Liu[1], Ming Feng[2], Renzhi Wang[2], Wenming Yang[3], and Jianhua Yao[1(✉)]

[1] Tencent AI Laboratory, Shenzhen, China
jianhuayao@tencent.com
[2] Peking Union Medical College Hospital, Beijing, China
[3] Shenzhen International Graduate School, Tsinghua University, Shenzhen, China

Abstract. The ideal midsagittal plane (MSP) approximately bisects the human brain into two cerebral hemispheres, and its projection on the cranial surface serves as an important guideline for surgical navigation, which lays a foundation for its significant role in assisting neurosurgeons in planning surgical incisions during preoperative planning. However, the existing plane detection algorithms are generally based on iteration procedure, which have the disadvantages of low efficiency, poor accuracy, and unable to extract the non-local plane features. In this study, we propose an end-to-end deep Hough plane network (DHPN) for ideal MSP detection, which has four highlights. First, we introduce differentiable deep Hough transform (DHT) and inverse deep Hough transform (IDHT) to achieve the mutual transformation between semantic features and Hough features, which converts and simplifies the plane detection problem in the image space into a keypoint detection problem in the Hough space. Second, we design a sparse DHT strategy to increase the sparsity of features, improving inference speed and greatly reducing calculation cost in the voting process. Third, we propose a Hough pyramid attention network (HPAN) to further extract non-local features by aggregating Hough attention modules (HAM). Fourth, we introduce dual space supervision (DSS) mechanism to integrate training loss from both image and Hough spaces. Through extensive validations on a large inhouse dataset, our method outperforms state-of-the-art methods on the ideal MSP detection task.

Keywords: Midsagittal plane detection · Deep hough transform · Hough pyramid attention

C. Qin and W. Zhou–Contributed equally to this work

Supplementary Information The online version contains supplementary material available at https://doi.org/10.1007/978-3-031-16449-1_56.

L. Wang (Eds.): MICCAI 2022, LNCS 13437, pp. 585–593, 2022.
https://doi.org/10.1007/978-3-031-16449-1_56

1   Introduction

Normal human brain holds an anatomically ideal midsagittal plane (MSP) that approximately bisects it into two symmetrical cerebral hemispheres. The intersection line of the ideal MSP and cranial surface serves as an important guideline for surgical navigation. At present, the guideline is delineated manually in clinical practice according to the nearly symmetrical features on the cranial surface, such as nose tip and eyes, which do not reflect the real symmetry inside the head. Several researches [1] have found that the left and right hemispheres of the human brain are not completely symmetrical in size, shape, connectivity and many other aspects. Moreover, MSP is susceptible to pathological factors such as intracerebral hematoma, brain injury, tumor and so on [2], making it deviate from the position of bilateral symmetry. Therefore, the manually delineated guideline may not be sufficient to assist neurosurgeons in designing surgical incisions during preoperative planning (Fig. 1). Consequently, it is considerably crucial to develop a robust algorithm for extracting and visualizing the ideal MSP, which remains an inherent challenge.

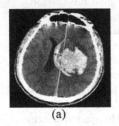 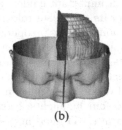 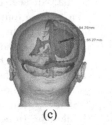

(a)   (b)   (c)

Fig. 1. Schematic diagram of planning surgical incision in case with intracerebral hematoma. (a) is an axial slice, where yellow line denotes the ideal MSP and red arrow indicates the midline offset. (b) denotes the 3D illustration of (a). (c) shows the surgical incision planning as the red dotted line. Green line is the guideline where the ideal MSP intersects the cranial surface. Black line indicates the planned entry direction of the puncture needle. (Color figure online)

In the field of standard plane detection and localization, there are typically three types of methods. The first type of method directly detects the plane. H. Chen et al. [3] and Z. Yu et al. [4] identified fetal facial standard plane by using a classification network based on CNN to judge whether 2D oblique slices with different angles and positions belong to the standard plane. However, these methods based on iteration procedure generally take a huge amount of computation due to the large searching space. H. Dou et al. [5] proposed a reinforcement learning (RL) framework to search for an optimal fetal MSP. There are also methods that directly estimate parameters of plane based on regression network [6]. The above methods generally regard plane detection as a special case of target detection. Consequently, they commonly have the disadvantages of low efficiency, poor accuracy, and unable to extract the non-local feature of plane, leading to unsatisfactory and suboptimal results.

The second type of method first detects landmarks and then fits a plane using them. H. Wu et al. [7] fitted the MSP through keypoint detection and iterative least-median of squares plane regression. However, landmarks in brain are vulnerable to the pressure caused by abnormalities such as intracerebral hematoma and tumor, resulting in uncertain deviation. Even we select landmarks with high probability located on the skull midline and have relatively constant locations, they are prone to be not detectable due to improper scanning and other factors. In addition, insufficient number of landmarks will also make the plane fitting unstable, which ultimately leads to the poor robustness and accuracy of the plane detection based on the location of landmarks.

The third type of method is based on the Hough transform. Qi Han et al. [8] incorporated the classical Hough transform into deeply learned representations to detect straight lines in photographic images. Nevertheless, when Hough transform is directly applied to plane detection task, it is necessary to vote and accumulate the probability of each feature point on the MSP in the Hough space, resulting in high computation cost. Moreover, it also causes many local pseudo-peaks in the Hough space so that it is ambiguous to find the optimal solution.

To address the aforementioned issues, we propose an end-to-end deep Hough plane network (DHPN) for ideal MSP detection, which combines the feature learning capability of CNN with Hough transform to obtain an efficient solution for semantic plane detection. Firstly, we introduce the differentiable deep Hough transform (DHT) and inverse deep Hough transform (IDHT) into CNN architecture, which realizes the mutual transformation of features between the image space and the Hough space, and simplifies the problem of detecting the ideal MSP of the sample in the image space as seeking the keypoint response in the Hough space. Secondly, we propose a strategy of sparse DHT, which increases the sparsity of features and greatly reduces the computation cost in the voting process of Hough space. Thirdly, we design a Hough pyramid attention network (HPAN) to further extract non-local features by aggregating Hough attention module (HAM), which can be combined with other network architectures as a plug-in component. Fourthly, we introduce dual space supervision (DSS) that utilizes both the semantic plane heatmap and Hough keypoint heatmap to constrain the planar expression forms in the image space and the Hough space respectively. All these components are unified in DHPN to obtain more accurate and more robust plane detection solutions.

2 Methodology

As shown in Fig. 2, the proposed DHPN framework consists of four main components: (1) a semantic feature extractor formed by cascading 3D ResNet [9] with FPN [10] to extract pixel-wise representations of MSP features from the original 3D images and to generate semantic plane heatmap in the image space. (2)

HPAN structure with multiple HAMs to enhance the model capability in non-local feature extraction. (3) a Hough feature extractor containing DHT modules to yield Hough keypoint heatmap in the Hough space. (4) a DSS module to integrate loss from both image and Hough spaces.

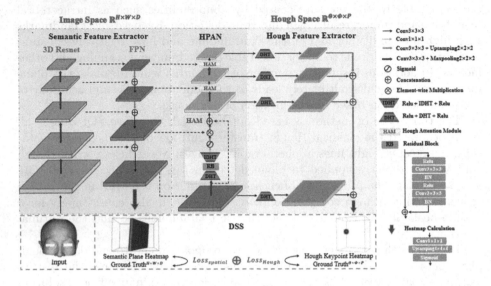

Fig. 2. The overall framework of the proposed DHPN for ideal MSP detection.

2.1 DHT and IDHT

We set the center of a given 3D image $I^{H \times W \times D}$ as the origin, where H, W and D are the image spatial size. In the polar coordinate system, the expression of a plane p is defined as:

$$x \sin \phi \cos \theta + y \sin \phi \sin \theta + z \cos \phi = \rho \qquad (1)$$

where θ and ϕ represent the angles between p and the x-axis and between p and the z-axis respectively, and ρ represents the vertical distance from the origin to p. $\theta \in [0°, 180°)$, $\phi \in [-90°, +90°)$, $\rho \in [-L/2, L/2]$. $L = \sqrt{(H^2 + W^2 + D^2)}$, represents the diagonal length of the volume. Then the size of the Hough space can be calculated by:

$$\Delta \alpha = arctan \frac{1}{max(H/2, W/2, D/2)} \qquad (2)$$

$$size(\Theta, \Phi, P) = (\frac{range(\theta)}{\Delta \alpha}, \frac{range(\phi)}{\Delta \alpha}, \frac{range(\rho)}{\Delta \rho}) \qquad (3)$$

where $\Delta\alpha$ and $\Delta\rho$ represent the optimal step sizes for angle and distance respectively. Thus, the discretized image space $R^{H \times W \times D}$ can match with the discretized Hough space $R^{\Theta \times \Phi \times P}$. DHT is responsible for allocating and voting the plane features extracted by CNN in the image space to the corresponding positions in the Hough space. That is, each feature point p_i in the spatial domain is destined to get a parameter set after DHT that defines planes where it may be located on. Then the voting score of all parameters in the set is increased by 1 respectively with an accumulator. Once all feature points have been voted, the peak point (θ, ϕ, ρ) corresponding to the ideal MSP is obtained. While IDHT converts features of keypoints in the Hough space into semantic plane features in the image space. By executing IDHT, a series of plane bundles are obtained, where the highlighted position is the ideal MSP. The process is illustrated in Fig. 3 and detailed steps are described in the supplementary material. The forward propagation of either DHT or IDHT is the reverse propagation process of the other. Consequently, DHT-IDHT pair ensures the mutual transformation of features in image space and Hough spaces, so that the problem of detecting semantic planes in the spatial domain is transformed into a keypoint identification problem in the Hough parametric domain, making the post-processing steps more efficient.

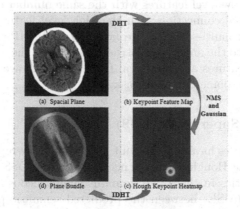

Fig. 3. Illustration of DHT and IDHT operations. The plane features are transformed into the keypoint features through DHT. Then the non-maximization suppression (NMS) is used to obtain the peak point, whose coordinates are the parameters of the ideal MSP, and applied with Gaussian filter to improve the searching range. IDHT converts the keypoint features into a series of plane bundles.

2.2 Sparse DHT Strategy

According to the voting principle of Hough transform, the processes of DHT and IDHT take multiple cycles, and their time complexities reach $O(n^3)$. Since the sparsity of features is crucial to the improvement of the speed, we propose a sparse DHT strategy. The specific improvements to the voting mechanism are described as follows: (1) Add Relu layers as the front layer of DHT and IDHT modules to increase the sparsity of features; (2) Pixels with value of 0 in the image space are excluded from the voting process. With this strategy, feature points become vastly sparse, which directly leads to a significant reduction in computation cost and inference time by nearly 20 times in our experiments.

2.3   Hough Pyramid Attention Network (HPAN)

Some existing methods have proposed a variety of attention modules to improve the performance of CNN in extracting non-local features [11,12]. Whereas, they all work in the image space, and fail to meet the requirement of considering global information due to the limitation of perception field. In this paper, we introduce attention mechanism into the Hough space as Hough attention Module (HAM). Specifically, the formula for calculating plane attention map is define as:

$$y = \epsilon(h^{-1}(f(h(x)))) \tag{4}$$

where h and h^{-1} denote the DHT and IDHT operation respectively. f and ϵ represent the residual block and sigmoid function. Plane attention map eventually weights on the low level features with the same number of channels, and its results makes residual connection with the high level features. By aggregating a series of HAMs in multi-scale feature layers of FPN, HPAN is established to put more attention to the anatomical structures around the MSP, and efficiently extract geometric features and non-local features, which are highly critical and suitable for accurate plane detection task.

2.4   Dual Space Supervision (DSS)

It is noted that features have different but complementary representations in the image space and the Hough space. Hence, we propose DSS strategy to supervise the characteristics in these two domains simultaneously. We compute the mean square error (MSE) between the predicted semantic plane heatmap and its smoothed ground truth which is obtained by using a Gaussian filter of $\sigma = 2$ to the binary standard image and then normalizing its pixels to $[0, 1]$. Similarly, we calculate the MSE between the predicted Hough keypoint heatmap and its smoothed ground truth. The loss function is defined as:

$$Loss_{spacial} = \frac{1}{N}\sum_{i=1}^{N}(P_{out} - P_{gt})^2 \tag{5}$$

$$Loss_{Hough} = \frac{1}{M}\sum_{i=1}^{M}(H_{out} - H_{gt})^2 \tag{6}$$

$$Loss_{total} = (1 - \lambda)Loss_{spacial} + \lambda Loss_{Hough} \tag{7}$$

where P_{out} and P_{gt} denote the prediction and its ground truth in the image space, while H_{out} and H_{gt} represent those in the Hough space. N and M are the number of voxels for these two volumes respectively. λ is the hyperparameter determining the weight, which is set to 0.1. $Loss_{total}$ denotes the total loss function, which is the weighted sum of $Loss_{spatial}$ and $Loss_{Hough}$.

3   Experiments

3.1   Dataset and Implementation Details

We validated the proposed DHPN framework on an in-house dataset consisting of 519 CT volumes collected from our partner hospitals, including all cases with intracranial haemorrhagic disease and 221 cases with midline shift greater than 5 mm. Specifically, the volume size of images in our dataset is $192 \times 192 \times 160$ with a unified voxel spacing of $1.0 \times 1.0 \times 1.0\,mm^3$. The corresponding size in the Hough space is $288 \times 288 \times 193$ according to formula (2–3). A clinician manually annotated 14 landmarks evenly distributed on the MSP (including the bony structural landmarks such as crista galli, internal occipital protuberance, the central point of foramen magnum, dorsum sellae, vertex, etc., as well as the intracerebral landmarks such as the brain stem, ventricle, etc.) (Fig. 4). The ground truth MSPs were fitted by using the least square method and further reviewed by an experienced medical expert. The images were randomly divided into training, validation and test sets at a ratio of $3 : 1 : 1$.

3.2   Quantitative and Qualitative Evaluation

In this paper, we adopt two evaluation indexes to assess the similarity between the MSP predicted by different methods and the ground truth, including distance deviation (DD) and angular deviation (AD). The formulas are as follows [13]:

$$DD = |d_p - d_g| \tag{8}$$

$$AD = arccos\frac{l_p \cdot l_g}{|l_p||l_g|} \tag{9}$$

where d_p and d_g denote the perpendicular distances from the origin to the predicted plane and the ground truth plane respectively, measured in mm. l_p and l_g represent the normal directions of the predicted and ground truth planes, and the AD is the angle between these two directions.

Table 1 shows the quantitative comparisons of the detection performance of ideal MSP between our DHPN and three competitive methods, including regression network (RN) based on 3D ResNet18 [9], RL framework [5] and keypoint detection network based on 3D hourglass [14]. It can be observed that our method achieves the superior performance in terms of both DD and AD (Table 1), especially for the cases with midline shift (MS) greater than 5 mm which are challenging for the MSP detection. Specifically, DHPN improves the best DD to 3.32 mm

Table 1. Quantitative comparisons of different methods.

Method	MS < 5 mm (288 cases)		MS ≥ 5 mm (221 cases)		Total (519 cases)	
	AD	DD	AD	DD	AD	DD
RN [9]	12.82(±9.24)	4.21(±4.74)	14.18(±10.56)	4.47(±3.65)	13.62(±9.91)	4.32(±4.27)
RL [5]	9.72(±7.42)	7.52(±7.96)	11.48(±10.27)	6.68(±7.05)	10.54(±8.91)	7.13(±7.56)
Hourglass [14]	4.25(±8.36)	3.10(±7.00)	5.08(±8.23)	5.66(±14.21)	4.64(±8.31)	4.30(±11.05)
DHPN(Ours)	**2.10(±0.98)**	**3.48(±2.34)**	**3.55(±1.10)**	**3.13(±4.61)**	**2.78(±1.04)**	**3.32(±3.62)**

Table 2. Quantitative comparisons of ablation experiments.

Method	Modules		Index	
	DSS	HPAN	AD	DD
Baseline	×	×	11.88(±9.36)	8.33(±7.76)
DHPN(Ours)	✓	×	5.71(±8.25)	3.54(±5.49)
DHPN(Ours)	✓	✓	**2.78(±1.04)**	**3.32(±3.62)**

and AD to 2.78°, which significantly outperforms the other competitors. Figure 4 shows the 3D visualization results among the three competitive methods and our method, which indicates that DHPN yields more accurate results.

We further conduct ablation study to verify the contribution of DSS and HPAN architecture. As illustrated in Table 2, the performance of MSP detection is significantly improved after introducing DSS strategy alone (6.03° reduction in AD and 4.79 mm in DD). The adoption of HPAN structure makes AD and DD indexes continuously optimized. It should be noted that the values of $\Delta\alpha$ and $\Delta\rho$ vary at different levels of the pyramid structure. Specifically, the AD and DD totally decrease by 9.10° and 5.01 mm respectively from the baseline.

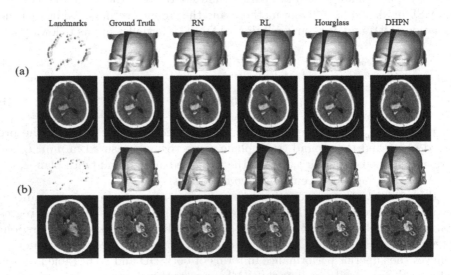

Fig. 4. Visual comparisons of different methods.

4   Conclusion

In this study, a novel and efficient framework for ideal brain midsagittal plane detection is proposed, which holds great potentials for assisting neurosurgeons in planning surgical incisions during preoperative planning. The quantitative results of experiments demonstrate the superiority of our method over the current state of the art.

References

1. Neubauer, S., Gunz, P., Scott, N.A., Hublin, J.-J., Mitteroecker, P.: Evolution of brain lateralization: a shared hominid pattern of endocranial asymmetry is much more variable in humans than in great apes. Sci. Adv. **6**(7), eaax9935 (2020)
2. Oertel-Knochel, V., Linden, D.E.J.: Cerebral asymmetry in schizophrenia. Neuroscci. **17**(5), 456–467 (2011)
3. Chen, H., et al.: Standard plane localization in fetal ultrasound via domain transferred deep neural networks. IEEE J. Biomed. Health Inform. **19**(5), 1627–1636 (2015)
4. Zhen, Yu., et al.: A deep convolutional neural network-based framework for automatic fetal facial standard plane recognition. IEEE J. Biomed. Health Inform. **22**(3), 874–885 (2017)
5. Dou, H., et al.: Agent with warm start and active termination for plane localization in 3D ultrasound. In: Shen, D., et al. (eds.) MICCAI 2019. LNCS, vol. 11768, pp. 290–298. Springer, Cham (2019). https://doi.org/10.1007/978-3-030-32254-0_33
6. Schmidt-Richberg, A., et al.: Offset regression networks for view plane estimation in 3D fetal ultrasound. In: Medical Imaging 2019: Image Processing, vol. 10949, pp. 907–912. SPIE (2019)
7. Wu, H., Wang, D., Shi, L., Wen, Z., Ming, Z.: Midsagittal plane extraction from brain images based on 3D sift. Phys. Med. Biol. **59**(6), 1367–1387 (2014)
8. Zhao, K., Han, Q., Zhang, C.-B., Xu, J., Cheng, M.-M.: Deep hough transform for semantic line detection. IEEE Trans. Patt. Anal. Mach. Intell. (2021)
9. He, K., Zhang, X., Ren, S., Sun, J.: Deep residual learning for image recognition. In: Proceedings of the IEEE Conference on Computer Vision and Pattern Recognition, pp. 770–778 (2016)
10. Lin, T.-Y., Dollár, P., Girshick, R., He, K., Hariharan, B., Belongie, S.: Feature pyramid networks for object detection. In: Proceedings of the IEEE Conference on Computer Vision and Pattern Recognition, pp. 2117–2125 (2017)
11. Li, H., Xiong, P., An, J., Wang, L.: Pyramid attention network for semantic segmentation. arXiv preprint arXiv:1805.10180 (2018)
12. Liu, S., Qi, L., Qin, H., Shi, J., Jia, J.: Path aggregation network for instance segmentation. In: Proceedings of the IEEE Conference on Computer Vision and Pattern Recognition, pp. 8759–8768 (2018)
13. Yang, X., et al.: Agent with warm start and adaptive dynamic termination for plane localization in 3D ultrasound. IEEE Trans. Med. Imaging **40**(7), 1950–1961 (2021)
14. Xu, J., et al.: Fetal pose estimation in volumetric MRI using a 3D convolution neural network. In: Shen, D., et al. (eds.) MICCAI 2019. LNCS, vol. 11767, pp. 403–410. Springer, Cham (2019). https://doi.org/10.1007/978-3-030-32251-9_44

Greedy Optimization of Electrode Arrangement for Epiretinal Prostheses

Ashley Bruce[1] and Michael Beyeler[1,2(✉)] (iD)

[1] Department of Computer Science, University of California, Santa Barbara, CA 93106, USA
{ashleybruce,mbeyeler}@ucsb.edu
[2] Department of Psychological and Brain Sciences, University of California, Santa Barbara, CA 93106, USA

Abstract. Visual neuroprostheses are the only FDA-approved technology for the treatment of retinal degenerative blindness. Although recent work has demonstrated a systematic relationship between electrode location and the shape of the elicited visual percept, this knowledge has yet to be incorporated into retinal prosthesis design, where electrodes are typically arranged on either a rectangular or hexagonal grid. Here we optimize the intraocular placement of epiretinal electrodes using dictionary learning. Importantly, the optimization process is informed by a previously established and psychophysically validated model of simulated prosthetic vision. We systematically evaluate three different electrode placement strategies across a wide range of possible phosphene shapes and recommend electrode arrangements that maximize visual subfield coverage. In the near future, our work may guide the prototyping of next-generation neuroprostheses.

Keywords: Retinal prosthesis · Implant design · Dictionary selection

1 Introduction

Current visual neuroprostheses consist of a microelectrode array (MEA) implanted into the eye or brain that is used to electrically stimulate surviving cells in the visual system in an effort to elicit visual percepts ("phosphenes"). Current epiretinal implant users perceive highly distorted percepts, which vary in shape not just across subjects, but also across electrodes [7,10], and may be caused by incidental stimulation of passing nerve fiber bundles (NFBs) in the retina [4,13]. However, this knowledge has yet to be incorporated into prosthesis design.

To address this challenge, we make the following contributions:

1. We optimize the intraocular placement of epiretinal electrodes using dictionary learning. Importantly, this optimization process is informed by a previously established and psychophysically validated model of simulated prosthetic vision [1,4].

L. Wang (Eds.): MICCAI 2022, LNCS 13437, pp. 594–603, 2022.
https://doi.org/10.1007/978-3-031-16449-1_57

2. We systematically evaluate three different electrode placement strategies across a wide range of possible phosphene shapes and recommend electrode arrangements that maximize visual subfield coverage.

2 Related Work

Sensory neuroprostheses such as retinal and cochlear implants are emerging as a promising technology to restore lost sensory function. These devices bypass the natural sensory transduction mechanism and provide direct electrical stimulation of (retinal or auditory) nerve fibers that the brain interprets as a (visual or auditory) percept. However, perceptual distortions can occur when multiple electrodes stimulate the same neural pathways [2,15], and electrode placement has been shown to impact the quality of visual and hearing outcomes [4,8].

However, this information has yet to be incorporated into neuroprosthetic device design and surgical placement, with most patients having less-than-optimal MEA placement [3,6]. Whereas previous studies have optimized the shape of individual electrodes [12], which electrodes to activate in order to produce a desired visual response [5,9,14], or the overall implant placement [3], we

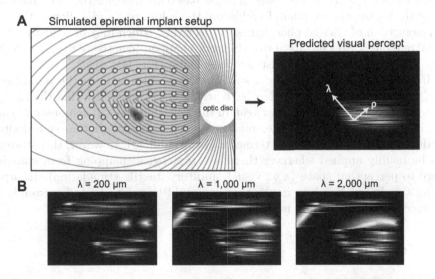

Fig. 1. A) Axon map model. *Left*: Electrical stimulation (red disc) of a NFB (gray lines) leads to tissue activation (dark-gray shaded region) elongated along the NFB trajectory away from the optic disc (white circle). The light-gray shaded region indicates the visual subfield that is being simulated. *Right*: The resulting visual percept appears elongated as well; its shape can be described by two parameters, λ (spatial extent along the NFB trajectory) and ρ (spatial extent perpendicular to the NFB). **B)** As λ increases, percepts become more elongated and start to overlap. (Color figure online)

are unaware of any studies that have attempted to optimize the placement of individual electrodes within an implant.

In the case of retinal implants, recent work has demonstrated that phosphene shape strongly depends on the retinal location of the stimulating electrode [4]. Because retinal ganglion cells (RGCs) send their axons on highly stereotyped pathways to the optic nerve, an electrode that stimulates nearby axonal fibers would be expected to antidromically activate RGC bodies located peripheral to the point of stimulation, leading to percepts that appear elongated in the direction of the underlying NFB trajectory (Fig. 1A, *right*). Using a simulated map of NFBs, Reference [4] was thus able to accurately predict phosphene shape for various users of the Argus Retinal Prosthesis System (Second Sight Medical Products, Inc.), by assuming that an axon's sensitivity to electrical stimulation:

i. decays exponentially with decay constant ρ as a function of distance from the stimulation site,
ii. decays exponentially with decay constant λ as a function of distance from the cell body, measured as axon path length.

As can be seen in Fig. 1B, electrodes near the horizontal meridian are predicted to elicit circular percepts, while other electrodes are predicted to produce elongated percepts that will differ in angle based on whether they fall above or below the horizontal meridian. In addition, the values of ρ and λ dictate the size and elongation of elicited phosphenes, respectively, which may drastically affect visual outcomes. Specifically, if two electrodes happen to activate the same NFB, they might not generate two distinct phosphenes (Fig. 1B, *right*).

Instead of arranging electrodes such that they efficiently tile the *retinal surface*, a better approach might thus be to arrange electrodes such that the *elicited percepts* effectively tile the *visual field*. In the following, we will demonstrate that this is equivalent to finding the smallest set of electrodes that cover a desired portion of the visual field (here termed a visual subfield). However, this strategy can be readily applied wherever there is a topological mapping from stimulus space to perceptual space (*e.g.*, visual, auditory, tactile stimulation). Incorporating this knowledge into implant design could therefore be indispensable to the success of future visual neuroprostheses.

3   Methods

3.1   Phosphene Model

Let $\mathcal{E} = \{e_1, \ldots, e_N\}$ be the set of N electrodes in a MEA, where the i-th electrode $e_i = (x_i, y_i, r_i)$ is described by its location on the retinal surface (x_i, y_i) and its radius $r_i > 0$. For example, Argus II can be described by $\mathcal{E}_{\text{ArgusII}}$, where $N = |\mathcal{E}_{\text{ArgusII}}| = 60$, $r_i = 122.5\,\mu\text{m}$ $\forall i$, and (x_i, y_i) are spaced $575\,\mu\text{m}$ apart on a rectangular grid. We do not assume any particular ordering of $\mathcal{E}$.

Furthermore, let $\mathcal{S} = \{(s_1, \ldots, s_k)\}$ be the set of stimuli sent to $k \leq N$ electrodes in the MEA, where the i-th stimulus $s_i = (e_i, a_i)$ is described by an electrode $e_i \in \mathcal{E}$ and its corresponding activation function a_i. In practice, a_i may be a biphasic pulse train of a given duration, pulse amplitude, and pulse frequency. However, for the purpose of this study, we limited ourselves to spatial activation values $a_i \in \mathbb{R}_{\geq 0}$, which did not contain a temporal component.

A phosphene model $\mathcal{M}$ then takes a set of stimuli $\mathcal{S}$ as input and outputs a visual percept $p \in \mathbb{R}_{\geq 0}^{H \times W}$, which is a height $(H) \times$ width (W) grayscale image. In general, $\mathcal{M}$ is a nonlinear function of $\mathcal{S}$ and depends on subject-specific parameters θ_{subject} such as ρ and λ; thus $p = \mathcal{M}(\mathcal{S}; \theta_{\text{subject}})$.

For the purpose of this study, we used *pulse2percept* 0.8.0, a Python-based simulation framework for bionic vision [1] that provides an open-source implementation of the axon map model [4], as described in the previous section. Constrained by electrophysiological and psychophysical data, this model predicts what a bionic eye user should "see" for any given set of stimuli $\mathcal{S}$.

3.2   Dictionary Selection

Problem Formulation. Let $\mathcal{T}$ now be the set of all nonoverlapping epiretinal electrodes, and $\mathcal{D}$ be a subset of those; that is, $\mathcal{D} \subset \mathcal{T}$, where $|\mathcal{D}| \ll |\mathcal{T}|$. In the dictionary selection problem, we are interested in finding the dictionary $\mathcal{D}^*$ that maximizes a utility function F (see next subsection):

$$\mathcal{D}^* = \arg\max_{|\mathcal{D}| \leq k} F(\mathcal{D}), \qquad (1)$$

where k is a constraint on the number of electrodes that the dictionary can be composed of.

This optimization problem presents combinatorial challenges, as we have to find the set $\mathcal{D}^*$ out of exponentially many options in $\mathcal{T}$. However, we will only consider utility functions F with the following properties:

i.   The empty set has zero utility; that is, $F(\emptyset) = 0$.
ii. F increases monotonically; that is, whenever $\mathcal{D} \subseteq \mathcal{D}'$, then $F(\mathcal{D}) \leq F(\mathcal{D}')$.
iii. F is approximately submodular; that is, there exists an ε such that whenever $\mathcal{D} \subseteq \mathcal{D}' \subseteq \mathcal{T}$ and an electrode $e \in \mathcal{T} \setminus \mathcal{D}'$, it holds that $F(\mathcal{D} \cup \{e\}) - F(\mathcal{D}) \geq F(\mathcal{D}' \cup \{e\}) - F(\mathcal{D}') - \varepsilon$. This property implies that adding a new electrode e to a larger dictionary $\mathcal{D}'$ helps at most ε more than adding e to a subset $\mathcal{D} \subseteq \mathcal{D}'$.

For utility functions with the above properties, Nemhauser et al. [11] proved that a simple greedy algorithm that starts with the empty set $\mathcal{D}_0 = \emptyset$, and at every iteration i adds the element

$$d_i = \arg\max_{d \in \mathcal{T} \setminus \mathcal{D}} F(\mathcal{D}_{i-1} \cup \{d\}), \qquad (2)$$

where $\mathcal{D}_i = \{d_1, \ldots, d_i\}$, is able to obtain a near-optimal solution.

Utility Function. Ideally, the utility function F would directly assess the quality of the generated artificial vision. As a first step towards such a quality measure, we considered the ability of a set of electrodes to lead to phosphenes that cover a specific visual subfield (i.e., the gray shaded region in Fig. 1). We would thus activate every electrode in $\mathcal{D}$, represented by $\mathcal{S}_\mathcal{D}$, and calculate the percept $p_\mathcal{D} = \mathcal{M}(\mathcal{S}_\mathcal{D}; \theta_{\text{subject}})$. Then F was given as the visual subfield coverage; that is:

$$F(\mathcal{D}) = \sum_{w=1}^{w=W} \sum_{h=1}^{h=H} p_{hw} \geq \varepsilon, \tag{3}$$

where W and H were the width and height of the percept, respectively, $\varepsilon = 0.1 \max_{\forall h,w}(p_{hw})$, and $F \in [0, HW]$.

Dictionary Selection Strategies. To find $\mathcal{D}^*$, we considered three different strategies:

- **No Overlap:** At each iteration i, d_i was chosen according to Eq. 2.
- **Not Too Close:** To consider manufacturing constraints, we modified Eq. 2 above to enforce that all electrodes were placed at least $c = 112.5\,\mu$m apart from each other:

$$d_i = \underset{d \in \mathcal{T}\backslash\mathcal{D} \text{ s.t. } ||d,d_j||_2 \geq c \,\forall d_j \in \mathcal{D}\backslash d}{\arg\max} F(\mathcal{D}_{i-1} \cup \{d\}). \tag{4}$$

- **Max Pairwise Distance:** As electrical crosstalk is one of the main causes of impaired spatial resolution in retinal implants [15], we further modified Eq. 4 to place electrodes as far away from each other as possible:

$$d_i = \underset{d \in \mathcal{T}\backslash\mathcal{D} \text{ s.t. } ||d,d_j||_2 \geq c \,\forall d_j \in \mathcal{D}\backslash d}{\arg\max} F(\mathcal{D}_{i-1} \cup \{d\}) + \alpha \sum_{d_i \in \mathcal{D}\backslash d} ||d, d_i||_2, \tag{5}$$

where $\alpha = 1 \times 10^{-4}$ was a scaling factor.

Implementation Details. For the sake of feasibility, we limited $\mathcal{T}$ to electrodes placed on a finely spaced search grid, $x_i \in [-3000, 3000]\,\mu$m and $y_i \in [-2000, 2000]\,\mu$m, sampled at $112.5\,\mu$m (i.e., the radius of an Argus II electrode). This led to a manageable set size ($|\mathcal{T}| \approx 2000$) while still allowing electrodes to be placed right next to each other (if desirable). To find the electrode d_i at each iteration in Eqs. 2, 4, and 5 above, we thus performed a grid search.

Stopping Criteria. The dictionary search was stopped when at least one of the following criteria were met:

- visual subfield coverage reached 99 %,
- the utility score F did not improve by $\geq 1 \times 10^{-6}$ on two consecutive runs,
- no more viable electrode locations were available (i.e., $\mathcal{T} \backslash \mathcal{D}$ s.t. $||d, d_i||_2 \geq c \,\forall d_i \in \mathcal{D} \backslash d = \emptyset$).

4   Results

4.1   Visual Subfield Coverage

The results of the greedy dictionary selection are shown in Fig. 2. For all three dictionary selection strategies, the number of electrodes required to cover at least 99 % of the visual subfield was inversely proportional to ρ and λ. As expected, the largest number was achieved with the smallest, most compact phosphene shape ($\rho = 100\,\mu m$, $\lambda = 200\,\mu m$). The required electrode number dropped rapidly with increasing ρ and λ, indicating that for large phosphenes, a prototype implant such as Argus I (4×4 electrodes) might be sufficient to cover the whole subfield.

It is interesting to note that, for any given ρ and λ combination, the Max Pairwise Distance strategy required a smaller number of electrodes than the No Overlap strategy. However, using the Not Too Close strategy, smaller ρ and λ combinations were no longer able to reach full coverage, as electrodes could no longer be placed too close to each other or to the visual subfield boundary. This issue was amplified with the Max Pairwise Distance strategy, with which coverage dropped for most ρ and λ combinations to 95%.

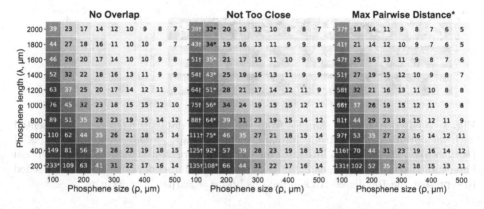

Fig. 2. Number of electrodes needed to reach maximum visual subfield coverage for the three dictionary selection strategies (†: 80−95% coverage, *: 95−99% coverage). Max Pairwise Distance always had *, unless otherwise noted by †.

With larger ρ and λ values, an implant required less than ten electrodes to cover the visual subfield. However, such a small number was not necessarily desirable, as it also reduced the number of distinct phosphenes that the implant can produce. Rather than focusing on visual subfield coverage alone, one might therefore ask what kinds of electrode arrangements the three dictionary selection strategies yield and what the resulting percepts look like.

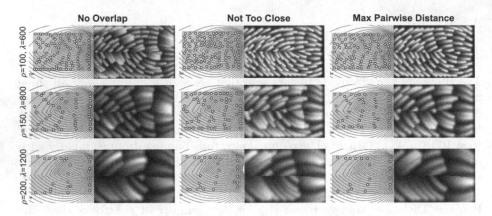

Fig. 3. Representative examples of the final electrode arrangements generated with the three dictionary selection strategies for different ρ and λ combinations. The left half of each panel shows the retinal location of all electrodes (small circles) in the implant, and the shaded region indicates the visual subfield (compare to Fig. 1). The visual percept that results from simultaneously activating every electrode in the implant is shown in the right half of each panel.

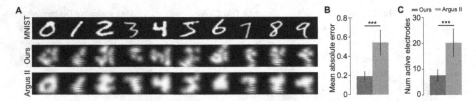

Fig. 4. A) Representative samples from the MNIST database (*top row*) represented with an optimized 60-electrode implant ("No Overlap", *middle row*) and Argus II (*bottom row*). **B)** Mean absolute error for 100 randomly selected MNIST digits (vertical bars: standard error of the mean, ***: $p < .001$). **C)** Number of active electrodes needed to represent the 100 digits. All simulations had $\rho = 200\,\mu\text{m}, \lambda = 400\,\mu\text{m}$.

4.2   Electrode Arrangement

Example electrode arrangements suggested by the three dictionary selection strategies are shown in Fig. 3. Here it is evident that, as λ increased, electrodes were preferentially placed on the top, bottom, and right boundaries of the visual subfield. This placement would often lead to the longest streaks, thus yielding the largest coverage. For small ρ values, electrodes aggregated mainly on NFBs, spaced λ apart, so that the streaks generated by different electrodes tiled the visual subfield. As ρ increased, electrodes tended to migrate away from the border and more inward, due to the outward spread of the generated percept.

4.3   Comparison with Argus II

To assess whether the optimized electrode arrangements provided an improvement over a rectangular MEA, we compared our results to Argus II. We thus modified our experiment such that the dictionary consisted of at most 60 electrodes (i.e., same as Argus II). We then considered the ability of the optimized implant to represent handwritten digits from the MNIST database (Fig. 4A). We found that the optimized implant not only led to smaller errors between predicted percepts and ground-truth digits (paired t-test, $p < .001$; Fig. 4B) but also required less active electrodes overall (paired t-test, $p < .001$; Fig. 4C).

Furthermore, we calculated the visual subfield coverage achieved with these 60 electrodes and compared the result to Argus II (Table 1). Given the fixed number of electrodes, our electrode selection strategies were always able to cover a larger portion of the visual subfield than Argus II.

Overall these results suggest that a rectangular grid might not be the best electrode arrangement for epiretinal prostheses.

Table 1. Visual subfield coverage achieved with 60 electrodes using different electrode placement strategies. AII: Argus II. NTC: Not Too Close. MPD: Max Pairwise Distance. Asterisk (*): Search terminated before utilizing 60 electrodes.

ρ	$100\,\mu m$			$150\,\mu m$			$200\,\mu m$		
λ	AII	NTC	MPD	AII	NTC	MPD	AII	NTC	MPD
$200\,\mu m$	39.2	43.2	42.9	74.1	75.6	75.9	86.2	>99*	95.2
$400\,\mu m$	55.0	66.8	66.8	80.6	91.3	91.4	87.5	>99*	95.1
$600\,\mu m$	62.4	81.8	81.3	81.9	96.9	95.3	88.4	>99*	95.2
$800\,\mu m$	66.7	88.8	88.8	82.8	97.7	95.2	89.0	>99*	95.4
$1000\,\mu m$	69.4	92.5	92.2	83.5	98.1*	95.2	89.6	>99*	95.8
$1200\,\mu m$	71.9	92.6	92.5	84.2	98.0*	95.3	90.1	>99*	95.7
$1400\,\mu m$	72.4	93.9*	93.8	84.6	98.1*	95.27	90.4	>99*	95.6
$1600\,\mu m$	73.4	93.9*	93.8	85.0	98.5*	95.41	90.7	>99*	95.8
$1800\,\mu m$	74.1	94.0*	93.9	85.4	98.6*	95.16	91.0	>99*	95.7
$2000\,\mu m$	74.7	94.4*	94.4	85.9	98.3*	95.08	91.4	>99*	95.2

5   Conclusion

We report epiretinal electrode arrangements that maximize visual subfield coverage as discovered by dictionary learning. We were able to obtain nearly full coverage across different ρ and λ values for a given implant size and electrode radius, and report the number of electrodes necessary to reach this coverage. Future work should extend the dictionary selection strategy to other implant sizes and explore all possible stimuli, not just the ones that may result in sufficient visual subfield coverage.

This preliminary study is a first step towards the use of computer simulations in the prototyping of novel neuroprostheses. To the best of our knowledge, this is the first study using a psychophysically validated phosphene model to optimize electrode arrangement. Even though we have focused on a specific implant technology, our strategy can be readily applied wherever there is a topological mapping from stimulus space to perceptual space (e.g., visual, auditory, tactile). This means that our approach could be extended to other neuromodulation technologies that include (but are not limited to) other electronic prostheses and optogenetic technologies. In the near future, our work may therefore guide the prototyping of next-generation epiretinal prostheses.

Acknowledgements. This work was supported by the National Institutes of Health (NIH R00 EY-029329 to MB). The authors would like to thank Madori Spiker and Alex Rasla for valuable contributions to an earlier version of the project.

References

1. Beyeler, M., Boynton, G.M., Fine, I., Rokem, A.: Pulse2percept: a Python-based simulation framework for bionic vision. In: Huff, K., Lippa, D., Niederhut, D., Pacer, M. (eds.) Proceedings of the 16th Science in Python Conference, pp. 81–88 (2017). https://doi.org/10.25080/shinma-7f4c6e7-00c
2. Beyeler, M., Rokem, A., Boynton, G.M., Fine, I.: Learning to see again: biological constraints on cortical plasticity and the implications for sight restoration technologies. J. Neural Eng. **14**(5), 051003 (2017). https://doi.org/10.1088/1741-2552/aa795e
3. Beyeler, M., Boynton, G.M., Fine, I., Rokem, A.: Model-based recommendations for optimal surgical placement of epiretinal implants. In: Shen, D., et al. (eds.) MICCAI 2019. LNCS, vol. 11768, pp. 394–402. Springer, Cham (2019). https://doi.org/10.1007/978-3-030-32254-0_44
4. Beyeler, M., Nanduri, D., Weiland, J.D., Rokem, A., Boynton, G.M., Fine, I.: A model of ganglion axon pathways accounts for percepts elicited by retinal implants. Sci. Rep. **9**(1), 1–16 (2019). https://doi.org/10.1038/s41598-019-45416-4
5. Bratu, E., Dwyer, R., Noble, J.: A graph-based method for optimal active electrode selection in cochlear implants. In: Martel, A.L., et al. (eds.) MICCAI 2020. LNCS, vol. 12263, pp. 34–43. Springer, Cham (2020). https://doi.org/10.1007/978-3-030-59716-0_4
6. Chakravorti, S., et al.: Further evidence of the relationship between cochlear implant electrode positioning and hearing outcomes. Otol. Neurotol. **40**(5), 617–624 (2019). https://doi.org/10.1097/MAO.0000000000002204
7. Erickson-Davis, C., Korzybska, H.: What do blind people "see" with retinal prostheses? Observations and qualitative reports of epiretinal implant users. PLOS ONE **16**(2), e0229189 (2021). https://doi.org/10.1371/journal.pone.0229189. publisher: Public Library of Science
8. Finley, C.C., et al.: Role of electrode placement as a contributor to variability in cochlear implant outcomes. Otol. Neurotol. **29**(7), 920–928 (2008). https://doi.org/10.1097/MAO.0b013e318184f492

9. Granley, J., Relic, L., Beyeler, M.: A hybrid neural autoencoder for sensory Neuroprostheses and Its Applications in Bionic Vision (2022). http://arxiv.org/abs/2205.13623,arXiv:2205.13623 [cs]

10. Luo, Y.H.L., Zhong, J.J., Clemo, M., da Cruz, L.: Long-term repeatability and reproducibility of phosphene characteristics in chronically implanted Argus II retinal prosthesis subjects. Am. J. Ophthalmol. **170**, 100–109 (2016). https://doi.org/10.1016/j.ajo.2016.07.021

11. Nemhauser, G.L., Wolsey, L.A., Fisher, M.L.: An analysis of approximations for maximizing submodular set functions-I. Math. Program. **14**(1), 265–294 (1978). https://doi.org/10.1007/BF01588971

12. Rattay, F., Resatz, S.: Effective electrode configuration for selective stimulation with inner eye prostheses. IEEE Trans. Biomed. Eng. **51**(9), 1659–1664 (2004). https://doi.org/10.1109/TBME.2004.828044

13. Rizzo, J.F., Wyatt, J., Loewenstein, J., Kelly, S., Shire, D.: Perceptual efficacy of electrical stimulation of human retina with a microelectrode array during short-term surgical trials. Invest. Ophthalmol. Vis. Sci. **44**(12), 5362–5369 (2003). https://doi.org/10.1167/iovs.02-0817

14. Shah, N.P., et al.: Optimization of electrical stimulation for a high-fidelity artificial retina. In: 2019 9th International IEEE/EMBS Conference on Neural Engineering (NER), pp. 714–718 (2019). https://doi.org/10.1109/NER.2019.8716987, ISSN: 1948-3546

15. Wilke, R.G.H., Moghadam, G.K., Lovell, N.H., Suaning, G.J., Dokos, S.: Electric crosstalk impairs spatial resolution of multi-electrode arrays in retinal implants. J. Neural Eng. **8**(4), 046016 (2011). https://doi.org/10.1088/1741-2560/8/4/046016

Stereo Depth Estimation via Self-supervised Contrastive Representation Learning

Samyakh Tukra[(✉)] and Stamatia Giannarou

The Hamlyn Centre for Robotic Surgery, Department of Surgery and Cancer,
Imperial College London, London SW7 2AZ, UK
{samyakh.tukra17,stamatia.giannarou}@imperial.ac.uk

Abstract. Accurate stereo depth estimation is crucial for 3D reconstruction in surgery. Self-supervised approaches are more preferable than supervised approaches when limited data is available for training but they can not learn clear discrete data representations. In this work, we propose a two-phase training procedure which entails: (1) Performing Contrastive Representation Learning (CRL) of left and right views to learn discrete stereo features (2) Utilising the trained CRL model to learn disparity via self-supervised training based on the photometric loss. For efficient and scalable CRL training on stereo images we introduce a momentum pseudo-supervised contrastive loss. Qualitative and quantitative performance evaluation on minimally invasive surgery and autonomous driving data shows that our approach achieves higher image reconstruction score and lower depth error when compared to state-of-the-art self-supervised models. This verifies that contrastive learning is effective in optimising stereo-depth estimation with self-supervised models.

Keywords: Deep learning · Depth · 3D · Stereo · Self-supervised · Contrastive representation learning

1 Introduction

Depth estimation holds substantial importance in a wide range of applications in Minimally Invasive Surgery (MIS) ranging from soft tissue 3D reconstruction, surgical robot navigation to augmented reality. One method of perceiving depth is with a stereo camera by estimating the horizontal displacement (disparity) from the left image pixels to the corresponding right. End-to-end deep learning stereo reconstruction methods, have become state of the art. They are typically categorised as supervised and self-supervised.

Stereo reconstruction models trained via supervised learning directly learn to regress the disparity values relying on ground truth data [7,10,16,21]. These models are typically trained on synthetic data [11] followed by fine-tuning on real scenes, to deal with limited ground truth availability. However, there is an

L. Wang (Eds.): MICCAI 2022, LNCS 13437, pp. 604–614, 2022.
https://doi.org/10.1007/978-3-031-16449-1_58

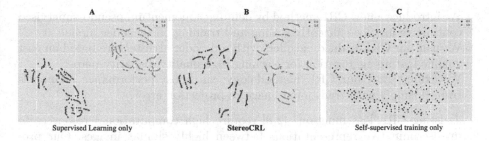

Fig. 1. T-SNE visualisation of model embeddings on Kitti2015 test data (200 images). Labels 0 and 1 signify embeddings of left and right views, respectively. (A) generated from pre-trained [10] (B) our proposed StereoCRL trained with Eq. 1 and 2 (C) our model variant trained only via self-supervision Eq. 2, on Kitti raw split (42,382 stereo frames).

inherent discrepancy between the two data sources, especially in surgical applications where scarce ground truth data, poorer image quality, limited space, shorter depth distance and a myriad of scene variations, can cause model performance to drop. For self-supervised methods, the training data is abundant since, only images and no ground truth is required. Therefore, an end-to-end self-supervised approach capable of harnessing diverse real data, can generate high-quality results for any scenario where ground truth data is limited. Self-supervised stereo reconstruction methods are trained to conduct novel view synthesis thus, optimising the photometric re-projection error [14,17,18,22]. This error however, can be optimised by a wide range of divergent disparity values resulting in inconsistent geometry. This is because self-supervised methods can not learn discrete feature representations between the stereo views. As shown in Fig. 1(A), supervised methods have a unique property of learning clear discrete feature representations for the stereo views (shown by the distance between left and right embeddings) whilst still learning similarities between the views (shown by the similar shape of the feature representations). Self-supervised methods are unable to reach that resolution in the feature space as shown in Fig. 1(C), learning a mixed embedding between left and right views and thus a confused disparity.

The question we address in this paper is, how can we make self-supervised approaches learn discrete representations similar to supervised methods, without relying on ground-truth? To achieve this, we propose *StereoCRL* a two-stage training process which entails (1) training an encoder via Contrastive Representation Learning (CRL) [2], to generate distinct feature embeddings for left and right views, followed by (2) self-supervised training for disparity estimation. Our contributions are the following:

- We design a procedure for training stereo depth models based on joint contrastive representation and self-supervised learning. To the best of our knowledge, this is the first CRL methodology applied for stereo depth estimation;

- We introduce a new CRL method by defining a momentum pseudo-supervised contrastive loss for efficiently scaling and training our encoder model;
- We pair our CRL-trained encoders with a decoder architecture based on cost volume feature aggregation for learning high quality disparity maps;
- We exhibit that our CRL methodology can boost performance of models trained simply with traditional image re-projection error.

Existing CRL loss functions [2,8] require high computational resources to learn discriminative representations between highly similar images. Our proposed CRL method is tailored for learning differences between highly similar images such as stereo pairs, and is scalable even on a single GPU hardware. The performance of our StereoCRLhas been evaluated on 2 MIS datasets *Hamlym* [5] and *SCARED* [1] and non-surgical scenes from Kitti2015 [13].

2 Methods

Stereo Contrastive Representation Learning: The premise of contrastive learning is to learn distinctiveness in the data representation. It involves training an encoder model to be capable of recognising similarities in scenes that are alike and differences in dissimilar scenes. The aim of CRL is to ensure the encoder learns general features such that its embeddings for similar objects (positive samples) are grouped closer together in the embedding space and separated from dissimilar objects (negative samples). In our proposed StereoCRL, positive and negative samples represent the left and right images, respectively [2]. Thus, our encoder model must learn discrete feature representations for the left and right images, such that for the same view the representations are close together and for the corresponding stereo view farther apart, as shown in Fig. 1.

To design an efficient CRL algorithm we must overcome the following challenges. *(1) Limited Resources:* CRL algorithms like SimCLR [2], require the encoder to observe vast number of negative samples at each training iteration (i.e. up to 4000 samples per batch). This is to ensure the embeddings for each positive sample are compared with enough divergent negative samples at each gradient update. *(2) High scene similarity:* most existing CRL algorithms are initially trained on classification tasks where, there is enough variation between objects in positive and negative samples. However, in the case of stereo images, the scene is extremely similar hence, optimising their embeddings to be far apart will require an additional learning signal.

Consider SimCLR [2] method adapted for stereo images is illustrated in Fig. 2(A). Each image is first parsed into a data augmentation module that performs the following set of transformations randomly: (1) horizontal flipping (2) colour distortion (3) Gaussian blurring and (4) variations in brightness, luminance (gamma parameter) and saturation. The augmented variants for each batch are processed by the encoder to generate a set of embeddings, h_i, for each image i in the batch. These embeddings are further parsed through a projection model which is a small Multi-Layer Perceptron (MLP) G_ϕ which re-projects them into a different latent space followed by L2 normalisation i.e. $z_i = G_\phi(h_i)$.

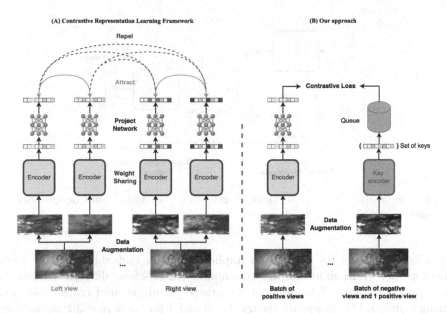

Fig. 2. The stereo contrastive learning framework. (*A*) The intuition of CRL and (B) Our proposed CRL approach.

The embeddings z_i for the same view, should maximise cosine similarity. G_ϕ is only used for CRL training and is discarded upon completion.

However, this standard CRL solution suffers from challenge (1) mentioned above. Hence, we advance the CRL process by formulating CRL as a dictionary look up problem as shown in Fig. 2(B). Instead of using large batch sizes to store negative samples for training, we use a memory bank (queue) to store the generated embeddings, same as [8]. In our case, the projection network G_ϕ, predicts $q_i = G_\phi(h_i)$ instead of z_i for each i image in the batch. Where q_i is a query embedding of h_i. Every time a batch is processed, a q_i is generated. At the same time, another encoder model entitled *key encoder* processes a batch of negative images and a single positive image to generate keys k_i^- and k_i^+, respectively. The *key encoder* is not trained, instead its weights are updated using the weights of the other encoder via momentum update [8]. The keys generated are stored in a memory bank, dynamically creating a dictionary of features on the fly. The intuition is that there is a single key k_i^+ that the q_i matches with. Thus, the contrastive loss value will be low when q is similar to k_i^+ and different from the other keys (k_i^-) in the queue. Ultimately, the encoder learns to make q for the same view highly similar, without needing large batches of images, since the loss can be scaled to observe negative samples as query-key vectors.

The above solution still suffers from challenge (2). To overcome this challenge, we define a new loss function to train our encoder based on a supervised variant of the contrastive loss [9]. This loss function leverages ground truth labels as

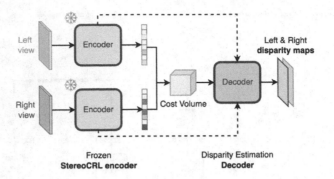

Fig. 3. StereoCRL model architecture. Solid black and dotted lines denote forward propagation and skip connections respectively.

an anchor to enable the encoder to implicitly cluster embeddings of the same class together whilst simultaneously distancing clusters from different classes. To ensure we maintain self-supervision, we create pseudo ground truth labels live during training. Our labels are binary, i.e. 0 and 1 for the left and right images, respectively. This forces the model to simply differentiate between left and right feature embeddings such that they are far apart from one another. We further augment this loss, to enable training with our proposed CRL approach, using query-key vectors. Hence, we introduce the momentum supervised contrastive loss:

$$L^{sup}_{mo:crl} = \Sigma^{2n}_{i=1} \frac{-1}{2|N_i|-1} \Sigma_{i \in N(y_i)} log \frac{exp(q_i \cdot k_i^+/\tau)}{\Sigma^K_{i=0} exp(q \cdot k_i/\tau)} \tag{1}$$

where, $N = (i, y_i)$ is the number of randomly sampled images in a batch and the corresponding label (y_i), i is the augmented image sample in the batch and τ is a temperature parameter, used for scaling the output in the range of -1 to 1 for calculating cosine similarity. k_i^+ is the single positive key in the memory bank. The sum is conducted over one positive and K negative samples. This loss is effectively the log loss of $K+1$ samples where a softmax based classifier (G_ϕ) is trying to classify q_i as y_i and k^+. Hence, phase (1) of our method, involves pre-training the encoder via Eq.(1), prior to the downstream task of disparity estimation. In phase (2) the key encoder is discarded, and a decoder model is trained for self-supervised disparity estimation. The final model is shown in Fig. 3 where the trained encoder is frozen and paired with a decoder.

Decoder Architecture: Our decoder comprises of 5 sequential blocks to mirror the encoder, schema shown in Fig. 4. Each decoder block utilises 3 convolution layers namely, *UpConv*, *iConv* and *DispPred*. All layers have a filter size of 3×3 except *DispPred* which is 1×1. All convolution layers are followed by normalisation (Batch) and activation (eLU) layers, except for *DispPred* since it outputs direct disparity. Each of the last 4 blocks outputs left and right disparity maps which are increasing spatially by a scale-factor of 2 in ascending order. Much

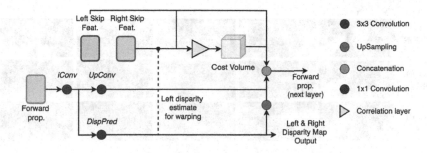

Fig. 4. Decoder block schema. The solid and dotted lines signify forward propagation and skip connections, respectively.

like a *DispNet* [12] architecture, our decoder takes forward propagating feature maps and skip connection feature maps from both left and right view encoders, shown in Fig. 3. We redesign the original *DispNet* decoder layer to ensure effective learning from both the incoming features and the skip connection features by generating a cost volume at each scale for multi-scale disparity estimation. The cost volume is used to encode features for disparity by discretising in disparity space and comparing the features along epipolar lines. This overcomes the issue of normalised disparity predicted by DispNet and other self-supervised models [14,18] which is affected by changes in image scale.

Typically, to process cost volumes, 3D convolution is used. However, this requires high computational cost and limits operations at low resolution. Given our loss is based on photometric error, we require iterative refinement. We opted for an efficient multi-scale cost volume generation same as PWCNet [15] at each decoder block. In particular, the right skip connection feature map is warped using the predicted left disparity, and the resulting warped feature map is correlated with the left skip connection feature map. For our cost volume, we set the search range to [-5, 4] pixels, allowing a sufficient radius for negative and positive disparities. Furthermore, we estimate the cost volume in each decoder block alongside disparity. At each block all the feature maps are increased by a scale factor of 2. Thus, StereoCRL iteratively refines its disparity estimate by adjusting its cost volume weights at each scale.

Self-Supervised Loss: In phase (2) of our method, only the decoder is trained via self-supervised photometric error using stereo video sequences, similar to [6] for disparity estimation. The output predicted left and right disparity maps (D_L^{pred} and D_R^{pred}) from the last 4 layers of the decoder are used to warp the input stereo frames I_L^{GT} and I_R^{GT}. The photometric warping error is defined by utilising a warping function F_{warp} on the input images for generating its corresponding stereo image via the opposite disparity map i.e. $F_{warp}(I_L^{GT}, D_R^{pred})$ generates the predicted right image I_R^{Pred} and vice versa. The loss function therefore, is based on the reconstruction error between the synthesised images and the original image which acts as ground truth. Standard loss components utilised

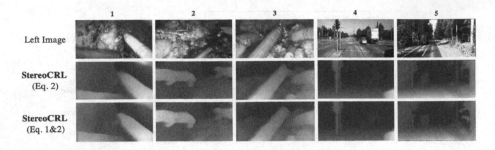

Fig. 5. Depth maps generated on the *Hamlyn* and the Kitti2015 test split. Inclusion of (Eq.) signifies the type of loss the model was trained with.

are image reconstruction L_{recon}, disparity smoothness L_{smooth} and left-right disparity consistency L_{LRcons} [6]. This is defined as:

$$L_{pe} = \delta_1 L_{recon} + \delta_2 L_{smooth} + \delta_3 L_{LRcons} \tag{2}$$

L_{pe} is performed on both the synthesised left and right images at each scale. L_{recon} is composed of multi-scale structural similarity index (MS-SSIM) together with a L1 loss term.

Implementation Details: The encoder architecture used is a ResNet50. For phase (1) of CRL training, the encoder was trained for a maximum of 200 epochs on a combination of the following 3 datasets: *Kitti raw data* containing 42,382 frames, *CityScapes* [3] containing 88,250 frames and *Hamlyn* MIS data containing 34,240 frames. For phase (2), only the decoder was trained via L_{pe} on *Hamlyn* MIS and Kitti raw data, independently. Our maximum disparity range is set to 192. The output dimension of query and key vectors q_i and k_i were 128. The temperature τ was set to 0.07 and the batch-size used in the study was 16 and maximum memory bank capacity was 3000. In L_{pe}, the weightings δ_1, δ_2 and δ_3 were 0.85, 1.0 and 1.0, respectively. The hyperparameters were selected using the original paper sources, for temperature τ [8] and for deltas δ_x [6], respectively. Testing was performed on the SCARED test dataset [1] comprising of 5,637 stereo image pairs without any finetuning (for testing generalisation), Kitti2015 comprising 200 frames (all of which were used only for testing) and *Hamlyn* comprising 7,191 frames. Images were processed at 256×512 resolution. Nvidia Tesla T4 was used for training. CRL and self-supervised training computation time was 3.5 and 1 GPU days, respectively. We also train a variant of StereoCRL only on Eq. 2 (no CRL pre-training) for comparison. For this process we randomly initialise weights of the encoder and jointly train it with the decoder.

3   Experimental Results and Discussion

The performance evaluation was conducted on the *Hamlyn*, SCARED and Kitti 2015 datasets. Due to lack of ground truth in *Hamlyn*, we compare the structural

Table 1. Performance evaluation on the *Hamlyn*. Best results highlighted in bold.

Methods	ELAS [4]	SPS [19]	V-Siamese [20]	StereoCRL Eq. 2	**StereoCRL** Eq. 1 and 2
Mean SSIM	47.3	54.7	60.4	80.8	**83.7**
Std. SSIM	0.079	0.092	0.066	0.033	**0.023**

Table 2. Mean absolute depth error (mm) on SCARED. Best self-supervised results are highlighted in bold.

Method	Method type	Test set 1	Test set 2
Trevor Zeffiro [1]	Supervised	3.60	3.47
Xiran Zhang [1]	Supervised	6.13	4.39
J.C. Rosenthal [1]	Supervised	3.44	4.05
Zhu Zhanshi [1]	Supervised	9.60	21.20
KeXue Fu [1]	Self-Supervised	20.94	17.22
Xiaohong Li [1]	Self-Supervised	22.77	20.52
StereoCRL Eq. 2	Self-Supervised	19.81	16.32
StereoCRL Eq. 1 and 2	Self-Supervised	**16.91**	**11.10**

Table 3. Evaluation Results on Kitti2015 Benchmark. Measuring % of disparity outliers (D1). Where 'all' signifies outliers averaged over all ground truth pixels image. Best results highlighted in bold.

Method	Method type	D1-all
SsSMnet [22]	Self-Supervised	3.57
MADNet [17]	Self-Supervised	4.66
UnOS [18]	Self-Supervised	5.58
StereoCRL Eq. 2	Self-Supervised	4.91
StereoCRL Eq. 1 and 2	Self-Supervised	**3.55**

similarity index (SSIM) between the original and the predicted stereo images, same as [20]. Since ground truth is provided in SCARED, we compare mean absolute depth error in mm on test sets 1 and 2 with other self-supervised methods submitted to the original challenge [1]. Since Kitti2015 contains ground truth disparities too, we show end-point-error benchmark.

Figure 5 compares depth estimated by StereoCRL model variants, i.e. one trained on the proposed approach with Eq. 1 and 2 and one trained only via self-supervision with Eq. 2. One can observe that StereoCRL Eq. 1 and 2 outperforms StereoCRL Eq. 2 in all scene types. This is visible from the sharper depth maps generated by StereoCRL Eq. 1 and 2. The model variant trained on Eq. 2 is unable to resolve depth for the anatomical structures in the background as shown in Fig. 5 samples 1 and 3. Furthermore, it results in blurry disparity maps around the edges of robotic tools and anatomical structures. Whereas, the Eq. 1 and 2 variant is exceptionally capable of learning different features for such anatomical structures in the background, creating smoother and sharper depth. Same performance is visible when observing non-surgical scene results in samples 4 and 5. The Eq. 1 and 2 variant estimates sharper depth around the leaves on the trees (sample 5) and the poles holding the traffic lights (sample 4). The full CRL approach overcomes the issue of mixed feature embeddings that pure self-supervised learning typically generates. This enables the model to learn independent discrete feature representations for both the input views, which are later combined by the decoder for prediction of sharper depth maps with iterative refinement.

The proposed model is also quantitatively validated on the *Hamlyn, SCARED* and Kitti2015 datasets in Tables 1, 2 and 3, respectively. In all three cases, our model outperforms traditional stereo depth estimation methods as well as, state-of-the-art self-supervised deep learning methods. In Table 1, the high SSIM score verifies that our model generates high fidelity disparity maps and yields better reconstruction accuracy. This exhibits the importance of explicitly modelling stereo features within self-supervised models. The same can be inferred from the results in Tables 2 and 3, which shows that our model improves on the current benchmark. Particularly, Table 2 also shows our model's capability in generalising to other laparoscopic stereo frames, since no fine-tuning was performed. In Table 3, we outperform other state-of-the-art methods. Additionally, to asses the impact of our proposed CRL loss defined in Eq. 1, we performed an ablation study by training 2 new variants of StereoCRL using (1) Vanilla CRL loss [2] and (2) Vanilla MoCo loss [8]. Both variants are later finetuned for disparity estimation via Eq. 2. Variant 1 achieved average depth error of 24.5 and 21.4 mm on SCARED datasets 1 and 2, respectively. Variant 2 achieves 23.2 and 23.6 mm on the same datasets. Both variants perform worse than the proposed StereoCRL method, which achieves 16.9 and 11.1 mm error, respectively. The results of the new variants are like a model trained only on photometric loss. This further verifies StereoCRL's ability in learning discriminative representations.

4   Conclusion

Supervised learning based depth estimation approaches typically outperform self-supervised counterparts. However, training a model to descritise the stereo input images into distinct feature representations is certainly the way forward to improve self-supervised training. We show that our approach despite being end-to-end self-supervised, is capable of learning generalisable feature representations without reliance on expensive ground truth on surgical data. This is due to contrastive representation learning boosting the model's feature representations. Our future work will entail experimenting with additional loss terms such as occlusion masking, and experimenting our contrastive learning approach paired with a supervised model to explore if generalisation can be improved.

Acknowledgements. Supported by the Royal Society (URF\R\201014 and RGF\EA\180084) and the NIHR Imperial Biomedical Research Centre.

References

1. Allan, M., McLeod, A.J., Wang, C.C., et al. J.R.: Stereo correspondence and reconstruction of endoscopic data challenge. CoRR arXiv:2101.01133 (2021)

2. Chen, T., Kornblith, S., Norouzi, M., Hinton, G.: A simple framework for contrastive learning of visual representations. In: Proceedings of the 37th International Conference on Machine Learning. Proceedings of Machine Learning Research, 13–18 Jul 2020, vol. 119, pp. 1597–1607 (2020)
3. Cordts, M., et al.: The cityscapes dataset for semantic urban scene understanding. In: Proceedings of the IEEE Conference on Computer Vision and Pattern Recognition (CVPR) (2016)
4. Geiger, A., Roser, M., Urtasun, R.: Efficient large-scale stereo matching. In: ACCV (2010)
5. Giannarou, S., Visentini-Scarzanella, M., Yang, G.Z.: Probabilistic tracking of affine-invariant anisotropic regions. IEEE Trans. Pattern Anal. Mach. Intell. **35**(1), 130–143 (2012)
6. Godard, C., Mac Aodha, O., Brostow, G.J.: Unsupervised monocular depth estimation with left-right consistency. In: IEEE Conference on Computer Vision and Pattern Recognition (CVPR) (2017)
7. Guo, X., Yang, K., Yang, W., Wang, X., Li, H.: Group-wise correlation stereo network. In: Proceedings of the IEEE Conference on Computer Vision and Pattern Recognition, pp. 3273–3282 (2019)
8. He, K., Fan, H., Wu, Y., Xie, S., Girshick, R.: Momentum contrast for unsupervised visual representation learning. arXiv preprint arXiv:1911.05722 (2019)
9. Khosla, P., et al.: Supervised contrastive learning. In: Larochelle, H., Ranzato, M., Hadsell, R., Balcan, M.F., Lin, H. (eds.) Advances in Neural Information Processing Systems, vol. 33, pp. 18661–18673. Curran Associates, Inc. (2020). https://proceedings.neurips.cc/paper/2020/file/d89a66c7c80a29b1bdbab0f2a1a94af8-Paper.pdf
10. Lipson, L., Teed, Z., Deng, J.: RAFT-Stereo: multilevel recurrent field transforms for stereo matching. arXiv preprint arXiv:2109.07547 (2021)
11. Mayer, N., Ilg, E., Häusser, P., Fischer, P., Cremers, D., Dosovitskiy, A., Brox, T.: A large dataset to train convolutional networks for disparity, optical flow, and scene flow estimation. In: IEEE International Conference on Computer Vision and Pattern Recognition (CVPR) (2016). http://lmb.informatik.uni-freiburg.de/Publications/2016/MIFDB16, arXiv:1512.02134
12. Mayer, N., Ilg, E., Häusser, P., Fischer, P., Cremers, D., Dosovitskiy, A., Brox, T.: A large dataset to train convolutional networks for disparity, optical flow, and scene flow estimation. In: CVPR, pp. 4040–4048 (2016)
13. Menze, M., Heipke, C., Geiger, A.: Object scene flow. ISPRS J. Photogrammetry Remote Sens. (JPRS) **140**, 60–76 (2018)
14. Pilzer, A., Lathuilière, S., Xu, D., Puscas, M.M., Ricci, E., Sebe, N.: Progressive fusion for unsupervised binocular depth estimation using cycled networks. IEEE Trans. Pattern Anal. Mach. Intell. **42**(10), 2380–2395 (2019)
15. Sun, D., Yang, X., Liu, M.Y., Kautz, J.: PWC-Net: CNNs for optical flow using pyramid, warping, and cost volume. In: IEEE Conference on Computer Vision and Pattern Recognition (2018)
16. Tankovich, V., Häne, C., Fanello, S., Zhang, Y., Izadi, S., Bouaziz, S.: Hitnet: hierarchical iterative tile refinement network for real-time stereo matching. In: IEEE/CVF Conference on Computer Vision and Pattern Recognition (CVPR), pp. 14357–14367 (2021)
17. Tonioni, A., Tosi, F., Poggi, M., Mattoccia, S., Di Stefano, L.: Real-time self-adaptive deep stereo. In: The IEEE Conference on Computer Vision and Pattern Recognition (CVPR) (2019)

18. Wang, Y., Wang, P., Yang, Z., Luo, C., Yang, Y., Xu, W.: Unos: unified unsupervised optical-flow and stereo-depth estimation by watching videos. In: Proceedings of the IEEE/CVF Conference on Computer Vision and Pattern Recognition (CVPR) (2019)
19. Yamaguchi, K., McAllester, D.A., Urtasun, R.: Efficient joint segmentation, occlusion labeling, stereo and flow estimation. In: ECCV (2014)
20. Ye, M., Johns, E., Handa, A., Zhang, L., Pratt, P., Yang, G.: Self-supervised siamese learning on stereo image pairs for depth estimation in robotic surgery. arXiv preprint arXiv:1705.08260 (2017)
21. Zhang, F., Prisacariu, V., Yang, R., Torr, P.H.: Ga-net: guided aggregation net for end-to-end stereo matching. In: IEEE Conference on Computer Vision and Pattern Recognition, pp. 185–194 (2019)
22. Zhong, Y., Dai, Y., Li, H.: Self-supervised learning for stereo matching with self-improving ability. arXiv preprint arXiv:1709.00930 (2017)

Deep Geometric Supervision Improves Spatial Generalization in Orthopedic Surgery Planning

Florian Kordon[1,2,3]($\boxtimes$) (iD), Andreas Maier[1,2] (iD), Benedict Swartman[4],
Maxim Privalov[4], Jan S. El Barbari[4], and Holger Kunze[3] (iD)

[1] Pattern Recognition Lab, Friedrich-Alexander-Universität
Erlangen-Nürnberg (FAU), Erlangen, Germany
florian.kordon@fau.de

[2] Erlangen Graduate School in Advanced Optical Technologies (SAOT),
Friedrich-Alexander-Universität Erlangen-Nürnberg (FAU), Erlangen, Germany

[3] Siemens Healthcare GmbH, Forchheim, Germany

[4] Department for Trauma and Orthopaedic Surgery, BG Trauma Center
Ludwigshafen, Ludwigshafen, Germany

Abstract. Careful surgical planning facilitates the precise and safe placement of implants and grafts in reconstructive orthopedics. Current attempts to (semi-)automatic planning separate the extraction of relevant anatomical structures on X-ray images and perform the actual positioning step using geometric post-processing. Such separation requires optimization of a proxy objective different from the actual planning target, limiting generalization to complex image impressions and the positioning accuracy that can be achieved. We address this problem by translating the geometric steps to a continuously differentiable function, enabling end-to-end gradient flow. Combining this companion objective function with the original proxy formulation improves target positioning directly while preserving the geometric relation of the underlying anatomical structures. We name this concept Deep Geometric Supervision. The developed method is evaluated for graft fixation site identification in medial patellofemoral ligament (MPFL) reconstruction surgery on (1) 221 diagnostic and (2) 89 intra-operative knee radiographs. Using the companion objective reduces the median Euclidean Distance error for MPFL insertion site localization from (1) 2.29 mm to 1.58 mm and (2) 8.70 px to 3.44 px, respectively. Furthermore, we empirically show that our method improves spatial generalization for strongly truncated anatomy.

The authors gratefully acknowledge funding of the Erlangen Graduate School in Advanced Optical Technologies (SAOT) by the Bavarian State Ministry for Science and Art.

Supplementary Information The online version contains supplementary material available at https://doi.org/10.1007/978-3-031-16449-1_59.

Keywords: Surgical planning · Orthopedics · Landmark detection

1   Introduction

Careful planning of the individual surgical steps is an indispensable tool for the orthopedic surgeon, elevating the procedure's safety and ensuring high levels of surgical precision [4,6,14,15]. A surgical plan for routine interventions like ligament reconstruction describes several salient landmarks on a 2D X-ray image and relates them in a geometric construction [5,8,22]. Previous attempts to automate this planning type typically separate automatic feature localization with a learning algorithm and geometric post-processing [11–13]. The separation allows to mimic the manual step-wise workflow and enables granular control over each planning step. However, this approach comes with the drawbacks of optimizing a proxy criterion. While this surrogate has shown to be a low-error approximation of the actual planning target for well-aligned anatomy, the strength of correlation depends on the level of image truncation and the visibility of the contained radiographic landmarks [2,24,27]. In manual planning, the user compensates for these effects by extrapolating visual cues and using prior anatomical knowledge. As the learning algorithm has no direct access to this knowledge, the variance in correlation limits spatial generalization to unseen data with a broad range of image characteristics. In this work, we develop and analyze a companion objective function that optimizes the planning target directly. We exploit that the planning geometry can be formulated as a continuously differentiable function, enabling end-to-end gradient flow. Through the combination with the original optimization of anatomical feature localization, the relations of the planning geometry can be retained. We name this concept Deep Geometric Supervision (DGS). We test its effectiveness by studying the following research questions.

RQ 1. How does DGS affect the overall positioning accuracy?
RQ 2. Does DGS improve spatial generalisation on truncated images?
RQ 3. Can the potential improvements be applied to more complex imaging data in an intra-operative setting?

The developed method is evaluated for medial patellofemoral ligament (MPFL) reconstruction planning on diagnostic and intra-operative knee radiographs. This planning involves calculating the Schoettle Point (SP) [22], which determines the physiologically correct insertion point of the replacement graft and ensures long-term joint stability. We demonstrate that DGS significantly improves localization accuracy, increases the success rate of plannings to be within the required precision range of 2.5 mm, and enables generalization to severely truncated images.

2   Materials and Methods

2.1   Automatic Approach to Orthopedic Surgical Planning

We build on the two-stage planning method proposed by Kordon et al. [12,13]. First, the positions of salient anatomical landmarks are automatically extracted using a multitask learning (MTL) approach. Then, the landmarks are interrelated through geometric post-processing to locate the actual planning target.

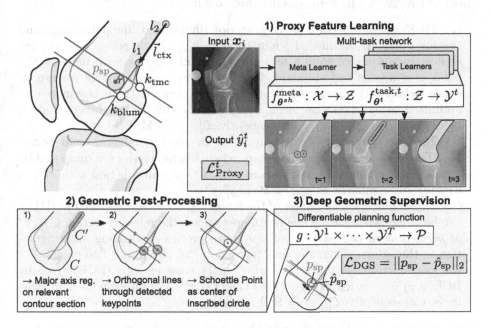

Fig. 1. Overview of the proposed method to automatic surgical planning with DGS.

In the first stage (Fig. 1-1), we want to optimize a mapping $f_\theta^t : \mathcal{X} \to \mathcal{Y}^t$ from the input domain $\mathcal{X}$ to several task solution spaces $\{\mathcal{Y}^t\}_{t \in [T]}$. θ marks a set of trainable function parameters, and T is the number of parallel tasks to solve. The function f_θ^t is optimized in a supervised manner using M datapoints $\{x_i, y_i^1, \ldots, y_i^T\}_{i \in [M]}$ with ground truth y_i^t. To exploit similarities between tasks and maintain task-specific complexity at the same time, we employ hard parameter-sharing [3,21]. Therefore, the model capacity θ is separated into disjoint sets of shared parameters θ^{sh} and task-specific parameters $\{\theta^t\}_{t \in [T]}$. According to Baxter [1], this separation can be interpreted as a subdivision of function f_θ^t into a meta learner $f_{\theta^{sh}}^{\mathrm{meta}} : \mathcal{X} \to \mathcal{Z}$ and task-specific learners $f_{\theta^t}^{\mathrm{task},t} : \mathcal{Z} \to \mathcal{Y}^t$, such that the composition $(f_{\theta^t}^{\mathrm{task},t} \circ f_{\theta^{sh}}^{\mathrm{meta}}) = f_{\theta^{sh},\theta^t}^t : \mathcal{X} \to \mathcal{Y}^t$. The task-specific parameters can be trained by minimizing the loss function $\mathcal{L}_{\mathrm{Proxy}}^t(\cdot, \cdot) : \mathcal{Y}^t \times \mathcal{Y}^t \to \mathbb{R}_0^+$. Following common practice in MTL literature, the shared parameters of the meta learner are optimized using a linear combination of all task losses [23]. Using this rationale, we arrive at the empirical risk minimization (ERM) objective

$$\min_{\theta^{sh};\theta^1,\ldots,\theta^T} \sum_{t=1}^{T} \hat{\mathcal{L}}_{\mathrm{Proxy}}^t(\theta^{sh}, \theta^t), \tag{1}$$

where $\hat{\mathcal{L}}_{\text{Proxy}}^t(\boldsymbol{\theta}^{sh}, \boldsymbol{\theta}^t) \triangleq \frac{1}{M} \sum_{i=1}^{M} \mathcal{L}_{\text{Proxy}}^t \left(f_{\boldsymbol{\theta}^{sh}, \boldsymbol{\theta}^t}^t(\boldsymbol{x}_i), y_i^t \right)$ is the task-specific empirical risk estimated on the training data. For the specific example of SP construction [22], we have to consider three distinct tasks (Fig. 1).

1. ($t = 1$): *Keypoint detection* of the posterior Blumensaat line point k_{blum} and turning point on the medial femur condyle k_{tmc}. Both points are encoded as heatmaps sampled from a bivariate Gaussian function with mean at the point coordinates and standard deviation of $\sigma_{\text{hm}} = 6\,\text{px}$. The correspondence between the predicted heatmap $\hat{y}$ and ground truth heatmap y is optimized using pixel-wise mean squared error (MSE) loss function given by $\text{MSE}(\hat{y}, y) = \mathbb{E}\,\|\hat{y} - y\|_2^2$. Consequently, $\mathcal{L}_{\text{Proxy}}^1 := \text{MSE}(\hat{y}_i^1, y_i^1)$.
2. ($t = 2$): *Line regression* of the tangent to the posterior femur shaft cortex l_{ctx}. The line is encoded as a heatmap, where the intensities are computed by evaluating the point-to-line distances with a Gaussian function with $\sigma_{\text{hm}} = 6\,\text{px}$ [11,13]. Similarly, $\mathcal{L}_{\text{Proxy}}^2 := \text{MSE}(\hat{y}_i^2, y_i^2)$.
3. ($t = 3$): *Semantic segmentation* of the femur region S. As described in [11], S can be combined with the line heatmap to mask the relevant section $C' \subseteq C$ of segmentation contour $C \subseteq S$ for more precise positioning and angulation in the subsequent major axis regression [26] of relevant line points. For the loss function, we use a pixel-wise binary cross entropy (BCE) given by $\text{BCE}(\hat{y}, y) := -[y \log(\sigma(\hat{y})) + (1 - y) \log(1 - \sigma(\hat{y}))]$ with sigmoid nonlinearity σ. Consequently, $\mathcal{L}_{\text{Proxy}}^3 := \text{BCE}(\hat{y}_i^3, y_i^3)$.

After extracting the relevant features, they are converted to geometric primitives and interrelated according to the planning geometry (Fig. 1-2). This geometry describes consecutive calculations to localize the planning target relevant to the surgeon. For MPFL planning, the cortex tangent line is determined by major axis regression [26] on relevant contour points. The SP can be approximated by the center of the inscribed circle bounded by the tangent and two orthogonal lines intersecting both detected keypoints (Fig. 1).

2.2 Deep Geometric Supervision (DGS)

The disconnect between the proxy function and the actual planning target limits generalization to unfavorable but common image characteristics. We approach this issue by introducing the concept of Deep Geometric Supervision. To this end, we add a companion objective function to the original ERM term (Eq. 1) that directly minimizes positioning errors of the planning target while retaining the relations of the planning geometry (Fig. 1-3).

Mathematically, we start by combining all geometric steps in a single non-parametric function $g : \mathcal{Y}^1 \times \cdots \times \mathcal{Y}^T \to \mathcal{P}$ that operates on the outputs of the anatomical feature extractor $f_{\boldsymbol{\theta}^{sh}, \boldsymbol{\theta}^t}^t$. The output of g is the desired planning target, e.g., a keypoint. Next, we calculate the positioning error of the planning target with the loss function $\mathcal{L}_{\text{DGS}}(\cdot, \cdot) : \mathcal{P} \times \mathcal{P} \to \mathbb{R}_0^+$. The empirical risk is given by

$$\hat{\mathcal{L}}_{\text{DGS}}(\boldsymbol{\theta}^{sh}, \boldsymbol{\theta}^1, \dots, \boldsymbol{\theta}^T) \triangleq \frac{1}{M} \sum_{i=1}^{M} \mathcal{L}_{\text{DGS}} \left(g \left(f_{\boldsymbol{\theta}^{sh}, \boldsymbol{\theta}^1}^1(\boldsymbol{x}_i), \dots, f_{\boldsymbol{\theta}^{sh}, \boldsymbol{\theta}^T}^T(\boldsymbol{x}_i) \right), p_i \right).$$

Finally, adding this term to the original ERM formulation yields

$$\min_{\theta^{sh};\theta^1,\dots,\theta^T} \lambda \sum_{t=1}^{T} \underbrace{\hat{\mathcal{L}}_{\text{Proxy}}^t(\theta^{sh},\theta^t)}_{\text{Anatomical Features}} + (1-\lambda)\underbrace{\hat{\mathcal{L}}_{\text{DGS}}(\theta^{sh},\theta^1,\dots,\theta^T)}_{\text{Surgical Target}}, \quad (2)$$

where $\lambda \in \mathbb{R}$ is a multiplicative risk-weighting term. Here, $\mathcal{L}_{\text{DGS}} := \|\hat{p}_i - p_i\|_2$. Since the planning function g is not subject to trainable parameters, the minimization of the additional risk term $\hat{\mathcal{L}}_{\text{DGS}}$ should directly contribute to updates of θ^{sh} and θ^t. For that purpose, $(g \circ f_{\theta^{sh},\theta^t}^t)$ must be a continuously differentiable function, such that $(g \circ f_{\theta^{sh},\theta^t}^t) \in C^1(\mathcal{X},\mathcal{P})$. To fulfill this constraint, the representations and objective functions for keypoint and line regression need to be changed from their original formulation in [13]. Matching the keypoint heatmaps with MSE is not feasible as it is typically followed by a subsequent non-differentiable argmax operation to extract the intensity peak's x and y coordinates. We instead use the regularized spatial-to-numerical transform (DSNT) [18]. For that purpose, the predicted heatmap $\hat{y}$ is rectified applying a spatial softmax and normalized with the L1 norm. The result of this standardization $\hat{y}'$ is transformed to the numerical coordinate $\hat{c} = \text{DSNT}(\hat{y}')$ in the range of $[-1,1]$ exploiting a probabilistic interpretation of $\hat{y}'$. This allows the cost function to operate directly on numerical coordinates and optimize the heatmaps implicitly. Finally, the keypoint loss function is updated to $\mathcal{L}_{\text{Proxy}}^1 := \|\hat{c}_i^1 - c_i^1\|_2 + D_{\text{JS}}(p(\hat{y}_i^1)\,\|\,\mathcal{N}(y_i^1,\sigma_{\text{hm}}^2 I_2))$. $p(\cdot)$ is a probability mass function under the interpretation of the predicted coordinates as discrete bivariate random vectors, and $D_{\text{JS}}(\cdot\|\cdot)$ is the Jensen-Shannon divergence which encourages similarity of the heatmaps to a Gaussian prior [18].

A differentiable representation of the line's position and orientation is obtained by calculating raw image moments M_{pq} and second order central moments μ_{pq}' on the line heatmap [7,20]. The centroid $c_{x,y}$ and orientation angle γ are given by $c_{x,y} = (M_{10}/M_{00}, M_{01}/M_{00})$ and $\gamma = \frac{1}{2}\arctan(2\mu_{11}'/(\mu_{20}' - \mu_{02}')) + \frac{\pi}{2}[\mu_{20}' < \mu_{02}']$. $[\cdot]$ marks the Iverson bracket.

2.3 Model Variants

We define three model variants to evaluate the effect of DGS on planning accuracy and spatial generalization. *A) Proxy*, which uses all three anatomical detection tasks and uses the original ERM term Eq. 1 without DGS. The planning target is calculated with geometric post-processing. *B) Proxy−Seg+DGS*, which utilizes the updated ERM term Eq. 2. Here, g is used to calculate the planning target, and the segmentation task is omitted. *C) Proxy+DGS*, which uses ERM term Eq. 2 but keeps the segmentation task to allow direct comparison without the external factor of different task and parameter counts.

2.4 Datasets and Training Protocol

Method evaluation was done on two radiographic image cohorts of the lateral knee joint. *Cohort 1)* contains 221 diagnostic radiographs collected retrospectively from anonymized databases. For each image, the SP geometry was annotated by an expert trauma surgeon with a proprietary tool by Siemens Healthcare GmbH. The femur polygon was labeled by a medical engineer using labelme software [25]. The images were split into three sets for training (167), validation (16), and testing (38) with stratified sampling [10]. There, all data showing a steel sphere of 30 mm diameter was assigned to the test split, allowing conversion from pixel to mm space. *Cohort 2)* contains 89 intra-operative X-ray images from 43 patients acquired with mobile C-arm systems. Most images show severely truncated bone shafts and instrumented anatomy. The images were annotated by a medical engineer using a custom extension of labelme [25]. The data was divided into training (61), validation (9), and test (19) with no patient overlap. During optimization, the training data was augmented using horizontal flipping ($p = 0.5$), rotation ($\alpha \in [-45°, 45°]$, $p = 1$), and scaling ($s \in [0.8, 1.2]$, $p = 1$) [13]. After min-max normalization of the images to the intensity range of $[0, 1]$, the data were standardized to the dimensions [H:256 × W:256] px by bicubic sampling and zero-padding to preserve the original aspect ratios. For each cohort and variant, an MTL hourglass network [13,17] (128 feature root) was trained for 450 epochs using an Adam optimizer, a learning rate of 0.001/0.0006 (cohort 1/2), a batch size of 2, and multiplicative learning rate decay of 0.1 after 350 epochs. The risk-weighting with $\lambda = 0.99$ was re-balanced [9] by decreasing $\lambda \in [0.01, 0.99]$ by 0.01 every forth epoch. Implementation was done in PyTorch v1.8 [19] (Python v3.8.12, CUDA v11.0) and reproducibility was confirmed.

3   Results

Effects on Feature Extraction and Positioning Accuracy (RQ 1). The results of model evaluation on Cohort 1) are summarized in Fig. 2. DGS reduces the median SP Euclidean Distance (ED) error from $2.29 [1.84, 2.82]_{CI95}$ mm (A) to $1.68 [1.19, 2.17]_{CI95}$ mm (B) and $1.58 [1.15 2.09]_{CI95}$ mm (C), respectively (Fig. 2-a). Between the two DGS variants, we observe no drop in performance when additionally solving the segmentation task. This understanding lets us reject insufficient model capacity or conflicting task configurations as a reason for inferior performance of the proxy optimization. Furthermore, DGS increases the number of predictions that fall within the clinically relevant precision range of 2.5 mm [22] from 63.2% (Model A) to 76.3% (Model B&C) (Fig. 2-b). Interestingly, DGS slightly changes the spatial appearance of the line heatmaps, increasing activations in the posterior aspect of the Blumensaat line (Fig. 2-c). Since this area resides on the tangential extension of the shaft cortex, we argue that activation in this region allows to compensate for small errors in line alignment.

Effects on Spatial Generalization (RQ2). To evaluate potential effects of DGS on spatial generalization, we constructed a secondary test set with different levels of shaft truncation. For this purpose, multiple crops per image were created such that the visible bone shaft corresponds to a fixed ratio $t \in [0.8, 2.7]$ between bone axis length and femur head width. The results are summarized in

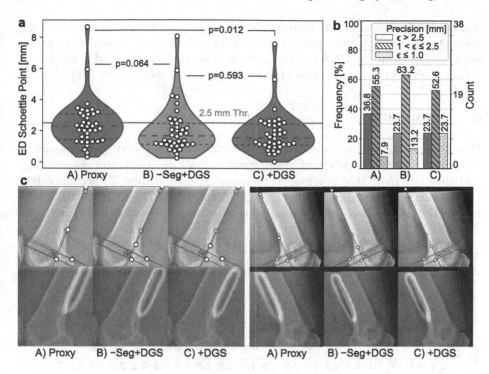

Fig. 2. Summary of model evaluation on Cohort 1). **a**, violin plots of the positioning errors on the test set (ED: Euclidean Distance; white dots mark individual planning samples). Statistical significance was evaluated with a two-sided Mann-Whitney U rank test. **b**, error distribution w.r.t. clinically relevant precision ranges. **c**, planning geometry (red: prediction, green: ground truth) and spatial appearance of line heatmaps. (Color figure online)

Fig. 3. We see considerable improvements for severely truncated shafts for both variants with DGS. This observation is underlined by a strong Spearman's rank correlation of $r_s = -0.79$ for Model A compared to moderate correlations of $r_s = -0.48$ and $r_s = -0.53$ for the DGS Models B and C, respectively ($p \ll 0.01$ for all correlations). Visual inspection shows that the tangent determined via the bone contour experiences a systematic angular offset in the postero-distal direction for very short shaft lengths. Optimization with DGS can recover the optimal tangent direction in most of these cases despite insufficient image information.

Application to Complex Intra-Operative Data (RQ 3). The evaluation on the intra-operative Cohort 2) is summarized in Fig. 4. Similar to Cohort 1), DGS reduces the positioning error significantly, yielding median ED scores of $3.50\,[2.47, 6.92]_{CI95}$ px and $3.44\,[2.09, 6.67]_{CI95}$ px for Model A and B, respectively. The improvements over the original proxy formulation with a median error of $8.70\,[5.06, 16.60]_{CI95}$ px can be explained by generally shorter shaft lengths caused by a smaller field of view of the mobile C-arm imaging device and less standardized acquisition. As seen before, this characteristic leads to misaligned tangent predictions during proxy optimization. The compensation effect previously identified in the DGS variants, which is characterized by additional activation peaks in the distal region of the femur, is clearly enhanced (Fig. 4-c).

4   Discussion

Directly optimizing the planning target position while preserving the geometric relation of the anatomical structures promises a more precise, better generalizing, and clinically motivated planning automation in orthopedics. To accommodate this rationale, we developed and analyzed the concept of Deep Geometric Supervision. By interpreting the planning geometry as a differentiable function, the

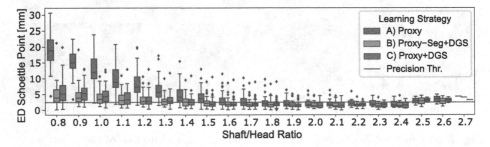

Fig. 3. Dependencies of the model variants to different levels of bone shaft truncation.

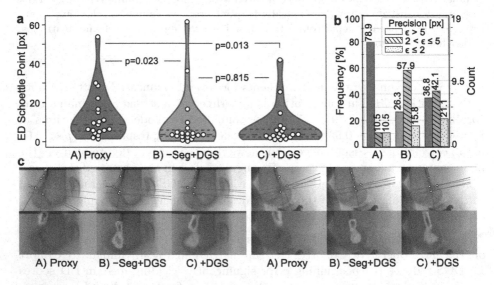

Fig. 4. Evaluation on Cohort 2). **a**, test set positioning errors. **b**, relevant precision classes. **c**, planning geometry (red: prediction, green: ground truth) and line heatmaps. (Color figure online)

planning target and the anatomical feature extractor can be optimized jointly. Improving target positioning accuracy while maintaining the core idea of stepwise geometric planning is a critical design decision that fosters clinical acceptance. Intriguingly, integrating the planning function into the computation graph can be interpreted as learning with a known operator [16], allowing end-to-end training and effectively reducing the upper error bound. In this context, it should be noted that minimizing only the DGS term yields a trivial solution where the extracted landmarks collapse to a single point at the planning target position. While these solutions offer competitive precision, they are undesirable because they do not mimic the clinically established planning workflow and cannot be easily verified for anatomic fidelity. An important trait of DGS is the improvement in spatial generalization. Especially in the intra-operative environment with constrained patient and device positioning, we cannot always expect standard acquisitions with sufficiently large bone shafts. There, DGS successfully bridges the semantic gaps present in the current proxy optimization strategy, reducing malpositioning when landmark visibility is limited. Besides these advantages, our current implementation of the planning function imposes little geometric constraints and, in theory, allows for different anatomical feature configurations that arrive at the same planning target. Reducing the space of possible solutions could ensure planning fidelity and smooth the optimization landscape.

Despite this limitation, our method effectively improves positioning accuracy and spatial generalization in orthopedic surgical planning. At the same time, it allows maintaining the clinically established planning geometry. We believe that these aspects facilitate the translation of planning automation concepts to the field and will ultimately motivate the development of new planning guidelines.

Disclaimer. The methods and information presented here are based on research and are not commercially available.

References

1. Baxter, J.: A model of inductive bias learning. J. Artif. Int. Res. **12**(1), 149–198 (2000)
2. Burrus, M.T., Werner, B.C., Cancienne, J.M., Diduch, D.R.: Correct positioning of the medial patellofemoral ligament: troubleshooting in the operating room. Am. J. Orthop. (Belle Mead N.J.) **46**(2), 76–81 (2017)
3. Caruana, R.: Multitask learning. In: Thrun, S., Pratt, L. (eds.) Learning to Learn, pp. 95–133. Springer, Boston (1998). https://doi.org/10.1007/978-1-4615-5529-2_5
4. Casari, F.A., et al.: Augmented reality in orthopedic surgery is emerging from proof of concept towards clinical studies: a literature review explaining the technology and current state of the art. Curr. Rev. Musculoskelet. Med. **14**(2), 192–203 (2021). https://doi.org/10.1007/s12178-021-09699-3
5. Colombet, P., Robinson, J., Christel, P., Franceschi, J.P., Djian, P., Bellier, G., et al.: Morphology of anterior cruciate ligament attachments for anatomic reconstruction: a cadaveric dissection and radiographic study. Arthroscopy **22**(9), 984–992 (2006). https://doi.org/10.1016/j.arthro.2006.04.102

6. Essert, C., Joskowicz, L.: Image-based surgery planning. In: Zhou, S.K., Rueckert, D., Fichtinger, G. (eds.) Handbook of Medical Image Computing and Computer Assisted Intervention, pp. 795–815. The Elsevier and MICCAI Society Book Series. Academic Press (2020). https://doi.org/10.1016/B978-0-12-816176-0.00037-5

7. Hu, M.K.: Visual pattern recognition by moment invariants. IEEE Trans. Inf. Theory **8**(2), 179–187 (1962). https://doi.org/10.1109/TIT.1962.1057692

8. Johannsen, A.M., Anderson, C.J., Wijdicks, C.A., Engebretsen, L., LaPrade, R.F.: Radiographic landmarks for tunnel positioning in posterior cruciate ligament reconstructions. Am. J. Sports Med. **41**(1), 35–42 (2013). https://doi.org/10.1177/0363546512465072

9. Kervadec, H., Bouchtiba, J., Desrosiers, C., Granger, E., Dolz, J., Ben Ayed, I.: Boundary loss for highly unbalanced segmentation. In: Cardoso, M.J., et al. (eds.) International Conference on Medical Imaging with Deep Learning. Proceedings of Machine Learning Research, vol. 102, pp. 285–296. PMLR (2019)

10. Kordon, F., Lasowski, R., Swartman, B., Franke, J., Fischer, P., Kunze, H.: Improved X-ray bone segmentation by normalization and augmentation strategies. In: Handels, H., Deserno, T., Maier, A., Maier-Hein, K., Palm, C., Tolxdorff, T. (eds.) Bildverarbeitung für die Medizin 2019. I, pp. 104–109. Springer, Wiesbaden (2019). https://doi.org/10.1007/978-3-658-25326-4_24

11. Kordon, F., Maier, A., Swartman, B., Privalov, M., El Barbari, J.S., Kunze, H.: Contour-based bone axis detection for X-Ray guided surgery on the knee. In: Martel, A.L., et al. (eds.) MICCAI 2020. LNCS, vol. 12266, pp. 671–680. Springer, Cham (2020). https://doi.org/10.1007/978-3-030-59725-2_65

12. Kordon, F., Maier, A., Swartman, B., Privalov, M., El Barbari, J.S., Kunze, H.: Multi-stage platform for (semi-)automatic planning in reconstructive orthopedic surgery. J. Imaging **8**(4), 108 (2022). https://doi.org/10.3390/jimaging8040108

13. Kordon, F., et al.: Multi-task localization and segmentation for X-Ray guided planning in knee surgery. In: Shen, D., et al. (eds.) MICCAI 2019. LNCS, vol. 11769, pp. 622–630. Springer, Cham (2019). https://doi.org/10.1007/978-3-030-32226-7_69

14. Kubicek, J., Tomanec, F., Cerny, M., Vilimek, D., Kalova, M., Oczka, D.: Recent trends, technical concepts and components of computer-assisted orthopedic surgery systems: a comprehensive review. Sensors **19**(23), 5199 (2019). https://doi.org/10.3390/s19235199

15. Linte, C.A., Moore, J.T., Chen, E.C., Peters, T.M.: Image-guided procedures. Bioeng. Surg. 59–90 (2016). https://doi.org/10.1016/b978-0-08-100123-3.00004-x

16. Maier, A., Syben, C., Stimpel, B., Würfl, T., Hoffmann, M., Schebesch, F., et al.: Learning with known operators reduces maximum error bounds. Nat. Mach. Intell. **2019**(1), 373–380 (2019). https://doi.org/10.1038/s42256-019-0077-5

17. Newell, A., Yang, K., Deng, J.: Stacked hourglass networks for human pose estimation. In: Leibe, B., Matas, J., Sebe, N., Welling, M. (eds.) ECCV 2016. LNCS, vol. 9912, pp. 483–499. Springer, Cham (2016). https://doi.org/10.1007/978-3-319-46484-8_29

18. Nibali, A., He, Z., Morgan, S., Prendergast, L.: Numerical coordinate regression with convolutional neural networks. arXiv e-prints arXiv:1801.07372, January 2018

19. Paszke, A., et al.: PyTorch: an imperative style, high-performance deep learning library. In: Wallach, H., Larochelle, H., Beygelzimer, A., Alché-Buc, F.D., Fox, E., Garnett, R. (eds.) Advances in Neural Information Processing Systems, pp. 8026–8037. Curran Associates, Inc. (2019)

20. Rocha, L., Velho, L., Carvalho, P.: Image moments-based structuring and tracking of objects. In: XV Brazilian Symposium on Computer Graphics and Image Processing, pp. 99–105 (2002). https://doi.org/10.1109/SIBGRA.2002.1167130
21. Ruder, S.: An overview of multi-task learning in deep neural networks. arXiv e-prints arXiv:1706.05098, June 2017
22. Schöttle, P.B., Schmeling, A., Rosenstiel, N., Weiler, A.: Radiographic landmarks for femoral tunnel placement in medial patellofemoral ligament reconstruction. Am. J. Sports Med. **35**(5), 801–804 (2007). https://doi.org/10.1177/0363546506296415
23. Sener, O., Koltun, V.: Multi-task learning as multi-objective optimization. In: Bengio, S., Wallach, H., Larochelle, H., Grauman, K., Cesa-Bianchi, N., Garnett, R. (eds.) Advances in Neural Information Processing Systems, vol. 31. Curran Associates, Inc. (2018)
24. Tanaka, M.J., Chahla, J., Farr, J., LaPrade, R.F., Arendt, E.A., Sanchis-Alfonso, V., et al.: Recognition of evolving medial patellofemoral anatomy provides insight for reconstruction. Knee Surg. Sports Traumatol. Arthrosc. (2018). https://doi.org/10.1007/s00167-018-5266-y
25. Wada, K.: Labelme: image polygonal annotation with Python (2016)
26. Warton, D.I., Wright, I.J., Falster, D.S., Westoby, M.: Bivariate line-fitting methods for allometry. Biol. Rev. Biol. Proc. Camb. Philos. Soc. **81**(2), 259–291 (2006). https://doi.org/10.1017/S1464793106007007
27. Ziegler, C.G., Fulkerson, J.P., Edgar, C.: Radiographic reference points are inaccurate with and without a true lateral radiograph: the importance of anatomy in medial patellofemoral ligament reconstruction. Am. J. Sports Med. **44**(1), 133–142 (2016). https://doi.org/10.1177/0363546515611652

On Surgical Planning of Percutaneous Nephrolithotomy with Patient-Specific CTRs

Filipe C. Pedrosa[1,2]([✉]), Navid Feizi[1,2], Ruisi Zhang[3,4], Remi Delaunay[3,4], Dianne Sacco[4,5], Jayender Jagadeesan[3,4], and Rajni Patel[1,2]

[1] Western University, London, ON, Canada
fpedrosa@uwo.ca
[2] Canadian Surgical Technologies and Advanced Robotics, London, ON, Canada
[3] Brigham and Women's Hospital, Boston, MA, USA
[4] Harvard Medical School, Boston, MA, USA
[5] Massachusetts General Hospital, Boston, MA, USA

Abstract. Percutaneous nephrolithotomy (PCNL) is considered a first-choice minimally invasive procedure for treating kidney stones larger than 2 cm. It yields higher stone-free rates than other minimally invasive techniques and is employed when extracorporeal shock wave lithotripsy or uteroscopy are, for instance, infeasible. Using this technique, surgeons create a tract through which a scope is inserted for gaining access to the stones. Traditional PCNL tools, however, present limited maneuverability, may require multiple punctures and often lead to excessive torquing of the instruments which can damage the kidney parenchyma and thus increase the risk of hemorrhage. We approach this problem by proposing a nested optimization-driven scheme for determining a single tract surgical plan along which a patient-specific concentric-tube robot (CTR) is deployed so as to enhance manipulability along the most dominant directions of the stone presentations. The approach is illustrated with seven sets of clinical data from patients who underwent PCNL. The simulated results may set the stage for achieving higher stone-free rates through single tract PCNL interventions while decreasing blood loss.

Keywords: Percutaneous nephrolithotomy · Surgical planning · Continuum robotics · Nonlinear optimization

1 Introduction

Percutaneous nephrolithotomy (PCNL) is a preferred treatment method for removing renal calculi (kidney stones) larger than 2 cm [7]. In PCNL, a urologist or radiologist determines the skin position and the primary calyx to puncture based on diagnostic CT scans for removing the stone, while considering smaller stones situated in harder to reach calyces [7]. In fact, the placement of percutaneous access is the most critical aspect of PCNL inasmuch as the clinical outcome

© The Author(s), under exclusive license to Springer Nature Switzerland AG 2022
L. Wang (Eds.): MICCAI 2022, LNCS 13437, pp. 626–635, 2022.
https://doi.org/10.1007/978-3-031-16449-1_60

is highly dependent on the accuracy of the preoperative image-based planning [17,19]. Largely due to difficulties in obtaining a percutaneuous access, a 2007 survey [11], found that only 27% of urologists who were trained in percutaneous access during residency continued to gain their own PCNL renal access with this percentage reducing to 11% amongst those who were untrained. In this setting, the best plan is one which allows the removal of the entire stone via a single tract while safeguarding surrounding tissues, organs, and blood vessels.

Owing to the structure of rigid nephroscopes and lithotripsy tools, accessing all stones from a single incision can be challenging. Once the tool is inserted into the renal collecting ducts, gentle torques must be applied to maintain skin and tissue integrity and decrease potential damage to the surrounding tissue. For many stone presentations, multiple punctures may be required to success-fully break and clear all stone fragments. However, this can lead to increased hemorrhagic complications [10]. Additionally, multiple punctures may lead to increased radiation exposure due to the need for including fluoroscopic guid-ance [23], increased operating time, increased anesthesia use, and increased risk of damage to surrounding tissue, kidney, and skin. Kyriazis *et al.* [10] com-pared multiple punctures to upper or lower kidney pole access, and showed that multiple-puncture patient groups required significantly more blood transfusion than their single puncture counterparts. In [8], single-tract patient groups had shorter hospital stay with no patient requiring blood transfusion. Lastly, the PCNL procedure for stones larger than 2 cm is reported to be 89.4% successful with an associated 13% complication rate due to the challenges of navigating the entire stone [15].

Against this background, the use of flexible robotic tools in PCNL may reduce the number of punctures and improve the stone-free rate while reducing the risk of tissue damage and hemorrhagic complications.

1.1 Concentric-Tube Robots (CTRs)

Our approach hinges on the use of CTRs as a means of gaining access to the renal collecting system. These special manipulators consist of a set of precurved and superelastic tubes that are concentrically inserted one inside another. Under independent and relative translational and rotational actuation of its component tubes, elastic interactions between overlapping sections of the tubes take place to yield the final equilibrium shape of the backbone [20]. Given their slender size and kinematic capability of achieving complex curved shapes, CTRs are particularly attractive for applications such as PCNL where the robot must navigate tortuous paths from the port of entry to the clinical targets while avoiding structures such as the lung, liver, spleen, colon and renal vessels.

For the purposes of this work, we consider a three-tube CTR whereby the innermost and intermediate tubes are composed of an initial straight transmis-sion section followed by a curved section with constant curvature. The outermost tube, however, is restricted to be straight.

1.2   Related Work

The number of tubes in the concentric arrangement and their corresponding kinematic parameters govern the kinematic behavior of CTRs and, thus, provides this class of robotic manipulators considerable freedom with respect to the number of possible designs that can be considered. As such, the selection of suitable designs for minimally invasive medical interventions must be guided by both the application and the patient-specific anatomy.

Several approaches have been reported in the literature in which methods are devised to design task-oriented, patient-tailored CTRs. Burgner *et al.* [4,5] proposed a grid search, voxel-based algorithm for maximizing the volume coverage of the clinical target while accounting for anatomical constraints. Bergeles *et al.* [3] proposed a framework for CTR design that optimizes the reachability of a set of targets while minimizing the lengths and curvatures of the tubes for kinematic stability while satisfying anatomical constraints. Morimoto *et al.* [13,14] introduced a surgeon-in-the-loop approach in which the tube parameters are calculated based on via points created manually by a medical expert in a virtual reality environment. Torres *et al.* [22] goes further and proposes a method for designing CTRs accounting for collision-free deployment from the port of entry to the targets by integrating a sampling-based motion planner. On a similar note, [1] attains asymptotic global optimality in the design of a variety of robots, CTR included, by integrating a sampling-based motion planner and stochastic optimization.

1.3   Contribution of the Paper

In this paper, we report a set of results on the issues of surgical planning and the design of patient-specific CTRs for PCNL. In contrast to [13,14], we account for the constraint at the renal pyramids and calyces for all posible configurations in which the CTR is deployed in the kidney. As opposed to [2], our method returns one single CTR design, thus not requiring change of tools and consequent redeployment of the robot into the anatomy during the PCNL procedure. The main contributions of this paper consist of a comprehensive framework, independent of any user's inputs, for devising surgical plans for percutaneous and renal access by virtue of patient-specific CTR designs. The algorithms developed can be applied seamlessly to different clinical cases. Once an optimized design is obtained, the proposed constrained inverse kinematics guides the navigation of the anatomically constrained manipulator within the kidney. In what follows, we briefly describe our method and demonstrate its usefulness by applying our algorithms to real clinical data of seven patients who underwent PCNL.

2   Surgical Planning

An appropriate percutaneous access is paramount to an effective PCNL as critical anatomical structures such as the bowel, spleen, liver, pleura and lungs must be avoided while allowing direct access to the kidney calculi, see Fig. 1. In addition, a suitable renal pyramid must be identified for renal access. To minimize damage to renal vessels, the CTR backbone must be contained within the caliceal ducts throughout the duration of the procedure.

In our approach, the surgical planning phase starts from an initial arbitrary CTR design and its corresponding constrained inverse kinematics (C-IK). Computer tomography (CT) scans segmented in 3D Slicer [6] provide information on the geometry and location of the stones, kidney, renal pyramids, and a patch of skin in the patient's supracostal region. Given the initial CTR design, the percutaneous access is selected by means of a grid search across skin points and renal pyramids such that the distal end of the CTR is delivered to the centroid of the stone with the smallest possible deviation at the pyramid. A surgical plan is described by a bijective transformation $H \in SE(3)$ which defines a supracostal puncture location as well as an insertion angle for the CTR and surgical tools. Any remaining position deviations, either at the end-effector or renal calyx, will be dealt with subsequently during the CTR design optimization.

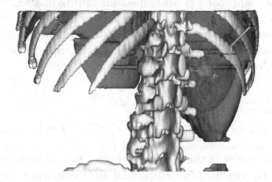

Fig. 1. Skeleton (gray), liver (amber), pleura (blue), kidney (magenta), kidney stones (white), renal pyramid (red marker), distal target (green marker) (Color figure online)

Constrained Inverse Kinematics (C-IK) for CTRs. Classically, the inverse kinematics (IK) problem for redundant manipulators has been addressed with by means of a composition of a minimum norm solution with local gradient projection techniques on the kernel of the corresponding Jacobians [18]. Here, we cast the IK for CTRs as a sequential quadratic programming (SQP) problem where the constraints and secondary goals are addressed in a more explicit manner.

Let $\dot{x}^{\text{aug}} := \begin{bmatrix} \dot{x}_{\text{ee}}^T & \dot{x}_{\text{clx}}^T \end{bmatrix}^T$ and $J^{\text{aug}} := \begin{bmatrix} J_{\text{ee}} & J_{\text{clx}} \end{bmatrix}$ denote the augmented vector and augmented Jacobian of spatial velocities at the end-effector and renal calyx, respectively. Then the C-IK is cast as

$$\underset{\dot{q},\,\delta}{\text{minimize}} \quad \frac{1}{2}\left(\dot{q}^T \Omega \dot{q} + \delta^T \Lambda \delta\right)$$

$$\text{subject to} \quad \dot{x}^{\text{aug}} + \delta = J^{\text{aug}}\dot{q}$$

$$\frac{1}{\alpha}(q^- - q) \le \dot{q} \le \frac{1}{\alpha}(q^+ - q)$$

$$\delta^- \le \delta \le \delta^+$$

(1)

where $\dot{q} \in \mathbb{R}^6$ is the vector of joint velocities of the CTR, $\delta \in \mathbb{R}^6$ is a slack variable used to relax the kinematic constraints on spatial velocities at both the end-efector and the renal calices, Ω, Λ are positive definite matrices of appropriate dimensions, and $q^{+,-}, \delta^{+,-}$ impose upper and lower bounds on the joint positions (resolved-rate collision avoidance) and spatial velocities relaxation.

3   Patient-Specific Design of CTRs

A key aspect of our approach is an ellipsoidal approximation to a point cloud $A = \{a_i\}_{1 \le i \le m}$, $a_i \in \mathbb{R}^3$ describing the geometry of the kidney stone. Simply put, the stone presentation is modeled as the minimum-volume covering ellipsoid (MVCE) which envelops all stone fragments. The computation of the MVCE is formulated as the following optimization problem [24]:

$$\underset{Q,\,c}{\text{minimize}} \quad \det\left(Q^{-\frac{1}{2}}\right), \quad Q \succ 0$$

$$\text{subject to} \quad (a_i - c)^T Q(a_i - c) \le 1, \quad i = 1, \ldots, m.$$

(2)

where $c \in \mathbb{R}^3$ is the center of the ellipsoid whose shape is defined by the positive definite matrix Q. By virtue of this ellipsoidal approximation, we lay the groundwork for the objective patient-specific design of CTRs across multiple patients with a single framework. More importantly, this approach allows us to claim that the semi-axes of the MVCE identify what we term as the dominant directions of the stone presentation, i.e., the most frequent directions that surgeons will find themselves navigating along during the stone ablation procedure.

3.1   Problem Formulation

Let us denote the MVCE in (2) by $E_{Q,c}$ and the extreme points of each of its semi-axes by p_i, where each p_i lies on the surface of $E_{Q,c}$. We seek a set of tube kinematic parameters that allows the CTR to reach the points p_i for $i = 1, \ldots, 6$. As alluded to in Subsect. 1.1, the structure of the CTR we consider for this problem is that of a three-tube CTR in which the outermost tube is straight whereas the remaining two consist of a straight section followed by a curved one. In view of this, we characterize the design space for the CTR as comprising the lengths of the straight and curved sections as well as the radii of curvature for tubes one and two (innermost and intermediate tubes). As a result, the design space is represented by a 6-dimensional vector of parameters, namely $p = [ls_1 \ lc_1 \ ls_2 \ lc_2 \ r_1 \ r_2]^{\mathsf{T}}$, where ls_j, lc_j, and r_j represent the lengths of the straight and curved sections, and the radius of curvature of the jth tube respectively. For the sake of tractability, we limit the design space $\mathcal{D} \subset \mathbb{R}^6_+$ to the six aforementioned parameters by fixing the remaining kinematic parameters such as the inner/outer diameters of the tubes and their Young's/shear modulus.

The shape and pose of the CTR backbone in the the patient's coordinate frame is defined by means of an access port which dictates the skin puncture location as the position $x_0 \in \mathbb{R}^3$ from which the CTR is deployed as well as the insertion angle (orientation) $h_{\text{port}} \in SO(3)$. For computing the CTR kinematics (see [20]), $\beta_i < \beta_{i+1} \leq 0$ defines the length of linear insertion/retraction of the ith CTR component tube and $\alpha_i \in [0, 2\pi)$, its rotation. As such, any CTR configuration can be expressed as $q = \{\beta_i, \alpha_i\}_{i=1,2,3}$. Given any such configuration q, the CTR shape $\Gamma(q, s) : \mathbb{R}^6 \times \mathbb{R} \to \mathbb{R}^3$ is defined by a parametric curved, via the arc-length variable $s \in [0, L]$, such that $\Gamma(q, L)$ yields the end-effector position x_{ee} while $\Gamma(q, 0)$, the skin puncture position x_0.

Define a vector valued distance function $f(q, \mathcal{D}) : \mathbb{R}^6 \times \mathcal{D} \to \mathbb{R}^6_+$ mapping the design space $\mathcal{D} \subset \mathbb{R}^6$ into a corresponding vector of accumulated minimum deviations at the distal end and renal calyx with respect to their desired positions. Namely, for any CTR design, the distance function $f(\cdot, \cdot)$ yields the minimum accumulated deviation with respect to the end-effector and the renal calices after running the constrained inversed kinematics (1):

$$f(q, \mathcal{D})_i := \min_q \left\{ \|p_i - x_{\text{ee}}\| + \|p_{s_i}^\perp - x_{\text{clx}}\| \right\}, \qquad i = 1, \dots, 6 \qquad (3)$$

where $p_{s_i}^\perp$ is the point on the CTR backbone which is closest to the renal pyramid, computed via orthogonal projection, when the CTR's end-effector is steered to the point p_i of $E_{Q,c}$.

From this background, the problem of determining a viable CTR design becomes a root-finding problem for the distance-like function (3), i.e., we seek a set of kinematic parameters which evaluate the $\mathcal{L}_2$ norm of $f(q, \mathcal{D})$ to zero within a specified numerical tolerance $\varepsilon > 0$. In this pursuit, we employ the iterative Nelder-Mead simplex algorithm, which is a direct search method for nonlinear multidimensional unconstrained minimization [16].

So that a box-constrained optimization problem can be considered with accepteable bounds on the tubes' lengths and pre-curvatures, the Nelder-Mead algorithm is instantiated as a subsidiary optimization algorithm to an augmented Lagrangian method implemented in C++ using the NLopt library [9]. In other words, the design optimization of the CTR tubes' is cast as the following nonlinear constrained optimization problem

$$
\begin{aligned}
\underset{\mathcal{D} \subset \mathbb{R}^6_+}{\text{minimize}} \quad & \|f(q, \mathcal{D})\| \\
\text{subject to} \quad & \ell_{s_i}^- \leq \ell_{s_i} \leq \ell_{s_i}^+, \qquad i = 1, 2 \\
& \ell_{c_i}^- \leq \ell_{c_i} \leq \ell_{c_i}^+ \\
& L_i \geq L_{i+1} + \gamma_i \\
& \|u_i^-\| \leq \|u_i\| \leq \|u_i^+\|
\end{aligned}
\qquad (4)
$$

where lower and upper bounds are imposed on the lengths of the straight ℓ_{s_i} and curved ℓ_{c_i} sections of the tubes as well as on the pre-curvature vectors u_i. Additionally, for any tube pair, the overall length of the innermost L_i tube must exceed that of the outer tube L_{i+1} with a minimum prescribed tolerance $\gamma_i > 0$.

4   Study Cases

We apply the optimization algorithm (4) on the clinical data of seven patients who underwent PCNL at the Massachusetts General Hospital under an institutional review board (IRB) approved protocol. In this context, the clinical data consists of preoperative CT scans from which a digital segmentation on 3D Slicer [6] yields a point cloud representing the shape and pose of the stones in the patient coordinate frame. The seven cases considered in this paper, see Fig. 2, were drawn from a larger database of clinical patient data. Across all cases considered, a CTR design is deemed apropriate if it can deliver the distal end to the points p_i, $i = 1, \dots, 6$ along the kidney stones' main directions while maintaining a radial position deviation of $\varepsilon = 5$ mm at the chosen renal calyx.

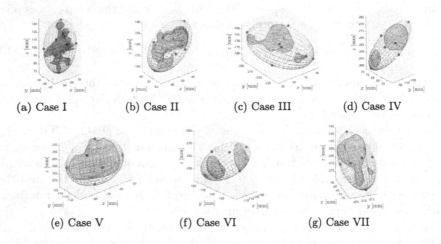

(a) Case I	(b) Case II	(c) Case III	(d) Case IV

(e) Case V	(f) Case VI	(g) Case VII

Fig. 2. MVCE is shown with kidney stones (black) and the critical points p_i which define the dominant directions of the kidney stone presentation (red markers). (Color figure online)

5   Results

We applied the optimization algorithms to the seven clinical cases, as depicted in Fig. 2. The simulation results were obtained by running all seven cases in parallel on a Intel i5-7500, 3.40 GHz Linux-based machine and took 5 h. For all cases, the algorithm (4) was run three successive times for the purpose of refining the solutions obtained. With the aim of investigating the kinematic capabilities of the optimized CTR in each case, the corresponding CTRs were steered along the direction defined by the points $p_i \in E_{Q,c}$ by means of the C-IK in (1). The error distributions at the calyx and distal ends are shown by the boxplots in Fig. 3.

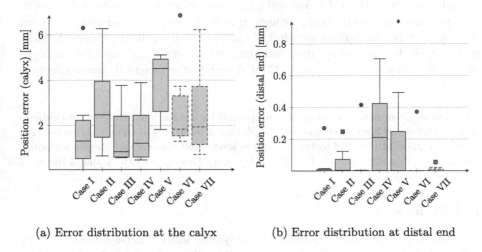

(a) Error distribution at the calyx   (b) Error distribution at distal end

Fig. 3. Position error distributions of the anatomically constrained CTR

It is worthwhile mentioning that, anatomically, in order to minimize vascular injury while obtaining the largest endoscopy angle, the ideal route for renal access passes through the axis of the target renal calyx [12,21]. However, given the diameter of the calyceal tracts, a 5 mm radial deviation with respect to the longitudinal axis of the targeted calyx during stone ablation is commonly deemed acceptable urologists. On the other hand, the distal end is also amenable to small position errors (usually <3 mm) as the emitted shockpulse lithotripsy waves affect neighboring regions of each point in the stone.

6   Conclusion

In this paper, we described an ellipsoidal approximation to kidney stones which underpins our approach to surgical planning and patient-specific design of CTRs for PCNL. By optimizing a set of kinematic tube parameters and the reachability of 6 critical points of the MVCE enclosing the stone presentation, our approach strives to enhance the manipulability of the robot along the dominant directions of the stone presentation. The approach is illustrated by seven clinical cases which demonstrates the usefulness of the algorithms presented.

While other existing approaches (see Subsect. 1.2) can be used, *a priori*, for the purpose of designing patient-specific CTRs, our approach can be applied seamlessly, as illustrated by the clinical cases of Sect. 4, to a myriad of cases of distinct clinical complexities while addressing the important constraint at the renal pyramid/calyces explicitly. Moreover, once an optimized CTR design is obtained, the proposed constrained inverse kinematics (C-IK) in (1) can be employed for guiding the navigation of the anatomically constrained CTR within the collecting system with clinically accepted tolerances as shown in Fig. 3.

In our future work, we plan to address the limitations in the current approach whereby intraoperative CT scans and ultrasound imaging modalites will allow us to depart from the static anatomy assumption in this paper and consider tissue deformation, fragmentation and shift of the stone burden in the kidney. We also plan to consider distinct optimization metrics and conduct a sensitivity analysis of the surgical plan and optimized CTR design in the face of dynamic effects in the anatomy.

Acknowledgements. This work was supported by the National Institute of Diabetes and Digestive and Kidney Diseases of the National Institutes of Health through Grant Number R01DK119269 and by the Natural Sciences and Engineering Research Council (NSERC) of Canada grant RGPIN1345 (Discovery Grant), and the Canada Research Chairs Program.

References

1. Baykal, C., Bowen, C., Alterovitz, R.: Asymptotically optimal kinematic design of robots using motion planning. Auton. Rob. **43**(2), 345–357 (2018). https://doi.org/10.1007/s10514-018-9766-x
2. Baykal, C., Torres, L.G., Alterovitz, R.: Optimizing design parameters for sets of concentric tube robots using sampling-based motion planning. In: 2015 IEEE/RSJ International Conference on Intelligent Robots and Systems (IROS), pp. 4381–4387 (2015)
3. Bergeles, C., Gosline, A.H., Vasilyev, N.V., Codd, P.J., del Nido, P.J., Dupont, P.E.: Concentric tube robot design and optimization based on task and anatomical constraints. IEEE Trans. Rob. **31**(1), 67–84 (2015)
4. Burgner, J., Gilbert, H.B., Webster, R.J.: On the computational design of concentric tube robots: Incorporating volume-based objectives. In: 2013 IEEE International Conference on Robotics and Automation, pp. 1193–1198 (2013)
5. Burgner, J., et al.: A telerobotic system for transnasal surgery. IEEE/ASME Trans. Mechatron. **19**(3), 996–1006 (2014)
6. Fedorov, A., et al.: 3D Slicer as an image computing platform for the quantitative imaging network. Magn. Reson. Imaging **30**(9), 1323–1341 (2012)
7. Ganpule, A.P., Vijayakumar, M., Malpani, A., Desai, M.R.: Percutaneous nephrolithotomy (PCNL) a critical review. Int. J. Surg. **36**, 660–664 (2016)
8. Hegarty, N.J., Desai, M.M.: Percutaneous nephrolithotomy requiring multiple tracts: Comparison of morbidity with single-tract procedures. J. Endourol. **20**(10), 753–760 (2006). pMID: 17094750
9. Johnson, S.G., Schueller, J.: Nlopt: Nonlinear optimization library. Astrophysics Source Code Library, pp. ascl-2111 (2021)
10. Kyriazis, I., Panagopoulos, V., Kallidonis, P., Özsoy, M., Vasilas, M., Liatsikos, E.: Complications in percutaneous nephrolithotomy. World J. Urol. **33**(8), 1069–1077 (2014). https://doi.org/10.1007/s00345-014-1400-8

11. Lee, C.L., Anderson, J.K., Monga, M.: Residency training in percutaneous renal access: does it affect urological practice? J. Urol. **171**(2), 592–595 (2004)
12. Miller, N.L., Matlaga, B.R., Lingeman, J.E.: Techniques for fluoroscopic percutaneous renal access. J. Urol. **178**(1), 15–23 (2007)
13. Morimoto, T.K., Cerrolaza, J.J., Hsieh, M.H., Cleary, K., Okamura, A.M., Linguraru, M.G.: Design of patient-specific concentric tube robots using path planning from 3-d ultrasound. In: 2017 39th Annual International Conference of the IEEE Engineering in Medicine and Biology Society (EMBC), pp. 165–168 (2017)
14. Morimoto, T.K., Greer, J.D., Hawkes, E.W., Hsieh, M.H., Okamura, A.M.: Toward the design of personalized continuum surgical robots. Ann. Biomed. Eng. **46**(10), 1522–1533 (2018). https://doi.org/10.1007/s10439-018-2062-2
15. Mousavi-Bahar, S.H., Mehrabi, S., Moslemi, M.K.: The safety and efficacy of PCNL with supracostal approach in the treatment of renal stones. Int. Urol. Nephrol. **43**(4), 983–987 (2011)
16. Nelder, J.A., Mead, R.: A simplex method for function minimization. Comput. J. **7**(4), 308–313 (1965)
17. Netto, N.R., Jr., Ikonomidis, J., Ikari, O., Claro, J.A.: Comparative study of percutaneous access for staghorn calculi. Urology **65**(4), 659–662 (2005)
18. Ott, C., Dietrich, A., Albu-Schäffer, A.: Prioritized multi-task compliance control of redundant manipulators. Automatica **53**, 416–423 (2015)
19. Rais-Bahrami, S., Friedlander, J.I., Duty, B.D., Okeke, Z., Smith, A.D.: Difficulties with access in percutaneous renal surgery. Therap. Adv. Urol. **3**(2), 59–68 (2011)
20. Rucker, D.C., Jones, B.A., Webster, R.J., III.: A geometrically exact model for externally loaded concentric-tube continuum robots. IEEE Trans. Rob. **26**(5), 769–780 (2010)
21. Seitz, C., Desai, M., Häcker, A., Hakenberg, O.W., Liatsikos, E., Nagele, U., Tolley, D.: Incidence, prevention, and management of complications following percutaneous nephrolitholapaxy. Eur. Urol. **61**(1), 146–158 (2012)
22. Torres, L.G., Webster, R.J., Alterovitz, R.: Task-oriented design of concentric tube robots using mechanics-based models. In: 2012 IEEE/RSJ International Conference on Intelligent Robots and Systems, pp. 4449–4455 (2012)
23. Yang, Y.-H., Wen, Y.-C., Chen, K.-C., Chen, C.: Ultrasound-guided versus fluoroscopy-guided percutaneous nephrolithotomy: a systematic review and meta-analysis. World J. Urol. **37**(5), 777–788 (2018). https://doi.org/10.1007/s00345-018-2443-z
24. Yildirim, E.A.: On the minimum volume covering ellipsoid of ellipsoids. SIAM J. Optim. **17**(3), 621–641 (2006)

Machine Learning – Domain Adaptation and Generalization

Low-Resource Adversarial Domain Adaptation for Cross-modality Nucleus Detection

Fuyong Xing[1](✉) and Toby C. Cornish[2]

[1] Department of Biostatistics and Informatics, University of Colorado Anschutz Medical Campus, Aurora, USA
`fuyong.xing@cuanschutz.edu`
[2] Department of Pathology, University of Colorado Anschutz Medical Campus, Aurora, USA

Abstract. Due to domain shifts, deep cell/nucleus detection models trained on one microscopy image dataset might not be applicable to other datasets acquired with different imaging modalities. Unsupervised domain adaptation (UDA) based on generative adversarial networks (GANs) has recently been exploited to close domain gaps and has achieved excellent nucleus detection performance. However, current GAN-based UDA model training often requires a large amount of unannotated target data, which may be prohibitively expensive to obtain in real practice. Additionally, these methods have significant performance degradation when using limited target training data. In this paper, we study a more realistic yet challenging UDA scenario, where (unannotated) target training data is very scarce, a low-resource case rarely explored for nucleus detection in previous work. Specifically, we augment a dual GAN network by leveraging a task-specific model to supplement the target-domain discriminator and facilitate generator learning with limited data. The task model is constrained by cross-domain prediction consistency to encourage semantic content preservation for image-to-image translation. Next, we incorporate a stochastic, differentiable data augmentation module into the task-augmented GAN network to further improve model training by alleviating discriminator overfitting. This data augmentation module is a plug-and-play component, requiring no modification of network architectures or loss functions. We evaluate the proposed low-resource UDA method for nucleus detection on multiple public cross-modality microscopy image datasets. With a single training image in the target domain, our method significantly outperforms recent state-of-the-art UDA approaches and delivers very competitive or superior performance over fully supervised models trained with real labeled target data.

Keywords: Nucleus detection · GAN · Domain adaptation

Supplementary Information The online version contains supplementary material available at https://doi.org/10.1007/978-3-031-16449-1_61.

L. Wang (Eds.): MICCAI 2022, LNCS 13437, pp. 639–649, 2022.
https://doi.org/10.1007/978-3-031-16449-1_61

1   Introduction

Because of domain shifts, deep cell/nucleus detection models trained on an image dataset acquired with one microscopic imaging modality (e.g., bright-field) will suffer from performance degradation when deployed to another dataset with a different modality (e.g., fluorescence). Meanwhile, image datasets with the same modality can also have different distributions due to inconsistent imaging protocols [11]. Many domain adaptation methods [11] addressing domain shifts in medical imaging align distributions in the feature or image space. Unsupervised domain adaptation (UDA) [5,9,20,23,24,26,39], particularly those based on generative neural networks (GANs) [10,41], has drawn much attention due to no need of target data labels for model training.

GAN-based UDA has been recently applied to various medical imaging tasks [2,4,15,25,37]. However, these methods do not address domain adaptation in a low-resource scenario, where the scarcity of unlabeled target training data can pose serious challenges for GAN training [3,21,38] and thus domain adaptation. GAN discriminators often suffer from overfitting when using limited training data, leading to significant performance degradation. Self-supervised learning, including image rotation prediction [6] and instance discrimination [35], has been adopted to improve data efficiency for GANs; however, these approaches require an auxiliary task which introduces additional computational burden. Some regularization techniques have been used to improve GAN training, such as adversarial defense [40], LeCam-divergence [31], consistency regularization [36], etc. These techniques have often exhibited inferior performance to data augmentation-based methods [21,38], which have been recently employed to mitigate discriminator overfitting without augmentation leakage so that GANs learn the original data distribution instead of augmented image distribution.

In this paper, we propose a novel GAN-based UDA method (see Fig. 1) for cross-modality cell/nucleus detection using limited unlabeled target data, which we refer to as low resource here. We first augment a dual GAN network by using a task-specific nucleus detection model to supplement the target-domain discriminator and provide additional gradient information to the source-to-target generator for training with limited target data. Meanwhile, we introduce cross-domain prediction consistency to preserve semantic content for image translation and facilitate learning of the target-to-source generator and the nucleus detection model. Next, we incorporate a differentiable data augmentation module into the GAN network, which stochastically applies a series of image transformations to *both real and translated images* to further improve GAN training by alleviating model overfitting without augmentation leakage. This data augmentation module allows the gradients to be backward propagated through the image transformations to the generators, requiring no modification of network architectures or loss functions. With an annotated source dataset and limited unlabeled target training data (e.g., one image), our UDA method can effectively identify nuclei in the target domain and produce superior performance over state-of-the-art UDA approaches in a low-resource setting.

2   Methodology

Figure 1 shows an overview of the proposed low-resource UDA framework. Given N^S annotated training images $(\boldsymbol{X}^S, \boldsymbol{Y}^S) = \{(\boldsymbol{x}_i^S, \boldsymbol{y}_i^S)\}_{i=1}^{N^S}$ from a data-rich source domain $\mathcal{S}$ and a limited, unlabeled training set $\boldsymbol{X}^T = \{\boldsymbol{x}_i^T\}_{i=1}^{N^T}$ from a target domain $\mathcal{T}$, the framework learns a source/target nucleus detector R^S/R^T and a source/target generator-discriminator pair $(G^{TS}, D^S)/(G^{ST}, D^T)$. The target nucleus detector R^T is trained via image-to-image translation constrained by cross-domain prediction consistency. Meanwhile, stochastic data augmentation including spatial and visual image transformations is applied on both real and translated images before feeding them to the discriminators, and the gradients are propagated through the augmentation back to the generators for model training. Here image $\boldsymbol{x}_i^S \in \mathbb{R}^{H \times W \times C}$, where H, W and C denote the image's height, width and channel respectively, and its associated label $\boldsymbol{y}_i^S \in [0,1]^{H \times W}$ is a continuous-valued proximity map [19,33]: $\boldsymbol{y}_i^S(u,v) = (e^{\beta(1-\frac{d_i(u,v)}{\epsilon})} - 1)/(e^\beta - 1)$ if $d_i(u,v) \le \epsilon$, and 0 otherwise, where $\beta = 3$ and $\epsilon = 16$ control the shape of the exponential function and $d_i(u,v)$ is the Euclidean distance between pixel (u,v) and its closest nucleus center in image $\boldsymbol{x}_i^S$.

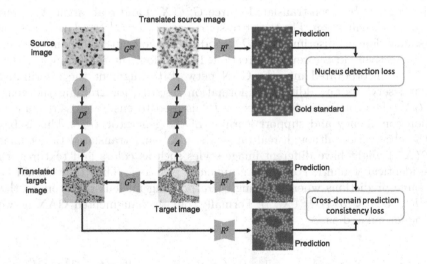

Fig. 1. The overview of the proposed low-resource UDA method for cross-modality nucleus detection. G^{ST}/G^{TS} and D^T/D^S are the source-to-target/target-to-source generator and its associated target-/source-domain discriminator, respectively. R^S/R^T is the source-/target-domain nucleus detector, and A represents the stochastic data augmentation. The R^T is learned with both translated, gold-standard-labeled source data via image-to-image translation and real target images via cross-domain prediction consistency. Source/target: Ki67 immunohistochemistry/hematoxylin & eosin staining.

2.1   Target Task-Augmented Bidirectional Adversarial Learning

In GAN-based UDA, limited target training data may not provide sufficient support so that the target-domain discriminator cannot effectively capture the data distribution, thus sending useless information to the corresponding generator and potentially failing the entire GAN training [3,21,38]. Instead of relying on self-supervised learning that introduces an additional, auxiliary task to improve data efficiency [6,35], we address the problem from a different perspective: directly using the target task, i.e., nucleus detection, to assist with GAN training using limited data. In addition, different from other task-specific GAN models [4,15,34] that train a single-directional task model, we use both source and target nucleus detectors to augment the GAN in a bidirectional manner. Furthermore, we adopt a relaxed, cross-domain prediction consistency constraint [7] to encourage semantic content preservation during image translation, instead of using the image reconstruction-based cycle consistency [41], which might be too strict when the target training data is scarce [8,13] due to the contrastive learning dynamics between the adversarial objective and the reconstruction intention.

In our UDA framework, the source-to-target generator G^{ST} learns to translate source images $\boldsymbol{X}^S$ to target-style data $G^{ST}(\boldsymbol{X}^S)$, which has an identical or similar distribution to $\boldsymbol{X}^T$, while the target-domain discriminator D^T aims to differentiate between translated source $G^{ST}(\boldsymbol{X}^S)$ and real target $\boldsymbol{X}^T$ images via binary classification. The target-to-source generator G^{TS} and its corresponding source-domain discriminator D^S conduct image translation and classification respectively in the reverse direction. In addition, the target nucleus detector R^T is incorporated into the GAN network to augment the discriminator D^T and serve as an additional information resource for training the generator G^{ST}. The source nucleus detector R^S is used to ensure cross-domain prediction consistency and support learning of the generator G^{TS}. This is based on the observation although real target data $\boldsymbol{X}^T$ and translated target images $G^{TS}(\boldsymbol{X}^T)$ might have different image styles such as colors and textures, they have identical semantic content, e.g., nucleus positions. Thus they should have the same predictions when applying corresponding nucleus detectors to them, i.e., $R^T(\boldsymbol{X}^T) \approx R^S(G^{TS}(\boldsymbol{X}^T))$. Formally, our task-augmented GAN network can be formulated as

$$\mathcal{L}^T(G^{ST}, D^T, R^T) = \mathbb{E}_{\boldsymbol{x}^T \sim \boldsymbol{X}^T}[\log D^T(\boldsymbol{x}^T)] + \mathbb{E}_{\boldsymbol{x}^S \sim \boldsymbol{X}^S}[\log(1 - D^T(G^{ST}(\boldsymbol{x}^S)))]$$
$$+ \mathbb{E}_{(\boldsymbol{x}^S, \boldsymbol{y}^S) \sim (\boldsymbol{X}^S, \boldsymbol{Y}^S)}[\||(\boldsymbol{y}^S + \alpha \bar{\boldsymbol{y}}^S \boldsymbol{1})^{1/2} \odot (R^T(G^{ST}(\boldsymbol{x}^S)) - \boldsymbol{y}^S)\|_F^2, \quad (1)$$

$$\mathcal{L}^S(G^{TS}, D^S, R^S) = \mathbb{E}_{\boldsymbol{x}^S \sim \boldsymbol{X}^S}[\log D^S(\boldsymbol{x}^S)] + \mathbb{E}_{\boldsymbol{x}^T \sim \boldsymbol{X}^T}[\log(1 - D^S(G^{TS}(\boldsymbol{x}^T)))]$$
$$+ \mathbb{E}_{(\boldsymbol{x}^S, \boldsymbol{y}^S) \sim (\boldsymbol{X}^S, \boldsymbol{Y}^S)}[\||(\boldsymbol{y}^S + \alpha \bar{\boldsymbol{y}}^S \boldsymbol{1})^{1/2} \odot (R^S(\boldsymbol{x}^S) - \boldsymbol{y}^S)\|_F^2, \quad (2)$$

$$\mathcal{L}_{con}(G^{TS}, R^S, R^T) = \mathbb{E}_{\boldsymbol{x}^T \sim \boldsymbol{X}^T}[\||R^T(\boldsymbol{x}^T) - R^S(G^{TS}(\boldsymbol{x}^T))\|_1], \quad (3)$$

where $\mathbb{E}$ means the expectation operator and the third term with the squared Frobenius norm $||\cdot||_F^2$ in Eqs. (1,2) is the nucleus detection loss, which is chosen as a weighted mean squared error to avoid trivial solutions [33,34]. The $\bar{y}^S$ is the mean value of $\boldsymbol{y}^S$, 1 is a matrix with all elements being 1, and α weights the contributions of non-zero regions. The $\odot$ denotes the element-wise multiplication. With the target nucleus detector R^T, the generator G^{ST} is enforced to learn meaningful data distributions with respective to predicting the labels, i.e., preserve the semantic content while changing the image style. Furthermore, the cross-domain prediction consistency loss $\mathcal{L}_{con}$ imposes an explicit constraint on the prediction for target data so that the generator G^{TS} is encouraged to produce content-consistent images. Another significant benefit of this loss is to assist with learning of the target nucleus detector R^T by directly taking advantage of real, high-quality target images $\boldsymbol{X}^T$ in Eq. (3), in addition to usage of the translated source data $G^{ST}(\boldsymbol{X}^S)$ in Eq. (1). Unlike pseudo-labeling-based domain adaptation [16,22,34], the cross-domain prediction consistency is differentiable and can be trained end-to-end via standard backpropagation.

2.2 Stochastic Data Augmentation

In addition to using a task specific model to augment the GAN network, we further improve the GAN training by explicitly alleviating discriminator overfitting, which often arises when training data is limited. Specifically, we design and incorporate a plug-and-play, differentiable data augmentation module into the GAN network to increase the size and diversity of training data. Note that directly applying spatial and visual image transformations to only real training data for GAN training is not helpful for image-to-image translation due to augmentation leakage [3,21,38], i.e., the generators would learn to match the distribution of the augmented data instead of the original images. For instance, an augmentation of adding noise to images would make the generators synthesize noisy images. In this way, translated images may be generated from a shifted distribution and deviate from real target data, thus potentially misleading the nucleus detectors. To address this issue, we apply identical data augmentation to both real and translated images before feeding them to the discriminators and conduct the augmentation when training both generators and discriminators (see Fig. 1). With invertible image transformations for data augmentation, the GAN will learn to capture the original data distribution instead of the augmented one [30]. Thus the GAN modeling, Eqs. (1, 2), can be rewritten as

$$\mathcal{L}_A^T = \mathbb{E}_{\boldsymbol{x}^T \sim \boldsymbol{X}^T}[\log D^T(A(\boldsymbol{x}^T))] + \mathbb{E}_{\boldsymbol{x}^S \sim \boldsymbol{X}^S}[\log(1 - D^T(A(G^{ST}(\boldsymbol{x}^S))))]$$
$$+ \mathbb{E}_{(\boldsymbol{x}^S, \boldsymbol{y}^S) \sim (\boldsymbol{X}^S, \boldsymbol{Y}^S)}[||(\boldsymbol{y}^S + \alpha \bar{y}^S 1)^{1/2} \odot (R^T(G^{ST}(\boldsymbol{x}^S)) - \boldsymbol{y}^S)||_F^2, \quad (4)$$

$$\mathcal{L}_A^S = \mathbb{E}_{\boldsymbol{x}^S \sim \boldsymbol{X}^S}[\log D^S(A(\boldsymbol{x}^S))] + \mathbb{E}_{\boldsymbol{x}^T \sim \boldsymbol{X}^T}[\log(1 - D^S(A(G^{TS}(\boldsymbol{x}^T))))]$$
$$+ \mathbb{E}_{(\boldsymbol{x}^S, \boldsymbol{y}^S) \sim (\boldsymbol{X}^S, \boldsymbol{Y}^S)}[||(\boldsymbol{y}^S + \alpha \bar{y}^S 1)^{1/2} \odot (R^S(\boldsymbol{x}^S) - \boldsymbol{y}^S)||_F^2, \quad (5)$$

where $A(\cdot)$ represents the data augmentation operation. Data augmentation in Eqs. (4, 5) can significantly increase the overlap of support between the distributions of real and translated image data [1, 28], thus boosting the image translation and nucleus detection performance. In order to propagate the gradients through the augmentation back to the generators during training, we implement all the image transformations with only differentiable primitive operations.

The requirement of invertible image transformations in Eqs. (4, 5) might be too rigorous in real applications, because many commonly used data augmentation techniques including image translation are not invertible. Inspired by [21], we combat this problem by performing the data augmentation with a certain probability p strictly less than 1, i.e., stochastic data augmentation. In this way, we increase the relative occurrence of non-augmented images among the discriminator input so that the generated distribution from the GAN will match that of original real training data, regardless of whether the augmentation is invertible. Therefore, we can use a large set of image transformations for data augmentation during training. Different from [21], we extend stochastic data augmentation to image-to-image translation for domain adaptation and use the same set of image transformations for the pair of real and translated images in the same batch to facilitate discriminator learning. Specifically, we compose 3 groups of 13 different image transformations into a data augmentation pipeline in a fixed order: 4 pixel blitting operations (horizontal flipping, vertical flipping, random $0°/90°/180°/270°$ rotation and integer translation), 4 general geometric transformations (random isotropic scaling, rotation, anisotropic scaling and fractional translation), and 5 color transformations (randomly adjusting image brightness, contrast and saturation, luma flipping, and hue rotation). Note that each image transformation is applied independently with probability $p < 1$. One important advantage of this stochastic data augmentation module is that it does not need to modify network architectures or loss functions.

With stochastic data augmentation, our low-resource UDA minimizes the following full objective, $\mathcal{L}$, for G^{ST}, G^{TS}, R^S, R^T and maximizes it for D^S, D^T:

$$\mathcal{L} = \mathcal{L}_A^T + \mathcal{L}_A^S + \lambda \mathcal{L}_{con}, \tag{6}$$

where λ represents a weighting parameter.

2.3 Implementation Details

We implement the generators using a fully convolutional residual network [18], the discriminators using a convolutional 70×70 PatchGAN [17], and the nucleus detectors using a U-Net-based regression network [33]. We empirically set $\alpha = 5$ in Eqs. (4, 5), $\lambda = 10^{-4}$ in Eq. (6), and $p = 0.8$ for stochastic data augmentation. We use the Adam optimizer for the GAN with exponential decay rates of 0.5 and 0.999, a learning rate of 2×10^{-4}, a batch size of 1 and a maximum epoch number of 100. We train first the source detector R^S with stochastic gradient descent and then all the other networks with Eq. (6). During testing, the detector R^T is applied to new real target images for nucleus detection by seeking local maxima in prediction maps [33].

3   Experiments

Datasets. We extensively evaluate the proposed method for nucleus detection on 4 public and 1 in-house microscopy image datasets, which are acquired with different imaging protocols, staining techniques and/or tissue preparations. The Ki67 immunohistochemistry (IHC)-stained bright-field microscopy image dataset [14] contains 803 training, 133 validation and 402 testing breast cancer images. The bone marrow dataset [19] has 11 hematoxylin and eosin (H&E)-stained bright-field microscopy images (H&E-Bone), and another H&E dataset [27] consists of 100 colon cancer images (H&E-Colon). The DAPI-stained fluorescence image dataset [29] is composed of 120 colon tissue images (Fluo-Colon). The in-house dataset has 111 Ki67 IHC-stained pancreatic neuroendocrine tumor images (IHC-Pancreas). We use the IHC breast cancer dataset as the source data due to its sufficient annotation, with the others as target datasets. We follow [27,33,34] to randomly split each target dataset into two halves for training and testing respectively. We further randomly choose 20% of training data as the validation set, and stop model training if the performance of the target nucleus detector on the validation set does not improve for 2×10^4 successive iterations. We use precision, recall and F_1 score as the evaluation metrics [27,33,34].

Table 1. Unsupervised domain adaptation evaluation for different target datasets. For each method, the mean and standard deviation (std) of 5 runs with different random seeds are reported: $\frac{mean}{(std)}$. The * indicates there is a statistically significant difference (p-value < 0.05) between our method and others in terms of F_1. The highest F_1 score for each dataset is highlighted with bold. P=Precision, R=Recall, and F_1=F_1 score.

	H&E-Bone			H&E-Colon			Fluo-Colon			IHC-Pancreas		
	P	R	F_1	P	R	F_1	P	R	F_1	P	R	F_1
ADDA [32]	49.0	94.0	64.4*	51.4	69.9	59.1*	37.1	39.1	36.8*	50.7	94.8	66.1*
	(0.9)	(1.9)	(0.7)	(0.7)	(5.8)	(2.2)	(7.3)	(15.6)	(9.6)	(2.0)	(0.5)	(1.7)
CyCADA [12]	64.5	72.3	66.4*	63.9	73.8	68.5*	69.1	85.5	75.8	66.6	51.7	58.2*
	(5.5)	(16.3)	(7.2)	(1.6)	(0.8)	(1.0)	(1.4)	(12.6)	(5.1)	(3.3)	(2.0)	(2.1)
ADAPL [34]	65.3	81.5	71.8	39.2	69.8	49.8*	53.0	79.3	63.2*	66.1	89.6	76.0*
	(3.0)	(13.7)	(6.2)	(11.2)	(21.6)	(14.2)	(12.1)	(11.0)	(11.3)	(2.6)	(1.2)	(1.3)
IUDA [22]	60.8	97.6	74.8	29.0	98.4	44.6*	79.6	68.4	73.6*	65.1	78.0	70.9*
	(5.9)	(1.0)	(4.4)	(4.6)	(2.3)	(5.7)	(0.7)	(0.5)	(0.3)	(6.2)	(4.7)	(5.4)
Source	46.9	99.1	63.6*	51.3	87.8	64.7*	54.0	56.4	54.6*	71.8	86.2	78.4
	(1.4)	(0.3)	(1.2)	(1.4)	(2.6)	(1.1)	(15.1)	(11.2)	(12.0)	(0.8)	(1.3)	(0.4)
CycleGAN	52.7	59.2	55.6*	30.7	29.4	29.6*	43.1	48.6	44.0*	66.3	83.5	73.8*
	(13.6)	(18.6)	(15.8)	(16.3)	(22.1)	(19.2)	(12.3)	(13.9)	(10.4)	(3.3)	(3.0)	(2.3)
Task-aug	76.5	64.3	67.7*	77.8	62.8	69.2	83.9	59.2	69.0*	78.8	77.9	78.3
	(11.5)	(14.5)	(8.1)	(3.9)	(4.4)	(1.5)	(7.0)	(4.0)	(2.0)	(0.6)	(1.2)	(0.7)
Data-aug	57.9	93.5	71.3*	59.0	73.8	65.5*	80.1	72.8	75.6*	67.9	76.2	71.6
	(4.2)	(4.1)	(2.2)	(2.6)	(1.2)	(1.5)	(6.9)	(7.5)	(2.2)	(8.8)	(7.1)	(7.5)
Ours	70.4	90.7	**79.2**	64.3	78.1	**70.4**	77.9	81.3	**79.5**	75.5	85.2	**79.9**
	(2.4)	(5.1)	(2.2)	(3.0)	(2.9)	(1.0)	(2.1)	(1.5)	(0.8)	(3.3)	(3.2)	(1.2)
Target	75.7	83.4	79.4	77.3	60.8	68.1*	70.9	92.3	80.2	81.0	79.8	80.4
	(1.3)	(2.9)	(1.9)	(1.4)	(1.4)	(0.7)	(0.4)	(0.6)	(0.4)	(0.7)	(1.1)	(0.4)

Comparison with State-of-the-Art Methods. Table 1 shows the nucleus detection performance of different methods using only 1 unlabeled target training image on different datasets. We compare our method with several recent state-of-the-art UDA approaches, including adversarial discriminative domain adaptation (ADDA) [32], cycle-consistent adversarial domain adaptation (CyCADA) [12], adversarial domain adaptation followed by pseudo-labeling (ADAPL) [34], and iterative unsupervised domain adaptation (IUDA) [22]. We note that in a low-resource setting, our method outperforms other approaches by large margins in terms of F_1 score for all the datasets and is significantly better than the others for most cases with p-value < 0.05 in Student's t-test. This demonstrates the effectiveness of our low-resource UDA method, which is trained with limited unlabeled target data. Qualitative results of nucleus detection are shown in the Supplementary Material.

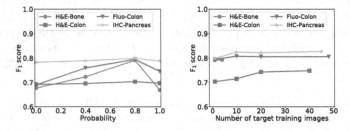

Fig. 2. The F_1 score of the proposed method with different probability values of stochastic data augmentation (left) and different numbers of unannotated target training images (right). Note the maximum number of bone target training images is 4.

Ablation Study. In Table 1 (lower panel), the *Source* denotes a nucleus detector trained with only source data and tested on target data without domain adaptation, and *CycleGAN* means using a cycle-consistent GAN [41] for image translation and then training a nucleus detector with translated source data. We also evaluate different variants of our method: 1) target task-augmented (*Task-aug*) GAN training with Eqs. (1–3), 2) data-augmented (*Data-aug*) GAN learning with the proposed stochastic data augmentation for image translation and then training a nucleus detector with translated source images, and 3) the proposed method (*ours*) trained with Eq. (6), i.e., with both task and data augmentation. The *Target* means directly training a nucleus detector with annotated target data. Compared with *CycleGAN*, incorporating a specific task (*Task-aug*) or stochastic data augmentation (*Data-aug*) into image translation GAN training is generally beneficial when target training data is limited. Furthermore, the proposed method provides a significantly higher F_1 score than others for most cases (p-value < 0.05) and produces very competitive or superior results over the *Target* model trained with labeled target data.

Analysis of Parameter Sensitivity. Figure 2 (left) shows the nucleus detection performance of our method using one target training image with different probability values of stochastic data augmentation. For all the datasets, our method with a probability strictly less than 1 (e.g., 0.8) produces higher accuracy than other cases, e.g., probability is 1 or 0 (no data augmentation). This suggests the importance of stochastic data augmentation for GAN training with limited target data. Figure 2 (right) displays the effects of the number of unlabeled target training images on nucleus detection. We see that using more training images is generally helpful for performance improvement, especially for the challenging H&E-Colon dataset that has high variability in image appearance.

4 Conclusion

We present a novel low-resource UDA method for cross-modality nucleus detection, which augments a dual GAN network with a target task-specific model and stochastic data augmentation. With limited unlabeled target training data, our method significantly outperforms several recent state-of-the-art UDA nucleus detection approaches and produces very competitive or superior performance over fully supervised models trained with annotated target data. One potential limitation of the method is its inapplicability to bioimages with invisible nuclei.

Acknowledgement. This research was supported by the National Cancer Institute of the National Institutes of Health under Award Number R21CA237493.

References

1. Arjovsky, M., Bottou, L.: Towards principled methods for training generative adversarial networks. In: ICLR, pp. 1–14 (2017)
2. Bentaieb, A., Hamarneh, G.: Adversarial stain transfer for histopathology image analysis. IEEE TMI **37**(3), 792–802 (2018)
3. Cao, J., Hou, L., Yang, M.H., He, R., Sun, Z.: Remix: towards image-to-image translation with limited data. In: CVPR, pp. 15013–15022 (2021)
4. Chen, C., Dou, Q., Chen, H., Heng, P.A.: Semantic-aware generative adversarial nets for unsupervised domain adaptation in chest x-ray segmentation. In: MLMI, pp. 143–151 (2018)
5. Chen, C., Liu, Q., Jin, Y., Dou, Q., Heng, P.-A.: Source-free domain adaptive fundus image segmentation with denoised pseudo-labeling. In: de Bruijne, M., et al. (eds.) MICCAI 2021. LNCS, vol. 12905, pp. 225–235. Springer, Cham (2021). https://doi.org/10.1007/978-3-030-87240-3_22
6. Chen, T., Zhai, X., Ritter, M., Lucic, M., Houlsby, N.: Self-supervised gans via auxiliary rotation loss. In: CVPR, pp. 12146–12155 (2019)
7. Chen, Y.C., Lin, Y.Y., Yang, M.H., Huang, J.B.: Crdoco: pixel-level domain transfer with cross-domain consistency. In: CVPR, pp. 1791–1800 (2019)
8. Choi, J., Kim, T., Kim, C.: Self-ensembling with gan-based data augmentation for domain adaptation in semantic segmentation. In: ICCV, pp. 6829–6839 (2019)

9. Gadermayr, M., et al.: Generative adversarial networks for facilitating stain-independent supervised and unsupervised segmentation: a study on kidney histology. IEEE TMI **38**(10), 2293–2302 (2019)

10. Goodfellow, I., et al.: Generative adversarial nets. In: NeurIPS (2014)

11. Guan, H., Liu, M.: Domain adaptation for medical image analysis: a survey. IEEE TBME **69**(3), 1173–1185 (2022)

12. Hoffman, J., et al.: CyCADA: cycle-consistent adversarial domain adaptation. In: ICML, pp. 1989–1998 (2018)

13. Hosseini-Asl, E., Zhou, Y., Xiong, C., Socher, R.: Augmented cyclic adversarial learning for low resource domain adaptation. In: ICLR, pp. 1–14 (2019)

14. Huang, Z., et al.: BCData: a large-scale dataset and benchmark for cell detection and counting. In: Martel, A.L., et al. (eds.) MICCAI 2020. LNCS, vol. 12265, pp. 289–298. Springer, Cham (2020). https://doi.org/10.1007/978-3-030-59722-1_28

15. Huo, Y., et al.: Synseg-net: synthetic segmentation without target modality ground truth. IEEE TMI **38**(4), 1016–1025 (2019)

16. Inoue, N., et al.: Cross-domain weakly-supervised object detection through progressive domain adaptation. In: CVPR, pp. 5001–5009 (2018)

17. Isola, P., Zhu, J., Zhou, T., Efros, A.A.: Image-to-image translation with conditional adversarial networks. In: CVPR, pp. 5967–5976 (2017)

18. Johnson, J., Alahi, A., Li, L.F.: Perceptual losses for real-time style transfer and super-resolution. In: ECCV, pp. 694–711 (2016)

19. Kainz, P., Urschler, M., Schulter, S., Wohlhart, P., Lepetit, V.: You should use regression to detect cells. In: Navab, N., Hornegger, J., Wells, W.M., Frangi, A.F. (eds.) MICCAI 2015. LNCS, vol. 9351, pp. 276–283. Springer, Cham (2015). https://doi.org/10.1007/978-3-319-24574-4_33

20. Kamnitsas, K., et al.: Unsupervised domain adaptation in brain lesion segmentation with adversarial networks. In: IPMI, pp. 597–609 (2017)

21. Karras, T., et al.: Training generative adversarial networks with limited data. In: NeurIPS, pp. 12104–12114 (2020)

22. Liimatainen, K., et al.: Iterative unsupervised domain adaptation for generalized cell detection from brightfield z-stacks. BMC Bioinf. **20**(1), 80 (2019)

23. Mahmood, F., et al.: Unsupervised reverse domain adaptation for synthetic medical images via adversarial training. IEEE TMI **37**(12), 2572–2581 (2018)

24. Ouyang, C., Kamnitsas, K., Biffi, C., Duan, J., Rueckert, D.: Data efficient unsupervised domain adaptation for cross-modality image segmentation. In: Shen, D., et al. (eds.) MICCAI 2019. LNCS, vol. 11765, pp. 669–677. Springer, Cham (2019). https://doi.org/10.1007/978-3-030-32245-8_74

25. Shaban, M.T., Baur, C., Navab, N., Albarqouni, S.: StainGAN: stain style transfer for digital histological images. In: ISBI, pp. 953–956 (2019)

26. Shin, S.Y., Lee, S., Summers, R.M.: Unsupervised domain adaptation for small bowel segmentation using disentangled representation. In: de Bruijne, M., et al. (eds.) MICCAI 2021. LNCS, vol. 12903, pp. 282–292. Springer, Cham (2021). https://doi.org/10.1007/978-3-030-87199-4_27

27. Sirinukunwattana, K., Raza, S.E.A., Tsang, Y.W., Snead, D.R.J., Cree, I.A., Rajpoot, N.M.: Locality sensitive deep learning for detection and classification of nuclei in routine colon cancer histology images. IEEE TMI **35**(5), 1196–1206 (2016)

28. Sønderby, C.K., Caballero, J., Theis, L., Shi, W., Huszár, F.: Amortised map inference for image super-resolution. In: ICLR, pp. 1–11 (2017)

29. Tofighi, M., Guo, T., Vanamala, J.K.P., Monga, V.: Prior information guided regularized deep learning for cell nucleus detection. IEEE TMI **38**(9), 2047–2058 (2019)

30. Tran, N.T., Tran, V.H., Nguyen, N.B., Nguyen, T.K., Cheung, N.M.: On data augmentation for GAN training. IEEE TIP **30**, 1882–1897 (2021)
31. Tseng, H.Y., Jiang, L., Liu, C., Yang, M.H., Yang, W.: Regularizing generative adversarial networks under limited data. In: CVPR, pp. 7917–7927 (2021)
32. Tzeng, E., Hoffman, J., Saenko, K., Darrell, T.: Adversarial discriminative domain adaptation. In: CVPR, pp. 2962–2971 (2017)
33. Xie, Y., Xing, F., Shi, X., Kong, X., Su, H., Yang, L.: Efficient and robust cell detection: a structured regression approach. MIA **44**, 245–254 (2018)
34. Xing, F., Bennett, T., Ghosh, D.: Adversarial domain adaptation and pseudo-labeling for cross-modality microscopy image quantification. In: Shen, D., et al. (eds.) MICCAI 2019. LNCS, vol. 11764, pp. 740–749. Springer, Cham (2019). https://doi.org/10.1007/978-3-030-32239-7_82
35. Yang, C., Shen, Y., Xu, Y., Zhou, B.: Data-efficient instance generation from instance discrimination. In: NeurIPS, pp. 9378–9390 (2021)
36. Zhang, H., Zhang, Z., Odena, A., Lee, H.: Consistency regularization for generative adversarial networks. In: ICLR, pp. 1–10 (2020)
37. Zhang, Y., Miao, S., Mansi, T., Liao, R.: Task driven generative modeling for unsupervised domain adaptation: application to x-ray image segmentation. In: Frangi, A.F., Schnabel, J.A., Davatzikos, C., Alberola-López, C., Fichtinger, G. (eds.) MICCAI 2018. LNCS, vol. 11071, pp. 599–607. Springer, Cham (2018). https://doi.org/10.1007/978-3-030-00934-2_67
38. Zhao, S., Liu, Z., Lin, J., Zhu, J.Y., Han, S.: Differentiable augmentation for data-efficient gan training. In: NeurIPS, pp. 7559–7570 (2020)
39. Zhao, Z., Xu, K., Li, S., Zeng, Z., Guan, C.: MT-UDA: towards unsupervised cross-modality medical image segmentation with limited source labels. In: MICCAI 2021. LNCS, vol. 12901, pp. 293–303. Springer, Cham (2021). https://doi.org/10.1007/978-3-030-87193-2_28
40. Zhou, B., Krähenbühl, P.: Don't let your discriminator be fooled. In: ICLR, pp. 1–10 (2019)
41. Zhu, J.Y., Park, T., Isola, P., Efros, A.A.: Unpaired image-to-image translation using cycle-consistent adversarial networks. In: ICCV. pp. 2223–2232 (2017)

Domain Specific Convolution and High Frequency Reconstruction Based Unsupervised Domain Adaptation for Medical Image Segmentation

Shishuai Hu[1], Zehui Liao[1], and Yong Xia[1,2,3](✉)

[1] National Engineering Laboratory for Integrated Aero-Space-Ground-Ocean Big Data Application Technology, School of Computer Science and Engineering, Northwestern Polytechnical University, Xi'an 710072, China
yxia@nwpu.edu.cn

[2] Ningbo Institute of Northwestern Polytechnical University, Ningbo 315048, China

[3] Research and Development Institute of Northwestern Polytechnical University in Shenzhen, Shenzhen 518057, China

Abstract. Although deep learning models have achieved remarkable success in medical image segmentation, the domain shift issue caused mainly by the highly variable quality of medical images is a major hurdle that prevents these models from being deployed for real clinical practices, since no one can predict the performance of a 'well-trained' model on a set of unseen clinical data. Previously, many methods have been proposed based on, for instance, CycleGAN or the Fourier transform to address this issue, which, however, suffer from either an inadequate ability to preserve anatomical structures or unexpectedly introduced artifacts. In this paper, we propose a multi-source-domain unsupervised domain adaptation (UDA) method called **Do**main specific **C**onvolution and high frequency **R**econstruction (**DoCR**) for medical image segmentation. We design an auxiliary high frequency reconstruction (HFR) task to facilitate UDA, and hence avoid the interference of the artifacts generated by the low-frequency component replacement. We also construct the domain specific convolution (DSC) module to boost the segmentation model's ability to domain-invariant features extraction. We evaluate DoCR on a benchmark fundus image dataset. Our results indicate that the proposed DoCR achieves superior performance over other UDA methods in multidomain joint optic cup and optic disc segmentation. Code is available at: https://github.com/ShishuaiHu/DoCR.

S. Hu, Z. Liao—Equal contribution.

Supplementary Information The online version contains supplementary material available at https://doi.org/10.1007/978-3-031-16449-1_62.

L. Wang (Eds.): MICCAI 2022, LNCS 13437, pp. 650–659, 2022.
https://doi.org/10.1007/978-3-031-16449-1_62

Keywords: Unsupervised domain adaptation · Medical image segmentation · Domain specific convolution · High frequency reconstruction

1   Introduction

Recent years have witnessed the tremendous success of deep learning in medical image segmentation [4,9,11,20,23]. Like any supervised learning approach, deep learning models reply their success on the assumption that training images and test ones share similar, if not identical, distributions. This assumption, however, is less likely to be held on medical image segmentation tasks, since the range of image quality and structure visibility can be considerable, depending on characteristics of the imaging equipment, skill of the operator, and compromises with factors such as patient radiation exposure and imaging time [15]. To relieve the performance degradation on test samples caused by the distribution discrepancy, tremendous research endeavors have been devoted to unsupervised domain adaptation (UDA), the setting where labeled training data is available on a source domain or source domains, but the goal is to have good performance on a target domain with only unlabeled data [2,5,14,16–18,21,24].

Currently, data-level UDA methods [7,13,16,21,24] attempt to address the distribution discrepancy issue by transforming source-domain images to the target domain using image synthesis. CycleGAN-based UDA methods [7,16,24] impose cycle consistency constraints between the source-domain image and corresponding synthetic target-domain image, and hence can be trained on the unpaired source- and target-domain images. Although they can transform image style well, these methods suffer from the poor preservation of structural information, which is unacceptable for segmentation applications. To preserve the anatomical structures in images, Fourier transform-based UDA methods [13,21] translate source-domain images to the target domain by replacing the low frequency components of the images in both domains, based on the assumption that the style information is embedded in low frequency components and structural information is embedded in high frequency components. Despite their prevalence, these approaches may unexpectedly introduce artifacts to synthetic images, since low frequency component replacement works like an ideal filter that introduces ringing artifacts in the filtered image. Moreover, it is less likely to estimate the cut-off between low and high frequency components. To address this issue, we suggest using high frequency image reconstruction (HFR), which is insusceptible to artifacts, to substitute for the low frequency component replacement. Similar to a high-pass filter, HFR filters out the low frequency components where most style information locates and hence improves the performance of UDA.

With the recent advance of adversarial training and feature normalization, feature-level UDA methods [2,12,14,19] boost their ability to domain-invariant feature extraction. Among them, feature disentangling based models [14] attempt to decouple domain-variant and domain-invariant features under the guidance of the game theory. Most of these models contain a pair of alternately trained

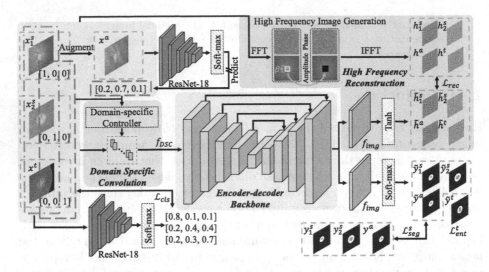

Fig. 1. Overview of DoCR for medical image segmentation. FFT: Fast Fourier Transform; IFFT: Inverse FFT. The box in yellow represents the domain-specific convolutional head. (Color figure online)

feature extractor and domain discriminator. The feature extractor is trained to fool the discriminator, and the discriminator is expected to identify the domain in which the features are extracted. After training, these models can well disentangle the discriminative features for domain classification. However, the equilibrium between the feature extractor and domain discriminator is hard to maintain, and hence it is difficult to train these models well enough to improve UDA. To avoid this issue, domain-specific batch normalization (DSBN) was developed to improve the ability to domain-invariant feature extraction [2,12]. Although DSBN can reduce the distribution discrepancy between source and target domain images in the feature space, its feature normalization ability is limited by the few learnable parameters and batch-level operations [19]. Therefore, we designed a domain-specific convolution (DSC) module based on dynamic convolutions [6,8,22] to extract domain-insensitive features.

In this paper, we incorporate both HFR and DSC into an encoder-decoder backbone, and thus propose a multi-source-domain UDA method called **Do**main specific **C**onvolution and high frequency **R**econstruction (**DoCR**) for medical image segmentation, with a particular focus on addressing the issue of training-test data distribution discrepancy caused by image quality variation. This method has been evaluated against other UDA methods on the task of the joint optic cup (OC) and optic disc (OD) segmentation on fundus images. Our main contributions are three-fold: (1) we use HFR as an auxiliary task to facilitate UDA in medical image segmentation, and hence avoid the interference of the artifacts generated by the low-frequency component replacement; (2) we design the DSC module to improve the segmentation model's ability to domain-invariant features extraction; and (3) as indicated by our results on OC/OD segmentation, the proposed DoCR model achieves superior performance over other state-of-the-art UDA methods.

2 Method

2.1 Problem Definition and Method Overview

Let a set of K source domains be represented as $\mathcal{D}^s = \{(x_{ki}^s, y_{ki}^s)_{i=1}^{N_k}\}_{k=1}^K$, where x_{ki}^s is the i-th image in the k-th source domain, and y_{ki}^s is the segmentation mask of x_{ki}^s. The target domain with unlabeled images can be denoted as $\mathcal{D}_u^t = (x_i^t)_{i=1}^{N_u}$. Our goal is to train a segmentation model $F_\theta : x \to y$ on $\mathcal{D}^s$ and $\mathcal{D}_u^t$, which can perform well on the target domain $\mathcal{D}^t$.

The proposed DoCR model has a U-shape architecture [4], equipped with a DSC module, an HFR head, and a segmentation head (see Fig. 1). The DSC module is used as the first convolutional layer to remove as much domain information as possible from the input images x and extract domain-insensitive features f_{DSC}. Then the encoder-decoder backbone takes f_{DSC} as its input and produces further processed features f_{img}. Based on f_{img}, the HFR head reconstructs the high frequency components h of x, and the segmentation head generates the segmentation prediction $\tilde{y}$. We now delve into the details of each part.

2.2 DSC Module

The DSC module consists of a DSC head and a domain-specific controller. The DSC head has a light-weighted design, containing a 3×3 convolutional layer, a ReLU layer, and a batch normalization layer. It takes the C-channel input image x as input and outputs a 32-channel domain-insensitive feature map f_{DSC}, which will be fed to the encoder-decoder backbone. Consequently, there are totally $C \times 32 \times 3 \times 3 + 32$ parameters in this head, denoted by ω_D. These parameters are dynamically generated by the controller F_C, which is also a convolutional layer, conditioned on the domain code $\mathscr{D}$ of the input images x.

Since there are K source domains and a target domain, $\mathscr{D}$ is a one-hot $K+1$ dimension vector. To relieve the insufficient optimization issue caused by one-hot $\mathscr{D}$, we conduct low frequency component replacement [21] to increase the diversity of source-domain data. Specifically, the low frequency components in a source-domain image can be replaced by the same number of low frequency components in either a source-domain image or a target-domain image. Then, we train ResNet-18 as a domain predictor using original images and their one-hot domain codes, and then use the trained ResNet-18 to predict the domain code $\mathscr{D}^a$ of each augmented image x^a (see the bottom left part of Fig. 1).

2.3 Encoder-decoder Backbone

We tailor ResNet-34 as the encoder by replacing its average pooling layer and fully connected (FC) layer with a ReLU layer and setting the channel number of the first convolutional layer to 32. The encoder contains a convolutional block and four residual blocks, which gradually convert the input feature map f_{DSC} into a lower resolution and more abstract one. Symmetrically, the decoder is also composed of five convolutional blocks, which gradually upsample the feature map to its original resolution. Skip connections are made from the first convolutional block and first three residual blocks in the encoder to the corresponding locations in the decoder. In each of the first four decoder blocks, the feature map is processed by a 2×2 transposed convolutional layer with a stride of 2 and a 1×1 convolutional layer with a stride of 1. After that, the feature map is concatenated with the feature map skipped from the encoder, and then fed to a ReLU layer and a batch normalization layer. The last decoder block only upsamples the feature map using a transposed convolutional layer and a batch normalization layer and produces a 32-channel feature map f_{img} with the same size $H \times W$ as the input image.

2.4   HFR and Segmentation

The HFR head F_{RC} contains a convolutional layer and a tanh layer to reconstruct the high frequency image $\tilde{h}$ based on the input feature map f_{img}. This can be expressed as

$$\tilde{h} = Tanh(F_{RC}(f_{img}; \theta_{RC})) \tag{1}$$

where θ_{RC} is the parameters of F_{RC}, and $Tanh(\cdot)$ represents the tanh layer. The ground truth of high frequency image h is generated by setting low frequency components to zero in the frequency domain (see the top right part of Fig. 1). The size of the replaced area is $\beta \times (H \times W)$, where β is a hyper-parameter.

The segmentation head F_S is composed of a convolutional layer and a soft-max layer to convert f_{img} to segmentation map $\tilde{y}$. This can be expressed as

$$\tilde{y} = SM(F_S(f_{img}; \theta_S)) \tag{2}$$

where θ_S is the parameters of F_S, and $SM(\cdot)$ is a soft-max layer.

2.5   Training and Inference

Training. We pretrain the domain predictor for 20 epochs using the one-hot domain code and corresponding images without augmentation. The objective of the domain prediction task is the cross-entropy loss

$$\mathcal{L}_{cls} = -\sum_{k=1}^{K+1} d_k \log(d_k^p) \tag{3}$$

where d_k is the domain label, and d_k^p is the soft-max probability of belonging to the k-th domain.

After that, the augmented image, the source image, and the target image batches are fed to the model randomly. For the reconstruction task, the objective is the $L1$ loss

$$\mathcal{L}_{rec} = \sum \left| h - \tilde{h} \right| \tag{4}$$

For the segmentation task, the cross-entropy loss is adopted as the objective for the supervised source domain data, and entropy loss [21] is used for the unlabeled target domain images

$$\mathcal{L}_{seg} = -(y^s \log \widetilde{y}^s + (1 - y^s) \log(1 - \widetilde{y}^s)) + \lambda_{ent}((-\widetilde{y}^t \log_2 \widetilde{y}^t)^2 + 0.001^2)^2 \quad (5)$$

where y^s and $\widetilde{y}^s$ represent the ground truth and prediction of source domain image, $\widetilde{y}^t$ is the prediction of target domain image, and λ_{ent} is a hyper-parameter, which is set to 0.005 as suggested in [21].

The total loss can be expressed as

$$\mathcal{L} = I(x^a \notin x)\mathcal{L}_{cls} + \mathcal{L}_{seg} + \lambda_{rec}\mathcal{L}_{rec} \quad (6)$$

where $I(\cdot)$ is an indicator function, and λ_{rec} is a hyper-parameter.

Inference. During inference, given a test image and its domain code, the DSC block produces a 32-channel feature map f_{DSC}, which is then fed to the backbone network and converted into a 32-channel feature map f_{img}. Next, the segmentation head takes f_{img} as input and generates the segmentation prediction $\widetilde{y}$.

3 Experiments and Results

Materials and Evaluation Metrics. To reduce the annotator bias [10] among different datasets, we use a multi-domain joint OC/OD segmentation dataset annotated by the same group of ophthalmologists, namely RIGA+ [1,3], and only use rater 1's annotations to train and evaluate all the algorithms. It contains 195 labeled data from BinRushed, 95 labeled data from Magrabia, 454 labeled, and 717 unlabeled data from the MESSIDOR database (including data collected from 3 medical centers, i.e., BASE1, BASE2, and BASE3). The details of the RIGA+ were summarized in Table 1. Dice similarity coefficient ($D_{disc}(\%)$, $D_{cup}(\%)$) is adopted as evaluation metrics.

Table 1. Details of the dataset used for this study. The 20% test cases in the target domain were selected randomly.

Domain	Dataset	Labeled images (training/test)	Unlabeled images
Source	BinRushed	195 (195/0)	0
	Magrabia	95 (95/0)	0
Target	BASE1	173 (138/35)	227
Target	BASE2	148 (118/30)	238
Target	BASE3	133 (106/27)	252

Implementation Details. Each input image was normalized by subtracting its mean and dividing by its standard deviation. For all experiments, the mini-batch size was set to 8, and all the images were center-cropped [17] and resized to

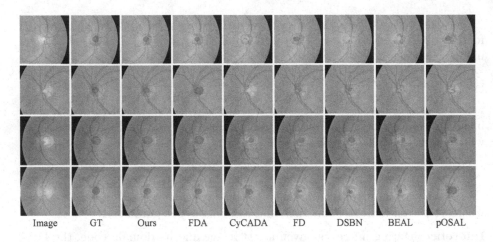

| Image | GT | Ours | FDA | CyCADA | FD | DSBN | BEAL | pOSAL |

Fig. 2. Visualization of segmentation masks predicted by our DoCR and six competing methods, together with ground truth.

512×512. The SGD algorithm with a momentum of 0.99 and an initial learning rate lr_0 of 0.01 was adopted as the optimizer. The learning rate was decayed according to the polynomial rule $lr = lr_0 \times (1 - t/T)^{0.9}$, where t is the current epoch and T is the maximum epoch, which was set to 100. The hyper-parameters λ_{rec} and β were set to 0.1 and 0.01 according to the ablation experiments. All experiments were implemented using the PyTorch framework. It takes about 5 h to train our DoCR model on one NVIDIA 2080Ti GPU.

Comparative Experiments. We compare our DoCR with three baselines: 'Intra-Domain' setting (*i.e.*, training and test on the data from the same

Table 2. Performance of our DoCR and nine competing methods in OC/OD segmentation. The best results except for 'Intra-Domain' are highlighted with **bold**.

Methods		Target domain		
		BASE1	BASE2	BASE3
	Intra-Domain	(95.87, <u>87.27</u>)	(95.92, <u>88.83</u>)	(95.73, <u>88.42</u>)
Baseline	w/o DA	(94.30, 84.01)	(94.58, 79.22)	(93.82, 81.40)
	Self-Training	(95.66, 85.47)	(94.97, 78.69)	(94.98, 82.76)
Data level	FDA [21]	(**95.91**, 85.69)	(96.22, 83.19)	(95.93, 84.36)
	CyCADA [7]	(95.04, 83.50)	(93.86, 80.64)	(94.77, 82.55)
Feature level	FD [14]	(95.58, 84.29)	(95.26, 82.72)	(95.40, 85.26)
	DSBN [2]	(95.60, 85.30)	(94.19, 82.43)	(95.08, 85.14)
Decision level	BEAL [17]	(94.96, 83.60)	(95.29, 80.12)	(94.78, 82.86)
	pOSAL [18]	(93.96, 83.07)	(95.36, 84.36)	(94.73, 84.36)
	DoCR (ours)	(95.89, **85.72**)	(**96.39**, **86.17**)	(**96.19**, **85.96**)

target domain), 'w/o DA' (*i.e.*, training on the data from the source domains and test on the target domain data), and 'Self-Training' (*i.e.*, using source domain data and pseudo labeled target domain data for iterative training and test on the target domain data); two data level UDA methods, including (1) FDA: a Fourier transform based method [21], and (2) CyCADA: a CycleGAN based method [7]; two feature level UDA methods, including (1) FD: a feature disentanglement based approach [14], and (2) DSBN: a feature normalization based method [2]; and two decision level UDA methods, including (1) pOSAL: an output adversarial training based method [18], and (2) BEAL: a boundary and entropy adversarial training based method [17]. Each target domain is chosen as the test domain alternately to simulate the deployment scenario, *i.e.*, only images from one target domain can be accessed per time.

The results of DoCR and its competitors were given in Table 2. It can be seen that the overall performance of our DoCR is not only higher than the baseline methods but also better than that of the other UDA approaches. We also visualize the segmentation maps predicted by our DoCR and other six competing methods, as shown in Fig. 2. It reveals that our DoCR can produce the most accurate segmentation map compared to the ground truth.

Table 3. Performance of DoCR, its four variants, and two baseline methods.

Methods		Target domain		
		BASE1	BASE2	BASE3
Baseline	w/o DA	(94.30, 84.01)	(94.58, 79.22)	(93.82, 81.40)
	Self-Training	(95.66, 85.47)	(94.97, 78.69)	(94.98, 82.76)
Analysis of DSC	HFR	(95.81, 85.03)	(95.85, 84.88)	(95.43, 85.67)
	HFR + Multi-Input-Head	(57.92, 36.74)	(52.96, 23.67)	(14.82, 00.00)
	HFR + DSC (DoCR)	(**95.89, 85.72**)	(**96.39, 86.17**)	(**96.19, 85.96**)
Analysis of HFR	DSC (Self-Training)	(95.94, 84.65)	(95.46, 82.82)	(95.92, 84.39)
	DSC + Image Rec.	(**96.24**, 85.45)	(95.90, 82.51)	(95.59, 84.54)
	DSC + HFR (DoCR)	(95.89, **85.72**)	(**96.39, 86.17**)	(**96.19, 85.96**)

Ablation Analysis. To evaluate the effectiveness of HFR and DSC, we conducted a series of ablation experiments, as shown in Table 3. The results of 'w/o DA', 'Self-Training', 'HFR', 'DSC', and 'DoCR' reveal that both HFR and DSC contribute to the final results. We compare the performances of reconstructing the high frequency image and reconstructing the image itself. The results are presented in 'Analysis of HFR' part of Table 3. It can be seen that reconstructing the image itself can not bring much performance gain and the OD segmentation results on 'BASE2' even get worse. It can be attributed to the fact that reconstructing the image itself forces the network to recover the details of the image, most of which are redundant for the segmentation task. Different from that, high frequency image reconstruction facilitates the network filter out domain-sensitive low frequency features and hence improves the segmentation performance on the target domain data.

We also compare the results of adopting dynamic convolution (*i.e.*, HFR + DSC) and using multi-input-head (*i.e.*, HFR + Multi-Input-Head). It can be seen that using multi-input-head is hardly to perform segmentation since the target domain-specific parameters are mainly optimized under the supervision of reconstruction. Even though, adopting dynamic convolution can achieve improved performance than HFR, owing to its flexible generated parameters and diverse domain codes.

4   Conclusion

In this paper, we propose a DoCR based unsupervised domain adaptation model for medical image segmentation. It is composed of a DSC module, an encoder-decoder backbone, an HFR head, and a segmentation head. The DSC module improves the model's domain-invariant feature extraction ability by adopting domain-specific parameters generated conditioned on the domain code. The HFR head performs reconstruction under the supervision of high frequency image, and hence forces the model to behave like a high-pass filter to filter out low frequency domain-sensitive features. Experimental results on the multi-domain joint OC/OD segmentation task suggest the proposed DoCR can achieve superior performance over the baselines and other state-of-the-art UDA methods.

Acknowledgment. This work was supported in part by the National Natural Science Foundation of China under Grants 62171377, in part by the Key Research and Development Program of Shaanxi Province under Grant 2022GY-084, and in part by the Natural Science Foundation of Ningbo City, China, under Grant 2021J052.

References

1. Almazroa, A., et al.: Retinal fundus images for glaucoma analysis: the Riga dataset. In: Medical Imaging 2018: Imaging Informatics for Healthcare, Research, and Applications, vol. 10579, p. 105790B. International Society for Optics and Photonics (2018)
2. Chang, W.G., You, T., Seo, S., Kwak, S., Han, B.: Domain-specific batch normalization for unsupervised domain adaptation. In: IEEE Conference on Computer Vision and Pattern Recognition (CVPR), pp. 7354–7362 (2019)
3. Decencière, E., et al.: Feedback on a publicly distributed image database: the Messidor database. Image Anal. Stereology **33**(3), 231–234 (2014)
4. Falk, T., et al.: U-Net: deep learning for cell counting, detection, and morphometry. Nat. Methods **16**(1), 67–70 (2019). https://doi.org/10.1038/s41592-018-0261-2
5. Guan, H., Liu, M.: Domain adaptation for medical image analysis: a survey. IEEE Trans. Biomed. Eng. **69**(3), 1173–1185. https://doi.org/10.1109/TBME.2021.3117407
6. Han, Y., Huang, G., Song, S., Yang, L., Wang, H., Wang, Y.: Dynamic neural networks: a survey. arXiv:2102.04906 [cs], February 2021
7. Hoffman, J., et al.: Cycada: cycle-consistent adversarial domain adaptation. In: International Conference on Machine Learning (ICML), pp. 1989–1998. PMLR (2018)

8. Hu, S., Liao, Z., Zhang, J., Xia, Y.: Domain and content adaptive convolution for domain generalization in medical image segmentation. arXiv preprint arXiv:2109.05676 (2021)
9. Isensee, F., Jaeger, P.F., Kohl, S.A.A., Petersen, J., Maier-Hein, K.H.: nnU-Net: a self-configuring method for deep learning-based biomedical image segmentation. Nat. Methods **18**(2), 203–211 (2020). https://doi.org/10.1038/s41592-020-01008-z
10. Liao, Z., Hu, S., Xie, Y., Xia, Y.: Modeling human preference and stochastic error for medical image segmentation with multiple annotators. arXiv preprint arXiv:2111.13410 (2021)
11. Litjens, G., et al.: A survey on deep learning in medical image analysis. Med. Image Anal. **42**, 60–88 (2017). https://doi.org/10.1016/j.media.2017.07.005
12. Liu, Q., Dou, Q., Yu, L., Heng, P.A.: MS-Net: multi-site network for improving prostate segmentation with heterogeneous MRI data. IEEE Trans. Med. Imaging **39**(9), 2713–2724 (2020). https://doi.org/10.1109/TMI.2020.2974574
13. Liu, Q., Chen, C., Qin, J., Dou, Q., Heng, P.A.: FedDG: federated domain generalization on medical image segmentation via episodic learning in continuous frequency space. In: IEEE Conference on Computer Vision and Pattern Recognition (CVPR), pp. 1013–1023, June 2021
14. Shin, S.Y., Lee, S., Summers, R.M.: Unsupervised domain adaptation for small bowel segmentation using disentangled representation. In: de Bruijne, M., et al. (eds.) MICCAI 2021. LNCS, vol. 12903, pp. 282–292. Springer, Cham (2021). https://doi.org/10.1007/978-3-030-87199-4_27
15. Sprawls, P.: Image characteristics and quality. In: Physical Principles of Medical Imaging, pp. 1–16. Aspen Gaithersburg (1993)
16. Wang, R., Zheng, G.: CyCMIS: cycle-consistent cross-domain medical image segmentation via diverse image augmentation. Med. Image Anal. **76**, 102328 (2022)
17. Wang, S., Yu, L., Li, K., Yang, X., Fu, C.-W., Heng, P.-A.: Boundary and entropy-driven adversarial learning for fundus image segmentation. In: Shen, D., et al. (eds.) MICCAI 2019. LNCS, vol. 11764, pp. 102–110. Springer, Cham (2019). https://doi.org/10.1007/978-3-030-32239-7_12
18. Wang, S., Yu, L., Yang, X., Fu, C.W., Heng, P.A.: Patch-based output space adversarial learning for joint optic disc and cup segmentation. IEEE Trans. Med. Imaging **38**(11), 2485–2495 (2019)
19. Xiao, J., Yu, L., Xing, L., Yuille, A., Zhou, Y.: Dualnorm-unet: incorporating global and local statistics for robust medical image segmentation. arXiv preprint arXiv:2103.15858 (2021)
20. Xie, X., Niu, J., Liu, X., Chen, Z., Tang, S., Yu, S.: A survey on incorporating domain knowledge into deep learning for medical image analysis. Med. Image Anal. **69**, 101985 (2021). https://doi.org/10.1016/j.media.2021.101985
21. Yang, Y., Soatto, S.: FDA: fourier domain adaptation for semantic segmentation. In: IEEE Conference on Computer Vision and Pattern Recognition (CVPR), pp. 4085–4095, June 2020
22. Zhang, J., Xie, Y., Xia, Y., Shen, C.: DoDNet: learning to segment multi-organ and tumors from multiple partially labeled datasets. In: IEEE Conference on Computer Vision and Pattern Recognition (CVPR), pp. 1195–1204, June 2021
23. Zhou, Z., Siddiquee, M.M.R., Tajbakhsh, N., Liang, J.: UNet++: redesigning skip connections to exploit multiscale features in image segmentation. IEEE Trans. Med. Imaging **39**(6), 1856–1867 (2020). https://doi.org/10.1109/TMI.2019.2959609
24. Zhu, J.Y., Park, T., Isola, P., Efros, A.A.: Unpaired image-to-image translation using cycle-consistent adversarial networks. In: International Conference on Computer Vision (ICCV), pp. 2223–2232 (2017)

Unsupervised Cross-disease Domain Adaptation by Lesion Scale Matching

Jun Gao[1,4], Qicheng Lao[3,5], Qingbo Kang[2], Paul Liu[4], Le Zhang[1(✉)], and Kang Li[2,5(✉)]

[1] College of Computer Science, Sichuan University, Chengdu, China
[2] West China Biomedical Big Data Center, West China Hospital, Sichuan University, Chengdu, China
likang@wchscu.cn
[3] School of Artificial Intelligence, BUPT, Beijing, China
[4] Stork Healthcare, Chengdu, China
[5] Shanghai Artificial Intelligence Laboratory, Shanghai, China

Abstract. Breast and thyroid lesions share many similarities in the feature representations of ultrasound images. However, there is a huge lesion scale gap between the two diseases, making it difficult to transfer knowledge between them through current methods for unsupervised domain adaptation. To address this problem, we propose a *lesion scale matching* approach where we employ a framework of latent space search for bounding box size to re-scale the source domain images, and then the Monte Carlo Expectation Maximization algorithm is used for optimization to match the lesion scales between the two disease domains. Extensive experimental results demonstrate the feasibility of cross-disease knowledge transfer, and our proposed method substantially improves the performance of unsupervised cross-disease domain adaptation models, with the Accuracy, Recall, Precision, and F1-score improved by 8.29%, 6.41%, 11.25%, and 9.14% on average in the three sets of ablation experiments.

Keywords: Cross-disease · Unsupervised domain adaptation · Scale matching · Classification · Ultrasound

1 Introduction

Breast cancer and thyroid cancer are the most common and top five most common cancers among women worldwide, respectively [1]. These two types of cancer are similar to each other in many aspects including molecular mechanisms [2–4], and radiomic feature representations (e.g., texture, shape, etc.) in various imaging modalities including the ultrasound (US) [5], which is now widely used for clinical screening of breast and thyroid lesions due to its easy accessibility, cost-effectiveness, non-invasive and real-time visualization [6].

Motivated by the similarities in the feature representations of both breast and thyroid lesions in US images, in this work, we investigate the feasibility

---

J. Gao and Q. Lao—Equal contribution.

© The Author(s), under exclusive license to Springer Nature Switzerland AG 2022
L. Wang (Eds.): MICCAI 2022, LNCS 13437, pp. 660–670, 2022.
https://doi.org/10.1007/978-3-031-16449-1_63

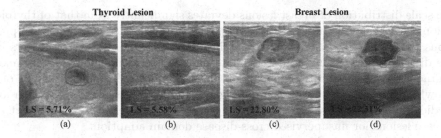

Fig. 1. Representative images of thyroid and breast lesions. (a) and (b) are thyroid US images with benign and malignant lesions respectively; (b) and (c) are breast US images with benign and malignant lesions respectively.

Table 1. Quantitative statistics of breast and thyroid lesion scales. The groups of lesions (small, big) are obtained by the K-means clustering.

Lesion type	Lesion scale		
	Small lesion group	Big lesion group	Overall
Thyroid	2.43 %	21.15%	5.55%
Breast	8.85 %	46.14%	22.45%

of transferring knowledge between the two diseases, e.g., whether the learned knowledge from thyroid nodule classification (i.e., benign or malignant) in US images could be transferred to breast nodule classification. Unlike previous deep learning approaches for breast and thyroid lesion classifications [5,7–11] that typically require a large amount of labeled US images, we aim to transfer the knowledge in an unsupervised domain adaptation (UDA) manner, i.e., unlabeled data in the target domain. This could tremendously increase the efficiency of the use of already labeled ultrasound image collected in one disease domain while transferring knowledge to another disease domain without further annotating the new disease domain. Nevertheless, since these two diseases are with different organs and have distinct physiological structures, to some extent, there still exists domain gap between them. Unexpectedly, although achieved great success in general computer vision tasks, our first attempt of direct applying current state-of-the-art UDA methods [12–16] for cross-disease domain adaptation still yield unsatisfactory performance.

In this paper, we identify the key problem that causes the failure of UDA methods for unsupervised cross-disease domain adaptation between thyroid and breast diseases, which we term as the *lesion scale gap* problem. The lesion scale (LS) means the proportion of the lesion in the image, and for different types of lesions, the LS distribution can be quite different. For example, it is known that the nodules in the thyroid are proportionally much smaller than those in the breast, as shown in Fig. 1 the representative images of thyroid lesion (left) and breast lesion (right). Note that we use the lesion scale instead of the lesion size for measurement because the whole size of the US images can be varying depending on the scanned tissues and the acquisition devices. We show quantitative statistics of breast and thyroid lesion scales in Table 1. Statistically, the scale distribution of breast lesions deviates significantly from that of thyroid lesions in both small and big lesion groups. This finding also coincides with previous study [17] where it is found that inconsistencies in the size of objects in the distribution of training and test data can lead to tremendous performance degradation of classification models on natural images. To the extent of our knowledge, there is few work conducting research on the scale gap problem for medical images, and we are the first to tackle this challenge in the context of disease lesions for unsupervised cross-disease domain adaption.

To address this problem, in this work, we propose a *lesion scale matching* approach where we employ a framework of latent space search for bounding box size to re-scale the source domain images based on our prior knowledge on the diseases. Then, the Monte Carlo Expectation Maximization algorithm is used for optimizing the UDA model and the latent variable of bounding box size to match the lesion scales between the two disease domains. We show in our experiments that the proposed approach can effectively decrease the gap of lesion scales between the two diseases, and as a result dramatically improves the adaptation performance. The contributions of our work can be summarized as follows: (1) For the first time, we demonstrate the feasibility of knowledge transfer from thyroid lesions to breast lesions in an unsupervised cross-disease domain adaptation manner, which may enlighten other similar scenarios for cross-disease knowledge transfer; (2) We propose a lesion scale matching approach which contains a framework of latent space search for bounding box size towards re-scaling, and together with the Monte Carlo Expectation Maximization algorithm, we effectively address the lesion scale gap problem; (3) Our proposed framework substantially improves the performance of unsupervised cross-disease domain adaptation models on the classification of thyroid and breast lesions.

2   Methodology

In this section, we present our proposed framework which is illustrated in Fig. 2. For the lesion scale matching, we first propose a latent space search for bounding box (BBox) size z where the proportion of lesions to the BBox (i.e., lesion scales) are matched for both the source and target disease domains based on prior medical knowledge. Then we randomly sample z in the latent space and crop the source domain image x_s at the centroid of the lesion with the BBox at size z. Finally, the cropped and labeled source domain image (x'_s, y_s) together with the unlabeled target domain image (x_t) are used to train the domain adaptation model, for which we simply adopt the *Margin Disparity Discrepancy* (MDD) method [14], with the Monte Carlo Expectation Maximization (MCEM) algorithm [18]. In the following, we review the MDD model as preliminaries in Subsect. 2.1, then the details of latent space search for BBox size and the MCEM algorithm are described in Subsect. 2.2 and Subsect. 2.3, respectively.

2.1   Preliminary: The Margin Disparity Discrepancy Method

We adopt the MDD method [14] as the backbone for unsupervised domain adaptation in our proposed framework. Note that we will show later in the

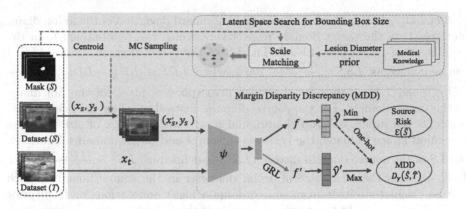

Fig. 2. Our proposed framework for unsupervised cross-disease domain adaptation.

experiments that our proposed framework is orthogonal to the backbone choices, although MDD gives the best performance. The network structure of MDD is shown in the bottom right of Fig. 2 which has a feature extractor ψ and two classifiers f and f', and its minimax objective function can be described as:

$$\min_{f,\psi} \max_{f'} \mathcal{E}(\widehat{S}) + \lambda \mathcal{D}_\gamma(\widehat{S}, \widehat{T}), \tag{1}$$

where $\widehat{S} = \left\{ \left(x_s^i, y_s^i \right) \right\}_{i=1}^n$ represents labelled samples drawn from source domain and $\widehat{T} = \{x_t^i\}_{i=1}^m$ denotes unlabelled samples drawn from target domain. $\mathcal{E}(\widehat{S}) = \mathbb{E}_{(x_s,y_s) \sim \widehat{S}} L\left(f\left(\psi\left(x_s \right) \right), y_s \right)$ is the source domain risk, λ is the trade-off coefficient, and $\mathcal{D}_\gamma(\widehat{S}, \widehat{T})$ is the margin disparity discrepancy between source and target domains given by:

$$\mathcal{D}_\gamma(\widehat{S}, \widehat{T}) = \mathbb{E}_{x_t \sim \widehat{T}} L'\left(f'\left(\psi\left(x_t \right) \right), f\left(\psi\left(x_t \right) \right) \right) - \gamma \mathbb{E}_{x_s \sim \widehat{S}} L\left(f'\left(\psi\left(x_s \right) \right), f\left(\psi\left(x_s \right) \right) \right), \tag{2}$$

where γ is the margin factor. The discrepancy loss term consists of L and L', with L being a standard cross-entropy loss for source domain, and L' being a modified cross-entropy loss for target domain which was introduced in [14]. In the training phase, our feature extractor ψ is directly trained to minimize the discrepancy loss term through a gradient reversal layer (GRL) [12].

2.2   Latent Space Search for Bounding Box Size

In order to match the lesion scale of the source domain disease to that of the target domain disease, we exploit prior medical knowledge to determine the range of the proper BBox size for the source domain (illustrated in the upper right of Fig. 2). Note that the prior knowledge of the tumor size range for each disease is usually not difficult to obtain by literature. However, when such knowledge is indeed unavailable, we can instead use the pseudo labels of the target data generated by existing tumor segmentation methods and extract the pseudo knowledge on the tumor size as a replacement for the ground truth.

Let LD_M^T, LD_{SD}^T denote the mean and standard deviation of the lesion diameter (pixel) in the target domain respectively, which can be obtained from the prior medical knowledge [19–23]. The LS of the target domain disease (LS^T) can be defined as: $LS^T = \frac{(LD^T)^2}{IA_M^T} \times 100\%$, where $LD^T \in [LD_M^T - LD_{SD}^T, LD_M^T + LD_{SD}^T]$, and IA_M^T stands for the mean area (pixel $\times$ pixel) of the full image in the target domain. For the source domain, instead of using the original full image as the input, we crop the original image with a BBox of size z as the new input image to match the LS between source and target domains. Similarly, the LS of the source domain disease (LS^S) is defined as: $LS^S = \frac{(LD_M^S)^2}{z \times z}$, where LD_M^S denotes the mean size of lesion diameter in the source domain, which can be calculated either from source domain mask annotations or from prior medical knowledge. In this study, we obtain it through the source domain mask annotations.

Finally, we define the mean LS distance (LSD) between the source and target domains as the l_1 distance between the LS in the source (LS^S) and target (LS^T) domains: $LSD = |LS^S - LS^T|$. By setting LSD to 0 (i.e., $LS^S = LS^T$), we obtain the latent space for the proper BBox size in the source domain:

$$z \in [\frac{\sqrt{IA_M^T \times LD_M^S}}{LD_M^T + LD_{SD}^T}, \frac{\sqrt{IA_M^T \times LD_M^S}}{LD_M^T - LD_{SD}^T}]. \qquad (3)$$

2.3   Monte Carlo Expectation Maximization

With the above-defined latent space for the BBox size z in Eq. 3, the paradigm of the MDD model in our proposed framework is then updated to:

$$\min_{f,\psi} \max_{f'} \mathcal{E}(\widehat{S}(z)) + \lambda \mathcal{D}_\gamma(\widehat{S}(z), \widehat{T}), \qquad z \sim P(z), \qquad (4)$$

where two types of parameters are to be optimized: the MDD network parameters (f, f', and ψ) and the hidden variable z. In this study, we assume that the BBox size follows a uniform distribution in the latent space, i.e., $z \sim U(z)$. Consequently, we apply the classical Monte Carlo Expectation Maximization algorithm (MCEM) [18] for optimization. Specifically, during the training phase, the BBox size z is randomly sampled in the latent space by the Monte Carlo (MC) method, which is then used to crop the source domain image centered at the lesion. Finally, we optimize the network parameters in the maximization step. By using the MCEM algorithm, we iteratively train the MDD model, thus ensuring that the LS of the source and target domains are matched for better cross-disease domain adaptation performance.

It is worth mentioning that the latent variable z 1) is not equivalent to obtaining scale invariance via data augmentation and 2) cannot be chosen completely randomly out of the desired space. Data augmentation with scaling ensures network output that is invariant with respect to scaling, but in our cross-disease domain adaptation case, we do not want scale invariance. This is because lesion size is a criterion in classifying malignant vs benign, with malignant lesions being larger. Breast lesions are typically twice as large as thyroid lesions (Table 1), and if scale zoom is too small for example, a malignant lesion in the target breast domain is shrunk and may potentially be erroneously classified as benign. We also demonstrate in the experiments that random scaling will only lead to worse results. Thus, correcting for scale mismatch is fundamental to domain adaptation and is not merely an artifact of the data acquisition process that we wish to achieve invariance with respect to, i.e., camera distance in computer vision.

3   Experiments

3.1   Experimental Settings

Datasets. The thyroid US dataset used in this study was collected from West China hospital, which contains 778 images with benign lesions and 946 images with malignant lesions. All images were acquired using GE LOGIQ E9 ultrasound system. The labels were either directly from the fine-needle aspiration results for malignant lesions or from clinical diagnosis by senior radiologists for benign lesions. Additionally, the lesion areas were annotated by three senior radiologists, checked and refined by another senior radiologist who has more than 25-year clinical experiences. The breast US dataset used in this study is a publicly available dataset (BUSI [24]), collected from the Baheya Hospital for Early Detection and Treatment of Women's Cancer. It consists of 210 images with benign lesions and 437 images with malignant lesions. The corresponding lesion masks are also available for the BUSI dataset.

Implementation Details and Evaluation Metrics. We implement our proposed framework with PyTorch [25]. ResNet-50 [26] is adopted as the feature extractor ψ with parameters pre-trained on ImageNet [27]. The two classifiers f and f' are both 2-layer neural networks with width 1024, and the bottleneck width in MDD is also set to 1024. For the optimization, we use the mini-batch SGD with the Nesterov momentum 0.9. Following [14], we set the learning rate for the bottleneck of the two classifiers to 0.004, which is 10 times to that for the feature extractor. All models are trained for 30 epochs with the batch size of 32. The hyper-parameters γ and λ are set to 4 and 0.25 respectively. For performance evaluation metrics, we adopt four commonly used metrics for image classification, including Accuracy, Recall, Precision and F1-score.

3.2   Experimental Results

State-of-the-art Performance on Cross-Disease Domain Adaptation. We compare our proposed method with three representative UDA methods including DANN [12], CDAN [13], and MDD [14], as well as the batch normalization based method DSBN [28] and MDD with random scale data augmentation (RSDA). The corresponding quantitative performance comparisons are shown in

Table 2. Quantitative performance comparison between different unsupervised domain adaptation methods and our method (Source domain: Thyroid; Target domain: Breast)

Method	Accuracy(%)	Recall(%)	Precision(%)	F1-score(%)
ResNet-50	61.85 ± 2.40	33.97 ± 8.87	40.71 ± 2.64	36.99 ± 5.75
Upper Bound	91.21 ± 0.98	87.93 ± 1.89	85.08 ± 2.24	86.42 ± 1.65
DANN [12]	65.22 ± 0.44	**72.62 ± 2.36**	47.46 ± 0.63	57.17 ± 0.04
CDAN [13]	65.43 ± 1.29	58.57 ± 6.55	47.43 ± 1.41	52.28 ± 2.57
DSBN [28]	69.67 ± 0.26	59.72 ± 0.98	53.73 ± 3.82	59.84 ± 2.25
MDD [14]	73.42 ± 0.80	54.29 ± 2.08	60.72 ± 1.78	57.36 ± 2.32
MDD+RSDA	76.77 ± 1.05	53.93 ± 5.71	67.98 ± 1.35	60.02 ± 3.44
Our method	**79.41 ± 0.49**	63.81 ± 2.95	**70.25 ± 2.52**	**66.42 ± 1.82**

Table 2, in which the thyroid disease is considered as the source domain and the breast disease as target domain. The first row in Table 2 represents the ResNet-50 baseline without any domain adaptation methods. The second row in Table 2 gives the upper bound, where both training and testing are conducted on breast US images. From the results in Table 2, we can clearly observe that our proposed method achieves the best performance in terms of Accuracy, Precision and F1-score compared with other UDA methods, and also outperforms the ResNet-50 baseline by a big margin, suggesting that there is knowledge transfer between thyroid lesion classification and breast lesion classification. To this end, we show a successful use case for unsupervised cross-disease domain adaptation, which may shed lights on other similar cross-disease knowledge transfer scenarios.

Correlation Between Lesion Scale Matching and Adaptation Performance. Furthermore, in order to verify the effect of different LSD between the source and target domains (i.e., lesion scale gap) on domain adaption performance, we train the model with different BBox sizes (e.g., $z = 168, 224, ..., 616$) in the source domain images which are correlated to different LSD. The quantitative results are shown in Table 3 and the corresponding line graph is visualized in Fig. 3. Both results reveal a strong correlation between the adaptation performance and lesion scale matching measured by LSD. Specifically, the domain adaption performance increases when the LSD decreases, and there exists a critical point where the LSD has the lowest value and corresponds to the best target accuracy. To conclude, these experimental results demonstrate that the model can achieve the best performance when the lesion scales between the source and target domains are matched. In addition, combining Eq. 3 and statistical prior anatomical information, we obtain a range of $[336, 414]$ for z, and this theoretical range is experimentally verified with highest accuracy at $z = 392$ (see Fig. 3). As indicated in Fig. 3 Table 3, choosing sizes smaller or larger than this range only decreases the accuracy. Thus scales out of our desired range not only is theoretically unjustified but also empirically performs worse.

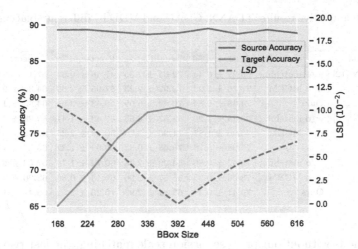

Fig. 3. Correlation between the adaptation performance (left) and lesion scale matching measured by LSD (right). Lower LSD means the lesion scales are close to matched.

Table 3. The effect of LSD on the domain adaption performance.

BBox size	LSD (10^{-2})	Accuracy(%)	Recall(%)	Precision(%)	F1-score(%)
168	10.60	65.07 ± 1.61	60.63 ± 1.53	46.81 ± 2.29	52.82 ± 1.92
224	8.59	69.35 ± 2.25	57.78 ± 2.44	52.57 ± 3.20	55.05 ± 2.86
280	5.58	74.39 ± 2.10	54.92 ± 1.45	62.00 ± 5.10	57.75 ± 2.04
336	2.50	77.90 ± 0.66	63.33 ± 2.47	68.52 ± 2.47	65.60 ± 2.43
392	0.04	78.41 ± 0.64	60.32 ± 0.27	68.91 ± 2.73	64.65 ± 0.87
448	2.31	77.41 ± 0.90	58.64 ± 3.08	68.08 ± 2.21	62.46 ± 2.51
504	4.25	77.23 ± 1.87	58.41 ± 1.53	68.19 ± 2.41	62.46 ± 2.51
560	5.52	75.79 ± 1.56	57.62 ± 3.12	64.48 ± 1.77	60.74 ± 2.86
616	6.63	75.06 ± 0.85	55.56 ± 6.21	64.06 ± 1.03	58.52 ± 2.67

Ablation Studies. In order to verify the effectiveness of each component in our proposed framework, we perform different ablation studies on our approach with three UDA methods (i.e., DANN [12], CDAN [13] and MDD [14]). The corresponding three groups of quantitative performance results are presented in Table 4. Specifically, different ablation settings are performed for each group. Firstly, 'w/o matching' indicates direct adoption of the whole images in both source (thyroid) and target (breast) domains for training UDA models (i.e., pure UDA methods without our proposed lesion scale matching). Secondly, 'w/o MCEM' means using a fixed size (e.g., 375, which is the mean of [336, 414]) for the BBox in the source domain images, that is without random sampling in the latent space for z using the MCEM algorithm. From the results summarized in Table 4, it can be clearly seen that, compared with the first setting which is the pure

Table 4. Quantitative results of DANN, CDAN and MDD by different ablation settings

Method		Accuracy(%)	Recall(%)	Precision(%)	F1-score(%)
DANN [12]	w/o matching	65.22 ± 0.44	72.62 ± 2.36	47.46 ± 0.63	57.17 ± 0.04
	w/o MCEM	73.76 ± 2.23	72.92 ± 2.23	57.95 ± 2.69	64.36 ± 2.44
	Ours	74.48 ± 0.89	70.42 ± 6.09	59.43 ± 2.04	64.43 ± 1.71
CDAN [13]	w/o matching	65.43 ± 1.29	58.57 ± 6.55	47.43 ± 1.41	52.28 ± 2.57
	w/o MCEM	73.96 ± 1.87	71.90 ± 3.78	56.98 ± 2.25	63.46 ± 2.08
	Ours	75.04 ± 0.78	70.48 ± 0.82	59.69 ± 1.32	63.37 ± 1.82
MDD [14]	w/o matching	73.42 ± 0.80	54.29 ± 2.08	60.72 ± 1.78	57.36 ± 2.32
	w/o MCEM	78.05 ± 0.33	58.67 ± 2.06	69.13 ± 0.86	63.26 ± 1.32
	Ours	79.41 ± 0.49	63.81 ± 2.95	70.25 ± 2.52	66.42 ± 1.82

UDA methods without our proposed lesion scale matching, the last two settings can significantly improve the adaptation performance ($p < 0.05$, paired t-test). Specifically, using prior medical knowledge to match the lesion scales of the source and target domains can significantly improve the adaptation performance, with the Accuracy, Recall, Precision, and F1-score improved by 7.23%, 6.00%, 9.48%, and 8.09% on average in the three sets of experiments. Similarly, the use of the MCEM training strategy can further improve the adaptation performance by an average of 1.05% in Accuracy, 0.51% in Recall, 1.77% in Precision, and 1.05% in F1-score, respectively. The ablation results demonstrate that our proposed approach is effective, robust and orthogonal to different UDA methods.

4   Conclusion and Future Work

This paper attempts to transfer knowledge between thyroid and breast lesions in an unsupervised domain adaptation manner given the similarities shared by these two diseases in ultrasound images. We identify in our work the key problem that causes the failure of current unsupervised domain adaptation methods directly used for cross-disease adaptation, i.e., the lesion scale gap problem. To tackle this, we propose a lesion scale matching approach with a framework of latent space search for the re-scaling bounding box size. The experimental results show that our proposed framework can significantly improve the performance of cross-disease domain adaption between the thyroid and breast diseases, with the Accuracy, Recall, Precision, and F1-score improved by 8.29%, 6.41%, 11.25%, and 9.14% on average in the three sets of ablation experiments. For the future work, we will further explore alternative algorithms for scale matching, e.g., rescale both disease domains instead of only the source domain in the current work, as well as other cross-disease knowledge transfer scenarios.

References

1. Sung, H., et al.: Global cancer statistics 2020: Globocan estimates of incidence and mortality worldwide for 36 cancers in 185 countries. CA Can. J. Clin. **71**(3), 209–249 (2021)
2. Agarwal, D.P., Soni, T.P., Sharma, O.P., Sharma, S.: Synchronous malignancies of breast and thyroid gland: a case report and review of literature. J. Can. Res. Ther. **3**(3), 172–173 (2007)
3. Chen, J., et al.: Correlation analysis of breast and thyroid nodules: a cross-sectional study. Int. J. Gener. Med. **14**, 3999–4010 (2021)
4. An, J.H., et al.: A possible association between thyroid cancer and breast cancer. Thyroid **25**(12), 1330–1338 (2015). PMID: 26442580
5. Yi-Cheng, Z., et al.: A generic deep learning framework to classify thyroid and breast lesions in ultrasound images. Ultrasonics **110**, 106300 (2021)
6. Sahiner, B., et al.: Malignant and benign breast masses on 3d us volumetric images: effect of computer-aided diagnosis on radiologist accuracy. Radiology **242**(3), 716–724 (2007). PMID: 17244717
7. Chen, K., et al.: Enhanced breast lesion classification via knowledge guided cross-modal and semantic data augmentation. In: de Bruijne, M., et al. (eds.) MICCAI 2021. LNCS, vol. 12905, pp. 53–63. Springer, Cham (2021). https://doi.org/10.1007/978-3-030-87240-3_6
8. Liu, T., et al.: Automated detection and classification of thyroid nodules in ultrasound images using clinical-knowledge-guided convolutional neural networks. Med. Image Anal. **58**, 101555 (2019)
9. Qian, X., et al.: Prospective assessment of breast cancer risk from multimodal multiview ultrasound images via clinically applicable deep learning. Nat. Biomed. Eng. **5**, 1–11 (2021)
10. Sharifi, Y., Bakhshali, M.A., Dehghani, T., DanaiAshgzari, M., Sargolzaei, M., Eslami, S.: Deep learning on ultrasound images of thyroid nodules. Biocybern. Biomed. Eng. **41**(2), 636–655 (2021)
11. Nguyen, D.T., Kang, J.K., Pham, T.D., Batchuluun, G., Park, K.R.: Ultrasound image-based diagnosis of malignant thyroid nodule using artificial intelligence. Sensors **20**(7), 1822 (2020)
12. Ganin, Y., et al.: Domain-adversarial training of neural networks. J. Mach. Learn. Res. **17**(1), 2030–2096 (2016)
13. Long, M., Cao, Z., Wang, J., Jordan, M.I.: Conditional adversarial domain adaptation. In: Bengio, S., Wallach, H., Larochelle, H., Grauman, K., Cesa-Bianchi, N., Garnett, R. (eds.), Advances in Neural Information Processing Systems, vol. 31. Curran Associates Inc (2018)
14. Zhang, Y., Liu, T., Long, M., Jordan, M.: Bridging theory and algorithm for domain adaptation. In: Proceedings of the 36th International Conference on Machine Learning, pp. 7404–7413. PMLR (2019)
15. Yang, Y., Soatto, S.: Fda: fourier domain adaptation for semantic segmentation. In: CVPR, pp. 4084–4094 (2020)
16. Lao, Q., Jiang, X., Havaei, M.: Hypothesis disparity regularized mutual information maximization. In: Proceedings of the AAAI Conference on Artificial Intelligence, vol. 35, pp. 8243–8251 (2021)
17. Touvron, H., Vedaldi, A., Douze, M., Jegou, H.: Fixing the train-test resolution discrepancy. In: Wallach, H., Larochelle, H., Beygelzimer, A., d' Alché-Buc, F., Fox, E., Garnett, R. (eds.), Advances in Neural Information Processing Systems, vol. 32. Curran Associates Inc (2019)

18. McLachlan, G.J., Krishnan, T.: The EM Algorithm and Extensions. Wiley Series in Probability and Statistics. Wiley (2007)

19. Jiang, Y.X., Liu, H., Liu, J.B., Zhu, Q.L., Sun, Q., Chang, X.Y.: Breast tumor size assessment: comparison of conventional ultrasound and contrast-enhanced ultrasound. Ultrasound Med. Biol. **33**, 1873–1881 (2007)

20. Golshan, M., Fung, B.B., Wiley, E., Wolfman, J., Rademaker, A., Morrow, M.: Prediction of breast cancer size by ultrasound, mammography and core biopsy. Breast **13**(4), 265–271 (2004)

21. Shoma, A., Moutamed, A., Ameen, M., Abdelwahab, A.: Ultrasound for accurate measurement of invasive breast cancer tumor size. Breast J. **12**(3), 252–256 (2006)

22. Zheng, X., et al.: Deep learning radiomics can predict axillary lymph node status in early-stage breast cancer. Nat. Commun. **11**, 03 (2020)

23. Cavallo, A., et al.: Thyroid nodule size at ultrasound as a predictor of malignancy and final pathologic size. Thyroid **27**, 01 (2017)

24. Al-Dhabyani, W., Gomaa, M., Khaled, H., Fahmy, A.: Dataset of breast ultrasound images. Data Brief **28**, 104863 (2020)

25. Paszke, A., et al.: Pytorch: an imperative style, high-performance deep learning library. Adv. Neural Inf. Process. Syst. **32** (2019)

26. He, K., Zhang, X., Ren, S., Sun, J.: Deep residual learning for image recognition. In: 2016 IEEE Conference on Computer Vision and Pattern Recognition (CVPR), pp. 770–778. IEEE Computer Society, Los Alamitos, CA, USA, June 2016

27. Russakovsky, O., et al.: ImageNet large scale visual recognition challenge. Int. J. Comput. Vis. **115**(3), 211–252 (2015). https://doi.org/10.1007/s11263-015-0816-y

28. Woong-Gi, C., You, T., Seo, S., Kwak, S., Han, B.: Domain-specific batch normalization for unsupervised domain adaptation. In: 2019 IEEE/CVF Conference on Computer Vision and Pattern Recognition (CVPR) (2019)

Adversarial Consistency for Single Domain Generalization in Medical Image Segmentation

Yanwu Xu[1]([⊠]), Shaoan Xie[3], Maxwell Reynolds[1], Matthew Ragoza[1], Mingming Gong[2], and Kayhan Batmanghelich[1]

[1] Department of Biomedical Informatics, University of Pittsburgh, Pittsburgh, USA
yanwuxu@pitt.edu
[2] School of Mathematics and Statistics, The University of Melbourne, Melbourne, Australia
[3] Department of Philosophy, Carnegie Mellon University, Pittsburgh, USA

Abstract. An organ segmentation method that can generalize to unseen contrasts and scanner settings can significantly reduce the need for retraining of deep learning models. Domain Generalization (DG) aims to achieve this goal. However, most DG methods for segmentation require training data from multiple domains during training. We propose a novel adversarial domain generalization method for organ segmentation trained on data from a *single* domain. We synthesize the new domains via learning an adversarial domain synthesizer (ADS) and presume that the synthetic domains cover a large enough area of plausible distributions so that unseen domains can be interpolated from synthetic domains. We propose a mutual information regularizer to enforce the semantic consistency between images from the synthetic domains, which can be estimated by patch-level contrastive learning. We evaluate our method for various organ segmentation for unseen modalities, scanning protocols, and scanner sites.

Keywords: Medical image segmentation · Single domain generalization · Adversarial training · Mutual information

1 Introduction

Deep Learning-based methods for the segmentation of medical images hold state-of-the-art performance across various organs and anatomies [13,18,24]. The independent and identically distributed (*i.i.d.*) is the underlying assumption of most of those methods.However, the difference in the image acquisition, such as scanning protocol and image modality, introduces domain shifts, rendering the assumption impractical. Domain Adaptation [2,6,7,20] (DA) and Multi-source Domain generalization [11] (MDG) aim to alleviate the domain shift

---

M. Gong and K. Batmanghelich—Equal Contribution.

issue.However, those approaches are not data-efficient as they either require access to test distribution (i.e., DA) or need multiple labeled source domains during training (i.e., MDG). In this paper, we focus on *single-source* domain generalization (SDG), aiming to train a generalizable deep model on only one source domain.

The *SDG* does not require access to the test distribution or labeled data from multiple sources during the training. As a result, it reduces annotation costs and avoids repetitive adaptation for each new domain. In the literature, various SDG methods have been proposed that are based on augmentation of input image [15, 22] and meta learning [17]. The meta-learning techniques tend to be extremely slow during inference time. The augmentation methods synthesize new images using random initialization of convolution filter [15, 22]. However, those methods cannot avoid over-fitting to a regular pattern of synthetic data. Thus, we propose synthesizing the new domains via learning an adversarial framework.

We propose synthesizing the new domains via learning an adversarial domain synthesizer (ADS). The intuition is that the synthetic domains cover a large enough area of plausible distributions so that unseen domains can be interpolated from synthetic domains. Specifically, we design the synthesizer with a random style module, enabling ADS to synthesize random textures during adversarial training. Without a constraint, adversarial training may change the image semantics, making the synthetic domains irrelevant. To remedy this problem, we propose to keep the underlying semantic information between the source image and the synthetic image via a mutual information regularizer. As estimating mutual information is hard for high dimensional data, we utilize the patch-level contrastive loss [14] as a surrogate to maximize the mutual information between the original and synthesized images.

The main contributions of this work can be summarized as follows: 1) We propose an adversarial framework for single domain generalization of medical image segmentation. 2) We redesign the network structure of synthesizing new domains. 3) To constrain the adversarial training, we propose a regularization method for synthetic images. To evaluate our model, we conduct experiments of single domain generalization of medical image segmentation on cross-modality image segmentation (CT → MRI), -imaging protocol, and -organizations.

2   Related Works

Unsupervised Domain Adaptation and Domain Generalization. Unsupervised Domain Adaptation (UDA) is a proposed strategy for addressing domain shift between the training data and testing data of deployed applications. In the UDA framework, the model has access to labelled data from a source domain and unlabelled data from a target domain during training. Prior work on UDA can be classified as distribution alignment [2, 6] or self-supervised learning [4].

Multi-domain generalization (MDG) is an alternative framework where the goal is to learn domain-invariant features from multiple labelled source domains. Recent works [11,12] give a theoretical proof that domain-invariant features can be learned using MDG. The common goal of UDA and MDG is to learn domain-invariant features for downstream tasks (i.e. classification, segmentation and object detection) by training a model on data from multiple domains. Therefore, UDA and MDG both require access to data from multiple domains during training. For single domain generalization, M-ADA [17] and L2D [21] propose an adversarial training framework for SDG learning. M-ADA proposes a meta-scheme method to find the adversarial perturbation, which calculates the gradient direction for each specific input and is extremely slow for real-life applications. While similar to RandConv [22] and GIN [15], M-ADA also proposes to synthesize new domain but with an adversarial training. However, M-ADA constrains the perturbation in latent space and cannot guarantee the semantic consistency for segmentation task, i.e. object boundary and shape, which is especially crucial for medical image segmentation.

Data Augmentation. Data Augmentation is a low-level data sampling technique, which gains a free performance improvement with use of human prior. To further improve the generalizability of models, AdaTransform [20] proposes to augment the data by maximizing the entropy of the data. AdvBias [3] also apply the adversarial technique for learning the bias-field and deformation field respectively. Cutout [5] randomly samples masks and crops out part of images via the masks. Mixup [23] simply mixes the images and also forces the prediction of mixed images to be the interpolation of ground truth labels.

3 Proposed Method

In the single domain generalization setting, we aim at training a segmentation model S on a single source domain $\mathcal{S} = \{x^i, y^i\}_{i=1}^N$ such that the learned model can generalize well to multiple target domains $\mathcal{T}_k = \{x_k^j, y_k^j\}_{j=1}^M$, where k is the domain indicator. The target domains are not accessible during training.

We propose an adversarial training-based method to synthesize data from new domains $\hat{\mathcal{T}}_l = \{\hat{x}_l^i, y^i\}_{i=1}^N$ from the single source domain, where l is the synthetic domain indicator. An assumption of our method is that the space of the collections of the unseen domains belongs to the collections of synthetic domains, i.e. $\cup_k \mathcal{T}_k \subset \cup_l \hat{\mathcal{T}}_l$. Specifically, the unseen domains can be generated by interpolating the synthetic domains. Our adversarial framework can enlarge the diversity the synthetic domains and guarantee sufficient coverage of the unseen domains. To ensure the synthetic images adhere to the semantics of the source images, we also apply a regularizer that promotes the mutual information between synthetic images and source images.

Our method consists of four modules: two adversarial domain synthesizers paramaterized as $T(\cdot; \theta_1)$ and $T(\cdot; \theta_2)$, a segmentation network parameterized as

$S(; \theta_3)$ and the mutual information regularizer $MI(\cdot, \cdot; \theta_4)$. We train the models with a two-step min-max procedure that involves supervised, consistency, adversarial, and regularization loss terms. A schematic illustration of our proposed model is shown in Fig. 1.

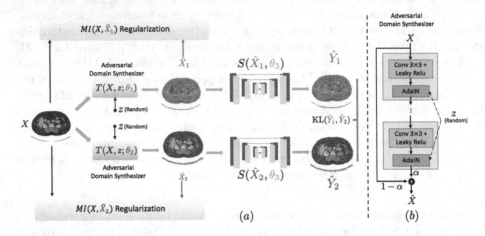

Fig. 1. Schematic of the proposed method. (a) shows the completed model structure, which consists of our adversarial domain synthesizer (ADS), the mutual information regularization between input and synthetic image, the segmentation network and the KL consistency loss between two predictions. (b) is the detailed structure of our proposed ADS. (Color figure online)

3.1 Adversarial Domain Synthesizer

To maximize the effectiveness of the synthetic domains, we utilize adversarial training to learn an Adversarial Domain Synthesizer (ADS) $T(X, z; \theta)$, which takes as input a source image and random noise sample z and outputs a synthetic image $\hat{X}$. We assume that T only changes the texture of the source images and not their segmentation annotation. As a result, the segmentation network S should generate the same segmentation mask for all $\hat{X}$ from a given X. For the adversarial training, T competes with S to generate the adversarial domain images. We construct the adversarial consistency via two differently parameterized T, and we can obtain two randomly synthesized images $\hat{X}_1 = T(X, z; \theta_1), \hat{X}_2 = T(X, z; \theta_2)$. The purpose of random variable z will be explained below. The adversarial competition can be formulated as follows:

$$\ell_{Con}(\hat{X}_1, \hat{X}_2) = \text{KL}(S(\hat{X}_1; \theta_3)||S(\hat{X}_2; \theta_3)), \tag{1}$$

where KL divergence is measured between the two softmax outputs. In Eq. 1, the two synthesizers T are trained to maximize $\ell_{Consistency}$ and S is trained to minimize it.

ADS Network Architecture. The goal of the ADS T is to generate the synthetic domains with hardest perturbation for S. Meanwhile, the semantics of the images, such as the boundary and shape of the organs, should keep consistent with source images and the changes are limited to the texture of the tissues. To achieve the effect mentioned above, we proposed a modified version of Global Intensity Non-Linear Augmentation module (GIN) [15]. GIN utilizes a shallow CNN structure with randomly initialized weights for the convolution filters plus a mixup with ratio of $\alpha \sim U(0,1)$ to generate multi-modalities from source domain. Unlike GIN, in our method, the synthesizer T is learned adversarially, so we cannot initialize T randomly to generalize multiple modalities. Thus, without randomness from the parameters of T, we can only interpolate the domains with the mixup parameter, which limits the variance of synthetic domains. To alleviate this issue, we incorporate the adaptive instance normalization (AdaIN) block [8] to introduce randomness to the generated domain. The design of the proposed T is shown in Fig. 1 (b) (Fig. 2).

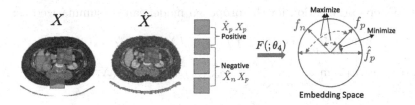

Fig. 2. Mutual information maximization.

3.2   MI Regularization

Without proper regularization, T can change semantics of the image arbitrarily. To ensure that the boundary and shape are shared underlying semantics between the synthetic image and the source image, we can maximize the mutual information between two images to keep the semantics. Thus, we propose to add the mutual information maximization (MIM) constraint to the synthesized images as maximizing $\mathrm{MI}(X, \hat{X})$, where θ_4 is the parameters of the mutual information estimator. Exact calculation of mutual information is computationally prohibitive. Inspired by the success of contrastive learning in image-to-image translation [16], we use patch-based contrastive loss as a surrogate for MI. Specifically, we utilize constrastive learning on the image patch level to maximize the MI between the source images and the transformed images, which exactly satisfies our goal. Thus, we can adapt [16] to our model and reformulate it as follows:

$$\ell_{MI}(X, \hat{X}) = \log \frac{\exp(\hat{f}_p \cdot f_p/\tau)}{\exp(\hat{f}_p \cdot f_p/\tau) + \sum_{n \neq p} \exp(f_p \cdot f_n/\tau)}. \tag{2}$$

In Eq. 2, the ℓ_{MI} is maximized and the numerator is the inner product of two corresponding patch features with exponential re-scaling, the $\hat{f}_p$ and f_p denote the features extracted by $F(\hat{X}_p; \theta_4), F(X_p; \theta_4)$, where $\hat{X}_p, X_p$ represent the patches of the synthesized image and the source image at the corresponding location respectively and they are the positive pair for contrastive learning. For the negative pair, we can see the same operation of the f_p and f_q in the denominator. In Eq. 2, we are trying to maximize the semantic correlation between the synthesized image and the input via pushing similarity of patches at the corresponding location and dissimilarity at different locations.

3.3   Model Optimization

As mentioned above, the segmentation masks are kept consistent from the source domain to synthesized domains. Thus we also have the supervised loss as below:

$$\ell_{Sup}(\hat{X}, Y) = \mathrm{KL}(S(\hat{X}; \theta_3) \| Y). \tag{3}$$

Our two-step optimization for the proposed model can be summarized as:

$$\text{update } \theta_3 : \min_{\theta_3} \mathbb{E}_{X,Y \sim P_{XY}} \ell_{Sup}(\hat{X}_1, Y) + \ell_{Sup}(\hat{X}_2, Y) + \ell_{Cons}(\hat{X}_1, \hat{X}_2)$$

$$\text{update } \theta_1, \theta_2, \theta_4 : \max_{\theta_1, \theta_2, \theta_4} \mathbb{E}_{X \sim P_X, z \sim P_z} \ell_{Con}(\hat{X}_1, \hat{X}_2) + \ell_{MI}(X, \hat{X}_1) + \ell_{MI}(X, \hat{X}_2).$$

4   Experiments

We evaluate our model on two experimental settings. For the first setting, we test the generalizability of the model on an unseen modality by using abdominal CT scans from [1] and MRI scans from [9] as the source domain and target domain respectively. For the second setting, we aim to test the generalizability of our model on unseen data distribution shift caused by different scanner machine settings. We obtain the data from six different organizations (RUNMC,BMC,HCRUDB,UCL,BIDMC,HK) and adopt the following cross-validation procedure: we select the data from one organization as the source domain for training and hold out the rest as the target domains for testing. Furthermore, we conduct a detailed ablation study to analyze the effectiveness of each component of the proposed method.

4.1   Training Configuration

We use Efficient-b2 [19] as the backbone for the segmentation network S and modify it with UNet-style skip connection. The synthesizer network T has 4 convolutional blocks, each block consisting of 3×3 convolutional kernel with the channel size of 2, Leaky-ReLU and AdaIN. For the MIM model, we use the encoder part of the generator with a fully connected layer as the feature extractor F from [16]. We choose Adam optimizer [10] for all of the models with initial learning rate 3×10^{-4} and $\beta = [0.5, 0.999]$. The learning rate will linearly decay to zero at the end of the training. The total training epochs are 2,000. We compare our method to Cutout, AdvBias, RandConv and GIN [15] under the same settings.

4.2   Data Prepossessing and Evaluation Metrics

Before training and testing, we first resize all of the 3D volumes at the axial plane to the size of 192×192. For the CT modality, we clip the intensity into the range of $[-275, 125]$. For the MRI modality, the top 99.5% of intensity values are cut out. We also apply all variations of predefined augmentations to the 3D volumes before we slice them into 2D images. These augmentations include random contrast via gamma transformation, Gaussian noise addition, affine transformation, 3D elastic transformation and intensity normalization of zero mean and unit variance. All of these predefined augmentations are performed with MONAI. More details of the data prepossessing and augmentation steps can be found in Appendix A.

To evaluate single source domain generalization, we do not access the target testing dataset during training. Thus, we split all source datasets into training and evaluation partitions with proportions 70% and 30%, respectively. For the final testing on the target datasets, we pick the saved model checkpoint which has the best predictive performance on the source evaluation split. We use the Dice score (DS) for evaluating the performance of the different methods.

4.3   Experimental Results and Empirical Analysis

We summarize our results in Table 1. Across three different single domain generalization settings, our proposed method achieves the best performance compared with the baseline models. Notably, compared with [15], the overall performance on Cardiac bSSFP-LGE is worse than their reported results, although we have tried our best to replicate the baseline results under the suggested settings of [15]. We also visualize the segmentation results in the Fig. 3. Our method succeeds in some difficult samples, where the partial spleen is missing or mis-classified in the Abdominal CT-MRI, or missing semantic structure in Cardiac bSSFP-LGE. We also observe the coarse segmentation mask in Prostate Cross-Centers for the baseline methods. The above quantitative and qualitative results show the effectiveness of our adversarial training framework. more visual results are shown in Appendix B.

4.4   Ablation Study

We conduct extra experiments to study the effectiveness of each component in our model. In Table 2, we compare the performance results of our model without adversarial training, our model without the mutual information regularizer,

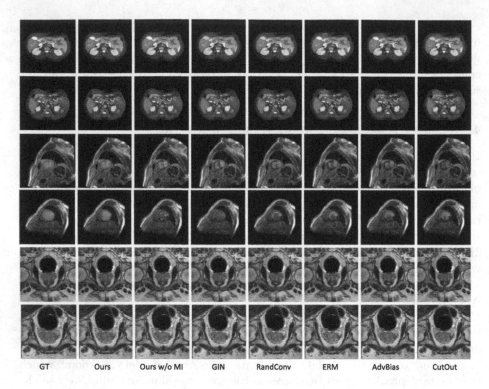

Fig. 3. Visualization of results.

Table 1. Results on three single source domain generalization settings.

Method	Abdominal CT-MRI					Cardiac bSSFP-LGE				Prostate Cross-center
	Liver	R-Kidney	L-Kidney	Spleen	Average	L-ventricle	Myocardium	R-ventricle	Average	Average
ERM	73.35	75.92	74.20	**82.93**	76.60	51.19	72.7	70.40	64.76	51.57
Cutout	76.57	80.06	77.96	78.90	78.37	65.13	77.58	70.44	71.05	58.79
AdvBias	74.97	86.16	78.39	72.45	77.99	60.10	77.52	72.29	69.97	60.47
RandConv	78.07	83.31	80.69	80.23	80.58	67.48	85.39	80.46	77.78	68.14
GIN	83.16	85.99	82.17	82.22	83.39	71.22	84.10	**82.06**	79.12	70.15
Ours	**85.24**	**89.87**	**86.92**	81.69	**85.93**	**73.37**	**86.10**	81.45	**80.31**	**71.42**

and our completed model. This experiment is conducted in the Abdominal CT-MRI under the same setups as described above. The model without adversarial training still achieves better results than the model without mutual information regularization by a small gap. However, we can see that both adversarial training and the MI regularizer can coorporate with each other to achieve superior performance.

Table 2. Ablation on Abdominal CT → MRI.

Abdominal CT-MRI					
Method	Liver	R-kidney	L-kidney	Spleen	Average
Ours w/o adversarial	84.59	85.52	**87.79**	79.21	84.28
Ours w/o MI	83.23	85.62	83.71	**82.79**	83.59
Ours	**85.24**	**89.87**	86.92	81.69	**85.93**

5   Conclusion

We successfully trained a model on a single source domain and deploy it on unseen target domains for the purpose of *single* domain generalization (SDG). Because of the limited training data and the unpredictable target domain shift, the generalizability of segmentation models is restricted, especially for medical images. Therefore, we propose a novel adversarial training framework to improve the generalizability of the SDG segmentation model. We synthesize the new domains via learning an adversarial domain synthesizer (ADS) in the proposed method. We assume that the unseen domains can be interpolated from synthetic domains, and the adversarial synthetic domains can guarantee sufficient coverage. To constrain the semantic consistency between synthetic images and the corresponding source images, we propose a mutual information regularizer, which can be estimated by patch-level contrastive learning. We evaluate our method for various organ segmentation for unseen modalities, scanning protocols, and scanner sites. The proposed method shows a consistent improvement over the baseline methods.

Acknowledge. This work was partially supported by NIH Award Number 1R01HL141813-01, NSF 1839332 Tripod+X, SAP SE, and Pennsylvania Department of Health. We are grateful for the computational resources provided by Pittsburgh Super-Computing grant number TG-ASC170024. MG is supported by Australian Research Council Project DE210101624. KZ would like to acknowledge the support by the National Institutes of Health (NIH) under Contract R01HL159805, by the NSF-Convergence Accelerator Track-D award #2134901, and by the United States Air Force under Contract No. FA8650-17-C7715.

References

1. (Author Name Not Available): Segmentation outside the cranial vault challenge (2015). https://doi.org/10.7303/SYN3193805, https://repo-prod.prod.sagebase.org/repo/v1/doi/locate?id=syn3193805&type=ENTITY
2. Bousmalis, K., Trigeorgis, G., Silberman, N., Krishnan, D., Erhan, D.: Domain separation networks. In: Lee, D., Sugiyama, M., Luxburg, U., Guyon, I., Garnett, R. (eds.) Advances in Neural Information Processing Systems, vol. 29. Curran Associates, Inc. (2016). https://proceedings.neurips.cc/paper/2016/file/45fbc6d3e05ebd93369ce542e8f2322d-Paper.pdf

3. Chen, C., et al.: Realistic adversarial data augmentation for MR image segmentation. In: Martel, A.L., et al. (eds.) MICCAI 2020. LNCS, vol. 12261, pp. 667–677. Springer, Cham (2020). https://doi.org/10.1007/978-3-030-59710-8_65

4. Chen, Y., Wei, C., Kumar, A., Ma, T.: Self-training avoids using spurious features under domain shift. CoRR abs/2006.10032 (2020). https://arxiv.org/abs/2006.10032

5. Devries, T., Taylor, G.W.: Improved regularization of convolutional neural networks with cutout. CoRR abs/1708.04552 (2017). http://arxiv.org/abs/1708.04552

6. Ganin, Y., et al.: Domain-adversarial training of neural networks. J. Mach. Learn. Res. **17**(59), 1–35 (2016). http://jmlr.org/papers/v17/15-239.html

7. Hoffman, J., et al.: CyCADA: cycle-consistent adversarial domain adaptation. In: Dy, J., Krause, A. (eds.) Proceedings of the 35th International Conference on Machine Learning. Proceedings of Machine Learning Research, vol. 80, pp. 1989–1998. PMLR, 10–15 Jul 2018. https://proceedings.mlr.press/v80/hoffman18a.html

8. Huang, X., Belongie, S.J.: Arbitrary style transfer in real-time with adaptive instance normalization. CoRR abs/1703.06868 (2017). http://arxiv.org/abs/1703.06868

9. Kavur, A.E., et al.: CHAOS challenge - combined (CT-MR) healthy abdominal organ segmentation. Med. Image Anal. **69**, 101950 (2021). https://doi.org/10.1016/j.media.2020.101950, http://www.sciencedirect.com/science/article/pii/S1361841520303145

10. Kingma, D.P., Ba, J.: Adam: a method for stochastic optimization (2014). http://arxiv.org/abs/1412.6980, cite arxiv:1412.6980Comment: Published as a conference paper at the 3rd International Conference for Learning Representations, San Diego (2015)

11. Li, Y., Tian, X., Gong, M., Liu, Y., Liu, T., Zhang, K., Tao, D.: Deep domain generalization via conditional invariant adversarial networks. In: Proceedings of the European Conference on Computer Vision (ECCV), September 2018

12. Liu, Q., Dou, Q., Heng, P.: Shape-aware meta-learning for generalizing prostate MRI segmentation to unseen domains. CoRR abs/2007.02035 (2020). https://arxiv.org/abs/2007.02035

13. Milletari, F., Navab, N., Ahmadi, S.A.: V-net: fully convolutional neural networks for volumetric medical image segmentation. In: 2016 Fourth International Conference on 3D Vision (3DV), pp. 565–571 (2016)

14. van den Oord, A., Li, Y., Vinyals, O.: Representation learning with contrastive predictive coding. CoRR abs/1807.03748 (2018). http://arxiv.org/abs/1807.03748

15. Ouyang, C., et al.: Causality-inspired single-source domain generalization for medical image segmentation. CoRR abs/2111.12525 (2021). https://arxiv.org/abs/2111.12525

16. Park, T., Efros, A.A., Zhang, R., Zhu, J.-Y.: Contrastive learning for unpaired image-to-image translation. In: Vedaldi, A., Bischof, H., Brox, T., Frahm, J.-M. (eds.) ECCV 2020. LNCS, vol. 12354, pp. 319–345. Springer, Cham (2020). https://doi.org/10.1007/978-3-030-58545-7_19

17. Qiao, F., Zhao, L., Peng, X.: Learning to learn single domain generalization. In: IEEE Conference on Computer Vision and Pattern Recognition (CVPR), pp. 12556–12565 (2020)

18. Ronneberger, O., Fischer, P., Brox, T.: U-Net: convolutional networks for biomedical image segmentation. In: Navab, N., Hornegger, J., Wells, W.M., Frangi, A.F. (eds.) MICCAI 2015. LNCS, vol. 9351, pp. 234–241. Springer, Cham (2015). https://doi.org/10.1007/978-3-319-24574-4_28

19. Tan, M., Le, Q.: EfficientNet: rethinking model scaling for convolutional neural networks. In: Chaudhuri, K., Salakhutdinov, R. (eds.) Proceedings of the 36th International Conference on Machine Learning. Proceedings of Machine Learning Research, vol. 97, pp. 6105–6114. PMLR, 09–15 June 2019. https://proceedings.mlr.press/v97/tan19a.html

20. Tang, Z., Peng, X., Li, T., Zhu, Y., Metaxas, D.: Adatransform: adaptive data transformation. In: 2019 IEEE/CVF International Conference on Computer Vision (ICCV), pp. 2998–3006 (2019). https://doi.org/10.1109/ICCV.2019.00309

21. Wang, Z., Luo, Y., Qiu, R., Huang, Z., Baktashmotlagh, M.: Learning to diversify for single domain generalization. In: Proceedings of the IEEE/CVF International Conference on Computer Vision (ICCV), pp. 834–843, October 2021

22. Xu, Z., Liu, D., Yang, J., Raffel, C., Niethammer, M.: Robust and generalizable visual representation learning via random convolutions. arXiv preprint arXiv:2007.13003 (2020)

23. Zhang, H., Cisse, M., Dauphin, Y.N., Lopez-Paz, D.: mixup: Beyond empirical risk minimization. In: International Conference on Learning Representations (2018). https://openreview.net/forum?id=r1Ddp1-Rb

24. Zhu, Q., Du, B., Yan, P.: Boundary-weighted domain adaptive neural network for prostate MR image segmentation. CoRR abs/1902.08128 (2019). http://arxiv.org/abs/1902.08128

Delving into Local Features for Open-Set Domain Adaptation in Fundus Image Analysis

Yi Zhou[1(✉)], Shaochen Bai[1], Tao Zhou[2], Yu Zhang[1], and Huazhu Fu[3]

[1] School of Computer Science and Engineering, Southeast University, Nanjing, China
yizhou.szcn@gmail.com
[2] Nanjing University of Science and Technology, Nanjing, China
[3] Agency for Science, Technology and Research (A*STAR), Singapore, Singapore

Abstract. Unsupervised domain adaptation (UDA) has received significant attention in medical image analysis when labels are only available for the source domain data but not for the target domain. Previous UDA methods mainly focused on the closed-set scenario, assuming that only the domain distribution shifts across domains while the label space is the same. However, in practice of medical imaging, the disease categories of training data in source domain are usually limited, and the open-world target domain data may have many *unknown* classes private to the source domain. Thus, open-set domain adaptation (OSDA) has great potential in this area. In this paper, we explore the OSDA problem by delving into local features for fundus disease recognition. We propose a collaborative regional clustering and alignment method to identify the common local feature patterns which are category-agnostic. Then, a cluster-aware contrastive adaptation loss is introduced to adapt the distributions based on the common local features. We also construct the first fundus image benchmark for OSDA to evaluate our methods and carry out extensive experiments for comparison. It shows that our model achieves consistent improvements over the state-of-the-art methods.

1 Introduction

Advanced vision and learning models have been largely investigated for early diagnosis of eye diseases [8,10,17,29,30]. In the past decade, efforts have been paid to develop models based on fundus images provided by separate datasets with limited disease categories. However, in the real world scenario, various data collected from multiple countries or sites usually have different domain distributions and disease categories (*i.e.* label space). Thus, how to build domain adaption models to simultaneously deal with domain distribution shift and label space shift is a crucial problem to develop models achieving good generalization performance, which is hugely neglected by the community.

Unsupervised domain adaptation (UDA) has attracted large attention in medical imaging [7]. Previous UDA methods [4,5,11,13,24,26] only address the

L. Wang (Eds.): MICCAI 2022, LNCS 13437, pp. 682–692, 2022.
https://doi.org/10.1007/978-3-031-16449-1_65

domain distribution shift across different datasets with the same disease category space (termed closed-set DA). Once the source and target domain data have different label spaces, they cannot perform properly since a huge negative transfer will be introduced by the misalignment of label spaces. Recently, in the computer vision community, open-set domain adaptation (OSDA [22]), has been proposed to investigate the label space shift in UDA tasks, where the target domain has private image classes (or termed *unknown* classes), besides the common classes same with the source domain. Previous OSDA methods [3,12,18,23,25,27] focus on identifying samples of common classes between the two domains and then performing domain adaptation on the shared class set, while disregarding the *unknown* classes. More recently, researchers attempt to improve robustness by exploiting latent structures of *unknown* classes. [21] proposes an additional branch that learns underlying information through the category-agnostic clustering in the target domain. [16] takes a further step by performing self-adapted category-level clustering and associating the common cluster centers.

Although existing OSDA methods obtain promising results on natural images, we did not satisfactorily apply them into fundus images due to **two main reasons**. First, although fundus images of different diseases have different class labels, they usually have similar global structures with uniformed appearances, such as the optic disks, vessels, and even common lesion patterns appeared in different diseases (In natural images, different object classes usually have significant visual appearance discrepancy). In other words, it is difficult to identify the common classes between the source and target domain in the global feature space due to such small inter-class variations. We also believe this challenge exists in most medical imaging areas, such as chest X-ray and CT. Second, because of the small inter-class variation, besides the samples of common classes, most of the regions without private lesion appearance of *unknown* classes in the target domain can be also exploited for domain adaptation, but none of previous works discover this point. Therefore, our motivation is to delve into detailed local regions whose features can be more precisely adopted for OSDA.

In this paper, rather than directly separating images of common classes from private classes in the target domain, we aim to cluster and align the common local features for OSDA. Specifically, we first propose a collaborative regional clustering and alignment (CRCA) method to identify the common local feature clusters which are category-agnostic. To mitigate the computational complexity and improve the performance of clustering, we introduce an informative region selection (IRS) operation on the dense local features before the clustering. Once the identified common local features are grouped into clusters, a cluster-aware contrastive adaptation (CCA) method is proposed to enforce the local features of aligned clusters across domains to remain close-by, while pushing those of the misaligned clusters far apart. Therefore, the proposed framework learns domain-invariant representations by exploring local features to guide the distribution adaptation. **The main contributions are highlighted as follows: 1)** We investigate the OSDA problem in the medical imaging scenario, specifically for fundus disease recognition, for the first time. We analyze the challenges in terms of the characteristics of medical images. **2)** A novel method is proposed to perform positive transfer more precisely while avoiding negative transfer across domains, through delving into local features. **3)** We construct the first fundus OSDA benchmark to study retinal disease recognition. Our proposed method achieves state-of-the-art performance compared to previous OSDA methods.

2 Proposed Methods

2.1 Problem Formulation

To address domain adaptation when both the domain space and label space shift, we propose to delve into local features and perform region-level clustering and adaptation between the source and target domain. An overview of our framework is illustrated in Fig. 1. Given the labeled samples from the source domain $\mathcal{D}^s = \{(X^s, Y^s)\}$ and the unlabeled samples from the target domain $\mathcal{D}^t = \{(X^t)\}$, we define the common label space $\mathcal{C}$ which is shared between $\mathcal{D}^s$ and $\mathcal{D}^t$ containing N common classes. In the OSDA setting, the source domain label space $\mathcal{C}^s = \mathcal{C}$, while additional *unknown* classes $\mathcal{C}^{unk}$ only exist in the target domain label space $\mathcal{C}^t = \mathcal{C} \cup \mathcal{C}^{unk}$, and make up of $N+1$ classes in total.

2.2 Region-Level Clustering and Adaptation

Rethinking Image-Level Clustering for OSDA. Part of our idea is originally inspired by DCC [16], which performs category-level (*i.e.* image-level) clustering based on global image features for OSDA. It proposes cycle-consistent matching to associate common cluster centers (*i.e.* common classes) across domains. Specifically, once two clusters in different domains act as each other's nearest center, this pair of clusters is considered as common clusters, whereas the unmatched clusters are rejected as *unknown* outliers. Moreover, it further computes the sample-level consensus to optimally search the number K of clusters and promote the effectiveness of cycle-consistent matching. However, this category-level feature clustering only works when there is a clear gap between a pair of category clusters, which is not suitable for fundus images. On the other hand, due to the uniformed global pattern of fundus images, many regions are similar even across the common and *unknown* classes, which motivates us to explore the knowledge transferability through learning region-level feature clusters.

Collaborative Regional Clustering and Alignment. We improve the clustering based idea by proposing category-agnostic regional clustering, and collaboratively optimize and align clusters on all sample across domains. In other words, we aim to match common local regions rather than identifying common classes, which is more fine-grained for learning domain-invariant features. Specifically,

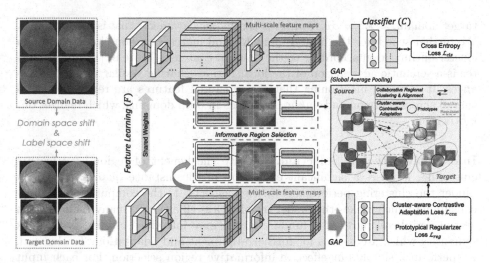

Fig. 1. An overview of our method. The feature learning part is adopted to extract multi-scale feature maps first, followed by an informative local region selection step. Then, we leverage the proposed collaborative regional clustering and alignment method to associate common clusters between the two domains. Finally, the cluster-aware contrastive adaptation loss is utilized to facilitate the domain adaptation.

for each input image $x \in \{X^s, X^t\}$, the local region features $f(x)$ are extracted from the $W \times H$ sized feature maps of a convolutional (*conv*) layer:

$$f(x) = \{F_{(w,h)}(x)\}_{w=1,h=1}^{W,H}, \tag{1}$$

where (w, h) denotes the spatial location of a grid and $F_{(w,h)}(x)$ is the normalized feature vector of the grid corresponding to the local representation of a patch on the original image. Ideally, these local segmented patches containing either lesion appearance or retinal tissue, are descriptive to contribute to more fine-grained clustering. However, various local patterns usually appear in different sizes (*e.g.* microaneurysms appear in tiny size, while optic disks appear in moderate size), clustering only on single-scale feature maps (*e.g.* the last *conv* layer) is coarse and less sensitive to small-size patterns due to oversized segments. Thus, we adopt multi-scale feature maps from the last l *conv* layers, and denote them as $f_1(x), f_2(x), ..., f_l(x)$. In this way, local features from different scales can be extracted and clustered, leading to multi-scale cluster centers. We apply the procedure on all the samples from the source domain and target domain, respectively, and obtain the region-level feature sets, denoted as $\mathcal{F}^s = [f_1(X^s), f_2(X^s), ..., f_l(X^s)]$ and $\mathcal{F}^t = [f_1(X^t), f_2(X^t), ..., f_l(X^t)]$.

Before clustering, informative region selection (next subsection) is operated to slim the $\mathcal{F}$ to $\hat{\mathcal{F}}$. Then, given the selected local feature sets $\hat{\mathcal{F}}^s$ and $\hat{\mathcal{F}}^t$, we perform K-means clustering on both domains, respectively. Since the target domain containing private *unknown* classes may appear more unseen patterns, we adjust the relationship of the initial cluster number between the source and target domain, as $K^t = \beta K^s$, where β is a hyper-parameter that is larger than 1. We further search for the best K^s and K^t and match common clusters through cluster optimization according to the domain consensus score in [16]. Each cluster is a semantic aggregation of the local regions sharing similar visual appearances. It is worth mentioning that the clustered features are region-level and shared across common and *unknown* classes in both domains, which realizes our motivation.

Informative Region Selection. Despite the benefit of region-level feature clustering, it raises a significant expansion of the instance quantity and may hinder the clustering performance in two aspects: 1) the computation is time-consuming which makes the model training less efficient; 2) many regions, such as the peripheral parts of a fundus image, may introduce clusters with weak semantics which are useless for further alignment and adaptation. To this end, we perform a simple but effective informative region selection. For each input x, we first compute the importance weight map $I^c_{(w,h)}(x)$ of each class c over location (w, h) using CAM [28]. Then, all the maps are merged along the class, by:

$$I_{(w,h)}(x) = \max_c I^c_{(w,h)}(x), \tag{2}$$

so that we can squeeze all these discriminative maps into a category-agnostic informative map. To complete the selection, we filter the standardized $I_{(w,h)}(x)$ with a threshold α and keep the local features by:

$$\hat{f}(x) = \{f(x) \quad if \ I_{(w,h)}(x) > \alpha\}_{w=1,h=1}^{W,H}. \tag{3}$$

In this way, the selected informative features can contribute to the distribution alignment and adaptation more effectively, while the remaining regions are left out. Please note that for different-scale feature maps, we simply resize the obtained importance weight map into corresponding scales for selection.

Cluster-Aware Contrastive Adaptation. Once the local features are grouped into clusters and aligned across domains, we propose a cluster-aware contrastive adaptation loss $\mathcal{L}_{cca}$ to facilitate the domain adaptation over the aligned common local features in a cluster-aware style. Specifically, for each cluster $i \in K^s$ and $j \in K^t$, we introduce the prototype vector p^s_i and p^t_j by computing the mean of feature vectors of the corresponding cluster. Then, we minimize $\mathcal{L}_{cca}$ as follows:

$$\mathcal{L}_{cca} = \sum_{i=1}^{K^s} \sum_{j=1}^{K^t} \ell(f^t_j, p^s_i) + \ell(f^s_i, p^t_j), \text{ where,} \tag{4}$$

$$\ell(f,p) = \begin{cases} d(f,p) & i \text{ and } j \text{ are aligned} \\ \max(0, \Delta - d(f,p)) & \text{otherwise,} \end{cases}$$

where $d(\cdot, \cdot)$ can be any distance function (we simply adopt L2 distance). Minimizing $\mathcal{L}_{cca}$ forces to learn a local feature vector f to have 0 distance to its aligned cross-domain cluster prototype p, and a distance greater than a margin Δ for mis-aligned cases. Moreover, as the end-to-end model training, the cluster prototype sets $\mathcal{P}$ of both domains are gradually updated, since their constituent local features evolve during training. To this end, we compute the new prototype sets $\mathcal{P}_{new}$ after every T_p iterations. Then, the existing $\mathcal{P}$ are updated by $\mathcal{P} \leftarrow \eta\mathcal{P} + (1 - \eta)\mathcal{P}_{new}$, where η is a momentum parameter to add a fraction of the previous cluster prototypes to the current ones. This CCA loss is adopted for optimization together with the standard classification entropy losses.

2.3   Optimization and Inference

Overall, we adopt three losses to optimize the whole model. The basic cross-entropy loss $\mathcal{L}_{cls}$ is used to train the classifier using the labeled data from the source domain. The cluster-aware contrastive adaptation loss $\mathcal{L}_{cca}$ aims to adapt the distributions based on the aligned common local clusters. Moreover, an auxiliary prototypical regularizer loss $\mathcal{L}_{reg}$ from [16] is imposed on target domain features to enhance clusters with more discriminativeness. Thus, the model is jointly optimized by minimizing the following objective:

$$\mathcal{L} = \mathcal{L}_{cls} + \lambda_1\mathcal{L}_{cca} + \lambda_2\mathcal{L}_{reg}, \tag{5}$$

where λ_1 and λ_2 are set to 0.1 and 0.01 to weight different loss terms.

In the inference phase, we simply apply the threshold determination to separate *known* and *unknown* samples by assigning a fixed value to the probability of the *unknown* class after the final Softmax classifier. If the probability of any *known* class is higher than the threshold, the sample is considered to belong to the common class, otherwise, it is considered as the private class.

3   Experiments and Results

3.1   Dataset Setup and Implementation Details

To evaluate OSDA methods for fundus disease recognition, we adopt three public datasets (TAOP [2], ODIR [1], and RFMiD [20]) to build two source and target domain pairs. TAOP, which contains 3,297 images with 5 retinal diseases, is set as the source domain data. ODIR and RFMiD, which consist of 6,576 and 2,451 images with more disease categories (covering 5 classes in source), are set as the target domain data, respectively. We call these two open-set fundus image benchmarks, OSF-T2O and OSF-T2R. Since the data from three datasets were collected from different hospitals, whose images were captured by various fundus cameras on people from different regions, noticeable shifts exist across domains. We further enlarge the domain gap by applying contrast and brightness adjustment to better evaluate the performance.

Following [3], we evaluate the OSDA performance based on four metrics, i.e. OS*, OS, UNK, and HM. OS* and OS represent the mean accuracy over common classes and the mean accuracy over all classes, respectively. UNK represents the accuracy to identify *unknown* classes, and HM is a harmonic metric calculated from OS* and UNK as a comprehensive evaluation.

We adopt ResNet-50 [9] pre-trained on ImageNet [15] as the backbone. Nesterov momentum SGD is adopted as the optimizer, with the momentum and weight decay set to 0.9 and 5×10^4. The input image size is 224×224. During training, the base learning rate and mini-batch size are set to 5×10^{-2} and 36, respectively. In fundus image analysis, we extract the last 2 *Conv* layer feature maps for our multi-scale strategy. For α and β, we fix them to 0.7 and 1.5. Moreover, we select the best-performed T_p and η as 1000 and 0.95 through experiments.

Table 1. Result comparison (%) of state-of-the-art OSDA methods and ablation studies. "**w/o**" and "**w**" are the abbreviation of without and with, respectively.

Datasets	OSF-T2O				OSF-T2R			
Methods	OS*	OS	UNK	HM	OS*	OS	UNK	HM
DANN [6]	46.859	48.702	57.918	51.805	53.057	53.403	55.132	54.075
OSBP [23]	48.276	50.482	61.510	54.095	52.566	54.519	64.286	57.838
ROS [3]	38.558	40.874	52.456	44.445	41.150	39.196	29.426	34.149
DAMC [25]	45.269	45.256	45.192	45.231	43.067	43.131	43.454	43.259
UAN [27]	47.805	47.923	48.513	48.156	48.846	50.037	55.993	52.176
DCC [16]	38.704	40.289	48.215	42.939	45.244	46.093	50.338	47.656
Source-only	38.075	40.825	55.172	45.056	50.067	51.116	56.361	53.028
w/o MS	47.081	48.212	53.867	50.246	47.397	50.356	65.151	54.873
w/o IRS	47.478	49.985	62.517	53.970	46.100	49.337	65.519	54.120
w/o $\mathcal{L}_{reg}$	50.125	51.897	60.755	54.931	52.337	54.052	62.077	56.857
Ours **w** CDD [14]	51.437	53.646	64.691	57.307	53.762	56.162	68.162	60.111
Ours	**52.538**	**54.891**	**65.945**	**58.483**	**55.004**	**57.391**	**69.874**	**61.554**

3.2   Comparison with State-of-the-Arts

We first compare our method with six state-of-the-art methods. As illustrated in Table 1, our model consistently outperforms other methods by clear gaps in both benchmarks, showing satisfactory improvements overall. DANN [6] is a classic closed-set DA method that aims to align the distribution across domains only, without any attempt to separate classes. Such a simple design favors its performance in medical images with small inter-class discrepancies. OSBP [23] achieves more than a 2% increase in HM, compared to DANN. However, for recent OSDA methods, ROS [3] utilizes self-supervision through image rotation to train a binary classifier to identify *unknown* samples, and DAMC [25] proposes a non-adversarial domain discriminator. These methods work well on natural vision benchmarks since there exist clear gaps across different classes, but fail for medical images due to the small inter-class discrepancy in global feature spaces, showing noticeable performance decreases. DCC [16] aims to simultaneously train category-level clusters and separate the *unknown* samples, but suffers a significant performance drop due to its poor image-level clustering, which further illustrates the superiority of our region-level clustering.

3.3   Ablation Studies

To further evaluate proposed modules, five more baselines are compared in Table 1. **Source-only** simply adopts the network trained on the source domain data with an additional threshold classifier to predict *known* and *unknown* classes. **w/o MS** (*i.e.* single-scale) represents the model which excludes multi-scale features but only selects the regional features from the last *Conv* layer feature maps. We also detach the informative region selection module and prototypical regularizer loss, as **w/o IRS** and **w/o** $\mathcal{L}_{reg}$, to evaluate their corresponding effectiveness, respectively. Moreover, to evaluate the proposed CCA loss, we substitute it with another SOTA adaptation loss, CDD [14], as **Ours w CDD**.

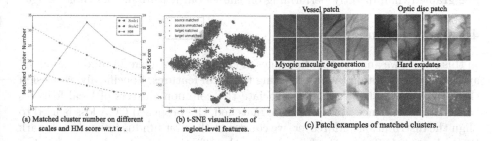

(a) Matched cluster number on different scales and HM score w.r.t α .

(b) t-SNE visualization of region-level features.

(c) Patch examples of matched clusters.

Fig. 2. Illustration of the effectiveness of local feature clustering performance.

As shown in Table 1 on OSF-T2O, we observed that the image-level clustering using global features even obtains lower scores than the source-only baseline due to the poor clustering performance. w/o MS further validates clustering on smaller regions is much better to associate common local patterns. It shows that the single-scale regional clustering already obtains an 7.31% increase in HM, compared to image-level clustering, and the multi-scale strategy achieves an extra 8.24% improvement. Moreover, the prototypical regularizer $\mathcal{L}_{reg}$ obtains 3.55% increase by enhancing clusters with more discriminativeness. Our proposed CCA is more effective than the CDD [14] method, showing an increase of 1.18%, since the learning of local representations and cluster prototypes is gradually updated. The results on OSF-T2R consistently show the superiority.

The IRS plays a significant role for the collaborative regional clustering. Once we detach it and perform clustering using all the local feature vectors, the computation complexity is large and the final HM score decreases by 4.51%, since many useless regions (*e.g.* peripheral areas) will be included to form a few noisy clusters which severely obstruct the following alignment and adaptation. Moreover, as illustrated in Fig. 2(a), the number of matched local feature clusters in both scales decreases with the increasing α. The best HM score is achieved when α reaches 0.7. The performance slightly drops when $\alpha \geq 0.8$ due to the over-filtering. We also visualize the learned local feature space using t-SNE [19] in Fig. 2(b). The matched features from both domains aggregating together prove the effectiveness of our local feature distribution alignment. Moreover, as to the computational cost of clustering in training, we perform the clustering every 300 epochs rather than running in every epoch which is unnecessary. Thus, the clustering time over all training time only takes approximately 10%.

Apart from the quantitative results, we also demonstrate the performance of the collaborative regional clustering and alignment by visualizing the image patch examples corresponding to the local features in the aligned clusters. As illustrated in Fig. 2(c) (In each pair, the left and right part shows examples from the source and target domain, respectively.), we clearly observed that the aligned clusters across domains have great semantic meanings, which capture appearances such as tissue structures and disease lesions. Thus, doing domain adaptation on these common local features is category-agnostic and much more fine-grained than directly performing on the global feature of the whole image.

4   Conclusion

In this paper, we analyze the challenges of the OSDA problem in fundus imaging, and propose a solution by investigating the local features. Specifically, we design the CRCA method with IRS to associate the common clusters of local features across domains. Once the clusters are matched, the proposed CCA loss is used to align the distributions. Moreover, we construct the first fundus image benchmark for evaluating OSDA methods, and prove the effectiveness of each component of our method and show clear improvements compared to other approaches.

Acknowledgement. This work was partially supported by the National Natural Science Foundation of China (Grants No 62106043), the Natural Science Foundation of Jiangsu Province (Grants No BK20210225), and the AME Programmatic Fund (A20H4b0141).

References

1. International competition on ocular disease intelligent recognition (2019). https://odir2019.grand-challenge.org
2. Tencent miying artificial intelligence competition for medical imaging (2021). https://contest.taop.qq.com/
3. Bucci, S., Loghmani, M.R., Tommasi, T.: On the effectiveness of image rotation for open set domain adaptation. In: Vedaldi, A., Bischof, H., Brox, T., Frahm, J.-M. (eds.) ECCV 2020. LNCS, vol. 12361, pp. 422–438. Springer, Cham (2020). https://doi.org/10.1007/978-3-030-58517-4_25

4. Chen, C., Dou, Q., Chen, H., Qin, J., Heng, P.A.: Synergistic image and feature adaptation: towards cross-modality domain adaptation for medical image segmentation. In: AAAI, vol. 33, pp. 865–872 (2019)
5. Dou, Q., et al.: PnP-AdaNet: plug-and-play adversarial domain adaptation network at unpaired cross-modality cardiac segmentation. IEEE Access **7**, 99065–99076 (2019)
6. Ganin, Y., et al.: Domain-adversarial training of neural networks. J. Mach. Learn. Res. **17**(1) (2016). ISSN 2096-2030
7. Guan, H., Liu, M.: Domain adaptation for medical image analysis: a survey. arXiv preprint arXiv:2102.09508 (2021)
8. Gulshan, V., et al.: Development and validation of a deep learning algorithm for detection of diabetic retinopathy in retinal fundus photographs. JAMA **316**(22), 2402–2410 (2016)
9. He, K., Zhang, X., Ren, S., Sun, J.: Deep residual learning for image recognition. In: CVPR, pp. 770–778. IEEE (2016)
10. He, X., Zhou, Y., Wang, B., Cui, S., Shao, L.: DME-Net: diabetic macular edema grading by auxiliary task learning. In: Shen, D., et al. (eds.) MICCAI 2019. LNCS, vol. 11764, pp. 788–796. Springer, Cham (2019). https://doi.org/10.1007/978-3-030-32239-7_87
11. Javanmardi, M., Tasdizen, T.: Domain adaptation for biomedical image segmentation using adversarial training. In: ISBI, pp. 554–558. IEEE (2018)
12. Jing, T., Liu, H., Ding, Z.: Towards novel target discovery through open-set domain adaptation. In: ICCV, pp. 9322–9331. IEEE (2021)
13. Kamnitsas, K., et al.: Unsupervised domain adaptation in brain lesion segmentation with adversarial networks. In: Niethammer, M., et al. (eds.) IPMI 2017. LNCS, vol. 10265, pp. 597–609. Springer, Cham (2017). https://doi.org/10.1007/978-3-319-59050-9_47
14. Kang, G., Jiang, L., Yang, Y., Hauptmann, A.G.: Contrastive adaptation network for unsupervised domain adaptation. In: CVPR, pp. 4893–4902. IEEE (2019)
15. Krizhevsky, A., Sutskever, I., Hinton, G.E.: Imagenet classification with deep convolutional neural networks. In: NeurIPS, vol. 25 (2012)
16. Li, G., Kang, G., Zhu, Y., Wei, Y., Yang, Y.: Domain consensus clustering for universal domain adaptation. In: CVPR, pp. 9757–9766. IEEE (2021)
17. Li, T., et al.: Applications of deep learning in fundus images: a review. Med. Image Anal. **69**, 101971 (2021)
18. Liu, H., Cao, Z., Long, M., Wang, J., Yang, Q.: Separate to adapt: open set domain adaptation via progressive separation. In: CVPR, pp. 2927–2936. IEEE (2019)
19. Van der Maaten, L., Hinton, G.: Visualizing data using t-SNE. J. Mach. Learn. Res. **9**(11) (2008)
20. Pachade, S., et al.: Retinal fundus multi-disease image dataset (RFMID): a dataset for multi-disease detection research. Data **6**(2), 14 (2021)
21. Pan, Y., Yao, T., Li, Y., Ngo, C.W., Mei, T.: Exploring category-agnostic clusters for open-set domain adaptation. In: CVPR, pp. 13867–13875. IEEE (2020)
22. Panareda Busto, P., Gall, J.: Open set domain adaptation. In: ICCV, pp. 754–763. IEEE (2017)
23. Saito, K., Yamamoto, S., Ushiku, Y., Harada, T.: Open set domain adaptation by backpropagation. In: ECCV, pp. 153–168 (2018)
24. Shen, Y., et al.: Domain-invariant interpretable fundus image quality assessment. Med. Image Anal. **61**, 101654 (2020)

25. Shermin, T., Lu, G., Teng, S.W., Murshed, M., Sohel, F.: Adversarial network with multiple classifiers for open set domain adaptation. IEEE Trans. Multimedia **23**, 2732–2744 (2020)
26. Yang, J., Dvornek, N.C., Zhang, F., Chapiro, J., Lin, M.D., Duncan, J.S.: Unsupervised domain adaptation via disentangled representations: application to cross-modality liver segmentation. In: Shen, D., et al. (eds.) MICCAI 2019. LNCS, vol. 11765, pp. 255–263. Springer, Cham (2019). https://doi.org/10.1007/978-3-030-32245-8_29
27. You, K., Long, M., Cao, Z., Wang, J., Jordan, M.I.: Universal domain adaptation. In: CVPR, pp. 2720–2729 (2019)
28. Zhou, B., Khosla, A., Lapedriza, A., Oliva, A., Torralba, A.: Learning deep features for discriminative localization. In: CVPR, pp. 2921–2929. IEEE (2016)
29. Zhou, Y., et al.: Collaborative learning of semi-supervised segmentation and classification for medical images. In: CVPR, pp. 2079–2088. IEEE (2019)
30. Zhou, Y., Wang, B., Huang, L., Cui, S., Shao, L.: A benchmark for studying diabetic retinopathy: segmentation, grading, and transferability. IEEE Trans. Med. Imaging **40**(3), 818–828 (2020)

Estimating Model Performance Under Domain Shifts with Class-Specific Confidence Scores

Zeju Li[1(✉)], Konstantinos Kamnitsas[1,2,3], Mobarakol Islam[1], Chen Chen[1], and Ben Glocker[1]

[1] BioMedIA Group, Department of Computing, Imperial College London, London, UK
zeju.li18@imperial.ac.uk
[2] Institute of Biomedical Engineering, Department of Engineering Science, University of Oxford, Oxford, UK
[3] School of Computer Science, University of Birmingham, Birmingham, UK

Abstract. Machine learning models are typically deployed in a test setting that differs from the training setting, potentially leading to decreased model performance because of domain shift. If we could estimate the performance that a pre-trained model would achieve on data from a specific deployment setting, for example a certain clinic, we could judge whether the model could safely be deployed or if its performance degrades unacceptably on the specific data. Existing approaches estimate this based on the confidence of predictions made on unlabeled test data from the deployment's domain. We find existing methods struggle with data that present class imbalance, because the methods used to calibrate confidence do not account for bias induced by class imbalance, consequently failing to estimate class-wise accuracy. Here, we introduce class-wise calibration within the framework of performance estimation for imbalanced datasets. Specifically, we derive class-specific modifications of state-of-the-art confidence-based model evaluation methods including temperature scaling (TS), difference of confidences (DoC), and average thresholded confidence (ATC). We also extend the methods to estimate Dice similarity coefficient (DSC) in image segmentation. We conduct experiments on four tasks and find the proposed modifications consistently improve the estimation accuracy for imbalanced datasets. Our methods improve accuracy estimation by 18% in classification under natural domain shifts, and double the estimation accuracy on segmentation tasks, when compared with prior methods (Code is available at https://github.com/ZerojumpLine/ModelEvaluationUnderClassImbalance).

Supplementary Information The online version contains supplementary material available at https://doi.org/10.1007/978-3-031-16449-1_66.

1 Introduction

Based on the independently and identically distributed (IID) assumption, one can estimate the model performance with labeled validation data which is collected from the same source domain as the training data. However, in many real-world scenarios where test data is typically sampled from a different distribution, the IID assumption does not hold anymore and the model's performance tends to degrade because of domain shifts [1]. It is difficult for the practitioner to assess the model reliability on the target domain as labeled data for a new test domain is usually not available. It would be of great practical value to develop a tool to estimate the performance of a trained model on an unseen test domain without access to ground truth [7]. For example, for a skin lesion classification model which is trained with training data from the source domain, we can decide whether to deploy it to a target domain based on the predicted accuracy of a few unlabeled test images. In this study we aim to estimate the model performance by only making use of unlabeled test data from the target domain.

Many methods have been proposed to tackle this problem and they can be categorized into three groups. The first line of research is based on the calculation of a distance metric between the training and test datasets [4,6]. The model performance can be obtained by re-weighting the validation accuracy [4] or training a regression model [6]. The second type of methods is based on reverse testing which trains a new reverse classifier using test images and predicted labels as pseudo ground truth [9,34]. The model performance is then determined by the prediction accuracy of the reverse classifier on the validation data. The third group of methods is based on model confidence scores [10–12]. The accuracy of a trained model can be estimated with the Average Confidence (AC) which is the mean value of the maximum softmax confidence on the target data. Previous studies have shown that confidence-based methods show better performance than the first two methods [8,10] and are efficient as they do not need to train additional models. We focus on confidence-based methods in this study.

The prediction results by AC are likely to be higher than the real accuracy as modern neural networks are known to be over-confident [12]. A few methods, such as Temperature Scaling (TS) [12], Difference of Confidences (DoC) [11], and Average Thresholded Confidence (ATC) [10], have been proposed to calibrate the probability of a trained model with validation data with the expectation to generalize well to unseen test data, as depicted in Fig. 1(a). We argue those methods do not generalize well to imbalanced datasets, which are common in medical image classification and segmentation. Specifically, previous work found that a model trained with imbalanced data is prone to predict a test sample from a minority class as a majority class, but not vice versa [21]. As a result, such models tend to have high classification accuracy but are under-confident on samples predicted as the minority classes, with opposite behaviour for the majority classes, as illustrated in Fig. 1(b). Current post-training calibration methods adopt the same parameter to adjust the probability, which would increase the estimation error for the minority class samples and thus, does not generalize well, as shown in Fig. 1(c). We introduce class-specific parameters into the performance estima-

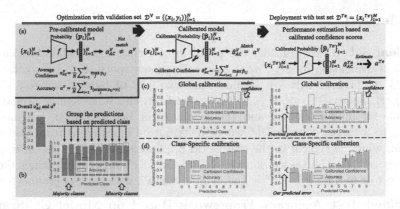

Fig. 1. Effect of class imbalance on confidence-based model evaluation methods. (a) The pipeline for model evaluation [10,11], which estimates the accuracy under domain shifts with Average Confidence (AC) after obtaining calibrated probabilities $\hat{p}_i$. (b) The confidence scores are biased towards the majority classes. (c) Prior methods reduce all probabilities in the same manner to match overall AC with the overall accuracy, but fail to calibrate class-wise probability. (d) Our proposed method with class-specific calibration parameters successfully reduces the bias and significantly increases the model estimation accuracy. The shown example is for imbalanced CIFAR-10.

tion framework using adaptive optimization, alleviating the confidence bias of neural networks against minor classes, as demonstrated in Fig. 1(d). It should be noted that our method differs from other TS methods with class-wise parameters such as vector scaling (VS) [12], class-distribution-aware TS [17] and normalized calibration (NORCAL) [29] as 1) our parameters are optimized so that class-specific confidence matches class-specific accuracy, whereas their parameters are optimized such that overall-confidence matches overall-accuracy; 2) we determine the calibrated class based on prediction instead of ground truth, therefore the model predictions would not be affected by the calibration process; 3) we apply the proposed methods to model evaluation; 4) we also consider other calibration methods than TS.

This study sheds new light on the problem of model performance estimation under domain shifts in the absence of ground truth with the following contributions: 1) We find existing solutions encounter difficulties in real-world datasets as they do not consider the confidence bias caused by class imbalance; 2) We propose class-specific variants for three state-of-the-art approaches including TS, DoC and ATC to obtain class-wise calibration; 3) We extend the confidence-based model performance framework to segmentation and prediction of Dice similarity coefficient (DSC); 4) Extensive experiments performed on four tasks with a total of 310 test settings show that the proposed methods can consistently and significantly improve the estimation accuracy compared to prior methods. Our methods can reduce the estimation error by 18% for skin lesion classification under natural domain shifts and double the estimation accuracy in segmentation.

2   Method

Problem Setup. We consider the problem of c-class image classification. We train a classification model f that maps the image space $\mathcal{X} \subset \mathbb{R}^d$ to the label space $\mathcal{Y} = \{1, 2, \ldots, c\}$. We assume that two sets of labeled images: $\mathcal{D}^{Tr} = \{(\boldsymbol{x}_i, y_i)\}_{i=1}^I$ and $\mathcal{D}^V = \{(\boldsymbol{x}_i, y_i)\}_{i=1}^N$ drawn from the *same* source domain are available for model training and validation. After being trained, the model is applied to a target domain $\mathcal{D}^{Te}$ whose distribution is *different* from the source domain, which is typical in the real-world applications. We aim to estimate its accuracy α^{Te} on the target domain to support decision making when there is only a set of test images $\mathcal{D}^{Te} = \{\boldsymbol{x}_i^{Te}\}_{i=1}^M$ available *without any ground-truth*.

A Unified Test Accuracy Framework Based on Model Calibration. To achieve this goal, we first provide a unified test accuracy estimation framework that is compatible with existing confidence-based methods. Let $\boldsymbol{p}_i$ be the predicted probability for input $\boldsymbol{x}_i$, which is obtained by applying the softmax function to the network output logit: $\boldsymbol{z}_i = f(\boldsymbol{x}_i)$. The predicted probability for the j'th class is computed as $p_{ij} = \sigma(\boldsymbol{z}_i)_j = \frac{e^{z_{ij}}}{\sum_{j=1}^c e^{z_{ij}}}$, and the predicted class y_i' is associated with the one of the largest probability: $y_i' = \mathrm{argmax}_j\, p_{ij}$. The test accuracy on the test set α^{Te} can be estimated via Average Confidence (AC) [12]: $\alpha_{AC}^{Te} = \frac{1}{N} \sum_{i=1}^N \max_j p_{ij}^{Te}$, which is an average of the probability of the predicted class j: $\max_j p_{ij}^{Te}$ for every test image $\boldsymbol{x}_i^{Te}$. Yet, directly computing over the softmax output p_{ij}^{Te} for estimation can be problematic as the predicted probability cannot reflect the real accuracy with modern neural networks [12]. It is essential to use calibrated probability $\hat{p}_{ij}^{Te}$ instead to estimate model accuracy with $\hat{\alpha}_{AC}^{Te} = \frac{1}{N} \sum_{i=1}^N \max_j \hat{p}_{ij}^{Te}$. Ideally, the calibrated probability $\hat{p}_{ij}^{Te}$ is supposed to perfectly reflect the expectation of the real accuracy. To adjust $\hat{p}_{ij}$, it is a common practice to use validation accuracy, which is defined as $\alpha^V = \frac{1}{N} \sum_{i=1}^N \mathbb{1}_{[y_i' = y_i]}$[1], as an surrogate objective to find the solution as there is no ground truth labels for test data [8,10,11]. On top of this pipeline which is also illustrated in Fig. 1(a), we will introduce state-of-the-art calibration methods including TS, DoC, and ATC and our proposed class-specific modifications of them (cf. Fig. 2), alleviating the calibration bias caused by the common class imbalance issue in medical imaging mentioned earlier.

Class-Specific Temperature Scaling. The TS method [12] rescales the logits using temperature parameter T. Calibrated probability is obtained with $\hat{p}_{ij} = \sigma(\boldsymbol{z}_i/T)_j$. To find T such that the estimated accuracy will approximate the real accuracy, we obtain optimal T^* using $\mathcal{D}^V$ with:

$$T^* = \underset{T}{\mathrm{argmin}} \left| \frac{1}{N} \sum_{i=1}^N \max_j \sigma(\frac{\boldsymbol{z}_i}{T})_j - \alpha^V \right|. \tag{1}$$

---

[1] $\mathbb{1}$ is an indicator function which is equal to 1 when the underlying criterion satisfies and 0 otherwise.

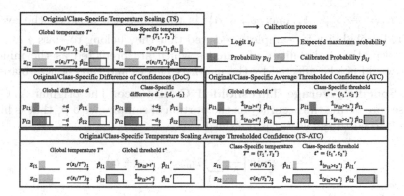

Fig. 2. Illustration of proposed Class-Specific modifications for four existing model evaluation methods. We show the calibration process of an under-confident prediction made for sample from minority class $c = 2$. Prior calibration methods use a global parameter for all classes, which leads to sub-optimal calibration and therefore bias for the minority class. The proposed variants adapt separate parameters per class, enabling improved, class-wise calibration.

To enable class-specific calibration, we extend temperature to vector $\boldsymbol{T}$ with c elements, where j'th element T_j corresponds to the j'th class. The calibrated probability is then written as $\hat{p}_{ij} = \sigma(\boldsymbol{z}_i/T^*_{y'_i})_j$, where T^*_j is optimized with:

$$T^*_j = \underset{T_j}{\operatorname{argmin}} \left| \frac{1}{N_j} \sum_{i=1}^{N} \mathbb{1}_{ij} \max_j \sigma\left(\frac{\boldsymbol{z}_i}{T_j}\right)_j - \alpha^V_j \right|, \tag{2}$$

where $\mathbb{1}_{ij} = \mathbb{1}_{[y'_i=j]}$ is 1 if the predicted most probable class is j and 0 otherwise, $N_j = \sum_{i=1}^{N} \mathbb{1}_{ij}$ and α^V_j is the accuracy for class j on the validation set which is calculated as $\alpha^V_j = \frac{1}{N_j} \sum_{i=1}^{N} \mathbb{1}_{ij} \mathbb{1}_{[y'_i=y_i]}$. This way we ensure optimized $\boldsymbol{T}^*$ leads to $\hat{p}_i$ matching the class-wise accuracy.

Class-Specific Difference of Confidences. The DoC method [11] adjusts p_i by subtracting the difference $d = \alpha^V_{AC} - \alpha^V$ between the real accuracy α^V and α^V_{AC} from the maximum probability:

$$\hat{p}_{ij} = \begin{cases} p_{ij} - d & \text{if } y'_i = j, \\ p_{ij} + \frac{d}{c-1} & \text{otherwise.} \end{cases} \tag{3}$$

The two parts of Eq. 3 ensure that the class-wise re-calibrated probabilities remain normalized. We propose a class-specific DoC method, where we define the difference vector $\boldsymbol{d}$ with c elements, with its j'th element d_j corresponding to the j'th class's difference. The class-wise calibrated probability $\hat{p}_{ij}$ is then obtained via:

$$\hat{p}_{ij} = \begin{cases} p_{ij} - d_j & \text{if } y'_i = j, \\ p_{ij} + \frac{d_{y'}}{c-1} & \text{otherwise,} \end{cases} \tag{4}$$

where $d_j = \frac{1}{N_j} \sum_{i=1}^{N} \mathbb{1}_{ij} \max_j p_{ij} - \alpha_j^V$ is the difference between the average predicted probability for samples predicted that they belong to class j and α_j^V. In this way, the probabilities of samples which are predicted as the minority classes would not be affected by the majority classes and thus be less biased.

Class-Specific Average Thresholded Confidence. The original ATC [10] proposes predicting the accuracy as the portion of samples with p_i that is higher than a learned threshold t. t is obtained via:

$$t^* = \operatorname*{argmin}_{t} \left| \frac{1}{N} \sum_{i=1}^{N} \mathbb{1}_{[\max_j p_{ij} > t]} - \alpha^V \right|. \tag{5}$$

We can then use t^* to compute $\hat{p}_{ij} = \mathbb{1}_{[p_{ij} > t^*]}$ for each test sample to calculate the estimated test accuracy, $\hat{\alpha}_{AC}^{Te}$, defined previously. We should note that for ATC, $\hat{p}_{ij}$ is no longer a calibrated probability but a discrete number $\in \{0, 1\}$. For class-specific calibration, we propose learning a set of thresholds given by vector t with c elements, one per class, where j'th element is obtained by:

$$t_j^* = \operatorname*{argmin}_{t_j} \left| \frac{1}{N_j} \sum_{i=1}^{N} \mathbb{1}_{ij} \mathbb{1}_{[max_k p_{ik} > t_j]} - \alpha_j^V \right|. \tag{6}$$

We can then calibrate probabilities as $\hat{p}_{ij} = \mathbb{1}_{[p_{ij} > t_{y_i'}^*]}$ based on predicted class y_i'. Hence the probabilities of samples that are predicted as different classes are discretized using different thresholds $t_{y_i'}^*$ to compensate the bias incurred by class imbalance. ATC can be combined with TS, by applying it to $\hat{p}_i$ which is calibrated by TS, and we denote this method as TS-ATC. By combining our class-specific modifications of TS and ATC, we similarly obtain the class-specific variant CS TS-ATC (cf. Fig. 2).

Extension to Segmentation. We propose extending confidence-based model evaluation methods to segmentation by predicting the DSC that the model would achieve on target data, instead of the accuracy metric that is more appropriate for classification. For this purpose, we slightly modify notation and write that the validation dataset consists of Z segmentation cases as $\mathcal{D}^V = \{\mathcal{S}_z\}_{z=1}^{Z}$, with $\mathcal{S}_z$ all the training pairs for the case z consisting of totally n pixels, defined as $\mathcal{S}_z = \{(\boldsymbol{x}_{zi}, y_{zi})\}_{i=1}^{n}$, where $\boldsymbol{x}_{zi}$ is an image patch sampled from case z and y_{zi} is the ground truth class of its central pixel. We then denote p_{zij} as the predicted probability of sample $\boldsymbol{x}_{zi}$ for the j'th class and its calibrated version as $\hat{p}_{zij}$. The estimated DSC can then be calculated using the predicted probabilities, in a fashion similar to the common soft DSC loss [26, 28]. Specifically, DSC_j, which is real DSC of $\mathcal{D}^V$ for the j'th class, is estimated with:

$$\text{sDSC}_j = \frac{1}{Z} \sum_{z=1}^{Z} \frac{2 \sum_{i=1}^{n} \mathbb{1}_{[\text{argmax}_k \, p_{zik} = j]} \hat{p}_{zij}}{\sum_{i=1}^{n} \mathbb{1}_{[\text{argmax}_k \, p_{zik} = j]} + \sum_{i=1}^{n} \hat{p}_{zij}}. \tag{7}$$

For the CS TS, CS ATC and CS ATC-CS, we first calibrate p_{zij} of the background class by matching class-wise average confidence and accuracy α_j^V, then calibrate p_{zij} of the foreground class j by matching sDSC$_j$ and DSC$_j$, in order to adjust the model to fit the metric of DSC. We calibrate the background class according to accuracy instead of DSC because the sDSC$_j$ is always very high and cannot be easily decrease to match DSC$_j$, which we found makes optimization to match them meaningless. For CS DoC, as d_j cannot be calculated with the difference between sDSC$_j$ and DSC$_j$, we keep $\hat{p}_{zij} = p_{zij}$ and estimate DSC on the target dataset $\mathcal{D}^{Te}$ on j'th class simply by DSC$^{Te}_{CSDoCj} = $ DSC$_j - $sDSC$_j + $sDSC$^{Te}_j$.

Table 1. Evaluation on different tasks under varied types of domain shifts based on Mean Absolute Error (MAE). Lower MAE is better. Best results with lowest MAE in **bold**. Class-specific calibration as proposed (CS methods) improves all baselines. This is most profound in segmentation tasks, which present extreme class imbalance.

Task	Classification			Segmentation		
Training dataset	CIFAR-10	HAM10000		ATLAS	Prostate	
Test domain shifts	Synthetic	Synthetic	Natural	Synthetic	Synthetic	Natural
AC	31.3 ± 8.2	12.3 ± 5.1	20.1 ± 13.4	35.6 ± 2.1	8.7 ± 4.9	18.7 ± 5.9
QC [30]	—	—	—	3.0 ± 1.7	5.2 ± 6.6	19.3 ± 7.1
TS [12]	5.7 ± 5.6	3.9 ± 4.3	12.1 ± 8.3	9.7 ± 2.5	3.7 ± 5.4	9.2 ± 4.9
VS [12]	3.8 ± 2.1	4.2 ± 4.2	13.6 ± 9.6	11.4 ± 2.5	4.8 ± 5.1	11.2 ± 4.9
NORCAL [29]	7.6 ± 3.8	4.2 ± 4.6	13.7 ± 9.6	6.7 ± 2.4	5.8 ± 5.7	7.3 ± 4.7
CS TS	5.5$^\sim$ ± 5.6	3.7$^\sim$ ± 4.0	11.9$^\sim$ ± 8.0	1.6** ± 1.8	3.0** ± 5.7	7.8** ± 4.8
DoC [11]	10.8 ± 8.2	4.6 ± 5.0	15.3 ± 9.7	4.2 ± 3.2	3.7 ± 5.8	13.9 ± 6.5
CS DoC	9.4** ± 7.2	4.5$^\sim$ ± 4.9	14.7* ± 9.2	1.3** ± 1.9	3.5* ± 6.1	12.1* ± 5.9
ATC [10]	4.6 ± 4.4	3.4 ± 3.9	7.1 ± 6.3	30.4 ± 1.8	8.6 ± 3.3	16.7 ± 5.3
CS ATC	2.8** ± 2.9	3.3$^\sim$ ± 4.8	5.8$^\sim$ ± 7.6	1.6** ± 1.5	1.1** ± 1.7	4.3** ± 2.2
TS-ATC [10,12]	5.3 ± 3.9	4.2 ± 4.2	7.3 ± 7.1	30.4 ± 1.8	8.5 ± 3.3	16.7 ± 5.3
CS TS-ATC	2.7** ± 2.3	4.2$^\sim$ ± 5.4	5.9** ± 8.4	1.3** ± 1.4	1.2** ± 1.7	4.2** ± 2.2

*p-value $<$ 0.05; **p-value $<$ 0.01; $^\sim$$p$-value $\geq$ 0.05 (compared with their class-agnostic counterparts)

3   Experiments

Setting. In our experiments, we first train a classifier f with $\mathcal{D}^{Tr}$, then we make predictions on $\mathcal{D}^V$ and optimize the calibration parameters based on model output and the validation labels. Finally, we evaluate our model on $\mathcal{D}^{Te}$ and estimate the quality of the predicted results based on $\hat{p}_i$. In addition to the confidence-based methods described in Sect. 2, for the segmentation tasks, we also compare with a neural network based quality control (QC) method [30]. Following [30], we train a 3D ResNet-50 that takes as input the image and the predicted segmentation mask and predicts the DSC of the predicted versus the manual segmentation. For fair comparison, we train the QC model with the same validation data we use for model calibration.

Datasets. *Imbalanced CIFAR-10:* We train a ResNet-32 [14] with imbalanced CIFAR-10 [19], using imbalance ratio of 100 following [3]. We employ synthetic domain shifts using CIFAR-10-C [15] that consists of 95 distinct corruptions.

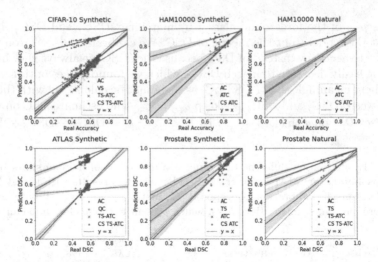

Fig. 3. Real versus predicted accuracy/DSC for classification/segmentation tasks respectively. Each point represents model performance on the unseen target data. We plot a baseline method (AC) in blue, the best among previous methods in orange, the best among the proposed methods in red and its class-agnostic counterpart in green. Best methods determined based on MAE. (Color figure online)

Skin lesion classification: We train ResNet-50 for skin lesion classification with $c = 7$ following [25,32]. We randomly select 70% data from HAM10000 [33] as the training source domain and leave the rest as validation. We also create synthetic domain shifts based on the validation data following [15], obtaining 38 synthetic test datasets with domain shifts. We additionally collect 6 publicly available skin lesion datasets which were collected in different institutions including BCN [5], VIE [31], MSK [13], UDA [13], D7P [18] and PH2 [27] following [35]. The detailed dataset statistics are summarized in supplementary material. *Brain lesion segmentation:* We apply nnU-Net [16] for all the segmentation tasks. We evaluate the proposed methods with brain lesion segmentation based on T1-weighted magnetic resonance (MR) images using data from Anatomical Tracings of Lesions After Stroke (ATLAS) [22] for which classes are highly imbalanced. We randomly select 145 cases for training and leave 75 cases for validation. We create synthetic domains shifts by applying 83 distinct transformations, including spatial, appearance and noise operations to the validation data. Details on implementation are in the supplementary material. *Prostate segmentation:* We also evaluate on the task of cross-site prostate segmentation based on T2-weighted MR images following [24]. We use 30 cases collected with 1.5T MRI machines from [2] as the source domain. We randomly select 20 cases for training and 10 cases for validation. We create synthetic domains shifts using the same 83 transformations as for brain lesions. As target domain, we use the 5 datasets described in [2,20,23], collected from different institutions using scanners of different field strength and manufacturers.

Results and Discussion. We calculate the Mean Absolute Error (MAE) and R2 Score between the predicted and the real performance (accuracy or DSC). We summarize MAE results in Table 1 and R2 Score results in supplementary material. We show scatter plots of predicted versus real performance in Fig. 3. Each reported result is the average of two experiments with different random seeds. We observe in Table 1 that our proposed modifications for class-specific calibration improve all baselines. Improvements are especially profound in segmentation tasks, where class imbalance is very high. It might be because previous model calibration methods bias towards the background class significantly as a result of class imbalance. These results clearly demonstrate the importance of accounting for class imbalance in the design of performance-prediction methods, to counter biases induced by the data. Figure 3 shows that the predicted performance by AC baseline is always higher than real performance because neural networks are commonly over-confident [12]. QC is effective in predicting real DSC under synthetic domain shifts but under-performs with natural domain shifts. Results in both Table 1 and Fig. 3 show that CS ATC (or CS TS-ATC) consistently performs best, outperforming prior methods by a large margin.

4   Conclusion

In this paper, we propose estimating model performance under domain shift for medical image classification and segmentation based on class-specific confidence scores. We derive class-specific variants for state-of-the-art methods to alleviate calibration bias incurred by class imbalance, which is profound in real-world medical imaging databases, and show very promising results. We expect the proposed methods to be useful for safe deployment of machine learning in real world settings, for example by facilitating decisions such as whether to deploy a given, pre-trained model based on whether estimated model performance on the target data of a deployment environment meets acceptance criteria.

Acknowledgements. ZL is grateful for a China Scholarship Council (CSC) Imperial Scholarship. This project has received funding from the ERC under the EU's Horizon 2020 research and innovation programme (grant No 757173) and the UKRI London Medical Imaging & Artificial Intelligence Centre for Value Based Healthcare, and a EPSRC Programme Grant (EP/P001009/1).

References

1. Ben-David, S., Blitzer, J., Crammer, K., Kulesza, A., Pereira, F., Vaughan, J.W.: A theory of learning from different domains. Mach. Learn. **79**(1), 151175 (2010)
2. Bloch, N., et al.: NCI-ISBI 2013 challenge: automated segmentation of prostate structures. The Cancer Imaging Archive, vol. 370 (2015)
3. Cao, K., Wei, C., Gaidon, A., Arechiga, N., Ma, T.: Learning imbalanced datasets with label-distribution-aware margin loss. In: NeurIPS, vol. 32 (2019)
4. Chen, M., Goel, K., Sohoni, N.S., Poms, F., Fatahalian, K., Ré, C.: Mandoline: model evaluation under distribution shift. In: ICML, pp. 1617–1629. PMLR (2021)

5. Combalia, M., et al.: Bcn20000: dermoscopic lesions in the wild. arXiv preprint arXiv:1908.02288 (2019)
6. Deng, W., Zheng, L.: Are labels always necessary for classifier accuracy evaluation? In: CVPR, pp. 15069–15078 (2021)
7. Eche, T., Schwartz, L.H., Mokrane, F.Z., Dercle, L.: Toward generalizability in the deployment of artificial intelligence in radiology: role of computation stress testing to overcome underspecification. Radiol. Artif. Intell. 3(6), e210097 (2021)
8. Elsahar, H., Gallé, M.: To annotate or not? Predicting performance drop under domain shift. In: EMNLP-IJCNLP, pp. 2163–2173 (2019)
9. Fan, W., Davidson, I.: Reverse testing: an efficient framework to select amongst classifiers under sample selection bias. In: ACM SIGKDD, pp. 147–156 (2006)
10. Garg, S., Balakrishnan, S., Lipton, Z.C., Neyshabur, B., Sedghi, H.: Leveraging unlabeled data to predict out-of-distribution performance. In: ICLR (2022). https://openreview.net/forum?id=o_HsiMPYh_x
11. Guillory, D., Shankar, V., Ebrahimi, S., Darrell, T., Schmidt, L.: Predicting with confidence on unseen distributions. In: ICCV, pp. 1134–1144 (2021)
12. Guo, C., Pleiss, G., Sun, Y., Weinberger, K.Q.: On calibration of modern neural networks. In: ICML, pp. 1321–1330. PMLR (2017)
13. Gutman, D., et al.: Skin lesion analysis toward melanoma detection: a challenge at the international symposium on biomedical imaging (ISBI) 2016, hosted by the international skin imaging collaboration (ISIC). arXiv preprint arXiv:1605.01397 (2016)
14. He, K., Zhang, X., Ren, S., Sun, J.: Deep residual learning for image recognition. In: CVPR, pp. 770–778 (2016)
15. Hendrycks, D., Dietterich, T.: Benchmarking neural network robustness to common corruptions and perturbations. In: Proceedings of the ICLR (2019)
16. Isensee, F., Jaeger, P.F., Kohl, S.A., Petersen, J., Maier-Hein, K.H.: nnU-Net: a self-configuring method for deep learning-based biomedical image segmentation. Nat. Methods 18(2), 203–211 (2021)
17. Islam, M., Seenivasan, L., Ren, H., Glocker, B.: Class-distribution-aware calibration for long-tailed visual recognition. arXiv preprint arXiv:2109.05263 (2021)
18. Kawahara, J., Daneshvar, S., Argenziano, G., Hamarneh, G.: Seven-point checklist and skin lesion classification using multitask multimodal neural nets. IEEE J. Biomed. Health Inform. 23(2), 538–546 (2018)
19. Krizhevsky, A., Hinton, G., et al.: Learning multiple layers of features from tiny images. Technical report (2009)
20. Lemaître, G., Martí, R., Freixenet, J., Vilanova, J.C., Walker, P.M., Meriaudeau, F.: Computer aided detection and diagnosis for prostate cancer based on mono and multi-parametric MRI: a review. Comput. Biol. Med. 60, 8–31 (2015)
21. Li, Z., Kamnitsas, K., Glocker, B.: Analyzing overfitting under class imbalance in neural networks for image segmentation. IEEE Trans. Med. Imaging 40(3), 1065–1077 (2020)
22. Liew, S.L., et al.: A large, open source dataset of stroke anatomical brain images and manual lesion segmentations. Sci. Data 5, 180011 (2018)
23. Litjens, G., et al.: Evaluation of prostate segmentation algorithms for MRI: the PROMISE12 challenge. Med. Image Anal. 18(2), 359–373 (2014)
24. Liu, Q., Dou, Q., Heng, P.-A.: Shape-aware meta-learning for generalizing prostate MRI segmentation to unseen domains. In: Martel, A.L., et al. (eds.) MICCAI 2020. LNCS, vol. 12262, pp. 475–485. Springer, Cham (2020). https://doi.org/10.1007/978-3-030-59713-9_46

25. Marrakchi, Y., Makansi, O., Brox, T.: Fighting class imbalance with contrastive learning. In: de Bruijne, M., et al. (eds.) MICCAI 2021. LNCS, vol. 12903, pp. 466–476. Springer, Cham (2021). https://doi.org/10.1007/978-3-030-87199-4_44
26. Mehrtash, A., Wells, W.M., Tempany, C.M., Abolmaesumi, P., Kapur, T.: Confidence calibration and predictive uncertainty estimation for deep medical image segmentation. IEEE Trans. Med. Imaging **39**(12), 3868–3878 (2020)
27. Mendonça, T., Ferreira, P.M., Marques, J.S., Marcal, A.R., Rozeira, J.: PH 2-a dermoscopic image database for research and benchmarking. In: EMBC, pp. 5437–5440. IEEE (2013)
28. Milletari, F., Navab, N., Ahmadi, S.A.: V-net: fully convolutional neural networks for volumetric medical image segmentation. In: 3DV, pp. 565–571. IEEE (2016)
29. Pan, T.Y., et al.: On model calibration for long-tailed object detection and instance segmentation. In: NeurIPS, vol. 34 (2021)
30. Robinson, R., et al.: Real-time prediction of segmentation quality. In: Frangi, A.F., Schnabel, J.A., Davatzikos, C., Alberola-López, C., Fichtinger, G. (eds.) MICCAI 2018. LNCS, vol. 11073, pp. 578–585. Springer, Cham (2018). https://doi.org/10.1007/978-3-030-00937-3_66
31. Rotemberg, V., et al.: A patient-centric dataset of images and metadata for identifying melanomas using clinical context. Sci. Data **8**(1), 1–8 (2021)
32. Russakovsky, O., et al.: Imagenet large scale visual recognition challenge. Int. J. Comput. Vision **115**(3), 211–252 (2015)
33. Tschandl, P., Rosendahl, C., Kittler, H.: The ham10000 dataset, a large collection of multi-source dermatoscopic images of common pigmented skin lesions. Sci. Data **5**(1), 1–9 (2018)
34. Valindria, V.V., et al.: Reverse classification accuracy: predicting segmentation performance in the absence of ground truth. IEEE Trans. Med. Imaging **36**(8), 1597–1606 (2017)
35. Yoon, C., Hamarneh, G., Garbi, R.: Generalizable feature learning in the presence of data bias and domain class imbalance with application to skin lesion classification. In: Shen, D., et al. (eds.) MICCAI 2019. LNCS, vol. 11767, pp. 365–373. Springer, Cham (2019). https://doi.org/10.1007/978-3-030-32251-9_40

vMFNet: Compositionality Meets Domain-Generalised Segmentation

Xiao Liu[1,4]($\boxtimes$), Spyridon Thermos[2], Pedro Sanchez[1,4], Alison Q. O'Neil[1,4], and Sotirios A. Tsaftaris[1,3,4]

[1] School of Engineering, University of Edinburgh, Edinburgh EH9 3FB, UK
Xiao.Liu@ed.ac.uk
[2] AC Codewheel Ltd, Larnaca, Cyprus
[3] The Alan Turing Institute, London, UK
[4] Canon Medical Research Europe Ltd., Edinburgh, UK

Abstract. Training medical image segmentation models usually requires a large amount of labeled data. By contrast, humans can quickly learn to accurately recognise anatomy of interest from medical (e.g. MRI and CT) images with some limited guidance. Such recognition ability can easily generalise to new images from different clinical centres. This rapid and generalisable learning ability is mostly due to the compositional structure of image patterns in the human brain, which is less incorporated in medical image segmentation. In this paper, we model the compositional components (i.e. patterns) of human anatomy as learnable von-Mises-Fisher (vMF) kernels, which are robust to images collected from different domains (e.g. clinical centres). The image features can be decomposed to (or composed by) the components with the composing operations, i.e. the vMF likelihoods. The vMF likelihoods tell how likely each anatomical part is at each position of the image. Hence, the segmentation mask can be predicted based on the vMF likelihoods. Moreover, with a reconstruction module, unlabeled data can also be used to learn the vMF kernels and likelihoods by recombining them to reconstruct the input image. Extensive experiments show that the proposed vMFNet achieves improved generalisation performance on two benchmarks, especially when annotations are limited. Code is publicly available at: https://github.com/vios-s/vMFNet.

Keywords: Compositionality · Domain generalisation · Semi-supervised learning · Test-time training · Medical image segmentation

1 Introduction

Deep learning approaches can achieve impressive performance on medical image segmentation when provided with a large amount of labeled training data [3,8].

Supplementary Information The online version contains supplementary material available at https://doi.org/10.1007/978-3-031-16449-1_67.

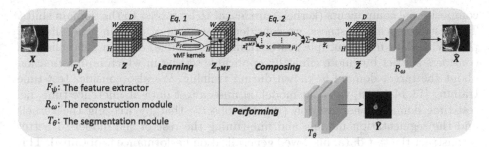

Fig. 1. The overall model design of vMFNet. The notations are specified in Sect. 3.

However, shifts in data statistics, i.e. *domain shifts* [41,42], can heavily degrade deep model performance on unseen data [4,24,33]. By contrast, humans can learn quickly with limited supervision and are less likely to be affected by such domain shifts, achieving accurate recognition on new images from different clinical centres. Recent studies [25,36] argue that this rapid and generalisable learning ability is mostly due to the compositional structure of concept components or image patterns in the human brain. For example, a clinical expert usually remembers the patterns (components) of human anatomy after seeing many medical images. When searching for the anatomy of interest in new images, patterns or combinations of patterns are used to find where and what the anatomy is at each position of the image. This process can be modeled as learning compositional components (*learning*), composing the components (*composing*), and performing downstream tasks (*performing*). This compositionality has been shown to improve the robustness and explainability of many computer vision tasks [16,21,36] but is less studied in medical image segmentation.

In this paper, we explore compositionality in domain-generalised medical image segmentation and propose vMFNet. The overall model design is shown in Fig. 1. Motivated by Compositional Networks [21] and considering medical images are first encoded by deep models as deep features, we model the compositional components of human anatomy as learnable von-Mises-Fisher (vMF) kernels (*learning*).[1] The features can be projected as vMF likelihoods that define how much each kernel is activated at each position. We then compose the kernels with the normalised vMF likelihoods to reconstruct the input image (*composing*). Hence, the kernels and the corresponding vMF likelihoods can also be learnt end-to-end with unlabeled data. In terms of a downstream segmentation task, the spatial vMF likelihoods are used as the input to a segmentation module to predict the segmentation mask (*performing*). In this case, the prediction is based on the activation of the components at each position of the image.

For data from different domains, the features of the same anatomical part of images from these domains will activate the same kernels. In other words, the

---

[1] vMF kernels are similar to prototypes in [35]. However, prototypes are usually calculated as the mean of the feature vectors for each class using the ground-truth masks. vMF kernels are learnt as the cluster centres of the feature vectors.

compositional components (kernels) are learnt to be robust to the domain shifts. To illustrate such robustness, we consider the setting of domain generalisation [22]. With data from multiple source domains available, domain generalisation considers a strict but more clinically applicable setting in which no information about the target domain is known during training. We also consider test-time training [13,14,19,38] with our model on unseen test data at inference time, i.e. test-time domain generalisation [18]. We observe that by freezing the kernels and the segmentation module and fine-tuning the rest of the model to better reconstruct the test data, improved generalisation performance is obtained. This is due to the fact that medical images from different domains are not significantly different; new components do not need to be learned, rather the composing operations need to be fine-tuned.

Contributions:

- Inspired by the human recognition process, we design a semi-supervised compositional model for domain-generalised medical image segmentation.
- We propose a reconstruction module to compose the vMF kernels with the vMF likelihoods to facilitate reconstruction of the input image, which allows the model to be trained also with unlabeled data.
- We apply the proposed method to two settings: semi-supervised domain generalisation and test-time domain generalisation.
- Extensive experiments on cardiac and gray matter datasets show improved performance over several baselines, especially when annotations are limited.

2   Related Work

Compositionality: Compositionality has been mostly incorporated for robust image classification [21,36] and recently for compositional image synthesis [2,25]. Among these work, Compositional Networks [21] originally designed for robust classification under object occlusion is easier to extend to pixel-wise tasks as it learns spatial and interpretable vMF likelihoods. Previous work integrates the vMF kernels and likelihoods [21] for object localisation [40] and recently for nuclei segmentation (with the bounding box as supervision) in a weakly supervised manner [43]. We rather use the vMF kernels and likelihoods for domain-generalised medical image segmentation and learn vMF kernels and likelihoods with unlabeled data in a semi-supervised manner.

Domain Generalisation: Several active research directions handle this problem by either augmenting the source domain data [7,42], regularising the feature space [5,15], aligning the source domain features or output distributions [23], carefully designing robust network modules [11], or using meta-learning to approximate the possible domain shifts across source and target domains [9,26,27,29]. Most of these methods consider a fully supervised setting. Recently, Liu et al. [29] proposed a gradient-based meta-learning model to facilitate semi-supervised domain generalisation by integrating disentanglement. Extensive results on a common backbone showed the benefits of meta-learning and disentanglement. Following [29], Yao et al. [39] adopted a pre-trained ResNet [12] as a backbone feature extractor and augmented the source data by mixing MRI images in the Fourier domain and employed pseudo-labelling to leverage the unlabelled data. Our proposed vMFNet aligns the image features to the same vMF distributions to handle the domain shifts. The reconstruction further allows the model to handle semi-supervised domain generalisation tasks.

3 Proposed Method

We propose vMFNet, a model consisting of three modules; the feature extractor F_ψ, the task network T_θ, and the reconstruction network R_ω, where ψ, θ and ω denote the network parameters. Overall, the compositional components are learned as vMF kernels by decomposing the features extracted by F_ψ. Then, we compose the vMF kernels to reconstruct the image with R_ω by using the vMF likelihoods as the composing operations. Finally, the vMF likelihoods that contain spatial information are used to predict the segmentation mask with T_θ. The decomposing and composing are shown in Fig. 1 and detailed below.

3.1 Learning Compositional Components

The primal goal of learning compositional components is to represent deep features in a compact low dimensional space. We denote the features extracted by F_ψ as $\mathbf{Z} \in \mathbb{R}^{H \times W \times D}$, where H and W are the spatial dimensions and D is the number of channels. The feature vector $\mathbf{z}_i \in \mathbb{R}^D$ is defined as a vector across channels at position i on the 2D lattice of the feature map. We follow Compositional Networks [21] to model $\mathbf{Z}$ with J vMF distributions, where the learnable mean of each distribution is defined as vMF kernel $\boldsymbol{\mu}_j \in \mathbb{R}^D$. To allow the modeling to be tractable, the variance σ of all distributions are fixed. In particular, the vMF likelihood for the jth distribution at each position i can be calculated as:

$$p(\mathbf{z}_i|\boldsymbol{\mu}_j) = C(\sigma)^{-1} \cdot e^{\sigma \boldsymbol{\mu}_j^T \mathbf{z}_i}, \text{ s.t. } \|\boldsymbol{\mu}_j\| = 1, \tag{1}$$

where $\|\mathbf{z}_i\| = 1$ and $C(\sigma)$ is a constant. After modeling the image features with J vMF distributions with Eq. 1, the vMF likelihoods $\mathbf{Z}_{vMF} \in \mathbb{R}^{H \times W \times J}$ can be obtained, which indicates how much each kernel is activated at each position. Here, the vMF loss $\mathcal{L}_{vMF}$ that forces the kernels to be the cluster centres of the features vectors is defined in [21] as $\mathcal{L}_{vMF}(\boldsymbol{\mu}, \mathbf{Z}) = -(HW)^{-1} \sum_i \max_j \boldsymbol{\mu}_j^T \mathbf{z}_i$. This avoids an iterative EM-type learning procedure. Overall, the feature vectors of different images corresponding to the same anatomical part will be clustered and activate the same kernels. In other words, the vMF kernels are learnt as the components or patterns of the anatomical parts. Hence, the vMF likelihoods $\mathbf{Z}_{vMF}$ for the features of different images will be aligned to follow the same distributions (with the same means). In this case, the vMF likelihoods can be considered as the content representation in the context of content-style disentanglement [6,30], where the vMF kernels contain the style information.

3.2 Composing Components for Reconstruction

After decomposing the image features with the vMF kernels and the likelihoods, we re-compose to reconstruct the input image. Reconstruction requires that complete information about the input image is captured [1]. However, the vMF likelihoods contain only spatial information as observed in [21], while the non-spatial information is compressed as varying kernels $\boldsymbol{\mu}_j, j \in \{1 \cdots J\}$, where the compression is not invertible. Consider that the vMF likelihood $p(\mathbf{z}_i|\boldsymbol{\mu}_j)$ denotes how much the kernel $\boldsymbol{\mu}_j$ is activated by the feature vector $\mathbf{z}_i$. We propose to construct a new feature space $\widetilde{\mathbf{Z}}$ with the vMF likelihoods and kernels. Let $\mathbf{z}_i^{vMF} \in \mathbb{R}^J$ be a normalised vector across $\mathbf{Z}_{vMF}$ channels at position i. We devise the new feature vector $\widetilde{\mathbf{z}}_i$ as the combination of the kernels with the normalised vMF likelihoods as the combination coefficients, namely:

$$\widetilde{\mathbf{z}}_i = \sum_{j=1}^{J} \mathbf{z}_{i,j}^{vMF} \boldsymbol{\mu}_j, \text{ where } ||\mathbf{z}_i^{vMF}|| = 1. \tag{2}$$

After obtaining $\widetilde{\mathbf{Z}}$ as the approximation of $\mathbf{Z}$, the reconstruction network $\boldsymbol{R}_\omega$ reconstructs the input image with $\widetilde{\mathbf{Z}}$ as the input, i.e. $\hat{\mathbf{X}} = \boldsymbol{R}_\omega(\widetilde{\mathbf{Z}})$.

3.3 Performing Downstream Task

As the vMF likelihoods contain only spatial information of the image that is highly correlated to the segmentation mask, we design a segmentation module, i.e. the task network $\boldsymbol{T}_\theta$, to predict the segmentation mask with the vMF likelihoods as input, i.e. $\hat{\mathbf{Y}} = \boldsymbol{T}_\theta(\mathbf{Z}_{vMF})$. Specifically, the segmentation mask tells what anatomical part the feature vector $\mathbf{z}_i$ corresponds to, which provides further guidance for the model to learn the vMF kernels as the components of the anatomical parts. Then the vMF likelihoods will be further aligned when trained with multi-domain data and hence perform well on domain generalisation tasks.

3.4 Learning Objective

Overall, the model contains trainable parameters ψ, θ, ω and the kernels $\boldsymbol{\mu}$. The model can be trained end-to-end with the following objective:

$$\underset{\psi,\theta,\omega,\mu}{\operatorname{argmin}} \lambda_{Dice}\mathcal{L}_{Dice}(\mathbf{Y}, \hat{\mathbf{Y}}) + \mathcal{L}_{rec}(\mathbf{X}, \hat{\mathbf{X}}) + \mathcal{L}_{vMF}(\boldsymbol{\mu}, \mathbf{Z}), \tag{3}$$

where $\lambda_{Dice} = 1$ when the ground-truth mask $\mathbf{Y}$ is available, otherwise $\lambda_{Dice} = 0$. $\mathcal{L}_{Dice}$ is the Dice loss [31], $\mathcal{L}_{rec}$ is the reconstruction loss (we use $L1$ distance).

4 Experiments

4.1 Datasets and Baseline Models

To make comparison easy, we adopt the datasets and baselines of [29], which we briefly summarise below. We also include [29] as a baseline.

Datasets: The **Multi-centre, multi-vendor & multi-disease cardiac image segmentation (M&Ms) dataset** [4] consists of 320 subjects scanned at 6 clinical centres using 4 different magnetic resonance scanner vendors i.e. domains A, B, C and D. For each subject, only the end-systole and end-diastole phases are annotated. Voxel resolutions range from $0.85 \times 0.85 \times 10$ mm to $1.45 \times 1.45 \times 9.9$ mm. Domain A contains 95 subjects, domain B contains 125 subjects, and domains C and D contain 50 subjects each. The **Spinal cord gray matter segmentation (SCGM) dataset** [33] images are collected from 4 different medical centres with different MRI systems i.e. domains 1, 2, 3 and 4. The voxel resolutions range from $0.25 \times 0.25 \times 2.5$ mm to $0.5 \times 0.5 \times 5$ mm. Each domain has 10 labeled subjects and 10 unlabelled subjects.

Baselines: For fair comparison, we compare all models with the same backbone feature extractor, i.e. UNet [34], without any pre-training. **nnUNet** [17] is a supervised baseline. It adapts its model design and searches the optimal hyperparameters to achieve the optimal performance. **SDNet+Aug.** [28] is a semi-supervised disentanglement model, which disentangles the input image to a spatial anatomy and a non-spatial modality factors. Augmenting the training data by mixing the anatomy and modality factors of different source domains, "SDNet+Aug." can potentially generalise to unseen domains. **LDDG** [23] is a fully-supervised domain generalisation model, in which low-rank regularisation is used and the features are aligned to Gaussian distributions. **SAML** [27] is a gradient-based meta-learning approach. It applies the compactness and smoothness constraints to learn domain-invariant features across meta-train and meta-test sets in a fully supervised setting. **DGNet** [29] is a semi-supervised gradient-based meta-learning approach. Combining meta-learning and disentanglement, the shifts between domains are captured in the disentangled representations. DGNet achieved the state-of-the-art (SOTA) domain generalisation performance on M&Ms and SCGM datasets.

4.2 Implementation Details

All models are trained using the Adam optimiser [20] with a learning rate of $1 \times e^{-4}$ for 50K iterations using batch size 4. Images are cropped to 288×288 for M&Ms and 144×144 for SCGM. F_ψ is a 2D UNet [34] without the last upsampling and output layers to extract features Z. Note that F_ψ can be easily replaced by other encoders such as a ResNet [12] and the feature vectors can be extracted from any layer of the encoder where performance may vary for different layers. T_θ and R_ω are two shallow convolutional networks that are detailed in Sect. 2 of Appendix. We follow [21] to set the variance of the vMF distributions as 30. The number of kernels is set to 12, as this number performed the best according to early experiments. For different medical datasets, the best number of kernels may be slightly different. All models are implemented in PyTorch [32] and are trained using an NVIDIA 2080 Ti GPU. In the semi-supervised setting, we use specific percentages of the subjects as labeled data and the rest as unlabeled data. We train the models with 3 source domains and treat the 4th

domain as the target one. We use Dice (%) and Hausdorff Distance (HD) [10] as the evaluation metrics.

Table 1. Average Dice (%) and Hausdorff Distance (HD) results and the standard deviations on M&Ms and SCGM datasets. For semi-supervised approaches, the training data contain all unlabeled data and different percentages of labeled data from source domains. The rest are trained with different percentages of labeled data only. Results of baseline models are taken from [29]. Bold numbers denote the best performance.

Percent	metrics	nnUNet	SDNet+Aug.	LDDG	SAML	DGNet	vMFNet
M&Ms 2%	Dice (↑)	$65.94_{8.3}$	$68.28_{8.6}$	$63.16_{5.4}$	$64.57_{8.5}$	$72.85_{4.3}$	$\mathbf{78.43_{3.6}}$
	HD (↓)	$20.96_{4.0}$	$20.17_{3.3}$	$22.02_{3.5}$	$21.22_{4.1}$	$19.32_{2.8}$	$\mathbf{16.56_{1.7}}$
M&Ms 5%	Dice (↑)	$76.09_{6.3}$	$77.47_{3.9}$	$71.29_{3.6}$	$74.88_{4.6}$	$79.75_{4.4}$	$\mathbf{82.12_{3.1}}$
	HD (↓)	$18.22_{3.0}$	$18.62_{3.1}$	$19.21_{3.0}$	$18.49_{2.9}$	$17.98_{3.2}$	$\mathbf{15.30_{1.8}}$
M&Ms 100%	Dice (↑)	$84.87_{2.5}$	$84.29_{1.6}$	$85.38_{1.6}$	$83.49_{1.3}$	$\mathbf{86.03_{1.7}}$	$85.92_{2.0}$
	HD (↓)	$14.80_{1.9}$	$15.06_{1.6}$	$14.88_{1.7}$	$15.52_{1.5}$	$14.53_{1.8}$	$\mathbf{14.05_{1.3}}$
SCGM 20%	Dice (↑)	$64.85_{5.2}$	76.73_{11}	63.31_{17}	73.50_{12}	79.58_{11}	$\mathbf{81.11_{8.8}}$
	HD (↓)	$3.49_{0.49}$	$2.07_{0.36}$	$2.38_{0.39}$	$2.11_{0.37}$	$1.97_{0.30}$	$\mathbf{1.96_{0.31}}$
SCGM 100%	Dice (↑)	$71.51_{5.4}$	81.37_{11}	79.29_{13}	80.95_{13}	82.25_{11}	$\mathbf{84.03_{8.0}}$
	HD (↓)	$3.53_{0.45}$	$1.93_{0.36}$	$2.11_{0.41}$	$1.95_{0.38}$	$1.92_{0.31}$	$\mathbf{1.84_{0.31}}$

4.3 Semi-supervised Domain Generalisation

Table 1 reports the average results over four leave-one-out experiments that treat each domain in turn as the target domain; more detailed results can be found in Sect. 1 of the Appendix. We highlight that the proposed vMFNet is **14 times faster to train** compared to the previous SOTA DGNet. Training vMFNet for one epoch takes 7 min, while DGNet needs 100 min for the M&Ms dataset due to the need to construct new computational graphs for the meta-test step in every iteration.

With limited annotations, vMFNet achieves 7.7% and 3.0% improvements (in Dice) for 2% and 5% cases compared to the previous SOTA DGNet on M&Ms dataset. For the 100% case, vMFNet and DGNet have similar performance of around 86% Dice and 14 HD. Overall, vMFNet has consistently better performance for almost all scenarios on the M&Ms dataset. Similar improvements are observed for the SCGM dataset.

Which Losses Help More? We ablate two key losses of vMFNet in the 2% of M&Ms setting. Note that both losses do not require the ground-truth masks. Removing $\mathcal{L}_{rec}$ results in 74.83% Dice and 18.57 HD, whereas removing $\mathcal{L}_{vMF}$ gives 75.45% Dice and 17.53 HD. Removing both gives 74.70% Dice and 18.25 HD. Compared with 78.43% Dice and 16.56 HD, training with both losses gives better generalisation results when the model is trained to learn better kernels and with unlabeled data. When removing the two losses, the model can still perform adequately compared to the baselines due to the decomposing mechanism.

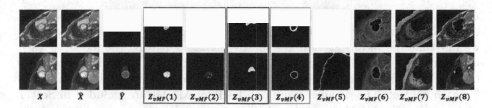

Fig. 2. Visualisation of images, reconstructions, predicted segmentation masks and 8 (of 12) most informative vMF likelihood channels for 2 examples from M&Ms dataset. (Color figure online)

Alignment Analysis: To show that the vMF likelihoods from different source domains are aligned, for M&Ms 100% cases, we first mask out the non-heart part of the images, features and vMF likelihoods. Then, we train classifiers to predict which domain the input is from with the masked images $\mathbf{X}$ or masked features $\mathbf{Z}$ or masked vMF likelihoods $\mathbf{Z}_{vMF}$ as input. The average cross-entropy errors are 0.718, 0.701 and 0.756, which means that it is harder to tell the domain class with the heart part of $\mathbf{Z}_{vMF}$, i.e. the vMF likelihoods for the downstream task are better aligned compared to the features $\mathbf{Z}$ from different source domains.

4.4 Visualisation of Compositionality

Overall, the segmentation prediction can be interpreted as the activation of corresponding kernels at each position, where false predictions occur when the wrong kernels are activated i.e. the wrong vMF likelihoods (composing operations) are used to predict the mask. We show example images, reconstructions, predicted segmentation masks, and 8 most informative vMF likelihoods channels in Fig. 2. As shown, vMF kernels 1 and 2 (red box) are mostly activated by the left ventricle (LV) feature vectors and kernels 3 (blue box) and 4 (green box) are mostly for right ventricle (RV) and myocardium (MYO) feature vectors. Interestingly, kernel 2 is mostly activated by papillary muscles in the left ventricle even though no supervision about the papillary muscles is provided during training. This supports that the model learns the kernels as the compositional components (patterns of papillary muscles, LV, RV and MYO) of the heart.

4.5 Test-Time Domain Generalisation

As discussed in Sect. 4.4, poor segmentation predictions are usually caused by the wrong kernels being activated. This results in the wrong vMF likelihoods being used to predict masks. Reconstruction quality is also affected by wrong vMF likelihoods. In fact, average reconstruction error is approximately 0.007 on the training set and 0.011 on the test set. Inspired by [13,37] we perform test-time training (TTT) to better reconstruct by fine-tuning the reconstruction loss

$\mathcal{L}_{rec}(\mathbf{X}, \hat{\mathbf{X}})$ to update $\boldsymbol{F}_\psi$ and $\boldsymbol{R}_\omega$ with the kernels and $\boldsymbol{T}_\theta$ fixed. This should in turn produce better vMF likelihoods. For images of each subject at test time, we fine-tune the reconstruction loss for 15 iterations (saving the model at each iteration) with a small learning rate of $1 \times e^{-6}$. Out of the 15 models, we choose the one with minimum reconstruction error to predict the segmentation masks for each subject. The detailed results of TTT for M&Ms are included in Sect. 1 of Appendix. For M&Ms 2%, 5% and 100% cases, TTT gives around 3.5%, 1.4% and 1% improvements in Dice compared to results (without TTT) in Table 1.

5 Conclusion

In this paper, inspired by the human recognition process, we have proposed a semi-supervised compositional model, vMFNet, that is effective for domain-generalised medical image segmentation. With extensive experiments, we showcased that vMFNet achieves consistently improved performance in terms of semi-supervised domain generalisation and highlighted the correlation between downstream segmentation task performance and the reconstruction. Based on the presented results, we are confident that freezing the kernels and fine-tuning the reconstruction to learn better composing operations (vMF likelihoods) at test time gives further improved generalisation performance.

Acknowledgement. This work was supported by the University of Edinburgh, the Royal Academy of Engineering and Canon Medical Research Europe by a PhD studentship to Xiao Liu. This work was partially supported by the Alan Turing Institute under the EPSRC grant EP/N510129/1. S.A. Tsaftaris acknowledges the support of Canon Medical and the Royal Academy of Engineering and the Research Chairs and Senior Research Fellowships scheme (grant RCSRF1819\8\25).

References

1. Achille, A., Soatto, S.: Emergence of invariance and disentanglement in deep representations. JMLR **19**(1), 1947–1980 (2018)
2. Arad Hudson, D., Zitnick, L.: Compositional transformers for scene generation. In: NeurIPS (2021)
3. Bernard, O., Lalande, A., Zotti, C., Cervenansky, F., Yang, X., et al.: Deep learning techniques for automatic MRI cardiac multi-structures segmentation and diagnosis: is the problem solved? IEEE TMI **37**(11), 2514–2525 (2018)
4. Campello, V.M., et al.: Multi-centre, multi-vendor and multi-disease cardiac segmentation: the M&Ms challenge. IEEE TMI **40**(12), 3543–3554 (2021)
5. Carlucci, F.M., D'Innocente, A., Bucci, S., Caputo, B., Tommasi, T.: Domain generalisation by solving jigsaw puzzles. In: CVPR, pp. 2229–2238 (2019)
6. Chartsias, A., Joyce, T., et al.: Disentangled representation learning in cardiac image analysis. Media **58**, 101535 (2019)
7. Chen, C., Hammernik, K., Ouyang, C., Qin, C., Bai, W., Rueckert, D.: Cooperative training and latent space data augmentation for robust medical image segmentation. In: de Bruijne, M., et al. (eds.) MICCAI 2021. LNCS, vol. 12903, pp. 149–159. Springer, Cham (2021). https://doi.org/10.1007/978-3-030-87199-4_14

8. Chen, C., Qin, C., Qiu, H., et al.: Deep learning for cardiac image segmentation: a review. Front. Cardiovasc. Med. **7**(25), 1–33 (2020)
9. Dou, Q., Castro, D.C., Kamnitsas, K., Glocker, B.: Domain generalisation via model-agnostic learning of semantic features. In: NeurIPS (2019)
10. Dubuisson, M.P., Jain, A.K.: A modified hausdorff distance for object matching. In: ICPR, vol. 1, pp. 566–568. IEEE (1994)
11. Gu, R., Zhang, J., Huang, R., Lei, W., Wang, G., Zhang, S.: Domain composition and attention for unseen-domain generalizable medical image segmentation. In: de Bruijne, M., et al. (eds.) MICCAI 2021. LNCS, vol. 12903, pp. 241–250. Springer, Cham (2021). https://doi.org/10.1007/978-3-030-87199-4_23
12. He, K., Zhang, X., Ren, S., Sun, J.: Deep residual learning for image recognition. In: CVPR, pp. 770–778 (2016)
13. He, Y., Carass, A., Zuo, L., et al.: Autoencoder based self-supervised test-time adaptation for medical image analysis. Media **72**, 102136 (2021)
14. Hu, M., et al.: Fully test-time adaptation for image segmentation. In: de Bruijne, M., et al. (eds.) MICCAI 2021. LNCS, vol. 12903, pp. 251–260. Springer, Cham (2021). https://doi.org/10.1007/978-3-030-87199-4_24
15. Huang, J., Guan, D., Xiao, A., Lu, S.: FSDR: frequency space domain randomization for domain generalization. In: CVPR (2021)
16. Huynh, D., Elhamifar, E.: Compositional zero-shot learning via fine-grained dense feature composition. In: NeurIPS, vol. 33, pp. 19849–19860 (2020)
17. Isensee, F., Jaeger, P.F., et al.: nnUNet: a self-configuring method for deep learning-based biomedical image segmentation. Nat. Methods **18**(2), 203–211 (2021)
18. Iwasawa, Y., Matsuo, Y.: Test-time classifier adjustment module for model-agnostic domain generalization. In: NeurIPS, vol. 34 (2021)
19. Karani, N., Erdil, E., Chaitanya, K., Konukoglu, E.: Test-time adaptable neural networks for robust medical image segmentation. Media **68**, 101907 (2021)
20. Kingma, D.P., Ba, J.: Adam: a method for stochastic optimization. In: ICLR (2015)
21. Kortylewski, A., He, J., Liu, Q., Yuille, A.L.: Compositional convolutional neural networks: a deep architecture with innate robustness to partial occlusion. In: CVPR, pp. 8940–8949 (2020)
22. Li, D., Yang, Y., Song, Y.Z., Hospedales, T.: Learning to generalise: meta-learning for domain generalisation. In: AAAI (2018)
23. Li, H., Wang, Y., Wan, R., Wang, S., et al.: Domain generalisation for medical imaging classification with linear-dependency regularization. In: NeurIPS (2020)
24. Li, L., Zimmer, V.A., Schnabel, J.A., Zhuang, X.: AtrialGeneral: domain generalization for left atrial segmentation of multi-center LGE MRIs. In: de Bruijne, M., et al. (eds.) MICCAI 2021. LNCS, vol. 12906, pp. 557–566. Springer, Cham (2021). https://doi.org/10.1007/978-3-030-87231-1_54
25. Liu, N., Li, S., Du, Y., Tenenbaum, J., Torralba, A.: Learning to compose visual relations. In: NeurIPS, vol. 34 (2021)
26. Liu, Q., Chen, C., Qin, J., Dou, Q., Heng, P.A.: FedDG: Federated domain generalization on medical image segmentation via episodic learning in continuous frequency space. In: CVPR, pp. 1013–1023 (2021)
27. Liu, Q., Dou, Q., Heng, P.-A.: Shape-aware meta-learning for generalizing prostate MRI segmentation to unseen domains. In: Martel, A.L., et al. (eds.) MICCAI 2020. LNCS, vol. 12262, pp. 475–485. Springer, Cham (2020). https://doi.org/10.1007/978-3-030-59713-9_46
28. Liu, X., Thermos, S., Chartsias, A., O'Neil, A., Tsaftaris, S.A.: Disentangled representations for domain-generalised cardiac segmentation. In: STACOM Workshop (2020)

29. Liu, X., Thermos, S., O'Neil, A., Tsaftaris, S.A.: Semi-supervised meta-learning with disentanglement for domain-generalised medical image segmentation. In: de Bruijne, M., et al. (eds.) MICCAI 2021. LNCS, vol. 12902, pp. 307–317. Springer, Cham (2021). https://doi.org/10.1007/978-3-030-87196-3_29

30. Liu, X., Thermos, S., Valvano, G., Chartsias, A., O'Neil, A., Tsaftaris, S.A.: Measuring the biases and effectiveness of content-style disentanglement. In: BMVC (2021)

31. Milletari, F., Navab, N., Ahmadi, S.A.: VNet: fully convolutional neural networks for volumetric medical image segmentation. In: 3DV, pp. 565–571. IEEE (2016)

32. Paszke, A., Gross, S., Massa, F., Lerer, A., et. al: PyTorch: an imperative style, high-performance deep learning library. In: NeurIPS, pp. 8026–8037 (2019)

33. Prados, F., Ashburner, J., Blaiotta, C., Brosch, T., et al.: Spinal cord grey matter segmentation challenge. Neuroimage **152**, 312–329 (2017)

34. Ronneberger, O., Fischer, P., Brox, T.: U-Net: convolutional networks for biomedical image segmentation. In: Navab, N., Hornegger, J., Wells, W.M., Frangi, A.F. (eds.) MICCAI 2015. LNCS, vol. 9351, pp. 234–241. Springer, Cham (2015). https://doi.org/10.1007/978-3-319-24574-4_28

35. Snell, J., Swersky, K., Zemel, R.: Prototypical networks for few-shot learning. In: NeurIPS, pp. 4080–4090 (2017)

36. Tokmakov, P., Wang, Y.X., Hebert, M.: Learning compositional representations for few-shot recognition. In: CVPR, pp. 6372–6381 (2019)

37. Valvano, G., Leo, A., Tsaftaris, S.A.: Re-using adversarial mask discriminators for test-time training under distribution shifts. arXiv preprint arXiv:2108.11926 (2021)

38. Valvano, G., Leo, A., Tsaftaris, S.A.: Stop throwing away discriminators! Re-using adversaries for test-time training. In: Albarqouni, S., et al. (eds.) DART/FAIR -2021. LNCS, vol. 12968, pp. 68–78. Springer, Cham (2021). https://doi.org/10.1007/978-3-030-87722-4_7

39. Yao, H., Hu, X., Li, X.: Enhancing pseudo label quality for semi-superviseddomain-generalized medical image segmentation. arXiv preprint arXiv:2201.08657 (2022)

40. Yuan, X., Kortylewski, A., et al.: Robust instance segmentation through reasoning about multi-object occlusion. In: CVPR, pp. 11141–11150 (2021)

41. Zakazov, I., Shirokikh, B., Chernyavskiy, A., Belyaev, M.: Anatomy of domain shift impact on U-Net layers in MRI segmentation. In: de Bruijne, M., et al. (eds.) MICCAI 2021. LNCS, vol. 12903, pp. 211–220. Springer, Cham (2021). https://doi.org/10.1007/978-3-030-87199-4_20

42. Zhang, L., Wang, X., Yang, D., Sanford, T., et al.: Generalising deep learning for medical image segmentation to unseen domains via deep stacked transformation. IEEE TMI **39**(7), 2531–2540 (2020)

43. Zhang, Y., Kortylewski, A., Liu, Q., et al.: A light-weight interpretable compositionalnetwork for nuclei detection and weakly-supervised segmentation. arXiv preprint arXiv:2110.13846 (2021)

Domain Adaptive Nuclei Instance Segmentation and Classification via Category-Aware Feature Alignment and Pseudo-Labelling

Canran Li[1], Dongnan Liu[1], Haoran Li[2,3], Zheng Zhang[4], Guangming Lu[4], Xiaojun Chang[2], and Weidong Cai[1(✉)]

[1] School of Computer Science, University of Sydney, Sydney, Australia
cali5184@uni.sydney.edu.au, tom.cai@sydney.edu.au
[2] ReLER, AAII, University of Technology Sydney, Sydney, Australia
[3] Department of Data Science and Artificial Intelligence, Monash University, Melbourne, Australia
[4] School of Computer Science and Technology, Harbin Institute of Technology (Shenzhen), Shenzhen, China

Abstract. Unsupervised domain adaptation (UDA) methods have been broadly utilized to improve the models' adaptation ability in general computer vision. However, different from the natural images, there exist huge semantic gaps for the nuclei from different categories in histopathology images. It is still under-explored how could we build generalized UDA models for precise segmentation or classification of nuclei instances across different datasets. In this work, we propose a novel deep neural network, namely Category-Aware feature alignment and Pseudo-Labelling Network (CAPL-Net) for UDA nuclei instance segmentation and classification. Specifically, we first propose a category-level feature alignment module with dynamic learnable trade-off weights. Second, we propose to facilitate the model performance on the target data via self-supervised training with pseudo labels based on nuclei-level prototype features. Comprehensive experiments on cross-domain nuclei instance segmentation and classification tasks demonstrate that our approach outperforms state-of-the-art UDA methods with a remarkable margin.

Keywords: Computational pathology · Nuclear segmentation · Nuclear classification · Unsupervised domain adaption · Deep learning

1 Introduction

Automatic nuclei instance segmentation and classification are crucial for digital pathology with various application scenarios, such as tumour classification

Supplementary Information The online version contains supplementary material available at https://doi.org/10.1007/978-3-031-16449-1_68.

and cancer grading [1]. However, manual labelling is limited by high subjective, low reproducibility, and resource-intensive [2,3]. Although deep learning-based methods can achieve appealing nuclei recognition performance, they require sufficient labelled data for training [3–7]. By directly adopting the off-the-shelf deep learning models to a new histopathology dataset with a distinct distribution, the models suffer from performance drop due to domain bias [8,9]. Recently, unsupervised domain adaption methods have been proposed to tackle this issue [10–12], and enable the learning models to transfer the knowledge from one labelled source domain to the other unlabelled target domain [13–15].

Several methods have recently been proposed for unsupervised domain adaptive nuclei instance segmentation in histopathology images [8,9]. Inspired by CyCADA [16], Liu et al. [8] first synthesize target-like images and use a nuclei inpainting mechanism to remove the incorrectly synthesized nuclei. The adversarial training strategies are then used separately at the image-, semantic- and instance-level with a task re-weighting mechanism. However, this work can only address the instance segmentation for nuclei within the same class. In Yang et al.'s work [9] , firstly, local features are aligned by an adversarial domain discriminator, and then a pseudo-labelling self-training approach is used to further induce the adaptation. However, the performance improvement of this work relies on weak labels and fails to get good training results without target domain labels. In addition, although this work is validated on the cross-domain nuclei segmentation and classification, their adaptation strategies are class-agnostic. In the real clinical, the nuclei objects in the histopathology images belong to various classes, and the characteristics of the nuclei in different classes are also distinct [17]. In addition, the number of objects within each category is also imbalanced in the histopathology datasets [18]. To this end, previous UDA methods are limited for cross-domain nuclei segmentation and classification due to the lack of analysis on the nuclei classes when transferring the knowledge.

To address the aforementioned issues, in this work, we study cross-domain nuclei instance segmentation and classification via a novel class-level adaptation framework. First, we propose a category-aware feature alignment module to facilitate the knowledge transfer for the cross-domain intra-class features while avoiding negative transfer for the inter-class ones. Second, a self-supervised learning stage via nuclei-level feature prototypes is further designed to improve the model performance on the unlabelled target data. Extensive experiments indicated the effectiveness of our proposed method by outperforming state-of-the-art UDA methods on nuclei instance segmentation and classification tasks. Furthermore, the performance of our UDA method is comparable or even better than the fully-supervised upper bound under various metrics.

2   Methods

2.1   Overview

Our proposed model is based on Hover-Net [6], a state-of-the-art method for fully-supervised nuclei instance segmentation and classification. The framework is constructed by three branches sharing the same encoder for feature extraction and using three decoders for different tasks: 1) Nuclear pixel (NP) branch to perform binary classification of a pixel (nuclei or background); 2) Hover (HV) branch to predict the horizontal and vertical distances of nuclei pixels to their centroid; 3) Nuclear classification (NC) branch to classify the nuclei types of pixels. The supervised Hover-Net loss function of our model is defined as $\mathcal{L}_F$:

$$\mathcal{L}_F = \mathcal{L}_{np} + \mathcal{L}_{hover} + \mathcal{L}_{nc} \tag{1}$$

The network architecture of our proposed model is shown in Fig. 1. Our proposed UDA framework is optimized in two stages. First, class-level feature alignment modules are proposed to alleviate the domain gap at the feature level. In the second stage, the pseudo-labelling process enhanced by the nuclei-level prototype is further proposed for self-supervised learning on the unlabelled target images.

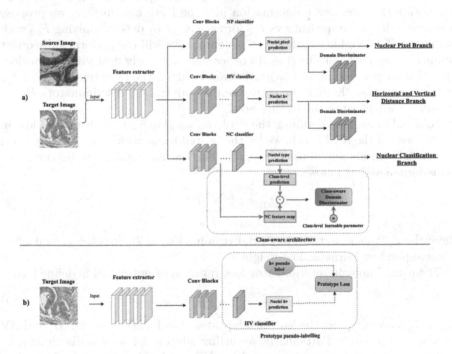

Fig. 1. Overview of the proposed category-aware prototype pseudo-labelling network.

2.2   Category-Aware Feature Alignment

In the histopathology datasets with multi-class nuclei, there is a very large gap in the number of nuclear in each category. This may lead to many images in the dataset having only a few nuclear categories, and some nuclear types are absent from an image. In addition, the features of the nuclei from different classes also vary. Under this situation, the typical class-agnostic feature alignment strategies may lead to the negative transfer of the features in different categories. To tackle this issue, for the NC branch on nuclei classification, we propose to conduct feature alignment for the cross-domain features within each class separately. Particularly, an adversarial domain discriminator D_c is introduced to adapt the features under the class c. Compared with directly employing a class-agnostic domain discriminator for the NC branch, the class-aware domain discriminators can avoid the misalignment across different classes and further encourage the knowledge transfer in classification learning.

The detailed paradigm is shown in Fig. 1a. First, we denote the features of the NC branch in the typical Hover-Net as F_{nc}, and the N-class prediction as P_{nc}. Note that P_{nc} contains N channels, and the predictions in each channel represent the classification results for each specific class. For each category c, we formulate the P_{nc} into a binary class prediction map P_{nc}^c, where the pixel value is set to 1 if it belongs to the nuclei in this category, otherwise set to 0. To incorporate the class-aware information into the feature alignment, we propose to generate the prototype features F_{pt}^c for the class c by dot-multiplying F_{nc} with the binary P_{nc}^c. In addition, if any P_{nc}^c is empty, we will not perform subsequent training on the nuclei in this class. In other words, we only deal with the nuclear types that exist in the input images. In each domain, the prototype features F_{pt}^c for the class c pass through the corresponding adversarial discriminators D_c for class-aware adaption at the feature level.

To avoid manually finetuning the trade-off weights for the adversarial loss in N categories of the NC branch, we let the overall framework automatically learn these weight parameters during training. The learnable weighted discriminator loss is formulated as follows:

$$\mathcal{L}_{NC}^{ca} = \omega_c^L \sum_{c=1}^L \mathcal{L}_c^{adv} \tag{2}$$

where the $\mathcal{L}_c^{adv}$ denotes the adversarial training loss of D_c for class c, and ω_c^L is its corresponding learnable loss weight.

The overall domain discriminator loss function of our model is defined as:

$$\mathcal{L}_{dis} = \mathcal{L}_{NC}^{ca} + \mathcal{L}_{NP}^{adv} + \mathcal{L}_{HV}^{adv} \tag{3}$$

where $\mathcal{L}_{NP}^{adv}$ and $\mathcal{L}_{HV}^{adv}$ are the feature adaptation loss functions in the NP and HV branches, respectively. Particularly, we utilize adversarial domain discriminators on the cross-domain output features of the NP and HV branches for adaptation. Details of the supervised Hover-Net loss and the adversarial loss are shown in the supplementary materials. With the above loss terms, the overall loss function of the first stage approach can be written as:

$$\mathcal{L}_{s1} = \mathcal{L}_F + \mathcal{L}_{dis} \tag{4}$$

2.3   Nuclei-Level Prototype Pseudo-Labelling

Although the class-aware feature alignment modules can narrow the domain gaps, the lack of supervised optimization on the target images still limits the model's segmentation and classification performance. Therefore, in the second stage, we use the output of the first stage model as pseudo labels for self-supervised learning to further improve the model performance on the target images. The detailed training process can be referred to Fig. 1b.

Different from the traditional self-training process with the pseudo labels, we only train the target domain of the HV branch during the second stage pseudo-labelling process. In the extensive experiments, we noticed that the performance of the classification predictions on the target images from the first stage model is limited. In addition, the binary segmentation predictions lack object-wise information. Moreover, the model trained with all pseudo labels may not perform well due to the low quality of some pseudo-labelling classes. To avoid the disturbance from the less accurate and representative pseudo labels, we no longer consider classification and binary segmentation branches in the second stage but particularly focus on the predictions from the HV branch, where the feature maps describe the distance from each pixel to the nuclei's centre point. Therefore, the features for each nuclear object can be regarded as a prototype at the object level, which contains morphological information such as the shape and size of the specific nuclear. By self-supervised learning with the pseudo labels on the predictions of the HV branch, the bias between nuclei objects can be further reduced. The overall loss function for the second stage is as follows:

$$\mathcal{L}_p = \frac{1}{N^p} \sum_{i=1}^{N^p} |x_i^p - \hat{y}_i^p|^2 \tag{5}$$

where x_i^p is the predicted features from the HV branch for each nuclear object p in the second stage and $\hat{y}_i^p$ is the object features generated by the pseudo labels.

3   Experiments

3.1   Datasets and Evaluation Metrics

We conduct experiments on two datasets from Lizard [18], a large-scale colon tissue histopathology database at the 20x objective magnification for nuclei instance segmentation and classification under six types: epithelial, connective tissue, lymphocytes, plasma, neutrophils, and eosinophils.

In this work, DigestPath (Dpath) and CRAG are employed, where the images in Dpath are extracted from histological samples from four different hospitals in China, and the CRAG dataset contains images extracted from whole-slide images (WSIs) from University Hospitals Coventry and Warwickshire (UHCW). For both two datasets, we select 2/3 of the whole images for training, and the remaining 1/3 for testing and validation. Specifically, we use Dpath as the source domain, with 46 images for training and the rest 22 for validation. CRAG dataset is used as the target domain, with 42 images for training and the remaining 21 for testing. The training images are randomly cropped to 256 × 256 patches, and the data augmentation methods are applied, including flip, rotate, Gaussian blur and median blur.

For evaluation, we choose the same metrics as Hover-Net [6]. Dice, Aggregated Jaccard Index (AJI), Detection Quality (DQ), Segmentation Quality (SQ), and Panoptic Quality (PQ) are for nuclei instance segmentation. In addition, the F1-scores at the detection and classification levels are employed to evaluate the nuclei detection and classification performance.

3.2 Implementation Details

We utilize the Hover-Net framework with ResNet50 [19] pre-trained weights on ImageNet as our base architecture. In the first stage of our adaptation process, the model is trained in two steps following the Hover-Net [6]. In the first step, only the decoders are trained 50 epochs. In the second step, all layers are trained for another 50 epochs. We use Adam optimization in both steps, with an initial learning rate of 1e–4, which is then reduced to 1e–5 after 25 epochs. In the self-training pseudo-labelling part, the Adam optimizer with a learning rate of 1e–4 was used to train 50 epochs with a batch size of 20. Experiments were conducted on one NVIDIA GeForce 3090 GPU and implemented using PyTorch.

3.3 Comparison Experiments

We conduct a series of comparative experiments to compare the performance. The details are as follows: (1) Source Only [6]: original Hover-Net without adaptation. (2) PDAM [8]: a UDA nuclei instance segmentation method with pixel-level and feature-level adaptation. Since it can only be used for binary classification for objects, we only compare the results of nuclei instance segmentation with this method. (3) Yang et al. [9]: a UDA framework is proposed based on global-level feature alignment for nuclei classification and nuclear instance segmentation. In addition, a weakly-supervised DA method is also proposed using weak labels for the target images such as nuclei centroid. We only compare with the UDA method in this work for a fair comparison. (4) Fully-supervised: fully supervised training on the labelled target images, as the upper bound.

Table 1. Experimental results on UDA nuclei instance segmentation.

Methods	$Dpath \rightarrow CRAG$				
	Dice	AJI	DQ	SQ	PQ
Source Only [6]	0.378	0.201	0.336	0.768	0.259
PDAM [8]	0.596	0.323	0.467	0.676	0.316
Yang et al. [9]	0.766	0.494	0.648	0.765	0.496
Baseline	0.750	0.455	0.604	0.759	0.458
Baseline+CA	0.772	0.502	0.661	**0.773**	0.510
Baseline+CA+PL	0.781	0.517	0.675	0.772	0.522
Proposed	**0.785**	**0.519**	**0.681**	0.769	**0.524**
Full-supervised	0.778	0.526	0.683	0.783	0.535

Evaluation on Nucleus Instance Segmentation. A comparison of the segmentation performance between our model and state-of-the-art methods is reported in Table 1. From the table, it can be observed that our instance segmentation effect is better than the two existing models. Compared with the source-only method, PDAM [8] has a performance improvement by aligning features at the panoptic level. The UDA method of Yang et al. [9] achieves good segmentation performance. Our method achieves the highest scores among all methods, with Dice and AJI being 2.3% and 1.6% higher than the previous methods, respectively. In comparison with the full-supervised model, the performance of our proposed UDA architecture is close to it. In particular, our Dice is higher than the results of the full-supervised model.

Table 2. Experimental results on UDA nuclei classification. F_c^1, F_c^2, F_c^3, F_c^4, F_c^5 and F_c^6 denote the F1 classification score for the Eosinophil, Epithelial, Lymphocyte, Plasma, Neutrophil and Connective tissue, respectively. F_{avg} denotes the average of all the F1-score for the classification under each category.

Methods	$Dpath \rightarrow CRAG$							
	Det	F_c^1	F_c^2	F_c^3	F_c^4	F_c^5	F_c^6	F_{avg}
Source Only [6]	0.490	0.022	0.389	0.324	0.195	0.038	0.161	0.188
Yang et al. [9]	0.736	0.037	0.670	0.330	0.371	0.017	0.428	0.309
Baseline	0.702	0.044	0.687	0.292	0.381	0.155	0.475	0.339
Baseline+CA	0.731	**0.128**	0.697	0.351	0.400	0.084	0.498	0.360
Baseline+CA+PL	0.772	0.110	**0.725**	0.327	0.383	0.352	**0.558**	0.409
Proposed	**0.775**	0.111	0.714	**0.389**	**0.402**	**0.377**	0.535	**0.421**
Full-supervised	0.748	0.167	0.724	0.388	0.419	0.428	0.545	0.445

Evaluation on Nucleus Classification. Table 2 reports the performance comparison of our method with other works on nuclei classification. Although Yang et al. [9] can achieve cross-domain nuclei classification and segmentation based on Hover-Net, their feature alignment modules are class-agnostic, which incurs the misalignment issues for the cross-domain features from different categories and further limits the performance. On the other hand, our proposed class-aware UDA framework has outperformed [9] by a large margin. Our model achieves a 4% improvement in the F1 detection score in the detection metric. In addition, our method improves the classification performance for all categories. Moreover, our method achieves significant improvements in classes with sparse samples like Neutrophil and Eosinophil, around 10% and 34%, respectively.

3.4   Ablation Studies

To test the validity of each component of our proposed model, we conducted ablation experiments. Firstly, based on our architecture, we kept only the feature-level domain discriminator on the three branches as our baseline method. Secondly, we kept only the class-aware structure, removing the pseudo labels based on the nuclei-level prototype (Baseline+CA). In addition, we also compared our nuclei-prototype pseudo-labelling process with the traditional one, which directly trains the model on the target images with all predictions in the first stage as the pseudo labels (Baseline+CA+PL).

Tables 1 and 2 show the instance segmentation and classification performance of all the ablation methods. From the tables, we can observe that the class-aware structure substantially improves the classification performance under categories with sparse samples (e.g. Neutrophil and Eosinophil), with a higher than 10% improvement in the classification F1-score. This phenomenon illustrates the effectiveness of class-aware adaptation in transferring the knowledge between the multi-class datasets. We note that the class-aware structure also has an approximate 2% improvement in nuclei segmentation.

Self-supervised training also improved both instance segmentation and classification performance. In addition, we note that prototype loss has a higher than 4% improvement on the F1-score for nuclei detection and achieves a better nuclei segmentation performance. In addition, our proposed nuclei prototype pseudo-labelling process also outperforms the typical pseudo-labelling. Due to the inferior classification performance of the first stage model, training models with all the pseudo labels might bring the noise to the network optimization, and limit the overall performance. Visualization examples of the ablation studies are shown in Fig. 2.

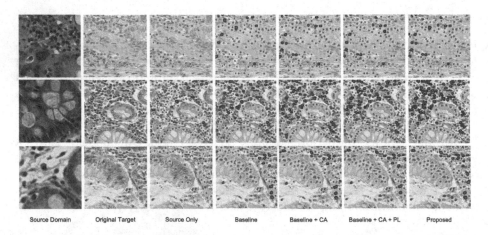

Fig. 2. Visualization predictions for the ablation experiments. Red: Eosinophil, Green: Epithelial, Yellow: Lymphocyte, Blue: Plasma, Magenta: Neutrophil, Cyan: Connective tissue. (Best to viewed in color and zoomed-in) (Color figure online)

4 Conclusion

In this paper, we proposed a category-aware prototype pseudo-labelling architecture for unsupervised domain adaptive nuclear instance segmentation and
classification. In our two-stage framework, category-aware feature alignment
with learnable trade-off loss weights is proposed to tackle the class-imbalance
issue and avoid misalignment during the cross-domain study. In addition, we
proposed a nuclei-level prototype loss to correct the deviation in the second
stage pseudo-labelling training, which further improves the segmentation and
classification performance on the target images by introducing auxiliary self-
supervision. Comprehensive results on various cross-domain nuclei instance segmentation and classification tasks demonstrate the prominent performance of
our approach. Given the appealing performance of our method on the UDA
nuclei instance segmentation tasks, we suggest that future directions can focus
on the cross-domain multi-class object recognition tasks for other medical and
general computer vision scenarios.

References

1. May, M.: A better lens on disease: computerized pathology slides may help doctors
   make faster and more accurate diagnoses. Sci. Am. **302**, 74–77 (2010)
2. Lee, H., Kim, J.: Segmentation of overlapping cervical cells in microscopic images
   with super-pixel partitioning and cell-wise contour refinement. In: Proceedings of
   the IEEE Conference on Computer Vision and Pattern Recognition Workshops,
   pp. 63–69 (2016)
3. Naylor, P., Laé, M., Reyal, F., Walter, T.: Segmentation of nuclei in histopathology
   images by deep regression of the distance map. IEEE Trans. Med. Imaging **38**(2),
   448–459 (2019)
4. Zhou, Y., Onder, O.F., Dou, Q., Tsougenis, E., Chen, H., Heng, P.-A.: CIA-Net:
   robust nuclei instance segmentation with contour-aware information aggregation.
   In: Chung, A.C.S., Gee, J.C., Yushkevich, P.A., Bao, S. (eds.) IPMI 2019. LNCS,
   vol. 11492, pp. 682–693. Springer, Cham (2019). https://doi.org/10.1007/978-3-
   030-20351-1_53
5. Chen, S., Ding, C., Tao, D.: Boundary-assisted region proposal networks for nucleus
   segmentation. In: Martel, A.L., et al. (eds.) MICCAI 2020. LNCS, vol. 12265, pp.
   279–288. Springer, Cham (2020). https://doi.org/10.1007/978-3-030-59722-1_27
6. Graham, S., et al.: Hover-Net: simultaneous segmentation and classification of
   nuclei in multi-tissue histology images. Med. Image Anal. **58**, 101563 (2019)
7. Liu, D., Zhang, D., Song, Y., Huang, H., Cai, W.: Panoptic feature fusion Net: a
   novel instance segmentation paradigm for biomedical and biological images. IEEE
   Trans. Image Process. **30**, 2045–2059 (2021)
8. Liu, D., et al.: PDAM: a panoptic-level feature alignment framework for unsupervised domain adaptive instance segmentation in microscopy images. IEEE Trans.
   Med. Imaging **40**(1), 154–165 (2021)
9. Yang, S., Zhang, J., Huang, J., Lovell, B.C., Han, X.: Minimizing labeling cost for
   nuclei instance segmentation and classification with cross-domain images and weak
   labels. Proc. AAAI Conf. Artif. Intell. **35**(1), 697–705 (2021)

10. Chen, C., Dou, Q., Chen, H., Qin, J., Heng, P.A.: Unsupervised bidirectional cross-modality adaptation via deeply synergistic image and feature alignment for medical image segmentation. IEEE Trans. Med. Imaging **39**(7), 2494–2505 (2020)
11. Zhou, Y., Huang, L., Zhou, T., Shao, L.: CCT-Net: category-invariant cross-domain transfer for medical single-to-multiple disease diagnosis. In Proceedings of the IEEE/CVF International Conference on Computer Vision, pp. 8260–8270 (2021)
12. Liu, D., et al.: Unsupervised instance segmentation in microscopy images via panoptic domain adaptation and task re-weighting. In: Proceedings of the IEEE/CVF Conference on Computer Vision and Pattern Recognition, pp. 4243–4252 (2020)
13. Kang, G., Jiang, L., Yang, Y., Hauptmann, A.G.: Contrastive adaptation network for unsupervised domain adaptation. In: Proceedings of the IEEE/CVF Conference on Computer Vision and Pattern Recognition, pp. 4893–4902 (2019)
14. Long, M., Cao, Y., Wang, J., Jordan, M.: Learning transferable features with deep adaptation networks. In: International Conference on Machine Learning, pp. 97–105 (2015)
15. Zhang, Q., Zhang, J., Liu, W., Tao, D.: Category anchor-guided unsupervised domain adaptation for semantic segmentation. In: Advances in Neural Information Processing Systems 32 (2019)
16. Hoffman, J., et al.: CyCADA: cycle-consistent adversarial domain adaptation. In: International Conference on Machine Learning, pp. 1989–1998 (2018)
17. Irshad, H., Veillard, A., Roux, L., Racoceanu, D.: Methods for nuclei detection, segmentation, and classification in digital histopathology: a review-current status and future potential. IEEE Rev. Biomed. Eng. **7**, 97–114 (2013)
18. Graham, S., et al.: Lizard: a Large-Scale Dataset for Colonic Nuclear Instance Segmentation and Classification. In: Proceedings of the IEEE/CVF International Conference on Computer Vision, pp. 684–693 (2021)
19. He, K., Zhang, X., Ren, S., Sun, J.: Deep residual learning for image recognition. In: Proceedings of the IEEE Conference on Computer Vision and Pattern Recognition, pp. 770–778 (2016)

Learn to Ignore: Domain Adaptation for Multi-site MRI Analysis

Julia Wolleb[1]([✉]), Robin Sandkühler[1], Florentin Bieder[1], Muhamed Barakovic[1],
Nouchine Hadjikhani[3,4], Athina Papadopoulou[1,2], Özgür Yaldizli[1,2],
Jens Kuhle[2], Cristina Granziera[1,2], and Philippe C. Cattin[1]

[1] Department of Biomedical Engineering, University of Basel, Allschwil, Switzerland
`julia.wolleb@unibas.ch`
[2] University Hospital Basel, Basel, Switzerland
[3] Massachusetts General Hospital, Harvard Medical School, Charlestown, MA, USA
[4] Gillberg Neuropsychiatry Center, Sahlgrenska Academy, University of Gothenburg,
Gothenburg, Sweden

Abstract. The limited availability of large image datasets, mainly due
to data privacy and differences in acquisition protocols or hardware,
is a significant issue in the development of accurate and generalizable
machine learning methods in medicine. This is especially the case for
Magnetic Resonance (MR) images, where different MR scanners introduce a bias that limits the performance of a machine learning model. We
present a novel method that learns to ignore the scanner-related features
present in MR images, by introducing specific additional constraints on
the latent space. We focus on a real-world classification scenario, where
only a small dataset provides images of all classes. Our method *Learn
to Ignore (L2I)* outperforms state-of-the-art domain adaptation methods on a multi-site MR dataset for a classification task between multiple
sclerosis patients and healthy controls.

Keywords: Domain adaptation · Scanner bias · MRI

1 Introduction

Due to its high soft-tissue contrast, Magnetic Resonance Imaging (MRI) is a
powerful diagnostic tool for many neurological disorders. However, compared to
other imaging modalities like computed tomography, MR images only provide
relative values for different tissue types. These relative values depend on the
scanner manufacturer, the scan protocol, or even the software version. We refer
to this problem as the scanner bias. While human medical experts can adapt
to these relative changes, they represent a major problem for machine learning

Supplementary Information The online version contains supplementary material
available at https://doi.org/10.1007/978-3-031-16449-1_69.

methods, leading to a low generalization quality of the model. By defining different scanner settings as different domains, we look at this problem from the perspective of domain adaptation (DA) [3], where the main task is learned on a *source domain*. The model then should perform well on a different *target domain*.

1.1 Related Work

An overview of DA in medical imaging can be found at [13]. One can generally distinguish between unsupervised domain adaptation (UDA) [1], where the target domain data is unlabeled, or supervised domain adaptation (SDA) [28], where the labels of the target domain are used during training.

The problem of scanner bias is widely known to disturb the automated analysis of MR images [22], and a lot of work already tackles the problem of multi-site MR harmonization [11]. Deepharmony [7] uses paired data to change the contrast of MRI from one scanner to another scanner with a modified U-Net. Generative Adversarial Networks aim to generate new images to overcome the domain shift [24]. These methods modify the intensities of each pixel before training for the main task. This approach is preferably avoided in medical applications, as it bears the risk of removing important pixel-level information required later for other tasks, such as segmentation or anomaly detection.

Domain-adversarial neural networks [12] can be used for multi-site brain lesion segmentation [18]. Unlearning the scanner bias [9] is an SDA method for MRI harmonization and improves the performance in age prediction from MR images. The introduction of contrastive loss terms [20,25,30] can also be used for domain generalization [10,19,23]. Disentangling the latent space has been done for MRI harmonization [4,8]. Recently, heterogeneous DA [2] was also of interest for lesion segmentation [5].

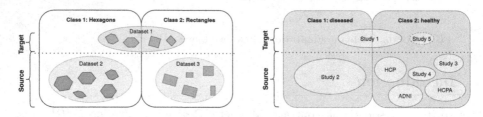

Fig. 1. Quantity charts for the datasets in the source and target domain. The chart on the left illustrates the problem, and the chart on the right shows the real-world application on the MS dataset. (Color figure online)

1.2 Problem Statement

All DA methods mentioned in Sect. 1.1 have in common that they must learn the main task on the source domain. However, it can happen that the bias present in datasets of various origins disturbs the learning of a specific task. Figure 1 on the left illustrates the problem and the relation of the different datasets on a toy example for the classification task between hexagons and rectangles. Due to the high variability in data, often only a small and specific dataset is at hand to learn the main task: Dataset 1 forms the target domain with only a small number of samples of hexagons and pentagons. Training on this dataset alone yields a low generalization quality of the model. To increase the number of training samples, we add Dataset 2 and Dataset 3. They form the source domain. As these additional datasets come from different origins, they differ from each other in color. Note that they only provide either rectangles (Dataset 3) or hexagons (Dataset 2). The challenge of such a setup is that during training on the source domain, the color is the dominant feature, and the model learns to distinguish between green and red rather than counting the number of vertices. Classical DA approaches then learn to overcome the domain shift between source and target domain. However, the model will show poor performance on the target domain: The learned features are not helpful, as all hexagons and rectangles are blue in Dataset 1.

This type of problem is highly common in the clinical environment, where different datasets are acquired with different settings, which corresponds to the colors in the toy example. In this project, the main task is to distinguish between multiple sclerosis (MS) patients and healthy controls. The quantity chart in Fig. 1 on the right visualizes the different allocations of the MS dataset. Only the small in-house Study 1 provides images of both MS patients and healthy subjects acquired with the same settings. To get more data, we collect images from other in-house studies. As in the hospital mostly data of patients are collected, we add healthy subjects from public datasets, resulting in the presented problem.

In this work, we present a new supervised DA method called *Learn to Ignore (L2I)*, which aims to ignore features related to the scanner bias while focusing on disease-related features for a classification task between healthy and diseased subjects. We exploit the fact that the target domain contains images of subjects of both classes with the same origin, and use this dataset to lead the model's attention to task-specific features. We developed specific constraints on the latent space and introduce two novel loss terms that can be added to any classification network. We evaluate our method on a multi-site MR dataset of MS patients and healthy subjects, compare it to state-of-the-art methods, and perform various ablation studies. The source code is available at https://gitlab.com/cian.unibas. ch/L2I.

2   Method

We developed a strategy that aims to ignore features that disturb the learning of a classification task between n classes. The building blocks of our setup are shown in Fig. 2. The input image $x_i \in \mathbb{R}^3$ of class $i \in \{1, ..., n\}$ is the input for the encoder network E with parameters θ_E, which follows the structure of Inception-ResNet-v1 [26]. However, we replaced the 2D convolutions with

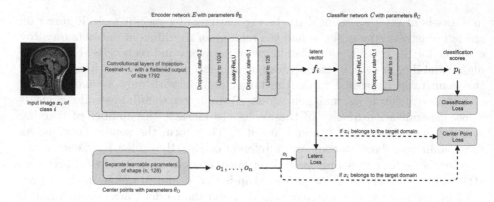

Fig. 2. Architecture of the classification network consisting of an encoder E with parameters θ_E, a fully connected classifier C with parameters θ_C, and separate learnable parameters θ_O. Here, $o_1, ..., o_n$ are learnable center points in the latent space, f_i is a vector in the latent space, and p_i is the classification score for class i. (Color figure online)

3D convolutions and changed the batch normalization layers to instance normalization layers. The output is the latent vector $f_i = E(x_i) \in \mathbb{R}^m$, where m denotes the dimension of the latent space. This latent vector is normalized to a length of 1 and forms the input for the classification network C with parameters θ_C. Finally, we get the classification scores $p_i = C(f_i) = C(E(x_i))$ for class $i \in \{1, ..., n\}$. To make the separation between the classes in the latent space learnable, we introduce additional parameters θ_O that learn normalized center points $\mathcal{O} = \{o_1, ..., o_n\} \subset \mathbb{R}^m$.

To suppress the scanner-related features, we embed the latent vectors in the latent space such that latent vectors from the same class are close to each other, and those from different classes are further apart, irrespective of the domain. We exploit the fact that the target domain contains images of all classes of the same origin. The model learns the separation of the embeddings using data of the target domain only. A schematic overview in 2D for the case of $n = 2$ classes is given in Fig. 3. We denote the latent vector of an image of the target domain of class i as $f_{i,t}$, where t denotes the affiliation to the target domain. The center points $\mathcal{O}$ are learned considering the latent vectors $f_{i,t}$ only, such that o_i is close to $f_{i,t}$, for $i \in \{1, ..., n\}$, and o_i is far from o_j for $i \neq j$. We force the latent vector f_i of an input image x_i into a hypersphere of radius r centered in o_i. For illustration, we use the toy example of Sect. 1.2: As all elements of the target domain are blue, the two learnable center points o_1 and o_2 are separated from each other based only on the number of vertices. The color is ignored. All latent vectors of hexagons f_1 of the source domain should lie in a ball around o_1, and all latent vectors of rectangles f_2 should lie in a ball around o_2.

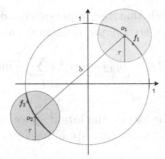

Fig. 3. The diagram shows a 2D sketch of the proposed latent space for $n = 2$ classes, with two learnable center points o_1 and o_2. The latent vectors are normalized and lie on the unit hypersphere. Latent vectors f_i of images of class i should lie within the circle around o_i, on the orange line or blue line respectively. (Color figure online)

2.1 Loss Functions

The overall objective function is given by

$$\mathcal{L}_{\text{total}} = \underbrace{\mathcal{L}_{\text{cls}}}_{\theta_C} + \underbrace{\lambda_{\text{cen}}\mathcal{L}_{\text{cen}}}_{\theta_O, \theta_E} + \underbrace{\lambda_{\text{latent}}\mathcal{L}_{\text{latent}}}_{\theta_E}. \tag{1}$$

It consists of three components: A classification loss $\mathcal{L}_{\text{cls}}$, a center point loss $\mathcal{L}_{\text{cen}}$ for learning $\mathcal{O} = \{o_1, ..., o_n\}$, and a loss $\mathcal{L}_{\text{latent}}$ on the latent space. Those components are weighted with the hyperparameters $\lambda_{\text{latent}}, \lambda_{\text{cen}} \in \mathbb{R}$. The parameters θ_E, θ_C and θ_O indicate which parameters of the network are updated with which components of the loss term. While the classification loss $\mathcal{L}_{\text{cls}}$ is separate and only responsible for the final score, it is the center point loss $\mathcal{L}_{\text{cen}}$ and the latent loss λ_{latent} that iteratively adapt the feature space to be scanner-invariant. With this total loss objective, any classification network can be extended by our method.

Classification Loss. The classification loss $\mathcal{L}_{\text{cls}, \theta_C}(f_i)$ is defined by the cross-entropy loss. The gradient is only calculated with respect to θ_C, as we do not want to disturb the parameters θ_E with the scanner bias.

Center Point Loss. To determine the center points, we designed a novel loss function defined by the distance from a latent vector $f_{i,t}$ of the target domain to its corresponding center point o_i. We define a radius $r > 0$ and force the latent vectors of the target domain $f_{i,t}$ to be within a hypersphere of radius r centered in o_i. Moreover, o_i and o_j should be far enough from each other for $i \neq j$. As o_i is normalized to a length of one, the maximal possible distance between o_i and o_j equals 2. We add a loss term forcing the distance between o_i and o_j to be larger than a distance d. The choice of $d < 2$ and $r > 0$ with $d > 2r$ is closely related to the choice of a margin in conventional contrastive loss terms [14,25]. The network is not penalized for not forcing o_i in the perfect position, but only to an acceptable region, such that the hyperspheres do not overlap. Then, the center point loss used to update the parameters θ_O and θ_E is given by

$$\mathcal{L}_{\text{cen},\theta_O,\theta_E}(f_{i,t},\mathcal{O}) = \max(\|f_{i,t}-o_i\|_2-r,0)^2 + \sum_{k\neq i} \frac{1}{2}\max(d-\|o_k-o_i\|_2,0)^2. \quad (2)$$

Latent Loss. We define the loss on the latent space, similar to the *Center Loss* [30], by the distance from f_i to its corresponding center point o_i

$$\mathcal{L}_{\text{latent},\theta_E}(f_i,\mathcal{O}) = \max(\|f_i - o_i\|_2 - r,0)^2. \quad (3)$$

With this loss, all latent vectors f_i of the training set of class i are forced to be within a hypersphere of radius r around the center point o_i. This loss is used to update the parameters θ_E of the encoder. By choosing $r > 0$, the network is given some leeway to force f_i to an acceptable region around o_i, denoted by the orange and blue lines in Fig. 3.

3   Experiments

For the MS dataset, we collected T1-weighted images acquired with 3T MR scanners with the MPRAGE sequence from five different in-house studies. For data privacy concerns, this patient data is not publicly available. Written informed consent was obtained from all subjects enrolled in the studies. All data were coded (i.e. pseudo-anonymized) at the time of the enrollment of the patients. To increase the number of healthy controls, we also randomly picked MPRAGE images from the Alzheimer's Disease Neuroimaging Initiative[1] (ADNI) dataset, the Young Adult Human Connectome Project (HCP) [29] and the Human Connectome Project - Aging (HCPA) [16]. More details of the different studies are given in Table 1 of the supplementary material, including the split into training, validation, and test set. An example of the scanner bias effect for two healthy control groups of the ADNI and HCP dataset can be found in Sect. 3 of the supplementary material.

All images were preprocessed using the same pipeline consisting of skull-stripping with HD-BET [17], N4 biasfield correction [27], resampling to a voxel size of 1 mm×1 mm×1 mm, cutting the top and lowest two percentiles of the pixel intensities, and finally an affine registration to the MNI 152 standard space[2]. All images were cropped to a size of (124, 120, 172). The dimension of the latent space is $m = 128$. This results in a total number of parameters of $36,431,842$. We use the Adam optimizer [21] with $\beta_1 = 0.9$, $\beta_2 = 0.999$, and a weight decay of

---

[1] Data used in preparation of this article were obtained from the Alzheimer's Disease Neuroimaging Initiative(ADNI) database (adni.loni.usc.edu). The investigators within the ADNI contributed to the design and implementation of ADNI and/or provided data but did not participate in analysis or writing of this report.

[2] Copyright (C) 1993–2009 Louis Collins, McConnell Brain Imaging Centre, Montreal Neurological Institute, McGill University.

$5 \cdot 10^{-5}$. The learning rate for the parameters θ_O is $lr_O = 10^{-4}$, and the learning rate for the parameters θ_C and θ_E is $lr_{E,C} = 5 \cdot 10^{-5}$. We manually choose the hyperparameters $\lambda_{latent} = 1$, $\lambda_{cen} = 100$, $d = 1.9$, and $r = 0.1$.

An early stopping criterion, with a patience value of 20, based on the validation loss on the target domain, is used. For data sampling in the training set, we use the scheme presented in Algorithm 1 in Sect. 2 of the supplementary material. Data augmentation includes rotation, gamma correction, flipping, scaling, and cropping. The training was performed on an NVIDIA Quadro RTX 6000 GPU, and took about 8 h on the MS dataset. As software framework, we used Pytorch 1.5.0.

4 Results and Discussion

To measure the classification performance, we calculate the classification accuracy, the Cohen's kappa score [6], and the area under the receiver operating characteristic curve (AUROC) [15]. We compare our approach against the methods listed below. Implementation details and the source code of all comparing methods can be found at https://gitlab.com/cian.unibas.ch/L2I.

- Vanilla classifier (*Vanilla*): Same architecture as *L2I*, but only $\mathcal{L}_{cls}$ is taken to update the parameters of both the encoder E and the classifier C.
- Class-aware Sampling (*Class-aware*): We train the *Vanilla* classifier with class-aware sampling. In every batch each class and domain is represented.
- Weighted Loss (*Weighted*): We train the *Vanilla* classifier, but the loss function $\mathcal{L}_{cls}$ is weighted to compensate for class and domain imbalances.
- Domain-Adversarial Neural Network (*DANN*) [12]: The classifier learns both to distinguish between the domains and between the classes. It includes a gradient reversal for the domain classification.
- Unlearning Scanner Bias (*Unlearning*) [9]: A confusion loss aims to maximally confuse a domain predictor, such that only features relevant for the main task are extracted.
- Supervised Contrastive Learning (*Contrastive*) [20]: The latent vectors are pushed into clusters far apart from each other, with a sampling scheme and a contrastive loss term allowing for multiple positives and negatives.
- Contrastive Adaptation Network (*CAN*) [19]: This state-of-the-art DA method combines the maximum mean discrepancy of the latent vectors as loss objective, class-aware sampling, and clustering.
- Fixed Center Points (*Fixed*): We train *L2I*, but instead of learning the centerpoints o_1 and o_2 using the target domain, we fix the center points at $o_i = \dfrac{v_i}{\|v_i\|_2}$ for $i \in \{1, 2\}$, with $v_1 = (1, ..., 1)$ and $v_2 = (-1, ..., -1)$.
- No margin (*No-margin*): We train *L2I* with $d = 2$ and $r = 0$, such that no margin is chosen in the contrastive loss term in Eqs. 2 and 3.

We report the mean and standard deviation of the scores on the test set for 10 runs. For each run the dataset is randomly divided into training, validation, and test set. In the first three lines of Table 1, the scores are shown when *Vanilla*, *Class-aware*, and *Weighted* are trained only on the target domain. The very poor performance is due to overfitting on such a small dataset. Therefore, the target domain needs to be supplemented with other datasets. In the remaining lines of Table 1, we summarize the classification results for all methods when trained on the source and the target domain. Our method *L2I* strongly outperforms all other methods on the target domain. Although we favor the target domain during training, we see that *L2I* still has a good performance on the source domain. Therefore, we claim that the model learned to distinguish the classes based on disease-related features that are present in both domains, rather than based on scanner-related features. The benefit of learning o_1 and o_2 by taking only the target domain into account can be seen when comparing our method against *Fixed*. Moreover, by comparing *L2I* to *No-margin*, we can see that choosing $d < 2$ and $r > 0$ brings an advantage. All methods perform well on the source domain, where scanner-related features can be taken into account for classification. However, on the target domain, where only disease-related features can be used, the *Vanilla*, *Class-aware* and *Weighted* methods show a poor performance. A visualization of the comparison between the *Vanilla* classifier and our method *L2I* can be found in the t-SNE plots in Sect. 4 of the supplementary material. *DANN* and *Contrastive*, as well as the state-of-the-art methods *CAN* and *Unlearning* fail to show the performance they achieve in classical DA tasks. Although *CAN* is an unsupervised method, we think that the comparison to our supervised method is fair, since *CAN* works very well on classical DA problems.

Table 1. Mean [standard deviation] of the scores on the test set for 10 runs.

Set		Target domain			Source domain		
		Accuracy	Kappa	AUROC	Accuracy	Kappa	AUROC
Target	Vanilla	50.0 [0.0]	0.0 [0.0]	59.5 [13.0]			
	· Class-aware	65.0 [5.0]	29.3 [9.5]	68.6 [12.4]			
	· Weighted	50.3 [1.1]	0.0 [0.0]	73.3 [13.5]			
Target and source	Vanilla	69.0 [6.1]	38.0 [12.1]	79.8 [7.3]	90.3 [5.7]	80.7 [11.3]	95.0 [5.9]
	· Class-aware	71.0 [8.3]	42.0 [16.6]	79.3 [13.3]	90.7[4.2]	81.3 [8.3]	95.3 [3.9]
	· Weighted	71.3 [6.3]	42.7 [12.7]	81.2 [7.8]	92.2 [3.1]	84.6 [6.3]	95.1 [3.4]
	DANN	67.0 [5.7]	34.0 [11.5]	74.7 [8.4]	93.1 [2.8]	86.3 [5.5]	98.7 [0.7]
	Unlearning	70.7 [6.6]	41.3 [13.3]	81.7 [9.2]	85.5 [3.2]	71.0 [6.5]	90.5 [3.3]
	Contrastive	76.3 [6.7]	52.6 [13.5]	86.7 [9.5]	**94.5 [2.4]**	**89.0 [4.7]**	**98.9 [0.7]**
	CAN	75.3 [6.9]	50.7 [13.8]	83.8 [8.8]	92.8 [2.5]	85.7 [5.0]	96.7 [1.8]
	L2I [Ours]	**89.0 [3.9]**	**78.0 [7.7]**	**89.7 [7.6]**	92.0 [2.5]	84.0 [4.9]	91.7 [4.9]
	·Fixed	71.7[7.2]	43.3 [14.5]	76.5 [11.4]	90.5 [3.9]	81.0 [7.9]	90.2 [5.8]
	·No-margin	82.0 [3.9]	64.0 [7.8]	75.3 9.1]	91.7 [4.8]	83.3 [9.7]	84.9 [3.9]

5 Conclusion

We presented a method that can ignore image features that are induced by different MR scanners. We designed specific constraints on the latent space and define two novel loss terms, which can be added to any classification network. The novelty lies in learning the center points in the latent space using images of the monocentric target domain only. Consequently, the separation of the latent space is learned based on task-specific features, also in cases where the main task cannot be learned from the source domain alone. Our problem therefore differs substantially from classical DA or contrastive learning problems. We apply our method *L2I* on a classification task between multiple sclerosis patients and healthy controls on a multi-site MR dataset. Due to the scanner bias in the images, a vanilla classification network and its variations, as well as classical DA and contrastive learning methods, show a weak performance. *L2I* strongly outperforms state-of-the-art methods on the target domain, without loss of performance on the source domain, improving the generalization quality of the model. Medical images acquired with different scanners are a common scenario in longterm or multi-center studies. Our method shows a major improvement for this scenario compared to state-of-the-art methods. We plan to investigate how other tasks like image segmentation will improve by integrating our approach.

Acknowledgements. This research was supported by the Novartis FreeNovation initiative and the Uniscientia Foundation (project #147–2018).

References

1. Ackaouy, A., Courty, N., Vallée, E., Commowick, O., Barillot, C., Galassi, F.: Unsupervised domain adaptation with optimal transport in multi-site segmentation of multiple sclerosis lesions from MRI data. Front. Comput. Neurosci. **14**, 19 (2020)
2. Alipour, N., Tahmoresnezhad, J.: Heterogeneous domain adaptation with statistical distribution alignment and progressive pseudo label selection. Appl. Intell. **52**, 8038–8055 (2021). https://doi.org/10.1007/s10489-021-02756-x
3. Ben-David, S., Blitzer, J., Crammer, K., Kulesza, A., Pereira, F., Vaughan, J.W.: A theory of learning from different domains. Mach. Learn. **79**(1), 151–175 (2010)
4. Chartsias, A., et al.: Disentangled representation learning in cardiac image analysis. Med. Image Anal. **58**, 101535 (2019)
5. Chiou, E., Giganti, F., Punwani, S., Kokkinos, I., Panagiotaki, E.: Unsupervised domain adaptation with semantic consistency across heterogeneous modalities for MRI prostate lesion segmentation. In: Albarqouni, S., et al. (eds.) DART/FAIR -2021. LNCS, vol. 12968, pp. 90–100. Springer, Cham (2021). https://doi.org/10.1007/978-3-030-87722-4_9
6. Cohen, J.: A coefficient of agreement for nominal scales. Educ. Psychol. Measur. **20**(1), 37–46 (1960)
7. Dewey, B.E., et al.: DeepHarmony: a deep learning approach to contrast harmonization across scanner changes. Magn. Reson. Imaging **64**, 160–170 (2019)
8. Dewey, B.E., et al.: A disentangled latent space for cross-site MRI harmonization. In: Martel, A.L., et al. (eds.) MICCAI 2020. LNCS, vol. 12267, pp. 720–729. Springer, Cham (2020). https://doi.org/10.1007/978-3-030-59728-3_70

9. Dinsdale, N.K., Jenkinson, M., Namburete, A.I.L.: Unlearning scanner bias for MRI harmonisation. In: Martel, A.L., et al. (eds.) MICCAI 2020. LNCS, vol. 12262, pp. 369–378. Springer, Cham (2020). https://doi.org/10.1007/978-3-030-59713-9_36

10. Dou, Q., Coelho de Castro, D., Kamnitsas, K., Glocker, B.: Domain generalization via model-agnostic learning of semantic features. In: Advances in Neural Information Processing Systems 32 (2019)

11. Eshaghzadeh Torbati, M., et al.: A multi-scanner neuroimaging data harmonization using RAVEL and ComBat. Neuroimage **245**, 118703 (2021)

12. Ganin, Y., et al.: Domain-adversarial training of neural networks. J. Mach. Learn. Res. **17**(1), 2030–2096 (2016)

13. Guan, H., Liu, M.: Domain adaptation for medical image analysis: a survey. IEEE Trans. Biomed. Eng. **68**(1) (2021)

14. Hadsell, R., Chopra, S., LeCun, Y.: Dimensionality reduction by learning an invariant mapping. In: 2006 IEEE Computer Society Conference on Computer Vision and Pattern Recognition, vol. 2, pp. 1735–1742 (2006)

15. Hanley, J.A., McNeil, B.J.: The meaning and use of the area under a receiver operating characteristic (ROC) curve. Radiology **143**(1), 29–36 (1982)

16. Harms, M.P., et al.: Extending the Human Connectome Project across ages: imaging protocols for the Lifespan Development and Aging projects. Neuroimage **183**, 972–984 (2018)

17. Isensee, F., et al.: Automated brain extraction of multisequence MRI using artificial neural networks. Hum. Brain Mapp. **40**(17), 4952–4964 (2019)

18. Kamnitsas, K., et al.: Unsupervised domain adaptation in brain lesion segmentation with adversarial networks. In: Niethammer, M., et al. (eds.) IPMI 2017. LNCS, vol. 10265, pp. 597–609. Springer, Cham (2017). https://doi.org/10.1007/978-3-319-59050-9_47

19. Kang, G., Jiang, L., Yang, Y., Hauptmann, A.G.: Contrastive adaptation network for unsupervised domain adaptation. In: Proceedings of the IEEE Conference on Computer Vision and Pattern Recognition, pp. 4893–4902 (2019)

20. Khosla, P., et al.: Supervised contrastive learning. arXiv preprint arXiv:2004.11362 (2020)

21. Kingma, D.P., Ba, J.: Adam: a method for stochastic optimization. arXiv preprint arXiv:1412.6980 (2014)

22. Mårtensson, G., et al.: The reliability of a deep learning model in clinical out-of-distribution MRI data: a multicohort study. Med. Image Anal. **66**, 101714 (2020)

23. Motiian, S., Piccirilli, M., Adjeroh, D.A., Doretto, G.: Unified deep supervised domain adaptation and generalization. In: Proceedings of the IEEE International Conference on Computer Vision, pp. 5715–5725 (2017)

24. Sankaranarayanan, S., Balaji, Y., Castillo, C.D., Chellappa, R.: Generate to adapt: aligning domains using generative adversarial networks. In: Proceedings of the IEEE Conference on Computer Vision and Pattern Recognition, pp. 8503–8512 (2018)

25. Schroff, F., Kalenichenko, D., Philbin, J.: Facenet: A unified embedding for face recognition and clustering. In: Proceedings of the IEEE Conference on Computer Vision and Pattern Recognition, pp. 815–823 (2015)

26. Szegedy, C., Ioffe, S., Vanhoucke, V., Alemi, A.: Inception-v4, inception-ResNet and the impact of residual connections on learning. In: Proceedings of the AAAI Conference on Artificial Intelligence, vol. 31, 4278–4284 (2017)

27. Tustison, N.J., et al.: N4ITK: improved N3 bias correction. IEEE Trans. Med. Imaging **29**(6), 1310–1320 (2010)

28. Valverde, S., et al.: One-shot domain adaptation in multiple sclerosis lesion segmentation using convolutional neural networks. Neuroimage Clin. **21**, 101638 (2019)
29. Van Essen, D.C., et al.: The Human Connectome Project: a data acquisition perspective. Neuroimage **62**(4), 2222–2231 (2012)
30. Wen, Y., Zhang, K., Li, Z., Qiao, Y.: A discriminative feature learning approach for deep face recognition. In: Leibe, B., Matas, J., Sebe, N., Welling, M. (eds.) ECCV 2016. LNCS, vol. 9911, pp. 499–515. Springer, Cham (2016). https://doi.org/10.1007/978-3-319-46478-7_31

Enhancing Model Generalization for Substantia Nigra Segmentation Using a Test-time Normalization-Based Method

Tao Hu[1], Hayato Itoh[1], Masahiro Oda[1,2], Yuichiro Hayashi[1], Zhongyang Lu[1], Shinji Saiki[3], Nobutaka Hattori[3], Koji Kamagata[3], Shigeki Aoki[3], Kanako K. Kumamaru[3], Toshiaki Akashi[3], and Kensaku Mori[1,4(✉)]

[1] Graduate School of Informatics, Nagoya University, Nagoya, Japan
`hitoh@mori.m.is.nagoya-u.ac.jp`, `kensaku@is.nagoya-u.ac.jp`
[2] Information Strategy Office, Information and Communications, Nagoya University, Nagoya, Japan
[3] Department of Neurology, School of Medicine, Juntendo University, Tokyo, Japan
[4] Research Center for Medical Bigdata, National Institute of Informatics, Tokyo, Japan

Abstract. Automatic segmentation of substantia nigra (SN), which is Parkinson's disease-related tissue, is an important step toward accurate computer-aided diagnosis systems. Conventional methods for SN segmentation depend heavily on limited magnetic resonance imaging (MRI) modalities such as neuromelanin and quantitative susceptibility mapping, which require longer imaging times and are rare in public datasets. To enable a multi-modal investigation for SN anatomic alterations based on medical bigdata researches, the need for automated SN segmentation arises from commonly investigated T2-weighted MRIs. To improve the performance of the automated SN segmentation from a T2-weighted MRI and enhance the model generalization for cross-center researches, this paper proposes a novel test-time normalization (TTN) method to increase the geometric and intensity similarity between the query data and the model's trained data. Our proposed method requires no additional training procedure or extra annotation for the unseen data. Our results showed that our proposed TTN achieved a mean Dice score of 71.08% in comparison with the baseline model's 69.87% score with in-house dataset. Additionally, improved SN segmentation performance was observed from the unseen and unlabeled datasets.

Keywords: Magnetic resonance imaging · Substantia nigra · Segmentation · Test-time normalization · Deep learning

1 Introduction

As a typical progressive neurodegenerative disorder, Parkinson's disease (PD) is characterized by continuous damage to motor functions. The most commonly reported anatomic alteration related to motor dysfunction is the progressive loss of dopaminergic neurons in the substantia nigra (SN) [1]. To further investigate

L. Wang (Eds.): MICCAI 2022, LNCS 13437, pp. 736–744, 2022.
https://doi.org/10.1007/978-3-031-16449-1_70

the underlying PD pathogenesis and search for an efficient boimarker for PD diagnosis, SN must be precisely segmented from MRI volumes.

Fully automated SN segmentation from MRI is a challenging task due to SN's tiny size, its ambiguous boundaries, and its large morphometric variability. Up to now, researchers have developed several semi-automated SN segmentation methods based on manually cropped regions of interest (ROI) and intensity threshold or region growing algorithms [2,3]. The most widely used fully automated method is the atlas-based method that uses input MRI and a labeled atlas pair [4]. Conventionally, the atlas-based method involves whole brain registration between the input MRI and the atlas. Since the registration process tends to pay more attention to larger structures like white matter, the atlas-based method does not perform so well in small structures like SN.

On the other hand, deep learning techniques have achieved significant improvements in various medical image scenarios, including organ and lesion segmentation of various sizes [5–7]. Inspired by a previous work [8], which segmented SN from neuromelanin MRI with a fully convolutional network (FCN), we propose a deep-learning-based method to improve the SN segmentation accuracy from a T2-weighted MRI. The major difficulty of SN segmentation from a T2-weighted MRI is the low contrast between SN and non-SN voxels. In addition, deep-learning-based segmentation methods tend to depend heavily on large-scaled annotated datasets, which are rather labor-intensive and time-consuming. Unfortunately, a model trained with small datasets carries relatively higher risks for over-fitting and might lack generalization in practice. For SN segmentation from a T2-weighted MRI, annotated bigdata are quite rare, greatly complicating training a model that has both acceptable accuracy and reliable generalization.

To tackle these problems, we first used a coarse-to-fine cascaded FCN to balance the memory requirement and local information preservation. The first FCN is trained to segment SN on a down-sampled MRI volume, and the predicted result is used for locating the ROI. The second FCN is trained to segment the SN at full resolution from a concatenation of the upsampled segmentation result of the first FCN and the original MRI in ROI.

Furthermore, we proposed the test-time normalization (TTN) method in the inference phase. By aligning the input MRI to the reference MRI in the support set, we generated a series of transformed input volumes that resemble the reference MRI volumes in latent space. Then the segmentation results from the normalized MRI volumes are combined for determining the final segmentation and estimating the voxel-wise uncertainty, which locates the ambiguous regions for SN identification.

Our paper's main contributions can be summarized as follows: (1) We propose a novel TTN for boosting SN segmentation accuracy, model generalization, and estimating voxel-wise model uncertainty. (2) We propose a prior-atlas-based metric normalized likelihood for estimating the SN segmentation performance in unlabeled datasets. (3) We investigate the feasibility and performance of fully automated SN segmentation using only the T2-weighted MRI and deep learning techniques.

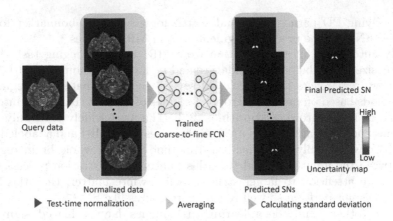

Fig. 1. Workflow in inference phase: Query MRI is first normalized to multiple transformed MRIs by our proposed TTN. Normalized MRIs are then fed to the trained, coarse-to-fine FCN to obtain multiple SN segmentation results. Final SN segmentation result is produced by averaging predicted SN probability maps. Voxel-wise uncertainty is calculated as standard deviation of predicted SN probability.

2   Methods

The cascaded FCN is comprised of two 3D U-Nets [9]. Coarse and fine 3D U-Nets share the same architecture and are trained sequentially, which means the coarse network is trained before the fine network. To encourage the model's attention on SN regions, we adopt the recently proposed loss function asymmetric loss (ASL) [10] to train the model.

2.1   Test-time Normalization

To further boost the SN segmentation accuracy and estimate the voxel-wise uncertainty as a reference for the users, we propose TTN[1] to refine the segmentation results by the coarse-to-fine FCN. Figure 1 shows the workflow in the inference phase with our proposed TTN. No additional training procedure is required to introduce our proposed TTN, which is only performed in the inference phase.

According to a recent study [11], if we combine (in the inference phase) the predictions of several geometrically transformed copies of a test image, higher segmentation accuracy might be achieved. Wang et al. proposed test-time augmentation (TTA) that employed an aleatoric uncertainty estimation method for medical image segmentation by applying affine transformation to test images [11]. The original aleatoric uncertainty estimation method mainly deals with geometrical variations by performing the same augmentation procedures in the training and testing phases.

---

[1] https://github.com/MoriLabNU/TTN_for_SN_segmentation.

However, in the scenario of a small dataset, since the training set might only be distributed in a small space of the overall distribution, the same augmentation procedure cannot guarantee the similar feature distribution of the query and training samples in latent space. Furthermore, the proposed TTA [11] does not take the samples' differences of intensity distribution into account, which is also not uncommon in the case of small datasets and may further limit the model's generalization.

Alternatively, we chose to use a support set to cross the gap between a test MRI and a fine-segmented MRI. We expect this process to normalize the former in both the spatial location and intensity distribution, allowing the normalized MRI to share more semantic information with the MRIs in the support set.

As shown in Fig. 2, the proposed TTN sequentially includes an affine transformation and a histogram-matching step [12]. The support set consists of N MRIs from the training set with the highest segmentation Dice coefficient. For the i-th reference MRI in the support set, the corresponding normalized test MRI of the j-th query case $\hat{\mathbf{I}}_i^{(j)}$ is defined by Eq. 1:

$$\hat{\mathbf{I}}_i^{(j)} = \mathcal{T}_{\text{Hist}}^{(i)}(\mathcal{T}_{\text{Aff}}^{(i)}(\mathbf{I}_j, \mathbf{I}_i), \mathbf{I}_i), \tag{1}$$

where $\mathcal{T}_{\text{Aff}}^{(i)}$ is the affine transformation, which moves query MRI $\mathbf{I}_j$ to the space of reference MRI $\mathbf{I}_i$, and $\mathcal{T}_{\text{Hist}}^{(i)}$ is a histogram-mapping function, which moves the histogram of the registered MRI to that of $\mathbf{I}_i$ [13]. The corresponding SN probability map in the reference space by FCN is then defined as $f(\hat{\mathbf{I}}_j^{(i)})$, where f represents the trained FCN. The final predicted SN probability map $\mathbf{S}_j$ of the j-th case in the original space is the averaged probability maps of the back-transformed MRIs, as shown in Eq. 2:

$$\mathbf{S}_j = \frac{1}{N} \sum_{i=1}^{N} \mathcal{T}_{\text{Aff}}^{(i)^{-1}}(f(\hat{\mathbf{I}}_j^{(i)})). \tag{2}$$

The uncertainty of the j-th test case in the k-th voxel is defined by Eq. 3 as the standard deviation of the predicted SN probability:

$$\mathbf{U}_{j,k} = \sqrt{\frac{1}{N-1} \sum_{i=1}^{N} (\mathbf{S}_{j,k} - \mathcal{T}_{\text{Aff}}^{(i)^{-1}}(f(\hat{\mathbf{I}}_{j,k}^{(i)})))^2}. \tag{3}$$

Note that we use the cumulative distribution function (CPD) based on the histogram-matching program not only for simplicity but also as a feature of the monotonicity preservation of pixel intensity, which might reduce the introduced artifact. In the present study, N is set to 10 to balance the computation time and the stable performance [11].

2.2   Prior-atlas-based Likelihood Estimation

To verify the improved generalization of our proposed TTN and to investigate the performance on unseen datasets, we collected T2-weighted MRI volumes from

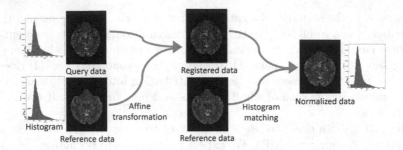

Fig. 2. Workflow of proposed TTN: Affine transformation first co-registers query and reference MRI. Then we performed histogram matching to normalize the registered MRI's histogram. Red circles indicate histogram range which shows significant difference between query and reference MRI before TTN and enhanced similarity after it.

two public datasets. Since these two datasets are not primarily designed for the research of brain tissue segmentation, there is no manually annotated ground truth of SN. To quantitatively evaluate the segmentation results on these public datasets, we propose a prior-atlas-based metric to estimate the segmentation accuracy using the annotated data of the in-house dataset.

We first selected a case from the training set of the in-house dataset as the MRI template. Then its remaining training cases were aligned to the MRI template using affine transformation. By counting the frequency of the transformed SN label in every voxel, we acquired the voxel-wise prior probability of SN in the space of the MRI template, and our study used this probability map as the SN atlas $\mathbf{A}$.

For a given query case, we can estimate the prior probability of SN in the original MRI space by registering the MRI template to the query case's MRI and applying the same transformation to atlas $\mathbf{A}$. Then for binary mask $\mathbf{M}$, we can estimate the likelihood of $\mathbf{M}$:

$$H(\mathbf{M}) = \sum_k ((1 - M_k)log(1 - \hat{A}_k) + M_i log \hat{A}_k), \qquad (4)$$

where $M_k \in \{0, 1\}$ is the predicted label in the k-th voxel and $\hat{A}_k$ is the prior probability of SN in the k-th voxel of the registered atlas. Normalized likelihood $\hat{H}$ of $\mathbf{M}$ is then defined as

$$\hat{H}(\mathbf{M}) = 1 - \frac{H(\mathbf{M})}{V log 0.5}, \qquad (5)$$

where V is the total number of voxels in $\mathbf{M}$. The normalized likelihood value measures the extent to which $\mathbf{M}$ is likely drawn from the distribution of the true SN masks in the training set. The higher $\hat{H}$ is, the more spatial similarity $\mathbf{M}$ shares with the true SN masks in the training set. Therefore, normalized likelihood $\hat{H}$ is used for approaching segmentation accuracy on public datasets in which the ground truth is unavailable.

Based on the normalized likelihood measurement, we propose a post-processing procedure called *re-threshold* to grid search the probability threshold to maximize $\hat{H}$. Specifically, for every query case, we compute the $\hat{H}$ values under 40 candidate thresholds, and select the threshold with the highest $\hat{H}$ as the final threshold.

3   Experiments

3.1   Datasets

In-House Dataset: This paper includes 156 cases (73 healthy cases (HC) and 83 PD cases) with T2-weighted MRI volumes and manually annotated SN masks. Among them, we randomly selected 84 for training, 52 for testing, and the remaining 20 for validation. The following are the major scan parameters: 3200/564 ms repetition time/echo time; echo train length of 303; 3.86-mm spacing between slices; 256 × 240 mm field of view; 0.80-mm slice spacing/thickness. The manual annotations of SN were performed by a board-certified radiologist with ten years of experience specializing in neuroradiology and checked and confirmed by two other experts. Note that the models are trained using only the in-house dataset, which means public datasets are only used for the models' generalization evaluation and totally unseen in the training phase.

Unseen Dataset 1: This dataset, which is a part of an ongoing longitudinal study (R01MH114870 to Anna Manelis), includes 39 unipolar depression patients and 47 HCs with T1- and T2-weighted MRIs.

Unseen Dataset 2: The ALLEN BRAIN ATLAS dataset released 8 healthy cases with T1- and T2-weighted MRI.

3.2   Implementation Details

As for pre-processing procedures, automated skull betting was first applied to remove the structures outside the brain. Then the intensities of the volumes were linearly normalized to zero to one intervals.

For the segmentation module, the number of convolution layers was 4 in every convolution block. Batch normalization and ReLU activation function were used. Specifically, the coarse U-Net was trained with cross entropy loss. The ASL was only used to train the fine U-Net.

We used random rotation in 3D for data augmentation. The angles rotating around the 3D axes are drawn from the same uniform distribution with a maximum of 30 °C and a minimum of minus 30 °C. Besides the rotation augmentation method, to verify the effectiveness of histogram matching in data augmentation, we also generated a mixed augmented MRI training set, including both 3D-rotated and histogram-matched MRI volumes. For the single histogram-matching-based augmentation procedure, we randomly selected one case other than the augmented case itself from the training set as a reference case.

Table 1. Dice ratio (%, mean ± standard deviation) in HCs, PD patients, and overall cases of in-house dataset by different methods

	ASL	ASL+HistAug	ASL+TTA [11]	ASL+TTN	ASL+TTN +re-threshold
HC	68.65 ± 11.07	67.56 ± 11.87	65.24 ± 9.70	**70.16 ± 11.15**	69.27 ± 12.08
PD	71.09 ± 9.01	71.27 ± 8.85	62.96 ± 9.75	**72.01 ± 6.27**	71.16 ± 6.31
Overall	69.87 ± 10.16	69.41 ± 10.63	64.10 ± 9.79	**71.08 ± 9.09**	70.21 ± 9.69

The Adam optimizer with an initial learning rate of 0.001 was used for the ASL training. We trained all the segmentation models until the loss converged on the validation set, and the models acquiring the highest Dice score in the validation set were selected for further evaluation.

4   Results

4.1   Quantitative Evaluation

We computed the overall Dice ratio (mean±standard deviation) and the Dice ratio of the PD cases and the healthy cases. From Table 1, the highest overall Dice ratio and the Dice ratios in the healthy cases were achieved by combining the proposed TTN and ASL: 71.08 ± 9.79%. Compared with the proposed TTN, the histogram-matching-based augmentation in the training phase and TTA did not improve the segmentation performance.

Table 2 shows the normalized likelihood values for evaluating the segmentation accuracy of the unlabeled datasets. The results indicated that the proposed TTN improved the normalized likelihood in the two public datasets, and the introduction of *re-threshold* indeed promoted the normalized likelihood of the segmented SN masks.

Table 2. Normalized likelihood (%, mean±standard deviation) in unseen datasets by different methods

	ASL	ASL+TTN	ASL+TTN+re-threshold
Unseen dataset 1	68.43 ± 5.70	69.28 ± 5.78	**70.36 ± 4.53**
Unseen dataset 2	58.82 ± 0.19	63.48 ± 5.51	**68.86 ± 5.29**

4.2   Qualitative Evaluation

Figure 3 presents the SN segmentation examples from the unlabeled datasets. The introduction of TTN improved the model's identification of the SN regions. In addition, the regions with vague boundaries tended to present higher uncertainty, a result in accordance with the view of human inspections.

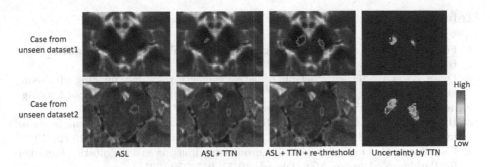

Fig. 3. SN segmentation results from unseen datasets: First row shows results of a case from unseen dataset 2. Second row shows results of a case from unseen dataset 1.

5 Discussion

As shown in the results, our proposed TTN improved the segmentation performance in both the in-house and unseen datasets, indicating enhanced model generalization. Different from a previous study [11], we introduced a histogram-matching step to normalize the initial intensity distribution of the test MRI. This step is based on the fact that the real MRI volumes not only differ from overall spatial or anatomical information such as shapes and sizes but also the intensity distribution which can be easily affected by diseases, personal variations, and even scan protocols. The histogram-matching procedure normalizes the intensity distribution of the test MRI to a seen case of the model. This process might reduce the potential gap between the test and training MRIs, which might be introduced by over-fitting. In addition, the proposed TTN is based on a support set selected from a training set, thus avoiding the latent parameter estimation which may be largely biased in small datasets. Furthermore, the uncertainty map of the proposed TTN shows reasonable distribution over the regions of vague appearances and may provide a reference for further human-computer interactions.

The following are the limitations of our study. First, we used a limited number of recruited cases. In the future, we plan to include more annotated cases. The relatively long inference time for TTN might impose a heavy computation burden with big datasets.

6 Conclusions

We proposed a novel TTN to enhance the model generalization for SN segmentation. Our proposed TTN promoted segmentation performance in both in-house and unseen datasets, indicating its potential for real-world applications.

Acknowledgements. We appreciate the help and advice from the members of the Mori laboratory. A part of this research was supported by the AMED Grant Numbers 22dm0307101h0004 and JSPS KAKENHI 21k19898, 17K20099.

References

1. Fearnley, J.M., Lees, A.J.: Ageing and Parkinson's disease: substantia Nigra regional selectivity. Brain **114**(5), 2283–2301 (1991)
2. Ogisu, K., et al.: 3D neuromelanin-sensitive magnetic resonance imaging with semi-automated volume measurement of the substantia nigra pars compacta for diagnosis of Parkinson's disease. Neuroradiology **55**(6), 719–724 (2013)
3. Hatano, T., et al.: Neuromelanin MRI is useful for monitoring motor complications in Parkinson's and PARK2 disease. J. Neural Transm. **124**(4), 407–415 (2017)
4. Castellanos, G., et al.: Automated neuromelanin imaging as a diagnostic biomarker for Parkinson's Disease. Mov. Disord. **30**(7), 945–952 (2015)
5. Zhao, X., Wu, Y., Song, G., Li, Z., Zhang, Y., Fan, Y.: A deep learning model integrating FCNNs and CRFs for brain tumor segmentation. Med. Image Anal. **43**, 98–111 (2018)
6. Shen, C., et al.: Spatial information-embedded fully convolutional networks for multi-organ segmentation with improved data augmentation and instance normalization. In: Medical Imaging 2020: Image Processing, pp. 261–267. SPIE, Houston (2020)
7. Hu, T., et al.: Aorta-aware GAN for non-contrast to artery contrasted CT translation and its application to abdominal aortic aneurysm detection. Int. J. Comput. Assist. Radiol. Surg. **17**(1), 97–105 (2021). https://doi.org/10.1007/s11548-021-02492-0
8. Le Berre, A., et al.: Convolutional neural network-based segmentation can help in assessing the substantia Nigra in neuromelanin MRI. Neuroradiology **61**(12), 1387–1395 (2019). https://doi.org/10.1007/s00234-019-02279-w
9. Ronneberger, O., Fischer, P., Brox, T.: U-net: convolutional networks for biomedical image segmentation. In: Navab, N., Hornegger, J., Wells, W.M., Frangi, A.F. (eds.) MICCAI 2015. LNCS, vol. 9351, pp. 234–241. Springer, Cham (2015). https://doi.org/10.1007/978-3-319-24574-4_28
10. Hashemi, S.R., Salehi, S.S., Erdogmus, D., Prabhu, S.P., Warfield, S.K., Gholipour, A.: Asymmetric loss functions and deep densely-connected networks for highly-imbalanced medical image segmentation: Application to multiple sclerosis lesion detection. IEEE Access **7**, 1721–1735 (2018)
11. Wang, G., Li, W., Aertsen, M., Deprest, J., Ourselin, S., Vercauteren, T.: Aleatoric uncertainty estimation with test-time augmentation for medical image segmentation with convolutional neural networks. Neurocomputing **338**, 34–45 (2019)
12. Bourke, P.: Histogram Matching. http://paulbourke.net/miscellaneous/equalisation/. (2011)
13. Milletari, F., Navab, N., Ahmadi, S.A.: V-Net: Fully convolutional neural networks for volumetric medical image segmentation. In: 2016 fourth international conference on 3D vision (3DV), pp. 565–571. IEEE, Stanford (2016)

Attention-Enhanced Disentangled Representation Learning for Unsupervised Domain Adaptation in Cardiac Segmentation

Xiaoyi Sun[1,2], Zhizhe Liu[1,2], Shuai Zheng[1,2], Chen Lin[1,2], Zhenfeng Zhu[1,2(✉)], and Yao Zhao[1,2]

[1] Institute of Information Science, Beijing Jiaotong University, Beijing, China
zhfzhu@bjtu.edu.cn
[2] Beijing Key Laboratory of Advanced Information Science and Network Technology, Beijing, China

Abstract. To overcome the barriers of multimodality and scarcity of annotations in medical image segmentation, many unsupervised domain adaptation (UDA) methods have been proposed, especially in cardiac segmentation. However, these methods may not completely avoid the interference of domain-specific information. To tackle this problem, we propose a novel **A**ttention-enhanced **D**isentangled **R**epresentation (ADR) learning model for UDA in cardiac segmentation. To sufficiently remove domain shift and mine more precise domain-invariant features, we first put forward a strategy from image-level coarse alignment to fine removal of remaining domain shift. Unlike previous dual path disentanglement methods, we present channel-wise disentangled representation learning to promote mutual guidance between domain-invariant and domain-specific features. Meanwhile, Hilbert-Schmidt independence criterion (HSIC) is adopted to establish the independence between the disentangled features. Furthermore, we propose an attention bias for adversarial learning in the output space to enhance the learning of task-relevant domain-invariant features. To obtain more accurate predictions during inference, an information fusion calibration (IFC) is also proposed. Extensive experiments on the MMWHS 2017 dataset demonstrate the superiority of our method. Code is available at https://github.com/Sunxy11/ADR.

Keywords: Unsupervised domain adaptation · Disentanglement · Cardiac segmentation

1 Introduction

Recently, the success of convolutional neural networks (CNNs) has enabled computer-aided diagnosis to develop significantly [13,22,23], especially in medical image segmentation. For instance, complete renal structures segmentation

L. Wang (Eds.): MICCAI 2022, LNCS 13437, pp. 745–754, 2022.
https://doi.org/10.1007/978-3-031-16449-1_71

can provide positional information for laparoscopic partial nephrectomy [6]. Nevertheless, most current studies highly rely on sufficient pixel-wise annotations and the assumption that the data satisfy independent identical distribution, which have significant limitations in clinical application. Due to the domain shift between medical images from different sources including but not limited to hospitals, imaging protocols and scanners, the performance of models trained on specific domain data (named as source) decreases significantly when inferring on new domain data (named as target). Moreover, retraining on new domain data is impractical, since manual labeling is expensive and time-consuming.

To bridge the gap between source and target domains, many unsupervised domain adaptation (UDA) methods have been proposed by considering domain alignment from both image and feature perspectives. The former [3,10] aims to address the domain shift issue by image-to-image transformation. Chen et al. [3] proposed to transform the appearance of images across domains and enhance domain invariance of the extracted features by adversarial learning. The latter [11,18,19] achieves domain alignment by adapting the structured output space in semantic segmentation, ensuring the similar spatial layout and local context between source and target domains. Additionally, a few recent studies [1,17,20,21] have attempted to tackle the domain shift problem by disentangled representation learning. Specially, from the frequency domain, Yang et al. [20] utilized Fourier Transform to extract high-frequency and low-frequency spectrums. In view of the feature-based disentanglement learning, some methods [1,16,21] focused on learning domain-invariant and domain-specific features by dual path encoding and pixel-wise reconstruction.

Despite the remarkable performances achieved by the above methods, there are still three key issues need to be further considered for disentanglement-based methods in UDA. First of all, the existing methods [17] generally directly disentangle the source and target representations into domain-invariant and domain-specific features. However, in the case of large domain shift, it will be difficult to obtain effective disentanglement. Additionally, in the previous works [21], dual path CNNs are usually used to extract disentangled features independently, but it is hard to learn effective domain-invariant features without mutual guidance between disentangled features. Finally, how to enhance the domain-invariant features to capture more task-relevant information are also rarely considered.

Motivated by the above observations, we propose a novel disentanglement-based framework to address the domain shift problem, aiming to eliminate cross-domain differences and assist doctors to make better decisions. The main contributions of this paper can be briefly summarized as follows:

- We propose a new Attention-enhanced Disentangled Representation (ADR) learning framework for cross-domain cardiac segmentation, which can capture more precise domain-invariant features simply and efficiently.
- To promote mutual guidance between domain-invariant and domain-specific features, a channel-wise disentanglement approach is proposed, in which the Hilbert-Schmidt independence criterion (HSIC) is adopted to restrict the independence and complementarity between disentangled features.

– An attention bias module is further proposed to motivate the model to focus more on the alignment of task-relevant regions, thus improving the discriminability of the domain-invariant features.

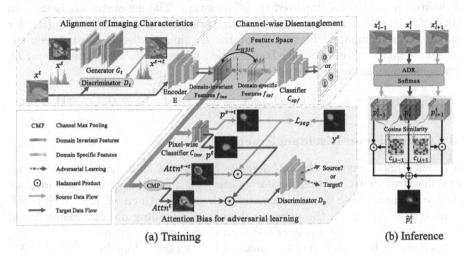

(a) Training

(b) Inference

Fig. 1. (a) An overview of our proposed ADR. (b) The information fusion calibration (IFC) during inference.

2   Methodology

2.1   Overall Framework

We focus on UDA for cardiac segmentation, aiming to accurately segment the cardiac of the target domain with a given labeled source dataset $\{X^s, Y^s\}$ and an unlabeled target dataset $\{X^t\}$. The overview of the proposed ADR model is shown in Fig. 1(a), which mainly consists of three parts: 1)**Alignment of Imaging Characteristics** employs generative adversarial networks to coarsely align cross-domain imaging disparity caused by scanners, sites, and so on; 2)**Channel-wise Disentanglement** is used to remove the interference of remaining domain-specific information, and HSIC is applied to restrict the independence and complementarity between disentangled features; 3)**Attention Bias for Adversarial Learning** aims to make the model focused more on the alignment of task-relevant regions. In addition, to obtain more accurate predictions during inference, an information fusion calibration (IFC) is also adopted as in Fig. 1(b).

2.2   Alignment of Imaging Characteristics

To achieve cross-domain alignment, recent disentanglement-based methods aim to directly encode source or target images into domain-invariant and domain-specific features. However, they may not be effective enough to address the domain shift problem due to the significant imaging disparity between the two domains. Therefore, we first attempt to perform coarse alignment by image transformation before disentanglement to remove obvious eye-discernible domain shift between MR and CT such as brightness. Like [3,7,24], we employ generative adversarial networks for unpaired transformation. The generator G_t is used to transform the source image $x^s \in X^s$ to the target-like image $x^{s \to t} = G_t(x^s)$ for fooling the discriminator. The discriminator D_t aims to distinguish between the real source image x^s and the fake transformed image $x^{s \to t}$. Thus, the G_t and D_t form a minimax game and are optimized by:

$$\mathcal{L}_{adv}^t(G_t, D_t) = \mathbb{E}_{x^t \sim X^t}\left[log D_t(x^t)\right] + \mathbb{E}_{x^s \sim X^s}\left[log\left(1 - D_t(x^{s \to t})\right)\right] \quad (1)$$

2.3   Channel-wise Disentanglement

To further finely remove the interference of domain shift (such as the imaging differences, noise, or even artifacts, etc.) on domain-invariant features, we propose a channel-wise disentanglement with single path encoding, which is different significantly from the previous dual path encoding disentanglement and can promote the mutual guidance between disentangled features.

Specifically, for each target-like image $x^{s \to t}$ or target domain image $x^t \in X^t$, we use an encoder E to extract the visual feature $Z \in \mathbb{R}^{H \times W \times C}$ (C denotes the number of channels), and then the first half of the channels of Z are treated as the domain-invariant features $f_{inv} \in \mathbb{R}^{H \times W \times \frac{C}{2}}$ and the second half are treated as domain-specific features $f_{spf} \in \mathbb{R}^{H \times W \times \frac{C}{2}}$. In order to realize the independence and complementarity between f_{inv} and f_{spf}, the HSIC [12] is applied for disentangled representation learning, which is given as follows:

$$\mathcal{L}_{HSIC}(E) = HSIC\left(\mathcal{G}(f_{inv}), \mathcal{G}(f_{spf})\right) \quad (2)$$

where $\mathcal{G}(\cdot)$ stands for global average pooling, and for $HSIC(A_1, A_2)$, we have $HSIC(A_1, A_2) = (n-1)^{-2} tr(K_{A_1} M K_{A_2} M)$, $A_i \in \mathbb{R}^{n \times \frac{C}{2}}$, n is the number of samples, K_{A_i} indicates the Gaussian kernel function applied to A_i, and M is a matrix with zero mean.

Meanwhile, to ensure that f_{spf} is highly correlated with domain-specific information, a classifier C_{spf} is proposed to distinguish f_{spf}, so that the distributions of $f_{spf}^{s \to t}$ and f_{spf}^t are separated as much as possible. Specially, given f_{spf}, the classifier C_{spf} is optimized by the cross-entropy loss:

$$\mathcal{L}_{spf}(E, C_{spf}) = \mathcal{L}_{ce}(y_{spf}, p_{spf}) \quad (3)$$

where y_{spf} is one-hot vector about label and p_{spf} is the predicted probability.

2.4   Attention Bias for Adversarial Learning

Since the labels of the source domain are assumed to be available in UDA, we first conduct supervised learning on target-like images to train the pixel-wise classifier C_{inv}. Given a domain-invariant feature $f_{inv}^{s \to t}$, C_{inv} generates a pixel-level prediction $p^{s \to t} \in \mathbb{R}^{H \times W \times L}$ (L is the number of categories). Here, we adopt a hybrid loss of cross-entropy and dice to train C_{inv} which is the same as [5]:

$$\mathcal{L}_{seg}\left(E, C_{inv}\right) = \mathcal{L}_{ce}\left(y^s, p^{s \to t}\right) + \mathcal{L}_{Dice}\left(y^s, p^{s \to t}\right) \tag{4}$$

where y^s represents the ground-truth mask of the source data.

To adapt the structured output space for semantic segmentation in UDA, some recent works [3,18] have generally enforced adversarial learning on pixel-level predictions to achieve similar spatial layout and local context. However, for the above approaches, each region is treated equally, which is not sufficient to achieve effective alignment. Therefore, we propose an attention bias module to make the model focused more on the alignment of task-relevant regions.

Inspired by [14], the channel max pooling (CMP) can gather the significant features (i.e., task-relevant information). Thus, given the domain-invariant features f_{inv} of source or target domains, we attempt to obtain the attention bias $Attn \in \mathbb{R}^{H \times W \times 1}$ by CMP and the value of $Attn$ at position (i, j) is defined as:

$$Attn\left(i, j, m\right) = \max\left[\mathcal{N}_k\left(f_{inv}\left(i, j, e \times m\right)\right)\right] \tag{5}$$

where $1 \le i \le H$, $1 \le j \le W$, e is the stride of the CMP kernel movement, $\mathcal{N}_k$ represents the operation of taking k neighbors on the channel side. Considering that the channel dimension m of our attention bias $Attn$ is 1, we set $k = \frac{C}{2}$, $e = 1$ and no padding in the CMP layer.

Most importantly, we enforce the obtained attention bias $Attn$ on pixel-level predictions by the Hadamard product. Then, as shown in Fig. 1(a), the result is fed into the domain discriminator D_p to achieve cross-domain alignment at the output level and the model is optimized by the adversarial learning:

$$\mathcal{L}_{adv}^p\left(E, C_{inv}, D_p\right) = \mathbb{E}_{x^{s \to t} \sim X^{s \to t}}\left[logD_p\left(p^{s \to t} \odot Attn^{s \to t}\right)\right] \\ + \mathbb{E}_{x^t \sim X^t}\left[log\left(1 - D_p\left(p^t \odot Attn^t\right)\right)\right] \tag{6}$$

Although the adopted attention bias module is simple, it is effective just as demonstrated in the visualization experiments because it can effectively enhance the domain alignment of cardiac regions. Thus, in turn, it can constrain the model to learn more task-relevant domain-invariant features. Specially, referring to [18], we perform multi-level adversarial learning at different feature levels. Combining Eq. (1)–(4) and Eq. (6), the total optimization loss can be given by:

$$\mathcal{L} = \mathcal{L}_{adv}^t\left(G, D_t\right) + \lambda_{HSIC}\mathcal{L}_{HSIC}\left(E\right) + \lambda_{spf}\mathcal{L}_{spf}\left(E, C_{spf}\right) + \lambda_{seg}^1\mathcal{L}_{seg}^1\left(E, C_{inv}^1\right) \\ + \lambda_{seg}^2\mathcal{L}_{seg}^2\left(E, C_{inv}^2\right) + \lambda_{adv}^{p1}\mathcal{L}_{adv}^{p1}\left(E, C_{inv}, D_{p1}\right) + \lambda_{adv}^{p2}\mathcal{L}_{adv}^{p2}\left(E, C_{inv}, D_{p2}\right) \tag{7}$$

where $\left\{\lambda_{HSIC}, \lambda_{cls}, \lambda_{seg}^1, \lambda_{seg}^2, \lambda_{adv}^{p1}, \lambda_{adv}^{p2}\right\}$ are some hyperparameters and consistently set as $\{0.1, 0.05, 1.0, 0.1, 0.1, 0.01\}$ in all experiments.

Table 1. Quantitative comparison with different methods on Dice(%) and ASD(mm).

Methods	Cardiac MRI → Cardiac CT										Cardiac CT → Cardiac MRI									
	Dice↑					ASD↓					Dice↑					ASD↓				
	AA	LAC	LVC	MYO	Avg	AA	LAC	LVC	MYO	Avg.	AA	LAC	LVC	MYO	Avg.	AA	LAC	LVC	MYO	Avg.
Supervised training	92.7	91.1	91.9	87.7	90.9	1.5	3.5	1.7	2.1	2.2	82.8	80.5	92.4	78.8	83.6	3.6	3.9	2.1	1.9	2.9
W/o adaptation	28.4	27.7	4.0	8.7	17.2	20.6	16.2	N/A	48.4	N/A	5.4	30.2	24.6	2.7	15.7	15.4	16.8	13.0	10.8	14.0
SynSeg-Net [8]	71.6	69.0	51.6	40.8	58.2	11.7	7.8	7.0	9.2	8.9	41.3	57.5	63.6	36.5	49.7	8.6	10.7	5.4	5.9	7.6
AdaOutput [18]	65.1	76.6	54.4	43.6	59.9	17.9	5.5	5.9	8.9	9.6	60.8	39.8	71.5	35.5	51.9	**5.7**	8.0	4.6	4.6	5.7
CycleGAN [24]	73.8	75.7	52.3	28.7	57.6	11.5	13.6	9.2	8.8	10.8	64.3	30.7	65.0	43.0	50.7	5.8	9.8	6.0	5.0	6.6
CyCADA [7]	72.9	77.0	62.4	45.3	64.4	9.6	8.0	9.6	10.5	9.4	60.5	44.0	77.6	47.9	57.5	7.7	13.9	4.8	5.2	7.9
Prior SIFA [2]	81.1	76.4	75.7	58.7	73.0	10.6	7.4	6.7	7.8	8.1	67.0	60.7	75.1	45.8	62.1	6.2	9.8	4.4	4.4	6.2
SIFA [3]	81.3	79.5	73.8	61.6	74.1	7.9	6.2	5.5	8.5	7.0	65.3	62.3	78.9	47.3	63.4	7.3	7.4	3.8	4.4	5.7
DDA-GAN [4]	68.3	75.7	78.5	**77.8**	75.1	6.5	4.8	5.4	**5.2**	5.5	-	-	-	-	-	-	-	-	-	-
ADR(Ours)	87.6	86.6	81.8	64.0	80.0	6.4	4.4	5.7	5.7	5.5	66.8	68.9	**81.0**	48.3	66.2	7.0	5.6	**3.4**	**4.0**	5.0
ADR+IFC	**87.9**	**86.8**	**82.1**	64.2	**80.2**	**5.9**	**4.1**	**4.8**	5.7	**5.1**	**66.9**	**69.1**	**81.0**	48.3	**66.3**	6.9	**5.5**	**3.4**	**4.0**	**4.9**

2.5   Implementation Details

Training. For the generator G_t and target domain discriminator D_t in coarse alignment, we load the parameters pretrained by the cycle consistency loss from SIFA [3] and fine-tune them with a very small learning rate[1], so that the network tends to be fixed to avoid degradation of the ability in preserving the anatomical structure. All other networks in our framework are trained from scratch with a learning rate of 2e–4. Both the encoder and the pixel-wise classifier are same as the configuration in SIFA [3]. The classifier C_{spf} contains a convolutional layer with kernel size of 3 and an average pooling layer, followed by two fully connected layers. Similar to AdaOutput [18], we adopt multi-level adversarial learning for domain alignment and our domain discriminator is consistent with PatchGAN [9]. We use the Adam optimizer to optimize all models, and perform 20k iterations with a mini-batch size of 8. The input image size is set to 256×256×1 and the momentum parameters of Adam optimizer are set as 0.5 and 0.99 for encoder E, target domain discriminator D_t and classifier C_{spf}.

Inference. To fully utilize the inter-slice information and reduce the prediction error without increasing the network burden, an information fusion calibration (IFC) strategy is proposed, as shown in Fig. 1(b). Specially, for position (u, v), given the normalized pixel-level predictions of target image at $i-1$-th slice $p_{i-1}^{u,v}$ and i-th slice $p_i^{u,v}$, we calculate the cosine similarity of the two slices pixel-to-pixel to obtain the cosine similarity matrix $c_{i,i-1}^{u,v}$. Similarly, we get the cosine similarity matrix $c_{i,i+1}^{u,v}$. Then, the refined pixel-level prediction $\hat{p}_i^{u,v}$ is calculated by a fusion operation as follows:

$$\hat{p}_i^{u,v} = \alpha p_i^{u,v} + (1-\alpha)\left[\frac{c_{i,i-1}^{u,v}}{c_{i,i-1}^{u,v} + c_{i,i+1}^{u,v}} \odot p_{i-1}^{u,v} + \frac{c_{i,i+1}^{u,v}}{c_{i,i-1}^{u,v} + c_{i,i+1}^{u,v}} \odot p_{i+1}^{u,v} \right] \quad (8)$$

where α is a hyperparameter with a value of 0.5 and $\odot$ is Hadamard product.

---

[1] In fact, we simply make fine-tuning on the pre-trained weights without introducing additional complex reconstruction loss, thus facilitating the stable training of the overall model.

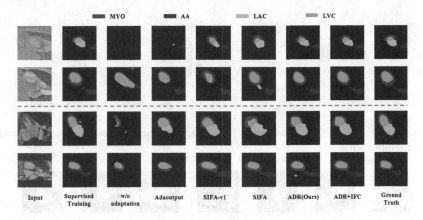

Fig. 2. Visual comparison of Cardiac segmentation results produced by different methods on "MR to CT" task (1st–2nd row) and "CT to MR" task (3rd-4th row).

3 Experiment Results and Analysis

3.1 Dataset and Evaluation Metrics

We conduct experiments on the Multi Modality Whole Heart Segmentation (MMWHS) challenge 2017 dataset [25] containing unpaired 20 CT and 20 MR volumes with ground-truth masks. To evaluate our model quantitatively, we perform segmentation on the following four cardiac structures: ascending aorta (AA), left atrium blood cavity (LAC), left ventricle blood cavity (LVC), and myocardium of the left ventricle (MYO). For a fair comparison, we adopt the preprocessed data published by SIFA [3], in particular, the preprocessing operations contain rotation, scaling, and affine transformation. To quantitatively evaluate the performance of our method, we adopt two widely used metrics, the Dice similarity coefficient (Dice) and the average symmetric surface distance (ASD), which cover two aspects: the overlap and the difference between the predictions and ground-truth masks. DBI is adopted to measure the degree of separation between disentangled features.

3.2 Effectiveness of ADR in Cardiac Segmentation

To fully validate the effectiveness of our ADR in addressing the domain shift problem, we conduct bidirectional domain adaptation experiments (i.e., MRI to CT and CT to MRI) and present the quantitative evaluation results in Table 1. It can be seen that: i) by comparison of the supervised training and without-adaptation model, there is indeed the serious domain shift problem between MR and CT domains; ii) remarkably, our ADR model achieves more significant performance improvements in terms of both Dice and ASD measurements. Specifically, we improve the average Dice to 80.0% over the four cardiac structures while the average ASD being reduced to 5.5 on the MRI to CT task; iii)

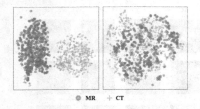

Fig. 3. t-SNE [15] w/o and with domain adaptation.

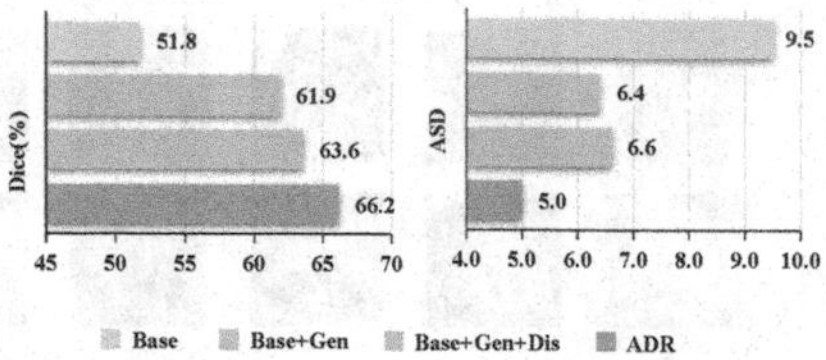

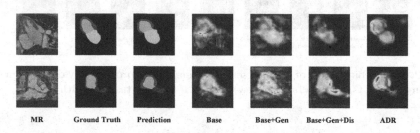

Fig. 4. Ablation results of our method.

Fig. 5. Visualization of attention maps from our method and its ablations.

additionally, the segmentation performance is further improved when the information fusion calibration strategy is adopted, and for CT images, it helps to achieve a improvement of 0.2% in the average Dice and a reduction of 0.4 in the average ASD. To further show the effectiveness of our method, the visualization of segmentation results is presented in Fig. 2 for qualitative comparison. It is obvious that the outputs of our ADR are more consistent with the ground truth for the slice images in both two directions. Meanwhile, we also provide the qualitative t-SNE [15] analysis in Fig. 3 to visualize the distribution of features without and with domain adaptation. The visualization results demonstrate on the one hand the significant domain shift problem between MR and CT, and on the other hand, our ADR significantly decreases the distribution discrepancies between the two domains. To summarize, the comparative experiments indicate that our method can successfully achieve domain alignment in cardiac segmentation.

3.3   Ablation Study

We also conduct comprehensive ablation experiments on CT to MR adaptation to demonstrate both quantitatively and qualitatively the effectiveness of each component. The experimental results are shown in Fig. 4 and Fig. 5. We take the network with only multi-level adversarial learning as the baseline, and add coarse alignment, channel-wise disentanglement and attention bias sequentially, named Base, Base+Gen, Base+Gen+Dis and ADR, respectively. As we can see, after adding the alignment of imaging characteristics, we achieve a clear improvement of 10.1% in the average Dice. In addition, the introduction of channel-wise disentanglement improved the average Dice to 63.6%, and the DBI is 0.29 in the MR to CT direction and 0.34 in the other direction, indicating that the disentangled features are well separated. Finally, with the attention bias for adversarial learning, we further achieve a clear improvement of 2.6% in the average Dice. Figure 5 provides more in-depth and intuitive validation of the attention maps from our ADR and its ablations. Compared to other ablation settings, our ADR indeed promotes the model to focus on more task-relevant regions, thus helping to extract more discriminative domain-invariant features.

4   Conclusions

In this paper, we propose an attention-enhanced disentanglement framework. The embedding space is disentangled into complementary domain-invariant and domain-specific subspaces. And attention bias is introduced to enhance the learning of task-relevant domain-invariant features. In addition, the information fusion calibration makes the predictions more accurate. Qualitative and quantitative comparative analyses demonstrate the effectiveness of our method.

Acknowledgement. This work was supported in part by the Science and Technology Innovation 2030 - New Generation Artificial Intelligence Major Project under Grant No. 2018AAA0102100, Beijing Natural Science Foundation under Grant No. 7222313, and National Natural Science Foundation of China under Grant No. 61976018.

References

1. Bercea, C.I., Wiestler, B., Rueckert, D., Albarqouni, S.: Feddis: disentangled federated learning for unsupervised brain pathology segmentation. arXiv preprint arXiv:2103.03705 (2021)
2. Chen, C., Dou, Q., Chen, H., Qin, J., Heng, P.A.: Synergistic image and feature adaptation: towards cross-modality domain adaptation for medical image segmentation. In: Proceedings of AAAI, vol. 33, pp. 865–872 (2019)
3. Chen, C., Dou, Q., Chen, H., Qin, J., Heng, P.A.: Unsupervised bidirectional cross-modality adaptation via deeply synergistic image and feature alignment for medical image segmentation. IEEE TMI **39**(7), 2494–2505 (2020)
4. Chen, X., et al.: Diverse data augmentation for learning image segmentation with cross-modality annotations. MedIA **71**, 102060 (2021)
5. Dou, Q., Ouyang, C., Chen, C., Chen, H., Heng, P.A.: Unsupervised cross-modality domain adaptation of convnets for biomedical image segmentations with adversarial loss. In: IJCAI, pp. 691–697 (2018)
6. He, Y., et al.: EnMcGAN: adversarial ensemble learning for 3d complete renal structures segmentation. In: Feragen, A., Sommer, S., Schnabel, J., Nielsen, M. (eds.) IPMI 2021. LNCS, vol. 12729, pp. 465–477. Springer, Cham (2021). https://doi.org/10.1007/978-3-030-78191-0_36
7. Hoffman, J., et al.: Cycada: cycle-consistent adversarial domain adaptation. In: ICML, pp. 1989–1998. PMLR (2018)
8. Huo, Y., et al.: Synseg-net: synthetic segmentation without target modality ground truth. IEEE TMI **38**(4), 1016–1025 (2018)

9. Isola, P., Zhu, J.Y., Zhou, T., Efros, A.A.: Image-to-image translation with conditional adversarial networks. In: Proceedings of CVPR, pp. 1125–1134 (2017)

10. Kim, M., Byun, H.: Learning texture invariant representation for domain adaptation of semantic segmentation. In: Proceedings of CVPR, pp. 12975–12984 (2020)

11. Li, H., Loehr, T., Sekuboyina, A., Zhang, J., Wiestler, B., Menze, B.: Domain adaptive medical image segmentation via adversarial learning of disease-specific spatial patterns. arXiv preprint arXiv:2001.09313 (2020)

12. Liu, X., Thermos, S., O'Neil, A., Tsaftaris, S.A.: Semi-supervised meta-learning with disentanglement for domain-generalised medical image segmentation. In: de Bruijne, M., et al. (eds.) MICCAI 2021. LNCS, vol. 12902, pp. 307–317. Springer, Cham (2021). https://doi.org/10.1007/978-3-030-87196-3_29

13. Liu, Z., Zhu, Z., Zheng, S., Liu, Y., Zhou, J., Zhao, Y.: Margin preserving self-paced contrastive learning towards domain adaptation for medical image segmentation. IEEE J. Biomed. Health Inf. **26**(2), 638–647 (2022)

14. Ma, Z., et al.: Fine-grained vehicle classification with channel max pooling modified CNNs. IEEE Trans. Veh. Technol. **68**(4), 3224–3233 (2019)

15. Van der Maaten, L., Hinton, G.: Visualizing data using t-sne. JMLR **9**(11) (2008)

16. Ning, M., et al.: A new bidirectional unsupervised domain adaptation segmentation framework. In: Feragen, A., Sommer, S., Schnabel, J., Nielsen, M. (eds.) IPMI 2021. LNCS, vol. 12729, pp. 492–503. Springer, Cham (2021). https://doi.org/10.1007/978-3-030-78191-0_38

17. Shin, S.Y., Lee, S., Summers, R.M.: Unsupervised domain adaptation for small bowel segmentation using disentangled representation. In: de Bruijne, M., et al. (eds.) MICCAI 2021. LNCS, vol. 12903, pp. 282–292. Springer, Cham (2021). https://doi.org/10.1007/978-3-030-87199-4_27

18. Tsai, Y.H., Hung, W.C., Schulter, S., Sohn, K., Yang, M.H., Chandraker, M.: Learning to adapt structured output space for semantic segmentation. In: Proceedings of CVPR, pp. 7472–7481 (2018)

19. Vu, T.H., Jain, H., Bucher, M., Cord, M., Pérez, P.: Advent: adversarial entropy minimization for domain adaptation in semantic segmentation. In: Proceedings of CVPR, pp. 2517–2526 (2019)

20. Yang, Y., Soatto, S.: Fda: fourier domain adaptation for semantic segmentation. In: Proceedings of CVPR, pp. 4085–4095 (2020)

21. You, C., Yang, J., Chapiro, J., Duncan, J.S.: Unsupervised Wasserstein distance guided domain adaptation for 3D multi-domain liver segmentation. In: Cardoso, J., et al. (eds.) IMIMIC/MIL3ID/LABELS -2020. LNCS, vol. 12446, pp. 155–163. Springer, Cham (2020). https://doi.org/10.1007/978-3-030-61166-8_17

22. Zhang, W., et al.: Deep learning based torsional nystagmus detection for dizziness and vertigo diagnosis. Biomed. Sig. Process. Control **68**, 102616 (2021)

23. Zheng, S., Zhu, Z., Liu, Z., Guo, Z., Liu, Y., Yang, Y., Zhao, Y.: Multi-modal graph learning for disease prediction. IEEE Trans. Med. Imaging **41**(9), 2207–2216 (2022)

24. Zhu, J.Y., Park, T., Isola, P., Efros, A.A.: Unpaired image-to-image translation using cycle-consistent adversarial networks. In: Proceedings of ICCV, pp. 2223–2232 (2017)

25. Zhuang, X., Shen, J.: Multi-scale patch and multi-modality atlases for whole heart segmentation of MRI. MedIA **31**, 77–87 (2016)

Histogram-Based Unsupervised Domain Adaptation for Medical Image Classification

Pengfei Diao[1,2], Akshay Pai[2,3(✉)], Christian Igel[1],
and Christian Hedeager Krag[4]

[1] Department of Computer Science, University of Copenhagen, Universitetsparken 1,
2100 Copenhagen, Denmark
[2] Cerebriu A/S, Copenhagen, Denmark
ap@cerebriu.com
[3] Rigshospitalet, Copenhagen, Denmark
[4] Department of Radiology, Herlev Hospital, Borgmester Ib Juuls Vej 1,
2730 Herlev, Denmark

Abstract. Domain shift is a common problem in machine learning and medical imaging. Currently one of the most popular domain adaptation approaches is the domain-invariant mapping method using generative adversarial networks (GANs). These methods deploy some variation of a GAN to learn target domain distributions which work on pixel level. However, they often produce too complicated or unnecessary transformations. This paper is based on the hypothesis that most domain shifts in medical images are variations of global intensity changes which can be captured by transforming histograms along with individual pixel intensities. We propose a histogram-based GAN methodology for domain adaptation that outperforms standard pixel-based GAN methods in classifying chest x-rays from various heterogeneous target domains.

Keywords: Unsupervised domain adaptation · Histogram layer · Lung disease classification

1 Introduction

One of the most ubiquitous application of deep learning has been in the classification of medical images to aid triage, diagnosis, and resource management. Even though several products have been developed, large scale deployment has been somewhat limited due to the sensitivity of large over-parameterized neural networks (NN) to domain shift. Domain shift is a commonly seen problem where the data distribution on which the NN has been trained has different statistics compared to the test data distribution. Most often domain shift manifest as covariate shifts where the marginal label distributions remain the same.

Supplementary Information The online version contains supplementary material available at https://doi.org/10.1007/978-3-031-16449-1_72.

In this study, we show that, as opposed to the existing domain adaptation approaches, addition of the pixel intensity histogram as a feature for discrimination (on top of raw intensities) and simplifying the generator to produce global intensity transformations have a positive effect on domain adaptation regardless of the site. Through experiments on a mix of publicly available datasets, namely *Chexpert* [12] and *NIH* [20], and an internal dataset referred to as *RH*, we show that orderless features along with a generator that allows global intensity transformations provided a better domain invariant mapping and thereby more stable generalization compared to the standard approaches using generative adversarial networks (GANs) at the level of complete images.

2 Literature Review

Several methods have been proposed to address domain shifts. A few of them are: out-of-distribution detection (OOD), subspace mapping [6], domain-invariant mapping [4,14,16], feature/data augmentation [15], or more expensively just supervised fine-tuning on new domains. For unsupervised domain adaptation, one of the most commonly used method is domain-invariant feature generation (or some modification of it). Previous works [4,14,16] employ GANs to train a classifier with domain-invariant features. These methods however require the primary training of the NN to happen with both target and source domain data available, and fine-tuning of the whole network when deploying to new domain. Here we set out with the assumption that the classifier remains unchanged and that the data for learning or fine-tuning any mapping is only possible at a deployment site. Unpaired image-to-image translation GAN methods [2,7,8,13,21] have been successfully applied in medical image tasks such as segmentation, data augmentation, and image synthesis. Few work has employed these methods for unsupervised domain adaptation in disease classification tasks.

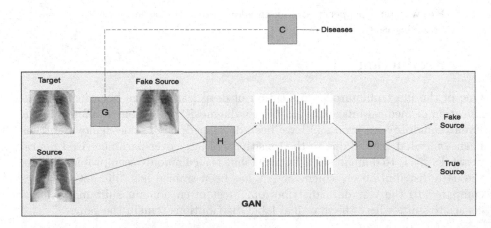

Fig. 1. Overview of our GAN based domain adaptation. The classifier (C) is trained on source image data for disease classification. The GAN is trained to translate images from target domain to source domain. The GAN is comprised of a generator (G), a histogram layer (H), and a discriminator (D). The input to the GAN is a batch of unpaired images from source and target domains. While images from source domain will directly proceed to the histogram layer, images from target domain will be passed through generator first.

3 Methods

3.1 Overview

An overview of the workflow is illustrated Fig. 1. Our model has two components: the classifier and the domain-transforming generator. The classifier is trained to classify five lung diseases and is an ensemble of five Densenet-121 [11] models. As explained earlier, one of the novelties of the methods is that the classifier remains fixed. What gets trained is a domain-transforming generator on both source and target domains. The domain-transforming generator essentially acts as a pre-processing step to the classifier. Similar to existing GAN-based approaches, the domain-transforming generator is trained in the standard adversarial fashion. However, the difference to existing approaches is that we feed histograms as the primary feature to the discriminator and not full images. This simple approach allows us to focus on global domain changes which we believe is the most common domain change in chest x-ray classification problems [3].

We propose two kinds of GAN methods – Graymap and Gamma-adjustment GAN. The Graymap (similar to Colormap GAN [19]) learns a global intensity transformation from target domain to source domain. In contrast, the Gamma-adjustment generator learns instance based intensity transformations formulated as gamma transformations. Both of these generators transform the image at a global intensity level and therefore maintain semantic consistency between the original and the generated image.

3.2 Histogram Layer

Inspired by the work of Sedighi et al. [18], our histogram layer is constructed with a set of Gaussian functions. Our histogram layer differs from the originally proposed one in two ways. First, the tails of the Gaussian-shaped kernels on the side are not replaced by a constant value of 1. So when the input intensity is outside of pre-defined range, its contribution to the bin will be close to zero. Second, the histogram is normalized by the sum of all bins instead of the total number of pixels. For a histogram with K, the frequency of intensity occurrences within the bin $k \in \{1, \ldots, K\}$ is approximated by

$$B(k) = \sum_{i=0}^{W} \sum_{j=0}^{H} e^{-\left(\frac{I_{ij} - \mu_k}{\sigma}\right)^2}, \tag{1}$$

where W and H are the width and height of input image I, respectively, I_{ij} is the intensity of pixel (i, j), μ_k is the center of k-th bin, and σ plays the role of the bin width which controls the spread of each bin. The normalized histogram is given by

$$B_{\text{norm}}(k) = \frac{B(k)}{\sum_{\hat{k}=1}^{K} B\left(\hat{k}\right)}. \tag{2}$$

This histogram layer does not have learnable weights. The bin centers μ_k are pre-computed. For a histogram with K bins, the k-th bin center μ_k is calculated as

$$\mu_k = \frac{(2k-1)\left(\max\left(L\right) - \min\left(L\right)\right)}{2K} + \min\left(L\right) , \tag{3}$$

where L is the intensity range, $\min\left(L\right)$ and $\max\left(L\right)$ are the minimum and maximum intensity levels accordingly. The bin width μ is determined by visually comparing the the output of histogram layer with the actual histogram. With a larger σ the bins overlap more, resulting in a smoother histogram. With smaller σ the histogram is closer to the actual histogram. However as $\sigma \to \frac{1}{\infty}$ the Gaussian function will eventually become Dirac delta function and the gradient can hardly flow across bins. In practice, we start with an relatively large σ and gradually decrease it until the layer output is visually close enough to the actual histogram. An illustration of two histogram layers with different bin widths can be found in Fig. 3 from Appendix A.2.

For back-propagation, we use auto differentiation of Tensorflow [1]. The derivatives of histogram and normalization functions are given in Eq. 11 and Eq. 12 in Appendix A.3.

3.3   Histogram Discriminator

The histogram discriminator consists of a histogram layer and a 1D CNN of ResNet type [10], see Appendix A.5 for the exact architecture. The discriminator takes the image as an input, computes the histogram through histogram layer and then discriminates between true source image and fake source image using the 1D CNN. This discriminator is used in both Graymap and Gamma-adjustment GAN.

3.4   Graymap GAN

The first generator we propose is a graymap generator, which is based on the generator in the Colormap GAN [19]. In Colormap GAN, the generator has together over 100 millions of weights ($256 \times 256 \times 256 \times 3$) and biases ($256 \times 256 \times 256 \times 3$) to translate RGB image from one domain to another. Our graymap generator has only 256 weights and 256 biases since we are dealing with 8-bit single channel gray-scale input. For an image I with intensity value normalized into the range $(0, 1)$ the graymap transformation is defined as

$$G\left(I_{i,j}\right) = \left(2I_{i,j} - 1\right) W_{\left(L\left(I_{i,j}\right)\right)} + B_{\left(L\left(I_{i,j}\right)\right)} , \tag{4}$$

where W and B are the weight and bias vectors of length 256, respectively. Assuming the element index of weight and bias vectors ranges between 0 and 255, $L\left(I_{i,j}\right)$ is defined as

$$L\left(I_{i,j}\right) = \lfloor I_{i,j} * 255 \rfloor \tag{5}$$

that computes the index of corresponding weight and bias to pixel $I_{i,j}$. We further clip the intensity value of generated images and re-scale the intensity level back to range $(0, 1)$ by

$$\hat{G}(I_{i,j}) = \begin{cases} 1 & \text{if } G(I_{i,j}) > 1 \\ 0 & \text{if } G(I_{i,j}) < -1 \\ 0.5 \cdot G(I_{i,j}) + 0.5 & \text{otherwise.} \end{cases} \quad (6)$$

Combining this graymap generator with the histogram discriminator gives the proposed Graymap. Because the generator has 256 degrees of freedom to transform each distinct intensity level, in the experiments we used 256 bins with $\sigma = 0.03$ for the histogram layer to capture the differences across intensity levels. For the generator, we initialize the weights with ones and the biases with zeros. For the discriminator, we used 256 convolutional filters in all convolutional layers including ones inside the residual blocks. Similar to Colormap GAN, we use the least square loss proposed by Mao et al. [17] for optimizing the generation of an image. The losses for the generator and discriminator are defined as

$$L_G = \mathbb{E}_{t \in T} \left[(D(G(t)) - 1)^2 \right] \quad (7)$$

and

$$L_D = 0.5 \left(\mathbb{E}_{s \in S} \left[(D(s) - 1)^2 \right] + \mathbb{E}_{t \in T} \left[D(G(t))^2 \right] \right) . \quad (8)$$

Here S and T are the sets of source and target images, respectively. The constant scalar 0.5 in Eq. (8) is used to balance the losses for generator and discriminator at each training step.

3.5 Gamma-Adjustment GAN

The second generator we propose does instance-based gamma adjustment to the image. For an image I with intensity value normalized into the range $(0, 1)$, the gamma adjustment is given as

$$G(I_{i,j}) = (I_{i,j} + \epsilon)^{\gamma(z)} , \quad (9)$$

where γ is a scalar regressed by a 1D CNN and ϵ is a small constant used for preventing undefined gradient during back-propagation. The construction of the γ regression is similar to the histogram discriminator. We first compute the histogram of input image using a histogram layer which is then fed to the 1D CNN (see Appendix A.5). The CNN outputs a scalar z which is scaled by activation function given as

$$\gamma(z) = (\alpha - \beta) \frac{1}{1 + e^{-z}} + \beta , \quad (10)$$

where α and β are two positive constants restricting γ in the range (β, α). The purpose of restricting the range of γ is to stabilize the training of GAN. Equation (9) has the property that for γ less than 1 the histogram will shift towards the right, resulting in a brighter image, and for γ greater than 1 the histogram will shift towards the left resulting in a darker image. Combining the gamma adjustment generator with the histogram discriminator we get the Gamma-adjustment GAN. Unlike Graymap that adjusts each intensity level independently, the gamma adjustment has only 1 degree of freedom. In our experiments we therefore used only 32 bins with $\sigma = 0.1$ for histogram layers in both generator and discriminator. The losses for generator and discriminator are the same as Eqs. (7) and (8).

4   Experiments

4.1   Datasets

We conducted studies on two publicly available large x-ray chest image datasets Chexpert [12] and NIH [20] as well as on a small internal dataset RH collected from Rigshospitalet, Denmark.

The **Chexpert** dataset consists of 223,414 images with 14 categories of observations, with 191,027 acquired in the frontal view and 32,387 acquired laterally. The separate test dataset comprises 202 frontal views and 32 lateral views. All images are provided in 8-bit JPG format and were originally post-processed with histogram equalization. The **NIH** dataset consists of 112,120 frontal view images with 15 categories of observations, from which 25,596 images are hold out for test. The images are provided in 8-bit PNG format without histogram equalization. The **RH** dataset consists of 884 frontal view images with 7 categories of observations. A separate test dataset which consists of 231 frontal view images is given. The images are provided in 8-bit PNG format without histogram equalization. For simplicity, we evaluated our methods on the classes atelectasis, cardiomegaly, consolidation, edema, and pleural effusion.

4.2   Training

For a fair comparison, we first reproduced the results presented in the Chexpert article [12]. The model's input size is 320×320 pixels. We used the Adam optimizer with $\beta_1 = 0.9$, $\beta_2 = 0.999$ and learning rate 10^{-4}. Random rotation (± 5 degrees) and random zoom-in/zoom-out (0.95, 1.05) were used for data augmentation. We held 6,886 (for Chexpert) images out for test. We trained the network for 10 epochs on 216,528 images with a batch size of 16 and 1% holdout for validation. We saved the checkpoint with best validation AUC. We shuffled the training data and repeated the experiment to get 5 checkpoints in total. The classifier combined these five networks by averaging their predictions. Following the same procedure, we trained another ensemble of 5 networks on the NIH dataset. We refer to these two classifiers as Chexpert-net and NIH-net accordingly.

In the **plain input** setting, we tested Chexpert-net and NIH-net on the test set of Chexpert, NIH and RH without translating the input. In **Cycle-GAN** [22] baseline experiments, we trained CycleGAN for translating images from NIH to Chexpert, from RH to Chexpert, from Chexpert to NIH, and from RH to NIH. For Chexpert and NIH, we used 5,000 unlabelled images from each dataset to train the GAN. For RH we used 800 images to train the GAN. We prepended the corresponding CycleGAN generator to Chexpert-net and NIH-net, and tested them on the corresponding datasets. In the **Colormap GAN** [19] setting, we replaced the generator of Colormap GAN with our graymap generator (Eq. (4)) for dealing with grayscale images. The discriminator and losses remained unchanged. We trained and tested Colormap GAN with the same dataset setup as CycleGAN. Our Graymap GAN and Gamma-Adjustment GAN were also trained and tested with this dataset setup, see Appendix A.4 for details.

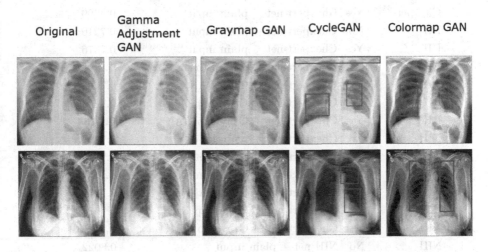

Fig. 2. Example of generated images. First row shows transformation from RH to NIH. Second row shows transformation from Chexpert to NIH. From left to right are, respectively, original, Gamma-adjustment GAN generated, Graymap gan generated, CycleGAN generated, and Colormap GAN generated. Red-boxes highlight where the artifacts are added or local details are lost. (Color figure online)

5 Results, Discussion, and Conclusions

Table 1 illustrates the AUCs generated by each methodology, the statistical evaluation based on DeLong tests [5] is summarized in Table 3 (found in the Appendix). While the newly proposed methods gave the highest average AUC values on the RH data, the individual differences in the AUC for the different classes are mostly not statistically significant, most likely due to the small test sample size. On NIH, the newly proposed methods gave the highest average AUC values and the individial differences on all five classes are highly signifi-cant ($p < 0.001$). On Chexpert, the new methods gave the highest average AUC values. For NIH-net + Gamma-adjustment GAN, the AUCs for Consolidation

Table 1. AUCs (area under the receiving operator curve) for different methods evaluated on the test data specified in the leftmost column. The AUC is the macro average over 5 classes. The dataset name refers to the dataset on which the classifier was trained (e.g., NIH-net was trained on NIH). The numbers of test images were 231, 25596, and 25523 for RH, NIH, and Chexpert, respectively, except for Chexpert[234] and Chexpert[6886], where 234 (standard Chexpert test set) and 6886 images were used. HE indicates whether the test images were histogram equalized. The results of DeLong significance tests comparing the AUC values against baselines can be found in Table 3 in the Appendix.

Test on	HE	Methods	Mean AUC
Chexpert[234]	Yes	Chexpert-net + plain input	0.8850
Chexpert[6886]	Yes	Chexpert-net + plain input	0.8409
RH	No	Chexpert-net + plain input	0.7210
RH	Yes	Chexpert-net + plain input[RH baseline]	0.7376
RH	Yes	Chexpert-net + γ-adjustment GAN	**0.7541**
RH	Yes	Chexpert-net + CycleGAN	0.7263
RH	Yes	Chexpert-net + Colormap GAN	0.7253
RH	Yes	Chexpert-net + Graymap GAN	**0.7434**
NIH	No	Chexpert-net + plain input	0.7737
NIH	Yes	Chexpert-net + plain input[NIH baseline]	0.7870
NIH	Yes	Chexpert-net + γ-adjustment GAN	**0.7993**
NIH	Yes	Chexpert-net + CycleGAN	0.7379
NIH	Yes	Chexpert-net + Colormap GAN	0.7671
NIH	Yes	Chexpert-net + Graymap GAN	**0.7986**
NIH	No	NIH-net + plain input	0.8022
RH	No	NIH-net + plain input[RH baseline]	0.6513
RH	No	NIH-net + γ-adjustment GAN	**0.6741**
RH	No	NIH-net + CycleGAN	0.6385
RH	No	NIH-net + Colormap GAN	0.6460
RH	No	NIH-net + Graymap GAN	**0.6619**
Chexpert	Yes	NIH-net + plain input[Chexpert baseline]	0.7458
Chexpert	Yes	NIH-net + γ-adjustment GAN	**0.7501**
Chexpert	Yes	NIH-net + CycleGAN	0.7274
Chexpert	Yes	NIH-net + Colormap GAN	0.7402
Chexpert	Yes	NIH-net + Graymap GAN	0.7458

and Edema are significantly better than the baselines ($p < 0.001$) and significantly better for the other classes ($p < 0.05$). For NIH-net + Graymap GAN, only the AUCs for Cardiomegaly and Edema were statistically significantly better ($p < 0.05$). Overall, the proposed histogram based methods gave considerably better AUCs when compared to either no domain adaptation or to domain adaptation using CycleGAN. For more ablation studies, we refer the reader to Appendix A.1. If the source and domain distributions are close by, like Chexpert and NIH, the performances without domain-adaptation were reasonably closer to our proposed method but still inferior. However, the performances of domain adaptation methods with discriminators based on the raw images (illustrated in Fig. 2) produced worse results compared to networks that do not have any in-built domain adaptation. We intend to look at this anomaly in the future and to explore various hyper-parameters.

In this paper, we have shown that in situations where the domain shift is due to global intensity changes (for instance different exposures on x-rays), over-parameterized pixel-level transformation/discrimination methods like Cycle-GAN or Colormap-GAN are unnecessary. This is consistent with a recent observation [3] that simple binary classifiers discriminating domains yield better results compared to more sophisticated distribution discriminators. Having said this, we would like to point out that in cases where domain shifts are characterized by more local changes (for instance, in brain magnetic resonance images), a combination of our proposed methodology and pixel/voxel-level transformations/discrimination may be the desired solution to account for domain shifts.

References

1. Abadi, M., et al.: TensorFlow: large-scale machine learning on heterogeneous systems (2015). https://www.tensorflow.org/
2. Bar-El, A., Cohen, D., Cahan, N., Greenspan, H.: Improved CycleGAN with application to COVID-19 classification. In: Išgum, I., Landman, B.A. (eds.) Medical Imaging 2021: Image Processing. vol. 11596, pp. 296–305. International Society for Optics and Photonics, SPIE (2021). https://doi.org/10.1117/12.2582162
3. Cao, T., Huang, C.W., Hui, D.Y.T., Cohen, J.P.: A benchmark of medical out of distribution detection. arXiv preprint arXiv:2007.04250 (2020)
4. Chen, H., Jiang, Y., Loew, M., Ko, H.: Unsupervised domain adaptation based COVID-19 CT infection segmentation network. Appl. Intell. **52**(6), 6340–6353 (2021). https://doi.org/10.1007/s10489-021-02691-x
5. DeLong, E.R., DeLong, D.M., Clarke-Pearson, D.L.: Comparing the areas under two or more correlated receiver operating characteristic curves: a nonparametric approach. Biometrics **44**(3), 837–845 (1988)
6. Fernando, B., Habrard, A., Sebban, M., Tuytelaars, T.: Unsupervised visual domain adaptation using subspace alignment. In: Proceedings of the IEEE International Conference on Computer Vision (CVPR), pp. 2960–2967 (2013). https://doi.org/10.1109/ICCV.2013.368
7. Gadermayr, M., et al.: Image-to-image translation for simplified MRI muscle segmentation. Front. Radiol. **1**, 664444 (2021). https://doi.org/10.3389/fradi.2021.664444. https://www.frontiersin.org/article/10.3389/fradi.2021.664444
8. Hammami, M., Friboulet, D., Kechichian, R.: Cycle GAN-based data augmentation for multi-organ detection in CT images via yolo. In: 2020 IEEE International Conference on Image Processing (ICIP), pp. 390–393 (2020). https://doi.org/10.1109/ICIP40778.2020.9191127
9. Harris, C.R., et al.: Array programming with NumPy. Nature **585**(7825), 357–362 (2020). https://doi.org/10.1038/s41586-020-2649-2

10. He, K., Zhang, X., Ren, S., Sun, J.: Deep residual learning for image recognition. In: 2016 IEEE Conference on Computer Vision and Pattern Recognition (CVPR), pp. 770–778 (2016). https://doi.org/10.1109/cvpr.2016.90

11. Huang, G., Liu, Z., Van Der Maaten, L., Weinberger, K.Q.: Densely connected convolutional networks. 2017 IEEE Conference on Computer Vision and Pattern Recognition (CVPR), pp. 2261–2269 (2017). https://doi.org/10.1109/CVPR.2017.243

12. Irvin, J., et al.: CheXpert: a large chest radiograph dataset with uncertainty labels and expert comparison. Proc. AAAI Conf. Artif. Intell. **33**, 590–597 (2019). https://doi.org/10.1609/aaai.v33i01.3301590

13. Jiang, J., et al.: Tumor-aware, adversarial domain adaptation from CT to MRI for lung cancer segmentation. In: Frangi, A.F., Schnabel, J.A., Davatzikos, C., Alberola-López, C., Fichtinger, G. (eds.) MICCAI 2018. LNCS, vol. 11071, pp. 777–785. Springer, Cham (2018). https://doi.org/10.1007/978-3-030-00934-2_86

14. Kamnitsas, K., et al.: Unsupervised domain adaptation in brain lesion segmentation with adversarial networks. In: Niethammer, M., et al. (eds.) IPMI 2017. LNCS, vol. 10265, pp. 597–609. Springer, Cham (2017). https://doi.org/10.1007/978-3-319-59050-9_47

15. Ma, J.: Histogram matching augmentation for domain adaptation with application to multi-centre, multi-vendor and multi-disease cardiac image segmentation. In: Puyol Anton, E., et al. (eds.) STACOM 2020. LNCS, vol. 12592, pp. 177–186. Springer, Cham (2021). https://doi.org/10.1007/978-3-030-68107-4_18

16. Madani, A., Moradi, M., Karargyris, A., Syeda-Mahmood, T.: Semi-supervised learning with generative adversarial networks for chest X-ray classification with ability of data domain adaptation. In: 2018 IEEE 15th International Symposium on Biomedical Imaging (ISBI 2018), pp. 1038–1042 (2018). https://doi.org/10.1109/ISBI.2018.8363749

17. Mao, X., Li, Q., Xie, H., Lau, R.Y., Wang, Z., Smolley, S.P.: Least squares generative adversarial networks. 2017 IEEE International Conference on Computer Vision (ICCV), pp. 2813–2821 (2017). https://doi.org/10.1109/iccv.2017.304

18. Sedighi, V., Fridrich, J.J.: Histogram layer, moving convolutional neural networks towards feature-based steganalysis. In: Media Watermarking, Security, and Forensics, pp. 50–55 (2017). https://doi.org/10.2352/ISSN.2470-1173.2017.7.MWSF-325

19. Tasar, O., Happy, S.L., Tarabalka, Y., Alliez, P.: ColorMapGAN: unsupervised domain adaptation for semantic segmentation using color mapping generative adversarial networks. IEEE Trans. Geosci. Remote Sens. **58**(10), 7178–7193 (2020). https://doi.org/10.1109/tgrs.2020.2980417

20. Wang, X., Peng, Y., Lu, L., Lu, Z., Bagheri, M., Summers, R.M.: ChestX-ray8: hospital-scale chest X-ray database and benchmarks on weakly-supervised classification and localization of common thorax diseases. In: 2017 IEEE Conference on Computer Vision and Pattern Recognition (CVPR), pp. 3462–3471 (2017)

21. Wolterink, J.M., Dinkla, A.M., Savenije, M.H.F., Seevinck, P.R., van den Berg, C.A.T., Išgum, I.: Deep MR to CT synthesis using unpaired data. In: Tsaftaris, S.A., Gooya, A., Frangi, A.F., Prince, J.L. (eds.) SASHIMI 2017. LNCS, vol. 10557, pp. 14–23. Springer, Cham (2017). https://doi.org/10.1007/978-3-319-68127-6_2

22. Zhu, J.Y., Park, T., Isola, P., Efros, A.A.: Unpaired image-to-image translation using cycle-consistent adversarial networks. In: 2017 IEEE International Conference on Computer Vision (ICCV), pp. 2242–2251 (2017). https://doi.org/10.1109/iccv.2017.244

Multi-institutional Investigation of Model Generalizability for Virtual Contrast-Enhanced MRI Synthesis

Wen Li[1], Saikit Lam[1], Tian Li[1], Andy Lai-Yin Cheung[1], Haonan Xiao[1], Chenyang Liu[1], Jiang Zhang[1], Xinzhi Teng[1], Shaohua Zhi[1], Ge Ren[1], Francis Kar-ho Lee[2], Kwok-hung Au[2], Victor Ho-fun Lee[3], Amy Tien Yee Chang[4], and Jing Cai[1(✉)]

[1] Department of Health Technology and Informatics, The Hong Kong Polytechnic University, Hong Kong SAR, China
jing.cai@polyu.edu.hk
[2] Department of Clinical Oncology, Queen Elizabeth Hospital, Hong Kong SAR, China
[3] Department of Clinical Oncology, The University of Hong Kong, Hong Kong SAR, China
[4] Comprehensive Oncology Centre, Hong Kong Sanatorium & Hospital, Hong Kong SAR, China

Abstract. The purpose of this study is to investigate the model generalizability using multi-institutional data for virtual contrast-enhanced MRI (VCE-MRI) synthesis. This study presented a retrospective analysis of contrast-free T1-weighted (T1w), T2-weighted (T2w), and gadolinium-based contrast-enhanced T1w MRI (CE-MRI) images of 231 NPC patients enrolled from four institutions. Data from three of the participating institutions were employed to generate a training and an internal testing set, while data from the remaining institution was employed as an independent external testing set. The multi-institutional data were trained separately (single-institutional model) and jointly (joint-institutional model) and tested using the internal and external sets. The synthetic VCE-MRI was quantitatively evaluated using MAE and SSIM. In addition, visual qualitative evaluation was performed to assess the quality of synthetic VCE-MRI compared to the ground-truth CE-MRI. Quantitative analyses showed that the joint-institutional models outperformed single-institutional models in both internal and external testing sets, and demonstrated high model generalizability, yielding top-ranked MAE, and SSIM of 71.69 ± 21.09 and 0.81 ± 0.04 respectively on the external testing set. Qualitative evaluation indicated that the joint-institutional model gave a closer visual approximation between the synthetic VCE-MRI and ground-truth CE-MRI on the external testing set, compared with single-institutional models. The model generalizability for VCE-MRI synthesis was enhanced, both quantitatively and qualitatively, when data from more institutions was involved during model development.

Keywords: Contrast-enhanced MRI · Model generalizability · Nasopharyngeal carcinoma

© The Author(s), under exclusive license to Springer Nature Switzerland AG 2022
L. Wang (Eds.): MICCAI 2022, LNCS 13437, pp. 765–773, 2022.
https://doi.org/10.1007/978-3-031-16449-1_73

1 Introduction

Nephrogenic systemic fibrosis (NSF), also known as nephrogenic fibrosing dermopathy, is a skin-related disease that occurs primarily in patients with end-stage renal disease [1]. NSF is a systemic disease involving multiple systems and organs, leading to widespread fibrosis of internal organs such as heart, lung, esophagus, skeletal muscle, and kidney, etc. [2]. In some cases, NSF leads to serious physical disability and even death [3]. The occurrence of NSF is highly related to the administration of gadolinium-based contrast agents (GBCAs) during MRI contrast enhancement [4, 5]. Marckmann et al. [1] analysed 13 patients with confirmed NSF, they found that all of the patients had been exposed to GBCAs before the first sign of NSF. GBCAs is mainly excreted by kidney. It was reported in healthy people, the average half-life of GBCAs is 1.3 h. In patients with end-stage renal disease, however, the half-life time is 34.3 h [1]. Gadolinium is highly toxic. Usually, large amounts of chelate are combined with gadolinium to avoid harm to the body. However, in some cases, trace amounts GBCA releases free Gd^{3+} [6]. By binding to endogenous ions in the body, the gadolinium will deposit in brain, bone, and skin, etc. [7]. Besides the patients with NSF, the gadolinium deposition has also been observed in healthy populations after CE-MRI scanning [5]. The long-term effects of gadolinium deposition remain unknown. For safety consideration, the use of GBCAs should be reduced or avoided.

From 2018, researchers were started exploring the feasibility of using deep learning to synthesize the virtual contrast enhanced MRI (VCE-MRI) from contrast-free MRI sequences, with the aim of finding a CE-MRI alternative to reduce or eliminate the use of GBCAs. Gong et al. [8] first reported that deep learning is possible to reduce the use of GBCAs by 90% in patients with brain tumor. The general idea of their work is applying a deep learning network to learn the mapping from contrast-free T1-weighted (T1w) MRI and 10% low dose CE-MRI to full-dose CE-MRI. Followed by their work, Kleesiek et al. [9] explored the feasibility to totally eliminate the use of GBCAs for brain tumor patients. They developed a 3D Bayesian model and input the model with multi-parametric MRI images, including T1w MRI, T2-weighted (T2w) MRI, T2w-FLAIR, DWI and SWI. The purpose of their work is to synthesize the full-dose CE-MRI from contrast-free multiparametric MRI images. In recent two years, more cancer types have been investigated, including liver cancer [10, 11], breast cancer [12], and nasopharyngeal carcinoma (NPC) [13]. These studies mainly focused on the feasibility studies or development of deep learning algorithms. However, the models they developed were based on dataset with specific conditions, *e.g.*, scanners, imaging protocols, and patient demographics. The developed models may not work well on data with other conditions since the image features differ among different conditions, especially for the multi-parametric MRI [14]. For example, in Fig. 1, although both institutional images belong to the same MRI sequence, the image quality of institution-2 images is relatively poor with low single-to-noise ratio for both T1w and CE-MRI, compared with the institution-1 images.

In this work, we aim at exploring the model generalizability using multi-institutional multiparametric MRI, and investigating the feasibility of using multi-institutional MRI to improve the model generalizability that specifically for the VCE-MRI synthesis task. To investigate the model performance on MRI data from unseen institutions, we first trained

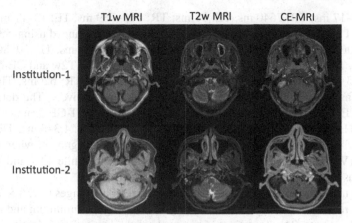

Fig. 1. Illustration of MRI difference between two institutions. The first and second row show the T1w MRI, T2w MRI, and CE-MRI images from institution-1 and institution-2 respectively.

three single-institutional models separately using data from three institutions. The single-institutional models were tested using internal (split from the same institution) and three external datasets. To investigate the feasibility of using multi-institutional MRI data for improving the model generalizability, we trained a joint-institutional model using data from three institutions. In order to keep the training samples of the joint-institutional model consistent with that of the single-institutional models, we randomly reduced the sample number of each institution to 1/3 of the original. The generalizability of the joint-institutional model was evaluated using an external dataset. The main contributions of this work are: (i) external evaluations were performed, which provided a research basis for a better understanding of inter-institutional MRI data biases; (ii) our model was trained with multi-institutional MRI data, which has the potential to provide a VCE-MRI synthesis model with high generalizability.

2 Materials and Methods

2.1 Data Description

The multi-parametric MRI of NPC patients from four institutions was labeled as institution-1, institution-2, institution-3, and institution-4, respectively. A total of 231 patients (71 from institution-1, 71 from institution-2, 71 from institution-3, and 18 from institution-4) were included in this study. All patients were accepted radiotherapy and T1w, T2w, and CE-MRI were retrospectively retrieved from each patient. For tumor delineation, all patients were administrated with GBCAs during CE-MRI imaging. The CE-MRI scanning was started immediately after injection of GBCAs. The T1w, T2w, and CE-MRI has been aligned by oncologists for radiotherapy purpose. The T1w, T2w, and CE-MRI images from the same institute were obtained using the same MRI scanner with the same pulse sequence. The MRI data from institution-1, institution-2, institution-3, and institution-4 were acquired using different scanners. The T1w, T2w, and CE-MRI images from institution-1 were scanned using a 1.5T-Siemens scanner with TR: 562–739 ms, TE: 13–17 ms; TR: 7640 ms, TE: 97 ms; TR: 562–739 ms, TE: 13–17 ms, respectively. The T1w, T2w, and CE-MRI from institution-2 were scanned using a 3T-Philips scanner with TR: 4.8–9.4 ms, TE: 2.4–8.0 ms; TR: 3500–4900 ms, TE: 50–80 ms; TR: 4.8–9.4 ms, TE: 2.4–8.0 ms, respectively. The images of T1w, T2w and CE-MRI from institution-3 were scanned using a 3T-Siemens scanner with TR: 620ms, TE: 9.8 ms; TR: 2500 ms, TE: 74 ms; TR: 3.42 ms, TE: 1.11 ms, respectively. The data of T1w, T2w and CE-MRI from institution-4 were scanned using a 3T-GE scanner with TR: 4.3–8 ms, TE: 3–7 ms; TR: 4200–4899 ms, TE: 60–80 ms; TR: 4.3–8 ms, TE: 3–7 ms, respectively. Rigid registration was applied to fine-tune the alignment when necessary by Elastix (V5.0.1) [15]. CE-MRI was used as fixed image, while T1w and T2w MRI were used as moving images.

Prior to model training, we first resampled the size of all images to 256×224. Then we linearly rescaled the image intensities to $[-1, 1]$ using the minimum and maximum value of all patients. The 71 patients from each of the institution-1, institution-2, and institution-3 were randomly split to 53 patients for model training and 18 patients for model testing, the 18 patients from institution-4 were used as an external testing set to evaluate the model generalizability. In total, we have three training datasets and four testing datasets (three internal testing datasets and one external dataset).

2.2 Deep Learning Network

A previous proposed multimodality-guided synergistic neural network (MMgSN-Net) [13] was used to train the models in this study. The MMgSN-Net is a two inputs network, which consist of five key components: multimodality learning module, synthesis network, self-attention module, multi-level module and discriminator. The network architecture is illustrated in Fig. 2. The multimodality learning module is designed to extract the modality-specific features from input sequences. The synthesis network is used to fuse the extracted features from multimodality learning module and automatically assemble the complementary features for VCE-MRI synthesis. The self-attention module is applied to capture the regions of interest, such as the contrast enhanced regions and surrounding high risk organs. The multi-level module is utilized to enhance the edge-detection performance of the model. To generate more realistic VCE-MRI, a discriminator is added to distinguish the VCE-MRI from real CE-MRI.

The T1w and T2w MRI were used as input, while the CE-MRI was utilized as the learning target. L1 loss [13] was adopted as the loss function of the synthesis network, while Adam was set as the optimizer. We trained the MMgSN-Net with 200 epochs with a fixed learning rate of 0.0002. The batch size was set as 1, no data augmentation was performed in this study. All experiments were performed with an RTX 3090 GPU card using PyTorch library (V1.8.1).

2.3 Study Design

The purpose of this study is to explore and improve the model generalizability by using multi-institutional data sets. It has been reported that it is possible to improve the model generalizability by enlarging the view of the deep learning models [16]. In this study, we utilized the multi-institutional data which generated from different scanners and patient

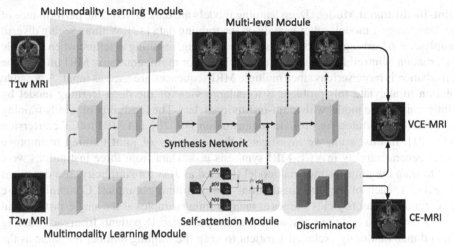

Fig. 2. Architecture of the MMgSN-Net. It consists of five key modules: multimodality learning module, synthesis network, multi-level module self-attention module and discriminator.

demographics to increase the diversity of training data. The overall design of this study is to train different deep learning models using datasets from three institutions, either separately or jointly, and then evaluate the intra-institutional performance and inter-institutional generalizability of these models using internal and external data sets. The overall relationship between training data and models is illustrated in Fig. 3.

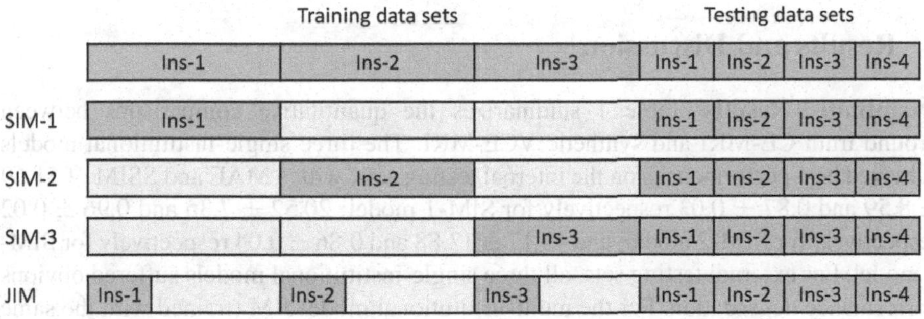

	Training data sets			Testing data sets			
	Ins-1	Ins-2	Ins-3	Ins-1	Ins-2	Ins-3	Ins-4
SIM-1	Ins-1			Ins-1	Ins-2	Ins-3	Ins-4
SIM-2		Ins-2		Ins-1	Ins-2	Ins-3	Ins-4
SIM-3			Ins-3	Ins-1	Ins-2	Ins-3	Ins-4
JIM	Ins-1	Ins-2	Ins-3	Ins-1	Ins-2	Ins-3	Ins-4

Fig. 3. Illustration of the training and testing data used for each model.

Single-Institutional Models. To evaluate the generalizability of models that trained with single-institutional data, we first trained three single-institutional models (single-institutional model-1, single-institutional model-2, and single-institutional model-3) separately using data from three institutions. For simplicity, these three models were labeled as SIM-1, SIM-2, and SIM-3 respectively. The three models were then evaluated using one internal and three external data sets.

Joint-Institutional Model. Deep learning models are data driven. The performance of the deep learning models relies heavily on the training data [17]. Without seeing diverse samples, deep learning models are easily to overfitting, resulting in failure when invisible perturbation is introduced to testing images [18]. For multiparametric MRI images, the perturbation is more serious since multiple MRI sequences are used as input. A possible solution to alleviate this problem is to enlarge view of the deep learning model by jointly training the model with multi-institutional data. The feasibility of jointly training has been demonstrated for prostate segmentation [19, 20] and MRI to CT conversion tasks [21]. In this work, we investigated the feasibility of joint training to improve model generalizability in VCE-MRI synthesis task. Data from three institutions were used to train the joint-institutional model (labeled as JIM for simplicity). As described in Sect. 2.1, each of the three institutions had 53 training samples. Combining these three training datasets yields 3 times as many training samples as the single-institutional models. To make a fair comparison, we randomly selected 18 patients from each training set, and then randomly excluded 1 patient to keep the training number the same as the single-institutional models.

Evaluation Methods. The synthetic VCE-MRI images were quantitatively and qualitatively compared with the ground truth CE-MRI images. For quantitatively comparison, the mean absolute error (MAE) and structural similarity index (SSIM) were calculated between VCE-MRI and CE-MRI of the patient. To visually assess the generalizability of these models, the T1w and T2w MRI of the external institution-4 data set was input to each model for generating VCE-MRI. The VCE-MRI generated from these models was visually compared and analyzed with ground truth CE-MRI.

3   Results and Discussion

Quantitative Results. Table 1 summarizes the quantitative comparisons between ground truth CE-MRI and synthetic VCE-MRI. The three single-institutional models achieved best performances on the internal testing sets, with a MAE and SSIM of 51.69 $\pm$ 8.59 and 0.87 $\pm$ 0.03 respectively for SIM-1 model; 20.52 $\pm$ 7.36 and 0.96 $\pm$ 0.02 respectively for SIM-2 model; and 48.17 $\pm$ 12.88 and 0.86 $\pm$ 0.04 respectively for SIM-3 model. For external testing sets, all three single-institutional models suffered obvious performance degradation. For the multi-institutional model JIM (trained with the same number of training samples as single-institutional models, but with more training institutions), comparable internal testing results were achieved on all three internal testing sets compared to the three single-institutional models, with a MAE and SSIM of 53.38 $\pm$ 7.91 and 0.87 $\pm$ 0.02 for Institution-1 testing set, 17.45 $\pm$ 6.16 and 0.86 $\pm$ 0.04 for institution-2 testing set, and 48.23 $\pm$ 13.27 and 0.85 $\pm$ 0.04 for institution-3 testing set, respectively. For external institution-4 testing set, JIM significantly outperforms SIM-1 and SIM-2 ($p < 0.05$) and obtained comparable results as SIM-3 ($p > 0.05$), with a MAE and SSIM of 71.69 $\pm$ 21.09 and 0.81 $\pm$ 0.04 respectively. The quantitative results from different institutions were quite different, which is reasonable since the MRI intensities that generated from different institutions or scanners are inconsistent. In our study, the MRI images from institution-2 had lower pixel intensities than those from the other three institutions. Compared with the other two training sets, the MRI data of institution-3 has the closest overall average intensity to the external institution-4. This may be the reason why institution-3 model outperformed other two models in institution-4 data. However, the JIM model with part of the institution-3 data also achieved comparable performance on institution-4 data, suggesting that enlarging the view of training sample does help improve the model generalizability, even when the sample is limited.

Table 1. Quantitative comparisons between synthetic VCE-MRI and ground truth CE-MRI. External results are shown in bold. MAE: mean absolute error; SSIM: structural similarity index; SD: standard deviation.

Model	Testing set	MAE ± SD	SSIM ± SD
SIM-1	Institution-1	51.69 ± 8.59	0.87 ± 0.03
	Institution-2	**134.21 ± 27.77**	**0.54 ± 0.08**
	Institution-3	**70.22 ± 11.63**	**0.77 ± 0.03**
	Institution-4	**95.98 ± 22.12**	**0.72 ± 0.05**
SIM-2	**Institution-1**	**125.89 ± 13.93**	**0.70 ± 0.03**
	Institution-2	20.52 ± 7.36	0.96 ± 0.02
	Institution-3	**84.49 ± 27.72**	**0.80 ± 0.06**
	Institution-4	**104.18 ± 39.18**	**0.76 ± 0.08**
SIM-3	**Institution-1**	**76.38 ± 8.86**	**0.78 ± 0.03**
	Institution-2	**78.97 ± 20.23**	**0.73 ± 0.07**
	Institution-3	48.17 ± 12.88	0.86 ± 0.04
	Institution-4	**72.14 ± 23.76**	**0.79 ± 0.05**
JIM	Institution-1	53.38 ± 7.91	0.87 ± 0.02
	Institution-2	17.45 ± 6.16	0.86 ± 0.04
	Institution-3	48.23 ± 13.27	0.85 ± 0.04
	Institution-4	**71.69 ± 21.09**	**0.81 ± 0.04**

Qualitative Results. Figure 4 shows a visual comparison between synthetic VCE-MRI and ground truth CE-MRI on the external institution-4 testing set. The synthetic VCE-MRI from both SIM-1 and SIM-2 models failed to accurately predict the contrast enhancement, especially the SIM-2 model. The JIM model correctly located and enhanced the tumor, and obtained a more realistic overall image than SIM-1 and SIM-2 model. The overall image generated from JIM outperforms SIM-3. Moreover, JIM was superior to SIM-3 in tumor texture. As shown in Fig. 4, the VCE-MRI from SIM-3 lost some texture details in tumor, while the synthetic image from JIM successfully reserved this information.

There are two main limitations of the current work: i) Only one hospital data was collected for external evaluation. In the future, data from more hospitals will be collected

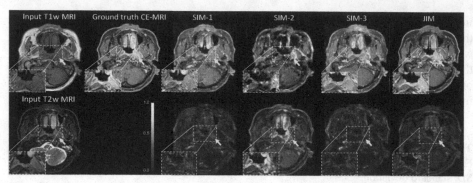

Fig. 4. Visual comparison between synthetic VCE-MRI and ground truth CE-MRI on the external institution-4 testing set. The first row from left to right are input T1w MRI, ground truth CE-MRI, and the synthetic VCE-MRI from four models. The second row from left to right are input T2w MRI and the difference map between CE-MRI and VCE-MRI of the four models. The yellow arrows show the position of tumor. (Color figure online)

for a more comprehensive evaluation. ii) Only one model was used to assess the generalizability. In the future, additional models will be applied to assess whether different models can also reach the same conclusions.

4   Conclusion

In this study, we investigated the model generalizability using multi-institutional multiparametric MRI data on VCE-MRI synthesis task. The results show that the model trained with single institutional data has poor model generalizability and failed to generate accurate VCE-MRI images on external dataset. When data from more institutions was involved during model development, the model generalizability for VCE-MRI synthesis was enhanced, both quantitatively and qualitatively.

Acknowledgment. This work was partly supported by funding GRF 151022/19M and ITS/080/19.

References

1. Marckmann, P., Skov, L., Rossen, K., et al.: Nephrogenic systemic fibrosis: suspected causative role of gadodiamide used for contrast-enhanced magnetic resonance imaging. J. Am. Soc. Nephrol. **17**(9), 2359–2362 (2006)
2. Saxena, S.K., Sharma, M., Patel, M., Oreopoulos, D.: Nephrogenic systemic fibrosis: an emerging entity. Int. Urol. Nephrol. **40**(3), 715–724 (2008)
3. Swaminathan, S., High, W., Ranville, J., et al.: Cardiac and vascular metal deposition with high mortality in nephrogenic systemic fibrosis. Kidney Int. **73**(12), 1413–1418 (2008)

4. Khawaja, A.Z., Cassidy, D.B., Al Shakarchi, J., McGrogan, D.G., Inston, N.G., Jones, R.G.: Revisiting the risks of MRI with Gadolinium based contrast agents-review of literature and guidelines. Insights Imaging **6**(5), 553–558 (2015). https://doi.org/10.1007/s13244-015-0420-2

5. Roberts, D.R., Chatterjee, A., Yazdani, M., et al.: Pediatric patients demonstrate progressive T1-weighted hyperintensity in the dentate nucleus following multiple doses of gadolinium-based contrast agent. Am. J. Neuroradiol. **37**(12), 2340–2347 (2016)

6. Grobner, T., Prischl, F.: Gadolinium and nephrogenic systemic fibrosis. Kidney Int. **72**(3), 260–264 (2007)

7. Aime, S., Caravan, P.: Biodistribution of gadolinium-based contrast agents, including gadolinium deposition. J. Magn. Reson. Imaging Official J. Int. Soc. Magn. Res. Med. **30**(6), 1259–1267 (2009)

8. Gong, E., Pauly, J.M., Wintermark, M., Zaharchuk, G.: Deep learning enables reduced gadolinium dose for contrast-enhanced brain MRI. J. Magn. Reson. Imaging **48**(2), 330–340 (2018)

9. Kleesiek, J., Morshuis, J.N., Isensee, F., et al.: Can virtual contrast enhancement in brain MRI replace gadolinium?: a feasibility study. Invest. Radiol. **54**(10), 653–660 (2019)

10. Xu, C., Zhang, D., Chong, J., Chen, B., Li, S.: Synthesis of gadolinium-enhanced liver tumors on nonenhanced liver MR images using pixel-level graph reinforcement learning. Med. Image Anal. **69**, 101976 (2021)

11. Zhao, J., Li, D., Kassam, Z., et al.: Tripartite-GAN: synthesizing liver contrast-enhanced MRI to improve tumor detection. Med. Image Anal. **63**, 101667 (2020)

12. Kim, E., Cho, H.-H., Ko, E., Park, H.: Generative adversarial network with local discriminator for synthesizing breast contrast-enhanced MRI. In: 2021 IEEE EMBS International Conference on Biomedical and Health Informatics (BHI) (2021)

13. Li, W., Xiao, H., Li, T., et al.: Virtual contrast-enhanced magnetic resonance images synthesis for patients with nasopharyngeal carcinoma using multimodality-guided synergistic neural network. Int. J. Radiat. Oncol. Biol. Phys. **112**(4), 1033–1044 (2022)

14. Cho, H., Nishimura, K., Watanabe, K., Bise, R.: Cell detection in domain shift problem using pseudo-cell-position heatmap. In: de Bruijne, M., et al. (eds.) Medical Image Computing and Computer Assisted Intervention – MICCAI 2021. LNCS, vol. 12908, pp. 384–394. Springer, Cham (2021). https://doi.org/10.1007/978-3-030-87237-3_37

15. Klein, S., Staring, M., Murphy, K., Viergever, M.A., Pluim, J.P.: Elastix: a toolbox for intensity-based medical image registration. IEEE Trans. Med. Imaging **29**(1), 196–205 (2009)

16. Onofrey, J.A., Casetti-Dinescu, D.I., Lauritzen, A.D., et al.: Generalizable multi-site training and testing of deep neural networks using image normalization. In: 2019 IEEE 16th International Symposium on Biomedical Imaging (ISBI 2019) (2019)

17. LeCun, Y., Bengio, Y., Hinton, G.: Deep learning. Nature **521**(7553), 436–444 (2015)

18. Kanbak, C., Moosavi-Dezfooli, S.-M., Frossard, P.: Geometric robustness of deep networks: analysis and improvement. In: Proceedings of the IEEE Conference on Computer Vision and Pattern Recognition (2018)

19. Sanford, T.H., Zhang, L., Harmon, S.A., et al.: Data augmentation and transfer learning to improve generalizability of an automated prostate segmentation model. Am. J. Roentgenol. **215**(6), 1403–1410 (2020)

20. Liu, Q., Dou, Q., Yu, L., Heng, P.A.: MS-Net: multi-site network for improving prostate segmentation with heterogeneous MRI data. IEEE Trans. Med. Imaging **39**(9), 2713–2724 (2020)

21. Li, W., Kazemifar, S., Bai, T., et al.: Synthesizing CT images from MR images with deep learning: model generalization for different datasets through transfer learning. Biomed. Phys. Eng. Express **7**(2), 025020 (2021)

Author Index

Printed in the United States
by Baker & Taylor Publisher Services